Developments in Petroleum Science, 3

PRODUCTION AND TRANSPORT OF OIL AND GAS

FURTHER TITLES IN THIS SERIES

1 A.G. COLLINS
Geochemistry of Oilfield Waters

2 W.H. FERTL
Abnormal Formation Pressures

Developments in Petroleum Science, 3

PRODUCTION AND TRANSPORT OF OIL AND GAS

A. P. SZILAS

Professor of Petroleum Engineering
Petroleum Engineering Department, Miskolc Technical University of Heavy Industries (Hungary)

ELSEVIER SCIENTIFIC PUBLISHING COMPANY

Amsterdam — Oxford — New York 1975

This work was originally published as

KŐOLAJ- ÉS FÖLDGÁZTERMELÉS
Tankönyvkiadó, Budapest

ENGLISH TRANSLATION BY

B. BALKAY

The distribution of this book is being handled by the following publishers:

for the U.S.A. and Canada

American Elsevier Publishing Company, Inc.
52 Vanderbilt Avenue
New York, New York 10017

for East Europe, China, Democratic People's Republic of Korea, Cuba, the Democratic Republic of Vietnam and of Mongolia

Akadémiai Kiadó, The Publishing House
of the Hungarian Academy of Sciences, Budapest (Hungary)

for all remaining areas

Elsevier Scientific Publishing Company
335 Jan van Galenstraat
P.O. Box 211, Amsterdam, The Netherlands

Library of Congress Catalog Card Number 75-1773

ISBN 0-444-99869-1

Joint edition published by
Elsevier Scientific Publishing Company, Amsterdam,
The Netherlands and Akadémiai Kiadó,
The Publishing House of the Hungarian Academy of Sciences, Budapest, Hungary

Printed in Hungary

Contents

Preface

Oil and gas production in the broad sense of the word can be subdivided into three more or less separate fields of science and technology, notably (1) production processes in the reservoir (reservoir engineering), (2) production of oil and gas from wells, and finally (3) surface gathering, separation and transportation. The present book deals with the second and the last of the three topics.

Chapter 1 reviews those calculations concerning flow in pipelines a knowledge of which is essential to the understanding and designing of single-phase and multi-phase flow in wells and in surface flow lines.

In compiling Chapters 2–5, which deal with oil and gas wells and in the treatment of those subjects, I have followed the principle that the main task of the production engineer is to ensure the production of that amount of liquid and/or gas prescribed for each well in the field's production plan, at the lowest feasible cost of production. The technical aim outlined above can often be attained by several different methods of production, with several types of production equipment and, within a given type, with various design and size of equipment; in fact, using a given type of equipment, several methods of operation are possible. Of the technically feasible solutions, there will be one that will be the most economical; this, of course, will be the one chosen.

I have attempted to cover the various subjects as fully as possible, but have nevertheless by-passed certain topics which are treated in other books, such as the dynamometry of sucker rod pumps and gas metering. A discussion of these topics in sufficient depth would have required too much space.

Chapter 6 deals with the main items of surface equipment used in oil and gas fields. In this case, I have also aimed at conveying a body of information setting out the choice of the technically and economically optimal equipment.

Equipment is not discussed in Chapters 7 and 8 which treat the flow of oil and gas in pipelines and pipeline systems. The reason for this is that comparatively short pipelines are encountered within the oil or gas field proper, and the relevant production equipment is discussed in Chapter 6; on the other hand, it seemed reasonable to emphasize the design conception which regards the series-connected hydraulic elements of wells, on-lease equipment and pipelines as a connected hydraulic system with an overall optimum that can be and must be determined. It should be emphasized, however, that this method of designing also requires a knowledge of rheology.

Naturally, in the treatment of each subject I have attempted to expose not only the 'hows' but also the 'whys' and 'wherefores' of the solutions outlined. It is a regrettable phenomenon, and one which I have often found during my own production experience and in my work at the University, that the logical consistency as well as the economy of the solution adopted will tend to suffer because the design or production engineer is just following 'cookbook rules' without understanding what he is actually trying to do. An understanding of the subject is a necessary critical foundation, and this is a prime reason of textbooks and handbooks.

In denoting physical quantities and in choosing physical units I have followed the SI nomenclature. In choosing the various suffixes to the symbols used in this book, the wide range of the subjects covered has necessitated some slight deviations from the principle of 'one concept — one symbol'. I sincerely hope that such compromises, adopted for the sake of simplicity, will not create any difficulties for the reader.

In compiling the present volume and in its preparation for publication I have been assisted by many of my co-workers at the Petroleum Engineering Department of the Miskolc Technical University of Heavy Industries. I am deeply grateful for their cooperation, without which the present book, a compendium of three decades' production and teaching experience, could hardly have been realized. Among them I wish to give special credit to Ferenc Patsch, Jr., who played a substantial part in the writing of Chapter 8, to Gábor Takács and Tibor Tóth, both of whom gave a great deal of help in the calculation and correction of the numerical examples in Chapters 1–7, and to Mrs. É. Szota for her painstaking work concerning the figures.

The Author

List of symbols

Note. — Items not entered here are: various infrequently used physical quantities (these are interpreted where they occur in the text); ordinal suffixes, whether letters or numbers; simplified forms of functions; symbols for constants and for expressions interpreted elsewhere and general symbols, such as, for example, pressure *difference* Δp, *mean* pressure $\bar{p}$, *maximum* pressure $p_{\max}$, *minimum* pressure $p_{\min}$, *optimal* fluid flow rate $q_{f\,\mathrm{opt}}$, *allowable* stress σ_{al}. Suffixes have been dropped for simplicity wherever the physical quantity in question is uniquely defined by the context, e.g. in a chapter on gas flow, standard gas density may be expressed simply by ϱ rather than ϱ_{gn}.

a	acceleration, m^2/s
$a_{\max}$	maximum acceleration of polished rod
a	temperature distribution or diffusivity factor, m^2/s
a_s	of soil
a_{sw}	of wet soil
b	weight reduction factor of sucker-rod string
c	specific heat
	per unit mass, J/(kg K)
	per mole mass, J/(kmole K)
c_g	of gas
c_l	of liquid
c_o	of oil
c_p	at constant pressure
c_{pa}	at atmospheric pressure
c_{pp}	at pressure p
c_{pL}	of liquid phase
c_s	of soil
c_{sd}	of dry soil
c_{sw}	of wet soil
c_{st}	of steel
c	compressibility, m^2/N
d	diameter, m
d_i	*ID*
d_{Ci}	of casing
d_o	*OD*
d_{To}	of tubing
d_e	equivalent

d_{ch} of choke
d_{chi} of choke in *i*th gas lift valve
d_{in} *OD* of insulation
d_p of sucker-rod pump plunger
d_r of sucker rod
$\left(-\frac{dv}{dr}\right)$ rate of shear by laminar flow in pipe, 1/s

e eccentricity
e figure of merit of pipe

f loss factor

g acceleration of gravity, m/s^2

h specific enthalpy, J/kg
h elevation (altitude difference), m
h_l of liquid column
h_{la} accumulated
h_{lo} produced
h_{lr} residual
h_{lu} moving up the tubing
h_p of sucker-rod pump barrel
h_v head loss of flow, m

k permeability, m^2
k_g effective, for gas
k_o effective, for oil
k heat transfer factor per unit length of pipe, W/(mK)
k^* heat transfer factor per unit surface, W/(m^{2}K)
k equivalent absolute roughness, m
k specific cost of transportation, Ft/(kgkm)
k specific cost of production, Ft/Mg

l length, distance from origin, m
l_c critical pipe length
l_g length of gelled oil plug

n total number of moles in liquid–gas system
n_L number of moles in liquid phase
n_V number of moles in gas phase
n stroke number (speed) of sucker-rod pump, 1/min
n exponent of productivity equation
n exponent of exponential law
n_c daily cycle number of intermittent gas lift, 1/d
n_o free oscillation number of sucker-rod string, 1/min

p — pressure, N/m^2
- p_c — critical
- p_{gl} — of injection gas
- p_{gL} — of static gas column of length L
- p_h — hydrostatic
- p_k — dome pressure of gas-lift valve
  - p_{kn} — at standard conditions
  - p_{kT} — at temperature T
- p_{k0} — surface casing pressure when gas lift valve closes
- p_n — at standard conditions
- p_n — at node
- p_{pc} — pseudocritical
- p_t — at instant t
- p_v — of vapour in tank
- p_{wf} — flowing BHP of well
- p_{wfi} — when injection gas flows through ith gas lift valve
- p_{ws} — static BHP of well
- p_C — of casing
  - p_{CO} — at wellhead, well producing
  - p_{CZ} — at wellhead, shut-in
  - p_{CL} — at tubing shoe
  - p_{Ci} — at ith gas lift valve
    - p_{Coi} — when valve opens
- p_T — of tubing
  - p_{TO} — at wellhead, well producing
  - p_{TZ} — at wellhead, shut-in
  - p_{TL} — at bottom of tubing string
  - p_{Ti} — at ith gas lift valve
    - p_{Toi} — when valve opens
  - p_{Tfi} — during continuous flow at ith gas lift valve

p_{pr} — pseudoreduced pressure

p_r — reduced pressure

p_s — pressure drop from friction, N/m^2
- p_{sL} — in laminar flow
- p_{sT} — in turbulent flow

Δp — pressure drop, N/m^2
- Δp_{ch} — across choke
- Δp_{cl} — closing pressure differential of gas-lift valve, N/m^2
- Δp_{gl} — spread of gas-lift valve, N/m^2
- Δp_{op} — opening pressure differential of gas-lift valve, N/m^2

q — flow rate, m^3/s
- q_g — of gas
  - q_i — standard-volume, flowing in ith connecting element
  - $q_{i,j}$ — standard-volume, flowing from ith to jth node
  - q_c — of compressor output
  - q_{gl} — of injection gas
  - q_{gn} — expressed in terms of standard volume

q_{in}	inflow rate
q_l	of liquid
q_o	outflow rate
q_o	oil flow rate
q_{ow}	flow rate of oil–water mixture
q_t	theoretical delivery of sucker-rod pump in terms of plunger stroke
q_{tot}	total volume production rate
q_t'	theoretical delivery of sucker-rod pump in terms of polished rod stroke
q_w	water flow rate
q_m	mass flow rate, kg/s
q_{gm}	of gas
q_{km}	of mixture
q_{om}	of oil
q_{oc}	oil produced by intermittent gas lift cycle, m^3/cycle
Δq	delivery correction, m^3/s
r	radius
r_e	of influence of well
r_i	of pipe, internal
r_w	of well
s	wall thickness of pipe, m
s	polished-rod stroke, m
s_p	sucker-rod pump plunger stroke length, m
s_v	valve travel, m
Δs	difference in stroke length of plunger and polished rod, m
t	time, duration, period, s
t_c	period of production cycle
t_p	period of temperature wave
u	flow velocity, m/s
v	flow velocity, m/s
v_g	of gas
v_{gs}	of gas slippage
v_k	of mixture
v_l	of liquid
v_o	of oil
v_{ow}	of oil–water mixture
v_{tot}	total flow velocity
v_w	of water
v_s	speed of sound, m/s
v_{mg}	mass velocity of gas, $kg/(m^2s)$
v_{ml}	mass velocity of liquid, $kg/(m^2s)$

w — well completion factor

x_i — mole fraction of ith component of liquid
x_i — measured data
  x_{ib} — base value
Δx_i — correction

y_i — mole fraction of ith component of gas

z — compressibility factor
  z_n — at standard temperature
  z_i — at temperature T_i
z — mole fraction in liquid-gas mixture
  z_i — of ith component
  z_L — of moles in liquid phase
  z_V — of moles in gas phase
z_h — mole fraction of hydrate-forming components
z — geodetic head, m
z_{ge} — critical length of deformation, m

A — cross-sectional area, m^2
  A_a — of gas anchor
  A_{ch} — of gas-lift-valve choke
  A_{chp} — of choke in pilot valve section
  A_i — internal, of pipeline
  A_k — of gas-lift valve dome
  A_p — of sucker-rod-pump plunger
  A_r — of sucker rod
  A_{ri} — internal, of hollow sucker rod
  A_t — of tubing wall
  A_v — of gas lift valve port
  A_{wk} — of well chamber
  A_T — internal, of tubing
  A_{TC} — of annulus
A — depreciation cost, Forint/year

B — isothermal speed of sound, m/s
B — cost, Forint/year
B — volume factor, m^3/m^3
  B_d — of well fluid on discharge from bottom-hole pump
  B_i — of well fluid on entry into bottom-hole pump
  B_o — of oil
  B_t — multiphase

C — coefficient of gas well's productivity equation, $\frac{m^{(3+n)}\,s^{(2n-1)}}{\mathrm{kg}^n}$
C — heat-transfer correction factor

D	rate of shear, 1/s
E	modulus of elasticity, N/m²
E_g	of gelled oil
E_r	of sucker rod
E_T	of tubing
F	force, load, weight, N
F_{fr}	friction force
F_h	buoyant force acting on sucker-rod pump
F_l	liquid load on sucker-rod pump in terms of full plunger cross-section
$F_{l'}$	liquid load on sucker-rod pump in terms of full plunger cross-section reduced by A_r
F_m	force of attraction of magnet
F_{op}	force required to open gas-lift valve
$F_{p\max}$	maximum polished-rod load
F_r	'dry' sucker-rod weight
F_s	spring force
F_s	static sucker-rod load
F_{sd}	downstroke
F_{su}	upstroke
ΔF_p	variable fluid load in sucker-rod pumping
G	unit weight of column, N/m
G_l	of liquid above plunger
$G_{l'}$	of liquid in tubing
G_r	of dry sucker-rod string
$G_{r'}$	of wet sucker-rod string
G_T	of tubing
H	head capacity of pump, m
H	enthalpy, J
J	productivity index of oil well, $\frac{m^{(3+n)}\, s^{(2n-1)}}{kg^n}$
K	Coberly factor
K	total transportation cost, Forint/year
K_i	equilibrium ratio in liquid–gas system
K_{hi}	equilibrium ratio of hydrate formation
L	length, depth, m
L_d	dynamic fluid level
L_i	depth of ith gas-lift valve
L_s	static fluid level
L_w	well depth

L_T length of tubing string
ΔL rod-string stretch, m
ΔL_{rd} under dynamic load
ΔL_{rp} under static load
ΔL_{Tv} variable stretch of tubing string, m

M torque, Nm
M molar mass, kg/kmole
M_g of gas
M_t multiphase mass factor, kg/m³

N dimensionless number
N_d pipe-diameter number
N_{lv} liquid velocity number
N_p pressure number
N_q flow rate number
N_t time number
N_{Fo} Fourier number
N_{Gr} Grashof number
N_{He} Hedström number
N_{Nu} Nusselt number
N_{Pr} Prandtl number
N_{Re} Reynolds number
$N_{\mathrm{Re}l}$ in liquid
$N_{\mathrm{Re}g}$ in gas
$N_{\mathrm{Re}c}$ critical
$N_{\mathrm{Re}ow}$ for flow of oil in water
$N_{\mathrm{Re}p}$ for plastic flow
$N_{\mathrm{Re}pp}$ for pseudoplastic flow
$N_{\mathrm{Re}ppc}$ critical
N_{W} Weber number
N_μ viscosity number
N_ϱ density number

P power, W
P_h active

Q heat, J
Q phase-transition heat, J/kg

R universal molar gas constant, J/(kmole K)
R volumetric ratio, m³/m³
R_{eff} effective *GOR*
R_{go} *GOR*
R_{gl} injection *GOR*
R_p production *GOR*
R_r specific gas production
R_s solution *GOR*

R_w water–liquid ratio
R_{wo} *WOR*
R_p pressure ratio, $\frac{N}{m^2} \Big/ \frac{N}{m^2}$
R_A tubing-to-annulus cross-sectional area ratio, m^2/m^2

S dimensionless slippage velocity
S (sign), flow direction indicator
S_c clay fraction of soil in percent of sand mass, kg/kg
S_w water content of soil in percent of sand mass, kg/kg

T temperature, K, °C
T_b of 'undisturbed' formation
T_c critical
T_i of *i*th gas-lift valve
T_i on inside of pipe
T_{in} on outside of insulation
T_f of fluid (axial)
T_{fk} corrected oil temperature
T_g of gas
T_n at standard conditions
T_o on outside of pipe
T_{pc} pseudocritical
T_s of soil
T_C in casing annulus
T_{CO} at wellhead
T_T of tubing
T_{TO} at wellhead
T_{TL} at bottom of tubing string
T' in transient heat flow
T_{pr} pseudoreduced temperature
T_r reduced temperature
ΔT_a amplitude of temperature wave, K
ΔT_h temperature difference at height h above well bottom, K

V volume m^3,
V_{gc} of injection gas required per cycle
V_{lo} evaporated from tank
V_n at standard conditions
V_o of oil
V specific volume, m^3/kg
V_k of mixture
V_{mol} molar volume, $m^3/kmole$

W work, energy, J
W' specific energy content, J/N
α angle of inclination, rad

α — throughput factor
α_T — temperature coefficient of oil density, kg/(m³ K)
α_p — pressure coefficient of oil density, s²/m²
α_1 — internal convection factor, W/(m² K)
α_2 — external heat-transfer factor, W/(m² K)
$\alpha_{\varrho r}$ — temperature coefficient of relative density, 1/K

β — dispersion factor
β_p — pressure coefficient of gas solubility, 1/N
β_p — pressure cubic shrinkage coefficient of oil, 1/N
β_T — thermal cubic expansion coefficient of oil, 1/K

γ — specific weight, N/m³
- γ_g — of gas
- γ_k — of mixture
- γ_l — of liquid
- γ_o — of oil
- γ_r — of sucker rods
- γ_w — of water

δ — dynamic factor of sucker-rod pumping

ε_g — cross-sectional fraction of the gas phase
ε_{ge} — relative critical deformation of oil gel
ε_l — cross-sectional fraction of the liquid phase
ε_p — specific quantity of separating paraffin wax, kg/(kg K)
ε_w — cross-sectional fraction of the water phase

η — efficiency
- η_v — volumetric, of sucker-rod pumping
  - η_{va} — filling of pump barrel
  - η_{vb} — tubing string
- $\eta_{v'}$ — referred to polished-rod stroke volume
- η_w — production efficiency of well producing gasless oil

$\varkappa$ — ratio of specific heats
$\varkappa$ — heating factor
$\varkappa$ — melting heat of paraffin, J/kg

λ — pipe friction factor
- λ_{sm} — for flow in smooth pipe
- $\lambda_{\sigma w}$ — for flow of oil-in-water emulsion

λ — thermal conductivity factor, W/(m K)
- λ_b — of rock
- λ_{in} — of insulator
- λ_l — of liquid
- λ_s — of soil
  - λ_{sd} — of dry soil
  - λ_{sw} — of wet soil

μ	outflow factor
μ	dynamic viscosity, Ns/m^2
μ_a	apparent viscosity
μ_g	of gas
μ_a	at atmospheric pressure
μ_p	at pressure p
μ_i	of oil at pipe-wall temperature
μ_l	of liquid
μ_{ow}	of oil-in water emulsion
μ_w	of water
μ_d	Joule–Thomson coefficient, Km^2/N
μ_{dL}	in liquid
μ_{dV}	in gas
ν	kinematic viscosity, m^2/s
ν_o	of oil at 0 °C
ξ	flow resistance factor
ξ	fluid-column pressure gradient, m/m
ξ_a	accelerating gradient
ξ_m	mass gradient
ξ_s	friction gradient
ϱ	density, kg/m^3
ϱ_g	of gas
ϱ_{gn}	at standard conditions
ϱ_k	of mixture
ϱ_l	of liquid
ϱ_o	of oil
ϱ_{on}	at standard conditions
ϱ_{oT}	at temperature T
ϱ_s	of soil
ϱ_{sd}	of dry soil
ϱ_{sw}	of wet soil
ϱ_{st}	of steel
ϱ_w	of water
ϱ_r	specific gravity
ϱ_{gr}	of gas
ϱ_i^j	of oil at temperature j referred to water at temperature i
σ	normal stress, N/m^2
σ_t	tangential stress
$\sigma_{\max}$	maximum tensile stress
σ_F	yield strength (of solid)
σ_B	tensile strength
σ	surface tension, N/m
σ_l	of liquid
σ_o	of oil

σ_w	of water
σ_{ow}	interface tension, N/m
σ_b	geothermal gradient, K/m
τ	shear stress, N/m^2
τ_e	static, or yield stress
$\tau_{e'}$	apparent yield stress
τ_i	next to pipe wall
Γ	liquid distribution factor
Φ	porosity
Φ	heat flow, W
Φ^*	specific heat flow, W/m
ω	angular velocity, cycle frequency, l/s

Chapter 1

Selected topics in flow mechanics

1.1. Fundamentals of flow in pipes

Pressure drop due to friction of an incompressible liquid flowing in a horizontal pipe is given by

$$p_s = \lambda \frac{v^2 l \varrho}{2d_i}, \qquad 1.1-1$$

where $v = q/A$. If the Reynolds number

$$N_{\mathrm{Re}} = \frac{vd_i}{\nu} \qquad 1.1-2$$

is less than about 2000—2300, then flow is laminar, and its friction factor λ is, after Hagen and Poiseuille,

$$\lambda = \frac{64}{N_{\mathrm{Re}}}. \qquad 1.1-3$$

For turbulent flow in a smooth pipe, for $N_{\mathrm{Re}} < 10^5$, the Blasius formula gives a fair approximation:

$$\lambda = \frac{0.316}{\sqrt[4]{N_{\mathrm{Re}}}}. \qquad 1.1-4$$

Likewise for a smooth pipe and for $N_{\mathrm{Re}} > 10^5$, the explicit Nikuradse formula is satisfactory:

$$\lambda = 0.0032 + 0.221 N_{\mathrm{Re}}^{-0.237}. \qquad 1.1-5$$

The Prandtl–Kármán formula

$$\frac{1}{\sqrt{\lambda}} = 2 \lg \frac{N_{\mathrm{Re}}\sqrt{\lambda}}{2.51} \qquad 1.1-6$$

is valid over the entire turbulent region but its implicit form makes it difficult to manipulate. In rough pipes, for the transition zone between the curve defined by Eq. 1.1—6 and the so-called boundary curve (cf. Eq. 1.1—12). the Colebrook formula gives

$$\frac{1}{\sqrt{\lambda}} = -2 \lg \left[\frac{2.51}{N_{\mathrm{Re}}\sqrt{\lambda}} + \frac{k}{3.71\, d_i}\right]. \qquad 1.1-7$$

Also for rough pipes, but for the zone beyond the boundary curve, Prandtl and Kármán give the relationship

$$\frac{1}{\sqrt{\lambda}} = 2 \lg \frac{3.71\, d_i}{k} . \qquad 1.1-8$$

Although Eqs 1.1—7 and 1.1—8 provide results sufficiently accurate for any practical purpose, other formulae are often used to determine the pressure drop of turbulent flow in rough pipes, in order to avoid the cumbersome implicit equations. Explicit formulae can be derived from the following consideration.

If we have an idea of the relative roughness to be expected, then we can characterize the relationship λ v. N_{Re} by a formula which differs from Eq. 1.1—6 only in its constants. Consider e.g. the formula of this type of Drew and Genereaux (Gyulay 1942):

$$\frac{1}{\sqrt{\lambda}} = 1.6 \lg \frac{N_{Re}\sqrt{\lambda}}{0.762} . \qquad 1.1-9$$

Short sections of the graph of this function can be approximated fairly well by an exponential function

$$\lambda = aN_{Re}^{-b} , \qquad 1.1-10$$

where a and b are constants characteristic of the actual value of relative roughness and of the N_{Re} range involved. — The drawback of formulae of type 1.1—10 is that they do not provide a satisfactory accuracy beyond a N_{Re} range broader than just two orders of magnitude. A relationship that is somewhat more complicated but provides a fair approximation over a broader N_{Re} range is the Supino formula

$$\lambda = \lambda_{sm} + 0.17 N_{Re} \lambda_{sm}^2 \frac{k}{d_i} , \qquad 1.1-11$$

where λ_{sm} is the friction factor of smooth pipe, to be calculated using Eq. 1.1—4 or 1.1—5. — The graphs of Eqs 1.1—3 and from 1.1—6 to 1.1—8 are illustrated in the Moody diagram shown as Fig. 1.1—1. The dashed curve in the diagram is the boundary curve that separates the transition zone from the region of full turbulence. In the transition zone λ depends both on the relative roughness k/d_i and on N_{Re}, whereas in the region of full turbulence it is a function of k/d_i alone. The equation of the boundary curve is

$$N_{Re}\sqrt{\lambda}\,\frac{k}{d_i} = 200 . \qquad 1.1-12$$

Example 1.1—1. Let us find the friction pressure drop of oil flowing in a horizontal pipeline $l = 25$ km long, if $d_i = 0.300$ m, $q = 270$ m³/h. At the temperature and pressure prevailing in the fluid, $\nu = 2.5$ cSt and $\varrho = 850$ kg/m³. The pipeline is made of seamless steel pipe, for which $k/d_i =$

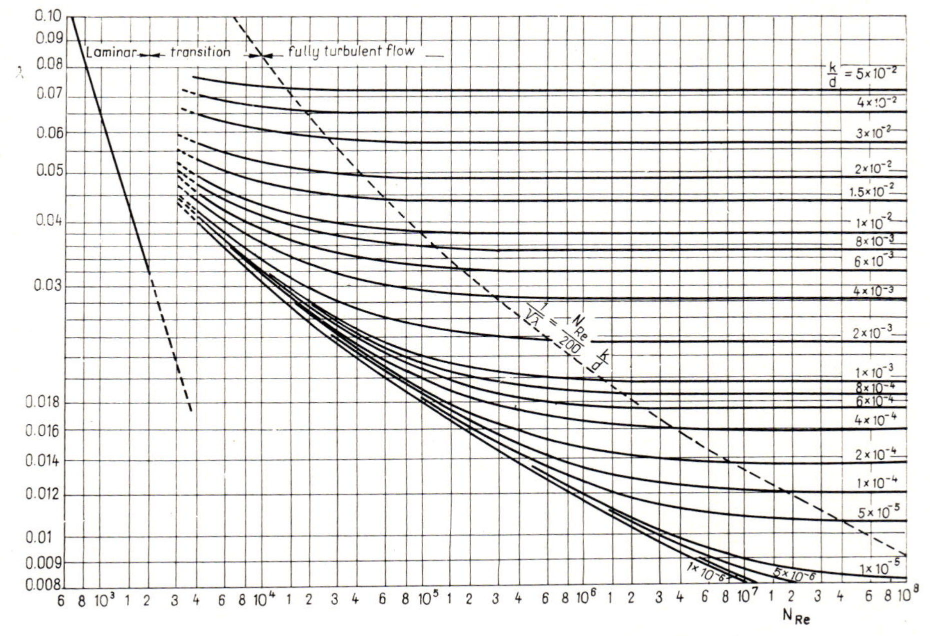

Fig. 1.1—1 Friction factor in pipes, according to Moody

$= 0.00017$. Converting the data of the problem to SI units, we have $l = 25{,}000$ m, $d_i = 0.3$ m, $q = 0.075$ m³/s, $\nu = 2.5 \times 10^{-6}$ m²/s, $\varrho = 850$ kg/m³, $k/d_i = 0.00017$. Flow velocity is

$$v = \frac{q}{\frac{d_i^2 \pi}{4}} = \frac{0.075}{0.7854 \times 0.3^2} = 1.06 \text{ m/s} ,$$

and

$$N_{\text{Re}} = \frac{1.06 \times 0.3}{2.5 \times 10^{-6}} = 1.27 \times 10^5 .$$

Flow is turbulent because 1.27×10^5 is greater than the critical Reynolds number, $N_{\text{Rec}} = 2300$. — The Moody diagram (Fig. 1.1—1) reveals that for $k/d_i = 0.00017$, flow is in the transition zone where Eq. 1.1—7 holds. It enables us to read off directly that, for the case in hand, $\lambda = 0.018$. If a more accurate value is required (which is, however, usually rendered superfluous by the difficulty of accurately determining relative roughness), the value of λ thus read off the diagram may be put into the right-hand side of Eq. 1.1—7 and the definitive value of λ can be found using that equation. The procedure is rather insensitive to the error of reading off the diagram. In the case in hand,

$$\frac{1}{\sqrt{\lambda}} = -2 \lg \left[\frac{2.51}{1.27 \times 10^5 \sqrt{0.018}} + \frac{0.00017}{3.71} \right] = 7.428$$

and hence,

$$\lambda = 0.0181 .$$

Let us calculate the friction factor also from Eq. 1.1—11, using a λ_{sm} furnished by Eq. 1.1—5:

$$\lambda_{sm} = 0.0032 + 0.221(1.27 \times 10^5)^{-0.237} = 0.0168 ;$$

$$\lambda = 0.0168 + 0.17 \times 1.27 \times 10^5 \times 0.0168^2 \times 0.00017 = 0.0178$$

Using in further computation the value $\lambda = 0.018$ we get by Eq. 1.1—1 for the flowing pressure drop

$$p_s = 0.018 \frac{1.06^2 \times 25{,}000 \times 850}{2 \times 0.300} = 0.72 \text{ MN/m}^2 = 7.2 \text{ bars} .$$

The pressure drop of flow in spaces of annular section can be determined as follows. In Eq. 1.1—1, substitute d_i by the equivalent pipe diameter, d_e. In a general way,

$$d_e = 4 \times \frac{\text{wetted cross-section}}{\text{wetted circumference}} .$$

For an annular space, then,

$$d_e = 4 \frac{\frac{d_1^2 \pi}{4} - \frac{d_2^2 \pi}{4}}{d_1 \pi + d_2 \pi} = d_1 - d_2 , \qquad 1.1\text{—}13$$

where d_1 is the *ID* of the outer pipe and d_2 is the *OD* of the inner pipe; Eq. 1.1—1 thus modifies to

$$p_s = \lambda \frac{v^2 l \varrho}{2(d_1 - d_2)} . \qquad 1.1\text{—}14$$

For laminar flow, the friction factor is given to a fair enough accuracy by

$$\lambda = \frac{64}{N_{\text{Re}}} \Phi \left(\frac{d_1}{d_2} \right) \qquad 1.1\text{—}15$$

(Knudsen and Katz 1958), where

$$\Phi \left(\frac{d_1}{d_2} \right) = \frac{(1 - d_1/d_2)^2}{1 + (d_1/d_2)^2 + \frac{1 - (d_1/d_2)^2}{\ln d_1/d_2}} .$$

For turbulent flow, no satisfactory result is to be expected save if the walls can be regarded as hydraulically smooth. In that case, according to Knudsen and Katz (1958):

$$\lambda = 0.304 N_{Re}^{-0.25} . \qquad 1.1-16$$

N_{Re} is to be computed using the hydraulic diameter $(d_1 - d_2)$. The limit between laminar and turbulent flow is at approximately $N_{Re} = 2{,}000$. Turbulent flow, however, will develop gradually, starting according to Prengle and Rothfus (Knudsen and Katz 1958) at the point of maximum velocity.The relationships derived by these authors imply for the $N_{Re'}$ belonging to maximum velocity the formula

$$N_{Re'} = \frac{d' v}{\nu} , \qquad 1.1-17$$

where

$$d' = \frac{2}{r_2} \left[r_2^2 - \frac{r_2^2 - r_1^2}{2 \ln \frac{r_2}{r_1}} \right] .$$

Even at $N_{Re'} = 700$ the actual friction factor will deviate from the value valid for laminar flow given by Eq. 1.1—15. Full turbulence sets in at $N_{Re'} = 2{,}200$.

Quite often the inner pipe is eccentrical within the outer pipe. According to Deyssler and Taylor, the friction factor decreases with increasing eccentricity (Knudsen 1958). Let us define eccentricity as the ratio of the distance between pipe centres to the difference between radii:

$$e = \frac{\vartheta}{r_1 - r_2} .$$

The decrease in friction factor may be appreciable. If for instance $r_2/r_1 = 3.5$ and $N_{Re} = 10^5$, then $\lambda = 0.019$ for $e = 0$, but $\lambda = 0.014$ for $e = 1$.

1.2. Gas flow in pipes

1.2.1. Fundamentals

The density and flow velocity of a gas flowing in a pipe will significantly vary along the pipeline as a result of temperature and pressure changes. The energy equation valid for steady flow will thus hold for infinitesimal lengths of pipe dl only when the pressure differential between the two ends of the infinitesimal section dl is dp. Then

$$\frac{dp}{\varrho} + v \, dv + g \, dh + \lambda \frac{v^2 \, dl}{2 d_i} = 0 \qquad 1.2-1$$

Let the pipe include an angle α with the horizontal. Then $dh = \sin \alpha dl$. The general gas law yields

$$\varrho = \frac{Mp}{RTz}$$

and

$$v = \frac{p_n}{T_n \pi/4} \frac{q_n Tz}{d_i^2 p} .$$

Most often, the energy spent in accelerating the gas flow is relatively small; it is therefore usual to assume that, in an approximation satisfactory for practical purposes, $vdv = 0$. Substituting the above expressions of λ and v into Eq. 1.2—1, and rearranging, we get

$$\frac{R}{M} \frac{Tz}{p} dp + \sin \alpha g dl + \left(\frac{p_n}{T_n \pi/4}\right)^2 \frac{q_n^2}{2d_i^5} \frac{\lambda T^2 z^2}{p^2} dl = 0 . \qquad 1.2\text{—}2$$

This equation has a variety of solutions. The flow is in most cases assumed to be isothermal, or to have a constant mean temperature, $T = \bar{T}$. The solutions of the equation will depend on the function used to describe the variation of z and λ v. p and T. In most formulae used to describe steady flow it is assumed in practice that, in addition to $T = \bar{T}$, also $z = \bar{z}$ and $\lambda = \bar{\lambda}$; i.e., the mean values in question are constant all along the pipeline. This assumption, together with the boundary conditions

$$p = p_1, \quad \text{if} \quad l = 0 \quad \text{and} \quad \sin \alpha = \frac{h}{l} = \text{const.}$$

leads to the following solution of Eq. 1.2—1:

$$p_1^2 = p_2^2 e^{\frac{2ghM}{R\bar{T}\bar{z}}} + \left(\frac{p_n}{T_n \pi/4}\right)^2 \frac{l}{2gh} \frac{\bar{\lambda} (q_n \bar{T} \bar{z})^2}{d_i^5} \left[e^{\frac{2ghM}{R\bar{T}\bar{z}}} - 1\right] .$$

R is 8315.1; let $g = 9.8067$; then

$$e^{\frac{2ghM}{R\bar{T}\bar{z}}} = e^{\frac{0.002359\, hM}{\bar{T}\bar{z}}} = e^m , \qquad 1.2\text{—}3$$

and hence,

$$p_1^2 = p_2^2 e^m + \left(\frac{p_n}{T_n \pi/4}\right)^2 \frac{l}{2gh} \frac{\bar{\lambda} (q_n \bar{T} \bar{z})^2}{d_i^5} (e^m - 1) . \qquad 1.2\text{—}4$$

$\bar{\lambda}$ is expressed in a variety of ways. One of the most widely used formulae was written up by Weymouth:

$$\lambda = \frac{0.009407}{\sqrt[3]{d_i}} . \qquad 1.2\text{—}5$$

It gives rather inaccurate results in most cases. Substituting this λ for $\bar{\lambda}$ in 1.2—4 we get

$$p_1^2 = p_2^2 e^m + 7.775 \times 10^{-4} \frac{l}{h} \left(\frac{p_n}{T_n}\right)^2 \left(\frac{q_n \bar{T} \bar{z}}{d_i^{8/3}}\right)^2 (e^m - 1) . \qquad 1.2\text{—}6$$

In a gas pipeline laid over terrain of gentle relief, the elevation difference h between the two ends of the pipeline can be neglected; Eq. 1.2—2 then yields for the horizontal pipeline, assuming, as in Eq. 1.2—4, $T = \bar{T}$, $z = \bar{z}$, $\lambda = \bar{\lambda}$ and $l = 0$ if $p = p_1$:

$$p_1^2 = p_2^2 + \frac{1}{\left(\frac{\pi}{4}\right)^2 R} \left(\frac{p_n}{T_n}\right)^2 \frac{l M q_n^2 \bar{T} \bar{z} \bar{\lambda}}{d_i^5} .$$

Substituting $R = 8315.1$ and the numerical value of $\pi/4$, we get

$$p_1^2 = p_2^2 + 1.950 \times 10^{-4} \left(\frac{p_n}{T_n}\right)^2 \frac{l M q_n^2 \bar{T} \bar{z} \bar{\lambda}}{d_i^5} . \qquad 1.2\text{—}7$$

Introducing the value of λ given by Weymouth's Eq. 1.2—5 we arrive at the widely used formula

$$p_1^2 = p_2^2 + 1.834 \times 10^{-6} \left(\frac{p_n}{T_n}\right)^2 \frac{l M q_n^2 \bar{T} \bar{z}}{d_i^{16/3}} . \qquad 1.2\text{—}8$$

Solving for gas flow rate, we get

$$q_n = 738.4 \frac{T_n}{p_n} d^{8/3} \left[\frac{p_1^2 - p_2^2}{l M \bar{T} \bar{z}}\right]^{0.5} . \qquad 1.2\text{—}9$$

Example 1.2—1. Using Eq. 1.2—9 let us find the gas flow rate in a horizontal pipeline if $T_n = 288.2$ K, $p_n = 1.013$ bars, $d_i = 0.1$ m, $p_1 = 44.1$ bars, $p_2 = 2.9$ bars, $\bar{T} = 275$ K, $M = 18.82$ kg/kmole, $l = 15$ kms. In order to find $\bar{z}$, let us first calculate by Eq. 1.2—26 an approximate mean pressure $\bar{p}$ in the pipeline:

$$\bar{p} = \frac{2}{3}\left[44.1 \times 10^5 + \frac{(2.9 \times 10^5)^2}{44.1 \times 10^5 + 2.9 \times 10^5}\right] = 29.5 \times 10^5 \text{ N/m}^2 .$$

According to Diagram 8.1—1 $p_c = 46.7$ bars, $T_c = 207$ K, and the reduced parameters $p_r = 0.63$ and $T_r = 1.33$ (cf. Eqs 8.1—3 and 8.1—4). Figure 8.1—2 yields $\bar{z} = 0.90$. The gas flow rate sought,

$$q_n = 738.4 \frac{288.2}{1.013 \times 10^5} 0.1^{8/3} \left[\frac{(44.1 \times 10^5)^2 - (2.9 \times 10^5)^2}{1.5 \times 10^4 \times 18.82 \times 275 \times 0.90}\right]^{0.5} = 2.383 \text{ m}^3\text{/s} .$$

Example 1.2—2. Using Eq. 1.2—6, find the input pressure in the pipeline of the foregoing Example provided the output end of the pipeline is situated higher by $h = 150$ m than its input end.

In Eq. 1.2—3,

$$m = 0.002359 \frac{150 \times 18.82}{275 \times 0.9} = 0.02691 ,$$

and hence

$$p_1^2 = (2.9 \times 10^5)^2 e^{0.02691} + 7.775 \times 10^{-4} \frac{1.5 \times 10^4}{1.5 \times 10^2} \left(\frac{1.013 \times 10^5}{288.2}\right)^2 \times$$

$$\times \left(\frac{2.383 \times 275 \times 0.9}{0.1^{8/3}}\right)^2 (e^{0.02691} - 1) = 1.974 \times 10^{13} \ \mathrm{N^2/m^4} .$$

Consequently,

$$p_1 = 4.44 \ \mathrm{MN/m^2} = 44.4 \ \mathrm{bars} .$$

In the foregoing Example we have had $p_1 = 44.1$ bars. An input pressure higher by 0.3 bar is thus required to overcome the elevation difference of 150 m if the gas flow rate of 2.383 m³/s is to be maintained.

Equation 1.2—7 becomes a more accurate tool of computation if λ is taken from Eq. 1.1—10 rather than from the Weymouth formula. The Reynolds number figuring in Eq. 1.1—10 is

$$N_{\mathrm{Re}} = \frac{d_i v \bar{\varrho}}{\bar{\mu}} ,$$

where

$$v = \frac{q}{\frac{d_i^2 \pi}{4}}$$

and the general gas law yields

$$\bar{\varrho} = \frac{M \bar{p}}{\bar{T} \bar{z} R} \quad \text{and} \quad q = \frac{p_n q_n \bar{z} \bar{T}}{\bar{p} z_n T_n} .$$

Substituting the expressions for v, $\bar{\varrho}$ and q into the fundamental equation and assuming that $z_n = 1$ in a fair approximation, we get

$$N_{\mathrm{Re}} = \frac{1}{\frac{\pi}{4} R} \frac{p_n q_n M}{d_i T_n \bar{\mu}} . \qquad 1.2\text{—}10$$

Substituting this into Eq. 1.1—10 and replacing the result into Eq. 1.2—7 we obtain the following general relationship for the calculation of q_n:

$$q_n = \left[\frac{R^{(1-b)}}{a}\right]^{\frac{1}{2-b}} \frac{T_n}{p_n} \frac{\pi}{4} \left[\frac{d_i^{(5-b)} (p_1^2 - p_2^2)}{l M^{(1-b)} \bar{T} \bar{z} \bar{\mu}^b}\right]^{\frac{1}{2-b}} . \qquad 1.2\text{—}11$$

The various formulae used in practice to express λ are all of the form 1.1—10. For a given roughness, the numerical values of the constants a

and b depend on the pipe diameter. A given set of constants will yield friction factors of acceptable accuracy for a given N_{Re} range only. For instance,

$$\lambda = 0.121 \left(\frac{1}{N_{Re}}\right)^{0.15}, \qquad 1.2-12$$

where, obviously, $a = 0.121$ and $b = 0.15$. Substitution into 1.2—11 yields

$$q_n = 156.3 \frac{T_n}{p_n} \left[\frac{d_i^{4.85}(p_1^2 - p_2^2)}{lM^{0.85}\,\bar{T}\bar{z}\,\bar{\mu}^{0.15}}\right]^{0.541}. \qquad 1.2-13$$

Example 1.2—3. Find the gas flow rate in a horizontal pipeline using Eq. 1.2—13 and the data of Example 1.2—1. Using the known values $p_r = 0.62$ and $T_r = 1.27$, we read off Diagrams 8.1—6 and 8.1—7:

$$\bar{\mu} = 10\ \mu\text{Ns/m}^2 .$$

Hence,

$$q_n = 156.3 \frac{288.2}{1.013 \times 10^5} \times$$

$$\times \left[\frac{0.1^{4.85}[(44.1 \times 10^5)^2 - (2.9 \times 10^5)^2]}{1.5 \times 10^4 \times 18.82^{0.85} \times 275 \times 0.9 \times (1 \times 10^{-5})^{0.15}}\right]^{0.541} = 3.00\ \text{m}^3/\text{s} .$$

The values furnished by the two formulae are seen to differ rather widely:

$$\varepsilon = \frac{3.00 - 2.38}{3.00} 100 = 20.7 \text{ per cent} .$$

A careful consideration of the suitability of any formula selected for use is essential. A useful basis for such considerations is a series of tests carried out at the Institute of Gas Technology (Uhl 1967, *1*). These tests have revealed a considerable difference between pipe in the laboratory and in the field. Its main cause is the considerable flow resistance due to pipe fittings, bends and breaks and weld seams in actual pipelines, which tend to bring about a modification of the Moody diagram.

The region of turbulent flow can be characterized by two types of equation. The first of these is the modified 'smooth-pipe' Equation 1.1—6, valid for relatively low Reynolds numbers:

$$\frac{1}{\sqrt{\lambda}} = 2\xi \lg \frac{N_{Re}\sqrt{\lambda_{sm}}}{2.8}, \qquad 1.2-14$$

where ξ is a resistance factor accounting for the fittings, bends, breaks and weld seams per unit length of pipe, and λ_{sm} is the friction factor for smooth pipe, which can be calculated for any given value of N_{Re}. At high Reynolds numbers, the relative roughness k/d has a decisive influence on the friction factor. The latter can be calculated to a satisfactory degree of accuracy using Eq. 1.1—8. The two equations respectively characterize the transition and fully turbulent regions. The transition between them is appreciably shorter, more abrupt than in the case of the curves illustrating the Colebrook Formula 1.1—7. The question as to which of the two equations (1.2—14 or 1.1—8) is to be used in any given case can be decided by finding the value of N_{Re} that satisfies both equations simultaneously:

$$N_{\mathrm{Re}} = 5.65 \left(\frac{3.7d}{k}\right)^{\frac{1}{\xi}} \lg \left(\frac{3.7d_i}{k}\right) . \qquad 1.2—15$$

Determining the value of ξ requires in-plant or field tests. Approximate values are given in a diagram by Uhl (1967, *3*).

We have so far assumed the mean values $\bar{T}$, $\bar{z}$ and $\bar{\lambda}$ to be constant all along the flow string. There are however, formulae that account also for changes of T, z and λ along the string (Aziz 1962). Among them, the calculation method of Cullender and Smith permits us to determine accurately the pressure drop of flow in a vertical string. The temperature is estimated from operational data. The calculation is based likewise on Eq. 1.2—2, which can be written to read

$$\frac{M}{R}\,\mathrm{d}l = \frac{-\frac{p}{Tz}\,\mathrm{d}p}{g\left(\frac{p}{Tz}\right)^2 + \left(\frac{p_n}{T_n \pi/4}\right)^2 \frac{\lambda q_n^2}{2d_i^5}} . \qquad 1.2—16$$

Integrating between the limits $l = 0$, $p = p_1$ and $l = L$, $p = p_2$ characterizing the vertical string (e.g. the tubing in a gas well) we get, formally,

$$\frac{M}{R} L = \int_{p_2}^{p_1} I \mathrm{d}p , \qquad 1.2—17$$

where

$$I = -\frac{\frac{p}{Tz}}{g\left(\frac{p}{Tz}\right)^2 + \left(\frac{p_n}{T_n \pi/4}\right)^2 \frac{\lambda q_n^2}{2d_i^5}} . \qquad 1.2—18$$

The integral can be evaluated by a successive approximation. In a general way,

$$\int_{p_1}^{p_n} I \mathrm{d}p = \frac{1}{2}[(p_2 - p_1)(I_1 + I_2) + (p_3 - p_2)(I_2 + I_3) + \ldots + (p_n - p_{n-1})(I_{n-1} + I_n)] . \qquad 1.2—19$$

To solve any practical problem it is usually sufficient to assume only one intermediate pressure p_k; then

$$\int_{p_2}^{p_1} I \mathrm{d}p = \frac{1}{2}[(p_k - p_2)(I_2 + I_k) + (p_1 - p_k)(I_k + I_1)] . \qquad 1.2—20$$

the pipe will be less than the static shear stress. The fluid filling out the pipe will not start flowing under a pressure gradient giving rise to τ_i. If $\tau_i \geqq \tau_e$, then the oil will flow 'as a liquid' only in annular space with outside radius r_i, and inside radius r_e, the latter being that radius for which the shear stress $\tau = \tau_e$. Within this radius, the plastic fluid flows as a solid plug, at a velocity equal to the liquid velocity v_e prevailing at the radius r_e (Longwell 1966):

$$v_e = \frac{\tau_i r_i}{2\mu''}\left[1 - \frac{2\tau_e}{\tau_i} + \left(\frac{\tau_e}{\tau_i}\right)^2\right]. \qquad 1.3-6$$

In a general way, the mean or bulk velocity is the mean height of the solid of rotation of radius r and 'height' v, that is,

$$\bar{v} = \frac{2}{r_i^2}\int_0^{r_i} vr\mathrm{d}r\ . \qquad 1.3-7$$

In a plastic fluid, the mean velocity can be characterized by the Buckingham equation (Reher 1967), which can be derived from Eq. 1.3—7. Written up after a slight formal modification, this equation reads

$$\bar{v} = \frac{\tau_i r_i}{4\mu''}\left[1 - \frac{4}{3}\frac{\tau_e}{\tau_i} + \frac{1}{3}\left(\frac{\tau_e}{\tau_i}\right)^4\right]. \qquad 1.3-8$$

Figure 1.3—5 illustrates the variation of the relative velocity $v/\bar{v}$ v. pipe radius (Longwell 1966). It is clear that for a given oil (τ_e = const.) the diameter of the solid plug moving along in the flow string will be the smaller, the greater the shear stress at the pipe wall, that is, the greater the pressure gradient that keeps the fluid flowing. On the other hand, at a given pressure gradient and τ_i engendered by it, the velocity distribution will approximate that of a Newtonian fluid the better, the 'less plastic' the fluid in flow, that is, the less the τ_e value characterizing it.

As regards *pseudoplastic fluids*, similar relationships exist in characterizing flow velocity v. pipe radius. These permit us to establish velocity distributions in much the same way as above. Because of the essential similarity between plastic and pseudoplastic flow, the parameters of flow in a pipe and hence also the velocity distributions are rather similar. Also in this case, an annular space with a rather steeply varying velocity distribution may develop next to the pipe wall; inside this annular region, we may likewise

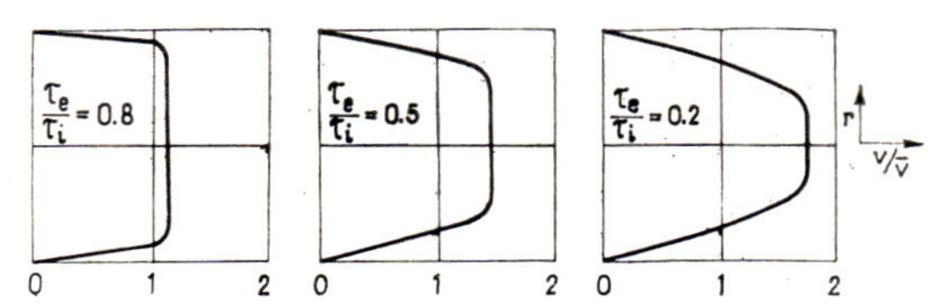

Fig. 1.3—5 Flow-velocity profiles of plastic fluids; after Longwell 1966, p. 377 (used with permission of McGraw-Hill Book Company)

find a central plug that flows at a velocity which, if not exactly the same, is only very slightly different. The calculated velocity distribution will depend to some extent on the mathematical model used to characterize pseudoplastic flow. Figure 1.3—6 shows flow velocity distributions for four fluids (Longwell 1966); sections (b), (c) and (d) refer to pseudoplastic fluids. The graphs for these pseudoplastic fluids are determined in two ways: one, using the power law 1.3—2 (dashed curves), and two, using the Ellis formula (given e.g. in Longwell 1966; full curves). In the Figures, the parameters $\tau_i/\tau_{1/2}$ refer to the Ellis formula, whereas the n's are the power-law exponents 1.3—2; section (a) shows the velocity distribution as a limiting case of a Newtonian fluid. The graphs of the Ellis formula and the power model coincide in sections (a) and (d), whereas they differ in sections (b) and (c). The Figure reveals, on the one hand, the way the velocity distribution is affected by various degrees of pseudoplasticity and, on the other, the non-negligible influence on the result of the mathematical model chosen. Notice how the velocity distribution approaches that of plastic flow as n increases.

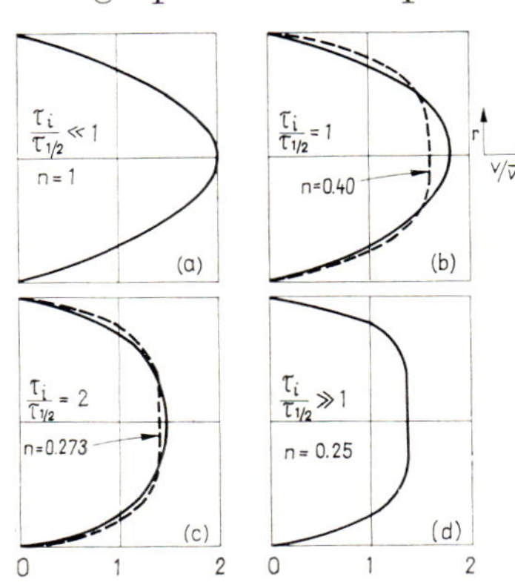

Fig. 1.3—6 Flow-velocity profiles of pseudoplastic fluids; after Longwell 1966, p. 383 (used with permission of McGraw-Hill Book Company)

1.3.3. The generalized Reynolds number

Let the shear stress in a fluid flowing in a pipe exceed the true or apparent static shear stress even in the centre line of the pipe; in this case, the flow will have no 'solid core', and the bulk velocity of flow will be correctly characterized by the general Eq. 1.3—7; from this equation the Wilkinson equation can be derived (Reher and Mylius 1967):

$$\frac{8\bar{v}}{d_i} = \frac{4}{\tau_i^2}\int_0^{\tau_i} \tau^2\left(-\frac{dv}{dr}\right) d\tau \,. \qquad 1.3\text{—}9$$

It is verifiable that the expression on the left-hand side is, in Newtonian fluids, equal to the deformation rate at the pipe wall, that is,

$$\frac{8\bar{v}}{d_i} = \left(-\frac{dv}{dr}\right)_i \,. \qquad 1.3\text{—}10$$

For pseudoplastic fluids, a formula describing the relation between the terms $8\bar{v}/d_i$ and $(-dv/dr)_i$ had been derived by Rabinowitsch and Mooney. This formula was written by Metzner and Reed (1955) in the form

$$\left(-\frac{dv}{dr}\right)_i = \frac{3n+1}{4n} \frac{8\bar{v}}{d_i}. \qquad 1.3-11$$

Substituting the expression for $(-dv/dr)$ into Eq. 1.3—2 we get

$$\tau_i = k\left(\frac{8\bar{v}}{d_i}\right)^n, \qquad 1.3-12$$

where

$$k = \mu'\left(\frac{3n+1}{4n}\right)^n. \qquad 1.3-13$$

For laminar flow in a horizontal pipeline Eqs 1.1—1 to 1.1—3 hold, provided $N_{Re} = N_{Re\,pp}$ and putting $v = \bar{v}$. The general relationships

$$\tau = \frac{d_i \Delta_p}{4l} \qquad 1.3-14$$

and

$$\nu = \frac{\mu}{\varrho} \qquad 1.3-15$$

are also valid. Using these equations, and the Eqs 1.3—2 and 1.3—11 of pseudoplastic flow we may write up the generalized Reynolds number, derived by Metzner and Reed, as

$$N_{Repp} = \frac{d_i^n \bar{v}^{(2-n)}}{\mu'} \frac{8}{(6+2/n)^n}. \qquad 1.3-16$$

By the above considerations, Eq. 1.3—15 is valid for pseudoplastic fluids obeying a power law, in which case μ' and n are constant and are given numerically in the equation of the flow curve. Replacing μ' in Eq. 1.3—15 by the expression in Eq. 1.3—13, we get

$$N_{Re\,pp} = \frac{d_i^n \bar{v}^{(2-n)} \varrho}{k\, 8^{(n-1)}}. \qquad 1.3-17$$

This formula is used if the rheological properties of the fluid have been determined by means of a capillary viscosimeter, or by field tests on a pipeline, or are known on the basis of a $\tau_i = f\,(8\bar{v}/d_i)$ curve. According to Eq. 1.3—12

$$\tau_i = k\left(\frac{8\bar{v}}{d_i}\right)^n;$$

if the fluid obeys the power law, then k and n are constants and their numerical values are known. The formula can, however, be used also if the fluid deviates from the power law. In this case it is sufficient to assume that Eq. 1.3—12 is the equation of the tangent to the $\tau_i = f(8\bar{v}/d_i)$ curve plotted in an orthogonal bilogarithmic system of co-ordinates; n is the slope of the tangent and k is the ordinate belonging to the value $(8\bar{v}/d_i) = 1$.

The tangent should touch the curve at the point whose abscissa $(8\bar{v}/d_i)$ corresponds to the actual values of q and d_i. If the flow behaviour is characterized by a $\tau_i = f(8\bar{v}/d_i)$ curve, then $N_{\mathrm{Re}pp}$ can be derived even more simply by the following consideration (LeBaron Bowen 1961).

On the basis of Eqs from 1.1—1 to 1.1—3 and the relationship $\nu = \mu/\varrho$,

$$\mu = \frac{\frac{d_i \Delta p}{4}}{\frac{8\bar{v}}{d_i}} = \frac{\tau_i}{\frac{8\bar{v}}{d_i}}.$$

Substituting this expression into Eq. 1.1—2, we get

$$N_{\mathrm{Re}} = \frac{d_i \bar{v} \varrho}{\tau_i / \frac{8\bar{v}}{d_i}}. \qquad 1.3\text{—}18$$

This relationship is of a general validity for all non-Newtonian fluids including pseudoplastic fluids deviating from the power law (where, obviously, $N_{\mathrm{Re}} = N_{\mathrm{Re}pp}$). To find the Reynolds number by this equation, read the τ_i belonging to the $(8\bar{v}/d_i)$ value corresponding to the intended $\bar{v}$ and d_i off an experimentally established $\tau_i = f(8\bar{v}/d_i)$ curve and substitute the appropriate data into Eq. 1.3—18.

1.3.4. Transition from laminar to turbulent flow

The transition from laminar to turbulent flow in non-Newtonian fluids depends, in addition to the Reynolds number, also on a number of other factors affected by the rheological properties of the fluid. No general equation has been derived so far, but individual research workers have published valuable partial results. Ryan and Johnson have introduced a stability parameter which has permitted them to write up the critical Reynolds number for pseudoplastic fluids obeying the power law as follows:

$$N_{\mathrm{Re}\,ppc} = \frac{d_i^n \bar{v}_c^{(2-n)}}{\mu'} \varrho \frac{8}{(6 + 2/n)^n} = \frac{6464}{\varphi(n)}, \qquad 1.3\text{—}19$$

where

$$\varphi(n) = \frac{(3n+1)^2}{n} \left[\frac{1}{n+2}\right]^{\frac{n+2}{n+1}}. \qquad 1.3\text{—}20$$

Assuming that $N_{\mathrm{Re}\,c} = 2100$ in Newtonian fluids, the critical Reynolds number of pseudoplastic fluids varies in the range of from 2100 to 2400, depending on the exponent n of the power law (Govier and Aziz 1972).

Dodge and Metzner (1959) have found $N_{\mathrm{Re}\,ppc}$ to fall more or less into the domain of transition of Newtonian fluids and to increase slightly as n decreases. According to these authors, e.g., $N_{\mathrm{Re}\,ppc} = 3100$ at $n = 0.38$.

According to Mirzadzhanzade et al. (1969), the development of turbulency depends to a significant extent on the particle size and concentration of the dispersed phase, as well as on the specific weight difference between the dispersing medium and the dispersed phase.

As the mathematical criteria established till now are far from unequivocal, it is expedient in doubtful cases to determine experimentally the type of flow prevailing under the intended flow conditions. The graphs in Fig. 1.3—7 have been determined experimentally (LeBaron Bowen 1961)

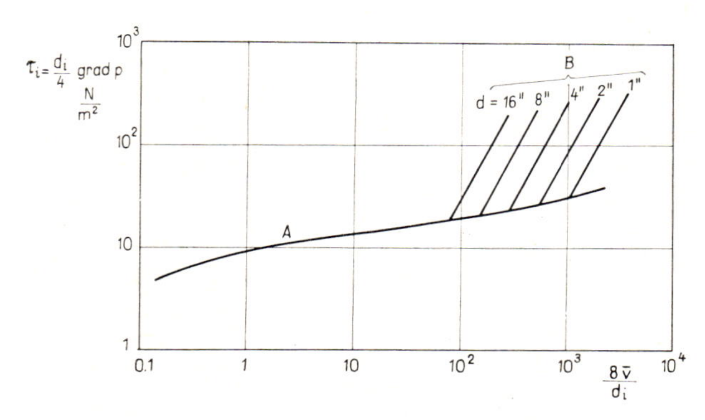

Fig. 1.3—7 Laminar and turbulent flows in pipelines; after LeBaron Bowen (1961)

Graph A characterizes laminar flow independently of pipe diameter. The set of Graphs B includes characteristic curves for turbulent flow in pipes of various diameters. The less the diameter, the greater the abscissa ($8\bar{v}/d_i$) at which flow becomes turbulent.

1.3.5. Calculation of friction loss

(a) Laminar flow of pseudoplastic fluids

By Eq. 1.3—12, the shear stress developing at the pipe wall in a fluid flowing in a pipe is at given values of k and n a function of $8\bar{v}/d_i$ only. This consideration permits, in the possession of experimental data obtained by means of a capillary extrusion viscosimeter of capillary diameter d_i, or in a flow test on a pipe, the direct calculation of the friction losses for any other pipe diameter.

Example 1.3—1. The variation of $\tau_i = \frac{d_i \text{ grad } p}{4}$ v. $8\bar{v}/d_i$ for a given oil is plotted in Fig. 1.3—8 on the basis of pipeline experiments. Find the

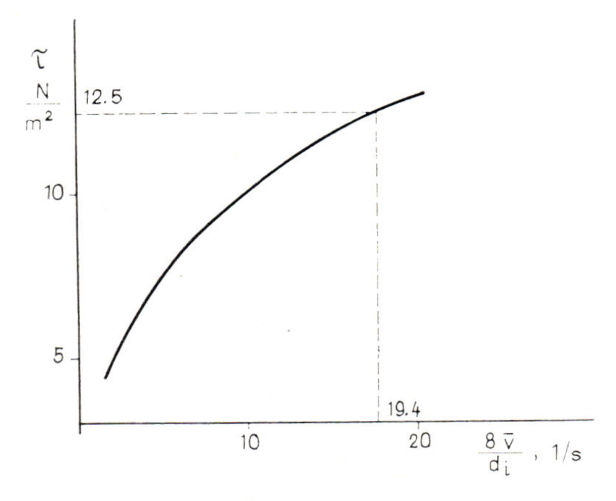

Fig. 1.3—8

pressure gradient in a pipe of $d_i = 0.308$ m, when $q = 200$ m³/h at the given flow parameters.

$$\bar{v} = \frac{q}{3600 \frac{d_i^2 \pi}{4}} = \frac{200}{3600 \frac{0.308^2 \pi}{4}} = 0.746 \text{ m/s} ;$$

$$\frac{8\bar{v}}{d_i} = \frac{8 \times 0.746}{0.308} = 19.4 \text{ 1/s} .$$

For this value of the abscissa, Fig. 1.3—8 gives

$$\tau_i = \frac{d_i \text{ grad } p}{4} = 12.5 \text{ N/m}^2 ,$$

and hence,

$$\text{grad } p = \frac{4\tau_i}{d_i} = \frac{4 \times 12.5}{0.308} = 162 \text{ N/m}^3 = 1.62 \text{ bar/km} .$$

If the fluid is a time-dependent, thixotropic, pseudoplastic one, it is often sufficient for the designer to know the parameters valid for steady-state flow. Then $\tau_i = f(8\bar{v}/d_i)$, and the flow curves permit us to find the pressure drop of steady-state flow in the pipe. The practical use of this idea is hindered by the fact that the flow curves of thixotropic pseudoplastic fluids are difficult and expensive to establish by field tests, and impossible to determine by the extrusion viscosimeter. Flow behaviour is thus characterized by flow curves, or sets of them, established by means of the rotation viscosimeter. Figure 1.3—9 shows flow curves of Algyő oil (Hungary) for steady-state flow at various temperatures (Szilas 1970). Pressure drop is then calculated in a way other than the above-outlined one. If *the flow curve obeys the power law*, then the constants μ' and n of Eq. 1.3—2 are known.

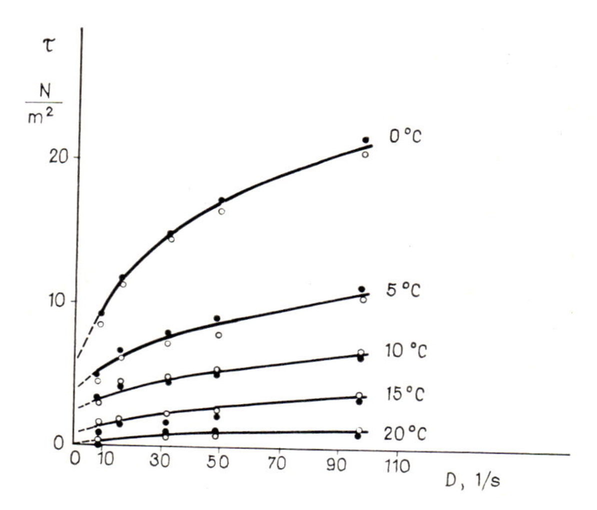

Fig. 1.3—9 Steady-state flow curves of thixotropic-pseudoplastic crude from Algyő

We compute $N_{\mathrm{Re}pp}$ for the intended oil flow rate q, flow velocity $\bar{v}$ and pipe diameter d_i using Eq. 1.3—16. On finding that flow is laminar, we compute λ by Eq. 1.1—3 and then the friction loss by Eq. 1.1—1.

Example 1.3—2. Find the flowing pressure gradient in a pipeline of $d_i = 0.308$ m if at the given parameters of flow $q = 200$ m³/h and $\varrho = 880$ kg/m³. The equation of the flow curve is

$$\tau = 4.08 D^{0.338},$$

that is, $\mu' = 4.08$ Ns/m² and $n = 0.338$; $\bar{v}$ is 0.746 m/s in agreement with the foregoing example. Substituting the figures obtained into Eq. 1.3—16, we get

$$N_{\mathrm{Re}pp} = \frac{0.308^{0.338} \times 0.746^{(2-0.338)} \times 880}{4.08} \; \frac{8}{(6 + 2/0.338)^{0.338}} = 308 \; .$$

Flow is laminar, it is therefore justified to use Eq. 1.1—3, which yields

$$\lambda = \frac{64}{308} = 0.208 \; .$$

Now by Eq. 1.1—1,

$$\mathrm{grad}\, p = \frac{\Delta p}{l} = \lambda \frac{\bar{v}^2 \varrho}{2 d_i} = 0.208 \frac{0.746^2 \times 880}{2 \times 0.308} = 165 \text{ N/m}^3 = 1.65 \text{ bar/km} \; .$$

If *the flow curve does not obey the power law*, then in Eq. 1.3—16 μ' means the ordinate intercept, at $D = 1$, of the tangent to the flow curve plotted in an orthogonal bilogarithmic system of coordinates, and n means the slope of said tangent. To find the deformation rate for which μ' and n hold at the intended velocity $\bar{v}$ and pipe diameter d_i, we may use Eq. 1.3—11 by Reed and Metzner (Govier and Ritter 1963). Calculation may proceed as follows: assuming several values of D, we determine the value of n belonging to each, using, on the one hand, Eq. 1.3—11 and, on the other, the tangents to the flow curves plotted in orthogonal bilogarithmic co-ordinates. The two functions are then plotted in a diagram. The desired value of n is given by their point of intersection. μ' is furnished by the ordinate intercept at $D = 1$ of the tangent to the flow curve valid for the deformation rate belonging to this particular value of n.

(b) Turbulent flow of pseudoplastic fluids

Theories and formulae describing pressure drop in turbulent flow are not yet sufficiently general or accurate to serve as a basis for designing installations. The relationships to be cited below are those considered most suitable by relevant literature. Dodge and Metzner were the first to develop a semi-empirical formula based on theoretical considerations to determine the pressure drop of a non-Newtonian fluid in turbulent flow. They carried out their experiments on aqueous solutions of a plastic (Dodge and Metzner 1959) which they regarded as pseudoplastic. The friction factor can, according to them, be calculated by the equation

$$\frac{1}{\sqrt{\lambda}} = \frac{2.0}{n^{0.75}} \lg \left[N_{\mathrm{Re}pp} \left(\frac{\lambda}{4} \right)^{\left(1 - \frac{n}{2}\right)} \right] - \frac{0.2}{n^{1.2}} . \qquad 1.3\text{—}21$$

For $n = 1$, this reduces to Eq. 1.1—6, valid for the flow of a Newtonian fluid in a smooth pipe.

Shaver and Merrill (1959) likewise used aqueous plastic dispersions in their experiments. They established the following relationship for the friction factor:

$$\lambda = \frac{0.316}{n^5 N_{\mathrm{Re}pp}^{\left(\frac{2.63}{10.5^n}\right)}} . \qquad 1.3\text{—}22$$

This equation cannot be used for fluids whose n is less than 0.4 . At $n = 1$ the equation reduces to Eq. 1.1—4. — Let us note that the above formulae have been determined by flow experiments in smooth tubes. The influence of pipe-wall roughness on friction losses is unknown.

(c) Thixotropic pseudoplastic fluids

Figure 1.3—2 shows that the flow curve belonging to any shear duration of a thixotropic-pseudoplastic oil looks like the flow curve of a time-independent pseudoplastic oil. The flow curve belonging to infinite shear duration is consequently suitable for determining the flow parameters of steady-state flow, and hence also the friction loss, in the way outlined in the previous paragraphs. In practice it is often found that after a shear duration on the order of 10 minutes the flow curve approximates quite closely the values to be expected at infinite duration of shear. In designing relatively long pipelines for pressure drop, accuracy is little influenced by the fact that the pressure gradient is slightly higher in a short section near the input end than the steady-state value in the rest of the pipeline. When, however, relatively short pipelines are to be designed for pressure drop, the error due to use of the steady-state pressure gradient may be quite considerable. A procedure for computing pressure drops under transient structural and flow conditions has been developed by Ritter and Batycky (1967).

(d) Plastic fluids

By the considerations in Section 1.3—5a, the pressure drop of a plastic fluid in laminar flow can also be determined in the way outlined in Example 1.3—1, provided the graph of the function $\tau_i = f(8\bar{v}/d_i)$ has been determined experimentally, by means of an extrusion viscosimeter or by field tests. In the possession of a flow curve characterizing the behaviour of the fluid, the pressure drop can be calculated by the following consideration. It has been shown by Hedström that the friction coefficient of plastic fluids is a function of two dimensionless numbers (Hedström 1952). One is a Reynolds number which involves the plastic viscosity μ'' in the place of the simple viscosity:

$$N_{\mathrm{Rep}} = \frac{d_i \bar{v} \varrho}{\mu''} . \qquad 1.3-23$$

The other dimensionless number is named the Hedström number

$$N_{\mathrm{He}} = \frac{d_i^2 \varrho \tau_e}{\mu''^2} . \qquad 1.3-24$$

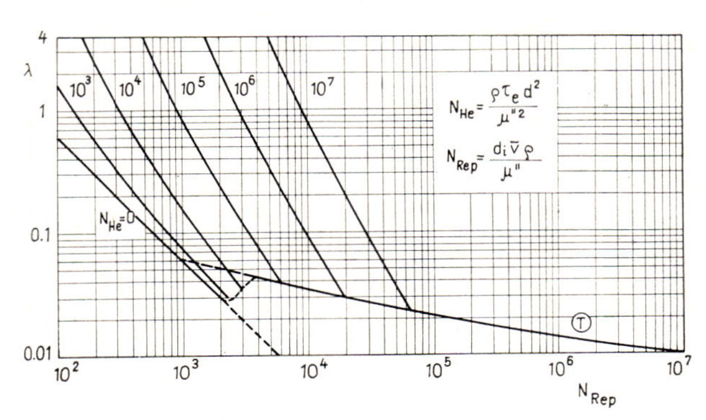

Fig. 1.3—10 Friction factor of plastic fluids, according to Hedström (1952)

This number accounts first and foremost for the fact that the 'solid core' of the flow reduces the cross-section free for 'liquid' flow (Le Fur 1966). — For flow in a pipeline, the friction factor can be read off Fig. 1.3—10 if N_{Rep} and N_{He} are known (API 1960). The curve marked T holds for turbulent flow; the rest hold for laminar flow. Let us add that the curve for turbulent flow refers to a smooth pipe and therefore gives approximate results only. Very little is known as yet on the pressure drop of flow in a rough pipe.

Example 1.3—3. Find the friction pressure gradient in the fluid characterized by the flow curve in Fig. 1.3—11, if $q = 200$ m³/h, $d_i = 0.308$ m and

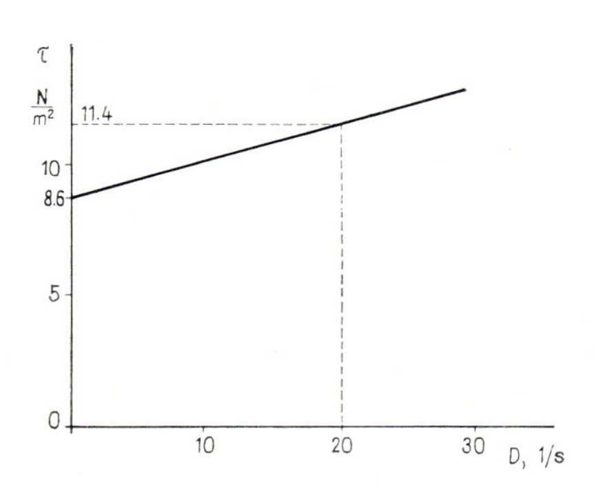

Fig. 1.3—11

$\varrho = 880$ kg/m³. — The flow curve yields $\tau_e = 8.6$ N/m² and, for instance at $(-dv/dr) = 20$ 1/s, $\tau = 11.4$ N/m². Hence

$$\mu' = \frac{11.4 - 8.6}{20} = 0.140 \text{ Ns/m}^2 ;$$

Now by Eq. 1.3—23,

$$N_{\mathrm{Rep}} = \frac{0.308 \times 0.746 \times 880}{0.140} = 1.44 \times 10^3 ;$$

and by Eq. 1.3—24,

$$N_{\mathrm{He}} = \frac{0.308^2 \times 880 \times 8.6}{0.140^2} = 3.66 \times 10^4 ;$$

Figure 1.3—10 yields $\lambda = 0.185$. Then, by Eq. 1.1—1,

$$\text{grad } p = \frac{\Delta p}{l} = 0.185 \frac{0.746^2 \times 880}{2 \times 0.308} = 147 \text{ N/m}^3 = 1.47 \text{ bar/km} .$$

1.4. Multiphase flow of liquids and gases

1.4.1. Flow in horizontal pipelines

Systematic research into the relationships governing the multiphase flow of liquids and gases in horizontal pipelines had started in 1939. The results were published by Lockhart and Martinelli in 1949 (Baker 1954). Kriegel (1967) derived and generalized the empirical formulae on the basis of theoretical considerations. These procedures were further improved by several authors including Baker, Hoogendorn, Chisholm, Laird, and Schlichting (Schlichting 1970), to mention only the better-known ones. Several other workers have followed other ways to derive formulae useful for practice. Let us cite here the theories of Baxendell (1955), Bertuzzi-Tek-Poettmann (1956) and Dukler, Wicks and Cleveland (1964B). The main target of research has been the two-phase flow of waterless oil and natural gas in horizontal pipelines. The influence of a hilly terrain upon pressure drop has also been investigated, however.

The effective friction loss of the two-phase flow of oil and gas in a horizontal pipeline is higher than the sum of the friction losses calculated separately for the two phases. This is due to the following circumstances. The friction loss of a liquid flowing in a pipe varies inversely as a power higher than unity of the flow cross-section. The gas phase occupies part of the cross-section and thus reduces the section available to the liquid phase. The friction loss of the gas phase is affected in a similar way by the presence of the liquid phase. In certain two-phase gas–liquid flow systems the gas–liquid interface is turbulent, 'rough'. This roughness increases friction losses in the same way as a rough pipe wall does. In any section of the pipeline, the height of the gas–liquid interface will vary rather frequently during flow, and this variation of height also consumes energy.

(a) Flow patterns

A gas and a liquid flowing together in a horizontal pipeline may assume a variety of geometrical arrangements with respect to each other. Typical arrangements are called flow patterns. According to Alves (Baker 1954), flow patterns include bubble or froth flow, plug flow, stratified flow, wave flow, slug flow, annular flow and spray or mist or dispersed flow (Fig. 1.4—1). To predict the flow pattern prevailing under any given conditions one may resort e.g. to the Baker diagram (Fig. 1.4—2). The abscissa is calibrated in the effective liquid–gas ratio in terms of the expression $v_{ml}\,\lambda\psi/v_{mg}$ whereas the ordinate is calibrated in gas mass velocity, given by the expression v_{mg}/λ; λ and ψ are pressure and temperature correction factors (after Holmes), by which the base

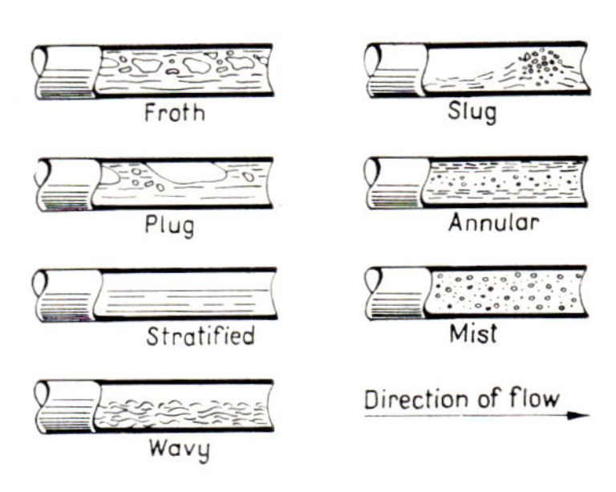

Fig. 1.4—1 Two-phase flow patterns in horizontal pipelines according to Alves, after Baker (1954)

factors derived for the flow of water and air at atmospheric pressure and 20 °C temperature can be adapted to the prevailing conditions (Baker 1954).

$$\lambda = \left[\frac{\varrho_g}{1.20} \frac{\varrho_l}{998} \right]^{0.5} , \qquad 1.4-1$$

where ϱ_g is the flowing density of the gas, and ϱ_l is the density of the liquid.

$$\psi = \frac{0.073}{\sigma_l} \left[10^3 \mu_l \left(\frac{998}{\varrho_l} \right)^{-2} \right]^{1/3} , \qquad 1.4-2$$

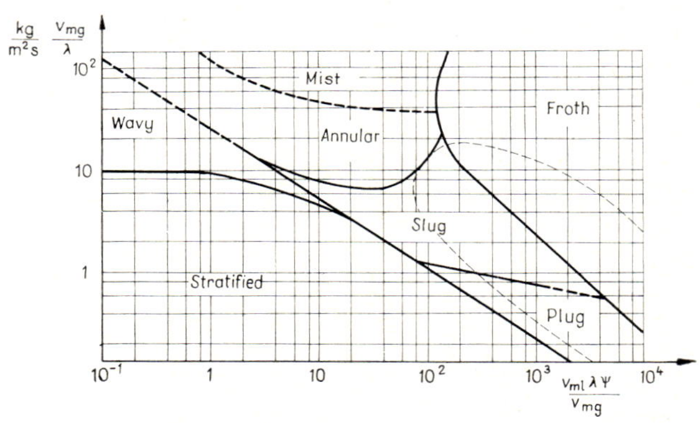

Fig. 1.4—2 Limits of the various flow patterns in horizontal two-phase flow, after Baker (1954)

where σ_l is the surface tension of the liquid, and μ_l is its viscosity. All factors are to be taken at the mean flowing pressure and at the temperature prevailing in the flow string or string section considered.

Example 1.4—1. What is the prevailing flow pattern when oil and gas flow together in a horizontal pipeline of $d_i = 0.257$ m, if $q_g = 3.89$ m³/s and $\varrho_{gn} = 0.722$ kg/m³ under standard conditions. At the mean pipeline pressure and temperature, $q_l = 0.0121$ m³/s, $\varrho_g = 53.2$ kg/m³, $\varrho_l = 777$ kg/m³, $\mu_l = 0.58$ mNs/m² and $\sigma_l = 1.67$ cN/m.

$$v_{ml} = \frac{q_l \varrho_l}{\frac{d_i^2 \pi}{4}} = \frac{0.0121 \times 777}{\frac{0.257^2 \pi}{4}} = 181 \text{ kg/(m}^2\text{s)} ;$$

$$v_{ml} = \frac{q_g \varrho_g}{\frac{d_i^2 \pi}{4}} = \frac{3.89 \times 0.722}{\frac{0.257^2 \pi}{4}} = 54.1 \text{ kg/(m}^2\text{s)} .$$

By Eqs 1.4—1 and 1.4—2, the correction factors are

$$\lambda = \left[\frac{53.1 \times 777}{1.20 \times 998} \right]^{0.5} = 5.87$$

and

$$\psi = \frac{0.073}{0.0167}\left[10^3 \times 5.8 \times 10^{-4}\left(\frac{998}{777}\right)^2\right]^{1/3} = 4.31 .$$

The value of the ordinate in the graph of Fig. 1.4—2 is, then,

$$\frac{v_{mg}}{\lambda} = \frac{54.1}{5.87} = 9.23 ,$$

and the value of the abscissa is

$$\frac{v_{ml}\,\lambda\psi}{v_{mg}} = \frac{181 \times 5.87 \times 4.31}{54.1} = 85 .$$

Plotting the calculated values in Fig. 1.4—1 reveals the flow pattern to be of the slug-type under these conditions.

(b) The calculation method of Lockhart and Martinelli

These authors have pointed out that (i) pressures are equal on both sides of the gas–liquid interface and that (ii) the sum of the cross-section areas occupied by the gas and the liquid, taken separately, is equal to the flow cross-section of the pipe. The fundamental relationships based on experimental data proposed by Lockhart and Martinelli are

$$\Delta p_{lg} = \Delta p_g \Phi_g^2 \qquad 1.4-3$$

and

$$\Delta p_{lg} = \Delta p_l \Phi_l^2 , \qquad 1.4-4$$

where Δp_g is the friction loss under the assumption that only gas is flowing in the pipe, and Δp_l is the friction loss under the corresponding assumption for the liquid (Baker 1954); Δp_l and Δp_g can be calculated by methods outlined in Chapters 1.1 and 1.2, respectively. For finding Φ_g and Φ_l,

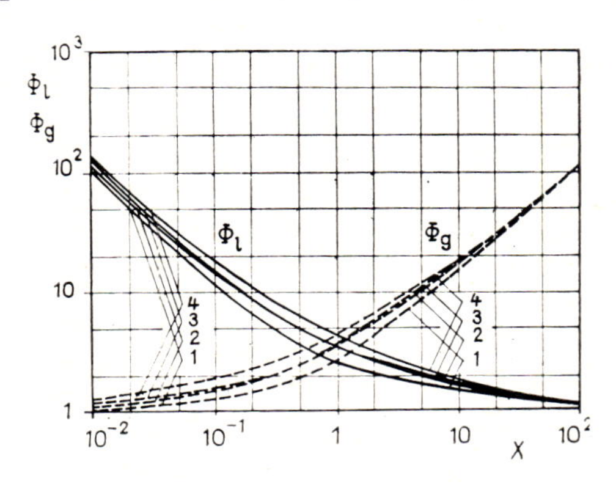

Fig. 1.4—3 Loss factors of horizontal two-phase flow according to Lockhart and Martinelli, after Schlichting (1970)

Lockhart and Martinelli have constructed the graph given as Fig. 1.4—3 here (Schlichting 1970). Either of the two equations 1.4—3 and 1.4—4 can be chosen to handle the problem. Either Φ_g or Φ_l plotted v. $X = (\Delta p_l/\Delta p_g)^{0.5}$ in Fig. 1.4—3 is read off, according to the choice made. The Figure shows four graphs for each of Φ_g and Φ_l. These are to be chosen according as the flow of the gas and liquid, taken separately, is laminar or turbulent. The Reynolds number is to be calculated for each phase as if the other phase were not present. The appropriate graph is then chosen as follows:

Flow of liquid	Flow of gas	Graph to be taken
Laminar	Laminar	No. 1
Turbulent	Laminar	No. 2
Laminar	Turbulent	No. 3
Turbulent	Turbulent	No. 4

The condition of laminar flow for both the liquid and the gas is that the respective Reynolds numbers, $N_{\mathrm{Re}l}$ and $N_{\mathrm{Re}g}$, be less than 1000. — The condition of turbulent flow is that these numbers be greater than 2000. — The method of Lockhart and Martinelli takes no account of the prevailing flow pattern. Subsequent investigations have shown, notwithstanding, that the calculation gives fairly good approximative results especially if the viscosity of the liquid is in the 50 cP range and if the liquid phase contains no free water (Schlichting 1970).

(c) Refinements of the Lockhart—Martinelli method

Several workers attempted to improve on the relationships defining the Φ_g factor in Eqs 1.4—3 and 1.4—4. The first rather widely known method was that of O. Baker (1954). This author developed a separate correction formula for each of the flow patterns shown in Fig. 1.4—2. His formulae were found, however, to give less accurate results than the original formulae of Lockart and Martinelli (Dukler, Wicks and Cleveland 1964A). Dukler presents also a table summarizing a multitude of experiments (Table 1.4—1) which lists the accuracy P as well as the standard deviation σ and the

Table 1.4—1

Flow pattern	Lockhart—Martinelli				Baker			
	P	σ	σ'	n	P	σ	σ'	n
Froth flow					15.6	168	30.0	960
Plug flow	9.4	36.3	20.0	270	116	92.9	100	69
Stratified flow	23.3	33.0	22.5	34	—	—	—	—
Wave flow	38.4	85.7	42.5	287	—91.0	2.5	2.5	98
Slug flow	2.9	31.2	17.5	974	61.0	218	135	251
Annular flow	—12.8	35.6	30.0	265	68.7	81.2	100	430
Mist flow	18.0	34.1	25.0	111	16.9	35.0	30.0	133

parameter σ_2 of the pressure drops calculated by the two methods. The accuracy P is presumably understood to mean the relative error

$$P = \frac{1}{n} \sum_{i=1}^{n} \frac{x_i - x_{ib}}{x_{ib}} 100\% . \qquad 1.4\text{—}5$$

As is well known, in the case of a normal (Gaussian) distribution, σ is a fair measure of the scatter of the measured values from the mean. 68 percent of the measured values fall within a band of width $\pm\sigma$ about the mean. Distributions are, however, quite often non-normal. For such cases, a measure of scatter σ' may be defined precisely as that deviation from the mean which includes 68 percent of the measured values. Let us add that in the formula above the x_i are the measured values and x_{ib} is the basis (e.g. the mean) to which the measurements are referred. — Hoogendorn, similarly to Baker, developed separate correction formulae for the individual flow patterns. These permit us to predict the pressure drop with a fair degree of accuracy, especially for liquids between 5 and 100 cP viscosity. — Chisholm and Laird, allowing for the relative roughness of the pipe, gave relationships suitable for calculation concerning the turbulent—turbulent flow of relatively low-viscosity liquids (from 0.5 to 2.0 cP; Schlichting 1970). — A method developed by Schlichting is applicable over a very broad liquid viscosity range (from 10 to 60,000 cP), provided the gas–liquid ratio computed with the standard volume of the gas does not significantly exceed 100. According to Schlichting, the standard deviation does not exceed ± 20 percent even if the free water content is as high as 80 percent. The suggested domain of application of the method is the area outlined by a dashed line in Fig. 1.4—2 (Schlichting 1970). The fundamental equation of the method is

$$\Delta p_{lg} = \left(C_1 + C_2 \sqrt{\frac{\Delta p_g}{\Delta p_l}}\right) \Delta p_l, \qquad 1.4\text{—}6$$

where

$$C_1 = \frac{1 + 0.65 \left(\frac{\mu_l}{10^4 \mu_g}\right)^{0.8}}{(1 - \varepsilon_w)^{1.3}}, \qquad 1.4\text{—}7$$

and

$$C_2 = 6 + 7.5 \left(\frac{\mu_l}{10^4 \mu_g}\right)^{0.5}. \qquad 1.4\text{—}8$$

The pressure drop of flow in a pipeline is calculated by subdividing the length l of the line into several sections of length Δl. The mean pressure drop Δp_{lgi} is calculated for each section. These values then furnish the pressure traverse of the line and the output end pressure.

Schlichting developed another relationship for calculating the temperature changes in a gas–liquid system flowing in a horizontal pipeline. His method accounts for the heat loss to the ground in which the line is laid, for the heat of evaporation of vapours escaping from the liquid, for cooling due to expansion and heating due to friction. If an appreciable change of temperature is to be expected along the pipeline, a temperature traverse has to be calculated prior to calculating the pressure drop. It is, however, often sufficient to use the mean line temperature provided it is close to the ground temperature.

(d) Baxendell's method

Equation 1.1—1 holds for infinitesimal sections of a horizontal pipeline in which a gas–liquid system is flowing:

$$dp = -\lambda \frac{\bar{v}^2 \varrho_k \, dl}{2d_i} . \qquad 1.4\text{—}9$$

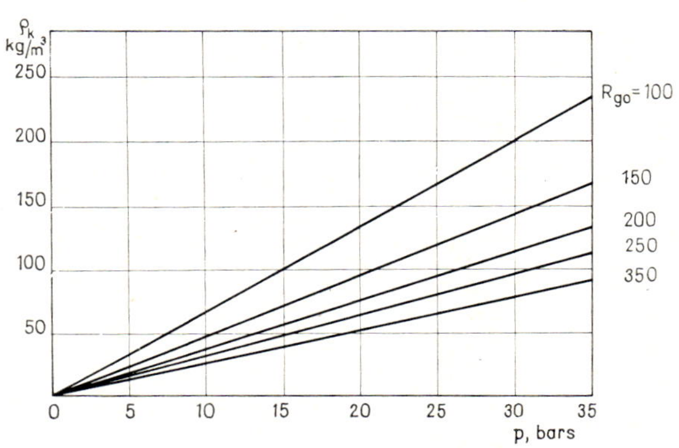

Fig. 1.4—4

If the flow rate through the pipe equals q_o m³ of stock-tank oil per unit of time and, in addition to it, liquids and gases in an amount per m³ of stock-tank oil, as given by the mass factor

$$M_t = \varrho_o + \varrho_g R_{go} + \varrho_w R_w , \qquad 1.4\text{—}10$$

then

$$v = \frac{q_o M_t}{\frac{d_i^2 \pi}{4} \varrho_k} . \qquad 1.4\text{—}11$$

Let — in agreement with Baxendell (1955) — $\lambda = 8f$. Baxendell experimentally determined the variation of density v. pressure for waterless La Paz oil for a variety of gas–oil ratios R_{go}. The results are shown in Fig. 1.4—4. It is apparent that, at a given R_{go}, density is a linear function of pressure. Let the slope of the straight lines be $n = \Delta\varrho_k/\Delta p$. Using these values of n, the variation of n v. R_{go} (Fig. 1.4—5) could be plotted; Fig. 1.4—4 thus implies

$$\varrho_k = np .$$

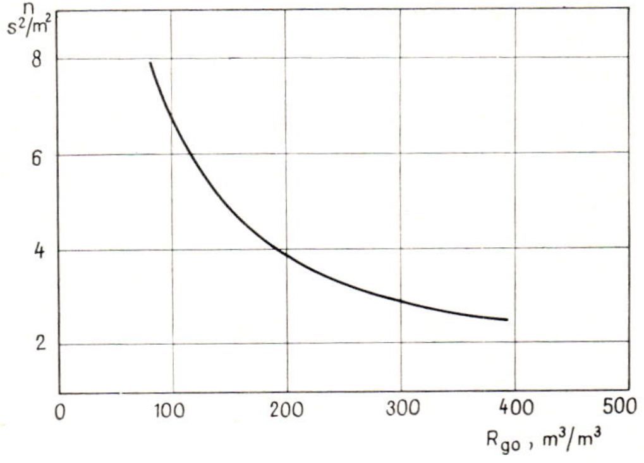

Fig. 1.4—5

Substituting appropriate values of v, λ and ϱ_k into Eq. 1.4—9 and solving the equation for pressure at the input end, we get

$$p_1 = \left[p_2^2 + 13.0 \frac{f l M_t^2 q_o^2}{d_i^5 n} \right]^{0.5} . \qquad 1.4\text{—}12$$

Baxendell determined the friction factor using experimental data concerning 4 in. and 6 in. line pipes. He assumed the friction factor f of the mixed flow of gas and liquid to be independent of viscosity, and plotted it v. the 'viscosity-less Reynolds number'

$$d_i \varrho v = 1.27 \frac{M_t q_o}{d_i} . \qquad 1.4\text{—}13$$

The graphs constructed from the experimental data are shown in Fig. 1.4—6. Baxendell has found flow to be turbulent at abscissa values greater

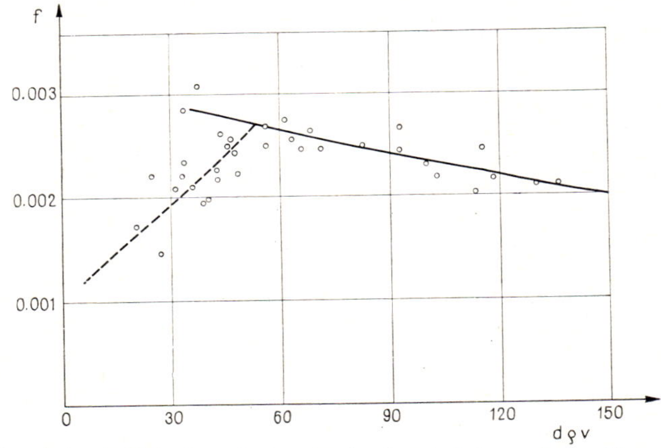

Fig. 1.4—6 Friction factor in horizontal two-phase flow, after Baxendell (1955)

than 54 and laminar at values less than 54. He states p_2 calculated from p_1 to be accurate within ± 5 percent in the first case and within ± 10 percent in the second.

Example 1.4—2. Find the pressure p_1 to be expected at the input end of a horizontal pipeline if $d_i = 0.1023$ m; $p_2 = 4.1$ bars; $l = 3300$ m; $q_o = 1.41 \times 10^{-3}$ m³/s; $R_{go} = 326$ m³/m³. The oil in question is the La Paz oil investigated by Baxendell, whose n v. R_{go} relationship is illustrated in

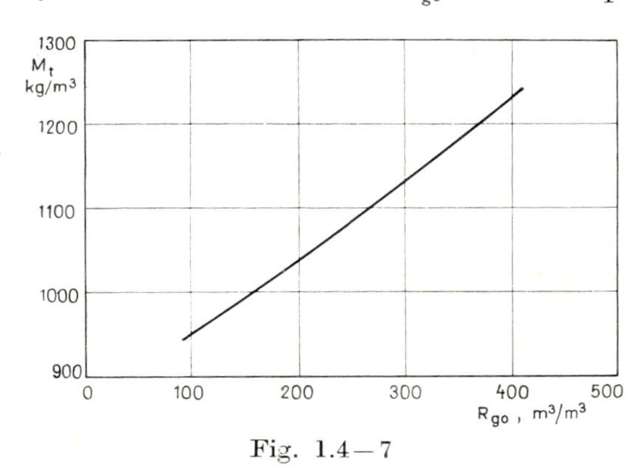

Fig. 1.4—7

Fig. 1.4—5 and whose M_t v. R_{go} relationship is shown as Fig. 1.4—7. — By Fig. 1.4—5, $n = 2.7 \times 10^{-5}$; by Fig. 1.4—4, $M_t = 1160$; by Eq. 1.4—13, the abscissa to be plotted in Fig. 1.4—6 is

$$1.27 \frac{1160 \times 1.41 \times 10^{-3}}{0.1023} = 20.3 .$$

Now from Fig. 1.4—6, $f = 0.00165$. — By Eq. 1.4—12,

$$p_1 = \left[(4.1 \times 10^5)^2 + 13.0 \frac{0.00165 \times 3300 \times 1160^2 (1.41 \times 10^{-3})^2}{0.1023^5 \times 2.7 \times 10^{-5}} \right]^{0.5} =$$

$$= 8.9 \times 10^5 \text{ N/m}^2 .$$

1.4.2. Flow in a pipeline laid over hilly terrains

In single-phase flow, the pressure drop is essentially composed of the friction loss plus the head difference equal to a liquid or gas column whose height equals the difference in elevation between the two ends of the line. In two-phase flow, however, the pressure drop is often appreciably higher than that which can be predicted by the above consideration. The deviation increases as the number and height of hills of terrain undulations increase. This is due to the circumstance that the two phases tend to separate after entrance into the pipeline. Gas flows at a higher velocity in the uphill sections and, by friction at the gas–liquid interface, tends to entrain the liquid. The amount of liquid left behind gradually increases, up to 80 percent of the pipe volume. The gas passing the liquid is accelerated because of the narrowing of cross-section free for the gas flow; this tends to increase the resistance to flow of the pipe and hence also the pressure drop between the two ends of the liquid plug. If this pressure drop is great enough, the accumulated liquid or part of it is entrained by the gas into the downhill section and from there into the next uphill section. As a consequence of the nature of this flow, the pressure drop between the ends of the pipeline depends on the aggregate head difference of the uphill pipe sections. The increase in pressure drop due to hilly terrains can according to Baker (1954) be calculated in a first approximation by the following formula:

$$\Delta p = K(h_1 + h_2 + \ldots + h_n)\gamma_l \,. \qquad 1.4-14$$

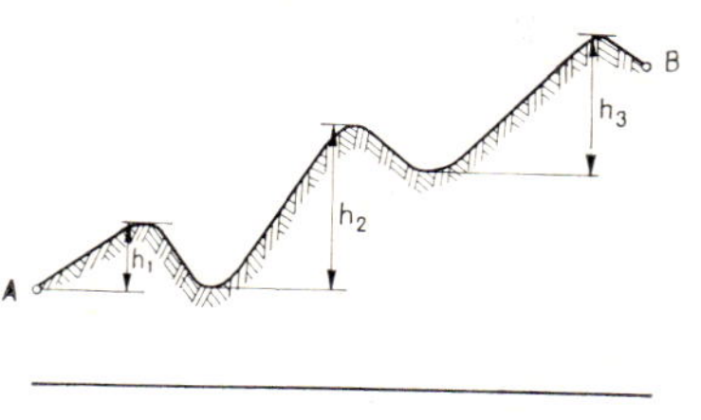

Fig. 1.4—8

K equals 0.5 for flow velocities $v < 3$ m/s and 0.38 for $v > 3$ m/s; $h_1 \ldots h_n$ are the heights of successive terrain elevations as identified in Fig. 1.4—8. The additional pressure drop due to hilly terrains may be appreciably greater than the pure flowing pressure drop.

Example 1.4—3. Consider a pipeline characterized by the same data as in the foregoing example, laid over hilly terrains, with the difference in elevation between line ends being zero but $h_i = 400$ m, and $\varrho_o = 830$ kg/m³. Let us estimate using Eq. 1.4—14 the additional pressure drop due to the elevations of the terrain. — Let us find ϱ_k figuring in Eq. 1.4—11 for the input pressure $p_1 = 8.9$ bars. From Fig. 1.4—4, $\varrho_k = 27$ kg/m³ at $R_{go} = 326$. — By Eq. 1.4—11,

$$v = \frac{1.41 \times 10^{-3} \times 1160}{\dfrac{0.1023^2 \pi}{4} \times 27} = 7.4 \text{ m/s} \,.$$

That is, the mean velocity of flow is greater than 3 m/s even in the most 'unfavourable', highest-pressure case. Now the additional pressure drop due to the hilly terrain is, by Eq. 1.4—14,

$$\Delta p = 0.38 \times 400 \times 8142 = 1.24 \times 10^6 \text{ N/m}^2 \,.$$

A series of experiments going on since 1958 under the direction of the AGA—API Multiphase Research Committee has been aimed, on the one hand, at improving the two-phase throughput capacity of pipelines and, on the other, at deriving more accurate relationships for computing the pressure drop in pipelines over hilly terrain (McDonald and Baker 1964). It has been found that by introducing into the pipeline at given intervals rubber balls of practically the same diameter as the *ID* of the pipe, it is possible to sweep out, so to speak, the liquid phase. The throughput capacity of pipelines was increased by from 30 to 70 percent by the use of the balls. In deriving the formulae describing this situation it was assumed that the sweeping effect of the balls gives rise to four types of flow; Fig. 1.4—9 shows after McDonald and Baker (1964) the flow patterns replacing the annular flow that would take place in the absence of the balls. In section (a) the flow is multiphase, annular. In section (b) gas flows within a wall wetted by oil; in section (c) liquid flows. In section (d) there is multiphase flow essentially identical with that in section (a). Separate formulae have been derived for each type of flow. Optimum operation is designed by computing optimum rubber-ball spacing, or, in other words, the optimum time interval between the insertion of two balls. Up-to-date field equipment required for ball sweeping operations has already been developed (Bean 1967)

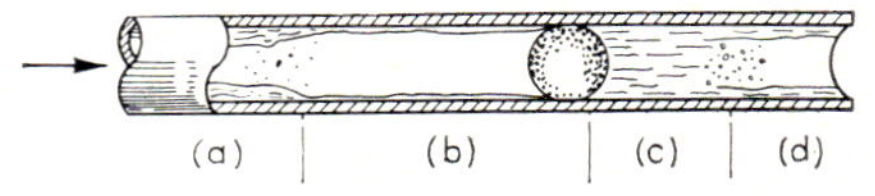

Fig. 1.4—9 Influence of rubber pipe ball upon horizontal two-phase flow, after McDonald and Baker (1964)

1.4.3. Flow in vertical pipe strings

Research into the laws governing two-phase vertical flows has long been pursued. Versluys (1930) was the first to give a general theory. Numerous new theories have been published since, and although research cannot be regarded as complete, we are in possession today of formulae of satisfactory accuracy for the flow regions of importance in practice. These formulae are based on various theoretical approaches, and in solving various problems or explaining various phenomena encountered in oil and gas production it is now one, now another that turns out most suitable. The reasons why the relationships describing two-phase vertical flow are so complicated are (i) that the specific volume of the flowing fluid varies appreaciably as a function of pressure and temperature, (ii) slippage losses arise in addition to friction losses, (iii) flow is affected by a great number of parameters and (iv) liquid and gas may assume a variety of flow patterns. — The problems of flow will be discussed below with the main emphasis on wells producing a gaseous liquid.

(i) The specific volume of gas gradually increases as it surges upward, owing to the decrease of the pressure acting on it. This is due on the one hand to the expansion of the free gas entering at the tubing shoe and, on the other, to the escape of more and more dissolved gas from the flowing oil under the decreasing pressure. Both effects are mitigated by the decrease in temperature. The specific volume of oil will slightly decrease owing to the escape of gas and to the decrease in temperature and slightly increase owing to the decrease in pressure.

(ii) Two-phase vertical flow involves two types of energy loss: friction loss and slippage loss. *Friction loss* is an energy loss similar to what arises in single-phase turbulent flows. The velocity distribution over the cross-section of the flow string, which considerably affects friction, is, however, significantly influenced by the flow pattern. In the course of upward flow, moreover, the friction factor may vary considerably because the relative gas content of the fluid in contact with the pipe wall and the flow velocity both increase monotonely. Hence, even if the only loss to be reckoned with were due to friction, we should be faced with much more of a problem than in the case of single-phase flow. The energy loss of flow is, however, further complicated by the phenomenon of slippage. *Slippage loss* is due primarily to the great difference in specific weight between the gas and the liquid. The gas bubble entering the flow string at the tubing shoe will leave the liquid element entering together with it far behind. As a consequence, the ratio of gas to oil in the fluid filling out a flow string (the so-called in-situ *GOR*) is less than the *GOR* calculated for the same temperature and pressure from the volumes of the gas and oil produced, the so-called effective producing *GOR*. The mean density of flow in the pipe is greater than it would be in the absence of slippage. To illustrate this fact suppose that there is a permeable, porous substance filling out the pipe and fixed to it. Let the aggregate section of the pores in any section of the pipe equal e.g. two thirds of the total pipe section. Let a gas be conveyed through this pipe. The in-situ 'gas–substance ratio' (the term 'substance' referring to the permeable-porous substance mentioned above) is 2/3 : 1/3 = 2. The effective producing 'gas–substance ratio' is, however, infinite, because only gas is produced without any 'substance'. If the 'substance' were not fixed to the pipe wall, it would start to rise at a given gas flow rate, and the effective producing 'gas–substance ratio' would begin to decrease. The same effect plays a role in the two-phase flow of gas and liquid. Slippage loss is greatest at relatively low gas flow rates in comparatively large pipes. — Let the liquid flow rate through the section A_T of the flow string situated at a given depth be q_l, and let the effective gas flow rate be q_g. Let a fraction ε_l of the section be filled with liquid and the rest, $1 - \varepsilon_l = \varepsilon_g$ by gas. The flow velocity of the liquid is

$$v_l^* = \frac{q_l}{A_T \varepsilon_l} = \frac{v_l}{\varepsilon_l}$$

and that of the gas is

$$v_g^* = \frac{q_g}{A_T(1 - \varepsilon_l)} = \frac{v_g}{(1 - \varepsilon_l)} .$$

The slippage velocity, proportional to the slippage loss is, then,

$$v_{gs} = \frac{q_g}{A_T(1 - \varepsilon_l)} - \frac{q_l}{A_T \varepsilon_l} \qquad 1.4-15$$

(iii) The pressure gradient of flow, also called the flow gradient, is affected by a host of independent variables. N. C. J. Ros (1961) has enumerated 12 such factors (*ID* of tubing; relative roughness of pipe wall; inclination of the tubing to the vertical; flow velocity, density, viscosity of both liquid and gas; interface tension; wetting angle; acceleration of gravity). There are further factors, e.g. flow temperature, whose influence can at best be estimated.

(iv) The spatial arrangement relative to each other of the liquid and gas, the filling-out of space by the individual phases, may take a variety of forms. The typical arrangements are called flow patterns. There is no full agreement in literature as to the classification of flow patterns. The following classification seems to be the most natural. (I) *Froth flow.* The continuous liquid phase contains dispersed gas bubbles. The number and size of gas bubbles may vary over a fairly wide range, but the gas–oil ratio is usually less than in the other flow patterns. If the gas bubbles move more or less parallel to the centre line of the pipe, the flow is called bubble flow, whereas in froth flow in the strict sense of the term the bubbles are in turbulent

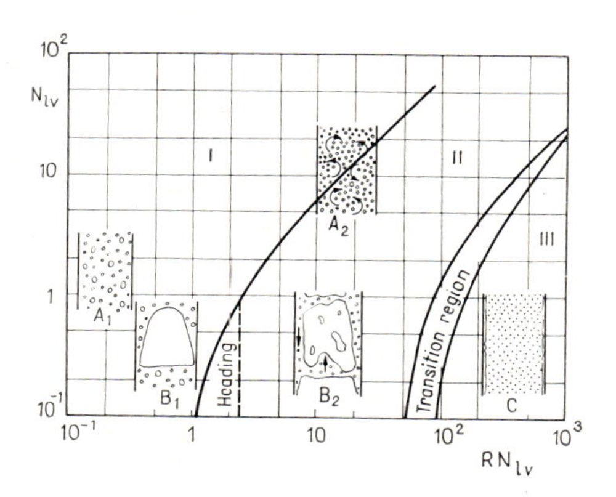

Fig. 1.4—10 Flow patterns of vertical two-phase flow, after Duns and Ros (1963)

motion. (II) *Plug flow.* The gas–oil ratio is higher; some of the coagulated gas bubbles fill out the entire pipe section in a certain height. Plug flow takes place in a succession of gas and liquid plugs. There may be an oil mist dispersed in the gas plugs and a gas froth dispersed in the oil plugs. (III) *Mist flow.* Here the gas is the continuous phase: it contains a finely dispersed mist of liquid. The gas–oil ratio is very high. In flowing and gas-lift wells, patterns (I) and (II) are of the greatest importance, whereas in condensate-gas wells pattern (III) comes into its own.

It is hard to accurately delimit the zones of the individual flow patterns. Gladkow for instance stated the upper limit of froth flow to be at a *GOR* of 20 m^3/m^3. According to Krylov, the upper limit of plug flow is at 100 m^3/m^3 (Muraviev and Krylov 1949). According to Duns and Ros (1963), flow velocity also plays an important part in addition to the effective gas–oil ratio. Figure 1.4—10 reproduced after these authors shows a diagram calibrated in dimensionless liquid velocity on the abscissa axis v. dimensionless gas velocity on the ordinate axis. Three flow regions marked by Roman numerals are distinguished in this diagram. Region (III) corresponds to the mist flow pattern described in paragraph (III) above. The froth and slug flow patterns do not, however, coincide with regions (I) and (II).

Present-day vertical two-phase flow theories of practical interest can be grouped whether the theory has an energy balance for its fundamental relationship; the fundamental relationship is a pressure balance; or the relationship is an experimentally established graph. — The first group includes two fundamental theories: that of Krylov and that of Poettmann and Carpenter. Both assume flow to be isothermal, or, more precisely, that the influence of changes in thermal energy content upon the energy balance is insignificant. For an infinitesimal length of tubing, the energy equation can be derived from Eq. 1.2—1, assuming that $\varrho = \varrho_k$ is the density and v is the velocity of the fluid mixture, and the fourth term, dh_v, is the energy loss. Dividing by g we have

$$\frac{dp}{\varrho_k g} + \frac{v dv}{g} + dh + dh_v = 0 . \qquad 1.4—16$$

Let us express density in terms of specific volume, that is, let $1/\varrho_k = V$, and let us put this into Eq. 1.4—16. The equation valid for the full length L of the flow string is obtained by integrating 1.4—16 between the bottom-hole and top-hole limits *1* and *2*:

$$\frac{1}{g} \int_{p_2}^{p_1} V dp = L + \frac{\overline{v}_1^2 - \overline{v}_2^2}{2g} + \int_1^2 dh_v . \qquad 1.4—17$$

The last term on the right-hand side is symbolic: it marks the total specific-energy loss between well bottom and wellhead.

(a) Krylov's theory. The q_g—q_l relationships in the operation of the flow string

Krylov's theory (Muraviev and Krylov 1949) is based on laboratory experiments. The experimental piping consisted of from 18 to 20 m long standard-size tubing pipes. Both ends of the pipes were fitted with quick-closing valves and pressure gauges. Krylov assumed that these lengths of pipe could be regarded as infinitesimal. The direct aim of his experiments was to establish the constants of Eq. 1.4—18. Krylov performed his experiments with water and air and corrected for the friction loss of the liquid by assuming the viscosity of oil to be five times that of water. The term $v dv/g$ in Eq. 1.4—16 is small enough to be negligible. Let $\varrho_k g = \gamma_k$. Let us multiply both sides of the equation by the expression $\gamma_k/\gamma_l dh$, where γ_l is the specific weight of the liquid, assumed to be constant irrespective of temperature and pressure. Then

$$\frac{dp}{\gamma_l dh} = \frac{\gamma_k}{\gamma_l} + \frac{\gamma_k dh_v}{\gamma_l dh} ,$$

where $(dp/\gamma_l dh)$ is the pressure gradient expressed in terms of liquid column height, which we shall in the following denote by ξ. The second term on the right-hand side is the friction-loss gradient, likewise expressed in terms of liquid column height, ξ_s. In the new notation,

$$\xi = \frac{\gamma_k}{\gamma_l} + \xi_s \, . \qquad 1.4\text{—}18$$

For various flow velocities of water and air through a variety of experimental pipe sizes, ξ, γ_k and γ_l could be unequivocally determined. From the data thus obtained, Krylov developed the relationship

$$\frac{\gamma_k}{\gamma_l} = \frac{q_l + 0.785\, d_i^2}{q_g + q_l + 0.785\, d_i^2}$$

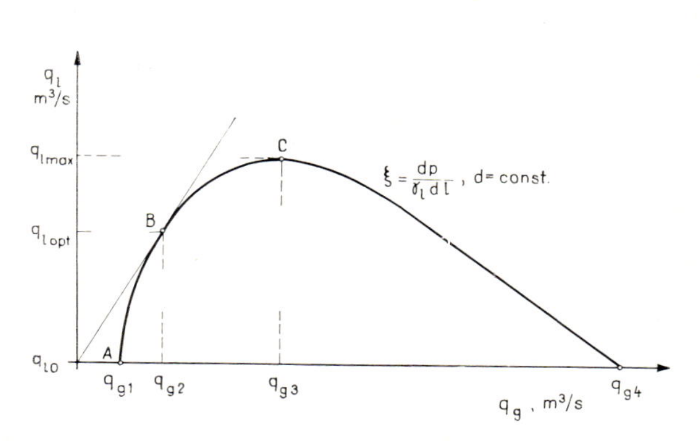

Fig. 1.4—11 Production characteristics of tubing string, according to Krylov

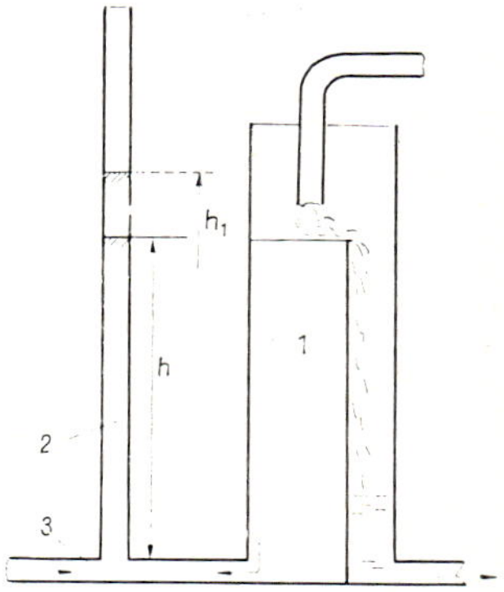

Fig. 1.4—12

This expression accounts in addition to the hydrostatic pressure of the liquid also for the slippage loss. For ξ_s, Krylov derived a three-term expression in which the second term is the friction loss that would arise if the gas flowed alone in the pipe, and the third term is the same for the liquid. The additional loss due to the flow being mixed-phase is taken care of by the first term. The fundamental formula obtained as a result of these operations is the pressure gradient for an infinitesimal length of pipe:

$$\xi = \frac{q_l + 0.785\, d_i^2}{q_g + q_l + 0.785\, d_i^2} + \frac{1.71 \times 10^{-3} q_g q_l}{d_i^5} + \frac{9.3 \times 10^{-7} q_g^2}{d_i^{5.33}} + \frac{1.17 \times 10^{-3} q_l^{1.75}}{d_i^{4.75}} \, . \qquad 1.4\text{—}19$$

This fundamental relation is seen to contain three independent variables or, better, five if ξ is regarded as defined by the relationship $dp = \xi dh \gamma_l$. Assuming ξ and d_i to be constant, Eq. 1.4—17 may be used to construct a $q_l = f(q_g)_{\xi; d_i}$ diagram characteristic of the operation of an infinitesimal length of tubing, shown as Fig. 1.4—11. The relationship $q_l = \mathrm{f}(q_g)_{\xi_i;d_i}$ for tubing strings of field length is of a similar form. The gas flow rate is in terms of standard cubic metres per unit of time. A curve of this type had earlier been described by Shaw who did not, however, succeed in giving a mathematical expression for it (Shaw 1939).

Consider now the simplified model shown in Fig. 1.4—12. Liquid flows from a constant-level reservoir tank *1* to the lower end of tubing *2*, where its pressure is p_1. The conduit *3* delivers gas likewise at a pressure p_1, but at an arbitrarily variable rate to the tubing *2*. The liquid rises in the flow string and flows out of the wellhead at a rate as high as can be carried by the gas flow. Since the pressures at the lower and upper end of the tubing string and the length of the tubing are constant, the mean pressure gradient will also be constant and so the operation of the tubing string will be characterized by Fig. 1.4—11. If gas is delivered to the tubing string at a rate lower than q_{g1}, then the gas will turn the liquid column of original height h in the tubing into a froth and raise its level to, say, h'. The level will reach the tubing head when the gas flow rate attains q_{g1}. The operating point q_{l0}, of the tubing, 'the operating point of kickoff', will be determined by the gas flow rate q_{g1} and the liquid flow rate $q_l = 0$. At gas flow rates below q_{g1}, the entire energy of the gas delivered into the tubing will be consumed by slippage. At gas flow rates above q_{g1}, the liquid flow rate will increase gradually until the gas flow rate attains a value q_{g2}. The specific gas consumption of the gas lift gradually decreases in the meanwhile. The operating point belonging to the gas flow rate q_{g2} is called 'the operating point of most economical gas lift'. The liquid flow rate belonging to this operating point is optimal, $q_{l\,\mathrm{opt}}$. The inverse slope of the position vector to any point of the graph equals q_g/q_l, the specific gas consumption. The slope of the position vector is greatest at $q_{l\,\mathrm{opt}}$; that is, specific gas consumption is least at that point. — A gas flow rate increased from q_{g2} to q_{g3} entails an increased liquid flow rate, but at the expense of an increased specific gas consumption, which renders operation less than optimally economical. The maximum liquid throughout capacity $q_{l\,\max}$ of the tubing is attained at a gas flow rate q_{g3}. The operating point belonging to this value is 'the operating point of maximum liquid throughput'. If the gas flow rate is increased above q_{g3}, both the liquid flow rate and the economy of operation will decline. At the gas flow rate q_{g4}, liquid transport ceases.

The shape of the $q_l = \mathrm{f}(q_g)_{\xi;d_i}$ diagram characterizing the liquid throughput capacity of the tubing string is determined by the relative magnitudes of the friction and slippage losses. This state of facts is illustrated by the approximate diagram of Fig. 1.4—13, showing P, the energy consumption rate v. the gas flow rate. The length of the flow string, the respective pressures p_1 and p_2 at its lower and upper ends, and the specific weight of the liquid γ_l, are considered constant. An isothermal change of state of a perfect gas delivers a total energy $P = p_n q_g \ln p_1/p_2$, that is, as shown by line *1*, energy varies directly as the gas flow rate. Curve *3* is geometrically similar to the $q_l = \mathrm{f}(q_g)_{\xi;d_i}$ diagram in Fig. 1.4—11, since the useful energy output is $P_h = h\gamma_l q_l$. Curve *2* is based on qualitative considerations. At a given v_g, the ordinate difference between the line *1* and the curve *2* represents friction loss; the ordinate difference between curves *2* and *3* represents slippage loss, whereas the ordinate of curve *3* represents the useful energy consumption. The Figure reveals the ratio of slippage loss to total energy consumption to be most important at comparatively low gas–oil ratios, whereas friction loss becomes predominant at comparatively high *GOR*s. — Figure 1.4—14 presents several $q_l = f(q_g)_{\xi;c_i}$ diagrams computed by using Eq. 1.4—19 for constant values of ξ usually encountered in practice. It is seen that the greater the value of ξ, (a) the less the gas flow rate at which

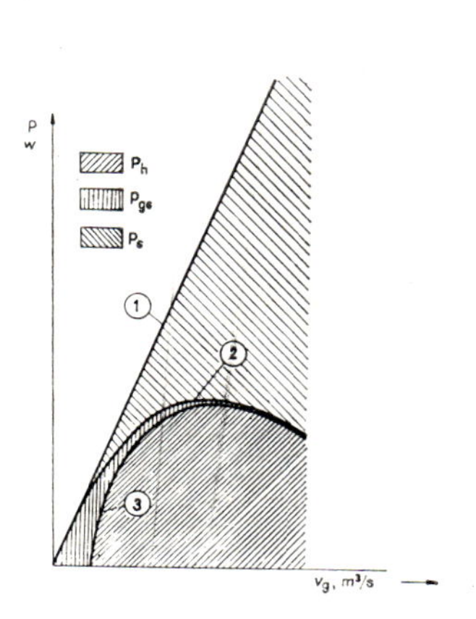

Fig. 1.4—13 Power consumption of vertical two-phase flow

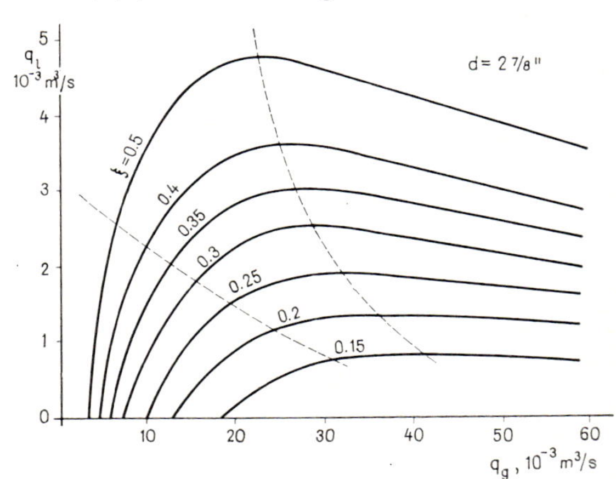

Fig. 1.4—14 Production characteristics of tubing string for various pressure gradients, according to Krylov

the well is kicked off, (b) the greater both the optimal and the maximal liquid flow rate, and (c) the less — in view of the slopes of the position vectors to the operating points — the specific gas consumption of a given liquid flow rate. — Figure 1.4—15 shows three graphs for standard tubing sizes at $\xi = 0.5$. The greater the tubing diameter d_i, (a) the greater the gas flow rate required for kickoff, (b) the greater the liquid throughput capacity of the tubing at both the optimal and the maximal liquid flow rate.

Equation 1.4—19 is valid for infinitesimal lengths of tubing. It can be extended to field lengths of tubing in two ways. Substituting this expression of ξ into the equation $dp = \xi \gamma_l dh$, we get an equation which — after certain simplifications — can be integrated between the limits set by the bottom and top ends of the tubing string. The relationship thus derived is, however, ungainly and unsuited for practical calculations. Krylov preferred certain approximate solutions of the differential equation, which he derived by substituting for ξ and q_g mean values valid for field lengths of tubing. When forming the means he assumed that the pressure traverse of the tubing string is linear; the flow and consequently the volume change of the gas is isothermal; flow temperature equals standard temperature; the volume changes of the gases can be characterized by the perfect gas laws; and gas solubility in produced liquids equals zero.

If the pressure traverse is linear, then the mean pressure gradient expressed in metres of liquid column is

$$\bar{\xi} = \frac{p_1 - p_2}{\gamma_l L} . \qquad 1.4-20$$

Let the gas flow rate through the tubing be q_{gn} m³/s of standard-state gas. It can be derived from the above assumptions that, at the mean pressure prevailing in the tubing string, the mean gas flow rate is

$$\bar{q}_g = \frac{q_{gn} p_n \ln p_1/p_2}{p_1 - p_2} . \qquad 1.4-21$$

Substituting the value of $\bar{\xi}$ as given by Eq. 1.4—20 and the value of q_g as given by 1.4—21 into Eq. 1.4—19, we obtain Krylov's formula for field lengths of tubing, which can, by certain rearrangements and simplifications, be brought to a form suitable for practical calculations. In Soviet oil production practice, however, relations derived for the three critical operating points of the tubing string are preferred to this formula.

At the operating point of kickoff $q_l = 0$. Putting this value into Eq. 1.4—19 we get as the kickoff criterion for an infinitesimal length of tubing

$$\xi = \frac{0.785\, d_i^2}{q_g + 0.785\, d_i^2} + \frac{9.3 \times 10^{-7} q_g^2}{d_i^{5.33}} .$$

Substituting into this equation the mean values $\bar{\xi}$ and $\bar{q}_g$ given by Eqs

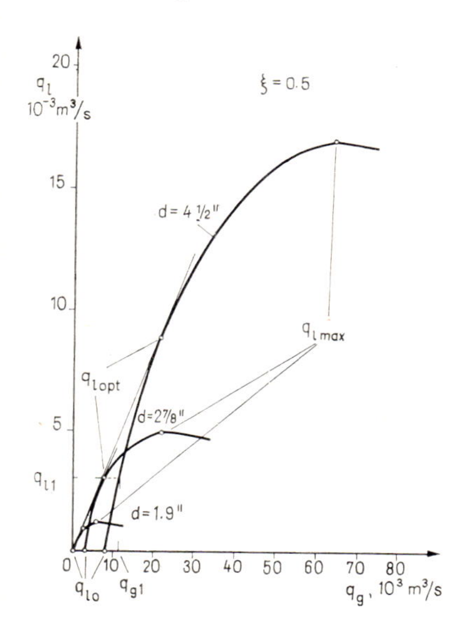

Fig. 1.4—15 Production characteristics of tubing strings for various tubing sizes, according to Krylov

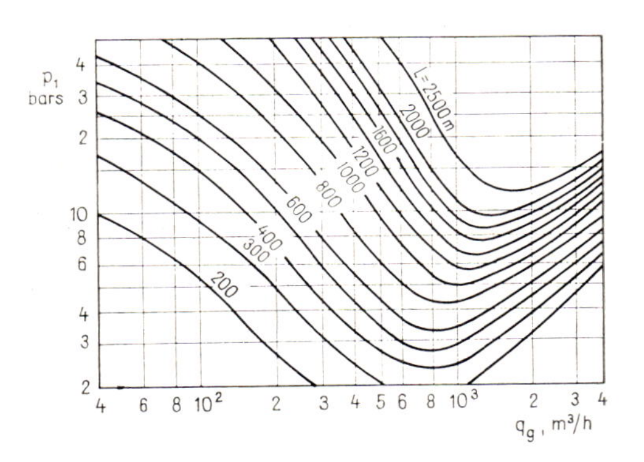

Fig. 1.4—16 Kickoff characteristics of long tubing string, according to Krylov

1.4—20 and 1.4—21, we see that the kickoff criterion for field lengths of tubing becomes

$$\frac{p_1 - p_2}{\gamma_l L} = \frac{0.785\, d_i^2}{\dfrac{q_{gn}\, p_n \ln p_1/p_2}{p_1 - p_2} + 0.785\, d_i^2} + \frac{9.3 \times 10^{-7}}{d_i^{5.33}} \left[\frac{q_{gn}\, p_n \ln p_1/p_2}{p_1 - p_2}\right]^2 . \qquad 1.4\text{—}22$$

Figure 1.4—16 illustrates this equation for $d = 2\ 7/8$ in., $p_2 = 1.4$ bars and $\gamma_l = 900$ kg/m³. It is seen that if the pressure prevailing at the lower end of a tubing string 1000 m long is 20 bars, then the gas flow rate required for kickoff is at least $q_{gn} = 290$ m³/h.

In order to establish formulae for the *operation points of optimum and maximum liquid flow rates*, Krylov connected the points $q_{l\,opt}$ and $q_{l\,max}$ respectively, of curves similar to the set in Fig. 1.4—14 referring to various tubing sizes (see dashed lines *1* and *2*), and wrote up his equations:

$$q_{l\,opt} = 55\, d_i^3 \xi^{1.5} (1 - \xi) ; \qquad 1.4\text{—}23$$

$$q_{l\,max} = 55\, d_i^3 \xi^{1.5} . \qquad 1.4\text{—}24$$

The gas flow rates required to assure liquid flow rates corresponding to these operating points are

$$q'_{g\,opt} = \frac{15.5\, d_i^{2.5} (1 - \xi)^2}{\xi^{0.5}} ; \qquad 1.4\text{—}25$$

$$q_{g\,max} = \frac{15.5\, d_i^{2.5}}{\xi^{0.5}} . \qquad 1.4\text{—}26$$

Equations 1.4—23 and 1.4—24 can also be used to derive the optimal and maximal liquid flow rates for field lengths of tubing, if ξ is replaced by the mean value $\bar{\xi}$ given by Eq. 1.4—20.

Soviet practice operates with flow rates expressed in terms of weight rather than volume per unit of time. To adapt the formulae to this viewpoint, both sides are multiplied by the specific weight of the liquid, γ_l. In a similar manner, a relationship for calculating the gas flow rates required for given liquid flow rates in field lengths of tubing can also be derived. For practical purposes, however, the specific gas consumption R (the amount of gas expressed in standard volume units, required to produce a unit volume or unit weight of liquid) is more suitable:

$$R = \frac{q_{gn}}{q_l} .$$

Using Eqs from 1.4—20 to 1.4—26 it can be shown that the specific gas consumption of liquid production through field lengths of tubing at the optimum and maximum liquid flow rates, respectively, is

$$R_{\rm opt} = 0.123\,\frac{L(1-\bar{\xi})\gamma_l}{d_i^{0.5}\bar{\xi}p_n \lg p_1/p_2}\,, \tag{1.4—27}$$

and

$$R_{\max} = 0.123\,\frac{L\,\gamma_l}{d_i^{0.5}\xi p_n \lg p_1/p_2}\,. \tag{1.4—28}$$

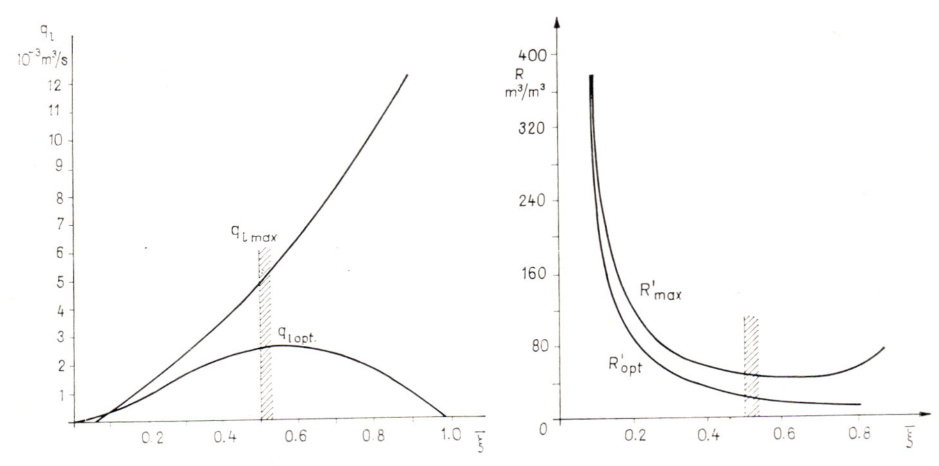

Fig. 1.4—17 Optimum and maximum flow rates in long tubing string, according to Krylov

Fig. 1.4—18 Optimum and maximum gas requirements in long tubing string, according to Krylov

Figure 1.4—17 shows the variation of $q_{l\rm opt}$ and $q_{l\max}$ for a tubing size of 2 7/8 in. and a liquid density of 900 kg/m³. It is clear that in the range of practical significance $\bar{\xi} < 0.5$, the liquid throughput capacity increases at both operating points as the hydraulic gradient increases. — Figure 1.4—18 shows the variation of $R_{\rm opt}$ and $R_{\max}$ v. $\bar{\xi}$, assuming that the dynamic level, that is, the effective height of lift as measured from the wellhead, is $h_d = L(1-\bar{\xi}) = 400$ m; and further, $p_2 = 1.0$ bar; $d = 2\ 7/8$ in. and $\varrho_l = 1000$ kg/m³. Obviously, in the range of practical significance, $\bar{\xi} < 0.5$; the specific gas consumption of liquid production decreases as the hydraulic gradient increases.

(b) The Poettmann—Carpenter theory. Pressure traverses for multiphase flow of gas, oil and water

The immediate aim of the theory and calculation method of Poettmann and Carpenter (1953) was to predict pressure traverses $p = f(h)$ in vertical flow strings, say between the lower and upper end of a tubing string.

Numerous problems connected with the operation of a well can be solved in the knowledge of the pressure traverses.

The salient features of the theory are as follows. (i) The basis of the calculation is an energy balance written up for the two ends of a long tubing string. Temperature is considered constant and equal to the mean temperature in the string. (ii) Friction and slippage losses are not separated, as the energy required to keep the fluid flowing is assumed to equal the difference in the energy content of the fluid entering and leaving the flow string, and to be independent of the in-situ gas–liquid ratio which depends on slippage loss. (iii) The energy loss of flow is characterized by the equation

$$h_v = 4f \frac{l\bar{v}^2}{2d_i g}, \qquad 1.4-29$$

called the Fanning equation in the English-speaking world, commonly used in hydraulics to calculate the friction loss of single-phase flow. This equation is derived from Eq. 1.1—1 by the substitutions $\lambda = 4f$, $p = h_v \varrho g$ and by putting $l = L$. (iv) Poettmann and Carpenter determined the total energy loss factor by an analysis of data from producing oil and gas wells. (v) The variation v. pressure of the specific volumes of various fluids of various gas–liquid ratios was taken into account by the introduction of multiphase factors.

Let $\int_1^2 dh_v$ in Eq. 1.4—17 be denoted h_v, and let us neglect the small kinetic-energy term. The energy balance between the two ends of the flow string then becomes

$$\frac{1}{g}\int_{p_2}^{p_1} V dp = L + h_v .$$

Let us replace h_v by Eq. 1.4—29, with $\bar{v}$, the mean velocity of the gas–liquid mixture flowing in the string. Substitution results in

$$\frac{1}{g}\int_{p_2}^{p_1} V dp = L + 4f \frac{L\bar{v}^2}{2d_i g} .$$

Expressing L from the above, we get

$$L = \frac{\frac{1}{g}\int_{p_2}^{p_2} V dp}{1 + \frac{4f\bar{v}^2}{2d_i g}} . \qquad 1.4-30$$

Let us introduce two factors, M_t and B_t. M_t is a mass factor defined in Eq. 1.4—10; B_t is the multiphase volume factor that indicates the volume, at the pressure p, of the gas, oil and water produced per m³ of stock-tank oil. Let $\bar{B}_t$ be the integral mean of B_t between the pressures p_1 and p_2. Then,

$$V_k = \frac{B_t}{M_t} \text{ and } \bar{v}^2 = \frac{q_o^2 \bar{B}_t^2}{0.785^2 d_i^4} .$$

Substitution of the expression for V_k and $\bar{v}$ into Eq. 1.4—30 results in

$$L = \frac{\frac{1}{M_t g} \int_{p_2}^{p_1} B_t \mathrm{d}p}{1 + \frac{0.331 f q_o^2 \bar{B}_t^2}{d_i^5}} \qquad 1.4-31$$

The integration written up in the numerator can most often be carried out to an accuracy satisfactory for practical purposes, as at high enough pressures B_t can be expressed as a simple function of p. In the knowledge of this function, $\bar{B}_t$ can be expressed in terms of B_t and the limits p_1 and p_2 as

$$\bar{B}_t = \frac{\int_{p_2}^{p_2} B_t \mathrm{d}p}{p_1 - p_2} .$$

Given the production data of an oil well, f can be calculated using Eq. 1.4—31. Poettmann and Carpenter determined values of f for numerous flowing and gas-lift wells and plotted them v. a 'viscosity-less Reynolds number' $N_{\mathrm{Re}''}$, that is, $(d_i \varrho_k v)$, because they considered the loss factor to be unaffected by viscosity at the high-velocity turbulence usual in two-phase flow regions. Replacing ϱ_k by $\varrho_k = M_t / B_t$ and v by $v = 4 q_o B_t / d_i^2 \pi$ we obtain $1.27 q_o M_t / d_i$ as the independent variable. The f v. $M_t q_o / d_i$ diagram of Poettmann and Carpenter, based on extensive experiments and calculations (cf. Eq. 1.4—13) is shown as Fig. 1.4—19.

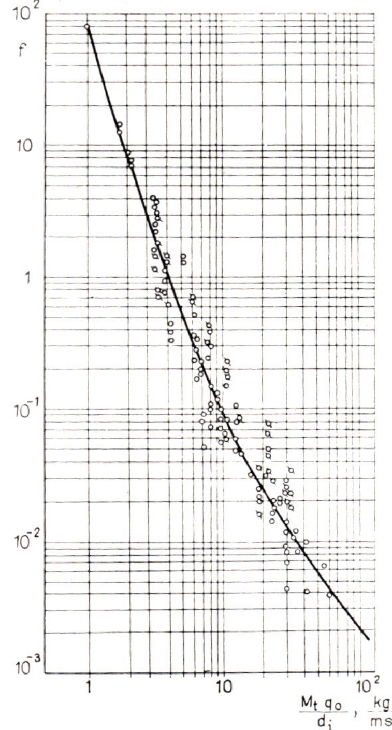

Fig. 1.4—19 Loss factor in vertical two-phase flow, according to Poettmann and Carpenter (1953)

The direct aim of the Poettmann—Carpenter theory was to give a procedure for establishing pressure traverses. The fundamental equation of the method is the differential form, valid for infinitesimal lengths of flow string, of Eq. 1.4—31:

$$\mathrm{d}h = \frac{\dfrac{B_t}{M_t g}\,\mathrm{d}p}{1 + \dfrac{0.331 f q_o^2 B_t^2}{d_i^5}}. \qquad 1.4\text{—}32$$

Let us express $1/\mathrm{d}h$ and let us multiply both sides of the equation thus obtained by $\mathrm{d}p$:

$$\frac{\mathrm{d}p}{\mathrm{d}h} = \frac{M_t g}{B_t} + \frac{M_t g}{B_t}\,\frac{0.331 f q_o^2 B_t^2}{d_i^5}. \qquad 1.4\text{—}33$$

Replacing $M_t g/B_t$ by γ_k and B^2 by $M_t^2 g^2/\gamma_k^2 = 96.23 M_t^2/\gamma_k^2$ results in

$$\frac{\mathrm{d}p}{\mathrm{d}h} = \gamma_k + \frac{C}{\gamma_k}, \qquad 1.4\text{—}34$$

where

$$C = \frac{31.8 f q_o^2 M_t^2}{d_i^5}. \qquad 1.4\text{—}35$$

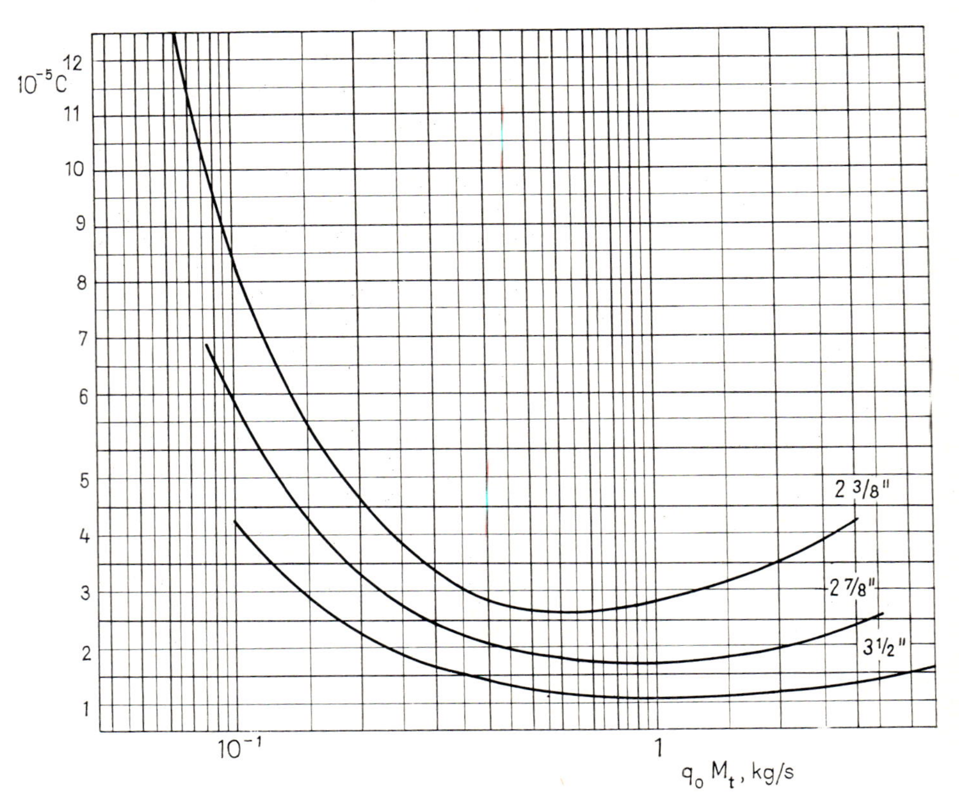

Fig. 1.4—20

Figure 1.4—19 shows the loss factor f to be a function of $q_o M_t$ and d_i. The factor C must therefore depend on these same quantities. The pressure gradient dp/dh depends in addition to these factors also on the flowing specific gravity γ_k. Poettmann and Carpenter constructed sets of $dp/dh = \Phi(q_o M_t, \gamma_k)_{di}$ nomograms on the basis of Eq. 1.4—33. When transposing their calculation procedure to SI units it seemed simpler to prepare just $C = \Phi(q_o M_t)$ diagrams using Eq. 1.4—35 (Fig. 1.4—20) and to calculate the pressure gradient using Eq. 1.4—34 in the knowledge of C. — The pressure traverse is constructed in the knowledge of the pressure at the lower end of the tubing string as follows: starting from said pressure p_1, one selects a sequence of decreasing pressures, for each of which one computes the actual flowing specific gravity $\gamma_k = M_t g / B_t$. The corresponding C is read off Diagram 1.4—20 and dp/dh is calculated using Eq. 1.4—34. The inverses of the values thus obtained, that is, the values of dh/dp, are plotted against pressure. Integrating graphically with respect to p, that is, constructing the graph $\Phi(p) = \int (dh/dp) dp$, we obtain the variation of lift height h v. pressure, or, conversely, the pressure traverse of the tubing string (Fig. 1.4—21). The pressure traverse can be constructed starting from a known wellhead pressure, too. — In order to determine γ_k, we have to know B_t for various pressures. The multiphase volume factor is the volume at pressure p of the oil, gas and water produced per m³ of stocktank oil, that is,

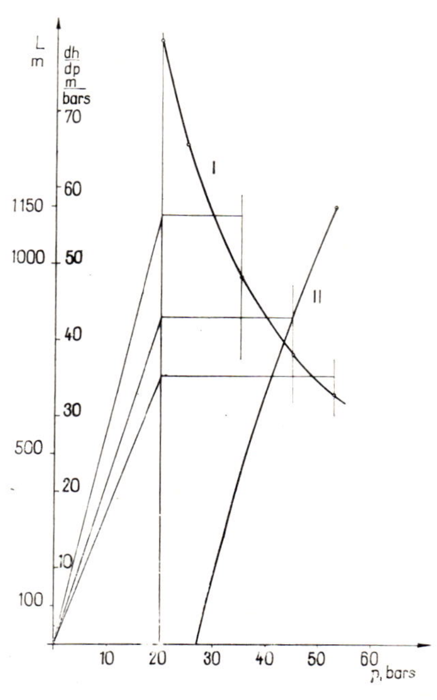

Fig. 1.4—21 Graphical construction of pressure traverse

$$B_t = B_o + \frac{p_n T z}{p T_n} (R_{go} - R_s) + R_{wo} \,. \qquad 1.4\text{—}36$$

The variation of B_o, the volume factor of oil, and of gas solubility R_s, the amount of gas dissolved in a unit volume of oil, against pressure at the mean flowing temperature can be established by means of diagrams based on laboratory experiments. The curves $B_o = \Phi(p)$ and $R_s = \Phi(p)$ can often be substituted in a fair approximation by straight lines (Figs 1.4—22 and 1.4—23). Deviations will not be significant except at low pressures. In the notation used in the graphs,

$$B_o = B_{ok} + n_B p$$

and

$$R_s = R_{sk} + n_R p \,.$$

The mass factor M_t, also required for computing γ_k, is independent of temperature and pressure and can be computed using Eq. 1.4—10, given farther above:

$$M_t = \varrho_o + \varrho_g R_{go} + \varrho_w R_{wo} .$$

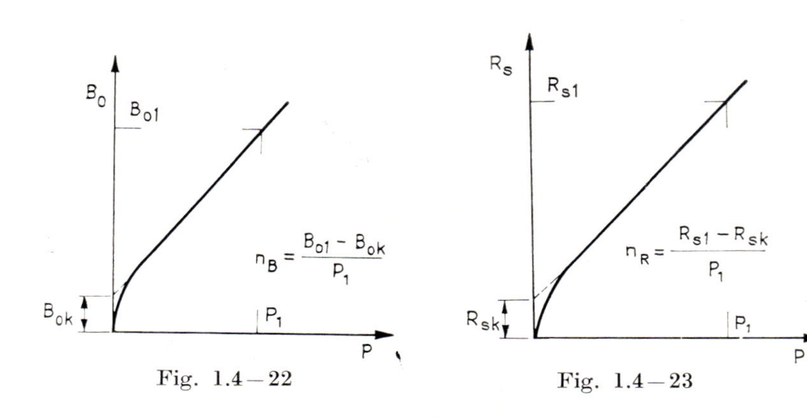

Fig. 1.4—22

Fig. 1.4—23

Example 1.4—4. Construct the pressure traverse if $q_o = 42.4$ m³/d; $R_{go} = 164$ m³/m³; $R_{wo} = 0.00$ m³/m³; $L = 1150$ m; $d = 2\ 7/8$ in.; $d_i = 0.062$ m; $p_1 = 53$ bars; $\bar{T} = 324$ K; $\varrho_{on} = 830$ kg/m³; $\varrho_{gn} = 1.1$ kg/m³; $\varrho_{wn} = 1000$ kg/m³; $R_{sk} = 10$ m³/m³; $n_R = 0.75$ (m³/m³)/bar; $B_{ok} = 1.14$ m³/m³ $n_B = 1.47 \times 10^{-3}$(m³/m³)/bar; $T_c = 232$ K; $p_c = 45.5$ bars; $p_n = 0.981$ bar and $T_n = 288$ K. — B_t is calculated using Eq. 1.4—34 at the pressure values 53, 45, 35, 25 and 20 bars. The principal intermediate results of the calculation are listed in Table 1.4—2. By Eq. 1.4—10,

$$M_t = 830 + 164 \times 1.1 = 1010 \text{ kg/m}^3 ,$$

and

$$q_o M_t = 4.91 \times 10^{-4} \times 1010 = 0.496 \text{ kg/s} .$$

Diagram 1.4—20 now yields $C = 1.45 \times 10^3$. — Graph I in Fig. 1.4—21 is the differential curve $dh/dp = \Phi(p)$. Graph II represents the function $h = \Phi'(p)$, resulting from the graphical integration of Graph I. The starting point of this graph is defined by the co-ordinates L and p_1 valid at the lower end of the tubing string. According to the diagram, the flowing wellhead pressure will be 26.8 bars. If the pressure required to transport the well fluids through the flow line is less than that, the well will flow. — Poettmann and Carpenter determined the loss factor f by measuring flow parameters in tubing-size pipes. In wells of very high producing rate, the casing annulus is often used to lift oil in order to better utilize the well cross-section. Experiments concerning flow in the casing annulus were carried out by Baxendell on the La Paz field, Venezuela.

Table 1.4—2

p	T_{pr}	p_{pr}	z	n_{RP}	R_s	$R-R_s$	B_g
bars	—	—	—	m³/m³	m³/m³	m³/m³	m³/m³
53	1.40	1.16	0.855	40.0	50.0	114	2.10
45	1.40	0.99	0.875	34.0	44.0	120	2.66
35	1.40	0.77	0.900	26.4	36.4	128	3.74
25	1.40	0.55	0.935	18.9	28.9	135	5.76
20	1.40	0.44	0.950	15.1	25.1	139	7.52

p	n_{BP}	B_o	B_t	ϱ_k	$\frac{C}{\varrho_k}$	$\frac{dp}{dh}$	$\frac{dh}{dp}$
bars	$10^{-2}\frac{m^3}{m^3}$	m³/m³	m³/m³	kg/m³	kg/m³	10^{-2} bar/m	m/bar
53	8.10	1.22	3.32	304	4.77	3.03	33.0
45	6.89	1.21	3.87	261	5.56	2.61	38.2
35	5.35	1.19	4.93	204	7.11	2.07	48.3
25	3.83	1.18	6.94	145	10.00	1.52	65.7
20	3.06	1.17	8.69	116	12.50	1.26	79.3

$M_t = 830 + 164 \times 1.1 = 1010$ kg/m³
$q_o M_t = 4.91 \times 10^{-4} \times 1010 = 0.496$ kg/s $\quad C = 1.45 \times 10^3$

(c) Refinements of the Poettmann—Carpenter method

According to Baxendell, Eq. 1.4—32 can be used to calculate the pressure gradient of flow in strings of annular section if rewritten in the form

$$dh = \frac{\dfrac{B_t}{M_t g}\,dp}{1 + \dfrac{0.331 f q_o^2 B_t^2}{(d_{Ci}^2 - d_{To}^2)^2 (d_{Ci} - d_{To})}} . \qquad 1.4\text{—}37$$

The f diagram, which has been determined experimentally, is section *3* of the diagram shown in Fig. 1.4—24. In the case of flow through an annular string, the d_i figuring in the expression $q_o M_t / d_i$ is to be replaced by $(d_{Ci} + d_{To})$. Section *1* of the curve is the original 'Poettmann—Carpenter f-curve' (Fig. 1.4—19). Baxendell suggests to join the two curves by the transitional section *2*. He states Eq. 1.4—37 to be suited also for determining the pressure drop in liquid and gas moving through horizontal flow lines. Baxendell established his 'f-curve' partly by experiments on horizontal flow lines.

In multiple and midi completions, small conduits of 1.66 in. and 1.9 in. diameter are increasingly gaining in importance. Hagedorn and Brown

(1964) checked the Poettmann–Carpenter theory on 1.66 in. tubing. As a first result they have found the 'f-curve' shown in Fig. 1.4–19 to require some modifications. If oil viscosity is less than 12 cP, viscosity effects are negligible, but the fact that oil and gas are in two-phase flow must not be left out of consideration. The abscissa of the 'f-curve' is calibrated in terms of

$$\left(a\frac{q_o M_t}{d_i} + bR\right),$$

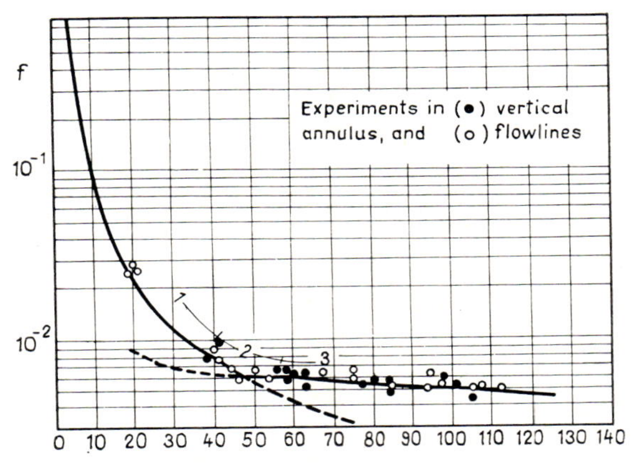

Fig. 1.4–24 Loss factor of horizontal and vertical two-phase flow, after Baxendell and Thomas (1961)

where a and b are given constants. If oil viscosity exceeds 12 cP, the abscissa is calibrated in terms of

$$\left(a'\frac{q_o M_t}{d_i \mu^{0.25}} + b'R\right).$$

The procedure is not to be applied at oil viscosities exceeding 25 cP.

Hagedorn and Brown have further found that the friction factor can simply be read off the Moody diagram (Fig. 1.1–1), if the Reynolds number is calculated from the data of the two-phase flow, using a relationship that accounts also for slippage loss. Hagedorn and Brown do not neglect the kinetic energy term, either. They consider their generalized procedure to apply to tubing of any size (Hagedorn and Brown 1965). Although the work of Hagedorn and Brown has considerably improved the accuracy and logical consistency of the Poettmann–Carpenter method, we shall not describe it in detail, as we prefer Ros' method both didactically and accuracywise (cf. Section 1.4.3e).

(d) Gilbert's theory. Gradient curves

According to Gilbert (1955) it is superfluous to determine pressure traverses by calculation. Gilbert has constructed sets of curves based on flow experiments in oil wells: it is sufficient to choose the flow curve suited to the problem in hand. He performed his experiments on wells producing oil of 825—964 kg/m³ density, but deemed his curves approximately suitable also for liquids of other densities. The Gilbert procedure presupposes, then, that the pressure gradient depends largely on the tubing diameter d_i, the flow rate q_l, the gas–oil ratio R, and the pressure p. Figures from A—1 to A—10 in the Appendix show Gilbert's pressure-gradient curves. The first five Figures refer to tubing of 2 3/8 in. diameter; the last five to tubing of 2 7/8 in. diameter. Each sheet carries a set of $h = \Phi(p)_{di:ql:R}$ curves for various flow rates. The curve with the least mean gradient is indicated by an arrow. — In addition to the gradient curves, which are shown here transposed into *SI* units, Gilbert has published similar sets of gradient curves also for the standard tubing sizes of 1.6, 1.9 and 3.5 in. The required pressure gradient curve can be selected as follows. From the set of sheets corresponding to the tubing size d in question one selects the two sheets with the q_l's bracketing the actual q_l figuring in the problem. If the sheets show no gradient curve for the desired R value, the curve should be constructed by interpolation on tracing paper placed into the sheet. The gradient curve for the desired value of q_l is interpolated between the two gradient curves belonging to the two bracketing values of q_l, both of which are traced on the same sheet of tracing paper. This operation results in the $h = \Phi(p)_{di:ql:R}$ curve shown in Fig. 1.4—25. The ordinate h' corresponding to a pressure p_1

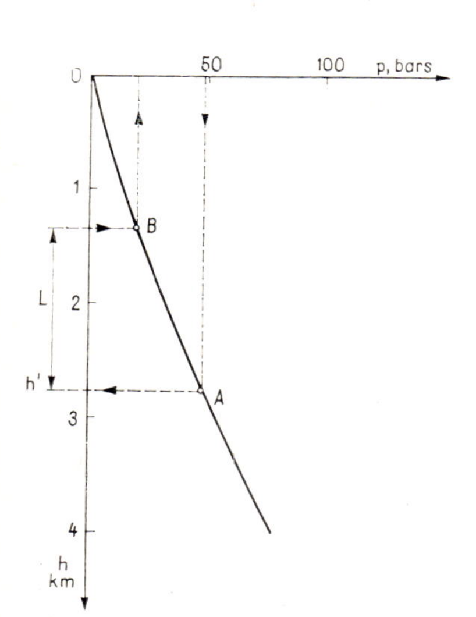

Fig. 1.4—25 Determining wellhead pressure using the pressure traverse

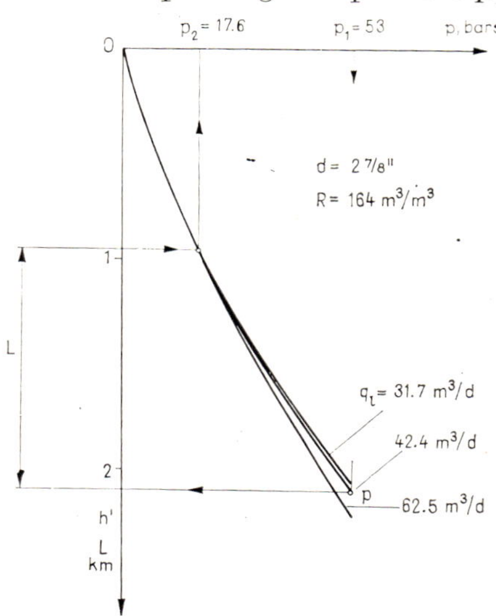

Fig. 1.4—26 Plotting the pressure traverse by interpolation

at the lower end of the tubing is read off this curve. At a distance L from that point one finds the wellhead pressure p_2; the curve section between the points A and B is the actual pressure gradient curve.

Example 1.4—5. Solve Example 1.4—4 by Gilbert's method. The data of the problem are: $p_1 = 35$ bars; $q_l = 42.4$ m³/d; $R = 164$ m³/m³; $L = 1150$ m; $d = 2\ 7/8$ in. — Select the sheets A—8 and A—9 corresponding to $d = 2\ 7/8$ in. and the bracketing q_ls 31.7 and 62.5 m³/d, respectively. On a sheet of tracing paper, interpolate curves for $R = 164$ between the curves for $R = 150$ and $R = 200$ on both sheets; then interpolate between the two traces so obtained the pressure gradient curve corresponding to $q_l = 31.7$ m³/d. According to the co-ordinate system on the sheet which is visible through or can be transferred to the tracing paper, the h' value belonging to 53.0 bars is 2100 m. The wellhead pressure is consequently 17.6 bars, read at 2100—1150 = 950 m. Flow in the tubing string will be characterized by the pressure traverse between 43.0 and 17.6 bars (Fig. 1.4—26). It was

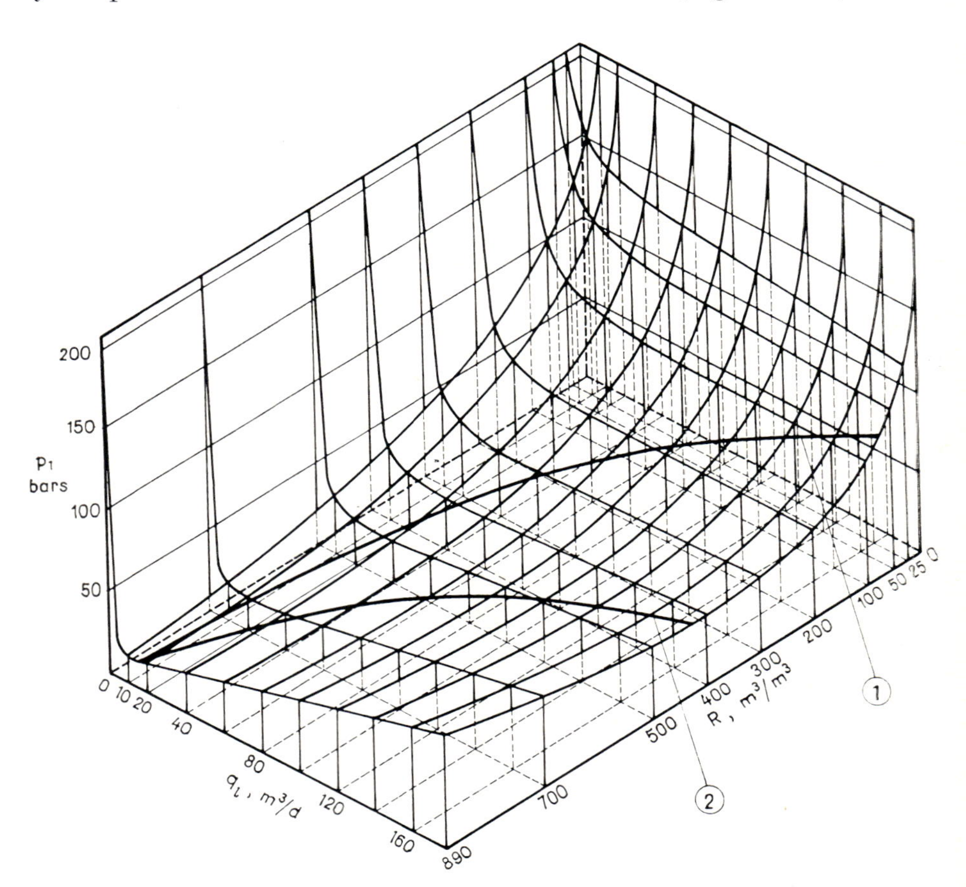

Fig. 1.4—27 Characteristic surface of vertical two-phase flow, after Gilbert (1954)

on the basis of the pressure gradient curves that Gilbert prepared his diagram which, transposed into *SI* units, is shown as Fig. 1.4—27 here. It shows the liquid flow rates of a 2 7/8 in. tubing string of 2438 m length at a wellhead pressure of 1 bar for a variety of gas–oil ratios. The diagram thus essentially shows the relationship between p_1, q_l and R; with L, d and p_2 kept constant. Thinking back to Krylov's theory we may visualize the vertical co-ordinate axis calibrated in the pressure gradient $\bar{\xi}$ rather than in p_1, since in the relationship $\bar{\xi} = (p_1 - p_2)/L\gamma_l$, p_1 is the only independent variable. The $p_1 = \mathrm{f}(q_l)_R$ curves reveal that to any value of specific gas consumption there is a liquid flow rate q_l at which the tubing-bottom pressure p_1 (or the pressure gradient $\bar{\xi}$) is a minimum. The total energy loss increases above this minimum by the increase in friction loss at higher flow rates and by the increase in slippage loss at lower ones, so that p_1 (or $\bar{\xi}$) will be greater than the minimum. The points $p_{1\,\mathrm{min}}$ are connected by the curve *1*. On the other hand, the curves $p_1 = \mathrm{f}'(R)_{ql}$ reveal that to any liquid flow rate q_l there belongs a specific gas consumption R at which p_1 (or $\bar{\xi}$) is least. The points $p'_{1\,\mathrm{min}}$ are connected by curve *2*. The total energy loss increases above this minimum owing to an increased slippage loss at lower, and to an increased friction loss at higher relative gas concentrations.

Gilbert's pressure gradient curves permit the rapid solution of practical problems. There is, however, the disadvantage that pressure changes

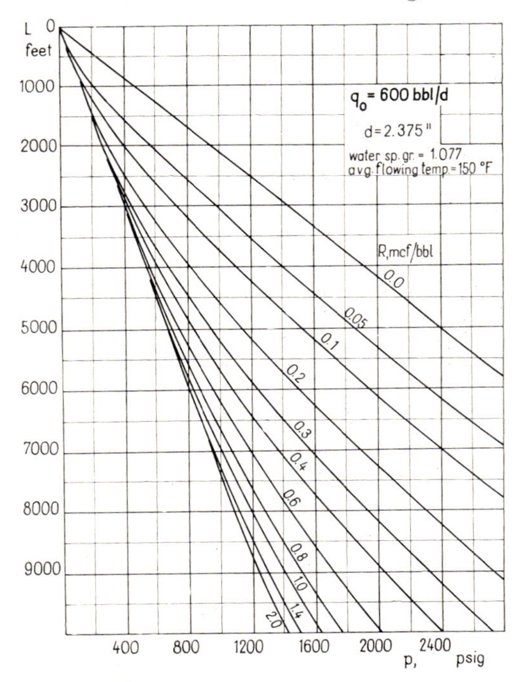

Fig. 1.4—28 Family of pressure traverses; from *USI Handbook of Gas Lift, 1959*, p. 825 (by permission of Axelson, Inc., Longview, Texas, USA)

cannot be established to a satisfactory accuracy except if the properties of the liquid and gas in hand closely resemble those studied by Gilbert. This circumstance has induced certain workers to compute sets of pressure gradient curves for liquids and gases of given properties, using the Poettmann—Carpenter method (McAfee 1961). The set of curves published by Garrett Oil covers for instance the production range from 8 to 1590 m³/d and the *GOR* range from 0 to 712 m³/m³, with numerous intermediate values in both ranges. Separate sets of curves were prepared for oil of density 865 kg/m³ and for water of density 1.074 kg/m³ (USI Handbook 1959). According to McAfee, the mean deviation of flowing bottom-hole pressure computed from the wellhead pressure did not exceed 3.9 percent under the conditions investigated by him. This cannot be generalized, however. For an example, consider Fig. 1.4—28 showing one curve of the Garretts' set, in the original American units. The pressure gradient is seen, in contrast to Gilbert's curves, to be the less, the greater the gas–oil ratio. This seems unjustified, however, in the light of both theoretical considerations and practical experience. Nind (1964) gave a comparative table which is reproduced here in metric units as Table 1.4—3. The Table gives flowing bottom-hole

Table 1.4—3

d	q_o	R	L	p_{wf} bars, overpressure		
in.	m³/d	m³/m³	m	From Gilbert curves	From Garrett curves	
					For oil	For water
1.66	63.5	107	3048	103	180.5	200.1
1.90	15.9	17.8	2134	113.8	163.8	188.3
	95.3	178	3048	96.5	138.3	152
2 3/8	15.9	178	3048	68.9	133.4	134.4
	95.3	17.8	1524	77.2	79.2	104.9
	95.3	17.8	3048	179.5	200	255
2 7/8	31.8	178	3048	86.2	85.5	255
	95.3	35.6	2438	114.7	112.8	141.2
3 1/2	63.5	17.8	1524	84.7	57.9	93
		17.8	3048	206.9	176.6	255
		178	1524	33	26.2	26.2
		178	3048	86.9	58.6	68.9

pressures for various tubing sizes, rate of flow and specific gas consumptions on the basis of Gilbert's and the Garretts' sets of curves. The differences between the bottom-hole pressures established by the two methods are seen to be quite considerable in some instances. This bids caution as regards the practical application of these curves.

One of the principal merits of the Poettmann—Carpenter and Gilbert methods is that the practical importance of gradient curves have been pointed out. These curves may be regarded as building blocks that play a highly useful role in buildings of widely different functions, i.e. in the solution of a great variety of practical problems.

(e) Ros' theory

Ros pointed out that pressure traverses established by the Poettmann—Carpenter method may be fraught with considerable error, for the following reasons: (i) Eq. 1.4—29 regards the total energy loss to be frictional in nature (with h_v varying as the square of flow velocity). Now if slippage losses play an important role, this may give rise to a significant deviation

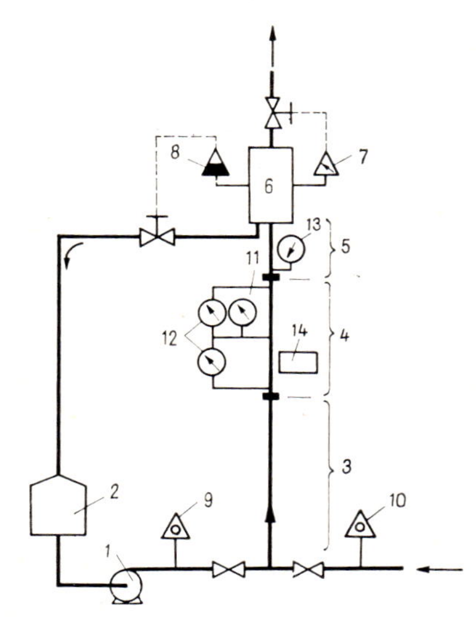

Fig. 1.4—29 Ros' experimental set-up, after Ros (1961)

from reality. (ii) In Eq. 1.4—29 the loss factor f is considered, at a given diameter d shown in Fig. 1.4—19, to depend exclusively on the mass flow rate $q_o M_t$, that is, to be constant all along the flow string. In reality, the changes in the flow velocities of both the gas and the liquid phase entail a change in f from the lower to the upper end of the tubing string. (iii) The energy loss is not independent of viscosity, although below 6 cP the effect of viscosity is very small indeed. (iv) It is more appropriate to choose the pressure rather than the energy balance as a fundamental equation, because the pressure balance permits the unambiguous separation of the friction loss from the slippage loss. (v) The total energy loss is significantly affected by the nature of the flow pattern. Losses are to be described by a different relationship for each type of flow pattern.

Ros' theory is based on laboratory experiments (Ros 1961, Duns and Ros 1963). The diagram of the experimental apparatus is shown as Fig. 1.4—29. Pump *1* delivers liquid from tank *2* to the experimental pipe *3*, *4*, *5*. The gas conduit on the right-hand side, indicated by an arrow, delivers air to this same pipe. The experimental pipe consists of three sections: the inflow section *3*, the measuring section *4* and the outflow section *5*. The liquid separated in separator *6* is recirculated to tank *2*; the air escapes in the direction of the upper arrow. Liquid level and pressure in the separator are regulated by devices *8* and *7*, respectively. Liquid throughput is measured by instrument *9*, gas throughput by instrument *10*. Pressures in pipe section *4* are measured by manometer *11* and differential pressure gauge *12*; flow temperature is measured by thermometer *13*. A radioactive tracer is admixed to the liquid: the counter *14* serves to determine the radiation level in the pipe section *4*; from its reading, the liquid–gas saturation affected by slippage can be calculated.

Ros established the very important fact that the pressure gradient is much greater in the initial section of an experimental pipe than higher up. He used an inflow section of 25 m length before his measuring pipe section of 10 m length. He found pressure gradients in the inflow section exceeding by as much as three times the gradients in the measuring section. The experimental data were put in a pressure-balance equation that was derived as follows. As a starting assumption, acceleration effects are considered to be negligible. By Fig. 1.4—30, the difference in compressive force between the two ends of an infinitesimal length of tubing dh is

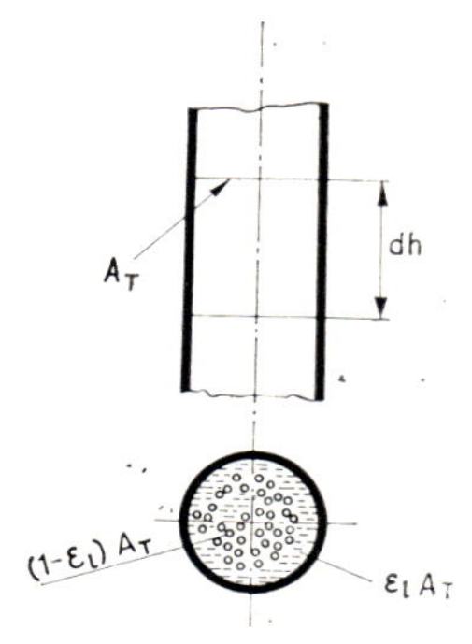

Fig. 1.4—30

$$\mathrm{d}p A_T = A_T \varepsilon_l \mathrm{d}h\, \gamma_l + A_T(1 - \varepsilon_l)\, \mathrm{d}h\, \gamma_g + A_T p_s\,;$$

that is, it equals the aggregate weight of the liquid plus gas contained in the tubing length considered, plus the pressure drop due to friction, $A_T\, p_s$. Division of both sides by $A_T\, \mathrm{d}h\, \gamma_l$ results in

$$\frac{\mathrm{d}p\, A_T}{A_T \mathrm{d}h\, \gamma_l} = \frac{A_T \varepsilon_l\, \mathrm{d}h\, \gamma_l}{A_T \mathrm{d}h\, \gamma_l} + \frac{A_T(1 - \varepsilon_l)\, \mathrm{d}h\, \gamma_g}{A_T \mathrm{d}h\, \gamma_l} + \frac{A_T p_s}{A_T \mathrm{d}h\, \gamma_l}. \qquad 1.4\text{—}38$$

The left-hand side now equals ξ, and the third term on the right-hand side equals ξ_s. The other two terms on the right-hand side can also be expressed more simply:

$$\xi = \varepsilon_l + (1 - \varepsilon_l)\frac{\varrho_g}{\varrho_l} + \xi_s\,, \qquad 1.4\text{—}38\text{a}$$

where ε_l is the portion occupied by the liquid of the cross-sectional area A_T of the tubing; γ_l is the effective specific weight of the liquid and γ_g is that of the gas.

Ros has shown that the flowing pressure gradient ξ is a function of twelve variables, as listed in Table 1.4—4. These variables do not include pressure and temperature, because the gradient is to be established at the actual values of these two parameters. Applying the rules of dimensional analysis, one can derive 10 dimensionless factors from the 12 variables. Of these, the following five play significant roles according to Ros:

Table 1.4—4

	Designation of variable	Symbol
(I) Pipe parameters	I. D. Roughness of wall Deflection of hole from vertical	d k Φ
(II) Liquid and gas parameters	Liquid density Gas density Liquid viscosity Gas viscosity Flow velocity of liquid and Flow velocity of gas, both over entire cross-sectional area A_T	ϱ_l ϱ_g μ_l μ_g v_l v_g
(III) Parameters of interaction between liquid and gas	Surface tension Wetting angle Acceleration of gravity	σ α g

N_{lv}, liquid flow velocity number, $N_{lv} = v_l \sqrt[4]{\varrho_l/g\sigma}$;

R, in-situ gas–oil ratio, $R = \dfrac{v_g}{v_l}$;

N_d, pipe diameter number, $N_d = d_i \sqrt{\varrho_l\, g/\sigma}$;

N_μ, liquid-viscosity number, $N_\mu = \mu_l \sqrt[4]{g/\varrho_l\, \sigma^3}$;

N_ϱ, gas–liquid density number, $N_\varrho = \varrho_g/\varrho_l$.

The terms on the right-hand side of Eq. 1.4—38 are to be computed in different ways for different flow regions. Three flow regions are distinguished, shown marked I, II and III in Fig. 1.4—10.

(e.1) *The mass gradient* ξ_m. — If in Eq. 1.4—38, $\xi_s = 0$, then

$$\xi = \xi_m = \varepsilon_l + (1 - \varepsilon_l)\frac{\varrho_g}{\varrho_l}. \qquad 1.4\text{—}39$$

Ros terms ξ_m the static gradient. It is more appropriate in the present author's opinion to call it the mass gradient because in the static state there cannot be two phases side by side, and the gas–oil ratio's being affected by the slippage loss likewise presupposes flow, i.e. a dynamic state. By Eq. 1.4—15, slippage velocity is the difference between the bulk flow velocities of the gas and the liquid in the given section. Some rewriting results in

$$v_{gs} = \frac{v_g}{1 - \varepsilon_l} - \frac{v_l}{\varepsilon_l}. \qquad 1.4\text{—}40$$

Slippage velocity can be expressed in terms of the flow parameters also in a dimensionless form:

$$S = v_{gs}\sqrt[4]{\frac{\varrho_l}{g\sigma}}\,. \qquad 1.4-41$$

If S is known, the v_{gs} can be calculated using Eq. 1.4—41; next, ε_l can be calculated by Eq. 1.4—39 and, finally, ξ_m by Eq. 1.4—38.

S in Flow Region I. — By Fig. 1.4—10, flow in Region I is of the slug or foam type, and

$$S = F_1 + F_2 N_{lv} + \left(F_3 - \frac{F_4}{N_d}\right)\left(\frac{RN_{lv}}{1 + N_{lv}}\right)^2. \qquad 1.4-42$$

By Figs 1.4—31 and 1.4—32, the quantities $F_{1,2,3,4}$ are parameters depending on N_μ, the properties of the liquid. Consequently,

$$S = \Phi(N_\mu;\ N_d;\ R;\ N_{lv});$$

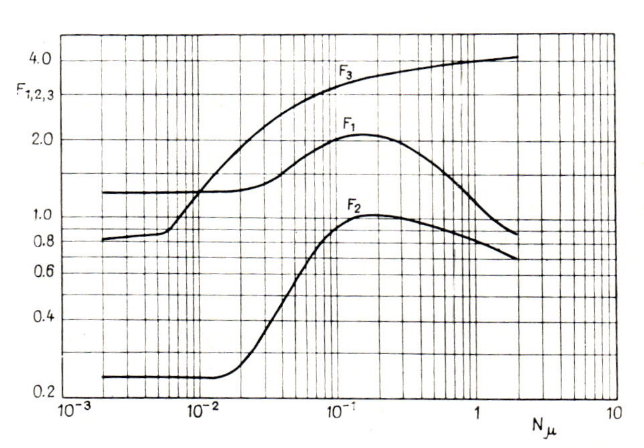

Fig. 1.4—31

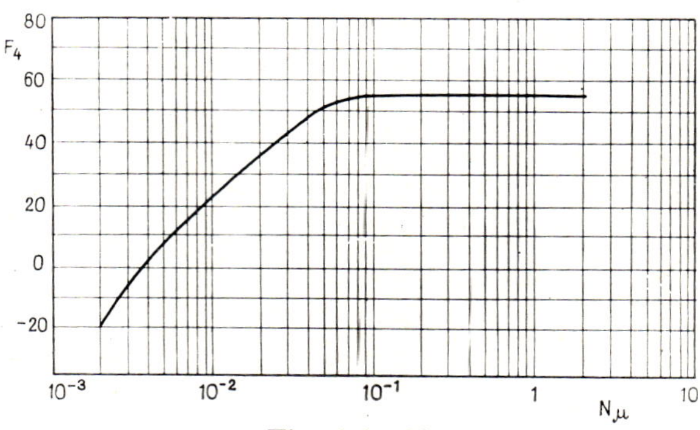

Fig. 1.4—32

that is, S is a function of four dimensionless quantities. In the case of annular flow N_d depends on the wetted circumference:

$$N_d = (d_{Ci} + d_{To}) \sqrt{\frac{\varrho_l g}{\sigma}} \,. \qquad 1.4\text{—}43$$

Region I extends between $RN_{lv} = 0$ and

$$RN_{lv} = L_1 + L_2 N_{lv} \,. \qquad 1.4\text{—}44$$

By Fig. 1.4—33, L_1 and L_2 both depend on N_d, the dimensionless tubing diameter. In the case of annular flow, Eq. 1.4—43 should be used also here.

S in Flow Region II. — This Region extends to medium gas velocities. Flow is of the slug or foam type.

$$S = (1 + F_5) \frac{(RN_{lv})^{0.982} + 0.029 N_d + F_6}{(1 + F_7 N_{lv})^2} \,. \qquad 1.4\text{—}45$$

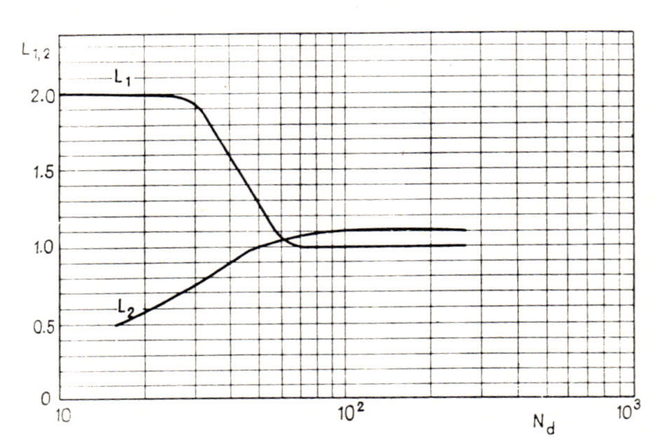

Fig. 1.4—33

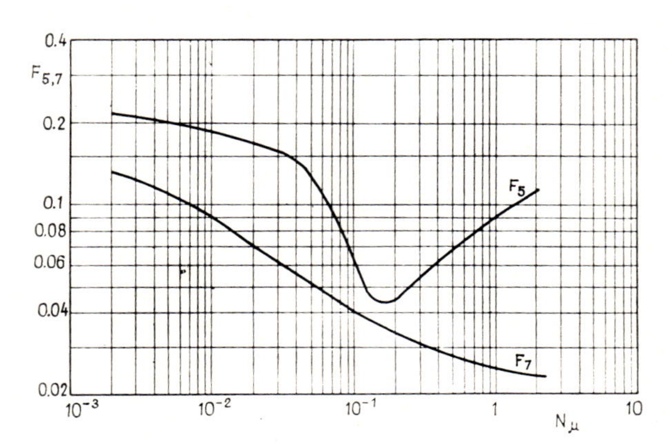

Fig. 1.4—34

Quantities $F_{5,7}$ depend by Figs 1.4—34 and 1.4—35 also in this case on the liquid properties; that is, S is, as above, a function of the four dimensionless factors. Flow Region II extends from the upper boundary of Region I to the boundary

$$RN_{lv} = 50 + 36\, N_{lv} \,.$$

Special attention should be paid to heading. The total pressure gradient is to be determined by interpolation between the gradients valid in Regions I and II, in the way described in the subsection on the 'Total pressure gradient'.

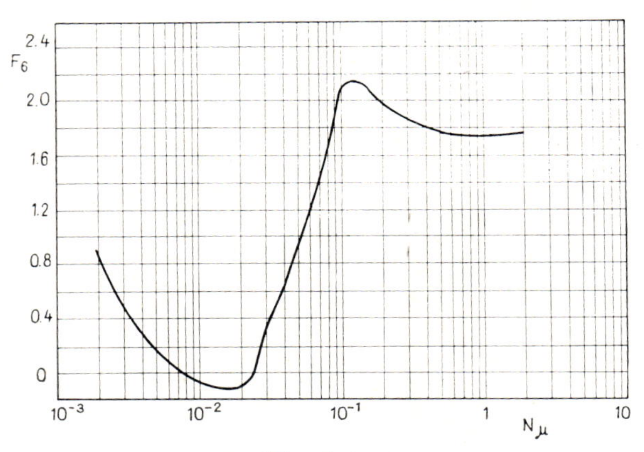

Fig. 1.4—35

S in Flow Region III. — In the case of mist flow, the high-velocity gas entrains the small drops of gas so fast that there is practically no velocity difference between the phases; that is, $v_{gs} = 0$. Substitution into Eq. 1.4—40 results in

$$\varepsilon_l = \frac{1}{1 + \frac{v_g}{v_l}} = \frac{1}{1 + R} \,. \qquad 1.4\text{—}46$$

Having determined ε_l, it is easy to calculate ξ_m using Eq. 1.4—39. There is a zone of transition between flow Regions II and III. The lower boundary of Region III does not coincide with the upper boundary of Region II: it is defined by the relationship

$$RN_{lv} = 75 + 84\, N_{lv}^{0.75} \,. \qquad 1.4\text{—}47$$

(e.2) *The friction gradient* ξ_s. — Friction loss in Flow Regions I and II is due to shear in the continuous liquid phase. A relationship suited for calculating this effect is the differential form of Eq. 1.4—29, with the interpretation $\mathrm{d}p = \mathrm{d}h_v\, \varrho_g$:

$$\mathrm{d}p = 4f \frac{v_k^2\, \bar{\varrho}_k}{2d_i}\, \mathrm{d}h \,. \qquad 1.4\text{—}48$$

The presence of gas results in an accelerated flow of the liquid phase. The bulk velocity through the cross-section A_T is

$$v_k = \frac{q_l + q_l R}{A_T} = \frac{q_l}{A_T}(1 + R) = v_l \left(1 + \frac{v_g}{v_l}\right). \qquad 1.4\text{—}49$$

Let the volume of an infinitesimal length of tubing be $(1 + R)$. Let the volume of the oil be unity, and that of the gas, R. The mass of liquid in this length of tubing is ϱ_l, that of the gas is $R\varrho_g$, and hence the total mass is $\varrho_l + R\varrho_g$. The total fluid density in the tubing is, then,

$$\bar{\varrho}_k = \frac{\varrho_l + R\,\varrho_g}{1 + R} = \frac{\varrho_l}{1 + \dfrac{v_g}{v_l}}, \qquad 1.4\text{—}50$$

because the mass of the gas is usually a negligible fraction of the total fluid mass. Substituting the above equations for v_k and $\bar{\varrho}_k$ into Eq. 1.4—48, we get

$$\left(\frac{dp}{dh}\right)_s = 4f\frac{v_l^2 \varrho_l}{2d_i}\left(1 + \frac{v_g}{v_l}\right). \qquad 1.4\text{—}51$$

Dividing both sides of this equation by the specific weight of the liquid phase, $\varrho_l g$, and substituting v_l, v_g and d_i by the appropriate dimensionless factors, we find that

$$\xi_s = \left(\frac{dp}{dh\varrho_l g}\right)_s = 2f\frac{N_{lv}(N_{lv} + RN_{lv})}{N_d}. \qquad 1.4\text{—}52$$

Experiments have furnished the friction factor formula

$$f = f_1 \frac{f_2}{f_3}. \qquad 1.4\text{—}53$$

Factor f depends largely on f_1, which can be read off a slightly modified Moody diagram as a function of the Reynolds number for the liquid (Fig. 1.4—36):

$$N_{Rel} = \frac{\varrho_l v_l d_i}{\mu_l}. \qquad 1.4\text{—}54$$

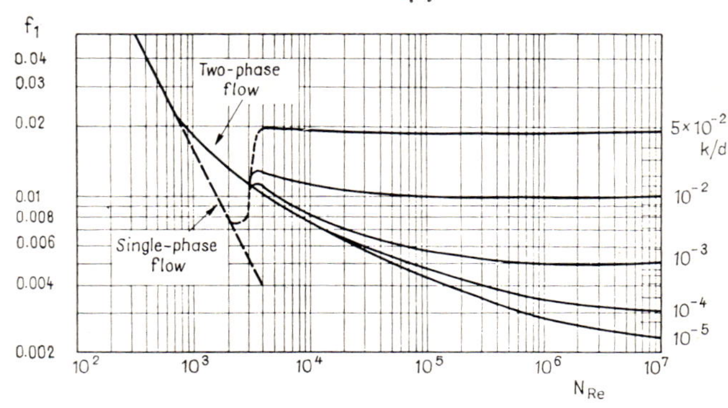

Fig. 1.4—36 Friction factor in vertical flow, after Duns and Ros (1963)

In the case of annular flow, d_i is to be substituted by $(d_{Ci} - d_{To})$, and

$$N_d = (d_{Ci} - d_{To})\sqrt{\frac{\varrho_l g}{\sigma}}. \qquad 1.4\text{—}55$$

The diagram shown as Fig. 1.4—36 differs from the Moody diagram for single-phase flow in that the transitional zone between the laminar and turbulent regions is different, simpler. Factor f_2 depends at a given tubing size primarily on the in-situ gas–liquid ratio and can be read off Fig. 1.4—37 as a function of the dimensionless expression $f_1RN_d^{2/3}$. Its value is close to unity at low values of R, but decreases appreciably at higher Rs. Factor

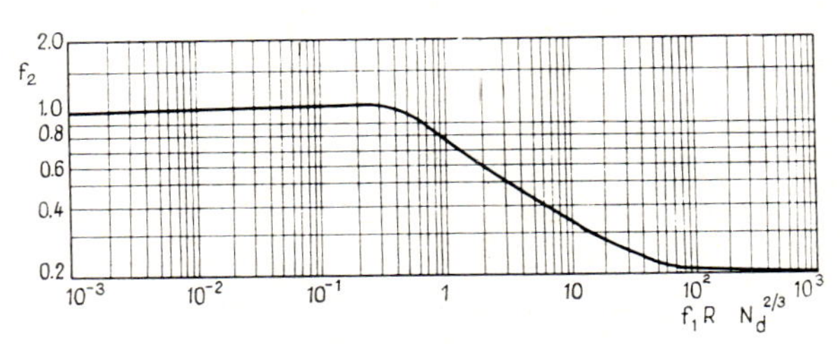

Fig. 1.4—37

f_3 is another correction factor, whose value depends on the viscosity of the liquid and, once again, on the gas–liquid ratio R. It can be calculated from the equation

$$f_3 = 1 + f_1\sqrt{\frac{R}{50}}. \qquad 1.4\text{—}56$$

At R values current in practice this factor has no particular significance except if liquid viscosity exceeds 5×10^{-5} m²/s. The relationship holds for Flow Regions I and II, that is, from $RN_{lv} = 0$ to $RN_{lv} = 50 + 36N_{lv}$. In Region III, the continuous phase is the gas: friction loss is therefore due to friction between the gas and the pipe wall. Since

$$\mathrm{d}p = 4f\frac{v_g^2\,\varrho_g}{2d_i}\mathrm{d}h \quad \text{and} \quad \xi_s = \frac{\mathrm{d}p}{\mathrm{d}h\,\varrho_l g},$$

the friction gradient, expressed in terms of liquid column height, is

$$\xi_s = 4f\frac{\varrho_g}{\varrho_l}\frac{v_g^2}{2gd_i}, \qquad 1.4\text{—}57$$

and, using dimensionless factors,

$$\xi_s = 2fN_\varrho\frac{(RN_{lv})^2}{N_d}. \qquad 1.4\text{—}58$$

Factor f equals f_1 and its value can be read off the Moody diagram at the Reynolds number referring to the gas, that is,

$$N_{\text{Reg}} = \frac{\varrho_g v_g d_i}{\mu_g} . \qquad 1.4-59$$

In mist flow, the roughness of the pipe does not enter into direct play: it is 'sensed' through the intermediary of a liquid film covering the pipe wall. This film can be strongly rippled and thus exert a considerable hydraulic resistance, being responsible for a major part of the friction loss. The intensity of this effect is a function of the roughness k. It used to be

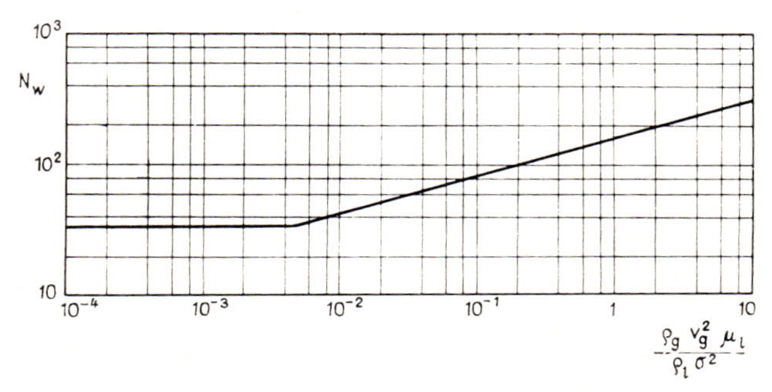

Fig. 1.4—38

assumed earlier that the liquid film is of constant thickness but we are aware today that the situation is much more complicated than that. The liquid ripples on the pipe wall are generated by the upward drag of the gas on the liquid film covering the wall. This phenomenon can be characterized by the Weber number

$$N_{\text{We}} = \frac{\varrho_g v_g^2 k}{\sigma} . \qquad 1.4-60$$

N_{We} can be computed from the operational parameters of the well. In its knowledge, k can be calculated using Eq. 1.4—60. In a first approximation, $N_{\text{We}} = 34$. More accurate values can be had by reading N_{We} off Fig. 1.4—38 as a function of the expression

$$\frac{\varrho_g v_g^2 \mu_l}{\varrho_l \sigma^2} .$$

In mist flow and at high values of v_g, the ripple effect is weak but never less than the roughness of the pipe, that is, about $10^{-3} \times d$. On the other hand, on transition to the Region II, in slug-type flow the waviness of the film may be quite considerable, owing to the breaking of each ripple as it collides with the next ripple above. k/d_i may then attain values up to 0.5.

If $k/d_i < 0.05$, then f_1 may be read off the Moody diagram. If, on the other hand, $k/d > 0.05$, then the following formula should be used:

$$f_1 = \frac{1}{4\left[\lg\left(0.27\,\frac{k}{d_i}\right)\right]^2} + 0.067\left(\frac{k}{d_i}\right)^{1.73} . \qquad 1.4\text{—}61$$

A strong rippling of the liquid adhering to the pipe wall may be an appreciable obstacle to gas flow. As a further refinement of the calculation method, d_i may be substituted by $(d_i - k)$ and v by $v_g d_i^2/(d_i - k)^2$ and k may be determined by iteration.

(e.3) *The accelerating gradient* ξ_a. — Acceleration is so slight as to be negligible in almost every practical application. It might be significant in the case of mist flow, though. There is practically no slippage in mist flow and flow velocity equals v_g. This value is so high that the thickness of the liquid film and its resistance to flow can be neglected. The mass flow rate is constant and equal to $(\varrho_l v_l + \varrho_g v_g)$. The acceleration gradient is

$$\left(\frac{\mathrm{d}p}{\mathrm{d}h}\right)_a = -(\varrho_l\, v_l + \varrho_g v_g)\,\frac{\mathrm{d}v_g}{\mathrm{d}h} . \qquad 1.4\text{—}62$$

Assuming an isothermal change of state, with $pv_g = C$,

$$\mathrm{d}(pv_g) = p\mathrm{d}v_g + v_g\mathrm{d}p = 0,$$

and hence,

$$\mathrm{d}v_g = -\frac{v_g}{p}\,\mathrm{d}p . \qquad 1.4\text{—}63$$

Let us divide both sides of Eq. 1.4—62 by $\varrho_l g$ and let us substitute $\mathrm{d}v_g$ by the expression in Eq. 1.4—63. Then

$$\xi_a = \left(\frac{\mathrm{d}p}{\mathrm{d}h}\right)_a \frac{1}{\varrho_l g} = -(\varrho_l\, v_l + \varrho_g\, v_g)\frac{-v_g \mathrm{d}p}{p\,\mathrm{d}h \varrho_l g} ; \qquad 1.4\text{—}64$$

and since

$$\frac{\mathrm{d}p}{\mathrm{d}h\,\varrho_l g} = \xi ,$$

we have

$$\xi_a = (\varrho_l v_l + \varrho_g v_g)\frac{v_g}{p}\,\xi . \qquad 1.4\text{—}65$$

(e.4) *The total pressure gradient.* — The total gradient adds up as follows:

$$\xi = \xi_m + \xi_s + \xi_a .$$

Replacing ξ_a by the expression in Eq. 1.4—65, and rearranging, we get

$$\xi\left[1 - (\varrho_l v_l + \varrho_g v_g)\frac{v_g}{p}\right] = \xi_m + \xi_s ,$$

and hence,

$$\xi = \frac{\xi_m + \xi_s}{1 - (\varrho_l v_l + \varrho_g v_g)\dfrac{v_g}{p}} . \qquad 1.4\text{—}66$$

Acceleration can thus be accounted for by a correcting factor. Its value may be significant at low pressures and high gas velocities, which conditions might, if anywhere, prevail in the upper reaches of a tubing string. In the zone of transition between Flow Regions II and III the pressure gradient is to be established by linear interpolation between the pressure gradients corresponding to the boundary values. The two boundary RN_{lv}'s corresponding to the value of N_{lv} are

$$RN_{lv} = 50 + 36 N_{lv} ,$$

and

$$RN_{lv} = 75 + 84\, N_{lv}^{0.75},$$

respectively. In mist flow, then, gas density by Eq. 1.4—58 significantly affects the friction gradient even at the boundary of Flow Region III. The value of ϱ_g' corresponding to the boundary RN_{lv} can be rather simply calculated by the following consideration: the mass flow rate $\varrho_g v_g$ is constant, wherefore $\varrho_g' v_g' = \varrho_g v_g$. From this equation, ϱ_g' can be expressed.

(e.5) *Three-phase flow of oil, gas and water.* — The above considerations have all referred to waterless mixtures of oil and gas. The same relationships can be applied also to the mixed flow of water and gas, although at a slightly impaired accuracy. The pressure gradient of water and oil cannot be determined if the two liquids form a stable emulsion. If, however, no such emulsion is to be expected to form, the pressure gradient can be calculated as follows. Pressure gradients are calculated for the entire flowing liquid throughput, first, as if the liquid were all oil [$(\mathrm{d}p/\mathrm{d}h)_0$], and second, as if it were all water [$(\mathrm{d}p/\mathrm{d}h)_w$]. Let the specific water content of the liquid throughput be R_w. Then the gradient for the actual flow can be computed from the above two gradient components as follows:

$$\left(\frac{\mathrm{d}p}{\mathrm{d}h}\right) = (1 - 7.3\, R_w)\left(\frac{\mathrm{d}p}{\mathrm{d}h}\right)_0 + 7.3\, R_w \left(\frac{\mathrm{d}p}{\mathrm{d}h}\right)_w . \qquad 1.4\text{—}67$$

The procedure is applicable to Flow Regions I and II if the water content is less than 10 percent. At the high velocities prevailing in Region III, the formation of an emulsion is always to be expected, especially in the liquid film on the pipe wall. The inaccuracy resulting therefrom may vary over a wide range, even if the oil and gas are easy to separate. A very small amount of oil ($\ll$ 1 percent) dispersed in water will turn the water 'milky'. The behaviour of this milky water differs rather appreciably from that of pure water. For instance, the region of heading is appreciably narrower.

In the region of intermittent flow and below, on the border of Flow Region I, deviations up to 30 percent were observed between the flow gradients of pure and milky water.

(e.6) *Applications.* — According to the authors, the standard deviation for waterless oil is 3 percent in Flow Region I, which equals the reproducibility of the laboratory experiments. The s.d. is 8 percent for slug-type flow in Region II, and 6 percent in Region III. For oil containing less than 10 percent water, the s.d. is 10 percent in Regions I and II. At higher water contents the equations may give results far off the actual conditions. In practice there are several other factors influencing the energy loss of flow, which cannot be taken into account in the laboratory, such as deposits of wax or scale in the tubing.

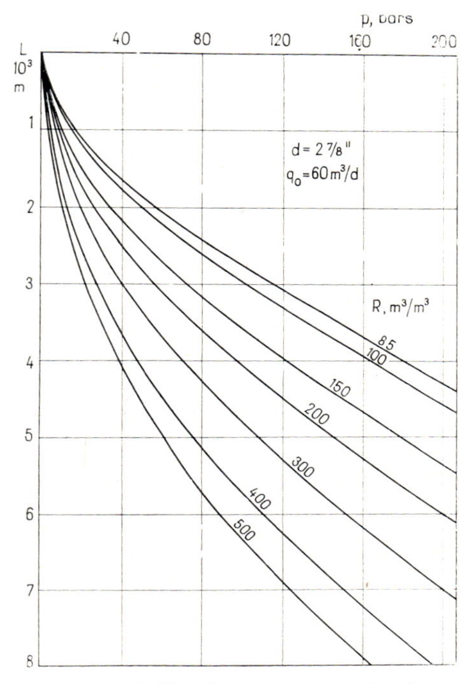

Fig. 1.4—39 Family of curves furnished for Algyő crude by the Ros method

The pressure-gradient curve can in principle be established from gradients calculated for a set of preselected pressures in much the same way as in the Poettmann—Carpenter method (Section 1.4—3). In practice, however, the calculation is so time-consuming that it is usually entrusted to a computer carrying out numerical integration. The result of such computation may be, e.g., a set of 'Gilbert's' pressure gradient curves resembling the set in Fig. 1.4—28. Figure 1.4—39 shows a set of curves calculated by Ros' method for the oil of the Algyő field, Hungary, for gas–oil ratios occurring at that field. The pressure gradient is calculated using Eqs 1.4—38 or 1.4—66. The pressure-dependent factors of the variables in these equations are ϱ_l, ϱ_g, v_l and v_g. These factors can be determined in a variety of ways, including the following procedure for waterless oil:

$$\varrho_l = \frac{\varrho_{on}}{B_o}; \qquad 1.4\text{—}68$$

$$\varrho_g = \frac{\varrho_{gn}}{\dfrac{p_n T z}{T_n p}}; \qquad 1.4\text{—}69$$

$$v_l = \frac{q_o B_o}{A_T};$$ 1.4—70

$$v_g = \frac{p_n \bar{T}(R - R_s) q_o z}{T_n A_T p}.$$ 1.4—71

Example. 1.4—6. Establish a pressure traverse on the basis of Ros' theory, for the following operating conditions: (I) $q_l = 42.4$ m³/day; $R = 164$ m³/m³; $\sigma = 0.030$ kg/m; $g = 9.81$ m/s²; $k = 5 \times 10^{-4} d$; $\mu_l = 5 \times 10^{-3}$ Ns/m²; $L = 1150$ m; $d_i = 0.062$ m; $p_1 = 53.0$ bars; $T = 324$ K; $\varrho_{on} = 830$ kg/m³; $\varrho_{gn} = 1.1$ kg/m³; $R_{sk} = 10$ m³/m³; $n_R = 0.726$ m³/m³ bar; $B_{ok} = 1.14$ m³/m³; $n_B = 1.47 \times 10^{-3}$ m³/m³/bar; $p_c = 45.5$ bars; $T_c = 232$ K; $p_n = 0.981$ bar; $T_n = 288$ K. — (II) $R = 400$ m³/m³; all other values as above. — The main intermediate results of the computation are given in Table 1.4—5 (see next pages). Inverse gradients are plotted v. pressure in Fig. 1.4—40 which also shows the pressure traverses $h = \Phi(p)$ obtained by graphical integration.

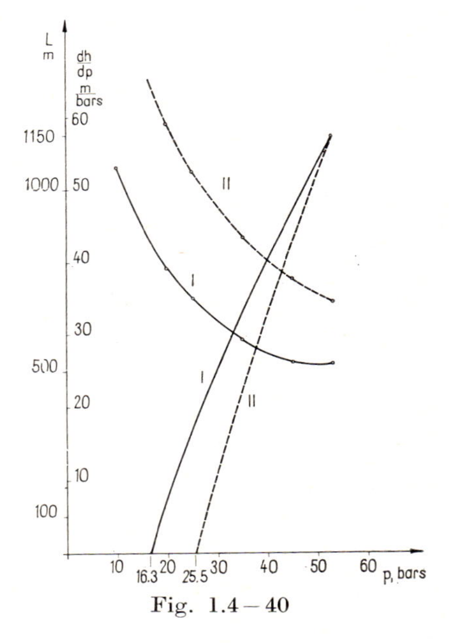

Fig. 1.4—40

(f) Orkiszewski's theory

A study of thirteen different theories of vertical multiphase flow has convinced Orkiszewski that none of these provides predictions of sufficient accuracy concerning the full ranges of all the flow patterns. — The Duns and Ros theory has the following drawbacks: (i) the error of prediction in slug flow is high for high rates of flow, but also for low rates if the crude is a high-viscosity one; (ii) a number of phenomenological relationships are employed for slug flow, so that the true nature of the flow phenomena taking place in the string are obscured; (iii) there is the further drawback, not pointed out by Orkiszewski, that the Duns—Ros theory is applicable only to a well fluid of comparatively low water content.

Orkiszewski divided flow theories in three groups according to whether in the calculation of densities they do or do not take into account the diversity of flow patterns and the slippage losses. For instance, the Poettmann—Carpenter theory accounts for none, whereas the Duns—Ros theory accounts for both.

The Orkiszewski theory (1967) exploits several preceding theories. For bubble and slug flow it is largely based on the Griffith—Wallis theory (Griffith and Wallis 1961, Griffith 1962) which, in Orkiszewski's evalua-

Table 1.4–5 (A)

p	ϱ_l	ϱ_g	$\bar{v}_l$	$\bar{v}_g$	R	N_{lv}	RN_{lv}	N_d	N_μ	L_1	L_2	Flow pattern
bars	kg/m³	kg/m³	m/s	m/s	—	—	—	—	10^{-2}	—	—	
53	679.8	59.78	0.1985	0.3411	1.72	1.376	2.365	29.23	2.404	1.91	0.751	I
45	686.6	49.65	0.1965	0.4323	2.20	1.365	3.004	29.38	2.398	1.92	0.748	II
35	695.4	37.55	0.1940	0.6076	3.13	1.353	4.236	29.57	2.390	1.92	0.747	II
25	704.4	25.81	0.1915	0.9360	4.89	1.340	6.547	29.76	2.383	1.92	0.744	II
20	709.0	20.32	0.1903	1.2220	6.42	1.333	8.561	29.85	2.379	1.92	0.741	II
10	717.5	11.04	0.1880	2.3590	12.55	1.321	16.580	30.05	2.371	1.93	0.739	II

Table 1.4–5 (B)

p	F_1	F_2	F_3	F_4	F_5	F_6	F_7	S	$\bar{v}_{gs}$	ε_l	$\left(\frac{dp}{dh}\right)_m$	N_{Rel}	f_1	f_2	f
bars	—	—	—	—	—	10^{-2}	10^{-2}	—	m/s	—	10^{-2} bar/m	—	10^{-2}	—	10^{-2}
53	1.31	0.312	2.02	39.0				2.415	0.3483	0.5287	3.802			1.067	1.502
45					0.163	5.66	6.62	3.768	0.5421	0.5274	3.783			1.071	1.506
35					0.163	5.23	6.63	4.932	0.7074	0.4613	3.345			1.030	1.448
25					0.163	4.79	6.64	7.100	1.0150	0.3825	2.800	1673	1.41	0.885	1.243
20					0.163	4.57	6.64	8.977	1.2810	0.3376	2.481			0.806	1.131
10					0.163	4.13	6.65	16.388	2.3320	0.2415	1.782			0.641	0.898

Table 1.4—5 (C)

p	$\left(\frac{dp}{dh}\right)_s$	$\frac{dp}{dh}$	$\frac{dh}{dp}$
bars	10^{-4} bar/m	10^{-2} bar/m	m/bar
53	3.5	3.837	26.06
45	4.1	3.824	26.15
35	5.1	3.396	29.45
25	6.1	2.861	34.95
20	7.0	2.551	39.20
10	9.9	1.881	53.16

Table 1.4—5 (D)

$\frac{dh}{dp}$
m/bar
34.64
37.85
43.44
52.55
59.12

Remark. $\frac{dp}{dh} = \xi \gamma_l$, $\left(\frac{dp}{dh}\right)_m = \xi_m \gamma_l$, and $\left(\frac{dp}{dh}\right)_s = \xi_s \gamma_l$. Parts A, B, and C of the Table list intermediate results of (I), Part D lists inverse gradients of (II) in Example 1.4—6.

tion, comes closest to the observation and interpretation of actual flow phenomena. Orkiszewski has, however, modified the original formulae so as to render them more general, as in their original form these proved suitable for predicting conditions in a slow and fully developed slug flow only. For transition and mist flow, Orkiszewski adopted the Duns—Ros method without change.

(f.1) *The fundamental equation.* Formula 1.4—16, derived from the energy equation for an infinitesimal length of flow string, can be slightly modified to read

$$dp = \varrho_k g\, dh + \varrho_k v\, dv + \left(\frac{dp_s}{dh}\right) dh \qquad 1.4-72$$

where (dp_s/dh) is the friction gradient, in N/m³. The kinetic-energy term $(\varrho_k v dv)$ has no significance except in mist flow. — By Eq. 1.4—62, the kinetic-energy term may be rewritten as

$$(dp)_a = -(\varrho_l v_l + \varrho_g v_g) dv_g\,.$$

Now in mist flow, $\varrho_l v_l \ll \varrho_g v_g$, and so, neglecting $\varrho_l v_l$,

$$(dp)_a = -\varrho_g v_g dv_g\,, \qquad 1.4-73$$

The factor dv_g figuring here may be expressed by Eq. 1.4—63 as

$$dv_g = -\frac{q_g\, dp}{A_T\, p}$$

and, making use of the concept of superficial gas velocity,

$$v_g = \frac{q_g}{A_T} = \frac{q_l R}{A_T} \qquad 1.4-74$$

Introducing the expression for mass flow, we find

$$q_m = q_l(\varrho_l + \varrho_g R)$$

whence

$$q_l = \frac{q_m}{\varrho_l + \varrho_g R}.$$

Substitution into Eq. 1.4—74 yields

$$v_g = \frac{q_m}{\varrho_l + \varrho_g R} \frac{R}{A_T}. \qquad 1.4-75$$

Now, summing up the above equations, we find

$$(\mathrm{d}p)_a = \frac{\varrho_g R}{A_T^2} q_g \frac{q_m \mathrm{d}p}{\varrho_l + \varrho_g Rp}.$$

Since in mist flow $\varrho_l \ll \varrho_g R$, one may neglect ϱ_l, and so the kinetic-energy term becomes

$$\varrho_k v \mathrm{d}v = \frac{q_m q_g \mathrm{d}p}{A_T^2 p}. \qquad 1.4-76$$

Introducing this expression into the fundamental Eq. 1.4—72, and rearranging, we obtain

$$\mathrm{d}p = \frac{\varrho_k g + (\mathrm{d}p_s/\mathrm{d}h)}{1 - \frac{q_m q_g}{A_T^2 p}} \mathrm{d}h.$$

This differential equation expresses change of pressure along the flow string Converting into a difference equation so as to make computation possible and solving for Δh, one finds for the ith step of pressure (or depth)

$$\Delta h = \frac{1 - \frac{q_m q_g}{A_T^2 \bar{p}}}{\bar{\varrho}_k g + (\mathrm{d}p_s/\mathrm{d}h)} \Delta p. \qquad 1.4-77$$

The individual terms of the equation are temperature-dependent. As temperature is a function of depth, the solution requires an interative procedure. In calculation, it is usual to fix the pressure step Δp and to find the corresponding Δh values. This approach has the advantage that, the mean fluid parameters being more affected by pressure than by temperature, convergence is faster. For each flow pattern, Orkiszewski provides a specific formula for mean density $\bar{\varrho}_k$ (in the sequel, $\bar{\varrho}_k = \bar{\varrho}$) and for the friction gradient $(\mathrm{d}p_s/\mathrm{d}h)$. In the knowledge of these expressions, Eq. 1.4—77 is suited for calculating the pressure traverse, that is, the pressure drop of two-phase flow.

Table 1.4 — 6

Flow pattern		
Designation	Physical nature	Equations of pattern boundaries
Bubble	Continuous liquid phase with dispersed gas bubbles	$q_g/q_t < \psi$
Slug	Liquid and gas slugs	$q_g/q_t > \psi$ $RN_{lv} < (50 + 36N_{lv})$
Transition	Continuous liquid phase gradually ceases	$RN_{lv} > (50 + 36Nl_v)$ $RN_{lv} < (75 + 84N_{lv}^{0.75})$
Mist	Continuous gas phase with liquid mist	$RN_{lv} > (75 + 84N_{lv}^{0.75})$

In contrast to the Duns—Ros approach, in Orkiszewski's theory the domains of the individual flow patterns and the different calculation procedures are closely interdependent. The unknown factors of Eq. 1.4—77 are to be determined in a different fashion for each of the four flow patterns.

The physical parameters and boundary values of the flow patterns distinguished by Orkiszewski are listed in Table 1.4—6. The value of ψ figuring in Table 1.4—6 is determined using the equation

$$\psi = 1.071 - 0.7277 \frac{v_t^2}{d}; \qquad \psi \geq 0.13 .$$

The boundary between mist flow and transition flow is the same as in the Duns—Ros theory (Fig. 1.4—10). The boundary between bubble and slug flow as specified by Orkiszewski is based, on the other hand, on the work of Griffith and Wallis, and expressed as a function of total flow velocity, flow-string diameter and the gas content of the well stream (Fig. 1.4—41).

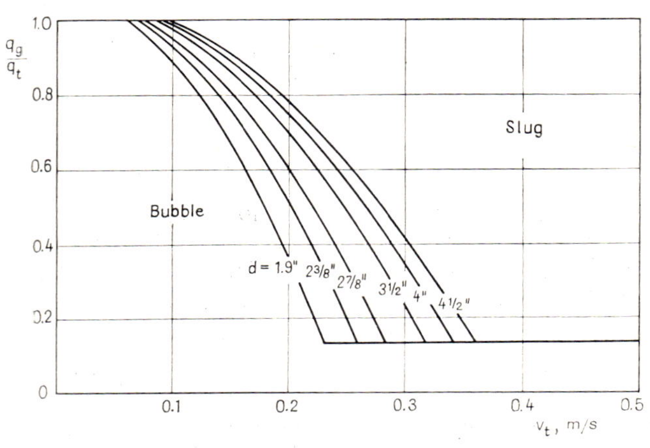

Fig. 1.4—41 Boundary of flow patterns, according to Orkiszewski (1967)

(f.2) *Determining $\bar{\varrho}$ and $(\mathrm{d}p_s/\mathrm{d}h)$ for the individual flow patterns.* — *Bubble flow.* Mean flow density as a function of cross-sectional fraction of the gas phase ε_g (cf. Section 1.4.3) can be calculated using

$$\bar{\varrho} = \varrho_l(1 - \varepsilon_g) + \varrho_g \varepsilon_g . \qquad 1.4-78$$

The unknown in the equation, ε_g, can be derived from the definition of slippage velocity (from a slightly modified Eq. 1.4—15):

$$v_{gs} = \frac{q_g}{A_T \varepsilon_g} - \frac{q_l}{A_T(1 - \varepsilon_g)} . \qquad 1.4-79$$

Solving this equation for ε_g, and introducing the substitution

$$q_l + q_g = q_t$$

one obtains the cross-sectional fraction of the gas phase as a function of slippage velocity:

$$\varepsilon_g = \frac{1}{2}\left[\left(1 + \frac{q_t}{A_T v_{gs}}\right) - \sqrt{\left(1 + \frac{q_t}{A_T v_{gs}}\right)^2 - 4\frac{q_g}{A_T v_{gs}}}\,\right] . \qquad 1.4-80$$

Adopting the considerations in Griffith (1962), Orkiszewski assumed the gas slippage velocity to be constant:

$$v_{gs} = 0.24 \text{ m/s}.$$

Introducing the ε_g value thus calculated into Eq. 1.4—78 one finds the value of $\bar{\varrho}$ valid for bubble flow. The liquid phase being continuous, the friction gradient can be determined using the expressions for single-phase liquid flow. The friction gradient is accordingly

$$(\mathrm{d}p_s/\mathrm{d}h) = \lambda \frac{\varrho_l (v_l^*)^2}{2d} \qquad 1.4-81$$

where v_l^* is the effective liquid velocity, expressed for this flow pattern using the definition of ε_g as

$$v_l^* = \frac{q_l}{A_T(1 - \varepsilon_g)} . \qquad 1.4-82$$

The friction factor λ can be derived in the knowledge of the liquid's Reynolds number out of the Moody diagram (Fig. 1.1—1). N_{Re} is calculated in this case using the effective flow velocity, that is,

$$N_{\mathrm{Re}} = \frac{\varrho_l d v_l^*}{\mu_l} . \qquad 1.4-83$$

Slug flow. The importance of this type of flow is in that most oil wells produce in this pattern over most of the time. The calculation procedure adopted by Orkiszewski was developed by Griffith and Wallis making use of research by Dumitrescu, Davies and Taylor. — By the Griffith—Wallis theory, the cylindrical control volume between $A-A$ and $B-B$ in Fig. 1.4—42 encloses a liquid slug of height h_l and a gas slug of height h_g. In the cylinder of height $(h_l + h_g)$ and cross-sectional area A_T, we have a mean density

$$\bar{\varrho} = \varrho_l \left(1 - \frac{V_g}{V_t}\right) + \varrho_g \frac{V_g}{V_t} \qquad 1.4\text{—}84$$

where $V_g = A_T h_g$ and $V_t = A_T(h_l + h_g)$. — V_g can be derived from the following considerations. The flow velocity of the liquid slug is, by Griffith and Wallis (1961)

$$v_l^* = \frac{q_l + q_g}{A_T}.$$

In a static liquid column, the gas slug will rise at a velocity v_{gs}. In a liquid column rising in the flow string, the rise velocity of the gas slug is

$$v_{gd} = v_l^* + v_{gs} = \frac{q_l + q_g}{A_T} + v_{gs}. \qquad 1.4\text{—}85$$

The gas slug rises to a height $(h_g + h_l)$ within a time interval Δt. Its rate of rise being v_{gd} we have

$$\Delta t = \frac{h_l + h_g}{v_{gd}} = \frac{(h_l + h_g)\, A_T}{q_l + q_g + v_{gs} A_T}. \qquad 1.4\text{—}86$$

The gas content of the control volume between $A-A$ and $B-B$ equals the amount of gas entering the volume during the time interval Δt, that is,

$$V_g = q_g \Delta t \qquad 1.4\text{—}87$$

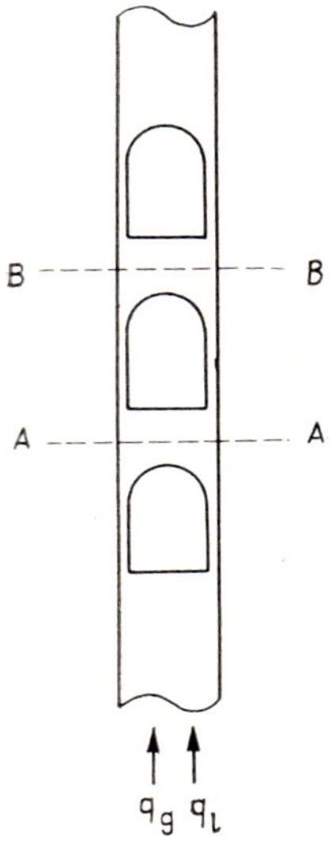

Fig. 1.4—42

Equations from 1.4—84 to 1.4—87 imply that

$$\bar{\varrho} = \varrho_l \frac{q_l + v_{gs} A_T}{q_l + q_g + v_{gs} A_T} + \varrho_g \frac{q_g}{q_l + q_g + v_{gs} A_T}$$

Defining the concepts of total volumetric and mass flow rate respectively by

$$q_t = q_l + q_g$$

and

$$q_m = \varrho_l q_l + \varrho_g q_g$$

we obtain the Griffith and Wallis formulation of the mean density of slug flow:

$$\bar{\varrho} = \frac{q_m + \varrho_l v_{gs} A_T}{q_t + v_{gs} A_T}. \qquad 1.4\text{—}88$$

Equation 1.4—88 was established by Griffith and Wallis for low flow velocities and regular, fully developed slug flow. In slug flow, however, velocities may well be quite high, and the liquid phase will be present not only as slugs separate from the gas phase, but also as a film covering the pipe wall and as a mist dispersed in the gas. It was to account for these phenomena that Orkiszewski complemented Eq. 1.4—88 with a liquid-distribution factor Γ, as follows:

$$\bar{\varrho} = \frac{q_m + \varrho_l v_{gs} A_T}{q_t + v_{gs} A_T} + \Gamma \varrho_l . \qquad 1.4\text{—}89$$

Tests on wells producing gaseous crudes gave rise to the following equations for the liquid-distribution factor. The continuous liquid phase is water:

$$\Gamma = a_1 \frac{\lg(10^3 \mu_l)}{d^n} + a_2 + a_3 \lg v_t + a_4 \lg d . \qquad 1.4\text{—}90$$

The continuous liquid phase is oil:

$$\Gamma = a_1 \frac{\lg(10^3 \mu_l + 1)}{d^n} + a_2 + a_3 \lg v_t + a_4 \lg d + a_5 . \qquad 1.4\text{—}91$$

The a_i and n can be read off Table 1.4—7, according as the liquid phase is water or oil, and as v_t is greater or smaller than 3 m/s. — Slip velocity v_{gs}, or, in the case in hand, the rising velocity of the gas plug is, according to the original equations of Griffith and Wallis (1961)

$$v_{gs} = C_1 C_2 \sqrt{gd} . \qquad 1.4\text{—}92$$

Table 1.4—7

v_t [m/s]	n	a_1	a_2	a_3	a_4	a_5	Limit of applicability
The continuous liquid phase is water							
<3	1.38	2.523×10^{-4}	−0.782	0.232	−0.428	0	$\Gamma \geq -0.213 \times v_t$
>3	0.799	1.742×10^{-4}	−1.251	−0.162	−0.888	0	$\Gamma \geq -\frac{v_{gs} A_T}{q_t + v_{gs} A_T}\left(1 - \frac{\bar{\varrho}}{\varrho_l}\right)$
The continuous liquid phase is oil							
<3	1.415	2.364×10^{-4}	−0.140	0.167	0.113	0	$\Gamma \geq -0.213 \times v_t$
>3	1.371	5.375×10^{-4}	0.455	0	0.569	*	$\Gamma \geq -\frac{v_{gs} A_T}{q_t + v_{gs} A_T}\left(1 - \frac{\bar{\varrho}}{\varrho_l}\right)$

* $a_5 = -(0.516 + \lg v_t)\left[1.547 \times 10^{-3} \frac{\lg(10^3 \mu_l + 1)}{d^{1.571}} + 0.722 + 0.63 \lg d\right]$

The coefficient C_1 can be read off Fig. 1.4—43 v. the Reynolds number of the gas slug:

$$N_{\mathrm{Re}gs} = \frac{\varrho_l d v_{gs}}{\mu_l}. \qquad 1.4-93$$

The coefficient C_2 is a function of $N_{\mathrm{Re}t}$ referring to total volumetric flow and also of $N_{\mathrm{Re}gs}$ (Fig. 1.4—44):

$$N_{\mathrm{Re}t} = \frac{\varrho_l d v_t}{\mu_l}. \qquad 1.4-94$$

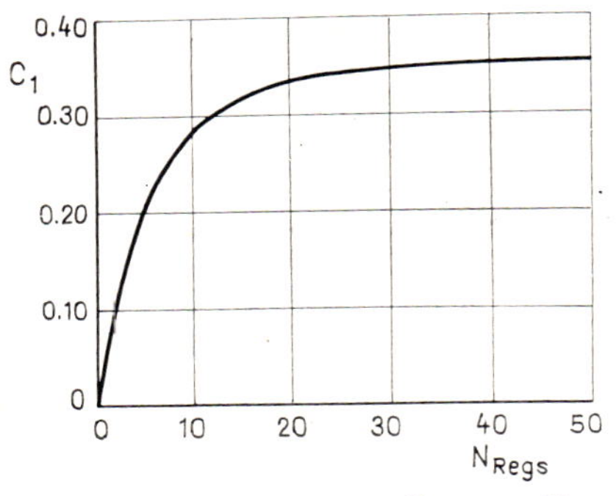

Fig. 1.4—43 Factor C_1 according to Orkiszewski (1967)

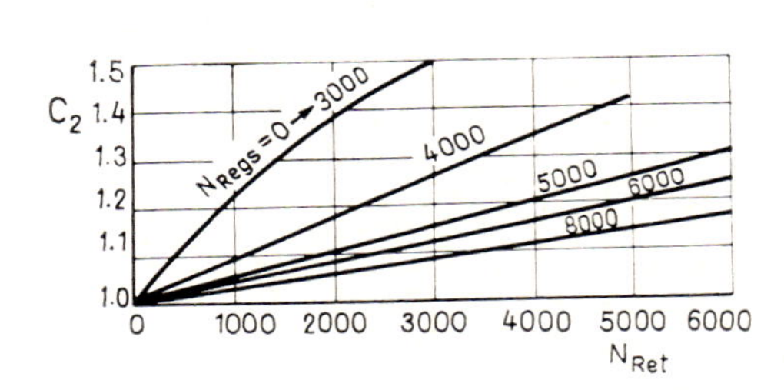

Fig. 1.4—44 Factor C_2 according to Orkiszewski (1967)

The above relationships permit the calculation of v_{gs} by means of an iterative procedure. Beyond the ranges of Figs 1.4—43 and 1.4—44, v_{gs} can be calculated using extrapolated values, by means of the equations:

if $N_{\mathrm{Re}gt} \leqslant 3000$

$$v_{gs} = (0.546 + 8.74 \times 10^{-6} N_{\mathrm{Re}t}) \sqrt{gd} \qquad 1.4-95$$

if $3000 < N_{\mathrm{Re}gs} < 8000$

$$v_{gs} = \frac{1}{2}\left[v_{gO} + \sqrt{v_{gO}^2 + \frac{11.17 \times 10^3 \mu_l}{\varrho_l \sqrt{d}}} \right] \qquad 1.4-96$$

with

$$v_{gO} = (0.251 + 8.74 \times 10^{-6} N_{\mathrm{Re}t}) \sqrt{gd}$$

if $N_{\mathrm{Re}gs} \geqslant 8000$

$$v_{gs} = (0.35 + 8.74 \times 10^{-6} N_{\mathrm{Re}t}) \sqrt{gd}. \qquad 1.4-97$$

According to Orkiszewski, in calculating the friction gradient the liquid distribution factor Γ should also be taken into account, because the distribution of the liquid among the slug, the film and the spray will significantly affect the friction head loss. Accordingly, the friction gradient is calculated by the formula

$$(\mathrm{d}p_s/\mathrm{d}h) = \lambda \frac{v^2}{2d}\left[\frac{q_l + v_{gs} A_T}{q_t + v_{gs} A_T} + \Gamma\right] \qquad 1.4-98$$

where $\lambda = \mathrm{f}(N_{\mathrm{Re}t})$; $N_{\mathrm{Re}t}$ is derived using Eq. 1.4—94.

The transition region. — For determining the $\bar{\varrho}$ and $(\mathrm{d}p_s/\mathrm{d}h)$ valid for the transition region, Orkiszewski employs essentially the weighted-average method presented in paragraph 1.4.3e.4. The relevant formulae are

$$\bar{\varrho} = \frac{(75 + 84N_{lv}^{0.75}) - RN_{lv}}{(75 + 84N_{lv}^{0.75}) - (50 + 36N_{lv})}[\bar{\varrho}]_{\mathrm{slug}} + \frac{RN_{lv} - (50 + 36N_{lv})}{(75 + 84N_{lv}^{0.75}) - (50 + 36N_{lv})}[\bar{\varrho}]_{\mathrm{mist}} \qquad 1.4-99$$

$$(\mathrm{d}p_s/\mathrm{d}h) = \frac{(75 + 84\,N_{lv}^{0.75}) - RN_{lv}}{(75 + 84\,N_{lv}^{0.75}) - (50 + 36\,N_{lv})}[(\mathrm{d}p_s/\mathrm{d}h)]_{\mathrm{slug}} + \frac{RN_{lv} - (50 + 36N_{lv})}{(75 + 84\,N_{lv}^{0.75}) - (50 + 36\,N_{lv})}[(\mathrm{d}p_s/\mathrm{d}h)]_{\mathrm{mist}}. \qquad 1.4-100$$

It was shown by Orkiszewski that in the transition region, the friction gradient can be established more accurately if the superficial velocity of the gas is calculated using the relationship

$$v_g = \frac{(75 + 84\,N_{lv}^{0.75})}{\sqrt[4]{\dfrac{\varrho_l}{g\sigma}}}. \qquad 1.4-101$$

Mist flow. — In this flow region the flow velocity of liquid equals the flow velocity of gas, that is, there is no slippage; Eq. 1.4—79 implies

$$v_{gs} = \frac{q_g}{A_T \varepsilon_g} - \frac{q_l}{A_T(1 - \varepsilon_g)} = 0,$$

whence

$$\varepsilon_g = \frac{1}{1 + \dfrac{q_l}{q_g}} = \frac{q_g}{q_t}.$$

By Eq. 1.4—78, the mean density is

$$\bar{\varrho} = \varrho_l \frac{q_l}{q_t} + \varrho_g \frac{q_g}{q_t}.$$

Friction loss is calculated, with due attention to the liquid film covering the pipe wall, as shown in paragraph 1.4.3e.2. All the modification introduced by Orkiszewski is restricted to a computative simplification. — From the graph shown as Fig. 1.4—38 Orkiszewski expressed the relative roughness k/d, in explicit form. If the abscissa is:

$N \leqslant 0.005$, then

$$\frac{k}{d} = \frac{34\,\sigma}{\varrho_g v_g^2 d} \qquad 1.4\text{—}103$$

$N > 0.005$, then

$$\frac{k}{d} = \frac{174.8\,\sigma\, N^{0.302}}{\varrho_g v_g^2 d}\,. \qquad 1.4\text{—}104$$

(f.3) *Determining pressure traverses.* — In order to determine a pressure traverse, one assumes a sequence of pressures p_i starting from the wellhead or the bottom-hole pressure, increasing or decreasing in steps of Δp_i. For satisfactory accuracy, Δp_i should equal about 10 percent of the given pressure p_i. Using the calculation procedure outlined, the depth increment Δh_i is established for each pressure value p_i. The summation of the depth steps Δh_i starting from the assumed starting level results in a depth sequence $L_i = L_O + \Sigma\Delta\, h_i$, and the corresponding values of p_i and L_i give the pressure traverse. The flow chart of calculating one step is shown in Fig. 1.4—45. Owing to the influence upon the 'in-situ' flow parameters of temperature (which is a predetermined function of depth), Δh_i is determined using an iterative procedure. If the estimated Δh_i^* falls within a prescribed accuracy

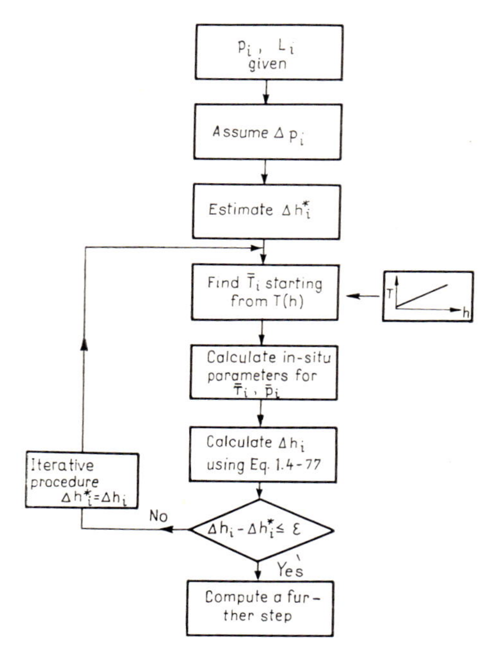

Fig. 1.4—45 Block diagram of calculating a pressure increment

about the Δh_i value supplied by Eq. 1.4—77, then the iterative procedure can be terminated and the next step can be attacked.

(f.4) *Accuracy of calculation.* — After the publication of Orkiszewski's flow theory, numerous authors have applied themselves to studying the accuracy attainable by the various theories. Their findings do not permit us to establish unequivocally whether Orkiszewski's theory is always and in each flow pattern more accurate than any other method. There is, however, agreement in the relevant literature as to the good comparative accuracy of the Orkiszewski method of calculating head loss for two-phase flow,

Table 1.4—8

q_o	q_w	R	Relative error %		
m³/d	m³/d	m³/m³	I	II	III
Well group 1, $q_0 < 16$ m³/d					
14.8	0	78.9	61.8	— 1.6	— 2.5
8.4	0	260.0	4.1	19.0	23.4
9.4	0	288.3	11.8	8.4	11.5
14.2	0	159.8	—13.3	—11.6	—10.1
13.0	0	198.0	— 3.0	— 8.0	— 5.4
6.9	0	1776.5	11.9	—63.7	—55.5
8.1	0	208.6	4.5	4.0	8.2
9.7	0	298.3	9.3	6.5	9.1
Well group 2, $q_0 > 32$ m³/d					
44.5	0	289.4	8.0	—24.7	—27.0
39.1	0	183.8	50.6	— 1.1	—23.8
38.5	0	190.2	30.1	—15.5	—35.0
36.7	0	228.0	—24.9	—29.0	—46.6
44.8	0	124.5	10.5	—15.7	—17.3
32.6	0	175.8	28.4	—11.4	—13.3
34.7	0	30.5	— 2.3	9.4	— 1.9
Well group 3, water-cut oil					
35.5	1.3	228.0	—24.9	—29.0	—46.6
31.2	7.9	183.8	50.6	— 1.1	—23.8
16.1	0.6	190.4	46.9	8.8	15.2
26.4	14.3	632.3	50.6	4.8	26.4
20.7	4.8	101.9	40.9	10.2	15.8
26.1	6.8	59.3	51.9	7.8	14.5
12.9	17.8	120.2	85.1	— 7.5	16.7
11.3	11.1	69.3	73.6	— 3.6	13.3

I Hagedorn—Brown }
II Orkiszewski } method
III Duns—Ros }

although the calculation is admitted to be fairly time-consuming. Espanol compared the accuracy of the Hagedorn—Brown, Orkiszewski, and Duns—Ros methods using data recorded on 44 wells. By a suitable grouping of the data, Espanol and his co-workers (1969) could show that the Orkiszewski method is particularly preferable in three-phase flow (water + oil+ gas), and also for bubble and slug flow. Some of the data and the percentual prediction errors are listed in Table 1.4—8. In another study (Aziz et al. 1972) an analysis of 48 test data revealed the absolute mean error of the Orkiszewski method to be 8.9 percent, whereas that of the Hagedorn—Brown, and Duns—Ros methods was established as 20.5 and 11.1 percent, respectively. — An examination of data on 78 wells, published by the authors proper who have put forward flow theories (McLeod et al. 1972) has found that in slug flow the predictions of the Orkiszewski method were not significantly better than those of the much simpler Poettmann and Carpenter method.

(g) Other theories; remarks

The deviation between measured data and the predictions of the various flow theories may be attributed to a number of causes. One possible factor is the deposition of wax in the production tubing. — In recent years, Soviet workers have studied intensely the problem that the crude rising in the well is often supersaturated, with the dissolved gas bubbling out at a lower pressure than the one predicted by equilibrium calculations. According to investigations by the Tuymazneft enterprise the bubble-point pressure may be different by as much as 10 bars according to whether the oil is stirred or at rest (Panteleev 1970). Short lengths of larger-bore inserts in the pipe strings, called turbulizers, have resulted in an increase of gas separation at any given depth. As a result, wellhead pressure likewise increased quite significantly. This phenomenon implies that, primarily in wells where the oil emerges from the reservoir at a pressure equal to or greater than the saturation pressure, one should expect the effective *GOR* at any depth to be lower than the calculated one in a pipe string of constant cross-section.

Krylov's differential Eq. 1.4—9 may be solved also by numeric integration, and if the real parameters of the liquid and gas are taken into account, the fundamental relationship will supply comparatively accurate pressure traverses (Patsch 1971).

1.5. Flow of compressible mediums through a choke

1.5.1. Flow of gases

The velocity of gas flowing through a choke can be calculated using the well-known de Saint Venant equation, assuming that the gas is perfect and that the flow is frictionless and adiabatic (see e.g. Binder 1958):

$$v_2 = \sqrt{2 \frac{\varkappa p_1}{\varkappa - 1\,\varrho_1}\left[1 - \left(\frac{p_2}{p_1}\right)^{\frac{\varkappa-1}{\varkappa}}\right]}. \qquad 1.5\text{—}1$$

Suffix *1* refers to the state on the upstream side of the choke, suffix *2* to the state on its downstream side. For an adiabatic change of state,

$$\varrho_1 = \varrho_2 \left(\frac{p_1}{p_2}\right)^{\frac{1}{\varkappa}}.$$

The mass of gas flowing through per unit of time is

$$q_{gm} = q_{g2}\varrho_2 = \frac{d_{\text{ch}}^2 \pi}{4} v_2 \varrho_2 ,$$

and the gas flow rate referred to the standard state is

$$q_g = \frac{q_{gm}}{\varrho_n}.$$

The combined gas law implies

$$\varrho_1 = \frac{p_1 M}{RT_1}, \text{ and } \varrho_n = \frac{p_n M}{RT_n}.$$

Substituting these expressions into Eq. 1.5—1, introducing the discharge coefficient α, and solving for the gas flow rate through the choke referred to standard conditions, we get

$$q_g = \sqrt{2R}\frac{\pi}{4} d_{\text{ch}}^2 \, p_1 \frac{T_n}{p_n} \alpha \sqrt{\frac{1}{MT_1} \frac{\varkappa}{\varkappa - 1}\left[\left(\frac{p_2}{p_1}\right)^{\frac{2}{\varkappa}} - \left(\frac{p_2}{p_1}\right)^{\frac{\varkappa+1}{\varkappa}}\right]} . \qquad 1.5\text{—}2$$

In *SI* units, $\sqrt{2R}\,\pi/4 = 101.3$. — The validity of Eqs 1.5—1 and 1.5—2 is limited by the critical pressure ratio $(p_2/p_1)_c$ at which the flow velocity attains the speed of sound. It is at this pressure ratio that the velocity (and the flow rate) of gas flowing through the choke is greatest:

$$\left(\frac{p_2}{p_1}\right)_c = \left(\frac{2}{\varkappa + 1}\right)^{\frac{\varkappa}{\varkappa-1}}. \qquad 1.5\text{—}3$$

For the purpose of calculating choke diameters, it is preferable to rewrite Eq. 1.5—2 in the form

$$d_{\text{ch}} = \sqrt{q_g \frac{p_n}{T_n} \frac{(MT_1)^{0.5}}{101.3\, p_1 C\alpha}} , \qquad 1.5\text{—}4$$

where

$$C = \sqrt{\frac{\varkappa}{\varkappa - 1}\left[\left(\frac{p_2}{p_1}\right)^{\frac{2}{\varkappa}} - \left(\frac{p_2}{p_1}\right)^{\frac{\varkappa+1}{\varkappa}}\right]}. \qquad 1.5\text{—}5$$

Equations from 1.5—2 to 1.5—5 are used as follows. First the critical pressure ratio is calculated by Eq. 1.5—3. If the given pressure ratio p_2/p_1 is less than that, it is necessary to replace p_2/p_1 in Eqs 1.5—2 and 1.5—5 by the critical ratio. If the given pressure ratio is greater, the formulae in question can be used without restriction. Let us add that the adiabatic gas exponent $\varkappa$ is relatively insensitive to temperature variations as well as to molecular weight within the range subtended by the gaseous hydrocarbons It is therefore satisfactory in practice to calculate with the constant value $\varkappa = 1.25$.

Example 1.5 — 1. Find the gas flow rate, referred to standard conditions. through a choke of diameter $d_{ch} = 10$ mm, if $M = 20$ kg/kmole; $\varkappa = 1.25$; $\alpha = 0.85$; $p_n = 1.00$ bar; $T_n = 288$ K; $p_1 = 35$ bars; $T_1 = 333$ K; let further (i) $p_2 = 28$ bars, and (ii) $p_1 = 14$ bars. — By Eq. 1.5 — 3,

$$\left(\frac{p_2}{p_1}\right)_c = \left(\frac{2}{1.25+1}\right)^{\frac{1.25}{1.25-1}} = 0.555 .$$

Now in the case of (i),

$$\frac{p_2}{p_1} = \frac{28\times10^5}{35\times10^5} = 0.80 ,$$

which is greater than the critical pressure ratio. Hence, by Eq. 1.5—2,

$$q_g = 101.3\times0.01^2\times35\times10^5\,\frac{288}{1\times10^5}\,0.85$$

$$\sqrt{\frac{1}{20\times333}\,\frac{1.25}{1.25-1}\left(0.8^{\frac{2}{1.25}} - 0.8^{\frac{1.25+1}{1.25}}\right)} = 0.415\ \text{m}^3/\text{s} .$$

In the case of (ii), however,

$$\frac{p_2}{p_1} = \frac{14\times10^5}{35\times10^5} = 0.4 ,$$

which is less than the critical value 0.555. By the above considerations,

$$q_g = 101.3\times0.01^2\times35\times10^5\,\frac{288}{1\times10^5}\,0.85$$

$$\sqrt{\frac{1}{20\times333}\,\frac{1.25}{1.25-1}\left(0.555^{\frac{2}{1.25}} - 0.555^{\frac{1.25+1}{1.25}}\right)} = 0.495\,\text{m}^3/\text{s}.$$

1.5.2. Two-phase flow of gases and liquids

The equation

$$p_1 = 3.59\times10^4\,\frac{q_o\,R^{0.546}}{d_{ch}^{1.89}} \qquad 1.5-6$$

was developed by Gilbert (1954) primarily to derive the upstream pressure of a fluid flowing through a choke mounted in a wellhead assembly. It is valid if

$$\frac{p_2}{p_1} \leq 0.59.$$

Example 1.5—2. Find the upstream pressure by Eq. 1.5—6 if $q_o = 98$ m³/d; $R = 115$ m³/m³; $d_{ch} = 8$ mm, and $p_2/p_1 = 0.59$.

$$p_1 = 3.59 \times 10^4 \frac{1.134 \times 10^{-3} \times 115^{0.546}}{0.008^{1.89}} = 5.00 \times 10^6 \text{ N/m}^2 = 50.0 \text{ bars.}$$

The Gilbert equation is based on the tacit assumption that at the pressure and temperature prevailing on the upstream side, q_o and R agree with the values referred to standard conditions. It further includes the assumption of a standard choke geometry whose discharge coefficient (in *SI* units) is incorporated in the constant 3.59×10^4. — The properties of the fluid flowing through the choke are taken into account to a more satisfactory degree by the Ros—Poettmann—Beck equation (Poettmann and Beck 1963), likewise valid at below-critical pressure ratios:

$$q_{km} = \frac{d_{ch}^2 \pi}{4} C \sqrt{\frac{2p_1}{V_1[1 + (1 - \beta)\,\varepsilon_{o1}]}} \cdot$$

$$\cdot \left[\frac{(0.4313 + 0.040\,\beta)\sqrt{R_1 + 0.7566 + 0.0188\,\beta}}{R_1 + 0.5553 + 0.0220}\right]. \qquad 1.5\text{—}7$$

Since $q_{km} = q_o M_t$, and the authors suggest the value of 0.5 for the dispersion coefficient β and 1.03 for the discharge coefficient C, the above equation can be used in the simpler form

$$q_o = \frac{0.516\, d_{ch}^2}{M_t(R_1 + 0.5663)} \sqrt{\frac{p_1(R_1 + 0.766)}{V_1(1 + 0.5\,\varepsilon_{o1})}}. \qquad 1.5\text{—}8$$

Here, $R_1 = v_{g1}/v_{o1}$, and the substitution of Eqs 1.4—70 and 1.4—71 results in

$$R_1 = \frac{p_n z_1 (R - R_{s1}) T_1}{T_n p_1 B_{o1}}. \qquad 1.5\text{—}9$$

Factor R is the gas–oil ratio referred to standard conditions. Suffix *1* refers to the temperature on the upstream side of the choke, estimated to equal 303 K by the authors.

$$\varepsilon_{o1} = \frac{\varrho_{o1}}{\varrho_{o1} + R_1 \varrho_{g1}} \qquad 1.5\text{—}10$$

is the oil fraction of the flowing mass. ϱ_{o1} and ϱ_{g1} can be calculated by means of the formulae 1.4—68 and 1.4—69. The specific volume of oil upstream of the choke is

$$V_1 = \frac{\varepsilon_{o1}}{\varrho_{o1}}. \qquad 1.5\text{—}11$$

Example 1.5—3. Find the rate of oil flow through a choke of diameter $d_{ch} = 10$ mm, if $R = 115$ m³/m³; $R_{s1} = 20$ m³/m³; $\varrho_{on} = 815$ kg/m³; $\varrho_{gn} = 0.910$ kg/m³; $p_1 = 50$ bars; $p_2 = 2.0$ bars; $p_n = 1.0$ bar; $T_1 = 303$ K; $T_n = 288$ K; $B_{o1} = 1.06$; $z_1 = 0.875$. — By Eqs 1.4—68 and 1.4—69,

$$\varrho_{o1} = \frac{815}{1.06} = 769 \text{ kg/m}^3 ;$$

$$\varrho_{g1} = \frac{0.910}{\dfrac{1.0\times 10^5 \times 303}{288} \dfrac{0.875}{50\times 10^5}} = 49.5 \text{ kg/m}^3 .$$

$$R_1 = \frac{1.0\times 10^5 \times 0.875(115 - 20)\, 303}{288\times 50\times 10^5 \times 1.06} = 1.65 \text{ m}^3/\text{m}^3 .$$

By Eq. 1.5—10,

$$\varepsilon_{o1} = \frac{769}{769 + 1.65\times 49.5} = 0.904 .$$

By Eq. 1.5—11,

$$V_1 = \frac{0.904}{769} = 1.18\times 10^{-3} \text{m}^3/\text{kg} .$$

By Eq. 1.4—10,

$$M_t = 815 + 115\times 0.910 = 920 \text{ kg/m}^3 .$$

Substitution of the values thus obtained into Eq. 1.5—8 yields

$$q_o = \frac{0.516\times 0.01^2}{920(1.65 + 0.5663)} \sqrt{\frac{50\times 10^5(1.65 + 0.766)}{1.18\times 10^{-3}(1 + 0.5\times 0.904)}} = 2.12\times 10^{-3} \text{ m}^3/\text{s} ,$$

that is, the oil flow rate through the choke is, according to the calculation method illustrated here, almost twice as high as the one found in Example 1.5—2, where Eq. 1.5—6 has been used.

An empirical relationship for the two-phase flow of water and gas has been developed that gives more accurate values than the preceding theory, especially in the case of small-bore chokes (1.6 mm $< d_{ch} <$ 5.6 mm), effective gas–liquid ratios exceeding unity and water flow rates less than 127 m³/day. The critical pressure ratio must exceed 0.546. In the pertinent experiments, the pressure upstream of the choke was varied over the range of from 28 to 69 bars (Omana et al. 1969). The relationship in question reads

$$N_{qw} = 0.263\, N_\varrho^{-3.49}\, N_{p_1}^{3.19} \left(\frac{1}{1 + R_1}\right)^{0.657} N_d^{1.8} , \qquad 1.5—12$$

where N_{qw} is the liquid volume rate number:

$$N_{qw} = q_w \left(\frac{\varrho_{w1}^5 g^3}{\sigma^5} \right)^{0.25};$$

N_ϱ is the density or mass ratio number:

$$N_\varrho = \frac{\varrho_{g1}}{\varrho_{w1}};$$

and N_{p1} is the upstream pressure number,

$$N_{p1} = p_1 \left(\frac{1}{\varrho_{w1} g \sigma} \right)^{0.5}.$$

R is the gas–liquid ratio

$$R = \frac{v_{g1}}{v_{w1}},$$

where v_{w1} and v_{g1} can be obtained from Eqs 1.4—70 and 1.4—71; N_d is the diameter number:

$$N_d = d_{\text{ch}} \left(\frac{\varrho_{w1} g}{\sigma} \right)^{0.5}.$$

According to the authors, the standard deviation (Eq. 1.4—5) at the flow parameters investigated by them was 44.4 percent for the Gilbert method, 29.0 percent for the Ros method and 1.15 percent for the method proposed by them.

Chapter 2

Producing oil wells – (1)

The fluid entering the well from the reservoir is endowed with an energy content composed among others of position, pressure, kinetic and thermal energies. If the available energy of the fluid is sufficient to lift it through the well to the surface, then the well will produce by flowing. Flowing wells are by far the most economical, because the well production equipment required is cheap and simple and keeping up the flow requires no extraneous source of energy. Economy therefore demands the selection of well completion and production parameters that will ensure flowing production as long as possible. In the case of continuous-flow production, this aim is achieved by a carefully designed combination of tubing size and wellhead pressure. This task looks simple enough but is complicated in reality by the fact that, in its passage to the stock tank, the fluid will traverse a number of hydraulical systems of differing parameters: the reservoir, the well, the choke and the flow line. Any change of flow parameters in any one of these components will affect all the other flow parameters, too. — It is desirable that the tubing string, run in on completion of the well, be optimally dimensioned. Changing the tubing in a flowing well is a time-consuming, costly operation. The production parameters of a well should therefore be ascertained even before completion. Relevant information may be gathered from geophysical well logs, the production maps of an already producing field, and, last but not least, the results of well testing operations.

2.1. Well testing; inflow performance curves

In steady-state flow the amount and composition of fluid entering and leaving the well per unit of time are identical. The reservoir and the well constitute a series-connected two-component hydraulical system. The interface between the two components, that portion of the wellbore surface where the fluid enters the wellbore through pores and perforations, is called the sand surface. The interface is most often situated at the bottom of the wellbore. On the interface the pressures prevailing in the two components of the hydraulic system are equal. The production capacity of the well is characterized by the relationship between various steady-state flowing bottom-hole pressures (*BHP*) p_{wf} and the corresponding oil production rates q_o. The graphical plot of this relationship is the inflow performance curve.

If the fluid flowing into the wellbore is pure oil, and if the oil is considered incompressible, then, in the laminar region of relatively low flow velocities, flow can be described by Darcy's law. For isothermal flow in a porous reservoir rock of homogeneous isotropic formation permeability k, the filtration rate is

$$v = \frac{q_0}{A} = -\frac{kdp}{\mu dl}, \qquad 2.1\text{—}1$$

where μ is the viscosity of the oil at the flowing pressure and temperature. — If the thickness h of the reservoir is very small as compared to the well's

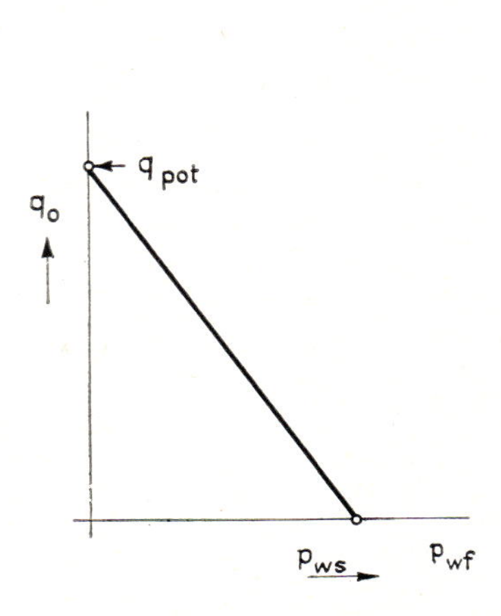

Fig. 2.1—1 Inflow performance curve

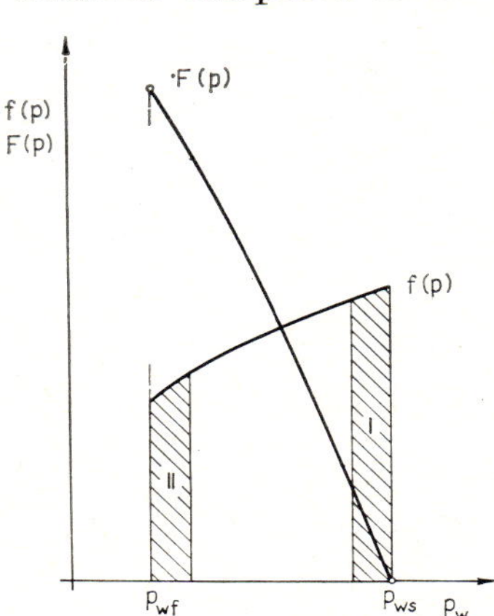

Fig. 2.1—2 Plotting the productivity curve for a gaseous fluid

influence radius r_e, that is, $h \ll r_e$, then the flow of oil from the reservoir into the wellbore of radius r_w will be plane-radial for all practical purposes. In the case of a horizontal reservoir and of steady-state flow, the solution of Eq. 2.1—1 is

$$q_o = \frac{2\pi hk(p_{ws} - p_{wf})}{\mu \ln \frac{r_e}{r_w}}. \qquad 2.1\text{—}2$$

Factor p_{ws} is the pressure prevailing on the circumference of the area of influence of the well, which in a very good approximation equals the steady-state *BHP* of the shut-in well, that is, the static *BHP*. If k, h, μ, r_e and r_w are constant, then

$$q_o = J(p_{ws} - p_{wf}). \qquad 2.1\text{—}3$$

In the course of production, p_{ws} has a tendency to vary rather slowly so that it may be regarded as constant over relatively short spans of time (a few weeks or months). The inflow performance curve corresponding to the productivity equation 2.1—3 is then a straight line, as shown in Fig. 2.1—1.

If $q_o = 0$, then $p_{wf} = p_{ws}$. The theoretically attainable production rate q_o at $p_{wf} = 0$ is the potential yield of the well. — If the reservoir is comparatively thick, flow may closely approach the spherical-radial type. In this case the productivity equation and the inflow performance curve are of the same form as Eq. 2.1—3 and the graph in Fig. 2.1—1, respectively, all other flow parameters being equal.

In wells producing a two-phase gas–oil mixture, the flowing *BHP* is often less than the saturation pressure. The pores of the formation will then contain some free gas as well as a liquid phase. The production of the well is then characterized, according to Muskat and Everdingen, instead of Eq. 2.1—2, by the following equation (Frick 1962):

$$q_o = \frac{2kh\pi}{\ln r_e/r_w} \int_{p_{wf}}^{p_{ws}} \frac{k_o/k}{\mu_o B_o} \mathrm{d}p \,. \qquad 2.1\text{—}4$$

$$\int_{p_{wf}}^{p_{ws}} \frac{k_o/k}{\mu_o B_o} \mathrm{d}p = \int_{p_{wf}}^{p_{ws}} \mathrm{f}(p)\mathrm{d}p \,. \qquad 2.1\text{—}5$$

The integrand is a function of pressure, because a decrease in pressure will liberate gas in the formation, and decrease the permeability with respect to oil, k_o. If the composition of the oil did not change, then its viscosity μ_o would decrease as the pressure decreases; however, the liberation of gas will result in an increase of liquid viscosity. On decrease of pressure, the oil formation volume factor B_o is decreased by degassing, but increased by the volume increase of the oil. The aggregate result of all these changes is a graph illustrating Eq. 2.1—5, which more or less closely resembles the Graph f(p) in Fig. 2.1—2. In the knowledge of this Graph, the inflow performance trend for depletion type reservoirs can be established by the following consideration. The integral curve of Graph f(p) is determined by graphic integration for initial pressure p_{ws} and various final pressures p_{wfi}. The integral curve defines Graph F(p), along which pressure p_{wf} varies at a given p_{ws}. While testing a given well, the expression

$$\frac{2\pi kh}{\ln r_e/r_w}$$

remains constant and can be denoted by the constant J'. The solution of Eq. 2.1—4 is, then,

$$q_o = J'F(p_{wf})_{pws} \,; \qquad 2.1\text{—}6$$

that is, the variation of the daily production of the well will have a curvature much like the heavy curve in Fig. 2.1—2. The indicator curve of a well producing gaseous oil from the reservoir described above will, then, be concave from below. Curves convex from below are bound to be the results of careless measurements, e.g. if in the course of well testing the stabilization of the individual flow rates is not given sufficient time. In the case of sandstone reservoirs, Eq. 2.1—6 can be approximated well enough by the formula

$$q_o = J(p_{ws} - p_{wf})^n \,, \qquad 2.1\text{—}7$$

where $n < 1$. In agreement with the notation used in Eq. 2.1—3, $J' = J$ here. — Comparison of Areas I and II below the Graph f(p) reveals the daily oil production at a given pressure difference $(p_{ws} - p_{wf})$ to be less if the reservoir pressure p_{ws} is lower. This implies that, even if all other parameters are equal, the decrease of reservoir pressure during the life of the well will entail a gradually decreasing production rate even if the draw-down remains constant. This decrease can be temporarily forestalled by increasing the drawdown $(p_{ws} - p_{wf})$, that is, by decreasing the flowing *BHP* faster than the reservoir pressure decreases.

Fractured limestone reservoirs are described by relationships other than Eq. 2.1—7. For instance, the inflow performance curve for a slightly compressible reservoir fluid becomes (Bán 1962)

$$q_o = a(p_{ws} - p_{wf}) + b(p_{ws} - p_{wf})^n \,, \qquad 2.1\text{—}8$$

where a and b are formation constants. — The inflow performance curves of wells are usually determined in practice by producing the wells against a variety of wellhead pressures by means of a number of production chokes of different sizes. To each wellhead pressure a different *BHP* will correspond. After the wellhead pressure has stabilized, the flowing *BHP* and the oil production rate are measured. Tests are usually carried out at three to five operating points. The static and flowing *BHP*s are usually measured by means of a down-hole, or reservoir, pressure gauge. The method itself belongs to the domain of reservoir engineering. If the casing annulus is not closed off at the tubing shoe and the flowing *BHP* is less than the saturation pressure of the fluid produced, it might be advantageous to calculate the pressure p_{wf} at the tubing shoe from the casinghead pressure p_{CO}. A relationship suitable for calculation can be derived from Eqs 1.2—3 and 1.2—4, putting $h = L_w$ and $q_n = 0$. Then

$$p_{wf} = p_{CO} e^{\frac{0.00118 L_w M}{T \bar{z}}} \,. \qquad 2.1\text{—}9$$

On well testing, steady-state gas production rates q_{gn} referred to standard conditions are also determined at each individual bottom-hole pressure. Figure 2.1—3 shows characteristic curves plotted from well-testing results. The producing gas–oil ratio curve $R_{go} = q_g/q_o$ has been plotted from pairs of q_{gn} and q_o belonging to each individual flowing bottom-hole pressure; R_{go} is seen to have a minimum in this particular case. Such minima do appear quite often. This is a result of certain reservoir mechanical phenomena. If there is no prescription as to the flowing *BHP* at which to produce the well, it may be expedient to operate it at R_{min} and the corresponding oil production rate q_o, because it is at this point that the amount of formation gas required to lift one m³ of oil is least. When changing chokes in the course of well testing, attention should be paid to the following. (i) One cannot attain lower wellhead pressures than the input pressure of the flow line. (ii) If in the course of well testing a choke is replaced with a smaller-size one, the restarting of production may trigger certain irregularities in wells with an open casing annulus. The annulus tends to fill up with gas during steady-state production. The pressure of the gas column at the level of the tubing shoe equals the shoe pressure p_{TL} of the fluid rising in the tubing string. When flow starts through the smaller-bore choke, the shoe pressure in-

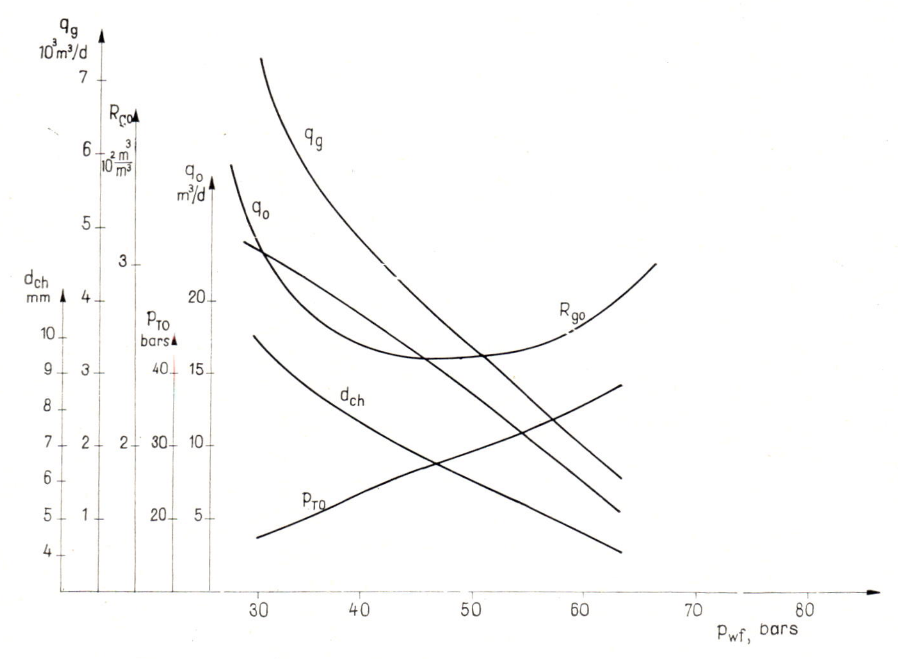

Fig. 2.1—3 Characteristics derived from well-test results

creases to a value $p'_{TL} > p_{TL}$. For the casing-annulus pressure to attain this increased value at shoe level, part of the fluid coming from the reservoir must flow into the annulus, and compress the gas column in it, with a liquid annulus developing above tubing bottom level. Subsequently, the fluid coming from the reservoir will start to let off gas into the casing annulus. This latter being closed on top, its gas pressure will increase, driving some of the oil from the above-mentioned liquid annulus into the tubing. The tubing will thus receive significantly more liquid and less gas than under steady-state conditions, so that the fluid produced will have a much lower *GOR*. Its gas content may be insufficient to ensure flowing production, so that the well may die. This type of trouble may often be prevented by gradually increasing the *BHP*; notably, by producing the well through some intermediate sizes of choke before introducing the small-bore one. — Well testing will, then, furnish a characteristic curve that can be approximated well enough by one of Eqs 2.1—7 or 2.1—8. In its possession, it is possible to determine (i) the optimum tubing size that will result in flowing production for the longest possible time, (ii) the most favourable conditions and means of artificial lifting if the well is to be produced at a rate impossible to achieve by natural means, and (iii) to predict several reservoir parameters affecting production, relying also on the results of other well tests.

2.2. Flowing wells producing gasless oil

The *GOR* of an oil-well fluid may be so low that no gas is liberated during flow through the well: flow is then single-phase. The operating parameters of the well are comparatively simple to calculate (Szilas 1955). The *BHP* is

$$p_{wf} = L_w \bar{\gamma} + p_s + p_s' + p_{TO} , \qquad 2.2-1$$

where p_s is friction loss in the tubing, and p_s' is friction loss in the casing between the well bottom and the tubing shoe. — Let us specify that well depth L_w will mean the depth below the wellhead on the top of the sand surface; P_{TO} is composed of the hydrostatic pressure of the fluid entering the flow line through the wellhead assembly, plus the energy losses of flow:

$$p_{TO} = \Delta p_{ch} + p_{wh} + p_l + p_h ,$$

where p_{wh} is the flowing energy loss in the wide-open (chokeless) wellhead assembly; p_l is the friction loss in the flow line and in the fittings of the tank station; p_h is the hydrostatic pressure acting upon the wellhead, composed of the hydrostatic pressures of the oil in the flow line and in the tank. — The resistance to flow of a chokeless wellhead assembly is usually negligible. Hence, in an approximation satisfactory for practical purposes,

$$p_{wf} = L_w \bar{\gamma} + p_s + p_s' + \Delta p_{ch} + p_{wh} + p_h . \qquad 2.2-2$$

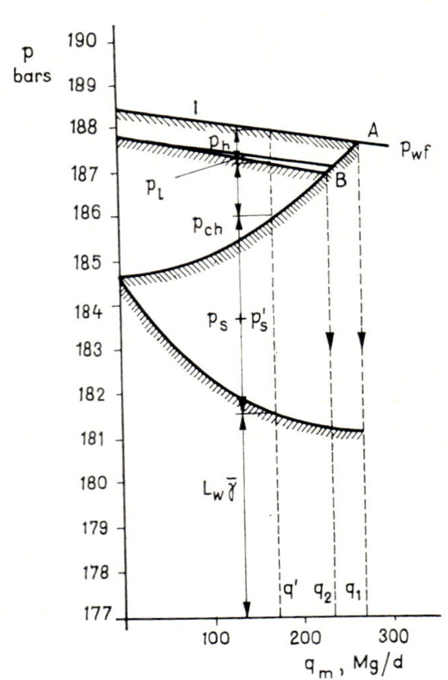

Fig. 2.2—1 Pressure utilization curves of well producing gasless crude, after Szilas (1955)

Figure 2.2—1 shows the variation of pressure as described by Eq. 2.2—2 applied to a well at Nagylengyel, Hungary. Graph I is an inflow performance curve characteristic of the inflow of oil into the wellbore, a plot of production rate v. flowing BHP, p_{wf}. Graph II is the internal energy-loss curve of the well, a plot of the hydrostatic pressure of the oil column flowing in the well plus the friction loss, v. rate of production. These two Graphs, representative of the interaction of reservoir and well, resemble the graphs of a centrifugal pump lifting fluid through a vertical flow string. The rate-of-production on graph, depending on the physical properties and saturation conditions of the formation, corresponds to the choke curve depending on structural and geometrical features of the centrifugal pump, whereas the internal energy-loss graph of the well corresponds to the resistance curve of the riser fed by the pump. In the case under consideration,

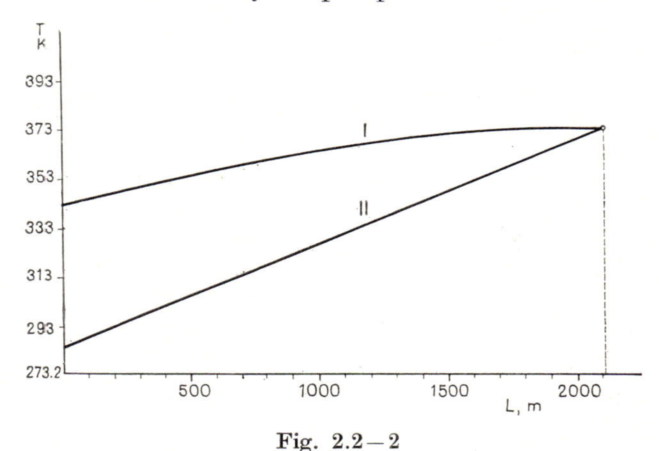

Fig. 2.2—2

the form of Graph II is considerably influenced by the fact that an increased production rate will increase the mean temperature of the oil flowing through the tubing, thus reducing its mean gravity and viscosity. This is why $L_w\bar{\gamma}$, the hydrostatic pressure of the liquid column in the well, decreases as the production rate increases, and friction losses do not increase as rapidly with the production rate as should be expected in isothermal flow. — Figure 2.2—2, based on operation parameters, provides some useful information concerning operation of the well: the intersection of Graphs I and II, marked A, is the operating point of the maximum liquid production rate q_1. This 'wide-open' flow rate will obtain at the gauge pressure $p_{TO} = 0$, that is, when the wellhead offers no resistance and oil flows from the tubing into the open through a wide-open valve. If, on the other hand, a given tank station is joined to the well by a given flow line, the maximum flow rate will be the lower value q_2. The corresponding operating point is B, with $\Delta p_{ch} = 0$, that is, with no choke in the wellhead assembly. The Figure reveals the increase in production rate that can be attained by minimizing the flow resistance of the surface equipment (increase of flowline diameter; heating or thermal insulation of flow line; sinking into the ground of separators, etc.). The diagram can often be used to predict performance at a future date. If the well produces for instance from a fractured, carbonic rock, the productivity index J will often have a tendency to stay approximately constant. This means that Graph I will have an unchanged slope; it will merely shift to a smaller intercept p_{ws}. Reservoir pressure will equal the p_{wf} valid at $q_o = 0$. If the energy loss curves are assumed to stay unchanged, then estimates will be on the safe side because the mean fluid gravity belonging to a given flow rate will decline in time owing to the warming up of the rock surrounding the wellbore. The energy-loss diagram can be constructed by calculating the factors in Eq. 2.2—2 for a given well at a number of different production rates, in the following way.

The expression $L_w \bar{\gamma}$ for the hydrostatic pressure of the fluid filling the well can be determined in a number of ways, one of which follows. We have to know the mean temperature of the oil flowing in the well. A formula for calculating the temperature of water flowing in wells has been derived by Boldizsár (1958). Most water wells have no tubing, and the water rises in the casing string. Boldizsár has assumed the hot water to lose heat to a thermally homogeneous surrounding. Oil flowing in the tubing of oil wells is surrounded by a jacket of oil filling the casing annulus. The present author has therefore added a heat-transfer correction factor C to Boldizsár's formula, to account for the heat passing from the flowing fluid to the host rock through piping, oil and cement. The C factor of a given well is constant over a rather wide flow rate range. Its value can be determined by field tests. Boldizsár's corrected formula is, then,

$$\Delta T_h = \frac{q_o c_o \sigma_b}{C k \lambda_b}\left[1 - e^{-\left(\frac{C k \lambda_b h}{q_o c_o}\right)}\right], \qquad 2.2\text{—}3$$

where ΔT_h is the temperature difference between the flowing fluid and the original temperature of the rock at a height h above the well bottom; λ_b is the thermal conductivity coefficient of the rock surrounding the well; and k is a dimensionless rock-heating coefficient, whose value is furnished by the integral

$$k_{(F)} = \frac{8}{\pi}\int_{u=0}^{-\infty} \frac{e^{-F_O u^2}}{J_0^2\, u + Y_0^2\, u}\,\frac{du}{u}\,. \qquad 2.2\text{—}4$$

J_0 and Y_0 are zero-order Bessel functions of the first and second kind, respectively; u is the integration variable; and F_O is the Fourier coefficient:

$$F_o = \frac{at}{r^2}, \qquad 2.2\text{—}5$$

where a is the theoretical thermal conductivity coefficient for liquid flowing through an uncased well; t is the time elapsed between the start-up of production to the data of testing; r is the radius of the flow string through which flow takes place. Values of the integral 2.2—5 are tabulated as a function of F_O in a paper by Jaeger and Clarke (1942). Equation 2.2—3 reveals that the oil entering the wellbore is at the height of the sand surface ($h = 0$) at precisely the same temperature as the rock surrounding the well, but at any other elevation the flowing oil is warmer; more specifically, it is the warmer, the higher the flow rate. Analysis of the formula makes it apparent that as production proceeds (as t increases), the flowing temperature of the oil will increase even if the flow rate remains unchanged. The reason for this is that some of the heat content of previously produced oil has already heated up some of the colder surroundings. Production gives rise to a 'thermal jacket', the geometry of which in the case of a given well drilled in a given rock is at any instant a function of previous production history. — If production preceding a well test was much longer than the duration of testing, F_O can be assumed to be constant in a fair approximation. In that case, a single test at a given production rate will directly furnish not only C but also the product $Ck = k'$.

Example 2.2—1. Given the data of a test on a well producing gasless oil, find (i) the expression $k' = Ck$ at the production rate q_1, (ii) the change of temperature of the oil as it rises through the flow string and its mean flowing temperature, both at the same production rate q_1, and (iii) the outflow temperature to be expected at a production rate q_2. — The tubing string reaches down to the well bottom; $L_w = 2108$ m; $d_i = 62$ mm; $q_1 = 1.082$ kg/s; $q_2 = 1.701$ kg/s. The outflow temperature is $T_{TO1} = 273.2 + 68.9 = 342.1$ K at the production rate q_1; $\varrho_{288} = 929$ kg/m^3; $\sigma_b = 4.24 \times 10^{-2}$ K/m; $\lambda_b = 1.838$ W/mK. Ground temperature next to the wellhead equals the annual mean temperature, $(273.2 + 11) = 284.2$ K.

(i) The mean flowing temperature $T = \bar{T}_{T1}$ required to determine the mean specific heat is estimated at $273.2 + 88 = 361.2$ K. — By Eq. 7.2—23,

$$c_0 = \frac{762.5 + 3.38 \times 361.2}{\sqrt{0.929}} = 2.06 \times 10^3.$$

The difference between the outflow temperature and the ground temperature next to the wellhead is

$$\Delta T_{O1} = 342.1 - 284.2 = 57.9 \text{ K}.$$

Substitution of the values obtained into Eq. 2.2—3 results in

$$57.9 = \frac{1.082 \times 2.06 \times 10^3 \times 4.24 \times 10^{-2}}{k'\,1.838}\left[1 - e^{-\left(\frac{k'1.838 \times 2108}{1.082 \times 2.06 \times 10^3}\right)}\right].$$

This resolves to

$$k' = 0.5418.$$

(ii) Substituting the above value of k' into Eq. 2.2—3 and assuming that the mean specific heat is approximately the same also at other flow rates, we get

$$\Delta T_{h1} = 95.1(1 - e^{-4.47 \times 10^{-4} h}).$$

The relationship will furnish the temperature differences ΔT_{h1} at various elevations h above the well bottom. Adding to this value the original rock temperature T_b at any elevation we get the flowing temperature at the height h. The variation of rock temperature v. depth can be calculated in the knowledge of $T_{bo} = 284.2$ K and $\sigma_b = 4.24 \times 10^{-2}$ K/m. Figure 2.2—2 is a plot v. depth of the rock and oil temperatures thus determined. By planimetering the surface under the curve $T_T = f(h)$, the mean flowing temperature is established as $T_{T1} = 361.2$ K.

(iii) By Eq. 2.2—3, the outflow temperature difference at the flow rate q_2 is

$$\Delta T_{O2} = \frac{1.701 \times 2.06 \times 10^3 \times 4.24 \times 10^{-2}}{0.5418 \times 1.838} \left[1 - e^{-\frac{0.5418 \times 1.838 \times 2108}{1.701 \times 2.0^6 \times 10^3}}\right] = 67.3 \text{ K}.$$

The outflow temperature of the oil is $T_{TO2} = 284.2 + 67.3 = 351.5$ K. If desired, the mean specific heat can be calculated more accurately, by means of an iteration procedure. Whether this is necessary is to be decided individually in each case.

The mean density can be calculated using Eq. 7.2—26, which, in the correct interpretation of the notation employed, gives

$$\bar{\varrho} = \varrho_{288} - \alpha_T(\bar{T} - 288.2) + \alpha_p \bar{p}.$$

α_T and α_p can be obtained from laboratory tests; α_p can also be determined from the shut-in data of the well (Szilas 1959). — The friction loss p_s can be determined either by field tests or by calculation. The latter is based on Eq. 1.1—1;

$$p_s = \lambda \frac{L \bar{v}^2 \bar{\varrho}}{2 d_i}.$$

Substituting

$$\bar{v} = q \frac{q_m}{d_i^2 \frac{\pi}{4} \bar{\varrho}},$$

and assuming the flow to be laminar, that is, Eqs 1.1—2 and 1.1—3 to hold, we obtain for pressure loss due to friction in the tubing the relationship

$$p_s = 40.7 \frac{L q_m \bar{\nu}}{d_i^4}. \qquad 2.2\text{—}6$$

If the tubing string is not run through to the well bottom, then the above relationships can be used to calculate also friction loss in the casing section between the well bottom and the tubing shoe. d_i then denotes the *ID* of the production casing.

p_s and p_s' can be determined quite accurately by a relatively simple well testing procedure. Notably, if the flowing well is abruptly shut in, the impulse content of the liquid column held by the flow string will give rise to a pressure surge in the wellhead, which will normally decay in a span of time on the order of 10 s (Fig. 2.2—3). The wellhead tubing pressure will increase from the steady-state flowing pressure p_{TO} to p_{TZ} and the casing head pressure from p_{CO} to p_{CZ}. Because of the abrupt shut-in, the *BHP* will not change from the steady-state flowing value and the temperature of the oil in the well will likewise remain unchanged. We may, then, write up that, before shut-in,

$$p_{wf} = (L_w - L_T)\,\bar{\gamma}_1 + L_T\,\bar{\gamma}_2 + p_s' + p_s + p_{TO}\,,$$

and after the decay of the pressure surge,

$$p_{wf} = (L_w - L_T)\,\bar{\gamma}_1 + L_T\bar{\gamma}_2 + p_{TZ}\,.$$

Subtracting the first equation from the second and rearranging, we get

$$\Delta p_T = p_{TZ} - p_{TO} = p_s' + p_s\,; \qquad 2.2\text{—}7$$

that is, the pressure surge in the tubing head equals the total pressure

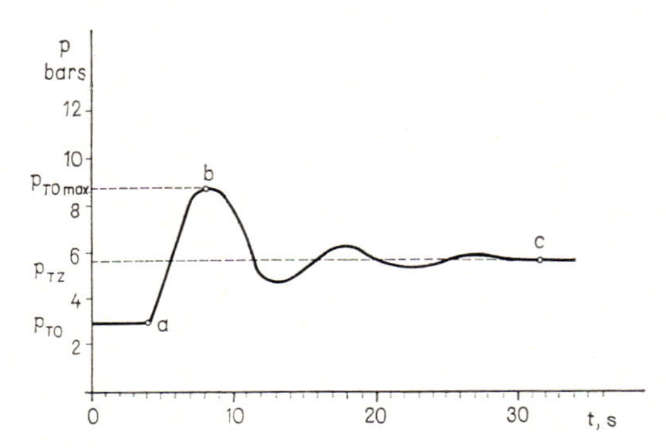

Fig. 2.2—3 Pressure buildup at wellhead after sudden shut-off, after Szilas (1959)

loss due to friction of oil flow in the well before shut-in. — The pressure balance between the well bottom and the casing-head gives, on the other hand,

$$p_{wf} = (L_w - L_T)\,\bar{\gamma}_1 + L\,\bar{\gamma}_3 + p_s' + p_{CO}\,,$$

and after the decay of the pressure surge,

$$p_{wf} = (L_w - L_T)\,\bar{\gamma}_1 + L\,\gamma_3 + p_{CZ}\,.$$

Subtraction of the first equation from the second gives

$$\Delta p_C = p_{CZ} - p_{CO} = p_s'\,; \qquad 2.2\text{—}8$$

that is, the pressure surge in the casing head equals that part of the friction loss of oil flow in the well before shut-in which arises between the well bottom and the tubing shoe. Let us note that $\bar{\gamma}_1$, $\bar{\gamma}_2$ and $\bar{\gamma}_3$ respectively denote the mean gravities of the oil in the casing between the well bottom and the bottom of the tubing string, in the tubing, and in the casing annulus. — Subtraction of Eq. 2.2—8 from Eq. 2.2—7 results in

$$\Delta p_T - \Delta p_C = p_s\,, \qquad 2.2\text{—}9$$

that is, the difference between the pressure surges in the tubing head and casing head equals that part of the friction loss of oil flow in the well before shut-in which is due to friction in the tubing string.

Example 2.2—2. Find the flowing pressure at the tubing shoe of a flowing well producing gasless oil. Given $L_w = 2016$ m; $d_i = 0.062$ m; $q_m = 1.505$ kg/s; $\varrho_{288} = 951$ kg/m³; $\alpha_T = 0.58$ kg/m³K; $\alpha_p = 9.384 \times 10^{-7}$ kg/mN; $\bar{T} = 367.0$ K; $\bar{\nu} = 4.10 \times 10^{-5}$ m²/s; $p_{TO} = 4.0$ bars; $p_{TZ} = 7.5$ bars; $p_{CO} = 6.4$ bars; $p_{CZ} = 6.5$ bars. — Equation 2.2—6 gives a friction loss

$$p_s = 40.7 \frac{2016 \times 1.505 \times 4.10 \times 10^{-5}}{0.062^4} = 3.4 \times 10^5 \text{ N/m}^2.$$

The friction loss as calculated from wellhead pressures measured during the abrupt shut-in test (Eq. 2.2—9) is

$$p_s = (7.5 \times 10^5 - 4.0 \times 10^5) - (6.5 \times 10^5 - 6.4 \times 10^5) = 3.4 \times 10^5 \text{ N/m}^2.$$

The mean flowing density required to calculate the hydrostatic pressure of the liquid column flowing through the tubing depends by Eq. 7.2—26 also on mean pressure; $\bar{\varrho}$ is therefore to be determined by successive approximation. Let us first assume that $\bar{p} = 0$. Then Eq. 7.2—26 gives the approximate mean density

$$\bar{\varrho} = 951 - 0.58(367.0 - 288.2) = 905.3 \text{ kg/m}^3.$$

The approximate mean pressure is

$$\bar{p} = \frac{p'_{wf} + p_{TO}}{2} = \frac{L\bar{\varrho}g + p_s + p_{TO} + p_{TO}}{2} =$$

$$= \frac{2016 \times 905.3 \times 9.81 + (3.4 + 4.0 + 4.0)10^5}{2} =$$

$$= 9.52 \times 10^6 \text{ N/m}^2.$$

Using this value, we get as the mean density

$$\bar{\varrho} = 905.3 + 9.834 \times 10^{-7} \times 9.52 \times 10^6 = 914.7 \text{ kg/m}^3;$$

$$p_{wf} = 2016 \times 914.7 \times 9.81 + 3.4 \times 10^5 + 4.0 \times 10^5 =$$

$$= 1.883 \times 10^7 \text{ N/m}^2 = 188.3 \text{ bars}.$$

The more accurate mean pressure calculated using this *BHP* does not appreciably affect the final result any more.

By the procedures outlined above, the energy loss, Graph II and the inflow performance curve I, characterizing flow through the sand surface (Fig. 2.1—1), can be determined for a given well. As established above, the ordinate difference between Graphs I and II equals at any production rate q_m the corresponding wellhead pressure p_{TO}. The pressures Δp_{ch}, p_l and p_h can be calculated.

The efficiency of production out of the well can be found by dividing the useful work W_2 expended in lifting one kg of oil from the well bottom to the surface by the total energy expenditure W_1, that is,

$$\eta_W = \frac{W_2}{W_1} = \frac{L_w}{L_w + \dfrac{p_s}{\bar{\gamma}}} .$$

We have assumed for simplicity that the tubing string reaches down to the well bottom. Replacing p_s by the expression in Eq. 2.2—6, we obtain

$$\eta_W = \frac{1}{1 + \dfrac{40.7\, q_m\, \bar{\nu}}{d_i^4\, \bar{\gamma}}} . \qquad 2.2\text{—}10$$

The specific energy available at the well bottom is, then, exploited the more efficiently, the greater the *ID* of the tubing string. By this consideration, it would be most expedient to have wells untubed and to produce them through the casing. An indisputable advantage of this solution is, of course, the saving due to dispensing with the tubing. The method is not, however, applicable if (i) the well produces sandy oil which may lead to casing erosion, (ii) produces a corrosive fluid, (iii) the well is to be produced at a relatively low rate. This last case may be justified e.g. by the following consideration. Figure 2.2—4 shows pressure utilization graphs for a variety of tubing sizes. One of the graphs refers to simultaneous production through a 2 7/8-in. tubing plus the casing annulus. Tangents to the pressure utilization curves parallel to the performance line define a set of operating points connected by the dashed line. One and the same

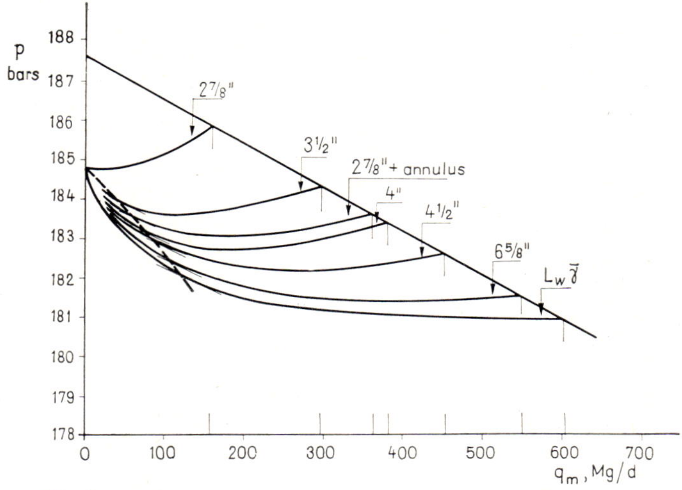

Fig. 2.2—4 Pressure utilization curves of well producing gasless crude, after Szilas (1955)

wellhead pressure p_{TO} may belong to two different production rates q_{m1} and q_{m2} on either side of the dashed line. (Cf. also Fig. 2.2—5.) The well in question cannot be produced at rates between $q_m = 0$ and the rate q'_m, belonging to the point of tangency, through the tubing in question. Production will be unsteady, fluctuating, and the well may even die. Running a smaller-size tubing will shift the upper limit of the unstable zone to the left in Fig. 2.2—4; that is, the well will permit also of a lower-rate production. Understanding this phenomenon of well behaviour is of considerable practical importance because reservoir engineering considerations (e.g. the prevention of water coning) may indeed demand a reduction of the production rate. Insertion of a smaller-bore choke may then result in fluctuation or eventual dying of the well. To remedy this, pumping is often started, although by the above considerations steady flow could be achieved simply by running in a smaller-size tubing. — In another possible case, a well with a relatively small-size tubing is to be produced above the operating point of maximum flow rate. The flowing cross-section can then be increased by bringing the casing annulus into play, provided the casing is not menaced by erosion and/or corrosion.

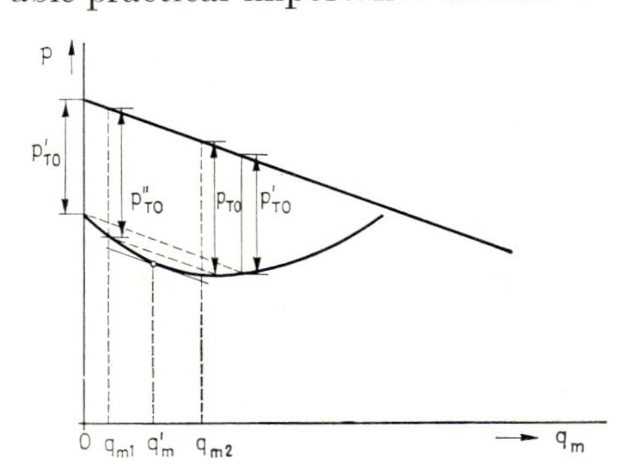

Fig. 2.2—5 Unstable operation interval of well producing gasless crude, after Szilas (1955)

Figure 2.2—4 has been plotted in a rather approximate fashion. The relationship $T = f(q_m)$ varies somewhat with the tubing size, and hence a slightly different $L_w \bar{\gamma} = f'(q_m)$ function should result for each tubing size. We have, however, refrained from an accurate calculation of these. The approximate curves have been calculated by extending also to other tubing sizes the temperature v. production-rate graph, plotted for the tubing size at which the well test has been carried out. — The friction loss of production through the casing annulus can be calculated by means of Eq. 1.1—14 or 1.1—15.

2.3. Flowing wells producing gaseous fluids

2.3.1. Interaction of well and formation

(a) Krylov's theory

Figure 2.3—1 is a plot of the oil production rate, q_o, and of the gas production rate referred to standard conditions, q_{gn}, v. the flowing *BHP* p_{wf}, as provided by a well test. The tubing in the well in question has been run to the well bottom; that is, $p_{wf} = p_{TL}$. In adapting the Figure we have used the approximation 1 at $\simeq$ 1 bar. Figure 2.3—2 shows pairs of values $q_g - q_{gn}$ belonging to given values of p_{wf} plotted in a coordinate system calibrated in q_o and q_{gn}. At a flowing *BHP* of 55 bars, for instance, $q_o = 1.16 \times 10^{-3}$ m³/s and $q_{gn} = 0.77$ m³/s. Krylov joined the corresponding pairs $q_o - q_{gn}$ by a curve calibrated in terms of flowing *BHP*. This curve is then the characteristic performance curve for inflow from one formation, plotted in a q_{gn} v. q_o coordinate system. — Krylov's general flow equation (Muraviev and Krylov 1949) for long tubing strings (not detailed in the present book) permits us to determine throughput curves for the tubing string. These curves resemble the

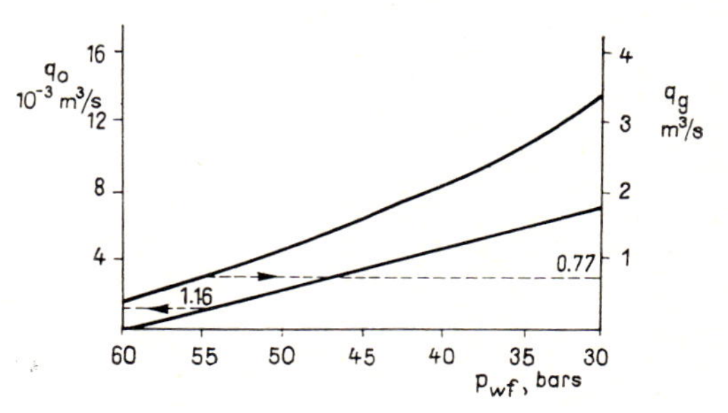

Fig. 2.3—1 Inflow curves from well tests

$q_{gn} - q_o$ curves referring to the throughput of infinitesimal lengths of tubing (cf. Fig. 1.4—11). The set of curves in Fig. 2.3—2 refers to the conditions $L_w = 1000$ m, $d = 2\ 7/8$ in., and $p_{TO} = 2$ bars. Instead of ξ, we use as a parameter p_{TL}, the only variable in its expression. The graphical solution of the set of equations represented by the inflow performance curve and a throughput capacity curve of the flow string furnished that flowing *BHP*, $p_{wf} = 49.3$ bars, at which the inflow rates into the well, $q_{gn} = 1.21$ m³/s for gas and $q_o = 2.45 \times 10^{-3}$ m³/s for oil, just equal the production rate which can be delivered to the surface through the given tubing, at the given wellhead pressure and at the tubing shoe pressure of 49.3 bars.

Performing the construction with L_w and d unchanged but for various values of wellhead pressure p_{TO}, we get the Graphs p_{TO} = const. on the left-hand side of Fig. 2.3—3. It is apparent that if the original wellhead

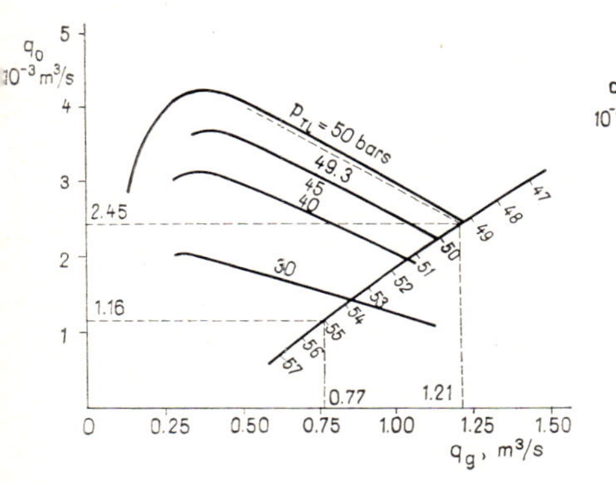

Fig. 2.3—2 Characteristics of well-formation interaction, for a given tubing size, according to Krylov

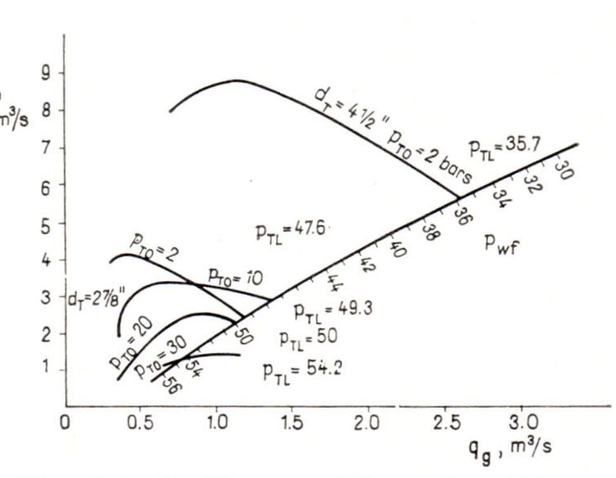

Fig. 2.3—3 Characteristics of well-formation interaction, for various tubing sizes, according to Krylov

pressure of 30 bars is decreased first to 20, then to 10 bars by the insertion of larger-bore chokes, then the flowing *BHP* will decrease accordingly, and the oil and gas flow rates will increase. The flowing *BHP* decreases by a smaller amount than the wellhead pressure, because the increased mass flow rate, mean specific volume and flow velocity will increase the total flowing energy loss and hence the mean flowing pressure gradient. Such situations are fairly frequent especially in medium-capacity wells, most of whose energy loss in the tubing is due to friction even at the least production rates, slippage being relatively insignificant. If the wellhead pressure is decreased to 2 bars, the flowing *BHP* will increase and the rate of production will decrease. This is due to the circumstance that, as outlined above, the increase in flowing pressure gradient entails an increase in *BHP* greater than the decrease in wellhead pressure. This situation is restricted to very high-capacity wells. — A different tubing size will have a different throughput capacity curve. In the example above, illustrated by Fig. 2.2—3, the flowing *BHP* will decrease to 35.7 bars for a wellhead pressure of 2 bars if the tubing size is changed to 4 1/2 in. It is thus apparent that when using a 2 7/8-in. size tubing, increasing the choke bore will not improve the production rate beyond a comparatively small increase (corresponding to $p_{wf} = 47.6$ bars at $p_{TO} = 10$ bars). If a greater production is required, a larger size tubing must be run in.

So far we have studied the interaction of well and reservoir at certain constant values of wellhead pressure. The system analysed was a series-connected two-component hydraulic system whose components had different flow characteristics. In production practice, the desired wellhead pressure is attained, as is well known, by the insertion of suitably dimensioned reduction orifices (chokes in common parlance). A method for establishing the common operating points of the three-component hydrodynamic system composed of the reservoir, the tubing and the choke was developed by Gilbert.

(b) Gilbert's theory

Graph I in Fig. 2.3—4 is an inflow performance graph characterizing the inflow of fluid from the reservoir. Let us consider various flow rates q_o and let us mark off the *BHP*'s p_{wf} belonging to each of these on the abscissa axis. Let us assume that the length of the tubing string is L_w and that the producing gas–oil ratio, R_{go}, is a constant independent of the production rate. Starting from various tubing-shoe pressures $p_{wf} = p_{TL}$, and using Gilbert's pressure-gradient curves, the curves in Fig. 2.3—4a, each illustrating pressure v. elevation in the tubing string at a given production rate, can be constructed (Gilbert 1955). The intercept $L = 0$ gives the wellhead pressure p_{TO} to be expected at the production rate considered. Transferring the pressures p_{TO} thus obtained to Fig. 2.3—4b, we get a set of corresponding pairs $p_{TO} - q_o$. Joining these pairs results in Graph II of wellhead pressure v. production rate. Flowing production is seen to be restricted to the interval between *A* and *B*. Let us determine now the production rate with a given choke diameter d_{ch} in place. — Assuming that pressures p_3 and p_2, respectively upstream and downstream of the choke, give a ratio less than critical, we can calculate the upstream pressure e.g. using Eq. 1.5—6. — If d_{ch} and R_{go} are constant, the wellhead pressure varies directly as the production rate. This relationship is illustrated by the straight-line, Graph III, in Fig. 2.3—4b. The actual choke graph is seen to deviate somewhat from a straight line near the origin of coordinates. Graph III intersects Graph II at the points D and E. These are in principle the operating points characterizing the reservoir-tubing-choke system. Are both operating points feasible?

Figure 2.3—5 shows Graphs II and III of the preceding Figure in somewhat more detail. Let us assume that the well will, at the operating point E, produce at a rate q_{oE} for a wellhead pressure p_{TOE}. In practice, flow will not be quite steady but quasi-steady, meaning that flow parameters will vary in time but their mean values will stay constant. If the production rate increases temporarily from q_{oE} to $(q_{oE} + \Delta q'_{oE})$, then the stability of flow requires that a pressure $(p_{TOE} - \Delta p_{TOE})$ should stabilize itself at the wellhead. At this increased flow rate, however, the pressure upstream of the choke will increase by Eq. 1.5—6 to $(p_{TOE} + \Delta p_{TOE})$. The higher wellhead pressure entails a higher BHP and hence a reduction of inflow. — If, on the other hand, the production rate decreases temporarily by a value $\Delta q''_{oE}$, then, in accordance with the operating point at E'', the pressure upstream of the choke will be less than the value needed to keep up flow at this reduced rate. The lower wellhead pressure entails a decrease

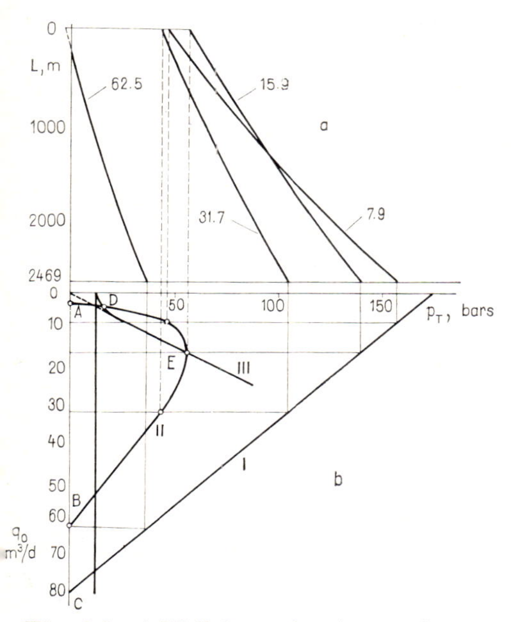

Fig. 2.3—4 Well-formation interaction, after Gilbert (1955)

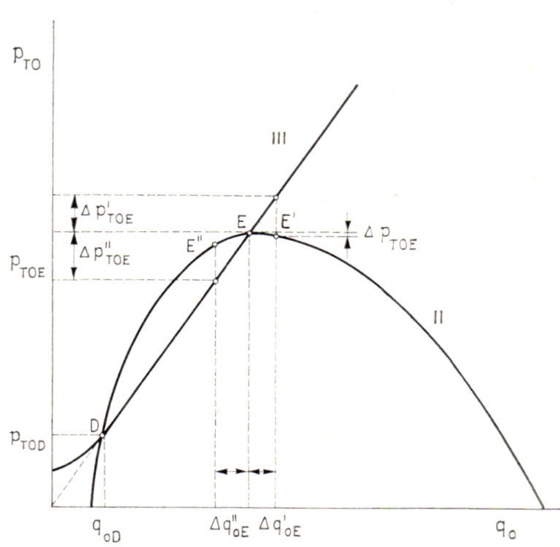

Fig. 2.3—5 Stable and unstable operating points of well producing gaseous crude, after Gilbert (1955)

in *BHP* and a consequent increase in inflow; E is thus revealed as a stable operating point, because after slight fluctuations of the production rate, the operating parameters will swing in spontaneously to their original values. — Now if the well produces at the rate q_{oD}, then a temporary increase of the flow rate will give rise to a pressure upstream of the choke that is insufficient to stabilize the flow. The decrease in wellhead pressure will reduce the *BHP* and increase the production rate until the value q_{oE} corresponding to the stable operating point is attained. If, on the other hand, the flow rate decreases below the original q_{oD}, then a process opposite

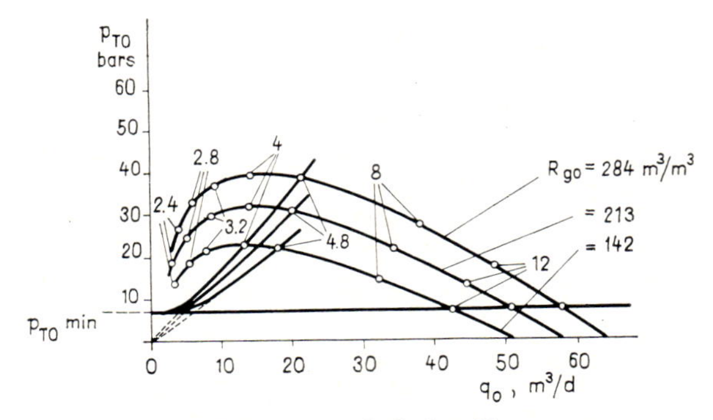

Fig. 2.3—6 Influence of choke diameter upon stability of operation of well producing gaseous crude, after Gilbert (1955)

to the one just described will become operative: flow will decrease to zero; the well will die. Operating point D is therefore unstable. By the above considerations, then, a given choke bore fixes one production rate at one stable operating point. The point in question is the common operating point of the formation-well-choke system. Let us now find out whether a stable production rate can be assigned to every choke size.

Figure 2.3—6 shows the operating points E, regarded as stable, determined for various values of R_{go} and various choke bores. Concentrating on the curve for $R_{go} = 284$ we find that a rather acute-angled intersection would develop, e.g., when using a 2.8-mm choke. The restoring pressure differential, forcing the system back to the stable operating point, is thus much less than in the case shown as Fig. 2.3—5. The restoring force being weak, the control process is sluggish and the flow rate will not yet have swung in to a stable value when it is deviated again by another fluctuation. The well will thus operate under the permanent risk of dying. In a general way, the restoring pressure differential will be insufficient unless the choke curve intersects the curves of formation plus well to the right of their respective peaks. Figure 2.3—6 shows that the choke diameter required to ensure stable flow is the greater, the greater the producing *GOR*, R_{go} and the production rate q_o.

By the above consideration, the peak of every Graph II will define a least production rate feasible in a given formation-plus-well system. If a lower production rate is desired, a smaller-size tubing string will have to be run in. It is further apparent that, given $p_{TO\,\min}$, the greater the producing GOR and R_{go}, the greater the rate at which a well can be produced through a given tubing size. — The common operating point of formation plus well at a prescribed wellhead pressure p_{TO} can be determined according to Nind in two ways, both simpler than the above procedure, already outlined in connection with Fig. 2.3—4 (Nind 1964). — In Fig. 2.3—7, Graph I is the inflow performance curve. Graph II is the $p_{TO} = f(q_o)$ curve introduced in connection with Fig. 2.3—4. This curve can be plotted also by selecting the pressure gradient curve belonging to a given triplet of d, q_o and R_{go}, finding the value of p_{TL} belonging to q_o read off Graph I, and then determining the wellhead pressure p_{TO} of the well of depth L_w in the manner discussed in Section 1.4—3d. In Fig. 2.3—7, this value of p_{TO} is plotted v. the assumed value of q_o; Graph II joins several points plotted in the same manner. The intersection with Graph II of a line parallel to the abscissa axis, passing through the prescribed wellhead pressure p_{TO}, provides the production rate q_{o1} looked for. — In the second procedure, the first step is likewise to find the pressure gradient curve belonging to the given triplet of d, q_o and R_{go}. Then, starting from the prescribed pressure p_{TO}, the tubing-shoe pressure to be expected at the depth L_w is read off. A plot of several pairs (q_o, p_{TO}) yields Graph III (the function $p_{TO} = f'(q_o)$) in Fig. 2.3—7. The production rate q_{o1} defined by the intersection of Graphs I and III is the same as that furnished by the foregoing procedure.

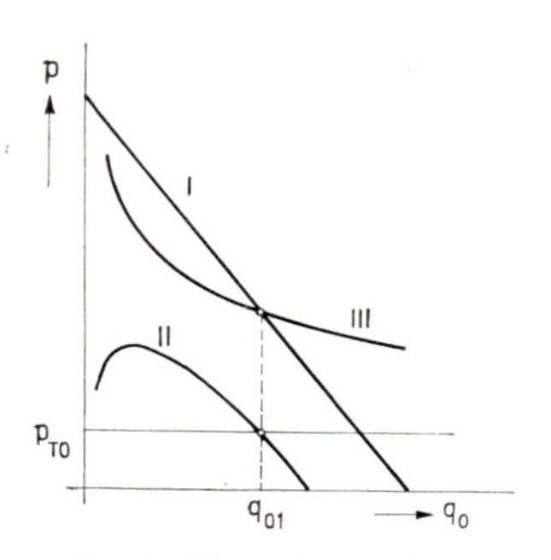

Fig. 2.3—7 Establishing well-formation interaction

(c) Influence of the flow line

Removing the wellhead choke and neglecting the pressure drop due to passage through the Christmas-tree assembly, we find that the least feasible wellhead pressure equals the least required pressure at the inflow end of the flow line. — If the flow line is laid over a level terrain, the pressure traverse of multiphase flow along the flow line can be characterized by gradient curve sets $p = f(l)_{d,qo,Rgo}$ constructed or calculated, similar to those used to characterize flow in vertical strings in Section 1.4—3. The set of curves characterizing the flow line may be determined in principle by any one of the calculation methods described in Section 1.4—1, provided the method in question is deemed to be accurate enough. Sets of curves of this type have been demonstrated by Brill et al. (1966). In possession of said curves, the common operating point of the formation-well-flowline system, and the maximum production rate of a given well through a given flow line, can respectively be determined as follows. Assuming R_{go} to be constant, one can plot the wellhead pressures of a well of given parameters v. the production rate in the manner of Graph II in Fig. 2.3—7. The resulting curve has likewise been marked Graph II in Fig. 2.3—8. Using Brill's curves, one can construct the curves $p_{TO} = f(q_o)$, valid for flow lines of a given length and a variety of pipe diameters (Graphs IVa—c). The intersection of Graphs II and IVa predicts the wellhead pressure $p_{TO\,min}$ and the production rate $q_{o\,max}$ for the corresponding size of line pipe. Line pipe sizes increase in the direction of the arrow. The greater the diameter of the horizontal flow line, the less the wellhead pressure, and the greater the production rate that can be realized in the given setup.

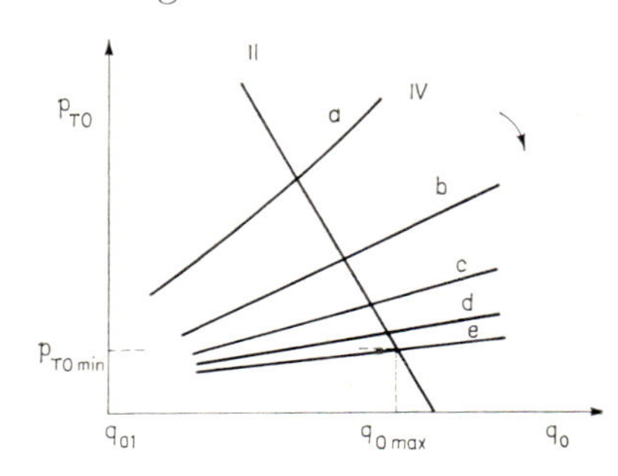

Fig. 2.3—8 Interaction of well and flowline, after Brill et al. (1966)

2.3.2 Time course of production parameters

(a) Calculating the flowing life of wells; Woodward's theory

The need to procure lifting equipment at the correct time and to predict the power requirements of artificial lifting makes it important to predict the flowing life of a well (Gilbert 1955). Reservoir engineering calculations permit us to determine the variation v. time of the total oil production V_o, the producing gas–oil ratio R_{go} and productivity index J of a well typical of a larger group of wells in a field, as well as the time variation of the reservoir pressure p_{ws}. A diagram prepared in this fashion is shown as Fig. 2.3—9. Figure 2.3—10 shows the inflow performance curves of the same well at seven instants of time marked in Fig. 2.3—9. The *IP* curves are straight lines here, as we have assumed the exponent n of the performance equation to equal unity (set of straight-line Graphs I). The set of Graphs II illustrates the function $p_{wf} = f(q_o)_{Rgo}$ characterizing the liquid throughput capacity of the tubing string. This set can be plotted by assuming a number of gas–oil ratios likely to occur sometime during the flowing life of the well. Keeping each R_{go} constant and using $p_{TO\,min}$, the flowing *BHP* is determined for various production rates by means of the pressure gradient curves (see Graph III in Fig. 2.3—7). The reservoir and the well will cooperate if the *GOR* as delivered by the reservoir to the flow string equals the producing *GOR* required to lift the fluid to the surface. The common operating points are joined by the curve marked with xs. The end point of said curve marks the end of the flowing life of the well. Tracing the *IP* line corresponding to said end point and reading off its intercept at $q_o = 0$,

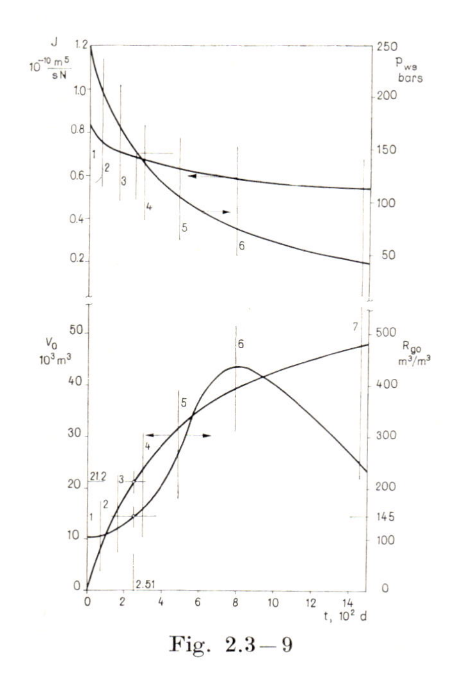

Fig. 2.3—9

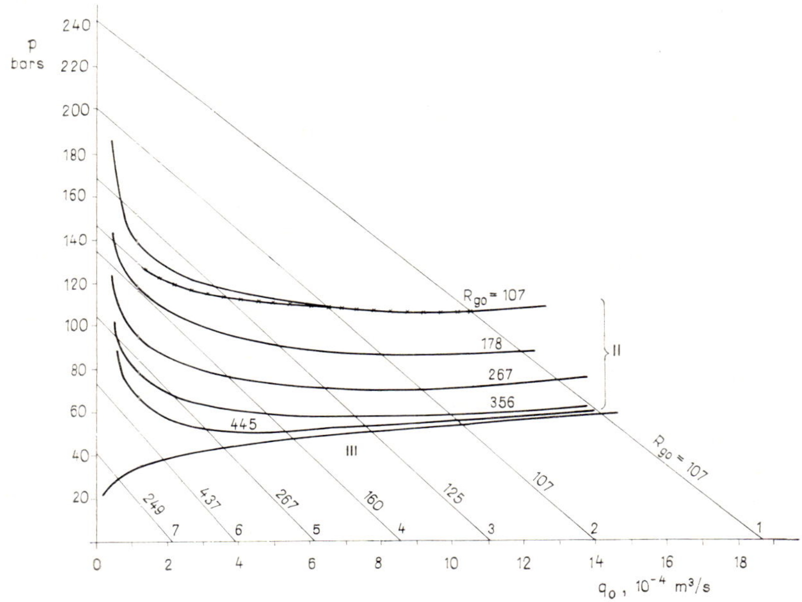

Fig. 2.3—10 Determination of flowing life, according to Woodward (Gilbert 1955)

we get the reservoir pressure to be expected at the end of the flowing life, 146.7 bars. Figure 2.3—9 shows that this pressure will be attained in approximately 251 days after start of production; this will be the flowing life of the well at the given tubing size.

(b) Prediction of production rates; Nind's theory

Reservoir-engineering estimations, similar to those described in connection with Fig. 2.3—9, permit the tracing of the set of inflow performance curves plotted in Fig. 2.3—11a—d. The first procedure described in connection with Fig. 2.3—7 can be used to construct tubing throughput capacity curves (*A*—*C*). Each curve marked with a capital letter corresponds to the line marked with the same lowercase letter. The points of intersection of the corresponding curves indicate the production rate of the well through a choke of diameter d_{ch} = 12.7 mm at the time when reservoir pressure assumes the value defined by the *IP* curve. — The figures will also furnish potential production rates at various values of reservoir pressure. In the knowledge of that value, the actual production rates q_o can be expressed as percentages of the potential production rates. The relevant parameters are listed in Table 2.3—1. It is apparent that a decline in reservoir pressure will entail a decrease in the ratio of feasible to potential production rates, because at a lower reservoir pressure the energy required to keep the fluid flowing through the tubing string is a greater fraction of

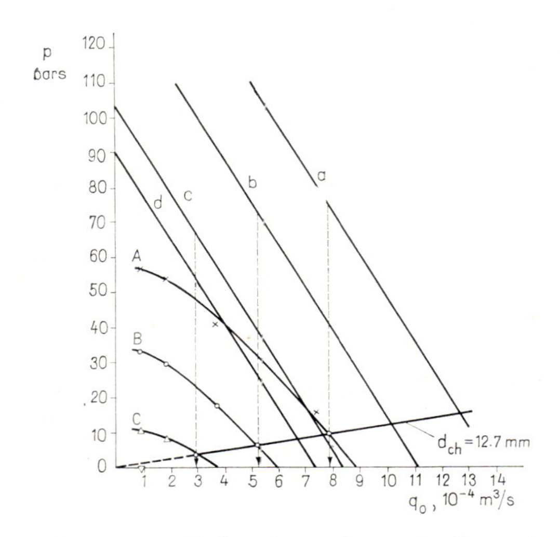

Fig. 2.3—11 Estimation of production-rate (after Nind 1964, p. 142; used with permission of McGraw-Hill Book Company)

Table 2.3 – 1

	p_{ws}	q_o	$q_o/q_{o\,pot}$
	bars	10^{-4} m³/s	%
At a,A	173	7.81	56.7
At b,B	138	5.15	46.7
At c,C	103	2.94	35.5
At d,D	90	0	0

the total pressure required to move the reservoir fluid from the periphery of the zone of influence to the wellhead (Nind 1964).

2.3.3. Designing flowing wells for optimum performance parameters

There is no consensus as to what the term optimum performance exactly means. However, once this or that interpretation of the term has been accepted, questions as to optimal well structure or production can be unambiguously answered. The parameters in question include the diameter d and length L of the tubing, and the well-head pressure p_{TO}. Of these, the choice of d and L is more critical, because once production has been started up, the tubing string can be changed only after shutting in the well; the change thus involves considerable cost and downtime. Let us assume, without justification for the time being, that the optimum tubing length equals the well depth L_w, with the tubing shoe flush with the upper limit of the sandface. We shall now concentrate on finding optimum tubing sizes under the following interpretations of the term optimum performance. (i) Tubing size is optimal if at a prescribed rate it delivers oil at a minimum producing GOR. (ii) According to Krylov (Muraviev and Krylov 1949, p. 270) the tubing size is optimal if it is large enough to let the start-up flow rate (which is usually maximal) pass, and assures flowing production for the longest possible time. (iii) According to Nind (1964, p. 84), the tubing size is optimal if it ensures a maximum flow rate out of the well at a given wellhead pressure p_{TO}. (iv) Tubing size is optimal if it ensures production at a minimum formation GOR.

The idea underlying each one of these interpretations is the choice of a tubing size that causes the least flowing pressure drop under the conditions envisaged. This is why the tubing should be run invariably down to the well bottom (which, in practice, means the top of the sandface). If the tubing were shorter, the well fluid would flow between the well bottom and the tubing shoe in the casing, that is, in a string of larger-than-optimum diameter, and the flowing pressure gradient would consequently be greater than optimal.

It would be a self-defeating attempt to try to uniformize the interpretation of optimum performance. Depending on the circumstances, now one, now another definition may turn out to be most useful. In the following we shall therefore outline methods for finding optimum tubing sizes for each of the above-mentioned criteria. We shall give both 'East-European' and 'Western' solutions to each of the problems.

(a) Dimensioning the tubing string for minimum GOR, with time-invariant flow parameters

Given a prescribed bottom-hole pressure $p_{TL} = p_{wf}$, the minimum realizable wellhead pressure $p_{TO\,min}$, the oil and gas flow rates q_o and q_g to be produced at the prescribed *BHP*, the length L_T of the tubing string, and the physical properties of the liquid and gas; let us find the tubing size which at the prescribed rate will produce at a minimum producing *GOR*.

(a.1) *Krylov's equations.* — Equation 1.4—23 provides the optimum tubing size:

$$d_{opt} = 0.263 \sqrt{\frac{1}{\xi}} \sqrt[3]{\frac{q_o}{1-\xi}} \, . \qquad 2.3\text{—}1$$

The standard tubing size closest to the calculated value is to be chosen. Substitution of this size into Eq. 1.4—27 yields the minimum producing *GOR* required for flowing production. If this is less than the effective *GOR* available, R_{eff}, the well will produce by flowing. In the contrary case, no flowing production is feasible through either the calculated tubing or tubing of any other size. The notion of effective *GOR* has to be introduced because part of the gas flow measured downstream of the separator is still dissolved in the oil while it rises in the tubing string. On the other hand, R_{go}, as suggested by its name 'gas–oil ratio', refers to oil, and the well may produce also some water, and the gas must lift this water, too. According to Krylov, the effective *GOR* is

$$R_{eff} = (R_{go} - R_s)\,(1 - R_w) \, , \qquad 2.3\text{—}2$$

where R_{go} is the *GOR* as determined after separation, referred to the temperature and pressure prevailing in the separator; R_s is the solution *GOR* in the tubing string at the mean pressure prevailing in it, provided that pressure in the string is a linear function of length:

$$R_s = \beta_p \left(\frac{p_{TL} + p_{TO}}{2} - p_n \right) ; \qquad 2.3\text{—}3$$

and R_w is the water–oil ratio of the liquid produced.

Example 2.3–1. Find (i) the tubing size requiring the least producing *GOR* and (ii) decide whether the well will flow if $q_o = 63.5$ m³/d, $R_{go} = 320$

m^3/m^3; $R_w = 0$; $\varrho_l = 830$ kg/m³; $\beta_p = 4.1 \times 10^{-6}$ m³/N; $L_T = 1400$ m; $p_{TL} = 23.0$ bars; $p_{TO\,\min} = 2.0$ bars; $p_n = 1.0$ bar. — Since $\gamma = \varrho g = 830 \times 9.81 = 8142$ N/m³, the mean pressure gradient is, by Eq. 1.4—20,

$$\bar{\xi} = \frac{23.0 \times 10^5 - 2.0 \times 10^5}{1400 \times 8142} = 0.184 .$$

The tubing size looked for is, by Eq. 2.3—1,

$$d_{\text{opt}} = 0.263 \sqrt{\frac{1}{0.184}} \sqrt[3]{\frac{7.35 \times 10^{-4}}{1 - 0.184}} = 0.059 \text{ m} .$$

Table 2.3—4a (see later) shows the next standard tubing size to be 2 7/8 in. nominal with an ID of 0.062 m. The least producing GOR required to keep the well flowing is, by Eq. 1.4—27,

$$R_{\text{opt}} = 0.123 \frac{1400(1 - 0.184)\, 8142}{0.062^{0.5} \times 0.184 \times 1.10^5 \lg \dfrac{23.0 \times 10^5}{2.0 \times 10^5}} = 236 .$$

The solution GOR at the mean tubing pressure is, by Eq. 2.3—3,

$$R_s = 4.1 \times 10^{-6} \left(\frac{23.0 \times 10^5 + 2.0 \times 10^5}{2} - 1.10^5 \right) = 4.7 \approx 5 \text{m}^3/\text{m}^3 .$$

The effective GOR is, by Eq. 2.3—2,

$$R_{\text{eff}} = (320 - 5)\,(1 - 0) = 315 \text{ m}^3/\text{m}^3 .$$

The well will consequently produce by flowing, since $R_{\text{eff}} > R_{\text{opt}}$.

(a.2) *Pressure-gradient curves.* — Of the set of curves valid for a given standard tubing size and a given oil flow rate q_o, let us select that curve along which the pressure increase from p_{TO} to p_{TL} takes precisely the length L_T. The R_{go} parameter of this curve furnishes the GOR at which the well will produce by flowing through the chosen size of tubing at the prescribed operation parameters. Of the several GORs corresponding to the various tubing sizes, one will be a minimum. The tubing size belonging to this R_{go} is optimal; this is the tubing that will, at the prescribed operation parameters, produce oil at the least producing GOR.

Example 2.3—2. Using Gilbert's pressure gradient curves find the tubing size giving the least producing GOR under the conditions stated in the previous example. — Figure A—4 (see Appendix) holds for a flow rate of $q_o = 63.5$ m³/day and at $d = 2\ 3/8$ in., while Fig. A—9 at $d = 2\ 7/8$ in —.

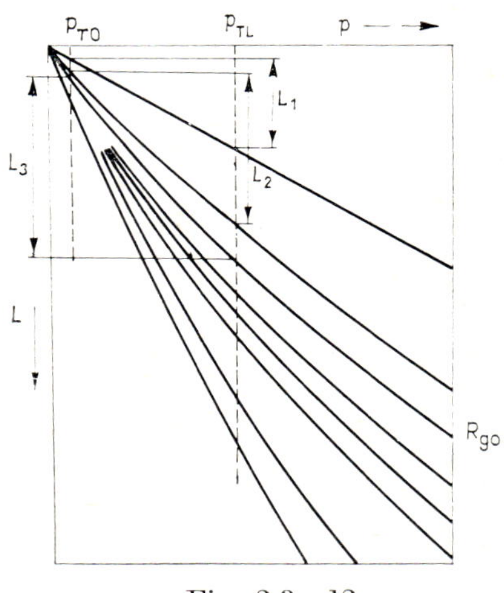

Fig. 2.3—12

Figure 2.3—12 shows one of the possible ways of choosing the suitable pressure-gradient curve. Let us place a sheet of tracing paper over the set of Gilbert curves, and draw parallels to the ordinate axis through the prescribed pressure values p_{TL} and p_{TO}. The two parallels will define sections of length L_i on the individual curves $L = f(p)_R$. Interpolation or the use of an auxiliary diagram will permit selection of the curve on which the prescribed pressure drop takes precisely a length L_T equal to the given tubing length. The dense set of steep curves at low pressures allows a rather approximate estimation only. The procedure permits us, notwithstanding, to pick out d_{opt}. Analysis based on the pressure gradient curves in Fig. A—4 (see Appendix) reveals that no flowing production is possible through a tubing string of 2 3/8 in. size. According to Fig. A—9 (see Appendix), however, a tubing string of 2 7/8 in. size will permit flowing production at a producing *GOR* of a round 300 m^3/m^3. A curve for 3 1/2-in. tubing published by Gilbert (1954), but not reproduced in the Appendix to this book, gives a specific gas consumption of a round 320 m^3/m^3.

Comparison of the solutions found respectively in paragraphs (a.1) and (a.2) reveals that although the results concerning the producing *GOR* are rather far apart, both procedures give the standard tubing of 2 7/8 in. as the optimum size.

(b) Dimensioning the tubing string for minimum GOR, with time-variable flow parameters

The composition of the fluid and the rate of production will vary over the flowing life of any well. The production will generally decline apart from certain fluctuations, and so will, most often, also the flowing *BHP*. Figure 1.4—15 shows how a decrease in oil flow rate will at a given producing *GOR* permit flowing production through a smaller-size tubing. The decline of the flowing *BHP* will entail an increase in producing *GOR* according to both Fig. 1.4—14 and Section 2.3—2b. The producing *GOR* of a well may vary in time according to a variety of functions. In the case of a dissolved-gas drive, for instance, the producing *GOR* of the well will at first increase rather sharply over a comparatively low initial value, and then gradually decline below it. Other flowing-life histories are also possible, though.

An interaction of the factors outlined above very often results in a situation where progressively smaller tubing diameters would be required in order to ensure a minimal producing *GOR*. The solution usually chosen is to run the least tubing size that still permits the initial, comparatively high production rate to flow through. Production efficiency will therefore be rather poor initially. This is, however, no particular disadvantage since a considerable specific-energy content remains available at the well bottom.

(b.1) *Krylov's equations.* — Let us substitute the initial *GOR* into Eq. 1.4—28 and find the wellhead pressure p_{TO} ensuring a maximum liquid flow rate for a number of standard tubing sizes d. Now we may calculate $q_{o\,max}$ by means of Eq. 1.4—24, using the chosen ds and the corresponding pressure gradients $\bar{\xi}$. The tubing size to be chosen is the least size of throughput capacity equal to or greater than the envisaged initial production rate.

Example 2.3—3. Given $q_o = 95.3$ m³/day $(= 1.10 \times 10^{-3}$ m³/s); $R_{go} = 255$ m³/m³; $R_w = 0$; $\varrho_o = 830$ kg/m³; $\beta_p = 4.1 \times 10^{-6}$ m²/N; $L_T = 1400$ m; $p_{TL} = 51.0$ bars; $p_{TO\,min} = 2.0$ bars; $p_n = 1.0$ bar; find the tubing size ensuring the longest flowing life of the well. — Assuming the mean wellhead pressure to be 10 bars, the effective *GOR* is stated by Eq. 2.3—2 to be

$$R_{eff} = 255 - 4.1 \times 10^{-6}\left(\frac{51.0 \times 10^5 + 10 \times 10^5}{2} - 1.0 \times 10^5\right) = 243 \text{ m}^3/\text{m}^3 .$$

Let us first find p_{TL} at the d values 0.04085 m, 0.0506 m and 0.062 m by successive approximations using Eq. 1.4—28, or by means of an auxiliary diagram, and then find $\bar{\xi}$ using Eq. 1.4—20 and $q_{o\,max}$ using Eq. 1.4—24. The results of the calculation are listed in Table 2.3—2. — The tubing to be chosen is 2 3/8-in. nominal whose *ID* is 0.0506 m, since $1.55 \times 10^{-3} > 1.1 \times 10^{-3}$. This is the least tubing size of throughput capacity greater than the initial production rate envisaged.

Table 2.3—2

d	p_{TO}	$\bar{\xi}$	$q_{l\,max}$
m	bars	—	10^{-3} m³/s
0.0409	8.3	0.374	0.86
0.0506	9.8	0.361	1.55
0.0620	10.9	0.353	2.76

(b.2) *Pressure gradient curves.* — Let us find the pressure gradient curve valid for the given q_o and R_{go} among the curve sets for various tubing sizes. Starting from the ordinate corresponding to the prescribed bottom-hole pressure p_{TL}, let us measure along the curve in question the tubing length L_T and read off the curve of the wellhead pressure p_{TO} to be expected. The well will produce by flowing through any size tubing for which $p_{TO} > p_{TO\,\min}$. The least of these sizes is to be chosen.

Example 2.3—4. Solve the preceding example using Gilbert's pressure gradient curves. The examination is to be extended to nominal tubing sizes of 1.9, 2 3/8 and 2 7/8 in. By Fig. A—10 (see Appendix), p_{TO} is 19.5 bars for the 2 7/8-in. size tubing (see Fig. A—5, Appendix), it is 17.5 bars for 2 3/8 in. and, by a set of Gilbert curves not reproduced in this book, no flowing production is possible through tubing of 1.9-in. size at a *GOR* of $R_{go} = 255$. Since $p_{TO\,\min}$ is less than the wellhead pressure of $p_{TO} = 17.5$ bars, to be expected with a tubing size of 2 3/8 in., the optimum tubing size is 2 3/8 in. — Comparison of the two solutions shows that, in the case considered, results are more or less identical: there is agreement in that 2 3/8-in. tubing will ensure the longest flowing life of the well, but procedure (b.2) furnishes a higher wellhead pressure than procedure (b.1).

(c) Dimensioning the tubing string for maximum liquid production rate, with time-invariant parameters

We have so far assumed that the flow rate to be attained is determined as a function of a prescribed flowing *BHP*, that is, that the liquid production rate is limited, e.g., by reservoir-engineering considerations. Quite often, however, the production rate may be permitted to vary over a comparatively wide range of flowing *BHP*s. In such cases, that tubing size is considered optimal which permits a maximum production rate.

(c.1) *Pressure gradient curves.* — Let us establish for various tubing sizes the characteristic $p_{TO} = f(q_o)$ curve of the interaction between reservoir and well by the procedure outlined in connection with Fig. 2.3—7. Let us determine for each case examined the maximum flow rates attainable at a wellhead pressure corresponding to $p_{TO\min}$. The optimal tubing size is that which permits a maximum production rate at a given $p_{TO} = p_{TO\min}$. Figure 2.3—13 shows the resultant curves, reproduced after Nind, transformed into metric units. Nind (1964) has solved the problem for a given well using Gilbert's sets of curves. The optimum tubing size in the case in question is seen to be 1.9 in., although the production rate through 2 3/8-in. size tubing is almost as high.

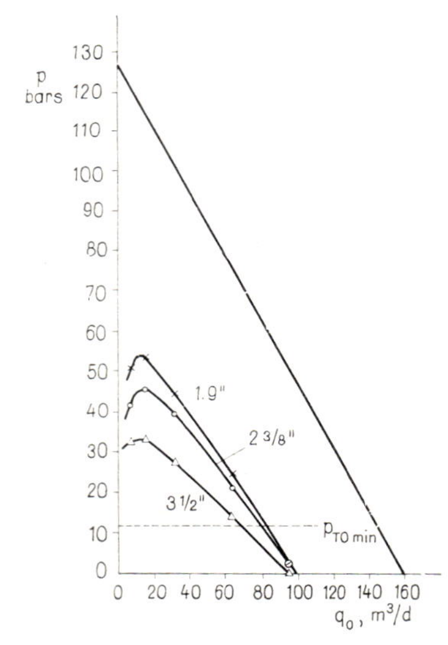

Fig. 2.3—13 Finding the tubing size of maximum production capacity (after Nind 1964, p. 129; used with permission of McGraw-Hill Book Company)

(c.2) *Krylov's equations.* — The problem can be solved in principle by the procedure given in connection with Fig. 2.3—3. The result looked for is the common operating point of the formation and the tubing string at various tubing sizes and wellhead pressures. The tubing size to be selected is that which ensures the maximum production rate. — The solution based on Krylov's general theory of the operation of a tubing string is rather cumbersome, it is therefore not discussed in the present book.

(d) Dimensioning the tubing string for minimum formation GOR, with time-invariant parameters

We have stated in connection with Fig. 2.1—3 that the curve describing the specific gas production of a well at different rates of total production possesses a minimum. Production at this operating point has the advantage of ensuring the most economical exploitation of the gas energy contained in the reservoir. The tubing is to be chosen so as to permit production at the operating point of R_{min}. This principle of dimensioning is of a more or less pure theoretical interest as the data available at the time of well completion do not usually provide the $R_{go} = f(q_o)$ function to the accuracy required by this mode of designing. If, on the other hand, the tubing size run into the well is too large or too small, the operating point of R_{min} will have to be established by extrapolation from a production test performed through a tubing unsuited for the purpose. The extrapolated data is rather unreliable, on the one hand and the advantages to be expected of a tubing of different size on the other, seldom exceed the drawbacks involved in exchanging the tubing string.

Note. — We have so far assumed that the entire string is made up of tubing of constant size. In earlier production practice, so-called telescopic strings, considered to cause less flowing pressure loss, were often employed. Such strings were composed of standard sizes of tubing, gradually increasing upward. The solution has, of course, a number of drawbacks: dewaxing, introduction of down-the-hole instruments, the unloading of the well by pumping become cumbersome if not impossible; in a selective completion, the casing-size requirement would depend on the maximum tubing size employed, etc. As far as the present author is aware, no telescopic tubing is employed anywhere today.

2.3.4. Well completions

The actual techniques of well completion involving a drilling or well completion rig will not be described here. Completions will be discussed only to such depth as is required for an understanding of production aspects.

(a) Wellhead equipment

The fluid entering the well across the sandface and rising through the tubing to the surface passes through the wellhead equipment on its way to the flow line. It is the wellhead equipment that holds in place the casing string(s) reaching to the surface, and the tubing string(s). Its three main parts are the casing head(s) *1*, the tubing head *2* and the Christmas-tree assembly *3*, shown in Fig. 2.3—14, which presents the wellhead equipment of a well producing a single zone, and incorporating two casing strings. The lowermost casing head is screwed onto the male thread of the largest-size casing string (the surface pipe) and supports the next casing string. The

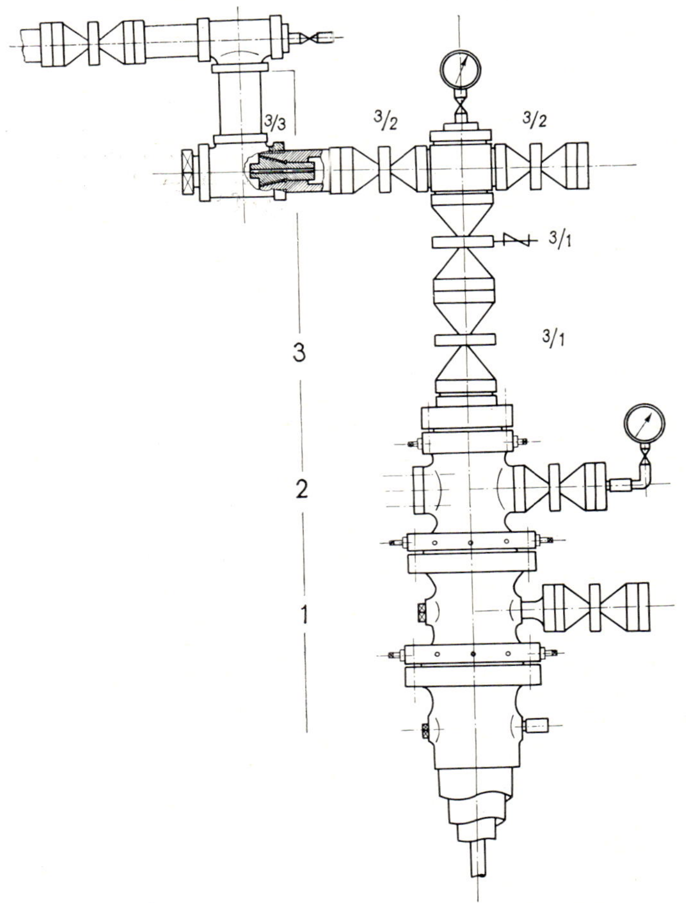

Fig. 2.3—14 Wellhead assembly of single completion

next casing head is connected with the lowermost casing head; it supports the third casing string. This is, in the present case, the so-called production casing. It is within this last casing string that the tubing string is run. The tubing head is connected with the so-called Christmas-tree assembly which incorporates all the valves and other equipment required to shut off, regulate and direct the flow of the well fluid. By API standards, wellhead equipment should be made of steel whose tensile strength is from 4830 to

6890 bars, and whose yield strength varies from 2480 to 5170 bars. The standard for the cold-working pressures of wellhead equipment proposes a scale of 66, 140, 207, 345, 690 and 1035 bars, with hydrostatic test pressures half as high again as these values (Frick 1962).

(a.1) *Casing heads.* — As already stated in connection with Fig. 2.3—14, the lowermost casing head is usually screwed onto the end of the outermost casing string. The second casing string is held in a tensioned state by the slips of the casing hanger. The annulus between the two casing strings can be packed off either by a resilient seal or by welding the casing top to the

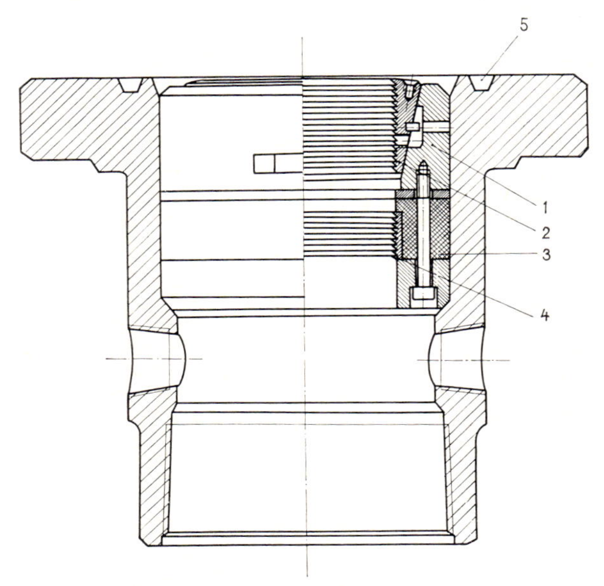

Fig. 2.3—15 OCT C-29 type casing head

casing head. Figure 2.3—15 shows a modern resilient-seal lowermost casing head, type C—29 of OCT Co. The casing, slightly slackened after drawing, is caught by slips *2* slipping into conical bowl *1*. The weight of the casing string energises the oil-resistant rubber packing to provide the positive packoff required. As a result, lower slips *4* will also engage the casing automatically. This solution with two sets of slips is preferable to the entire casing weight being supported by a single set, because casing deflection and deformation is less and the hanging capacity of the casing head is nevertheless higher. The next casing head is fixed with nuts and bolts to the flange of the first casing head, and the next casing string is hung in much the same way. The two casing heads are provided with a polished-in ring gasket usually of soft iron, fitting into groove *5*. Sections of API Std 6A seal rings are shown as Fig. 2.3—16. The polishing of the ring gasket and the subsequent transportation and assembly require great care. Even slight knocks may lead to deformation which entails the escape of gas, possibly throughout the life of the well. — The tubing head is fixed to the uppermost casing head.

(a.2) *Tubing heads.* — The permissible working pressures according to API Std 6A of standard tubing heads are round 69, 140, 207, 345, 690 and 1035 bars. In choosing a tubing head, the following criteria should be observed according to Foster (Frick 1962): (i) the geometry and permissible working pressure of the lower flange should equal those of the casing-head flange with which it is to be connected; (ii) the size and geometry of the tubing head should permit the passage of tools of size corresponding to the ID of the producing casing string, (iii) the seating of tubing hangers for single or multiple completions, (iv) the mounting of such valves as correspond to the pressure rating of the tubing head; (v) the top flange should ensure the required fit to the Christmas tree and should be provided by lock screws which ensure the sealing of the tubing hanger even with the Christmas tree not in place; (vi) the permitted working pressure should be equal to or greater than the maximum shut-in pressure to be expected at the wellhead.

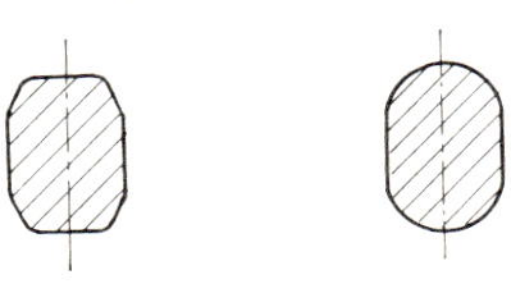

Fig. 2.3—16 Ring gaskets conformable to API Std 6A

No comment concerning (i), (ii), (v) and (vi) above seems indicated.

ad (iii). Several types of tubing hanger are in use. One type of latch-around hanger (the National H—7), shown as Fig. 2.3—17, is composed of two hinged halves. Each of the halves consists of one top and one bottom steel half-mandrel with a resilient half-ring seal sandwiched between the two. Once the hanger is seated in the tubing-head bowl, and secured in place by the tubing-head lock screws (*1* in Fig. 2.3—18), the top mandrel will compress the sealing element so as to provide pack-off between tubing and tubing head. — The advantage of the latch-around hanger is that if the well kicks off while the tubing is being run, it can be immediatly latched around the tubing and seated in the bowl of the tubing head, with the top upset of the landed tubing compressing the hanger. Circulation can then be started in the tubing. At comparatively low pressures, the tubing can even be stripped

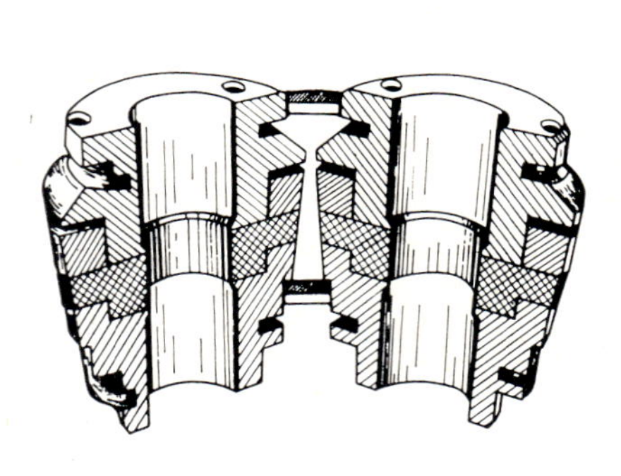

Fig. 2.3—17 National H-7 tubing hanger

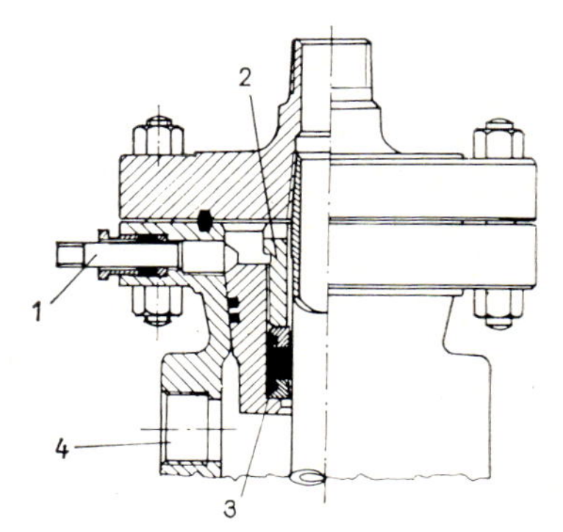

Fig. 2.3—18 National H-6 tubing hanger

through the hanger between upsets. After the well has quietened down and the entire tubing string has been run in, the Christmas tree (see p. 147) can be attached to the top tubing thread, in which case the hanger will merely function as a blow-out preventer without having to support the tubing weight. — One make of boll-weevil hanger of the stuffing-box type (National H—6) emplaced in a tubing head is shown as Fig. 2.3—18. Boll-weevil hangers are cheaper than latch-around ones. If the well kicks off, however, seating it is more complicated, as the mandrel *2* can be installed on disconnected tubing ends only. Packoff is provided by two O-ring seals or a plastic seal between the tubing head and the bowl, and by plastic seal *3* between the hanger and the tubing. The tubing can be moved under pressure in the hanger between upsets. The hanger is held in place by the lock-down screws *1*.

Mandrel hangers. The tubing string is hung from a mandrel attached to the top thread of the tubing string. The National H—1 type mandrel hanger is shown as Fig. 2.3—19. Packoff against the tubing-head bowl is ensured by two plastic O-ring seals (*1*). The tubing cannot be moved after the seating of the mandrel. If the well kicks off during the running or pulling of the tubing string, seating requires a fairly long time. This type of hanger is used in connection with 'quiet' wells. — A safe construction, used on high-pressure wells, is the Cameron LD tubing head. It functions as a blowout preventer during well completion and as a tubing head during production. It is fundamentally a split-packoff preventer which provides a seal towards the tubing and at the same time supports its upset. After seating, this upset reaches up into the Christmas tree where another safety seal of the high-pressure hydraulic grease-gun type is installed. — Other types of hanger used with comparatively low-pressure wells are also known. — Selective completion of wells producing several formations raise the following additional requirements as to tubing hangers: the tubing-head bowl should permit the hanging not only of the maximum designed number of tubing strings, but also of a smaller number; even only one if that be required. The orientated emplacement of hangers should be feasible. It should not be necessary to remove the blowout preventer before all the tubing strings have been run. — Several types are known. In the multiple-bore mandrel hanger, the mandrel is a cylindrical body with a separate bore for each tubing string. The individual strings are run with separate guides. This is the simplest and most easily installed hanger but has the drawback

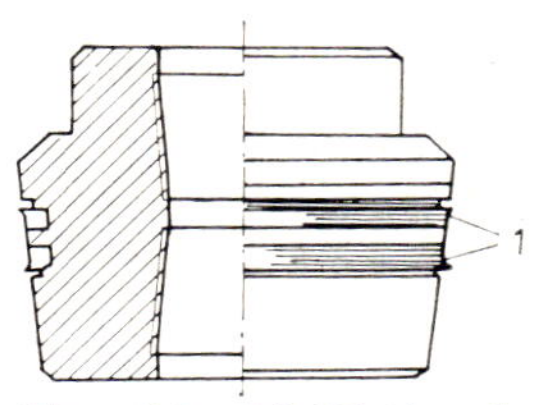

Fig. 2.3—19 National H-1 tubing hanger

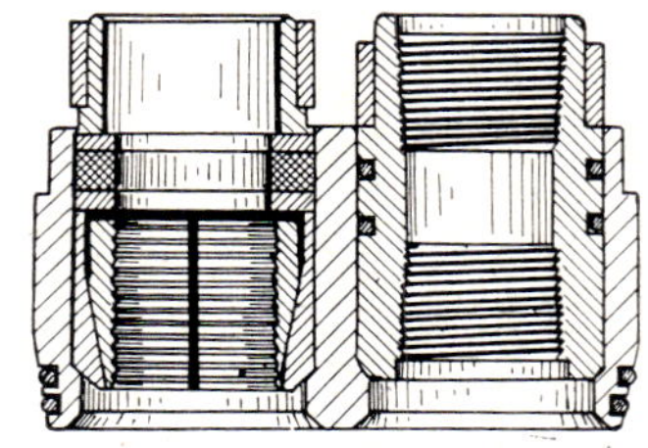
Fig. 2.3—20 OCT dual tubing hanger and seat

of being limited to applications where accessories jutting beyond the tubing-joint outline, such as gas-lift valves, are not required.

In the multiple-segment hanger, there is a separate hanger of cylinder-segment shape for each tubing string. Each segment occupies its part of the bowl when landed. This solution permits the use also of accessories jutting beyond the tubing-joint outline. Figure 2.3—20 shows an OCT make double hanger and seat of a combination type. The hanger on the left is of the mandrel type; that on the right is of the boll-weevil type. The hanger

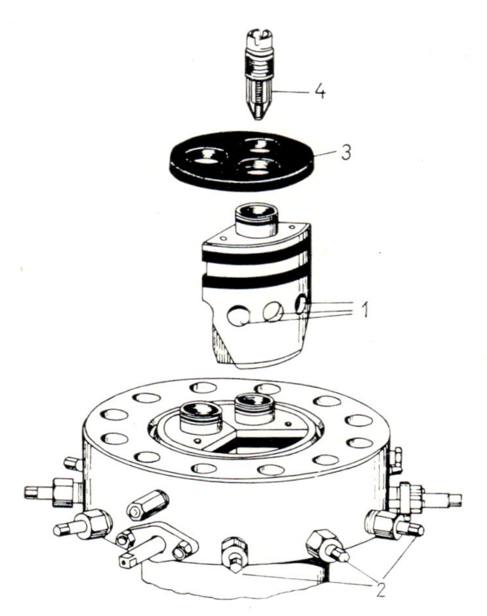

Fig. 2.3—21 Cameron tubing-head flange and hangers for triple completion

on the left has O-ring seals, that on the right has resilient rings. The arrangement of the tubing strings is readily visible in Fig. 2.3—21, showing a Cameron type tubing head flange with tubing hangers for a triple completion. Orientated seating and fixation are ensured by the holes *1* and screws *2*; *3* is a resilient seal disk; *4* is a back-pressure valve, of which one only is shown although three are used (see also the Christmas-tree assembly).

ad (iv). Side outlets on tubing heads permit access to the annulus between the production casing and the tubing. These outlets are to be provided with outlet valves. Modern types of equipment use full-opening gate or plug valves. The outlets are designed so as to permit the changing of leaky valves under pressure, using a valve-changing tool. The change is feasible only if the opening device of the valve is not damaged. The changing tool is fundamentally a length of pipe in whose interior a well-packed rod can be rotated. This pipe is fixed to the outlet of the valve. At the end of the rod there is a detachable male-threaded plug which fits into a female thread provided for this purpose in the side outlet. Changing a valve is performed as follows. The leaky valve is closed and the changing tool is installed. The valve is opened, and the rod is turned until the plug seats itself in the appropriate bore. The rod is then detached from the plug, and the changing tool and the leaky valve are both removed. A good valve is now installed with the changing tool mounted on it. The plug is retrieved; the valve is closed and finally the changing tool is removed. Side outlets are provided with threads (for working pressures of usually up to 140 bars), studs (usually above 207 bars) and extended flanges (for any, but usually high, pressures); *4* in Fig. 2.3—18 shows a threaded outlet.

(a.3) *Christmas-tree assembly.* — A Christmas tree is an assembly of valves, fittings and other accessories with a purpose of regulating production. The element in direct contact with the tubing string is the flange or bonnet. Numerous types are available. All have the common trait of being sealed together with the tubing head by a metal ring gasket. The main types differ in their connection with the master valve, which may be of the male or female thread or flange type. The bonnet of type National B—4T shown in Fig. 2.3—18 is of the male thread type. For master valves with a male lower thread, an adapter may be inserted. — Christmas-tree valves are also made of high-strength alloy steels. The two current types are gate and plug valves. Both have either flanged or threaded connections. Gate valves are more widely used. These may be of the wedging or non-wedging types. For simplicity, we shall concentrate on gate valves in the following. The master valve *3/1* in Fig. 2.3—14 should be of the full-opening type, with a clearance equal to or greater than the *ID* of the tubing. On high-pressure wells, two series-connected master valves are often used. Figure 2.3—14 shows such an assembly. The wing valve *3/2* in Fig. 2.3—14 may be of the restricted-opening type, with a nominal size somewhat less than that of the master valve, provided it does not significantly raise the flowing pressure drop of the fluid produced. Threaded connections are used up to working pressures of 345 bars, whereas flanged ones are unlimited as to pressure (the maximum API standard Christmas-tree work in pressure is of 1035 bars). Christmas-tree valves are required to close safely even if the well fluid is gaseous or pure gas. Recently, monoblock-type Christmas trees incorporating the functions of several valves and spools have become popular. Such blocks are, of course, much easier to install. Designs differ e.g. according to the number of tubing strings used in the completion, or to the rated working pressure. Figure 2.3—22 shows a Cameron B type valve block for a dual completion; *1* is the master valve of one tubing string; *2* is its wing valve and *3* is its swabbing valve. The corresponding items for the other tubing string are numbered *4*, *5* and *6*, respectively. Connections towards the flow lines, pressure gauges and lubricator flanges are of the stud type.

Christmas trees are sensitive to sand in the oil which may cause severe erosion especially in bends and deflections. The Soviet-type 1—AFT Christmas tree (Fig. 2.3—23) is used in producing sandy fluid. The normal outlet into the flow line is marked *1*. When the tree is eroded, wing valves *3* are opened and the master valve *2* is closed. Flow now continues in the direction marked *4*, while the eroded elements above valve *2* are changed. When that operation is finished, production is switched back to wing *1*. The choke mounted in the insert marked *5* is a wear-resistant ceramic-lined type.

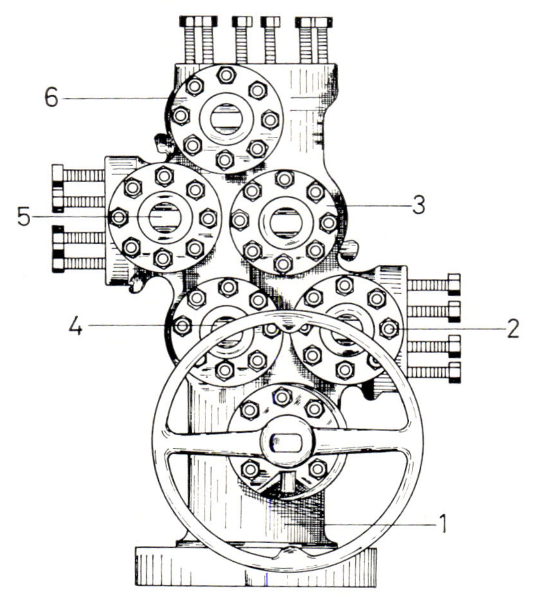

Fig. 2.3–22 Cameron dual-completion solid block wellhead equipment

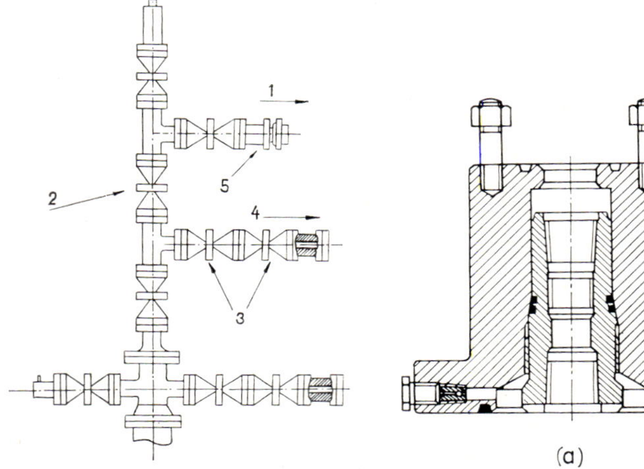

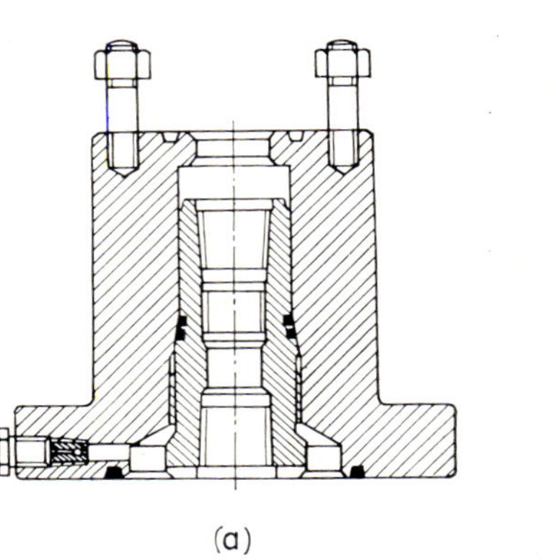

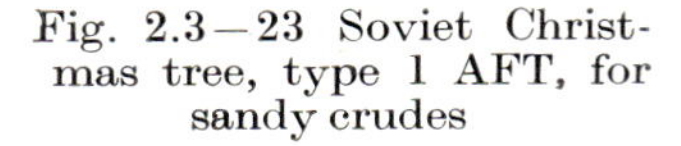

Fig. 2.3–23 Soviet Christmas tree, type 1 AFT, for sandy crudes

Fig. 2.3–24 National B-8M bonnet and back-pressure valve

(a.4) *Christmas-tree fittings.* — Back-pressure valve. This is a check valve installed in the tubing hanger or in the wellhead bonnet in order to seal the tubing bore while the blowout preventer is being removed and the Christmas tree is being mounted in place. Once the Christmas tree has been connected with the top end of the tubing and the tubing has been landed in the tubing head, the back-pressure valve can be retrieved through the open master valve, even under pressure, by means of a special tool. The tubing can even be circulated, if the need arises, through the back-pressure valve. Of the two current types of back-pressure valve, one is fixed in place with a thread, the other with a spring lock; (a) Fig. 2.3—24 shows a National

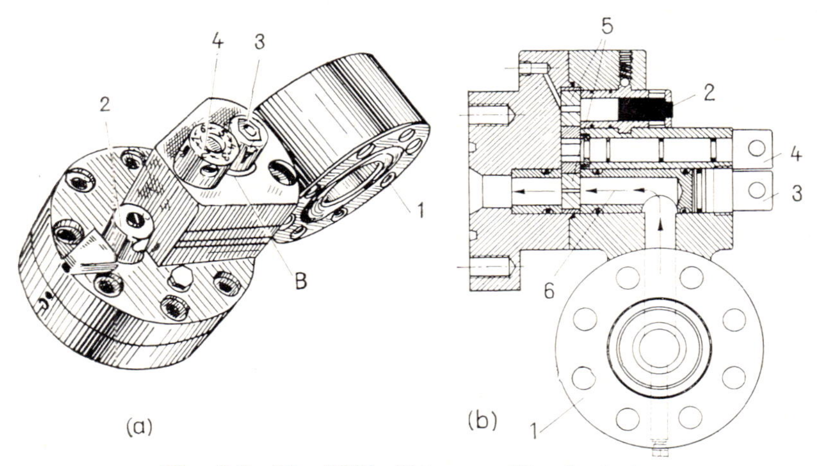

Fig. 2.3—25 Willis T-type wellhead choke

B—8M type bonnet whose top thread will accept a back-pressure valve of the first type. The valve in question is shown as (b) in Fig. 2.3—24.

Chokes. Figure 2.3—14 shows a choke-carrying insert with a choke mounted in a thread. When a change of choke is required, it is necessary either to shut in the well, or to have another choked outlet, similar to the one shown in Fig. 2.3—23. Modern wellhead assemblies include chokes which can be changed without shutting-in the well. There are several such designs. The Willis T-type shown as Fig. 2.3—25 is suited for working pressures of up to 690 bars. Flange *1* is connected with the master valve. The device functions at the same time as a wing valve. Section (b) of the Figure permits us to follow the path of the well fluid. If safety plug *2* is removed and screw *3* loosened, then by rotating choke cylinder *5* about pin *4* it is possible to bring any choke in the cylinder flush with channel *6*. A cylinder may incorporate five chokes and a blind plate. Any one of these can be installed without interrupting production.

Surface safety valve. This device is mounted on the Christmas tree or in the flow line. When installed on the Christmas tree, it is usually mounted between the wing valve and the choke. It will close when pressure builds up above or drops below a predetermined level. Overpressure protection is necessary if an increase in pressure may damage the flow line or other surface equipment. Such overpressure may be due e.g. to a hydrate plug or a closed valve in the flow line. Underpressure protection is necessary because a flow-line break, for instance, may deliver well fluid into the open

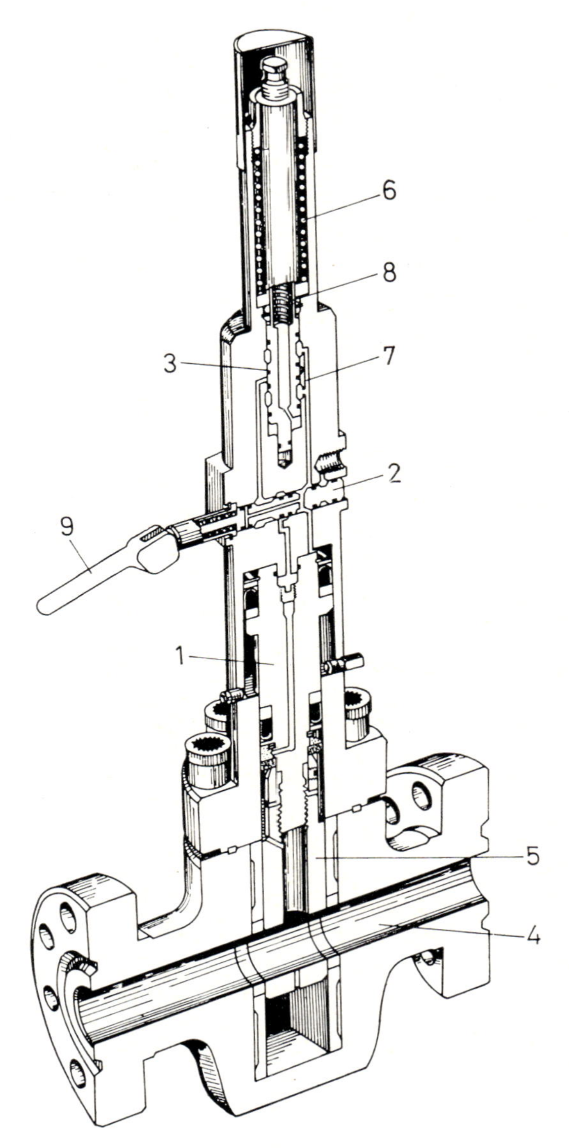

Fig. 2.3—26 Cameron B-type automatic safety valve

and constitute a fire and explosion hazard. — Safety valves are usually set so that the high-pressure limit is about 10 percent above normal flowing pressure and the low-pressure limit is from 10 to 15 percent below it. Safety valves may be arranged so as to be controlled by fluid levels and shut down e.g. when the level in a tank rises above a permissible maximum. They can also be used to cut the well off from the separator. In this latter arrangement, the closing of the valve at the output end of the flow line will entail a pressure surge in the line, which will close the valve. This solution may be favourable when producing a high-pressure intermittent well, because even when the well is shut-in, pressure in the flow line will be comparatively low and, moreover, it will not be necessary to take the trouble of going out to the well to shut it down. If the output-end valve is opened, the pressure will drop in the flow line and the safety valve will open. There are several current surface safety valve designs. The main types are: (i) actuated and controlled by pressure within the valve body; (ii) actuated by pressure within the valve body and controlled by pressure from an external source; (iii) actuated and controlled by pressure from an external source; (iv) actuated by pressure from an internal or external source and controlled by some electrical signal.

The Cameron B-type automatic safety valve shown as Fig. 2.3—26 belongs in type (i) above; that is, it is both actuated and controlled by pressure within the valve body. A substantial advantage of this design is that the device can be installed on a standard Cameron valve body. The valve operates as follows. The liquid or gas pressure prevailing in the valve body is communicated by power piston *1* through the manually operated piston valve *2* to the pilot valve *3*. If the pressure is between preset limits, the pressure in the valve body *4* tends to force gate *5* and piston *1* upward into the open position. If the internal pressure exceeds a preset maximum, then the pressure acting on the pilot valve *3* overcomes the spring force *6* and the pilot valve moves up. Channel *7* delivers pressure to the top of power piston *1*, and, depressing it, forces the valve gate to close. When the pressure drops again to the preset value, the pilot valve sinks to its previous position and the valve opens by a reverse of the above process. If the pressure in the valve drops below a preset minimum, then the upward pressure of the fluid acting on pilot valve *3* will be less than the force of spring *8*, and the pilot valve will move downward. Now again channel *7* will deliver pressure to the top end of power piston *1* which will thereupon close the gate. Opening is automatic once the pressure builds above the preset minimum. By moving handle *9* the automatic safety valve can be manually operated, too.

(b) Well safety equipment

Tubing safety valves (storm chokes). — Installed in the tubing string, the storm choke is open under normal operating conditions; it will shut the well in when damage to the wellhead permits flow above a predetermined rate or the pressure in the tubing drops below a predetermined value. Storm chokes were originally used on offshore and townsite locations, but they are to be recommended in any situation where the wellhead is liable to be damaged. Several types of storm choke are known. All can be run and retrieved by wire line. Some can be seated on a special landing nipple, others can be landed on slips at any point of the tubing. Some types are triggered by a pressure differential in excess of a predetermined value, others by a pressure drop below a predetermined value. The latter type includes the OTIS–H type storm choke shown as Fig. 2.3—27. Chamber *1* is charged with gas to a predetermined pressure prior to installation. If the pressure of the surrounding medium decreases below the pressure in the chamber plus the force of spring *2*, then cylinder *3* moves up and turns ball valve *4*. The latter then obstructs the aperture of the valve. This solution has the considerable advantage that the valve seat is not exposed to erosion because in the open state it is covered by the ball. This type of storm choke can be used up to pressure differentials of 700 bars.

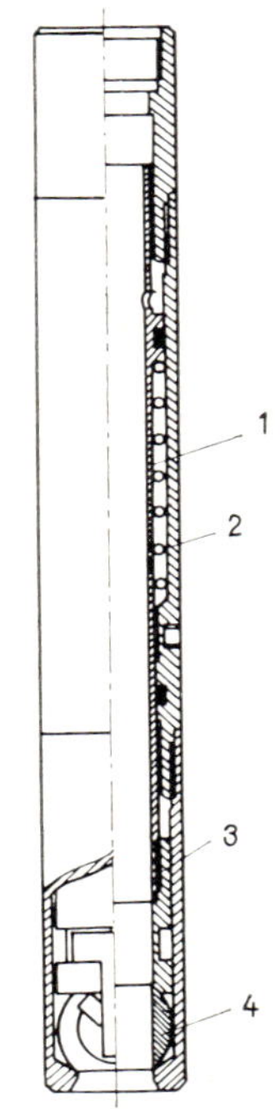

Fig. 2.3—27 OTIS-H type safety valve

The above-described type of storm choke is called the direct-control type because it is the pressure or the pressure gradient of the immediate surroundings that triggers closure. The situation may require, however, the use of a valve controlled by wellhead pressure, transmitted to the safety valve by a pressure conduit outside the tubing. The advantage of this solution is that the amount of well fluid that can escape into the atmosphere prior to closure is less than in the case of direct control.

Safety valve (storm choke) installed in the casing annulus. — A storm choke can be installed also in the casing annulus. The OTIS company, for instance, offers a device of this type for the protection of underground gas storage facilities but similar devices are widely used in numerous oil and gas wells, too, particularly where the wellhead is liable to be damaged by acts of sabotage or violence.

(c) Underground well equipment

Modern principles of production do not permit the exploitation of several zones through one and the same well, except if a multiple completion permits to produce the individual formations selectively, that is, with no communication between any two formations. Figure 2.3—28 shows equipment permitting the separate production of two zones in a dual completion. Table 2.3—3 summarizes the typical features of each piece of equipment after Turner (1954). The alphabetic order of the drawings is the order of increasing drawbacks on a relative scale. The feasibility or otherwise of using a solution is indicated by a plus or minus sign, respectively. Of the two signs in row 2.4, the upper one refers to the upper formation, and vice versa. Let us add a few explanatory details. Crossover piece *1* in part (a) of the Figure is shown in blow-up. The fluid rising from the lower zone passes plug *3* in piece *A*. The upper zone is produced through the casing annulus.

Table 2.3–3
Comparison of selectively producing dual well completions

	Well completion type					
	a	b	c	d	e	f
1 First cost	A	C	E	B	D	F
2 Maintenance cost						
2.1 Running and pulling	A	C	E	B	D	F
2.2 Formation treatment						
2.2.1 Acidizing	B	B	B	A	A	A
2.2.2 Fracturing	A	C	E	B	D	F
2.3 Downhole well testing	B	—	A	B	—	A
2.4 Dewaxing	±	±	++	±	±	++
3 Casing protection						
3.1 By liquid-column pressure	—	—	—	—	+	+
3.2 By inhibitor	—	+	+	—	+	+
4 Applicability						
4.1 Both formations flowing	B	B	A	B	B	B
4.2 Upper formation flowing, lower pumped	C	B	A	C	B	C
4.3 Lower formation flowing, upper pumped	—	D	C	A	B	A
4.4 Both formations pumped	—	D	A	—	C	B

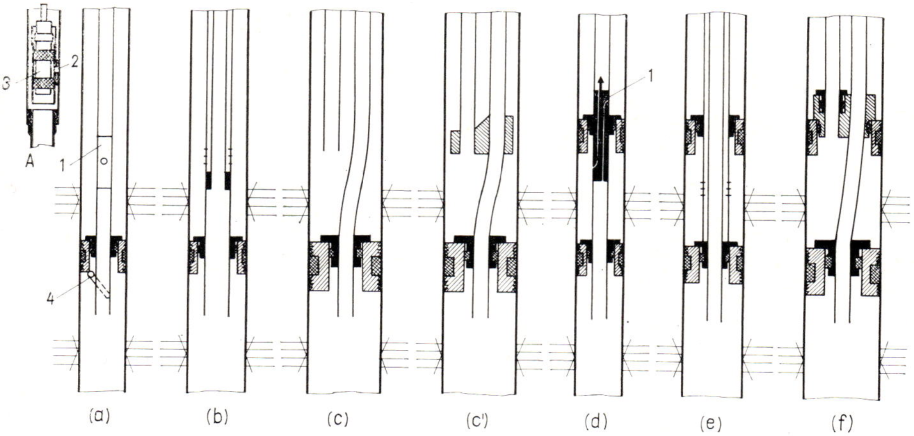

Fig. 2.3–28 Dual completions, after Turner (1954)

When the tubing is pulled, flapper valve *4* closes. Plug *3* can be pulled by means of a wire-line tool. If the flow of fluid through the casing annulus is shut in at the surface, the fluid from the upper zone will enter the tubing through port *2* and rise to the surface through the tubing. This solution permits the unloading by pumping of the upper formation as well as its acidizing and fracturing. Solution (c′) is a variant of type (c). If flowing production from the upper zone ceases, a pump can be built in. In order to improve the output of the pump, the tubing shoe can be anchored to the long string. The lower zone can be produced by means of a pump, too, if the need arises. Crossover choke *1* in solution (d) permits, as contrary to case (a), the production of the lower formation through the casing annulus and the upper one through the tubing. The choke can be changed by means of wire-line tools. It can be exchanged for piece permitting the same production

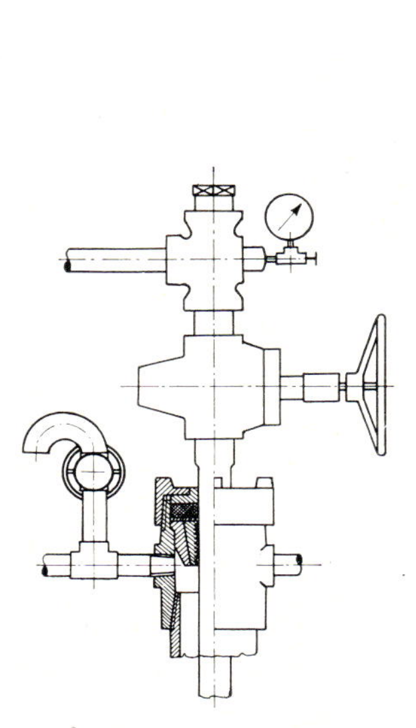

Fig. 2.3—29 Christmas tree of midi (slim-hole) completion, after Bonsall (1960)

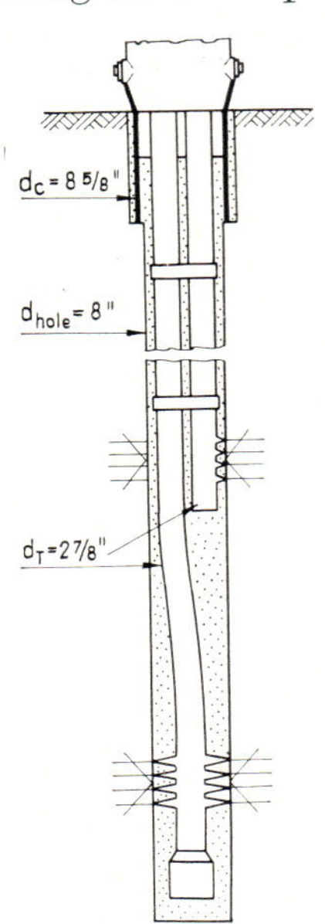

Fig. 2.3—30 Dual midi completion, after Bonsall (1960)

pattern as in (a). Also, it is possible to shut in one zone and produce the other through the casing. This flexibility is an advantage because the comparatively large cross-section of the casing annulus will permit flowing production of relatively high-capacity zones only. A zone must not be produced through the casing annulus if the well fluid is corrosive or if sand erosion is liable to occur. Paraffin removal is a problem in any casing annulus; often it can be performed only after pulling the tubing. Still, the dual completion requiring the slimmest well is the one with a single string of tubing. Coaxial double tubing requires a larger cross-section and two parallel strings of tubing an even larger one. But the adaptability of the well structure to various tasks of production increases in this same order. A dual completion usually costs from 66 to 83 percent of the cost of two single-completion wells producing the same two formations (Prutzman 1955). A greater saving is likely to be realized at greater depth. — In the last decade, equipment permitting the separate production of more and more formations has been designed and built. Equipment for octuple completions is available today.

Midi installations. The term midi as used here is an abbreviation for minimum diameter. (An alternative term is slim-hole structure.) The term refers to very small-diameter wells, usually cased with 2 7/8-in. tubing pipe. Quite often there is no separate tubing string; this is why these were originally called 'tubing-less completions'. This term, however, does not express the significant feature. A possible wellhead assembly for a midi well is shown as Fig. 2.3—29 (Bonsall 1960). A tubingless midi well is convenient provided the well fluid will not harm the casing. The well diameter puts a restriction on production capacity. At the designing stage, it is important to know the intended future production rates of the well and the periods over which these are supposed to be kept up, as well as the estimated variation with time of the *BHP* during the producing life of the well. In a given case, a midi well has turned out to be cheaper by 28 percent than a conventional well of the same depth (Bonsall 1960). Dual-completion midi wells are often used to advantage when two zones are to be produced selectively. The structure shown as Fig. 2.3—30 after Bonsall is a well drilled to a diameter of 200 mm to total depth, with two strings of 2 7/8 in. diameter tubing size cemented in. In the case considered by Bonsall, this solution was cheaper by 18 percent than a conventional dual completion.

(d) Tubing

From a production viewpoint, the tubing plays a prime role in a well completion. The tubing has to provide the most favourable flowing cross-section to the fluid rising from the well bottom. It must protect the casing from corrosive and erosive well fluids and, if a packer is used, also from excessive pressure. Accordingly, quality requirements facing tubing pipe are fairly stringent. It is often required to stand pressure differentials of several hundred bars, and its threads must provide a hermetic metal-to-metal seal. Quality requirements have been raised significantly by the spread of multiple completions. Because of the need to run several strings of tubing in a casing of the least possible diameter, tubing joints should have the least possible excess *OD* (flush joints). This has led to the use of higher-strength steels and to the devising of threaded joints providing a higher thread strength and a better seal. The risk of leaks and of breakdowns has been reduced by the use of integral joints to replace the coupling rings of the past. Table 2.3—4 shows the main dimensions, according to API Std 5A, of upset and plain tubing. Strength requirements according to the same standard are given in Table 2.3—5. — It is worth adding that nominal tubing size prior to 1950 meant the approximate *ID* in inches. Since 1950, the API standard regards as the nominal diameter the precise *OD* in inches. — Some standard and non-standard joints, the latter with special threads, are shown in Fig. 2.3—31.

Tubing is sometimes provided with an internal lining, most often as an anti-corrosion measure. Glass linings are most widespread in the Soviet Union

Table 2.3—4a
Main dimensions of API tubing (after API Std. 5A, 1963)

	Nomina size	*OD*	*ID*	Calculated mass with thread and joint
	in.	mm	mm	kg/m
External upset tubing	1.050	26.67	20.93	1.79
	1.315	33.40	26.64	2.60
	1.660	42.16	35.04	3.50
	1.900	48.26	40.90	4.20
	2 3/8	60.32	50.66	6.59
	2 3/8	60.32	47.42	8.83
	2 7/8	73.02	62.00	9.58
	2 7/8	73.02	57.38	12.93
	3 1/2	88.90	76.00	13.80
	3 1/2	88.90	69.86	19.24
	4	101.60	88.30	16.36
	4 1/2	114.30	100.54	19.20

Table 2.3—4b

Main dimensions of API tubing (after API Std. 5A, 1963)

	Nominal size	*OD*	*ID*	Calculated mass with thread and joint
	in.	mm	mm	kg/m
Plain tubing	1.050	26.67	20.93	1.70
	1.315	33.40	26.64	2.53
	1.660	42.16	35.04	3.44
	1.900	48.26	40.90	4.09
	2 3/8	60.32	51.84	5.98
	2 3/8	60.32	50.66	6.71
	2 3/8	60.32	47.42	8.66
	2 7/8	73.02	62.00	9.41
	2 7/8	73.02	57.38	12.75
	3 1/2	88.90	77.92	11.68
	3 1/2	88.90	76.00	13.48
	3 1/2	88.90	74.22	15.11
	3 1/2	88.90	69.86	18.93
	4	101.60	90.12	14.02
	4 1/2	114.30	100.54	18.66

(Zotov and Kand 1967); plastic is popular in the USA. The Hydril make tube shown in Fig. 2.3—31 is of this type. Little or no paraffin will settle on plastic- or glass-lined tubing walls. The lower roughness of the pipe wall results in an increased liquid and gas throughput capacity. Using this type of tubing it is necessary to ensure that the unlined thread at the joint be protected from contact with the well fluid. This can be ensured by correct design or by the insertion of a plastic seal ring. Tubing made entirely of plastic is also available. The Dowsmith Company, for instance, produces

Table 2.3—5
Strength of API tubing (after API Std. 5A—5AX)

Steel grade	Min. yield strength	Ultimate tensile strength	Min. elongation in 2 in.	
			%	
	bars	bars	Coupon	Hoop
H-40	27.6	41.4	27	32
J-55	38.0	51.7	20	25
N-80	55.1	69.0	16	18
P-105	72.4	82.8	15	17
	For wells producing gas containing H_2S			
N-80 R	49.1	78.5	16	18

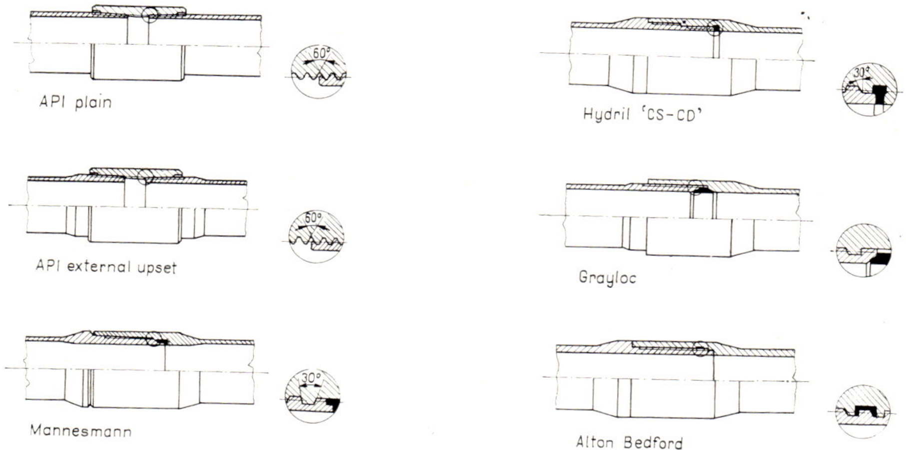

Fig. 2.3—31 Tubing joints

API standard tubing of fiberglass-reinforced epoxy resin. The unit mass of 2 3/8 in. tubing is 1.77 kg/m, round one-third of the weight of the lightest plain steel tubing of the same diameter. The maximum permissible internal-pressure differential is only round 82 bars, however.

2.3.5. Producing a well

(a) Starting up a well

A new well after perforating a zone or an older well that has been shut in for some time, will often fail to start flowing when switched to the tank battery. The liquid column filling the well must then be unloaded by means of an external energy source. Several suitable procedures are known; all have the aim of reducing the liquid column pressure acting on the well bottom to below the reservoir pressure, or, more precisely, to a value equal to or temporarily less than the required *BHP* for continuous flowing production. Unloading may be effected by swabbing, gas lift, the impression of liquid and gas together, and, especially in new wells, by exchanging the mud column in the well by a lower-gravity one.

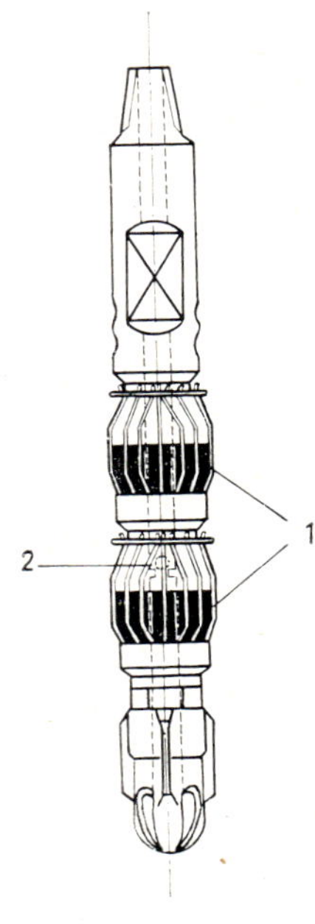

Fig. 2.3—32 Guiberson tubing swab

Swabbing. — One of the widely used types of tubing swabs is the Guiberson type shown as Fig. 2.3—32. During operation the swab depends from a wire line wound onto the relatively high-rpm winch of a hoist. The swab sinks easily into the fluid, because the *OD* of the wire-protected rubber cups *1* is less in the unloaded state than the *ID* of the tubing. Also, check valve *2* in the interior of the swab is open. The swab is lowered to a depth of 10—100 m below the fluid surface and then pulled up. While it is being pulled, the weight of the fluid makes the sealing cups press against the tubing wall and closes the check valve. Most of the liquid column above the swab is delivered to the surface, usually into a temporary storage tank placed beside the wellhead, or to the tank battery. At the beginning of swabbing, the casing annulus is open at the wellhead. Let us assume that there is no packer closing the annulus at the bottom, either. When the flow of air through the open valve into the annulus ceases, this means that the reservoir has started to deliver fluid to the well. The annulus is then shut down. Further swabbing will gradually decrease the *BHP*. — The fluid delivered by the reservoir would, if unhampered, flow into both the tubing and the annulus, in the proportion of their respective cross-sectional areas. Gas cannot now, however, escape from the annulus, as it is closed in on top. Volume and pressure of gas in the annulus will gradually increase; as a result, the liquid accumulated earlier in the annulus will be pushed into the tubing. During this period, then, only a fraction of the gas delivered by the reservoir will enter the tubing, roughly equal to the ratio of the cross-sectional areas of the tubing and the annulus, whereas the liquid flow rate in the tubing will be greater than the flow rate out of the reservoir. The *GOR* of the fluid in the tubing will therefore be significantly less than the *GOR* of the fluid delivered by the reservoir. At the instant when all the liquid has been pushed out of the annulus, this latter is occupied by a gas column whose pressure at tubing-shoe level is p_{TL}. The tubing-shoe pressure of the fluid column in the tubing is likewise p_{TL}. All further fluid delivered by the reservoir enters the tubing in the original composition. This is when flowing production usually starts. Now the tubing-shoe pressure in the tubing decreases and, as a result, some gas will pass from the annulus into the tubing via the tubing shoe. The annulus 'blows down'; its pressure decreases, and the GOR in the tubing increases. The decrease in BHP entails an increase in the rate of flow into the well, which in turn entails an increase in the gravity of the fluid in the tubing, and an increase in flowing BHP. Once more, fluid composed of both liquid and gas will enter the casing annulus. The liquid column will be rather short, however. The entire process repeats itself at a lower intensity, and may go on repeating itself a few times at declining pressure amplitudes; thereafter the well will produce at a uniform rate. In this state, the casing annulus will be filled with gas in its entire length, and the tubing will produce a fluid composed of gas and liquid. The flowing pressure at tubing-shoe level, p_{TL}, will equal the casing-head pressure p_{CO} plus the static pressure p_{gL} of the gas column in the annulus.

For unloading a well by gas lift see Section 2.4, where it is given full treatment.

The producing sandface may be sufficiently friable to cave in under a sudden decrease in BHP brought about by unloading processes. This may cause the well to sand up. It is therefore indicated to choose an unloading method which decreases the BHP slowly and gradually, such as the following. The casing annulus is first filled full of oil by means of a surface pump; then gas is added gradually to the liquid pumped in. The liquid load in the well is thus replaced by a gaseous fluid of decreasing gravity; the rate of decrease can be regulated at will. By the end of the procedure, pure gas is being pumped into the annulus.

The formation pressure of oil and gas reservoirs often stands close to the hydrostatic pressure of a water column whose height equals well depth. If such a well is filled at the time of completion with a mud of, say, 1200 kg/m^3 density, then there will be no inflow after perforating a productive zone. If, however, the mud is replaced by pure water, the BHP will decrease to round $L_w \gamma_w$. Some wells will start to flow at this BHP. Wells with a somewhat lower reservoir pressure will start to flow if the water is replaced by oil.

(b) Types and control of flowing wells

There are three types of flowing well: those producing at a steady uniform rate, those producing continuously but in surges and those producing intermittently. We have so far tacitly assumed all flowing wells to belong in the first group, that is, to produce a steady flow. Most flowing wells do indeed belong in this group, particularly in the first phases of their lives, when the specific energy content of the well fluid is still high. This type of well is characterized by casing- and tubing-head pressures that appear to be constant in the short term. Of course, these pressures are also subject to slow changes, as a result of the decline in reservoir energy and hydrocarbon reserves.

(b.1) *Surging wells.* These fall into two groups: one, whose tubing-head pressure fluctuates while the casing-head pressure stays constant, and two, whose casing-head pressure fluctuates as well as their tubing-head pressure. — Surging of the first type is a consequence of a slug pattern of flow. Figure 1.4—10 reveals that, intercalated between slug-flow regions B_1 and B_2, there is a zone of flow where surging will occur. Flow is quasi-steady here; tubing-head pressure fluctuates rather rapidly (at a period of a few minutes), but the average pressure over several periods is approximately constant. Surging is caused by the fact that the wellhead choke alternatingly passes now a gas, now a liquid slug, whose flow velocities through the tubing are nearly equal. Owing to the greater viscosity of the liquid, however, the resistance to liquid flow of the choke will be much greater than its resistance to gas flow. The pressure fluctuations observed at the tubing head do not reach deep; in fact, they cannot be perceived at the tubing shoe, nor, consequently, at the casing head.

Wells of the second type are comparatively low-capacity ones. If the tubing-shoe pressure temporarily declines for some reason during production, gas will flow from the annulus into the tubing through the tubing shoe. This will entail a decrease in fluid gravity and tubing-shoe pressure, and will permit a further amount of gas to enter the tubing from the annulus. During this process, the well produces fluid at a decreasing *GOR*. The process continues until the decrease in *BHP* and the consequent increase of liquid inflow rate into the well bring about an increase of fluid gravity in the tubing. The pressure of the gas that has stayed behind in the annulus is less than the now increasing tubing-shoe pressure in the tubing, wherefore a column of liquid will rise in the annulus. The gas delivered by the reservoir into the annulus cannot escape through the closed casing head: its pressure increases until it can push out all the liquid in the annulus through the tubing. In this phase the well will produce a fluid of low *GOR*. Hence, the well will continually produce both gas and oil, but the *GOR* is subject to appreciable fluctuations. The rates of change of the flow parameters are comparatively slow, a production cycle usually is of the order of some hours. — This second type of surging is harmful because the energy of gas delivered by the reservoir is utilized at a lower efficiency than in the case of continuous flow. This state of facts is illuminated by the following example (after Nind, transposed into SI units).

Example 2.3—5. Given $L_T = 1220$ m; $d = 2\ 3/8$ in. ($d_i = 0.0506$ m); flow of fluid into the well is represented by the *IP* curve *I* in Fig. 2.3—33; the well produces through a choke of diameter $d_{\mathrm{ch}} = 9.5$ mm; the production *GOR* of the well is $R_{go} = 20$ m³/m³ over 22 hours and then 350 m³/m³ over 2 hours. Find the rate of production of the well and the mean *GOR*; also, calculate the production rate of the well assuming, however, steady continuous flow through the same choke at the same daily gas consumption.

Using Gilbert's pressure gradient curves let us plot the $p_{TO} = \mathrm{f}(q_o)$ graphs describing the interaction of reservoir and well at $R_{go} = 20$ and $R_{go} = 350$ m³/m³, respectively, applying the procedure described in Section 2.3. Using Eq. 1.5—6 let us plot the $p_{TO} = \mathrm{f}'(q_o)$ graphs characterizing the operation of the choke at these same *GOR*s. The results of the construction are shown as Graphs II and III in Fig. 2.3—33. The intersections of the corresponding curves reveal flow rates of 4.51×10^{-4} m³/s at $R_{go} = 20$ and 5.27×10^{-4} m³/s at $R_{go} = 350$. Taking into account the respective durations of the two modes of production, the daily oil production is

$$V_o = 7.92 \times 10^4 \times 4.51 \times 10^{-4} + 7.2 \times 10^3 \times 5.27 \times 10^{-4} = 35.7 + 3.8 = 39.5 \text{ m}^3$$

and the daily gas production is

$$V_{gn} = 35.7 \times 20 + 3.8 \times 350 = 2044 \text{ m}^3 .$$

The mean *GOR* of surging production is

$$R_{go} = \frac{2044}{39.5} = 51.8 \text{ m}^3/\text{m}^3 .$$

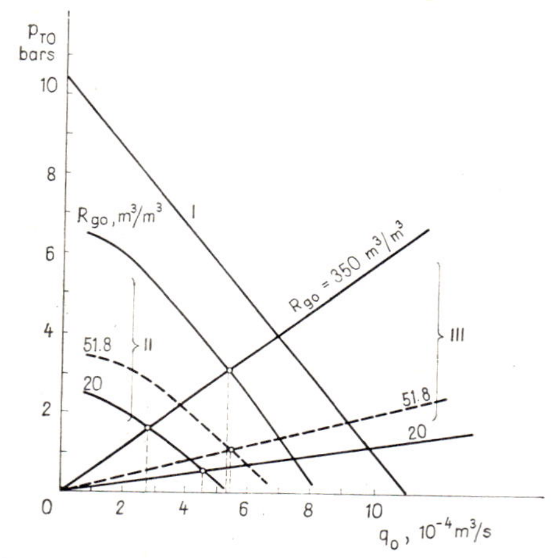

Fig. 2.3—33 Specific gas demand of continuous and intermittent flowing production after Nind 1964, p. 172 (used with permission of McGraw-Hill Book Company)

If the well produced steadily at this latter *GOR*, then the rate-of-production-v.-wellhead-pressure relationships would be represented by the graphs shown as dashed lines. The intersection of these two curves shows that the daily oil production of the well would then equal $5.44 \times 10^{-4} \times 86{,}400 = 47$ m³. Surging thus deprives the operator of a daily $47 - 39.5 = 7.5$ m³ of oil at the given gas flow rate.

(b.2) *Intermittent wells*. Intermittent production of a flowing well means that liquid flow out of the well will entirely cease periodically. Either the well is only periodically opened up to start with, so as to reduce its output, or else the well, although kept continually open, is incapable of delivering at a steady rate. This latter case is largely restricted to comparatively small-capacity wells producing from a low-pressure reservoir, in cases when the throughput capacity of the tubing is greater than the inflow capacity of the reservoir. The reservoir delivers liquid and gas to a tubing which is open at the surface. The gas present is insufficient to ensure flowing production. That part of it which enters the tubing bubbles through the liquid column without doing any useful work. Gas pressure in the casing annulus, closed on top, increases the while, until it is sufficient to push out the liquid accumulated in the annulus. Just as in the case of unloading the well (cf. Section 2.3.5a) the annulus at a given stage cannot hold more gas, so that all the gas delivered by the reservoir will enter the tubing. This gas will now be able to start liquid flow, and the tubing-shoe pressure will decrease accordingly. This effect is enhanced by the flow of higher-pressure gas from the annulus into the tubing. Since the liquid flow rate from the formation into the well is very low gas pressure in the annulus is free to decrease abruptly, and the well will 'blow off'. After a while the well will produce gas only; the well 'is empty'. Filling up the well with fluid then starts at a slow rate corresponding to the low productivity and low reservoir pressure of the well. This type of intermittent flowing production is harmful because an appreciable part of the gas is able to escape the well without doing any useful work.

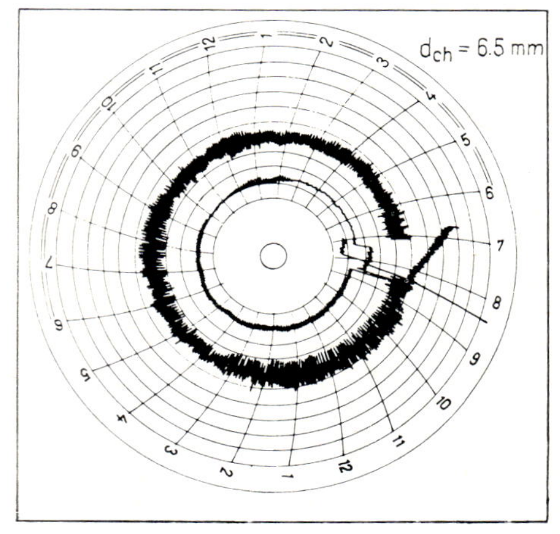

Fig. 2.3—34 Orifice meter charts of one well for two different size chokes

(b.3) *Flow regulation.* Surging production of the fluctuating-casing-pressure type and 'natural' intermittent production are comparatively inefficient ways of using formation gas to drive a well. There are several known ways to improve this efficiency. These fall into the following groups: (i) methods reducing the liquid throughput capacity of the well, (ii) methods preventing the abrupt entrance of large volumes of gas from the annulus into the tubing, and (iii) methods which, by periodical shut-in and unloading of the well, will prevent the production of gas without liquid. We shall discuss some of the more important solutions below.

(i) The wellhead choke is replaced by a smaller-bore one. The greater resistance to flow of the new choke will reduce the rate of flow of the well fluid. The continuous liquid throughput capacity of the well approaches, or indeed attains, the rate of continuous inflow from the reservoir. Figure 2.3—34 shows diagrams of the gas flow rates of a well at two different rates of liquid flow. Production is seen to have steadied considerably owing to the replacement of a 15.8 mm choke by a 6.5 mm one. The drawback of reducing the choke bore is that it entails a higher wellhead pressure and a consequent lower work potential of the gas; it might kill the well if the *GOR* is small. The risk of killing the well is considerably reduced if the choke is installed at the tubing shoe rather than in the wellhead. The choke is then called a bottom-hole choke. According to Soviet literature, the damping effect upon surging will be much the same, irrespective of whether the choking, i.e. the pressure drop, is brought about at the wellhead or at the tubing shoe. The work potential of the gas will be much greater, however. For instance, let the useful energy output per m³ of stock tank oil of gas in a flowing well approximately equal isothermal work, that is, $Rp_n \ln p_{TL}/p_{TO}$; this may be expressed as $c \lg p_{TL}/p_{TO}$ if the *GOR* is constant. Let the prescribed flowing *BHP* be 44 bars and the minimum feasible wellhead pressure, $p_{TO\min} = 2$ bars. Let us assume that a pressure drop of 9 bars is required to squelch surging. If this drop is brought about at the wellhead, then wellhead pressure will increase to $2 + 9 = 11$ bars. The isothermal work expended by the gas in the tubing will then equal $c \lg 4.4 \times 10^6/1.1 \times 10^6 = 0.6c$. Using a bottom-hole choke, on the other hand, we may bring about the required pressure drop at the tubing shoe, which reduces the *BHP* from 44 bars to 35 bars at the lower end of the tubing. Wellhead pressure will then be 2 bars and no further choke will be required in the wellhead. The work potential of the gas is, then, $c \lg 3.5 \times 10^6/2 \times 10^6 = 1.24c$, that is, the same relative amount of gas can do round twice as much work. The work potential of the gas is increased further by the fact that at the lower mean flowing pressure more gas will escape from solution and hence the effective *GOR* will increase. — A further advantage of a bottom-hole choke is that it reduces the pressure acting during production upon the wellhead assembly. Also, especially in high-pressure gas wells, the pressure drop at the wellhead choke reduces gas temperature below the hydrate point: gas hydrates will form and, obstructing the choke, will kill the well. At the depth where the bottom-hole choke is installed, on the other hand, the ambient temperature is likely to be much higher, so that no hydrate will form. If cooling due to expansion is still too great, it is expedient to install several bottom-hole chokes one above another, and so to distribute the required expansion.

There are several known types of bottom-hole choke. The non-removable type is practically a pressure-reducing insert in the tubing that can be removed only by pulling the tubing. Its operation is cumbersome and therefore not recommended. There are several types of removable bottom-hole choke that can be installed and retrieved by means of wire-line tools. Some have to be seated in a special landing nipple; others can be seated at any point of the tubing. Of the latter, some have seals energised by a pressure differential in the tubing; others provide packoff if triggered mechanically. Figure 2.3—35 shows the rather well-known OTIS–B type removable bottom-hole choke. It can be installed and retrieved under pressure. A lubricator is installed on the wellhead and the choke is lowered through it to the required depth by wire line with a suitable landing tool. If now the well is started up at a comparatively high production rate, then the well fluid will energise the resilient seal cups. Also, the slips will grab the tubing wall and thus fix the choke in place. The landing tool is then recovered by means of the wire line. Recovery is by a pulling tool run in likewise on a wire line: a jerk on the pulling tool engaging the fishing neck

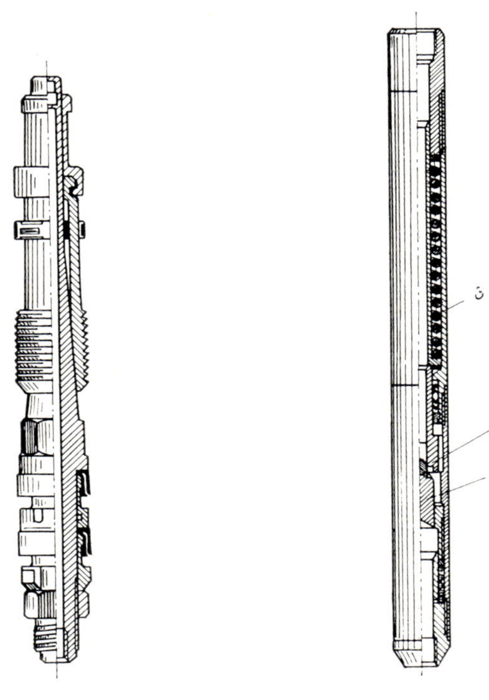

Fig. 2.3 – 35 OTIS B-type retrievable bottom-hole choke

Fig. 2.3 – 36 OTIS E-type bottom-hole regulator

of the choke will disengage the slips. In the absence of flow, the resilient cups will be slack, too, so that the bottom-hole choke can now be pulled. This type of choke is not recommended for pressure differentials in excess of 120 bars. For greater differentials, a bottom-hole choke of a design permitting mechanical locking should be used.

Removable bottom-hole regulator. — It serves much the same purpose as a removable bottom-hole choke. The difference is that the regulator maintains a constant pressure differential irrespective of the wellhead pressure. It is usually combined in use with a rather small-bore wellhead choke. The constant damping is provided by the regulator; the prescribed wellhead pressure can be set by means of the wellhead choke. Figure 2.3 – 36 shows the OTIS–E type bottom-hole regulator. Valve *1* is pressed by a rather weak spring load against the choke seat *2*, provided the latter is in the lower, that is, closed end position. If in the open state of the valve the pressure differential across the valve is greater (less) than the pressure represented by the spring *3*, then seat *2* will rise (fall). Thus the flow resistance between valve *1* and seat *2* will decrease (increase). The regulator can be used to maintain pressure differentials up to a round 100 bars. It may be seated at the desired depth in one of several landing devices not shown in the Figure.

As explained farther above, the flow resistance of a choke of conventional design is higher if the fluid flowing through is a liquid rather than a gas. In

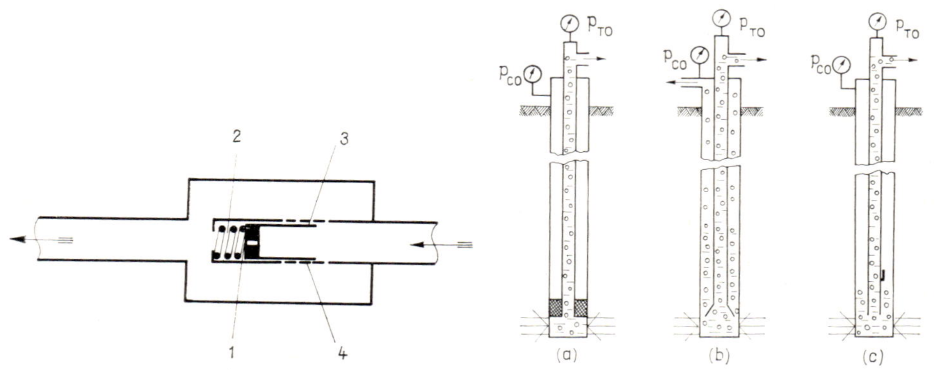

Fig. 2.3—37 Choke of self-adjusting aperture

Fig. 2.3—38 Surge-damping completions, after Muraviev and Krylov (1949)

order to provide a steadier flow regulation, a variable-resistance choke described by DeVerteuil (1953) and shown as Fig. 2.3—37, has been devised. If liquid flows through the choke, the pressure differential across orifice *1* will increase. The resultant pressure will displace sleeve *3* to the left, against the spring force *2*. The perforations *4* are thus opened to let the liquid pass and the flow resistance is thus lessened. The pressure differential across orifice *1* will, of course, be less in the case of gas flow. — As stated above, a surface choke providing the required damping may bring about a wellhead pressure high enough to kill the well. In such wells it may be recommended to change the tubing to a size suited to ensure cooperation between formation and well, optimal in the sense expounded in Section 2.3.3b.

(ii) If the casing annulus is shut off by a packer at tubing-shoe level, then the annulus will cease to function as a surge chamber, and the fluid in the tubing will have the original *GOR* as delivered by the formation. Installing such a packer may radically cure surging (Fig. 2.3—38a). If the *GOR* is comparatively high, then some of the gas may be bled off into the flow line through the partly opened casing valve provided the well has a packerless completion. The gas rising in the annulus will entrain a jacket of gaseous liquid. When the *BHP* declines, it is this 'reserve oil' that will enter the tubing first.

The pressure of fluid rising in the tubing will not, therefore, decrease any further, and the annulus will not blow down. — A Krylov funnel is an inverted funnel, fixed to the end of the tubing, of a rim diameter slightly less than the *ID* of the casing (Fig. 2.3—38b). In wells equipped with this device, the casing annulus is left to bleed as in the foregoing case. The small clearance between the funnel rim and the casing will damp the flow of gaseous liquid from the annulus into the tubing. — In a third solution, a gas-lift valve, set so as to pass gas from the annulus to the tubing at a pressure differential from 1 to 2 bars, is installed from 30 to 40 m above the tubing shoe. The casing valve is closed in this case. The annulus is packed off at the wellhead. Also in this case, a gaseous-liquid jacket will develop in the annulus below the gas-lift valve.

(iii) By the well-timed opening and shutting of the wellhead it may be achieved that the well starts producing only when a sufficient quantity of liquid has already accumulated, and is shut in as soon as the liquid is depleted, just before a blowoff of gas from the annulus occurs. In this case, a valve controlled by one of several possible signals is installed in the flow line. Control systems can, according to the signal used, be classified as follows: Opening controlled by a clockwork mechanism; shut-in controlled by a clockwork mechanism, a drop in casing pressure, a rise in tubing pressure, the arrival of a tubing plunger at the wellhead. Opening controlled by rise in casing pressure; shut-in controlled by a drop in casing pressure, a rise in tubing pressure, the arrival of a tubing plunger at the wellhead.

We shall now outline the operation of a Hungarian make of pneumatic control system, which, by its modular structure, can be adapted to several of the above-outlined combinations. The realization to be discussed here is clockwork-controlled as to opening and controlled by a rise in tubing pressure as to shut-in (see Figs 2.3—39a, b). The timer wheel *1*, rotated by the clockwork, bears cams in a number equal to the daily number of production cycles. When arm *2* is moved by one of the cams to the left of its position shown in the Figure, it permits the supply gas from the YES relay *3* (power amplifier) to bleed through nozzle *4*. The valve stem of relay *3* is forced upward by spring *5*. Pressure in the supply line *6* decreases, and so does the supply pressure acting upon the membrane of the control valve in the flow line: the control valve will open. As soon as the tubing pressure increases after the liquid production phase to a value set by spring *7*, valve stem *9* of pressure detector *8* will move upward; as a result, supply pressure above the membrane of the NO relay *10* will decrease. The valve stem of the relay rises and delivers supply gas to the pressure switch *11*. This displaces panel *12* to the left and, disengaging arm *2* from the cam on timer wheel *1*, shuts off nozzle *4*. Now the pressure of the supply gas will increase above the membrane of relay *3*. The valve stem of the relay will move downward and, by increasing the pressure of supply gas in conduit *6* and above the membrane of the control valve in the flow line, brings the control valve to

→

Fig. 2.3—39 Pneumatic control of flowing wells employed in Hungary; dashed line is high-pressure casing gas, broken line is low-pressure power gas, solid line is high/low pressure power gas

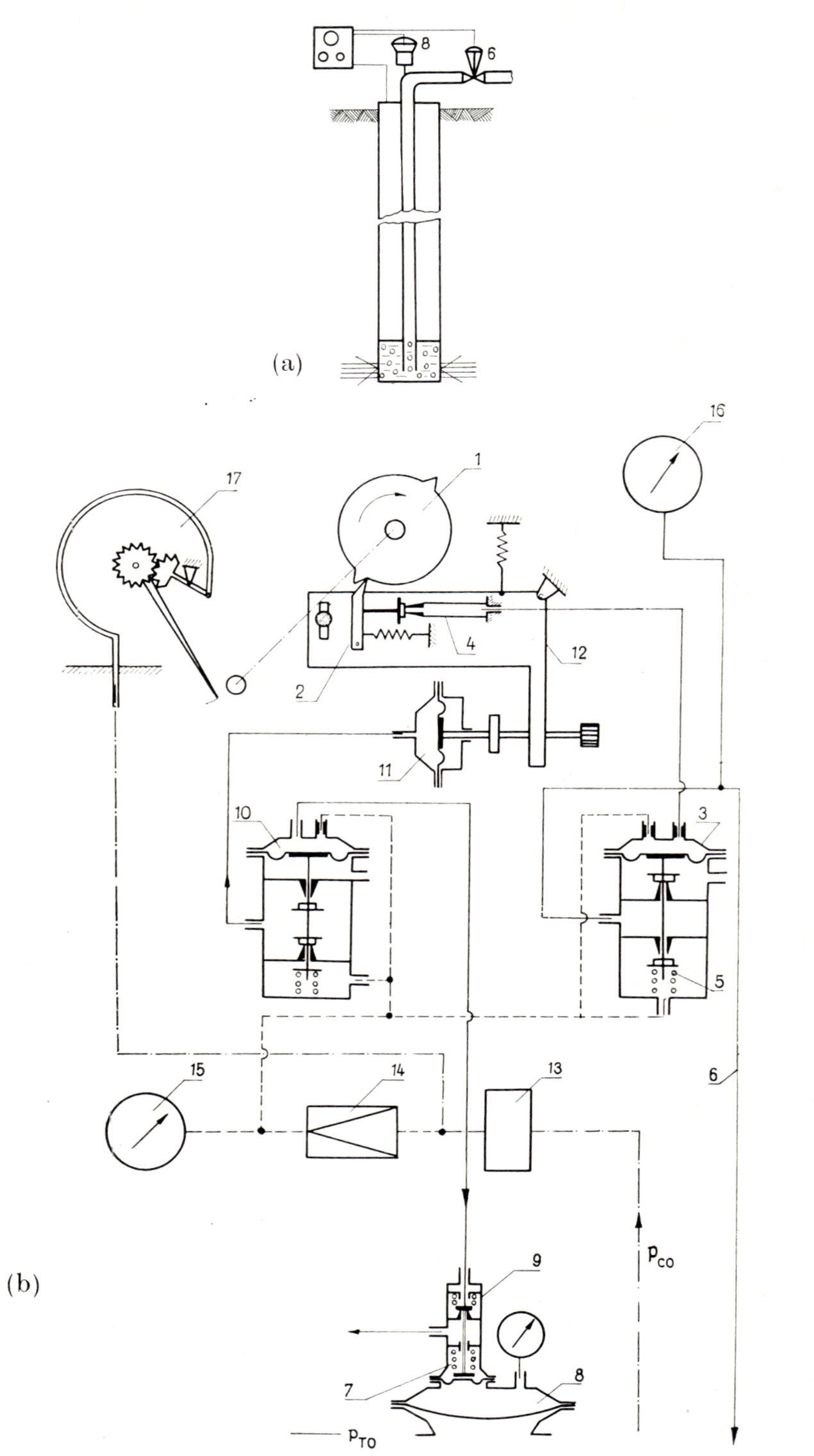

8
6
(a)
16
17
1
4
12
2
11
10
3
5
15
14
13
6
(b)
9
p_{co}
7
8
p_{TO}

close. Supply gas is taken from the casing annulus through filter *13*. Its pressure is reduced to the desired value in reductor *14*. Pressure gauges *15* and *16* respectively indicate input supply gas pressure and supply gas pressure over the membrane of the control valve. Bourdon-gauge recorder *17* records the pressure in the casing head. — The fact that the gas accumulated in the casing annulus is prevented from escaping will give rise to a rather high *BHP* and, in turn, to a relatively low inflow rate. The greater the permitted pressure drop in the casing annulus, the greater will be the amount of gas escaping without doing useful work, but the less will be the mean *BHP* and the greater the daily liquid production. Optimal operation parameters should be determined starting from reservoir engineering limitations and economic considerations.

(c) Well check-ups

In the course of production, it is imperative to run periodical checks on the condition of the well and also on the parameters of liquid and gas production. Checking the well should cover the following.

When the drilling contractor hands over the well, it is necessary to check (i) whether the well completion and the fittings conform to the prescriptions set down in the order, (ii) whether the tubing run in the well conforms to the agreement reached after well testing, (iii) whether the wellhead equipment provides the required packoff without a leak. The pressures as well as the internal gas and liquid contents in the wellhead assembly should be checked; (iv) whether the hand-over protocol contains all essential data concerning well completion and testing.

Checks on a producing well include the periodical or continuous recording and evaluation of certain operating parameters. The parameters fall into two groups: those that should be recorded at least once a day and those to be recorded at longer intervals. The first group includes the pressures in the tubing and casing head, and the daily duration of production. The wellhead assembly and the flow regulation equipment should be checked for good condition. The recording of wellhead pressures sheds light on a number of important circumstances. Any essential change in the rate of flow and composition of the well fluid is reflected by pressure changes at the tubing shoe (and hence, if the annulus is open below, at the casing head), and at the tubing head. The general rule is that if in a well having a casing annulus open below, p_{TO} and p_{CO} vary in the same direction, then the cause of the change is to be sought for outside the well completion. If both pressures increase, this may be due e.g. to an increase in reservoir energy, or an obstruction of the choke, or an accumulation of wax in the flow line. Decrease of both pressures may be due e.g. to a decrease in reservoir energy, or lowered flow resistance of the choke as a result of corrosion. If the two pressures vary in opposition, the cause will usually reside in the well completion. If p_{TO} decreases while p_{CO} increases, then the flow resistance of the tubing has increased, e.g. owing to deposits of scale or wax. The production of individual wells is measured only at rather long intervals. It is but the production of a group of wells producing e.g. into the same stock tank that is recorded. If the total production shows a drop, remedial operations cannot be started unless the individual well responsible for the drop has been pinpointed. Looking up the pressure records will at once reveal anomalous behaviour in the responsible well. It is this well that has to be subjected to a closer scrutiny.

The second group (that of the parameters to be checked at longer intervals) includes the following. (i) The oil, gas and water productions as well as the sand and/or mud production of the well are measured at intervals of 5—10 days. (ii) It is expedient to run well tests with several chokes at least once a year or after any significant change in well production. The results may indicate the desirability of exchanging the wellhead choke, or indeed, the tubing, for a more suitable size. (iii) Formation pressure is measured 3—6 months apart. The data recorded at the group of wells under consideration are processed into an isobar map which provides useful information as to the stage of depletion of the field. (iv) Usually in connection with item (iii), it is usual to record a *BHP* buildup curve, which permits the establishment of several parameters of the zone of influence of the well. (v) A fluid sample is taken under pressure from every well newly brought in, and periodically from key wells of the field, and tested for composition and physico-chemical properties in order to gather the information required for a satisfactory planning and management of production. (vi) On the occasion of every scraping operation, the position and thickness of the wax, sand and scale deposits in the tubing are to be determined. This is how the optimal frequency and method of scraping operations can be determined. Scale can often be eliminated only by changing the tubing. Observations permit us to predict optimum tubing-change intervals. (vii) At intervals depending on the composition and sand content of the well fluid, it is necessary to check whether the bore of the production choke has worn down beyond a certain limit, and to change it if need be.

2.4. Gas lifting

Gas lift is a means of artificial production by which the producing well is supplied from the surface with high-pressure injection gas, whose pressure energy is used to help lift the well fluid. There are two main types of gas lift, that is continuous-flow and intermittent. *Continuous-flow gas lift* can be regarded as an obvious continuation of flowing production. In order to supplement the formation-gas energy, insufficient in itself to ensure flowing production, a steady stream of injection gas is at the surface introduced into the casing annulus. The injection gas usually enters the tubing through a deeply installed gas lift valve, to aerate the well fluid derived from the formation. This means of production has the considerable advantage that it uses the total quantity of produced formation gas to lift liquid. It is most economical where flowing *BHP* and well capacity are both high, because this tends to keep specific injection gas requirement low. The Gilbert pressure-gradient curves given in the Appendix reveal, for example, that if a well 2000 m deep produces a daily 95.3 m^3 of liquid through 2 7/8 in. tubing, and the flowing *BHP* is 50 bars, then at 1 bar wellhead pressure the specific injection gas requirement is 140 m^3/m^3. If the flowing *BHP* were only 30 bars at the same rate of production, the specific injection gas requirement would be 350 m^3/m^3. At a flowing *BHP* of 30 bars and a daily liquid production rate of 7.9 m^3, the specific gas requirement would be 480 m^3/m^3. During production life, the production rate and the flowing *BHP* of gas-lifted wells will usually tend to decline gradually, which entails a gradual increase in specific gas requirement. — The simplest well completion for *intermittent gas lift* is the same as that for continuous-flow gas lift. The fundamental trait of intermittent gas lift is that the injection gas is introduced into the tubing slugwise rather than continously, notably at the instants when enough liquid has accumulated in the tubing. The gas slug, acting more or less like a piston, lifts the liquid slug to the surface. The specific gas requirement over a day of production equals in a fair approximation the specific gas requirement of each individual production cycle. The gas requirement of intermittent gas lift is lower at low rates and low *BHP*'s than that of continuous-flow gas lift; also, it hardly decreases with decreasing *BHP* as long as the amount of liquid lifted per cycle remains unchanged. For instance, the specific injection gas requirement of a 2000 m deep well of modern completion, producing a daily 7.9 m^3 of liquid at a *BHP* of 5 bars, is about 300 m^3/m^3. The drawback of intermittent gas lift is that—particularly in its modern forms of low specific injection gas requirement, which permit production also at low producing *BHP*s—the formation gas produced is not exploited in lifting the well fluid. A peculiar variety of intermittent gas lift is *plunger-lift* production. In this method, the liquid slug and the gas column are separated by a plunger made of metal or some plastic. Insertion of the plunger greatly reduces slippage. The main advantage of the method is that the specific injection gas requirement is low at medium *BHP*s, even if the rate is small; also, the formation gas produced will contribute to the lifting of the liquid.

2.4.1. Continuous-flow gas lift

(a) Theory of production; factors influencing operation

One of the possible well completions suitable for continuous flow gas lifting in a single-completion well is sketched up in Fig. 2.4—1. The casing annulus is packed off at tubing shoe level by packer *2*. Of the gas lift valves installed in the tubing wall, only operating valve *1* is normally open during steady production. The injection gas required to supplement the formation-gas energy, which is insufficient in itself to ensure flowing production, is led through injection gas line *3* into the casing annulus; the gas enters the tubing through operating valve *1*. Both injection gas flow and well flow are steady or quasi-steady. In the tubing, formation fluid and injection gas rise together. The pressure drop of flow in the tubing string can be determined by the methods discussed in Section 1.4, just as in the case of flowing production. Designing a continuous-flow gas lift operation for a single-completion well means to find the optimal size and length of tubing; the depth of continuous gas injection; the types, sizes and depths of the gas lift valves to be installed; and the decision whether to use a packerless (open) or a closed gas lift installation. — When choosing the tubing, the criterion of optimization is either to lift the formation fluid at the prescribed flowing *BHP* under the lowest possible injection *GLR*, or to make the well produce liquid at the greatest possible flow rate at a given consumption of injection gas. In both alternatives it is assumed that the wellhead pressure p_{TO} is the least possible value attainable with the given surface gathering and separation equipment.

In early production practice, lift gas was invariably introduced into the tubing at the tubing shoe. The annulus was not packed off. If, in such a well completion, injection gas pressure is less than the prescribed flowing *BHP*, then the length of the tubing string has to be less than well depth. In modern completions, the tubing string reaches down to the well bottom, and is provided farther up with an aperture to let injection gas enter. This aperture is usually fitted with a gas lift valve. The specific injection gas requirement of continuous-flow gas lift is less if the tubing string reaches down to the well bottom, because, by hypothesis, the tubing size chosen ensures either a minimum specific injection gas requirement or a minimum flowing pressure gradient. If the well fluid rises, on the first leg of its journey to the surface, in a casing which has of necessity a larger diameter than the tubing, then the specific gas requirement, or the flowing pressure gradient, will fail to be optimal. The point of gas injection is at the well bottom if injection gas pressure is equal to or greater than the flowing *BHP* to be realized. If it is less, then the point of injection has to be higher up the tubing. The correct depth of injection can be designed in the manner shown in Fig. 2.4—2. It is assumed first that the injection gas enters at the

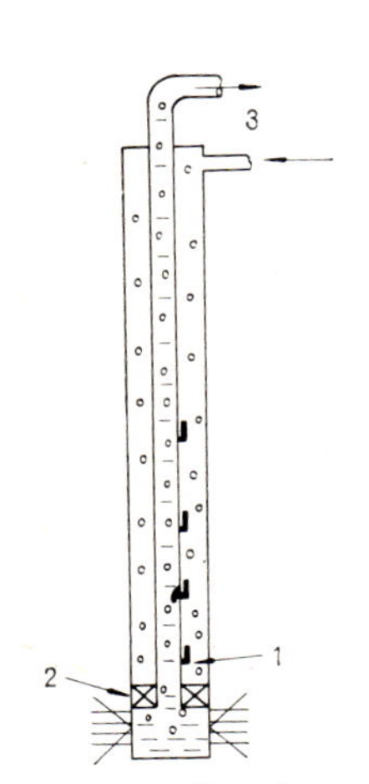

Fig. 2.4—1 Continuous gas-lift installation

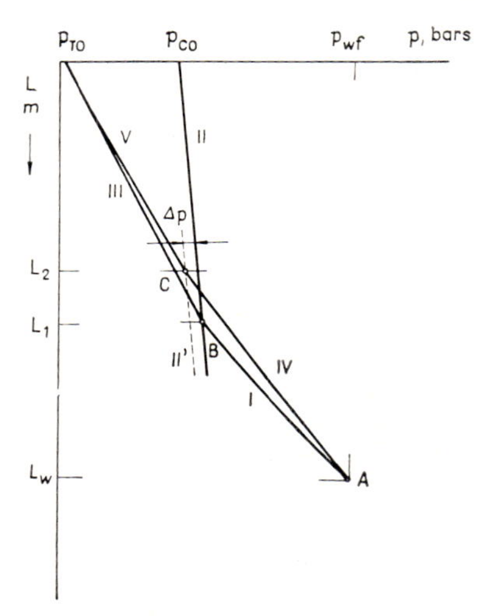

Fig. 2.4—2 Determining the point of gas injection

shoe of a tubing string not reaching down to the bottom of the well. Graph I is the pressure gradient curve, issuing from point A, determined by the flowing BHP p_{wf} prescribed for the well bottom situated at a depth L_w. This combination of data will prevail if the well fluid, characterized by a liquid inflow rate q_l and a formation GOR R_r, flows in a casing of diameter d_{Ci}; Graph II shows the injection gas pressure v. depth in the casing annulus. At the point of intersection B of Graphs I and II, the pressure of injection gas attains the pressure of the well fluid rising in the casing; the ordinate of this point determines the length L_1 of the tubing; Graph III is the pressure traverse in the tubing at a liquid flow rate q_l and a combined formation-plus-injection GOR $(R_r + R_{go})$. This graph is selected from a set of curves referring to the given q_l and the tubing size d_T by finding the curve along which pressure decreases from p_{CB} to p_{TO} over a length L_1. Subtraction of R_r from the R_{go} value belonging to this curve yields the specific injection gas requirement R_{gl}. Let the tubing string reach down to the well bottom and let the injection gas pass from the casing annulus into the tubing through a gas lift value. Graph IV is the pressure traverse of flow below the point of injection. Passage through the gas lift valve involves a pressure drop Δp. Hence, a Graph II' is to be traced parallel to, and at a distance Δp from, Graph II. The point of intersection C of Graphs IV and II' indicates the depth L_2 at which the injection valve is to be installed. The pressure traverse valid for the flow of formation fluid plus injection gas (Graph V) is chosen in the same way as the curve issuing from the point B in the previous case. Flowing pressure at depth L_2 in the tubing is thus less than what it would have been in the previous case, when the well fluid rose in the casing to a depth L_1. This means that, for rising to a depth L_2, the well fluid has consumed less pressure energy. Consequently, lifting it to the surface consumes less injection gas energy, too. In the following we shall analyse the influence of wellhead pressure and injection gas pressure upon the economy of continuous-flow gas lift (McAfee 1961).

The aim to be pursued is to make the wellhead pressure p_{TO} as low as possible. Causes for a high wellhead pressure may include a flow line too long or laid over a hilly terrain, or of a diameter too small or reduced by deposits; some fitting causing a significant local rise in flow resistance; an abrupt break of direction; or a separator installed far above the well; or too high a separator pressure. The influence of wellhead pressure upon the specific requirement of injection gas is shown in Fig. 2.4—3. Let wellhead pressures of 4.4, 7.9, 11.3 and 14.7 bars prevail in succession in the well characterized in the Figure. Starting from point C, let us select from the set of curves, in the manner described in connection with Fig. 2.4—2 above, the pressure gradient curves belonging to said wellhead pressures at the given oil flow rate q_o and tubing size d. Each curve is marked with the corresponding specific injection gas requirement. At the bottom of the Figure, the energy required to compress the injection gas is shown on a comparative scale; absolute values are not shown either here or in the next few figures. When calculating the compressors' power consumption, the intake pressure was taken to be 2.7 bars in each case. It is seen that the power consumption of compression is almost three times as high at a wellhead pressure of 14.7 bars than at one of 4.4 bars. Production is, then, the more economical, the less the flowing wellhead pressure.

The effective injection gas pressure at the surface (that is, the pressure to be ensured continuously during production) is to be chosen so that it should equal the flowing *BHP* at the well bottom. If the injection gas pressure is less than that, then injection should, by the above considerations, take place at a point above the well bottom. The compressor energy re-

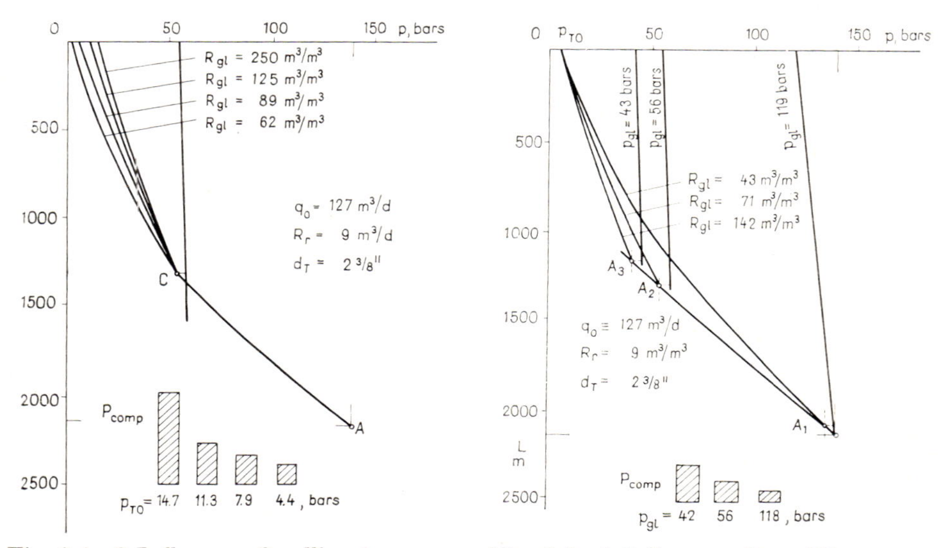

Fig. 2.4—3 Influence of wellhead pressure on power demand of injection-gas supply after McAfee (1961)

Fig. 2.4—4 Influence of gas-lift pressure on power demand of injection-gas supply, after McAfee (1961)

quired to compress the injection gas is the greater, the higher up the tubing the point of injection determined by lower injection gas pressure. Figure 2.4—4 refers to a well with pressure gradient curves for three values of injection gas pressure. Specific injection gas requirement is the parameter of the pressure gradient curves fitted between the point (p_{TO}, $L = 0$), corresponding to the wellhead, and the points A_1, A_2 and A_3, corresponding to the points of injection. In the given example, specific injection gas requirement rises from 43 to 142 m³/m³ if the injection gas pressure decreases from 119 to 43 bars. If the injection gas pressure is unnecessarily higher than the flowing *BHP* (cf. also paragraph a.2), then part of the pressure energy will escape without doing any useful work.

(a.1) *Selecting optimal tubing size.* — Two parameters of the tubing have to be optimized, viz. its length and its *ID*. In the foregoing paragraph we have found that length is optimal if the tubing string reaches down to the well bottom. The optimum diameter depends on what is regarded as the optimization criterion of the gas lift operation.

(i) *Dimensioning the tubing for a fixed rate of oil production and a minimum specific injection gas requirement with time-invariant parameters.* — Using pressure gradient curves, the problem can be solved regardless of whether the pressure of the injection gas is greater or less than the flowing *BHP* unequivocally defined by the oil production rate (McAfee 1961). In Fig. 2.4—5, the point of injection is likewise defined unambiguously by the point of intersection of the prescribed *BHP* and the given injection-gas pressure, or, retaining the notation of Fig. 2.4—2, of Graphs I and II′. Between this point, and the points $(p_{TO}, L = 0)$, pressure gradient curves referring to various tubing sizes have been fitted. It is clear from the Figure that, under the conditions stated, tubing sizes of 1.9, 2 3/8 and 2 7/8 in. will respectively result in specific injection gas requirements of 142, 44 and 36 m^3/m^3. Consequently, in the case considered, 2 7/8 in. size tubing is the optimal choice. The procedure is applicable also if injection gas pressure at the well bottom is equal to or greater than the prescribed flowing *BHP*, if it is noted that higher injection gas pressures have to be reduced to the desired value. If the injection gas pressure available, p_{gl}, is equal to or greater than the *BHP*, then the tubing size delivering the prescribed oil flow at the minimal injection gas pressure can rather simply be calculated using Krylov's relationships. The method has the advantage of not being conditional upon the accuracy of a set of pressure gradient curves. It is based on the consideration that $\bar{\xi}$ can be calculated from Eq. 1.4—20 if the given tubing-shoe pressure p_{TL} and the least feasible wellhead pressure p_{TO} are known. In the knowledge of the prescribed oil flow rate q_o, d_{opt} can be calculated using Eq. 2.3—1.

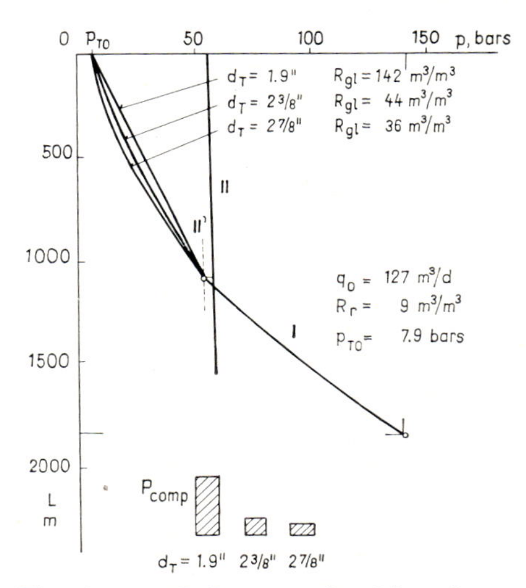

Fig. 2.4—5 Influence of tubing size on power demand of injection-gas supply, after McAfee (1961)

Example 2.4—1. Find the optimum tubing size if $L_w = 1153$ m; $p_{ws} = 79.5$ bars; $p_{wf} = 24.0$ bars; $p_{TO\min} = 1.2$ bar; $J = 2.23 \times 10^{-10} \frac{m^3}{s} \Big/ \frac{N}{m^2}$; $n = 1$; $\varrho_l = 900$ kg/m^3; $R_r = 15$ m^3/m^3. — The oil production rate corresponding to the prescribed flowing *BHP* is, by Eq. 2.1—3,

$$q_o = 2.23 \times 10^{-10} \ (79.5 \times 10^5 - 24.0 \times 10^5) = 1.24 \times 10^{-3} \ \text{m}^3/\text{s}.$$

The average pressure gradient is, by Eq. 1.4—20,

$$\bar{\xi} = \frac{24.0 \times 10^5 - 1.2 \times 10^5}{1153 \times 900 \times 9.81} = 0.224,$$

and the optimum tubing size is furnished by Eq. 2.3—1 as

$$d_{opt} = 0.263 \sqrt{\frac{1}{0.224}} \sqrt[3]{\frac{1.24 \times 10^{-3}}{1 - 0.224}} = 0.065 \text{ m}.$$

The *ID* of the next standard tubing size is 0.062 m: its nominal size is 2 7/8 in. — The total specific gas requirement is, by Eq. 1.4—27,

$$R_{opt} = 0.123 \frac{1153(1 - 0.224)900 \times 9.81}{0.062^{0,5} \times 0.224 \times 1.01 \times 10^5 \lg \dfrac{24.0 \times 10^5}{1.2 \times 10^5}} = 132 \text{ m}^3/\text{m}^3,$$

and the specific injection gas requirement is

$$R_{gl} = 132 - 15 = 117 \text{ m}^3/\text{m}^3.$$

(ii) *Dimensioning the tubing for the maximum oil flow rate feasible at a given specific injection gas consumption with time-invariant parameters.* — Let us solve the problem under the assumption that the injection-gas pressure at the well bottom equals the flowing *BHP*. Let us trace the inflow performance curve of the well in a bilinear orthogonal system of coordinates calibrated in q_o v. p_{wf}. Using a set of pressure gradient curves, let us plot in this Figure the throughput capacity v. pressure curves (q_o v. p_{wf}) of various tubing sizes. This is done by choosing from a set of curves valid for a given tubing size d and q_o = const. the curve belonging to the prescribed *GOR*, R_{go}. The value of p_{wf}, that is, the value to which tubing pressure increases from a prescribed wellhead pressure p_{TO} over a tubing length L_T, is then read off the curve. The values thus obtained are plotted against the oil production rate. The plots belonging to one and the same tubing size are connected with continuous lines. The points of intersection of the inflow performance curve, characterizing inflow into the well, with the throughput capacity curves of the individual tubing sizes, give the rate q_o at which the well will produce oil at the prescribed specific gas requirement and wellhead pressure. The tubing size to be chosen is that which ensures the highest oil flow rate.

Example 2.4—2. Find the standard tubing size at which a well 2000 m deep, producing an oil that can be characterized by one of Gilbert's set of curves, will produce at the highest rate, if p_{TO} = 4.0 bars and R_{gl} = 200 m³/m³. Let the formation gas production be negligible. Tubing available includes the sizes 2 3/8, 2 7/8, and 3 1/2 in., all three in sufficient length to reach down to the well bottom. The inflow performance curve is given as Graph *1* in Fig. 2.4—6a. The throughput capacity curves are the three top graphs in Fig. 2.4—6a, respectively marked 2 3/8, 2 7/8 and 3 1/2 in. at R_{gl} = 200. The graphs have been plotted using Columns 1, 2 and 3 of Table 2.4—1, whose values have in turn been read off Gilbert's pressure gradient curves. The data for 3 1/2-in. tubing have been taken from a set of curves presented in Gilbert's original work but not reproduced in this book. As revealed by the points of intersection of the inflow performance curve (Graph *1*) and of the throughput capacity graphs, the well will deliver a daily 73 m³ of oil through a tubing of 2 3/8 in. size and further 70 m³ through 2 7/8 and 67.5 m³ through 3 1/2 in. size tubing. That is, at a prescribed specific injection gas consumption $R_{gl} = 200$ m³/m³, it is the 2 3/8 in. tubing that ensures the highest rate of production.

(iii) *Dimensioning the tubing for the maximum feasible liquid production rate, with time-invariant parameters.* — It is assumed that injection gas pressure at the well bottom equals the flowing *BHP*. — Let us trace the inflow performance curve of the well in a bilinear orthogonal q_0 v. p_w

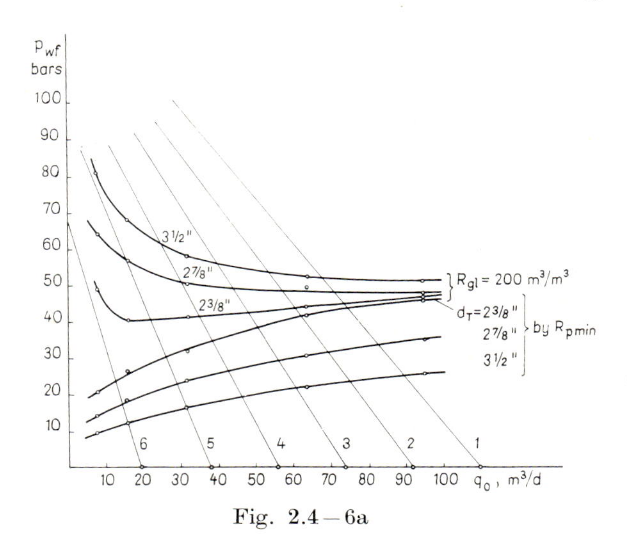

Fig. 2.4—6a

coordinate system. Using sets of pressure gradient curves, let us plot in this same Figure throughput capacity curves (q_o v. p_{wf}) corresponding to the least feasible flowing *BHP* for various tubing sizes. The values to be plotted are established by finding among the pressure gradient curves referring to a given tubing size d and various values of q_o = const. that curve which has the steepest slope, that is, the one which gives the least pressure at any depth below the surface. Reading off the pressures to which the prescribed wellhead pressure p_{TO} increases over a tubing length L_T (that is, the producing *BHP*s, p_{wf}), let us plot the values thus found against the production rate. The points belonging to a given tubing size are connected with a continuous line. The points of intersection of the inflow performance curve with the throughput capacity curves belonging to the various tubing sizes permit to find the maximum rates of production feasible by continuous flow gas lift through each of the tubing sizes examined. The pressure gradient curves permit us to determine the specific gas requirements of lifting the maximum rates of production through each tubing size. — Let us point out that this construction cannot be performed with the pressure-traverse curves based on the Poettmann—Carpenter equations, because in any set of curves belonging to a given pair of values

d and q_c, the slope of the curve invariably increases with the increase of R_{go}, the production *GOR*. This situation, which is contrary to observation, is due to certain lacks of the theory expounded in Section 1.4.3.

Example 2.4—3. For the well characterized in the foregoing example, let us find out, first, which of the tubing sizes 2 3/8, 2 7/8 and 3 1/2 in. will ensure the greatest rate of production, and second, the specific injection gas requirement of that rate of production. — The throughput capacity curves derived from the pressure gradient curves are shown as the lowermost three graphs marked respectively 2 3/8, 2 7/8 and 3 1/2 in. at $R_{p\,min}$ in Fig. 2.4—6a. The curves have been plotted using the data in Columns 1, 2 and 4 of Table 2.4—1, which have in turn been read off Gilbert's pressure gradient curves. Also in this case, the data for 3 1/2 in. tubing have been taken from Gilbert's original work. The points of intersection of the inflow performance curve with the throughput capacity curves give the maximum daily production by means of continuous flow gas lift as 75 m³ for 2 3/8 in., 82.5 m³ for 2 7/8 in., and 89.5 m³ for 3 1/2 in. tubing. The continuous curves in Fig. 2.4—6b illustrate the q_o v. $R_{p\,min}$ relationships for the three tubing sizes. These curves have been plotted using data in Columns 2 and 5 of Table 2.4—1. The curves can be used to read off the specific injection gas requirement $R_{p\,min}$ also for any intermediate value of q_o. Accordingly, the 75 m³ maximum daily production obtained in the foregoing paragraph requires 300 m³/m³ of injection gas. The 82.5 m³ produced through 2 7/8 in. tubing requires 460 m³/m³, and the 89.5 m³ through 3 1/2 in. tubing requires 670 m³/m³.

(iv) *Dimensioning of the tubing string, with time-variant parameters.* — The inflow performance curve of the well will vary during the life of the

Table 2.4—1

d	q_o	p_{wf200}	$p_{wf\,min}$	$R_{p\,min}$
in.	m³/d	bars	bars	m³/m³
1	2	3	4	5
	7.9	50	22	890
	15.9	41	28	632
2 3/8	31.7	42	32	445
	62.5	45	42	321
	95.3	48	47	262
	7.9	65	15	1570
	15.9	58	20	1120
2 7/8	31.7	51	25	765
	62.5	50	31	578
	95.3	49	36	421
	7.9	81	11	2320
	15.9	69	13	1590
3 1/2	31.7	59	78	1130
	62.5	53	23	805
	95.3	52	27	644

well, and it is usually possible to predict its variation by reservoir engineering considerations. Sclving the problems outlined above for various instants of time, each instant characterized by its own inflow performance curve, it is possible to derive the changes in time also of the production parameters.

Example 2.4—4. The inflow performance curves of a well 2000 m deep at intervals of 2.5, 5, 9, 14.5, 22.5 and 42.5 months after designing are

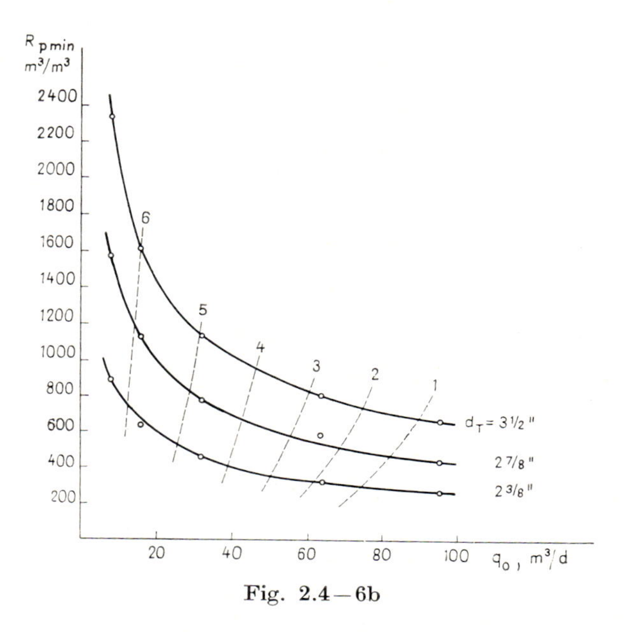

Fig. 2.4—6b

shown as Graphs *1—6* in Fig. 2.4—6a. The flow of the well fluid can be characterized by a set of Gilbert curves. Let the formation gas production of the well be negligible. The least feasible wellhead pressure is $p_{TO\,min} =$ $= 4.0$ bars. Tubing is available in the standard sizes 2 3/8, 2 7/8 and 3 1/2 in. Assuming continuous-flow gas lift over the period of time envisaged, let us answer the following questions: what is the maximum feasible daily oil production rate, $q_{o\,max}$; which of the tubing sizes will ensure the maximum oil flow rate $q_{o\,200}$ at $R_{gl} = 200$ m³/m³; what additional supply of injection gas is required to produce the additional amount of oil between $q_{o\,200}$ and $q_{o\,max}$ through any given size of tubing?

The throughput capacity curves in Fig. 2.4—6a have been plotted as explained in points (ii) and (iii); Fig. 2.4—6b provides the $R_{p\,min}$ values belonging to the various points of time, as explained in point (iii). The values of $q_{o\,max}$, $R_{p\,min}$ and $q_{o\,200}$, read off the graphs, are listed in Table 2.4—2. The greatest feasible oil production rate and the corresponding specific injection gas requirement are plotted v. time in Fig. 2.4—6c. It is apparent that 3 1/2 in. size tubing will permit the greatest oil production rate throughout the producing life of the well. However, this production, exceeding the maximum feasible production through 2 3/8 in. tubing by 30 percent, is realized at the considerable expense of a 120-percent increase in *GOR*. — Columns 1—4 of Table 2.4—3 reveal how far the maximum production rate $q_{o\,max}$ attainable through each of the tubing sizes available exceeds the production rate $q_{o\,200}$ through 2 3/8 in. tubing at a *GOR* of 200; Table 2.4—2 refers to 2 3/8 in. tubing. The differences between the cor-

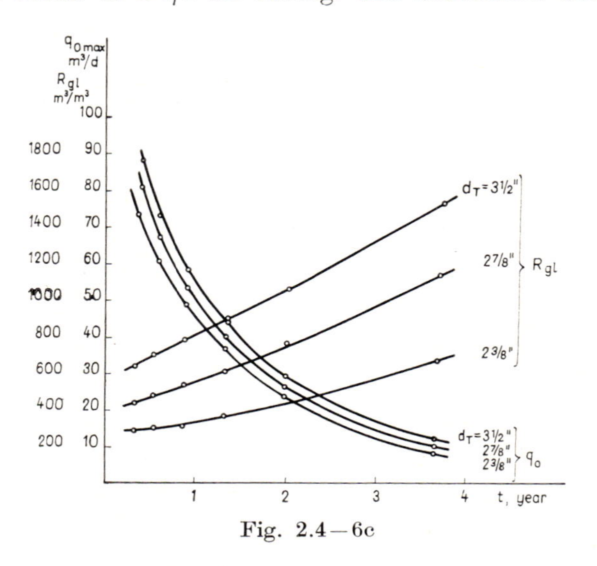

Fig. 2.4—6c

responding entries in Columns 1 and 7 of Table 2.4—2 are carried into Column 1 of Table 2.4—3. These values after graphical smoothing are carried into Column 2. In Columns 5—7, $q_{g\max}$ is the product of $q_{o\max}$ and R_{gl} taken from Table 2.4—2; Column 8 lists the values of $q_{g\,200}$, the products of the values $q_{o\,200}$ and $R_{gl} = 200$ taken from Table 2.4—2; Columns 9—11 list $\Delta q_g = q_{g\max} - q_{g\,200}$; Columns 13—15 give the ratios of the corresponding

Table 2.4—2

Serial number	$q_{o\,max}$			R_{gl}			$q_{o\,200}$
	2 3/8 in.	2 7/8 in.	3 1/2 in.	2 3/8 in.	2 7/8 in.	3 1/2 in.	2 3/8 in.
	m³/d			m³/m³			m³/d
	1	2	3	4	5	6	7
1	75.0	82.5	89.5	300	460	670	73.5
2	62.0	69.0	74.5	320	510	730	60.0
3	50.5	55.0	60.5	350	580	820	48.0
4	39.0	42.5	46.0	400	660	940	35.0
5	26.0	29.0	31.5	520	820	1120	22.0
6	12.0	14.0	16.0	750	1220	1620	—

Table 2.4 – 3

Serial number	Δq_0 2 3/8 in. read-off (m³/d)	Δq_0 2 3/8 in. corrected (m³/d)	Δq_0 2 7/8 in. (m³/d)	Δq_0 3 1/2 in. (m³/d)	$q_{g\,max}$ 2 3/8 in. (10³ m³/d)	$q_{g\,max}$ 2 7/8 in. (10³ m³/d)	$q_{g\,max}$ 3 1/2 in. (10³ m³/d)
	1	2	3	4	5	6	7
1	2.5	2.0	10.0	17.0	22.5	38.0	60.0
2	2.0	2.0	9.0	14.5	19.8	35.2	54.4
3	2.5	2.5	7.0	12.5	17.7	31.9	49.6
4	4.0	3.0	7.5	11.0	15.6	28.0	43.2
5	4.0	4.2	7.0	9.5	13.5	23.8	35.3
6	12.0	12.0	14.0	16.0	9.0	17.1	25.9

q_{g200} 2 3/8 in.	q_g 2 3/8 in. (10³ m³/d)	q_g 2 7/8 in. (10³ m³/d)	q_g 3 1/2 in. (10³ m³/d)	ΔR_{gl} 2 3/8 in. read-off (10³ m³/m³)	ΔR_{gl} 2 3/8 in. corrected (10³ m³/m³)	ΔR_{gl} 2 7/8 in. (10³ m³/m³)	ΔR_{gl} 3 1/2 in. (10³ m³/m³)
8	9	10	11	12	13	14	15
14.5	8.0	23.5	45.5	3.2	4.0	2.4	2.7
12.0	7.8	23.2	42.4	3.9	3.9	2.6	2.9
9.6	8.1	22.3	40.0	3.2	3.2	3.2	3.2
7.0	8.6	21.0	36.2	2.2	2.9	2.8	3.3
4.4	9.1	19.4	30.9	2.3	2.2	2.8	3.3
—	9.0	17.1	25.9	0.8	0.8	1.2	1.6

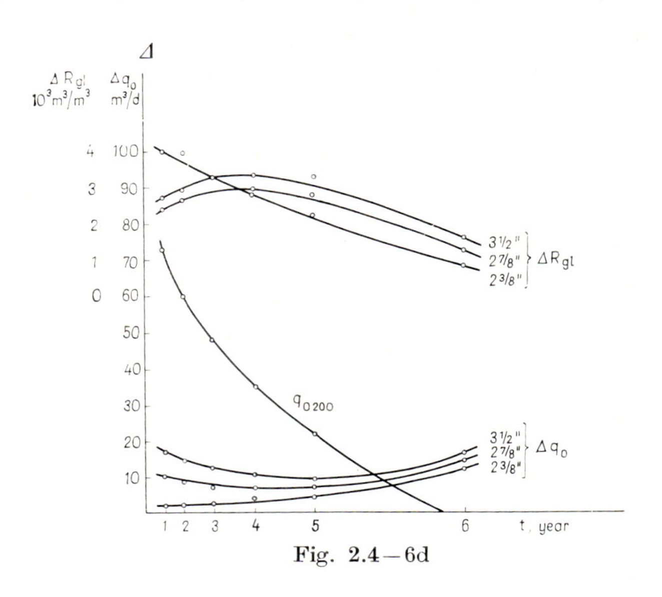

Fig. 2.4 – 6d

entries in Columns 9—11 and 2—4, respectively, that is, $\Delta R_{gl} = \Delta q_g/\Delta q_o$. These data indicate the specific amount of injection gas required to produce oil at any rate higher than what can be produced through 2 3/8 in. tubing at $R_{gl} = 200$. Figure 2.4—6d is a plot of the data in Columns 2, 3, 4, 13, 14 and 15 of Table 2.4—3 and Column 7 of Table 2.4—2. It is seen that the production of more oil requires an additional 2000—4000 m³/m³ of injection gas over a significant part of the life of the well.

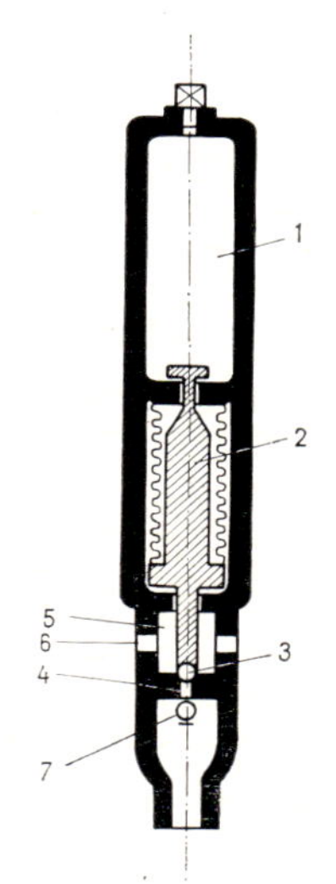

Fig. 2.4—7 Guiberson A-type gas-lift valve

(a.2) *Gas lift valve spacing.* — (i) *Operation and dimensioning of pressure-controlled gas lift valves.* Section 2.4—3 deals in detail with the valves used in modern gas lift operations. Designing production equipment presupposes, however, a knowledge of at least the fundamentals of valve operation. We shall therefore discuss at this point, slightly apart from our main trend of ideas, the operation of a gas lift valve permitting a continuous gas lift operation. Figure 2.4—7 shows a schematic section of a Guiberson A type gas lift valve. Dome *1* is charged with nitrogen gas. Its pressure p_k depresses the bellows *2* of effective cross-sectional area A_k and also the valve ball *3* fixed to it. If the valve is in the closed position, as shown in the Figure (for *7*, which is a check valve, see p. 211), then the pressure acting to open it is composed of the back pressure p_T acting upon the cross-sectional area A_{ch} of the valve port *4* and the pressure p_C acting upon the cross-sectional area $(A_k - A_{ch})$. The suffixes have been chosen in this way because the valve is usually installed so that the pressure in the tubing acts as back-pressure and the pressure in the casing annulus as gas pressure. The condition for the valve to open is

$$A_{ch}\, p_T + (A_k - A_{ch})\, p_C = A_k\, p_k. \qquad 2.4\text{—}1$$

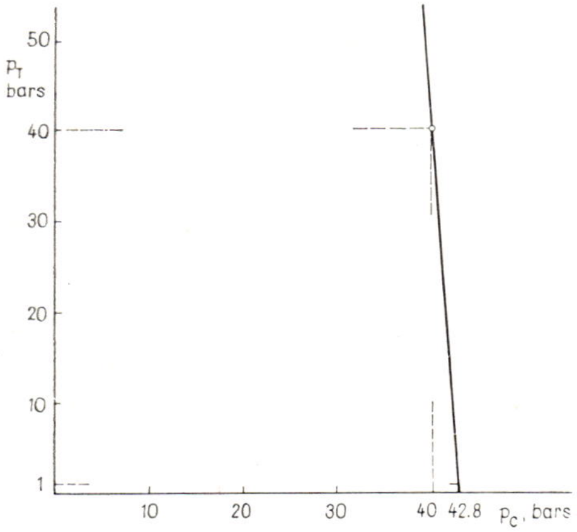

Let us express p_C from this equation; let e.g. $A_{ch} = 3.17 \times 10^{-5}$ m² and $A_k = 4.73 \times 10^{-4}$ m²; then

$$p_C = 1.072\, p_k - 0.072\, p_T.$$

Fig. 2.4—8

The relationship is illustrated for $p_k = 40$ bars in Fig. 2.4—8. The casing pressure required to open the valve is seen to depend on the pressure acting on the tubing side of the valve. If $p_T = 1$ bar, then the casing pressure required to open the valve is $p_C = 42.8$ bars. If $p_T = 40$ bars, then the casing pressure required to open the valve will also be 40 bars. The condition for gas to flow through the opened valve from the casing to the tubing is, of course, $p_C > p_T$. The valve can, then, be opened at any back-pressure $p_T < 40$ bars by bringing pressure in the casing to the appropriate value between 42.8 and 40 bars. If the valve is open, then pressure in cavity *5* (Fig. 2.4—7) approximately equals the casing pressure, because the area of the valve port *4* is much smaller than the combined sections of the inflow ports *6*. In the open position, the pressure in the tubing exerts no control on the moving system; the closing of the valve is controlled by casing pressure alone and the valve will close whenever casing pressure drops enough to equal dome pressure, that is, $p_C = p_k$.

The valve is at a higher temperature when installed in the well than when it has been charged. By the general gas law, the pressure of nitrogen in the dome at the temperature T_i prevailing at the depth of installation L_i is

$$p_{kT} = p_{kn} \frac{z_n T_i}{z_i T_n}.$$

The compressibility factor of nitrogen is small in most cases of practical interest. At 40 °C and 50 bars pressure, for instance, it equals 1.002. It should further be taken into account that establishing the valve temperature in a working valve involves some uncertainties, such as the cooling effect of the gas flowing through. It is therefore usually permissible to regard the ratio z_n/z_i as equal to unity, whence

$$p_{ki} = p_{kn} \frac{T_i}{T_n}. \qquad 2.4\text{—}2$$

In order to control the valve by regulating injection gas pressure on the surface, we have to know the injection gas pressure p_{Ci} at the depth of installation L_i of the valve, provided the casing-head pressure is p_{CO}. Equation 1.2—4 with the assumption $q_n = 0$ yields for the pressure at depth of the gas column

$$p_1 = p_2 e^{\frac{0.00188 h M}{\bar{T} z}}.$$

Let $p_1 = p_{Ci}$, $p_2 = p_{CO}$, $h = L_i$; the pressure of injection gas is p_{CO} on the surface; at the depth L_i where the gas lift valve is installed, it is

$$p_{Ci} = p_{CO} e^{\frac{0.00118 L_i M}{\bar{T} z}}. \qquad 2.4\text{—}3$$

Let us add that $\bar{T}$ can be regarded as the arithmetic average of mean annual temperature on the surface and of formation temperature at valve depth. $\bar{z}$ is the compressibility factor of the injection gas in the casing annulus, at the mean pressure $\bar{p}_C$ and temperature $\bar{T}$ prevailing there. Accuracy will not be unduly impaired by assuming $\bar{p}_C \approx p_{CO} \approx p_{CL}$.

Example 2.4—5. Find the dome charge pressure p_{kn} at 20 °C temperature, the closing pressure as measured on the casing head, p_{kO}, and the choke diameter d_{ch} of a pressure-controlled gas lift valve of opening equation $p_C = 1.072 p_k - 0.072 p_T$, provided the depth of installation of the valve is $L_i = 670$ m; temperature at that depth, $T_i = 30$ °C; pressure in the casing annulus at that depth, $p_{Ci} = 40.5$ bars; mean annual temperature at the surface, $T_{CO} = 11.0$ °C; maximum possible back-pressure during unloading after closing of valve, $p_{Ti\,\max} = 29.6$ bars; back-pressure during continuous production, $P_{Tfi} = 16.7$ bars; $M = 21.0$ kg/kmole; $\varkappa = 1.25$; gas flow rate through valve during continuous operation, $q_{gn} = 0.5$ m^3/s; discharge factor, $\mu' = 0.9$; $p_n = 1.01$ bar; $T_n = 288.2$ K. — Dome pressure in the valve installed in the well equals, by the opening equation,

$$p_{ki} = 0.933\,p_{Ci} + 0.067\,p_{Ti\,\max} = 0.933 \times 40.5 \times 10^5 + 0.067 \times 29.6 \times 10^5 = \\ = 39.7 \times 10^5 \text{ N/m}^2.$$

Dome pressure in the valve at 20 °C temperature is

$$p_{k\,20} = 39.7 \times 10^5 \frac{293.2}{303.2} = 38.4 \times 10^5 \text{ N/m}^2.$$

Closing pressure of the valve is equal to dome pressure. Its surface value can be calculated using Eq. 2.4—3. Calculation requires the knowledge of $\bar{T}$ and $\bar{z}$.

$$\bar{T} = \frac{T_{CO} + T_i}{2} = \frac{284.2 + 303.2}{2} = 293.7 \text{ K}.$$

By Figs 8.1—1 and 8.1—2 (see pp. 549–50) $\bar{z} = 0.87$ at $\bar{T} = 293.7$ K and $\bar{p} \approx p_{ki} = 39.7$ bars. Hence, casing-head pressure on closing of the valve is

$$p_{CO} = 39.7 \times 10^5 e^{-\frac{0.00118 \times 670 \times 21.0}{0.87 \times 293.7}} = 37.2 \times 10^5 \text{ N/m}^2.$$

The choke diameter of the valve can be calculated using Eq. 1.5—4. The pressure ratio during production is

$$\frac{p_2}{p_1} = \frac{p_{Tfi}}{p_{Ci}} = \frac{16.7 \times 10^5}{40.5 \times 10^5} = 0.412\,.$$

The critical pressure ratios for various values of $\varkappa$, calculated by means of Eq. 1.5—3, are listed in Table 2.4—4. Table 2.4—5, on the other hand, lists the C factors calculated using Eq. 1.5—5 for various pressure ratios p_2/p_1 on the assumption that $\varkappa = 1.25$. Table 2.4—4 gives $(p_2/p_1)_c = 0.555$ for $\varkappa = 1.25$. Since $0.412 < 0.555$, we may use $C = 0.465$ according to Table 2.4—5. By Eq. 1.5—4, where, by the logic of the situation, $T_c = T_i$ and $p_1 = p_{Ci}$, we may calculate the choke diameter:

$$d_{\text{ch}} = \sqrt{0.5 \frac{1.01 \times 10^5}{288.2} \frac{(21.0 \times 303.2)^{0.5}}{101.3 \times 40.5 \times 10^5 \times 0.9 \times 0.465}} = \\ = 0.00904 \text{ m; that is, about 9 mm.}$$

(ii) *Unloading a well.* In a well, shut in or dead for one of a variety of reasons, or intentionally filled up with liquid from the surface, production can be started up only by removing a substantial portion of the liquid column

Table 2.4 – 4

$\varkappa$	$\left(\frac{p_1}{p_2}\right)_c$	C
1.40	0.528	0.484
1.35	0.537	0.478
1.30	0.546	0.472
1.25	0.555	0.465
1.20	0.565	0.459
1.15	0.574	0.452

Table 2.4 – 5

$\frac{p_2}{p_1}$	C	$\frac{p_2}{p_1}$	C
1.000	0.000	0.900	0.297
0.990	0.099	0.850	0.351
0.980	0.140	0.800	0.391
0.970	0.170	0.750	0.420
0.960	0.195	0.700	0.441
0.950	0.217	0.650	0.455
0.940	0.236	0.600	0.463
0.930	0.253	0.555	0.465
0.920	0.269		
0.910	0.283		

from the well. This operation, called unloading, is carried out in gas lift wells by means of the injection gas. The simplest way of doing it would be to expel through the tubing the liquid filling the well by injecting gas into the casing annulus. This would, however, typically require quite a high injection gas pressure, with a number of attendant drawbacks such as the necessity of installing on the surface a separate high-pressure unloading network conduit; the abrupt pressure buildup during unloading may cause caving and a sand inrush at the sandface; etc. In modern practice the well is unloaded in several steps, by means of a number of gas lift valves installed along the tubing string. Figure 2.4 – 9 shows the main phases of unloading a well provided with three gas lift valves. The Figure shows a so-called semi-closed installation which has the casing annulus packed off but has no standing valve at the

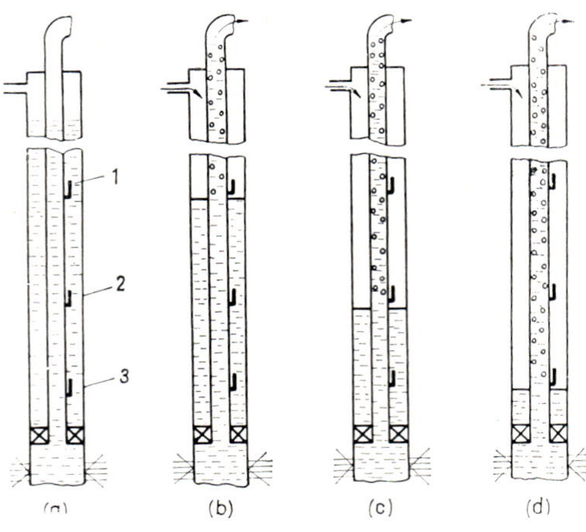

Fig. 2.4 – 9 Unloading with gas-lift valves

tubing shoe. In modern well completions, the gas lift valves are provided with reverse check valves, so that no oil might enter the casing annulus during production. This shortens the time required for unloading. After first installation, however, also the casing annulus will be filled with liquid. We shall in the following assume this to be the case. Our considerations can be transferred unchanged to so-called open completions, where the casing annulus is not packed off at the tubing shoe. Let us assume that the dome pressure of each valve is less by 1 bar than that of the next valve above. In the phase shown as Fig. 2.4—9a, the well is practically filled with liquid, e.g. to the static level determined by formation pressure.

Unloading begins by introducing into the casing annulus injection gas at the maximum pressure required to open the top gas lift valve. We have seen in connection with Fig. 2.4—8 that, at this gas pressure, the top valve will open even if the pressure in the tubing opposite it is zero. In the case considered, however, at the time when injection gas begins to flow into the annulus, there is a gasless liquid column of considerable pressure both in the annulus and in the tubing between the level of the first gas lift valve and the static-pressure level of the liquid. The top valve can consequently open even at the beginning of gas injection. The two lower valves will also have opened up, because their dome pressures are less than that of the uppermost valve, and the opening pressures acting upon them are greater than that acting upon the uppermost valve *1*. In the phase shown as Fig. 2.4—9b valve *1* is already uncovered. The depth of this valve is designed so that, at the time the gas attains the valve, there will be at valve depth a pressure differential of say 3.4 bars between annulus and tubing, in order to make gas flow through the valve. — After the beginning of gas flow through the valve, the liquid column in the tubing is gradually aerated and the well delivers a gaseous fluid through the tubing head. Tubing pressure opposite valve *1* decreases. As a result, the liquid level in the annulus keeps sinking, because at the levels of the lower gas lift valves the pressures in the annulus and in the tubing must be practically balanced. Tubing pressure opposite valve *1* decreases to a constant value, and the liquid level in the annulus sinks to a corresponding constant depth. Valve *2* is installed slightly above that depth, similarly to valve *1*, in such fashion that gas flow may be started by an initial pressure differential Δp. From this instant on, gas injected into the casing head may enter the tubing through two valves. If now the surface gas injection rate is less than the rate of gas flow into the tubing through the two valves, then pressure in the annulus will start to decrease. As soon as it attains the dome pressure p_{k1} of valve *1*, this valve will close. Gas will continue to flow into the tubing through valve *2* only. This initial stage is shown in Fig. 2.4—9c. — The liquid level sinks largely in the way just described to the depth of valve *3*. As soon as injection gas can enter the tubing through valve *3*, valve *2* closes under the influence of the annulus pressure's decrease to p_{k2}. The well will now be produced by continuous gas lift by the injection gas entering through valve *3*. — The opening and closing casing pressures of the gas lift valves are set so as to decrease down the hole, in order to ensure that, by applying the correct injection gas pressure, it may be possible to produce the well by injection through any one of the three valves. One thing to avoid is, however, that a valve should reopen once it has closed and gas flow has started through the next valve below. In order to prevent this, the maximum tubing pressure opposite any valve is calculated for the case that injection gas enters the tubing through the next valve below. In the knowledge of this tubing pressure and the casing pressure, already fixed, the dome pressure can be calculated using the opening equation of the valve.

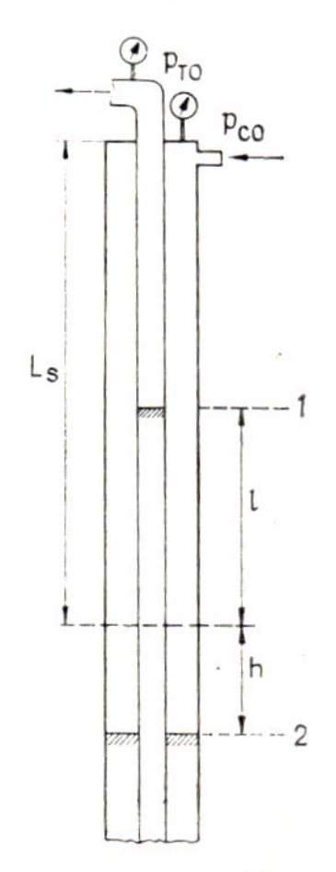

Fig. 2.4–10

Example 2.4–6. Find the back pressure at which the valve characterized in the previous example will open after closing. — If the valve is in the closed state, then injection gas will enter the tubing through one of the valves below. The most unfavourable case as far as the opening safety of the valve is concerned is when the lower valve in question is the next valve below, because it is in that case that tubing pressure may be highest. The casing pressure is already less than the initial value, say by a value $\Delta p = 0.7$ bar. Using the opening equation of the valve, we may calculate the tubing pressure at which the valve will open if its dome pressure is 39.7 bars and the casing pressure is $40.5 - 0.7 = 39.8$ bars:

$$p_{T1} = 14.88 \times 39.7 \times 10^5 - 13.88 \times 39.8 \times 10^5 =$$
$$= 38.3 \times 10^5 \text{ N/m}^2.$$

That is, a tubing pressure of at least 38.3 bars is required to open the valve. If the tubing pressure opposite the valve considered cannot exceed 29.6 bars after its closure, then the valve will stay closed, through whichever lower valve the gas will enter the tubing.

(iii) *Valve depths; choke sizes.* Calculating the depth of the top valve may involve one of two procedures, depending on whether the static liquid level prior to unloading is comparatively high or low. 'Comparatively high' means that the pressure of gas injected into the annulus will make liquid flow through the tubing head even before gas could enter from the annulus into the tubing. 'Comparatively low' means that the oil rising in the tubing has not yet attained the wellhead at the instant when injection starts into the tubing from the annulus through the top gas lift valve. Figure 2.4–10 is a schematic drawing of the upper part of a well, partly filled with a liquid of gravity γ_l. At the initial instant, liquid level is at a depth L_s below the surface in both the tubing and the annulus. If gas of pressure p_{CO} is injected into the annulus, then, provided no backflow into the formation takes place during the brief period of unloading, the liquid level will sink by a height h, and rise in the tubing to a level l. Let the cross-sectional area of the annulus be A_{TC}, that of the tubing, A_T. The relationship between the level changes in the annulus and in the tubing is $h(A_{TC}/A_T) = hR_A$. At the end of the U-tubing, the pressure at the liquid level (level 2) will be p_C in both the annulus and the tubing. We may write up for the tubing that, if wellhead pressure is p_{TO}, and the weight of the gas column is neglected, then

$$p_C = h\gamma_l + hR_A \gamma_l + p_{TO},$$

and hence,

$$h = \frac{p_C - p_{TO}}{(1 + R_A)\gamma_l}. \qquad 2.4\text{—}4$$

The tubing is called 'long' if at the initial instant its liquid-unfilled length L_s is longer than the column height hR_A of the liquid U-tubed from the annulus into the tubing, that is, if

$$L > hR_A.$$

Let us substitute h by the expression in Eq. 2.4—4. The condition for 'long' tubing is, then,

$$L_s > \frac{p_C - p_{TO}}{\left(\frac{1}{R_A} + 1\right)\gamma_l}. \qquad 2.4\text{—}5$$

If the tubing is 'long', then the depth of the top gas lift valve is given by the equation

$$L_1 = L_s + h - 10 = L_s + \frac{p_C - p_{TO}}{(1 + R_A)\gamma_l} - 10. \qquad 2.4\text{—}6$$

In order to provide the pressure differential necessary to make the valve pass gas from the annulus into the tubing, the 10 metres are subtracted. As in the case under consideration a static balance of pressures would require the sinking of the liquid level in the annulus by a further 10 m, a liquid column of height $10R_A$ is still missing from the tubing. The initial pressure differential is, then,

$$\Delta p = 10(1 + R_A)\gamma_l.$$

If the tubing is 'short', or the static level is uncertain, or the well has been filled with liquid from the surface, then the top valve is to be installed at depth

$$L_1 = \frac{p_C - \Delta p_C - p_{TO}}{2}. \qquad 2.4\text{—}7$$

This equation is based on the consideration that, at the instant when injection starts, the effective pressure in the tubing at valve depth ($p_C - \Delta p_C$) is equal to the hydrostatic pressure $L_1\gamma_l$ of the liquid column reaching to the wellhead in the tubing, plus the wellhead pressure p_{TO}; Δp_C is the pressure drop of injection gas across the gas lift valve. In the latter case, the depth of the top valve can be established also by a graphical construction. The depths of the remaining valves can also be established by several means. The procedure to be described below is a graphical one, modified after Winkler (Winkler and Smith 1962). It is assumed that, even at the initial stages of unloading, the *BHP* drops below the formation pressure, so that the formation will deliver fluid to the well at a gradually increasing rate. The steps of designing are as follows. (1) Starting from the point ($L = 0$, p_{TO}), the pressure gradient curve (Graph I), valid for the size of tubing under consideration, is plotted for a continuous-flow gas lift operation into a bilinear orthogonal system of coordinates (p v. L in Fig. 2.4—11). (2) Graph II (static pressure of gas column in annulus v. depth)

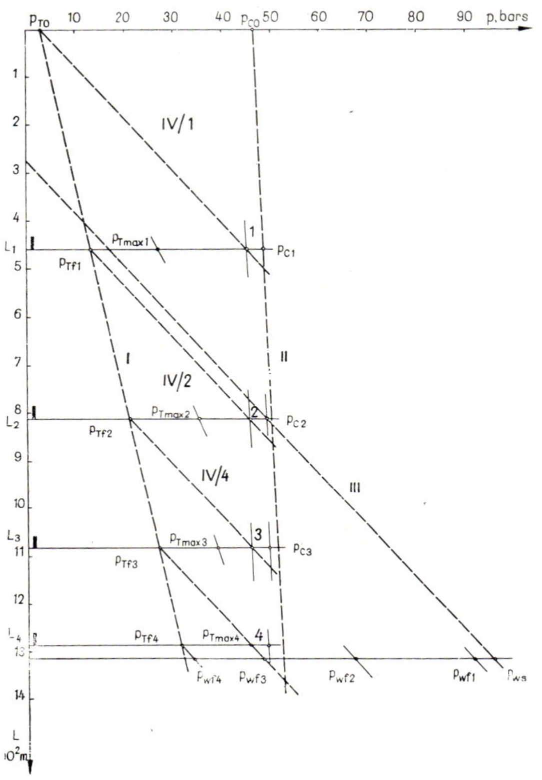

Fig. 2.4—11 Gas-lift valve spacing for continuous flow

is plotted, starting from the point ($L = 0$, p_{CO}). (3) Graph III (pressure of gasless oil v. depth) is plotted in the knowledge of $\bar{\gamma}_l$, starting from the point (L_w, p_{ws}). (4) A Graph IV/1, parallel to Graph III, is drawn through the point ($L = 0$, p_{TO}). At the point where it intersects Graph II, the gas pressure in the annulus will equal the pressure in a tubing filled with oil to the wellhead at a wellhead pressure of p_{TO}. Let us draw a parallel to Graph II, at a distance of $\Delta p = 3.4$ bars above the point of intersection. The point of intersection *1* of this last parallel with Graph IV/1 determines the depth L_1 of the top valve *1*. The pressure differential $\Delta p =$ $= 3.4$ bars ensures, on the one hand, that the valve will pass gas when uncovered and equals, on the other hand, the pressure drop to be expected across the valve when gas flows through it at the maximum designed rate. (5) In the case of continuous flow gas lift, a pressure p_{Tf1} will prevail in the tubing at depth L_1. During unloading, it is safe at any event to reduce tubing pressure to this value, because the well will at a flowing *BHP* of p_{wf1} still produce less liquid than during continuous flow. — Let us trace a parallel to Graph III through the point (L_1, p_{Tf1}) (Graph IV/2). Now let us draw parallels to the casing-pressure traverse, one at a distance of one bar, and another at a distance of a further 3.4 bars. The reason for the first one-bar reduction is to set the opening pressure of valve *2* one bar below that of valve *1*. The casing-pressure traverse reduced by 4.4 bars is intersected by Graph IV/2 at point *2*. It is the ordinate of this point that defines the depth of valve *2*. (6) The remaining valve depths are determined as for valve *2* above. The only difference is that casing pressure is reduced by a further one bar for each valve. (7) Let us determine for all valves but the last one the maximum tubing pressure that may arise during unloading. This value is obtained with a fair margin of safety by drawing for any valve a straight line connecting the point marked *2—4* of the next valve below and the point ($L = 0$, p_{TO}). $p_{T\max}$ is defined by the point of intersection of this line with a horizontal line drawn at the depth of the valve under consideration. (8) Using Eq. 2.4—3, we calculate the unloading tubing-head pressures p_{COi} corresponding to the pressures p_{Ci}. (9) Introducing the values p_{Ci} and p_{Tfi} into Eq. 1.5—4 we calculate the least possible choke diameters $d_{ch\,i}$ of the valves. (10) Using the catalogue of a manufacturer of gas lift valves (cf. e.g. the CAMCO Gas Lift Manual by Winkler and Smith 1962, p. A4—001) we choose the valve with the next greater port area than the value found in step (9). Taking from the catalogue the parameters A_{ch} and A_k we may write up the opening equation of the valve. (11) Using the opening equations of the valves, and the pressure data p_{Ci} and $p_{T\max i}$ we calculate for each valve the dome-pressures p_{ki} required at valve depth. By the logic of the situation, there is no '$p_{T\max}$' for the lowermost valve. Its actual dome pressure equals that of the next valve above. Using Eq. 2.4—2, we determine the charged pressures p_{kni} corresponding to the installed pressures p_{ki}.

Example 2.4—7. Find the depths of the gas lift valves required for the unloading and continuous flow gas lifting of a well of given capacity, as well as the surface unloading pressure, the filling pressures and choke diameters of the valves, if $L_w = 1320$ m, $d = 2\ 3/8$ in.; $p_{ws} = 92.4$ bars; $J = 1.79\times10^{-10} \dfrac{\text{m}^3}{s} \Big/ \dfrac{N}{\text{m}^2}$, $n = 1$; $q_o = 95.3$ m^3/d; $p_{CO} = 45.1$ bars; $p_{TO} = 2.9$ bars; $\mu' = 0.9$; mean oil density in the well, $\bar{\varrho}_o = 900$ kg/m^3; $M_g = 21.0$ kg/kmole; surface gas temperature, assumed to equal the mean annual temperature, $T_o = 11$ °C; geothermal gradient, $\sigma_b = 0.04$ K/m. Formation gas production negligible. Valves to be used are: CAMCO Type J–20. The suitable Gilbert pressure gradient curve set will yield the curve valid for a tubing size $d = 2\ 3/8$ in. at $q_o = 95.3$ m^3/d and $R_{go} = 262$ m^3/m^3. Let us copy this curve onto Fig. 2.4—11, so that it starts at $p_{TO} = 2.9$ bars.

This will be our Graph I. The flowing *BHP* will be 30.8 bars. — The results of the graphical construction in Fig. 2.4—11 and the calculation according to steps (1)—(8) above are listed in Table 2.4—6. — The results of the calculations according to steps (9)—(11) above are listed in Table 2.4—7. Let us add that the entries in Column 1, the least *BHP* (p_{wfi}) to be expected while the individual valves pass gas during unloading, have been determined graphically by the method illustrated in Fig. 2.4—11. The rates of production corresponding to these *BHP*s and listed in Column 2 have been calculated by means of the relationship

$$q_{oi} = 1.79 \times 10^{-10} (92.4 \times 10^5 - p_{wfi}).$$

Table 2.4—6

Number of valve	L	p_C	p_{Tf}	$p_{T\,max}$	T	p_{Co}
	m	bars	bars	bars	°C	bars
1	466	46.9	12.8	26.1	29.4	45.1
2	815	47.6	20.3	34.0	43.6	44.1
3	1088	47.8	25.9	37.6	54.5	43.4
4	1293	47.6	30.2	—	62.7	42.4

These rates of production are required only for calculating the injection gas requirement which in turn serves to calculate the choke diameter. The accuracy of the procedure will not be unduly impaired if the actual rate of production is replaced by the rate marked on the Gilbert gradient curve that stands closest to the actual rate. The values in question are entered in Column 3; Column 4 lists the *GOR*s required according to the pressure gradient curves to produce liquid at the rate in question if the tubing pressure at valve depth during continuous flow gas lift is p_{Tfi}; Column 5 gives the products of the corresponding entries of Columns 3 and 4. The p_{Tfi} and p_{Ci} data underlying the entries in Column 6 have been read off Fig. 2.4—11. The correction whose results are given in Column 7 is necessary because of overstepping the critical pressure ratio. The values of C have been taken from Tables 2.4—4 and 2.4—5. The $d_{ch\,i}$ values of Column 9 have been calculated by means of Eq. 1.5—4; Column 11 lists the choke diameters of the valves, type J–20, as established from the manufacturer's catalog. The opening equations of the valves chosen are (according to data found in the CAMCO Gas Lift Manual, by Winkler and Smith 1962, p. A4—001), for the 3/16 in. valve, $p_C = \frac{p_k}{0.962} - 0.040\, p_T$; for the 5/16 in. valve, $p_C = \frac{p_k}{0.896} - 0.116 p_T$. Columns 12 and 13 contain the charged pressures at depth and on the surface, corresponding to the values calculated in step (11).

Table 2.4 — 7

Number of valve	p_{wf}	q_o	q_o Gilbert	R_{go}	q_{gn}	p_{Tf}/P_C	p_{Tf}/P_C corr.	C	d_{ch} calc.	d_{ch}	d_{ch} chosen	p_k	p_{kn}
	bars	10^{-5} m³/s	10^{-5} m³/s	m³/m³	10^{-2} m³/s	—	—	—	mm	in.	in.	bars	bars
	1	2	3	4	5	6	7	8	9	10	11	12	13
1	88.4	7.04	9.14	250	1.76	0.272	0.555	0.465	1.57	1/16	3/16	46.1	43.9
2	64.7	49.6	72.3	150	7.44	0.427	0.555	0.465	3.24	5/32	3/16	47.1	42.8
3	46.6	82.2	72.3	150	12.33	0.543	0.555	0.465	4.21	3/16	3/16	48.4	42.5
4	32.9	106.8	110.3	262	27.98	0.635	0.635	0.457	6.45	9/32	5/16	48.4	41.5

2.4.2. Intermittent gas lift

(a) Theory of production; factors affecting operation

The simplest completion used in intermittent gas lift operation is shown in Fig. 2.4—12. The casing annulus is filled through regulating device *1* (called the surface controller) with injection gas at a predetermined pressure. Valve *2* is usually pressure-controlled; its operation is similar to that of the unloading valve shown in Fig. 2.4—7. The annulus is packed off at the tubing shoe by means of packer *3*. There is a standing valve *4* in the tubing string. By suitably regulating the casing pressure from the surface, it is possible to make the operating valve open only when a liquid column of sufficient height has built up above it. Three distinct phases of lift can be distinguished. We shall analyse them one by one, largely after White et al. (1963).

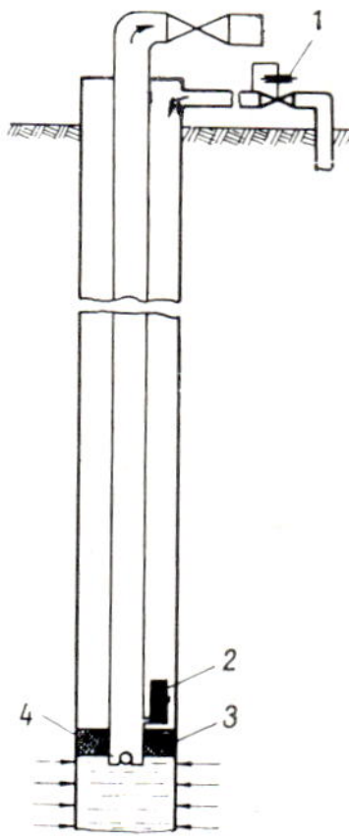

Fig. 2.4—12 Intermittent gas-lift installation

First phase. Start of flow. The pressure difference $(p_C - p_T)$ makes gas flow from the casing through the gas lift valve into the tubing. If the valve is a snap-acting one and the operating parameters are suitably chosen, then the lifted liquid slug will soon enough (in 10 sec or so) attain a constant (terminal) velocity. This is the end of the first phase. The opening pressure of the valve is

$$p_C = p_k + \Delta p_{gl}, \qquad 2.4\text{—}8$$

where Δp_{gl} is the valve spread, depending on the valve characteristic and the tubing pressure (cf. point (i) in paragraph 2.4.1a.2). At the same time, the tubing pressure opposite the valve is

$$p_T = h_{la}\,\gamma_l + p_{TO} + p_g, \qquad 2.4\text{—}9$$

where h_{la} is the length of the liquid column accumulated above the gas lift valve; p_g, the static pressure of the gas column above the liquid slug, is negligible in most cases.

Second phase. The liquid slug moves up the tubing at the practically constant velocity v_l. This phase ends when the top of the slug surfaces. The length of the liquid slug will gradually decrease during the lift period, because of the break-through of injection gas and the fallback (in the form of mist and of a liquid film covering the tubing wall) of the aerated tail of the liquid slug. According to White et al. (1963), in a tubing of a given size, the velocity of gas slippage v_{gs} is independent of the velocity v_l of the liquid slug, and is controlled only to some extent by the physical parameters of the liquid. Figure 2.4—13 shows gas slippage velocity in a 2 3/8-in. size tubing with a 12.6 mm bore choke to be about 0.6 m/s if the liquid is oil, and 1.1 m/s if it is salt water. The velocity of the liquid slug, v_l, depends on the velocity of the gas column lifting it.

The latter varies as shown in Fig. 2.4—13 v. the pressure ratio p_C/p_T for a given tubing size d_T and valve port size d_{ch}. Let the gas volume in the tubing below the liquid slug be V_1 at the instant t_1 when the injection into the tubing has already ceased. Then, at the instant $(t_1 + t)$, the gas volume will increase by expansion to

$$V_2 = V_1 + A_T v_l t.$$

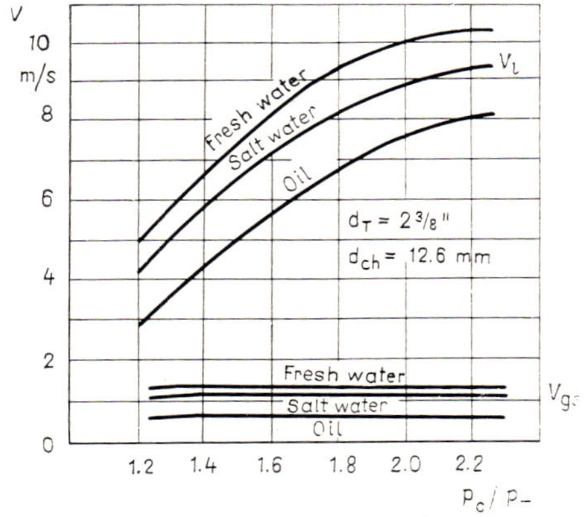

Fig. 2.4—13 Velocity of liquid slug and of slippage v. p_c/p_T, after White (1964)

Assuming that the weight of the gas column in the tubing is negligible, the volume change of the gas is isothermal, and the deviation from the perfect gas laws is evanescent, the pressure of the gas column will be

$$p_2 = \frac{p_1 V_1}{V_1 + A_T v_l t}.$$

The pressure change per unit of time can be determined by differentiating the above, that is,

$$\frac{dp}{dt} = -\frac{p_1 V_1 A_T v_l}{(V_1 + A_T v_l t)^2}. \qquad 2.4-10$$

During this same interval of time t, the volume of the liquid slug above the gas column will decrease owing to aeration from V_{l1} to

$$V_{l2} = V_{l1} - v_{gs} A_T t.$$

The volume change per unit of time can be obtained by differentiating the above:

$$\frac{dV_l}{dt_t} = -v_{gs} A_T . \qquad 2.4-11$$

Now since

$$\frac{dp}{dt} = \frac{dp}{dV_l}\frac{dV_l}{dt},$$

the introduction of Eqs 2.4—10 and 2.4—11 results in

$$\frac{dV_l}{dp} = \frac{(V_1 + v_l A_T t)^2 v_{gs}}{p_1 V_1 v_l}. \qquad 2.4-12$$

This equation is an approximate expression of volume loss of the liquid slug as a result of unity pressure decrease in the gas column.

Third phase. This starts when the top of the liquid slug surfaces and ends when the gasless liquid slug of length h_{lo} leaves the tubing. Let the volume of the gas below the liquid slug be V_1 at the instant t_1'. Assuming that no further gas is injected in the tubing after that instant, let the gas expand to a volume V_2 over a time span t. The pressure decrease per unit of time can be characterized by Eq. 2.4—10. Now, however, the liquid volume above the gas column is decreased not only owing to aeration but also because of outflow into the flow line. Hence, at an instant $(t_1'+t)$, the liquid volume will be

$$V_{l2} = V_{l1} - v_l A_T t - v_{gs} A_T t.$$

Differentiating and rearranging, we may write

$$\frac{dV_l}{dt} = - A_T(v_{gs} + v_l),$$

and by analogy to Eq. 2.4—12,

$$\frac{dV_l}{dp} = \frac{(V_1 + v_l A_T t)^2 (v_l + v_{gs})}{p_1 V_1 v_l}. \qquad 2.4\text{—}13$$

Comparison of Eqs 2.4—12 and 2.4—13 reveals that the volume decrease of the liquid slug per unit pressure drop, and hence, the specific decrease of its hydrostatic pressure, is greater in the third phase than in the second.

The above considerations permit us to formulate several criteria of modern intermittent gas lift production. In order to provide the greatest possible recovery h_{lo}/h_{la} against the starting slug length h_{la}, it is necessary that the liquid be lifted as fast as possible. Given a certain gas injection rate, this can be ensured in the first phase of production by accelerating the liquid column to its terminal velocity v_l within the shortest possible time. The valve should be snap-acting (see paragraph 2.4.3a.1), that is, after its opening, gas should be able to flow at once through the entire cross-section of the valve. In the second and third phases of production, the liquid velocity should not decrease. In the interest of this, (i) the injection of gas into the tubing should be shut off when the volume and pressure of gas in the tubing are already sufficient to keep lifting the liquid slug of decreasing hydrostatic pressure at unchanged speed, (ii) the wellhead should present the least possible resistance to flow.

(i) Comparison of Eqs 2.4—12 and 2.4—13 reveals that the decrease per unit pressure drop of both the volume and the hydrostatic pressure of the liquid slug may be much greater in the third than in the second phase. The probability that the expansion of the gas in the tubing will lift the liquid slug without decrease of velocity is thus greater in the third phase than in the second. It is therefore expedient to examine whether the injection of gas into the tubing can be stopped at the beginning of the third phase. Let a gas pressure p_T prevail below the liquid slug of hydrostatic pressure $h_{lu} \gamma_l = p_h$ at the instant when the gas lift valve closes. Let us assume that a pressure ratio p_h/p_T is sufficient to provide the required slug velocity. Subtracting the same pressure drop, Δp, from both the denominator and the numerator of this ratio, whose value is less than unity, we obtain another pressure ratio,

$$\frac{p_h - \Delta p}{p_T - \Delta p} = \frac{p_l'}{p_T'},$$

which is necessarily less than p_h/p_T. Thus the length per unit gas pressure of the liquid column lifted will decrease, or, in other words, the pressure gradient lifting the liquid slug will increase. The slug velocity cannot, therefore, decrease if its pressure decreases by at least as much as that of the gas column lifting it. — At the beginning of the third phase, let the tubing volume filled with gas of pressure p_1 be $V_1 = (L - h_{lu})A_T$. At the end of this phase, when the slug has just passed the wellhead, let the gas-filled volume be $V_2 = (L_T - h_{lr})A_T$; gas pressure will thus be (assuming a perfect gas for simplicity)

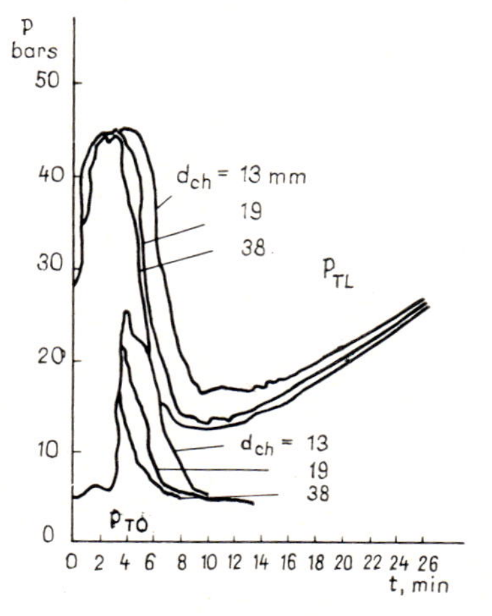

Fig. 2.4—14 Influence of wellhead choke size upon intermittent gas lift, after Beadle et al. (1963)

$$p_2 = p_1 \frac{(L_T - h_{lu})\, A_T}{(L_T - h_{lr})\, A_T}\,.$$

The pressure drop in the gas will be

$$\Delta p = p_1 - p_2 = p_1 \left(1 - \frac{L - h_{lu}}{L - h_{lr}}\right). \qquad 2.4\text{—}14$$

If this pressure drop is less than the hydrostatic pressure of the liquid slug of length h_{lO} lifted out of the well, that is, if $\Delta p \leqq h_{lO}\,\gamma_l$, then, by the above consideration, the slug velocity will not decrease. From this viewpoint, then, injection gas supply is correctly controlled if the gas lift valve closes at the instant when the top of the liquid slug surfaces. If, on the other hand, $\Delta p > h_{lO}\,\gamma_l$, then the gas lift valve should close after the onset of the third phase. This may be the case when the h_{la}/L_T ratio is comparatively great. Injection gas supply to the tubing is sometimes kept up too long, e.g. until the entire liquid slug has left the wellhead, or even longer. It is easy to see that this may increase the specific injection gas requirement rather far above the necessary minimum.

(ii) Except if unavoidable, do not install a production choke in the wellhead, nor any other wellhead equipment of high resistance to flow. Experiment has revealed (Beadle et al. 1963) that, the smaller the choke bore, the higher will be the maximum producing *BHP*, and the less liquid will be produced. The experimental data, plotted by the authors in graph form (Fig. 2.4—14) make it obvious that, the less the choke diameter, the greater will be the pressure at both ends of the tubing, and the slower will these pressures decrease. Some production parameters describing the case illustrated are compiled in Table 2.4—8.—Let us add that, if the liquid volume produced per cycle is high and the valve dome pressure p_k is great, and the separator is close to the producing well, it may be necessary for reasons of safety to install a production choke. This, however, should be at the separator station rather than in the wellhead. — If the separator station is at a higher elevation than the wellhead, then some of the liquid

Table 2.4—8

d_{ch} mm	h_{lo}/h_{la}	$p_{TO\,min}$	p_{TL}
		bars	
38	73	13.4	23.7
19	69	14.1	25.1
13	59	17.3	27.1

produced may have a tendency to flow back into the well. To prevent this, a reverse check valve of large liquid throughput capacity should be installed at the wellhead.

According to Muraviev and Krylov (1949), the prime consideration in designing an intermittent gas lift installation is the economical exploitation of pressure energy. Production is optimal if the specific injection gas requirement is a minimum. This occurs when slippage and friction losses are least as related to the total energy consumption. Total losses do not depend on the length of the liquid column to be lifted. True, a greater length h_{la} of the liquid slug entails a greater friction loss, but the liquid fallback will decrease by the same amount. Total loss of energy is significantly influenced, however, by the mean flow velocity of the liquid slug. There is an optimal rate of injection $q_{g\,opt}$ at which total loss is least. This is, according to Krylov, the goal to be attained.

(b) Intermittent gas-lift design

The complicated, transient nature of flow in an intermittent gas lift installation precludes an exact prediction of operating parameters. There are, however, several procedures permitting approximate estimations. We shall now discuss first the Winkler and Smith relationships (Winkler 1962), with certain modifications. These relationships provide a rapid first approach to a problem. If the conditions are not overly unfavourable (such as a small production choke or considerable emulsification), then the length of the liquid column lost owing to fallback is

$$1.6\times10^{-4}\, h_{la} L_T < h_{lr} < 2.3\times10^{-4}\, h_{la} L_T. \qquad 2.4-15$$

Let e.g. $h_{la} = 100$ m and $L_T = 1500$ m; then

$$h_{lr} = 1.6\times10^{-4}\times100\times1500 = 24 \text{ m},$$

and

$$h_{lr} = 2.3\times10^{-4}\times100\times1500 = 35 \text{ m}.$$

Of an accumulated liquid column of 100 m length, a column of length $h_{lo} = (h_{la} - h_{lr}) = 76$ (or 65) m can be lifted out of the well. The desirable lift velocity of the liquid slug is 5 m/s; the approximate lift duration is, consequently,

$$t_1 = 0.2L\,.$$

In a well 1500 m deep, for instance, $t_1 = 0.2 \times 1500 = 300$ s. The specific injection gas requirement is, for a conventional installation,

$$0.12L < R_{gl} < 0.24\,L \quad (\mathrm{m^3/m^3}).$$

The specific injection gas requirement may be greater than this if, at the given L_T, the casing pressure p_C is comparatively low and the tubing diameter d is comparatively great. In chamber installations,

$$0.12\,L < R_{gl} < 0.18\,L \quad (\mathrm{m^3/m^3}).$$

For instance, in a well 1500 m deep, a specific injection gas requirement between 180 and 360 $\mathrm{m^3/m^3}$ may be expected if the installation is conventional, and between 180 and 270 $\mathrm{m^3/m^3}$ in a chamber installation.

The surface closing pressure of the so-called operating valve, the valve controlling the intermittent lift in the well, is, for wells 900—2400 m deep,

$$p_{kO} = 2.3 \times 10^3 L_i \quad (\mathrm{N/m^2}),$$

and the available line pressure at the wellhead should exceed this by 7—10 bars. For example, the wellhead closing pressure of the operating valve in a well 1500 m deep should be

$$p_{kO} = 2.3 \times 10^3 \times 1500 = 3.45 \times 10^6 \ \mathrm{N/m^2} = 34.5 \ \text{bars},$$

and the available line pressure at the wellhead should be at least

$$34.5 \times 10^5 + 7 \times 10^5 = 41.5 \times 10^5 \ \mathrm{N/m^2} = 41.5 \ \text{bars}.$$

At the opening of the operating valve, the ratio of the casing and tubing pressures, p_{Ci} and p_{Ti}, is

$$1.3 < \frac{p_{Ci}}{p_{Ti}} < 2.0\,.$$

If we aim at realizing production parameters optimal in the sense of Muraviev and Krylov (1949), then the gas injection rate should be

$$q_{g\,\mathrm{opt}} = 0.474\,d^2 \sqrt[3]{L^2} \quad (\mathrm{m^3/s}). \qquad 2.4\text{—}16$$

In this case, the column length lost due to fallback is

$$h_{lr\,\mathrm{opt}} = \frac{80}{\sqrt{d}\,\varrho_l} \sqrt[3]{L^2} \ (\mathrm{m}), \qquad 2.4\text{—}17$$

and the pressure drop due to friction is

$$p_{s\,\mathrm{opt}} = \frac{100\,L}{\sqrt{d}} \ (\mathrm{N/m^2})\,. \qquad 2.4\text{—}18$$

Intermittent gas lift design means in principle the choice of a well completion that will lift oil at the desired rate at the lowest possible specific cost. In the case of a single completion, the installation may be either a conventional one, or one with a downhole chamber (Figs 2.4—12 and 2.4—40, respectively). We shall discuss below the design of a conventional intermittent gas lift installation. The statements to be made apply with slight modifications also to chamber installations. To find the most favourable tubing size, the procedure to be described below is to be performed for several tubing sizes; that size is to be chosen which gives the least specific cost of production or the least specific injection gas requirement for a given injection-gas line pressure. Taking the given gas line pressure and choosing a certain tubing size we design the unloading valves in much the same way as for a continuous-flow well. The differences in design principles are: (i) Valves should be instant-closing and opening (snap-action), that is, they are usually pilot-operated valves (see paragraph 2.4.3a.1). (ii) Their gas throughput capacity should be comparatively large. In order to prevent a significant fall-back of liquid, the following valve port sizes are recommended by Brown (1967): 10—14 mm for 1.9 in. tubing size, 13—17 mm for 2 3/8 in., 14—20 mm for 2 7/8 in. tubing size. — Figure 2.4—15, likewise after Brown, shows the recovery of the starting slug length v. the valve port area, under given conditions. (iii) The least back-pressure in the tubing, liable to come about during unloading, is determined by assuming that the well produces the desired rate by continuous flow. (iv) It is to be decided whether we want the upper valves to open once the liquid slug has passed them (multipoint injection) or not (single-point injection). In the latter case, the dome pressure of the operating valve is to be reduced accordingly. The choice of multipoint v. single-point injection shall be discussed later on.

If the valves are unbalanced (see later), then determining dome pressures of the unloading valves is usually based on one of the following alternatives (Brown 1967): surface closing pressures decreased downward by 0.7 bar per valve; all surface closing pressures are identical; surface closing pressures increased downward; the opening pressures at 1 bar back pressure, as measured in a valve tester, are equal for all valves; the surface opening pressures of all valves are equal: they decrease downward by 0.7—1.7 bar per valve. In the case when the surface closing pressures decrease downward by 0.7 bar per valve and injection is of the single-point type, design proceeds in the following steps.

(1) In a bilinear-orthogonal system of coordinates, p v. L, shown in Fig. 2.4—16, starting from a point ($L = 0$, p_{TO}), we trace the pressure gradient curve which would apply to the given tubing if the well were produced through that tubing at the prescribed rate, at the optimal *GOR* as defined by Gilbert, that is, at the least pressure at any given depth (Graph I). (2) Starting from the point ($L_T = 0$, p_{CO}) we trace an injection gas pressure traverse in the annulus for the instant when the top valve closes. Let p_{CO} be less by 7 bars than the injection gas pressure available at the surface (Graph II). (3) Starting from ($L_T = 0$, p_{TO}), we trace a line $p = L\gamma_l$ defined by the gravity γ_l of the gasless oil. Where this line, Graph IV/1, intersects Graph II, the gas pressure in the annulus is equal to or greater than the pressure in a tubing of wellhead pressure p_{TO}, filled up to the wellhead with gasless oil. When injection gas starts to flow through the valve, pressure in the annulus is higher by 7 bars than this pressure in the tubing.

The pressure differential required to start the gas flowing through the valve is thus assured. At the instant the valve closes, pressure in the tubing opposite the valve is p_{T1}. (4) Let us draw a parallel to Graph IV/1 through the point (L_1, p_{T1}) (Graph IV/2). Let us draw further a parallel to Graph II, the annulus pressure traverse, at a distance corresponding to a pressure difference of 0.7 bar, in order to ensure that the surface closing pressure of valve *2* be less by approximately 0.7 bar than the closing pressure of valve *1*. This latter line intersects Graph IV/2 at the point *2*. The ordinate

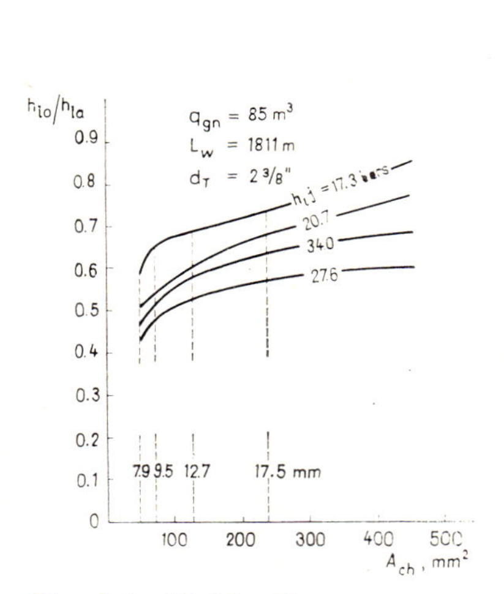

Fig. 2.4—15 Liquid recovery v. wellhead choke area, after Brown 1967 (by permission of Prentice-Hall, Inc., Englewood Cliffs, New Jersey, USA)

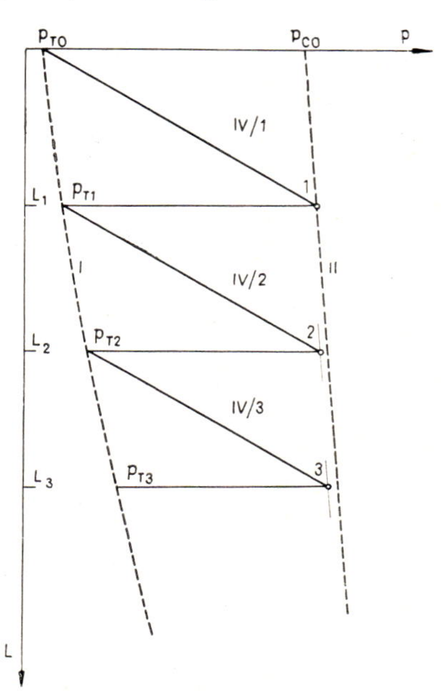

Fig. 2.4—16 Valve spacing for intermittent gas-lift well, after Brown (1967)

of this point defines the depth of valve *2*. (5) The depth of the remaining valves are established similarly to valve *2*. The annulus pressure is decreased by a further 0.7 bar per valve. (6) Let the dome pressure of the lowermost valve be less by 3.6 bars than that of the next valve above. The maximum pressure p_{Toi} permissible opposite the lowermost valve at the instant of its opening is established on the assumption that its value should be about two thirds of the dome pressure in said valve. (7) Using the opening equation of the lowermost valve, let us calculate the casing pressure p_{Coi} required to open it. As a result of the procedure under (6), the ratio p_{Coi}/p_{Toi} will be greater than 1.5. (8) If the upper valves are not to be opened by the rising liquid slug, we have to check whether the opening casing pressures of these valves are not, in the extreme case that is most conducive to their opening, less than the actual casing pressures opposite them. Let us assume that, in the said extreme case, the pressure in the tubing equals the opening casing pressure of the lowermost valve, minus a gas-flow pressure drop of, say, 3.4 bars. If under the influence of this tubing pressure p_T, and a casing pressure p_C equal to the opening pressure of the lowermost valve, the unloading valve directly above the operating valve will not open, then the dome pressure of the operating valve has been correctly chosen. It is moreover clear that the valves farther up will not open, either. However, if the lowermost unloading valve does open, then the dome pressure of the operating valve is to be reduced and the checking procedure repeated. (9) Using Eq. 2.4—2, we determine the charged dome pressures p_{kni} of each individual valve. (10) Using Eq. 2.4—3 we determine, on the assumption that $p_{kOi} = p_{CL}$, the surface closing pressures of the individual valves. (11) We determine the daily liquid production of the well. Flow in the drawdown area of an intermittent well is invariably transient. The production cannot in the general case be predicted with certainty. For purposes of an estimate, the liquid flowing into the well is considered incompressible. The productivity equation No. 2.1—7 valid for steady-state flow can then be applied. Let us further assume that the exponent n of the productivity equation equals unity; then

$$q_0 = J(p_{ws} - p_{wf}) = J\Delta p.$$

Assuming an installation similar to the one in Fig. 2.4—12, we may write up that liquid flowing into the well at a rate q_o over a period of time dt results in an increase of liquid volume by $A_T dh$ in a tubing of cross-sectional area A_T; that is,

$$A_T dh = q_o dt. \qquad 2.4\text{—}19$$

At the instant considered, bottom-hole pressure is

$$p_{wf} = h_l \gamma_l + p_{TO} + p_g, \qquad 2.4\text{—}20$$

where p_g is the pressure of the gas column above the liquid slug, and

$$h_l = \frac{p_w - p_{TO} - p_g}{\gamma_l}.$$

Since p_{TO} is constant, and p_g can also be considered approximately constant,

$$dh_l = \frac{dp_{wf}}{\gamma_l}.$$

Substitution of the expressions of q_o and dh_l into Eq. 2.4—19 yields, after rearranging and writing up an integration,

$$\int_{p_{wf_1}}^{p_{wf_2}} \frac{A_T dp_{wf}}{\gamma_l} = \int_{t_1}^{t_2} J(p_{ws} - p_{wf})\, dt.$$

Figure 2.4—17 shows the variation of p_{wf} v. time (full line). The limits of integration are identified in the Figure. Solution of this equation yields the increased bottom-hole pressure at the instant t_2' as

$$p_{wf2} = p_{ws} - (p_{ws} - p_{wf})e^{-\frac{J\gamma_l t_2'}{A_T}} .$$

The equation can be transformed to yield the drawdown

$$\Delta p_{w2} = (p_{ws} - p_{wf2})$$

by writing

$$\Delta p_{w2} = \Delta p_{w1}\, e^{-\frac{J\gamma_l t_2'}{A_T}} . \qquad 2.4-21$$

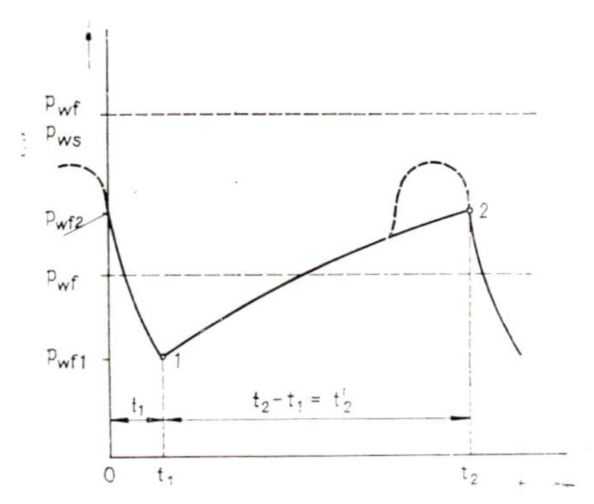

Fig. 2.4—17 Bottom-hole pressure v. time in intermittent gas-lift well

Clearly, the mean drawdown over the period of accumulation of the liquid is most expediently expressed as a logarithmic average, namely

$$\Delta p_w = \frac{\Delta p_{w1} - \Delta p_{w2}}{\ln \dfrac{\Delta p_{w1}}{\Delta p_{w2}}} . \qquad 2.4-22$$

If — as happens rather often — the production period t_1 is short as compared to the accumulation period t_2' (Fig. 2.4—17), then Δp_w applies in a fair enough approximation to the full intermittent cycle, and the daily liquid production turns out to be

$$q_o' = J\Delta p_w \quad (\mathrm{m^3/d}). \qquad 2.4-23$$

Drawdown at the beginning of the accumulation period is, if p_{wf1} is taken according to Eq. 2.4—20,

$$\Delta p_{w1} = p_{ws} - h_{lr}\gamma_l - p_{TO} - p_{g1} . \qquad 2.4-24$$

At the beginning of the accumulation period, the liquid column above the valve may be higher than h_{lr} because, e.g. in the case of an insert chamber installation, liquid will flow into the well even during the production period. This is not usually taken into account in calculations. The less t_1/t_2', the less the 'increment of h_{lr}'. Regardless of that, h_{lr} can be estimated only, e.g. using Eqs 2.4—15 and 2.4—17. Relationships for establishing $h_{lo} = (h_{la} - h_{lr})$ have been derived by White et al., (1963) who have also prepared nomograms based on their equations. These provide the ratio

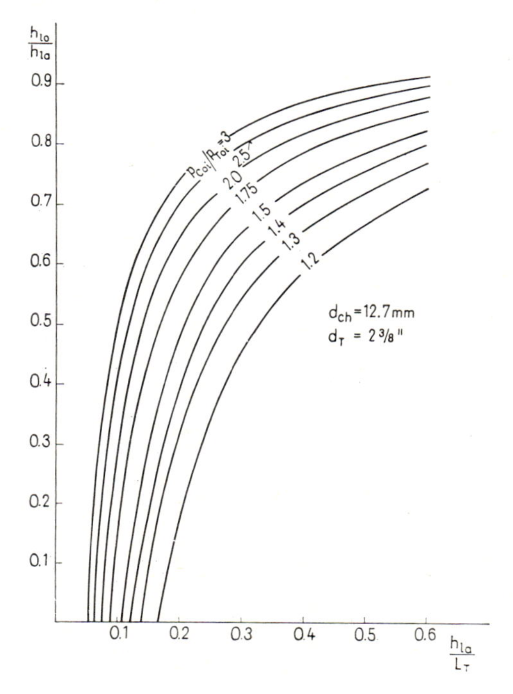

Fig. 2.4—18 Finding the liquid recovery of intermittent gas-lift wells, after White (1964)

h_{lo}/h_{la} for various values of p_{Cli}/p_{Tli} and h_{la}/L_T. One such diagram is shown as Fig. 2.4—18. Drawdown at the end of the accumulation period is estimated at

$$\Delta p_{w2} = p_{ws} - h_{la}\gamma_l - p_{TO} - p_{g2}. \qquad 2.4-25$$

So far we have tacitly assumed the operating valve to be installed at the bottom of the well. If it is higher up, then the bottom-hole pressure will be higher than the above-calculated value by the pressure of the liquid column between the well bottom and the valve. (12) The injection gas requirement can now be calculated. The gas used up in any production cycle equals the quantity of gas fed by the well to the flow line after the lifting of a liquid slug of length h_{lo}. By our considerations connected with Eq. 2.4—14, let us assume that the operating valve closes at the instant when the top of the liquid slug has surfaced. The pressure under the liquid slug then equals the wellhead pressure p_{TO} plus the hydrostatic pressure of the liquid column of length h_{lo} in a fair approximation. The tubing pressure opposite the operating valve is at the same instant equal to the dome (that is, closing) pressure of the valve, minus the pressure drop of gas flow through the valve, Δp. The mean pressure of the gas column is, then,

$$\bar{p}_T = \frac{p_{TO} + h_{lo}\gamma_l + p_{ki} - \Delta p}{2}$$

and the gas-filled volume of the tubing is

$$V_T = (L_T - h_{la})A_T.$$

By the combined gas laws, the standard-state volume of the gas in the tubing is

$$V_{gn} = \frac{V_T \bar{p}_T z_n T_n}{\bar{T}_T \bar{z}_T p_n}.$$

We have tacitly assumed above — in a tolerable approximation of the actual situation — that the gravity of the liquid is independent of temperature and pressure and that no gas dissolves in the liquid. The gas-filled volume in the tubing after the period of production is

$$V'_T = (L_T - h_{lr})A_T.$$

The mean pressure of the gas column is

$$\bar{p}_T = \frac{p_{TO} + p_{TO} + p'_g}{2}.$$

The standard volume of the remaining gas is

$$V'_{gn} = \frac{V'_T \bar{p}'_T z_n T_n}{\bar{T}'_T \bar{z}'_T p_n}.$$

Gas volume used per cycle is, then,

$$V_{gcn} = V_{gn} - V'_{gn}.$$

If — as is frequently the case — the wellhead pressure p_{TO} is comparatively low, then sufficient accuracy can be achieved by putting V'_{gn} equal to zero. Let further $z_n = 1$; then,

$$V_{gcn} = \frac{V_T \bar{p}_T T_n}{\bar{T}_T \bar{z}_T p_n}. \qquad 2.4\text{—}26$$

The specific injection gas requirement is the ratio of the gas volume used for one intermittent cycle and the liquid volume produced, that is,

$$R_{gl} = \frac{V_{gcn}}{h_{lo} A_T}. \qquad 2.4\text{—}27$$

The theoretical specific gas requirement of daily liquid production is equal to this value. In practice, the actual specific requirement may be greater, e.g. if the tubing and/or casing string is leaking or the operating valve closes only after the top of the liquid slug has left the well. Daily injection-gas requirement is

$$q_{gn} = R_{gl} q_o \,. \qquad 2.4-28$$

(13) Determining the daily number of cycles. Daily cycle number is

$$n = \frac{86{,}400}{t_1 + t_2'} \,. \qquad 2.4-29$$

By the considerations at the beginning of Section 2.4—2b, the duration of the lift period, estimated from the slug velocity, is

$$t_1 = 0.2 L_T \,.$$

By Eq. 2.4—21,

$$t_2' = \frac{A_T}{J \gamma_l} \ln \frac{\Delta p_{w1}}{\Delta p_{w2}} \,. \qquad 2.4-21a$$

(14) By the above line of thought, well production, injection gas requirement and daily cycle number can at best only be estimated. Another circumstance to be reckoned with is that the performances of well and formation will change in time. The unloading valves and especially the operating valve must therefore operate satisfactorily not only at the calculated production parameters but also at slightly different ones. It is expedient to consider the calculated production parameters as belonging to the maximum possible starting slug length accumulating prior to opening. If the actual length h_{la} of the liquid slug to be lifted is less than that, then the opening casing pressure of the operating valve will turn out higher. It may increase until it attains the opening pressure, valid during the rise of the liquid slug, of the next valve above. This is the opening pressure that determines the least length of the liquid column to be produced.

Example 2.4—8. Design an intermittent gas lift installation by the procedure just discussed for a well completion as shown in Fig. 2.4—12, if $L_T = 1640$ m; $d = 2\ 7/8$ in.; $p_{ws} = 122.1$ bars; $J = 3.7 \times 10^{-11} \frac{m^3}{s} \Big/ \frac{N}{m^2}$; $p_{TO} = 2.0$ bars; $p_{gl} = 45.1$ bars; estimated oil production rate, $q_o = 30$ m³/d; $M_g = 21.7$ kg/kmole; $\varrho_o = 850$ kg/m³; $T_{CO} = 11.0$ °C; $\sigma_b = 3.88 \times 10^{-2}$ K/m; $p_n = 1.01$ bar; $T_n = 15$ °C. The opening equation is $p_C = 1.17 p_k - 0.17 p_T$ for all gas lift valves.

In the order of the steps outlined above, the solution is found as follows. (1) We take as a basis that Gilbertian set of curves for 2 7/8-in. tubing which holds for the liquid production rate q_o closest to the expected production. The value in question is $q_o = 31.7$ m³/d. The least tubing pressure at any depth is ensured by a *GOR* of 765 m³/m³. The corresponding section of the curve having this parameter is copied onto Fig. 2.4—19 (Graph I). (2) Let

$$p_{kO} = p_{gl} - 7 \times 10^5 = 45.1 \times 10^5 - 7 \times 10^5 = 38.1 \times 10^5 \text{ N/m}^2 \,.$$

Starting from this pressure and using Eq. 2.4—3, we calculate the casing pressure v. depth curve (Graph II). (3) Using a pressure gradient of 850 kg/m³/m corresponding to a density of 850 kg/m³, we draw a pressure traverse for gasless oil in the tubing, starting from the point defined by the coordinates ($L = 0$, $p_{TO} = 2.0$ bars; Graph IV/1). (4) and (5) Let us determine graphically the depths of valves from *1* to *4*. Column 1 of Table 2.4—9 lists the valve depths read off Fig. 2.4—19; Column 2 gives the dome pressures at the respective depths of installation. (6):

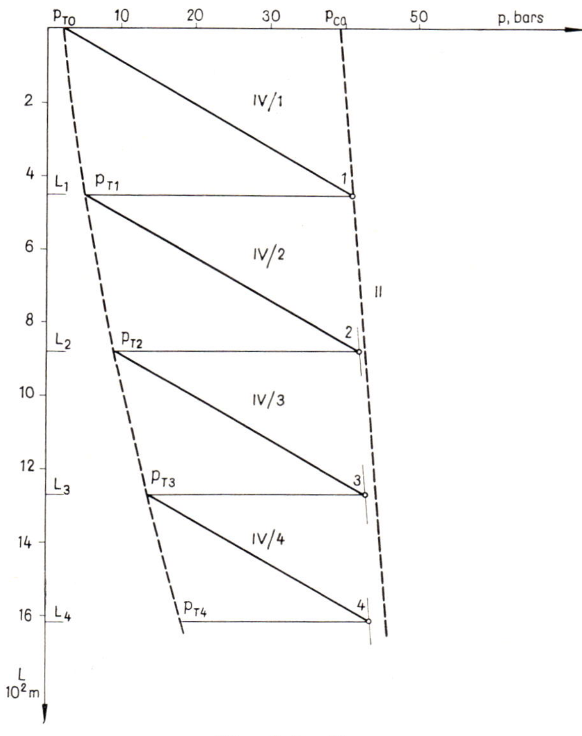

Fig. 2.4—19

$$p_{T4} = \frac{38.1 \times 10^5}{1.5} = 25.4 \times 10^5 \text{ N/m}^2 .$$

(7)

$$p_{C4} = 1.17 \times 38.1 \times 10^5 - 0.17 \times 25.4 \times 10^5 = 40.2 \times 10^5 \text{N/m}^2 .$$

(8) Let

$$p_{T3} = 40.2 \times 10^5 - 3.4 \times 10^5 = 36.8 \times 10^5 \text{N/m}^2 .$$

The least casing pressure required to open valve *3* is, then,

$$p_{C3} = 1.17 \times 41.7 \times 10^5 - 0.17 \times 36.8 \times 10^5 = 42.5 \times 10^5 \text{ N/m}^2 ,$$

and since

$$42.5 \times 10^5 > 40.2 \times 10^5 ,$$

valve *3* will not open when the liquid slug has passed it. (9) Column 6 of Table 2.4—9 gives the temperature at the depth of each valve; Column 7 lists the charged dome pressures calculated by means of Eq. 2.4—2. (10) The surface closing pressures calculated using Eq. 2.4—3 are shown in Column 8 of Table 2.4—9. (11) At the instant when the liquid slug starts to

Table 2.4—9

Serial number	L	p_k	p_T	p_C	$p_{T\,min}$	T	p_{kn}	p_{kO}
	m	bars				°C	bars	
	1	2	3	4	5	6	7	8
1	452	39.9				28.5	38.2	38.2
2	882	40.9				45.2	36.9	37.7
3	1275	41.7	36.8	42.5		60.5	36.0	36.9
4	1627	38.1	25.4	40.2	13.7	73.5	31.6	32.7

rise, tubing pressure opposite valve *4* is $p_{T4} = p_{Tf2} = 25.4$ bars. Using Eq. 2.4—20 and assuming $p_g = 0$, we have

$$h_{la} = \frac{25.4 \times 10^5 - 2.0 \times 10^5}{850 \times 9.81} = 281 \text{ m}.$$

The length of the liquid column lost by fallback is, by Eq. 2.4—15, and for a constant equal to 2×10^{-4},

$$h_{lr} = 2 \times 10^{-4} \times 281 \times 1610 = 90 \text{ m}.$$

By Eq. 2.4—17,

$$h_{lr} = \frac{80}{\sqrt{0.062 \quad 850}} \sqrt[3]{1610^2} = 52 \text{ m}.$$

Let us choose the less favourable $h_{lr} = 90$ m. By Eq. 2.4—20, and assuming $p_g = 0$, we have with reference to the bottom of the tubing

$$p_{Tf1} = 90 \times 850 \times 9.81 + 2.0 \times 10^5 = 9.5 \times 10^5 \text{ N/m}^2 .$$

The pressure of the liquid column between the well bottom and valve *4* is, provided density is 850 kg/m³ also here,

$$\Delta L\, \gamma_l = (1640 - 1627) 850 \times 9.81 = 1.1 \times 10^5 \text{ N/m}^2.$$

Referring this to the well bottom, we get

$$p_{w1} = p_{Tf1} + \Delta L\, \gamma_l = 9.5 \times 10^5 + 1.1 \times 10^5 = 10.6 \times 10^5 \text{ N/m}^2,$$

and

$$p_{w2} = p_{Tf2} + \Delta L\, \gamma_l = 25.4 \times 10^5 + 1.1 \times 10^5 = 26.5 \times 10^5 \text{ N/m}^2 ;$$

Hence

$$\Delta p_{w1} = 122.1 \times 10^5 - 10.6 \times 10^5 = 111.5 \text{ N/m}^2 ,$$

and

$$\Delta p_{w2} = 122.1 \times 10^5 - 26.5 \times 10^5 = 95.6 \times 10^5 \text{ N/m}^2 .$$

By Eq. 2.4—22, the mean drawdown is

$$\Delta \bar{p}_w = \frac{111.5 \times 10^5 - 95.6 \times 10^5}{\ln \dfrac{111.5 \times 10^5}{95.6 \times 10^5}} = 102.8 \times 10^5 \text{ N/m}^2 .$$

By Eq. 2.4—23, the daily liquid production is

$$q_0 = 86{,}400 \times 3.7 \times 10^{-11} \times 102.8 \times 10^5 = 32.8 \text{ m}^3\text{/d} .$$

(12) The value of $\bar{p}_T$ figuring in Eq. 2.4—26

$$\bar{p}_T = \frac{2 \times 10^5 + (281 - 90)850 \times 9.81 + 38.1 \times 10^5 - 3.4 \times 10^5}{2} =$$

$$= 26.2 \times 10^5 \text{ N/m}^2 ,$$

and

$$V_t = (1627 - 90)30.2 \times 10^{-4} = 4.64 \text{ m}^3;$$

hence,

$$V_{gcn} = \frac{4.64 \times 26.2 \times 10^5 \times 288.2}{315.5 \times 0.92 \times 1.01 \times 10^5} = 119 \text{ m}^3 .$$

By Eq. 2.4—27,

$$R_{gl} = \frac{119}{191 \times 30.2 \times 10^{-4}} = 206 \text{ m}^3\text{/m}^3 .$$

By Eq. 2.4—28,

$$q_{gn} = 206 \times 32.9 = 6777 \text{ m}^3\text{/d} .$$

(13) Determining the daily cycle number. We have first of all

$$t_1 = 0.2 \times 1627 = 325 \text{ s} ,$$

and by Eq. 2.4—21a,

$$t_2' = \frac{30.2 \times 10^{-4}}{3.7 \times 10^{-11} \times 850 \times 9.81} \ln \frac{111.5 \times 10^5}{95.6 \times 10^5} = 1506 \text{ s}.$$

Substitution of these values into Eq. 2.4—29 yields

$$n_c = \frac{86{,}400}{1506 + 325} = 47.2 \text{ l/d} .$$

(14) Let us find the least liquid column length $h_{l\,min}$ that can be lifted from valve *4* without the valves above opening: the critical valve to check is, of course, the lowermost unloading valve. The opening equation of valve *3* is

$$p_{C3} = 1.17 \times 41.7 \times 10^5 - 0.17 p_{T3} ,$$

and that of valve *4* is

$$p_{C4} = 1.17 \times 38.1 \times 10^5 - 0.17 p_{T4} .$$

If, in order to be on the safe side, we neglect the weight of the gas column in both the annulus and the tubing, it is clear that valve *3* will just not open if

$$p_{C3} = p_{C4} .$$

Further, by hypothesis,

$$p_{T3} = p_{C4} - 3.4 \times 10^5 .$$

The above four equations permit us to calculate the maximum permissible opening casing pressure:

$$p_{C4} = 42.2 \times 10^5 ,$$

and the corresponding minimum tubing pressure at opening is

$$p_{T4} = 13.7 \times 10^5 \ \mathrm{N/m^2} .$$

The minimum starting slug length is

$$h_{l\,min} = \frac{13.7 \times 10^5 - 2.0 \times 10^5}{850 \times 9.81} = 140.3 \ \mathrm{m} .$$

2.4.3. Gas-lift valves

(a) Pressure-controlled valves

Gas-lift valves used in modern practice can be controlled in a variety of ways. In the last few decades, valves operating on a variety of principles and incorporating various technological solutions have become available and widely used in the world oil industry. Since the Second World War, however, almost all other types of valves are being gradually replaced by pressure-operated valves. Their considerable advantages include simplicity of design, low cost of both the underground and surface equipment required for their operation, easy dimensioning and setting, together with adaptability to a wide range of operating conditions. Pressure-operated valves now in widespread use can be classified according to several criteria. In order to facilitate choice, the present author has introduced a system of six-digit tags (see below) which permits the characterization of any pressure-operated valve. Each digit of the tag is either a one or a two (except when two alternatives are possible; then the digit of the tag is a three). The meaning of a digit depends on its position within the tag. The tag code permits us to interpret e.g. the tag 211122 as: the valve in question is a non-instant-closing, bellows-type valve with gas-charged dome, whose opening is affected by the tubing pressure. It has no special opening or closing features, and is non-retrievable. — In our systematic description of the main valve types we shall invariable indicate the appropriate tags.

Tag Code

Position of digit from left to right	Value of digit	
	1	2
1 Is the valve of the instant-closing (snap-action) type?	yes	no
2 Is it of the metal bellows type?	yes	no
3 Is it a gas charged valve?	yes	no
4 Is its opening affected by the tubing pressure?	yes	no
5 Does it have special opening or closing features?	yes	no
6* Is it wire-line retrievable?	yes	no

* When both alternatives are possible, the value of digit will be: 3

(a.1) *Opening and closing of valve.* — The pressure-operated unloading valve shown in Fig. 2.4—7 can be used as an operating valve also. Its tag is 211123. Let us examine its opening and closing mechanism. — By Eq. 2.4—1, the opening of the valve is influenced at a given charged pressure by both the casing and the tubing pressure. If e.g. in a valve of a given make, $A_k = 6.2 \times 10^{-4}$ m^2 and $A_{ch} = 5.7 \times 10^{-5}$ m^2, then

$$p_T = 10.9\, p_k - 9.9 p_C \, .$$

Let $p_k = 24.0$ bars, then a back-pressure of $p_T = 5.0$ bars requires a casing pressure of 25.9 bars to balance the forces acting from above and below upon the inner valve. The opening takes in fact place when p_T or p_C rise slightly above the values just stated. The pressure acting upon the closing surface A_{ch} of the rising inner valve is $p_T = 5.0$ bars before opening and, in a fair approximation, $p_C = 25.9$ bars after opening. The force increment suddenly hitting the inner valve in the course of opening is, then,

$$\Delta F_1 = A_{ch}(p_C - p_T) = 5.7 \times 10^{-5}\,(25.9 \times 10^5 - 5.0 \times 10^5) = 119 \text{ N} \, .$$

This force makes the inner valve rise to a height h', that is, to open 'instantaneously'. The rise of the inner valve compresses the bellows. This generates, on the one hand, a spring force ΔF_s in the metal and, on the other, a pressure rise Δp_k in the gas dome, corresponding to a force increase ΔF_k. The inner valve will reach its highest possible position in the valve under consideration — that is, the valve will fully open — if the opening force ΔF_1 is greater than the force increase $\Delta F_2 = \Delta F_s + \Delta F_k$ resulting from the rise of the valve in the highest position, h. If $\Delta F_1 < \Delta F_2$, the valve does not open fully, and its gas throughput capacity will be less than in the fully open position.

Example 2.4.9. Let $A_k = 6.2$ cm² and $A_{ch} = 0.57$ cm². Let us assume a total valve stem travel of $h = 3$ mm, a force $\Delta F_r = 20$ N resulting from the load rate of the bellows at full opening and $\Delta F_k = 47$ N resulting from the pressure rise of the dome at 24 bars dome pressure. Let us write up the condition of full opening. — The relationship $\Delta F_1 = A_{ch}(p_C - p_T)$ is shown in Fig. 2.3—20 for dome pressures of 24 bars (Graph I), 28 bars

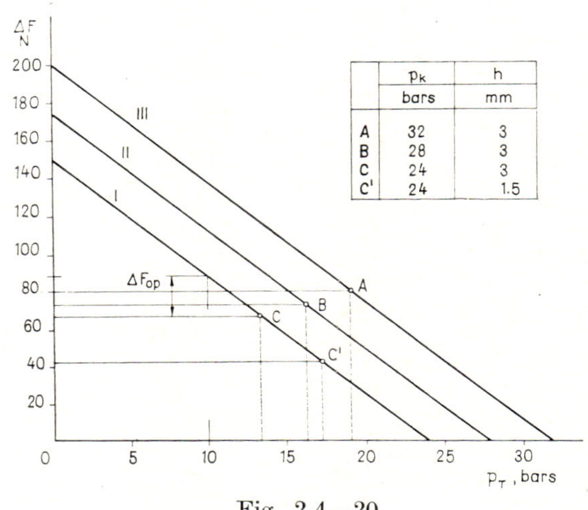

	p_k bars	h mm
A	32	3
B	28	3
C	24	3
C'	24	1.5

Fig. 2.4—20

(Graph II) and 32 bars (Graph III). Full opening at $p_k = 24$ bars gives rise to a $\Delta F_2 = 20 + 47 = 67$ N. Let us draw a parallel to the abscissa axis through this ordinate value. It intersects Graph I at C. Clearly, the valve will not open fully unless tubing pressure drops below the abscissa of C, that is, 13.5 bars. The corresponding points of intersection and limit pressures for 28 and 32 bars are B at 16.4 bars, and A at 19.0 bars, respectively. — Reduction of valve travel to, say, 1.5 mm, or increasing the dome volume of the valve entails a decrease of ΔF_k. For instance, at $p_k = 24$ bars, the limit pressure will be at C' instead of C. The full instant-opening pressure limit has consequently assumed a greater value.

The example reveals that full opening depends not only on valve design but also on dome pressure and, in the case of intermittent lift, also on the daily cycle number. If the cycle number in a given well is decreased, the pre-opening pressure p_T will increase, entailing a decrease of ΔF_1. It may thus happen that the gas throughput capacity of the valve, satisfactory as predicted at high cycle numbers, will be insufficient at a comparatively lower cycle number. Intermittent lift requires valves that will open fully under the operating conditions to be expected. Let us note that valve opening in testers is often checked at atmospheric back-pressure only.

This is the most favourable of all possible conditions as far as full instant opening is concerned. Closure is unaffected by tubing pressure. The valve will close when casing pressure decreases to equal dome pressure (cf. paragraph 2.4—1a.2). Keeping the valve stem in the upper (fully open) position requires an excess opening force $\Delta F_{op} = \Delta F_1 - \Delta F_2$.

Example 2.4—10. In continuation of the foregoing example let $p_k = 24$ bars and $h = 3$ mm. Then, e.g. at a pressure of $p_T = 10.0$ bars, $\Delta F_{op} = 87 - 67 = 20$ N. When the casing pressure starts to decrease as the liquid slug rises, the valve will stay in the upper limit position until the decreasing pressure attains the value $\Delta F_{op}/A_k$, that is, $20/6.2 \times 10^{-4} = 0.32 \times 10^5$ N/m² in the present example. Any further drop in casing

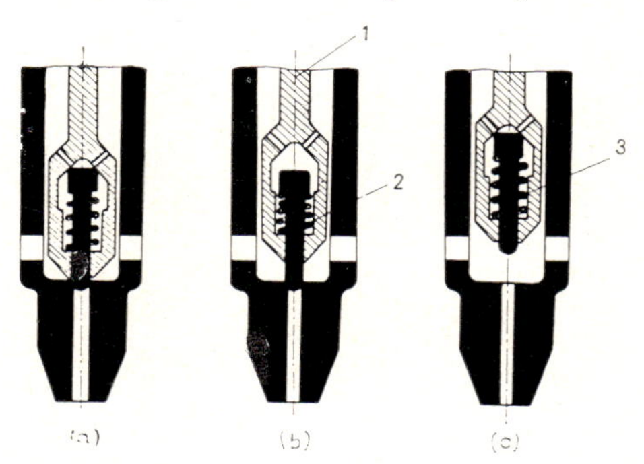

Fig. 2.4—21 McMurry-type inner valve

pressure will lower the valve stem with a consequent decrease in gas throughput cross-section. Closing occurs when the casing pressure has dropped to equal the dome pressure. The valve shown in Fig. 2.4—7 is, then, instant-opening but not instant-closing.

Instant opening is not, however, quite guaranteed by the play of forces just described, either. The valve stem may start to rise so smoothly that a small initial rise, h', may establish itself for a while, because the resistance to flow of the geometry $(d_{ch}\pi h')$ may be so great as to prevent any but the slightest pressure rise in the tubing below the valve. This transitory balance can of course be upset by a slight shock or vibration. During this aborted opening, injection gas will enter the tubing to no avail.

Instant (snap-action) opening can be ensured in several ways. (i) In the valve shown in Fig. 2.4—7, the check valve *7* can be pressed by a weak spring (not shown) against the valve port. The check valve will consequently open only when pressure below the valve *4* has risen to a higher value, boosting the instant-opening action. (ii) Instant opening is facilitated also by the McMurry-type valve (Fig. 2.4—21). Part (a) of the Figure shows the closed valve. An increase in casing pressure lifts the valve stem *1* to the position shown in part (b), without opening the valve closed by the needle *2*. Once the increase in casing pressure has lifted the valve stem above a limit position depending on the force arising in spring *3*, the spring will jerk the inner valve into the position shown in part (c) of Fig. 2.4—21.

In the Szilas-type gas lift valve (tag: **111123**), instant closure is ensured by a permanent magnet *1*, installed in the dome (Fig. 2.4—22; Szilas 1962). The magnet will let go of jacket *2* fixed to the valve stem only when the casing pressure has decreased to equal the closing pressure. The force needed to part the jacket from the magnet can be set by adjusting an air gap:

$$\Delta F_m = \Delta F_2 = \Delta F_s + \Delta F_k.$$

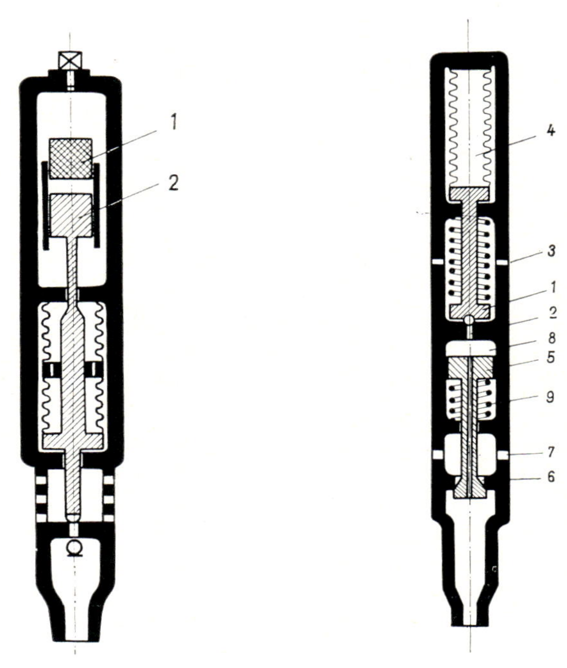

Fig. 2.4—22 Szilas-type magnetic snap-action valve

Fig. 2.4—23 Merla WFM-12-type gas-lift valve

Instant closing and opening of intermittent valves is most often ensured by pilot valves. As an example, consider the Merla make WMF-12-type operating valve (tag: 112122). When the main valve is closed, pilot *1* (see Fig. 2.4—23) is under the influence of a tubing pressure p_T acting on the area $A_{\mathrm{ch}\,p}$ of port *2* and a casing pressure p_C entering through the inlets *3*, acting on the effective area $(A_k - A_{\mathrm{ch}\,p})$ where A_k is dome area. The bellows is liquid-filled (cf. paragraph 2.4.3a.3). Operation of the pilot is thus the same as that of the valve in Fig. 2.4—7. The pilot will open if the algebraic sum of the forces acting on it from the direction of the tubing and the casing exceeds the 'bellows pressure force'. Once the pilot valve has opened, the casing pressure entering through port *2* will act on the top face of main valve *5*. The resultant force will depress and thereby open the main valve and permit injection gas to enter the tubing through inlets *7* and port *6*.

As soon as casing pressure decreases to equal the bellows pressure, the pilot will close. Gas of pressure p_k, trapped in space *8*, will escape towards the tubing through the bleed bore of the main valve stem, and pressure in space *8* will decrease to p_T. Spring *9* then lifts the main valve into the closed position. — In addition to safe instant opening and closure, pilot-operated intermitting valves have the considerable advantage that the opening equation and consequently the spread are independent of the valve-choke

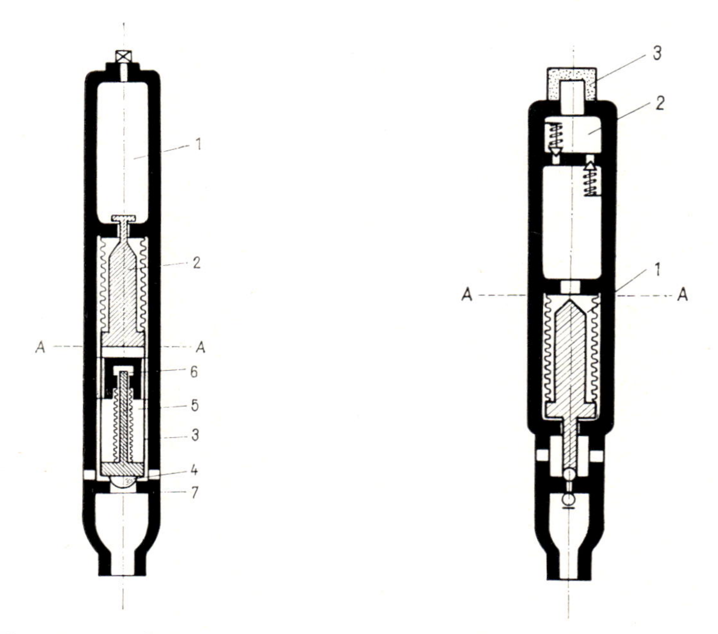

Fig. 2.4—24 Guiberson CR-type gas-lift valve

Fig. 2.4—25 Gas-lift valve with down-hole dome-pressure adjustment (Forsythe 1964)

area A_{ch}. In a pilotless valve an increase of valve port will increase the spread (the difference between the opening and closing pressure). Under certain conditions, this may entail an oversupply of gas to the tubing over the interval between opening and closing. This will in turn entail an excess injection gas requirement.

Instant-action intermitting valves are required first of all if there is no surface intermitter to control the cycles. If there is one, then non-instant-action valves may also be used, possibly at the expense of a slight increase in injection gas requirement (cf. also Section 2.4.5). Instant-action valves are not recommended for continuous-flow gas lift installations (cf. also paragraph 2.4.3a.4).

A valve with special opening and closing features is the Guiberson-CR (constant ratio)-type (Fig. 2.4—24; tag: 111113). Down to the line *A-A* this valve is essentially the same as the Guiberson-A-type valve (Fig.

2.4—7). Dome *1* is charged with nitrogen. In the position shown, rods *3* touch the bottom face of valve *2*, but are not fixed to it. The upper part of stem *5* of main valve *4* serves as a piston reaching into space *6*, filled with gas at a near-atmospheric pressure. When casing pressure exceeds the dome pressure in the valve, main valve *2* moves upwards, whereas the elements below it stay in place. For these latter, we may write

$$p_C \Delta A \geqq p_T A_{ch} ,$$

where ΔA is the area A_{ch} of valve port *7* minus the effective area of the lower bellows. Rearranging of the above equation results in

$$\frac{p_C}{p_T} = \frac{A_{ch}}{\Delta A}, \qquad 2.4\text{—}30$$

that is, whenever the casing pressure exceeds the dome pressure, the opening ratio of casing to tubing pressure depends exclusively on the design features of the valve, and is independent of the casing pressure p_C. The valve invariably closes at dome pressure. A considerable advantage of the CR type valve is that even if the length h of the liquid slug to be lifted varies in a given well, the ratio p_C/p_T determining the rising velocity of the slug remains constant. Changes in casing pressure p_C due to design errors or changes in well parameters hardly affect the optimal use of the injection gas.

Another special-feature valve (tag: 211113) can be charged in the well (Forsythe 1964). A sketch section of the valve is shown as Fig. 2.4—25. Below the line *A*-*A*, valve design is essentially the same as the corresponding part of the Guiberson-A valve (Fig. 2.4—7). Above dome *1* there is an antechamber *2*. The wall separating the two chambers has two ports closed by needle valves. The upper valve opens at a pressure differential of 1.4 bar; the lower one at a differential of about 7 bars. In a state of balance, the dome is under ambient pressure, e.g. the pressure in the casing. Filter *3* does not hamper the entry or exit of gas. The valve is installed in the well without a gas charge. Casing pressure at the depth of installation charges chamber *1* to casing pressure minus 7 bars. If the ambient pressure drops to a value less by 1.4 bar than the pressure in dome *1*, gas flows from the dome into the antechamber. By the above consideration, dome pressure does not change until the ambient pressure grows less than (p_k — 1.4) bars or greater than (p_k + 7.0) bars. This pressure-differential range of 8.4 bars is sufficient to make the valve operate e.g. like the Guiberson-A valve. The design has the following advantages. (i) The dome being charged to the required operating pressure at the actual depth of installation, the charged pressure is not affected by dimensioning uncertainties. (ii) The dome pressure can be changed without retrieving the valve, if changes in operating conditions require such a change to be made.

(a.2) *Metal-bellows-type and other valves.* — The metal-bellows-type valves described so far have the drawbacks that (i) dirt may settle in the convolution of the bellows, impeding its motion, so that it will not open or close as expected, (ii) if the well fills up with liquid, the resulting external pressure may be great enough to deform or burst the bellows.

In the Garrett-make valve of tag 211123, shown in Fig. 2.4—26a, dome pressure acts on the outside of the bellows. Valve stem *1* contains a cavity filled with high-viscosity lube oil. The space between the bellows and the outer surface of the perforated valve stem is filled with a similar oil. Mud, sand grains, scale or rust entering the valve can thus reach the lowermost convolution of the bellows at most. In normal operation, no well fluid rises higher than port *4*, being prevented from doing so by check valve *5*. Still, some inflow of well fluid is possible during installation, and the injection gas may

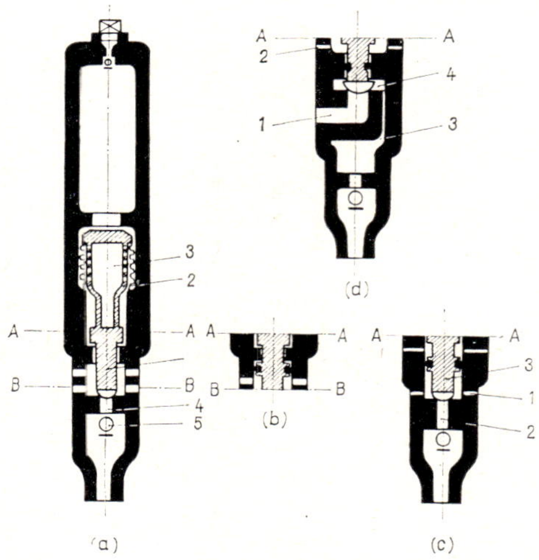

Fig. 2.4—26 Various Garrett gas-lift valves

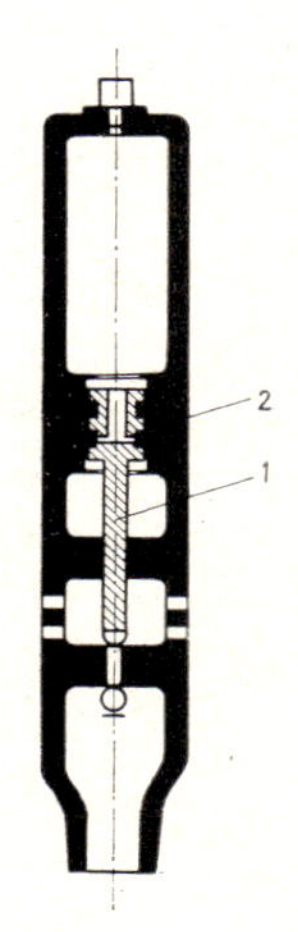

Fig. 2.4—27 Merla piston-type gas-lift valve

also contain some dirt. The bellows cannot be contaminated if the middle part of the valve is constructed as shown in part (b) of the Figure. The valve stem is provided with an O-ring type seal. The space above the seal is filled with high-viscosity oil, too. In some constructions, the bellows is made of several elements so as to minimize deformation (Fig. 2.4—22) or leans against a perforated jacket. — In the Merla-make valve (tag 221123) shown in Fig. 2.4—27, the dome is isolated by a piston *2*, integral with valve stem *1*, from the lower, open part of the valve. The top part of the piston is provided with O rings, and a Teflon seal ring below them. Above and between the seals there is a high-viscosity lube oil. This valve type is insensitive to both dirt and excess pressure. The piston does not provide a perfect seal, however, so that dome pressure in the valve installed in the well will drift upwards as high-pressure injection gas seeps in. — The OTIS-C valve of tag 221222 (Fig. 2.4—28) has a number of favourable properties. The valve is not mounted in the usual mandrel but threaded between two lengths of tubing as a sort of special sleeve. The nitrogen-charged dome *1* is of annular section. If dome pressure exceeds casing pressure, internal overpressure makes the elastic sleeve *2* close the injection gas inlet slots *3*. The similarly elastic reverse check valve *4* prevents the well fluid from flowing through or entering into the valve. If casing pressure exceeds dome pressure, sleeve *2* assumes the position shown in part (b) of the Figure, and injection gas can flow from the annulus through slots *3*, annular passage *5* and bores *6* into the tubing. The only moving parts are the two sleeves

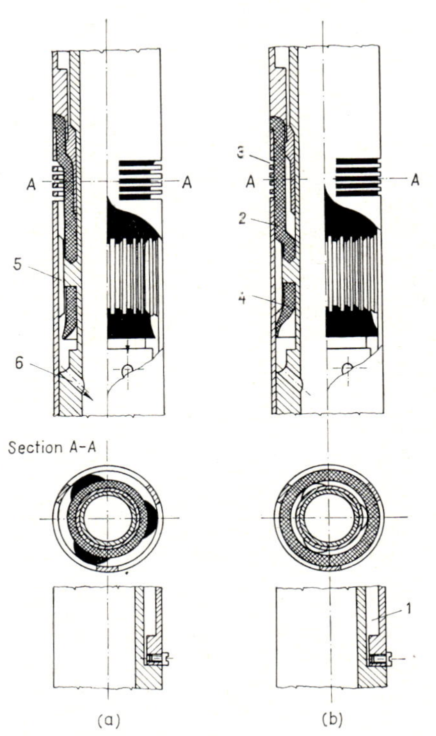

Fig. 2.4—28 OTIS-C type gas-lift valve

made of a rubber or plastic. The valve has a high safety of operation, being almost insensitive to dirt.

(a.3) *Temperature effects; spring-loaded valves.* — In paragraph (a.1) we have already considered the consequences of dome pressure changes in pressure-operated valves. Designing and operation control are seriously hampered by a lack of accurate knowledge of operating temperature of the valve prior to actual operation. This has made it desirable to design valves so that dome pressures are influenced little or not at all by temperature changes. In the CAMCO valve of tag 211123 (Fig. 2.4—29) the valve is pressed down on the seat in addition to dome pressure also by the force F_s of spring. The condition of opening is therefore

$$p_T A_{ch} + p_C(A_k - A_{ch}) \geqq p_k A_k + F_s , \qquad 2.4\text{—}31$$

and the condition of closing is

$$p_k A_k + F_s \geqq p_C A_k \,. \qquad 2.4-32$$

A spring force F_s is equivalent in a fair approximation to an increase $\Delta p_k = F_s/A_k$ in dome pressure. Hence, to obtain a valve response of the type shown in Fig. 2.4—8 it is sufficient to charge the dome to a lower pressure for a given A_k and A_{ch}. If the dome pressure is lower, then so is the dome pressure increase under a given increase of temperature. Component *2* is

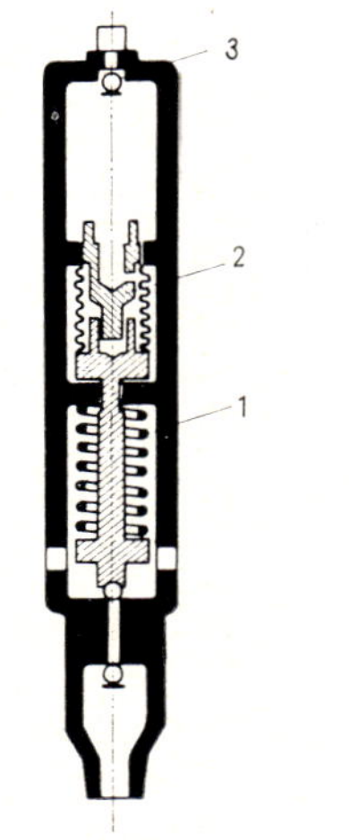

Fig. 2.4—29 CAMCO valve loaded by a spring plus dome-pressure

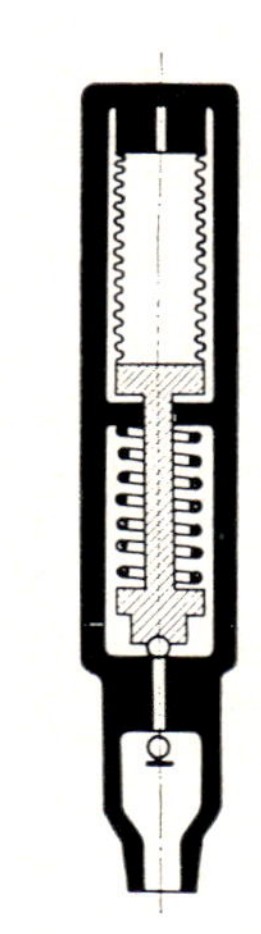

Fig. 2.4—30 Merla spring-loaded gas-lift valve

an anti-chatter device. The bellows space of the dome is filled, as usual in bellows-type valves, with high-viscosity lube oil. The retarded flow of oil through the broken passage makes the operation of the valve smoother, chatter-free. Check valve *3* helps to charge the chamber to an accurately controlled pressure. — The Merla-make valve of tag 212123 (Fig. 2.4—30) is a further exploitation of the above idea, with a liquid-filled chamber containing practically no gas at all. During filling, the bellows is in a compressed position. Valve operation is practically independent of ambient temperature, except for the slight temperature-dependence of the spring force. — In spring-loaded valves, especially in the last type mentioned, the term dome pressure changes its meaning; one may continue to speak of a dome pressure, but the term now means the expression $(p_k + F_s/A_k)$, or the expression F_s/A_k if $p_k = 0$.

(a.4) *Design features of continuous-flow and intermittent valves.* — Gas lift valves may be designed to perform one of three functions: (i) unloading, (ii) subsurface gas injection and regulation in a continuous-flow gas lift installation (continuous-flow valves), and (iii) subsurface injection-gas regulation in intermittent-flow installations (intermittent valves).

Unloading valves of essentially the same design are used in both types of installation. The requirements as to their opening and closing characteristics have been discussed in paragraph 2.4.1.a.2. — A continuous-flow valve is required to continuously pass injection gas from the 'gas conduit' of the well into the rising well fluid, while automatically regulating the rate of injection-gas flow. The rate of inflow of formation fluid and its composition may be subject to slight fluctuations. The fluid rising in the tubing may grow lighter or heavier, with a concomitant drop or rise in tubing-shoe pressure. Steady production may be promoted by a controlled addition of injection gas, that is, by installing a valve that will pass more gas at higher tubing-side pressures and vice versa. An obvious requirement is that the valve should not shut off under any pressure drop within the range of these fluctuations. It follows from the above that, in a continuous-flow valve, the A_{ch}/A_k ratio, and hence, the spread (the difference between opening and closing pressure) should be possibly great. As the valve is continually open in normal operation, no instant opening or closing is required. During continual operation, the forces acting upwards on the valve stem must be sufficient to position the valve stem so as to permit the desired gas flow. The relationship of valve stem travel to pressure increase in the space below the bellows depends on whether the valve is loaded by a spring alone or by gas pressure alone. In the former case, valve stem travel per, say, one bar of pressure rise is nearly equal at any stem position. If, on the other hand, the dome is gas charged, then the valve stem travel per one bar of pressure increase tends to decrease rather sharply as the pressure increases. The valve is thus a less efficient regulator at comparatively high operating pressures. The continuous-flow valve will automatically regulate its own injection gas throughput only if the space below the valve ball is under tubing pressure in the open state. This can be ensured in several ways. In the valve shown as Fig. 2.4—26c, the flow resistance of the bores *1* is substantially greater than that of the valve port *2*, and hence, pressure in space *3* equals tubing pressure to a fair approximation when the valve is open.

Requirements facing an intermittent valve may cover a fairly wide range depending on the conditions prevailing and the goal to be attained. It has been stated in paragraphs 2.4.3a.1 that instant-opening and -closing valves should as a rule be used whenever gas is supplied to the well by a line not provided with a surface cycle controller. If injection-gas flow is suitably controlled on the surface, no instant closing is required. Given the annulus volume, the spread of the intermittent valve will considerably affect the injection gas consumption per cycle. The annulus can be used to store gas between cycles. If the injection is controlled at the surface by an intermitter, and the available gas supply is sufficient, then the well annulus volume need not be used for gas storage. The process taking place will be as follows: injection gas starts to flow into the annulus at a sufficiently high rate; it enters the tubing through the gas lift valve and lifts the liquid slug at the desired velocity; gas supply from the surface into the annulus equals the gas injection rate, so that pressure in the annulus remains unchanged; as soon as the slug has risen high enough in the tubing, the surface controller shuts off the gas supply to the annulus. Casing pressure now decreases, making the gas lift valve close; expanding, the gas plug confined in the tubing lifts the liquid slug to the surface. The opening pressure of the valve may thus equal its closing pressure, that is, the opening of the valve may depend on casing pressure alone and not at all on tubing pressure. Valves of this type are called balanced. Figure 2.4—26d shows a detail of such a valve. The valve port of area A_{ch} is under casing pressure acting through inlet *1*, and so is the effective area $(A_k - A_{ch})$ through inlet *2*. The opening force thus equals $p_C A_k$, independent of tubing pressure. If the valve is open, then most of the pressure of injection gas passing through drops over the small-bore orifice *3*, so that pressure prevailing in space *4* equals casing pressure in a fair approximation. — Also in the OTIS-C valve shown as Fig. 2.4—28, the opening pressure is independent of tubing pressure.

When the injection gas supply is not under surface control, or the highest gas flow rate of which the controller is capable is less than that required to lift the liquid slug at the required velocity, balanced valves will not do the job. In such cases, intermittent valves of a great enough spread are required. These will permit the gas stored in the annulus to lift the liquid slug to the surface at the required velocity either by itself, or with the aid of further gas inflow from the surface gas supply line. Pressure-operated valves of this type are called unbalanced. It is important to match valve spread to the conditions prevailing in the well. If the spread is too small, then the fallback loss, $h_{lr} A_T$, will be too great. If the spread is too great, then too much gas will enter the tubing after the slug has surfaced. Most bellows-type valves are unbalanced. An unbalanced rubber sleeve valve has been designed at the Petroleum Engineering Department of the University of Heavy Industries in Miskolc (Hungary). Its special feature is that opening is controlled largely by the duration of the opening casing-pressure.

In production practice, tubing-pressure-operated (tubing-sensitive or fluid-operated) valves are also employed. These can be conventional unbalanced intermittent valves installed in mandrels providing a so-called reverse installation, with tubing pressure acting on the area $(A_k - A_{ch})$ and casing pressure acting on A_{ch}. The arrangement is illustrated in Fig. 2.4—31. If the valve is open, injection gas can pass into the tubing through check valve *1*, choke *2* and orifices *3*. By the properties of this arrangement, the valve is more sensitive to changes in tubing than in casing pressure. This setup is considered highly suited for multipoint injection (cf. Section 2.4.4a). Its drawbacks are that it will not operate suitably in wells whose flowing *BHP* is very low; cycle regulation is something of a problem; if tubing-

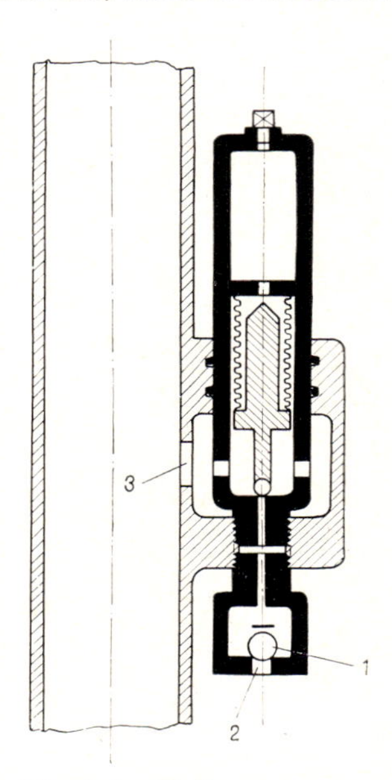

Fig. 2.4—31 Reverse-control gas-lift valve, from Winkler and Smith 1962 (used with permission of CAMCO Inc.)

head pressure rises considerably and then declines slowly during production, then the valve may stay open too long, entailing an overconsumption of injection gas; the closing casing pressure cannot be predicted, because pressure under the bellows in the open position may rather widely differ from the casing pressure; if the well fluid is sandy, the sand can easily deposit in the convolutions of the bellows.

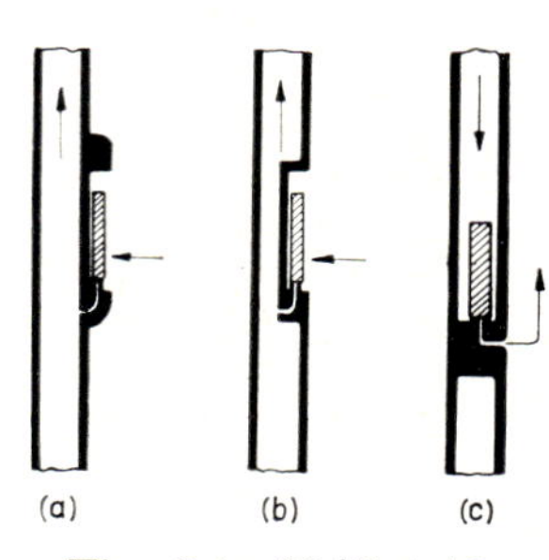

Fig. 2.4—32 Outside mounted gas-lift valve mandrels

(a.5) *Installation of gas lift valves; wireline-retrievable valves.* — Valves are installed in special mandrels run as parts of the tubing string. The valve is either threaded into the mandrel, or fixed with a small packer. In the first case, the valve is installed on the surface, prior to running the tubing. It can be retrieved only after pulling the tubing, or, in a more general formulation, after pulling the conduit which includes the valve mandrel. Figure 2.4—32 shows various gas lift valve mandrels. It has been assumed that the valve is installed in every case straight, that is, so as to be operated by casing pressure. In (a), the cross-sectional area of the mandrel equals that of the tubing. Case (b) differs in that the valve is installed in a concave mandrel. This arrangement requires less space for running, but does not provide the full-open-tubing feature. It is not usually possible to run instruments etc. below the valve. In (c), gas is conveyed to the valve through the tubing, which is usually of macaroni size, and the well is produced through either the casing annulus, or the annulus between the gas conduit and a production tubing.

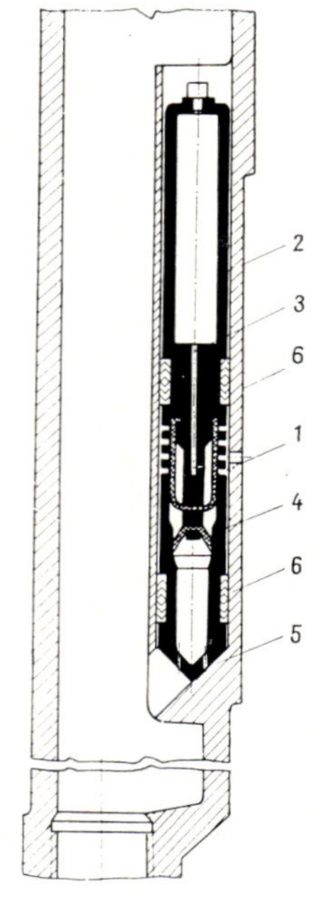

Fig. 2.4—33 OTIS wireline-retrievable gas-lift valve

Changing a wireline-retrievable valve is much less costly. Most of today's valves are manufactured in two variants: one wireline-retrievable and the other for outside mounting. A wireline-retrievable OTIS valve is shown in Fig. 2.4—33. Its operation is the same as that of the OTIS-C type valve described in paragraph 2.4.3a.2. Gas flows into the valve through inlets *1*. It is fixed in its seat by packing *6*, which also provides packoff between tubing and annulus. Running and retrieving a CAMCO type retrievable valve is illustrated in Fig. 2.4—34 (Wieland 1961a). Part (a) of the Figure shows the running operation. Valve *1* is installed by means of a wire-line tool, the main components of which are the running tool *5*, the knuckle joint *2*, and the kickover tool, which consists of three centring arms *3*, fixed to a sleeve at each end. On running, spring *4* pushes the centring arms up, so that the upper sleeve comes to rest against the knuckle joint. The valve is first run past the valve mandrel and then slightly pulled back. Pulling makes the upper sleeve slide down bar *6*, and get caught in the position shown in the Figure. On further lowering the valve is deflected by the centring arms in the eccentric mandrel, so as to slide precisely into the mandrel bore. A slight jerk will disengage the running tool. — Retrieval is shown in part (b) of the Figure. Retrieving tool *7* differs from the running tool in that its length is increased by spacer bar *8*. Also in this case, the tool is run past the valve first and then pulled up a small way. The kickover tool then gets caught in the lowermost possible position, and further lowering of the tool directs the pulling tool exactly towards the fishing neck of the valve. The valve gets caught and can be retrieved. — Figure 2.4—35 shows three types of mandrels for retrievable valves. In (a) injection gas flows from the annulus into the tubing. In (b) the gas enters a chamber (cf. Section 2.4.4b). In (c) the well is produced through the tubing by means of gas supplied through a separate gas conduit.

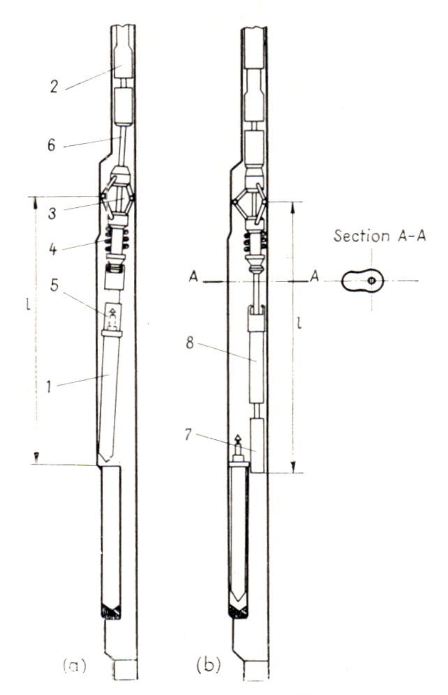

Fig. 2.4—34 Running and retrieving a CAMCO gas-lift valve, after Wieland (1961a)

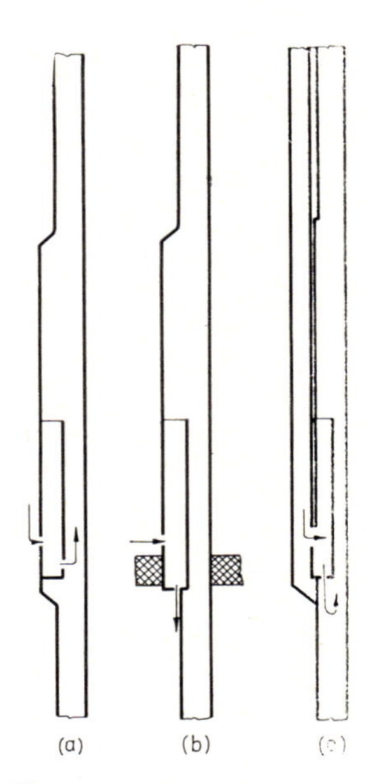

Fig. 2.4—35 Retrievable gas-lift valve mandrels

(b) Other types of gas-lift valves

As stated in Section 2.4.3a, gas lift valves operating on a variety of principles have been used in early production practice (Brown 1967). Of these, only the so-called differential type gas lift valve is used to any advantage today. Its main feature is that it will operate even if casing pressure remains constant. It opens when the pressure differential across it is small and closes when it is great. It can be used as an unloading valve, or in a chamber installation as a bleed valve (see Section 2.4.4b).

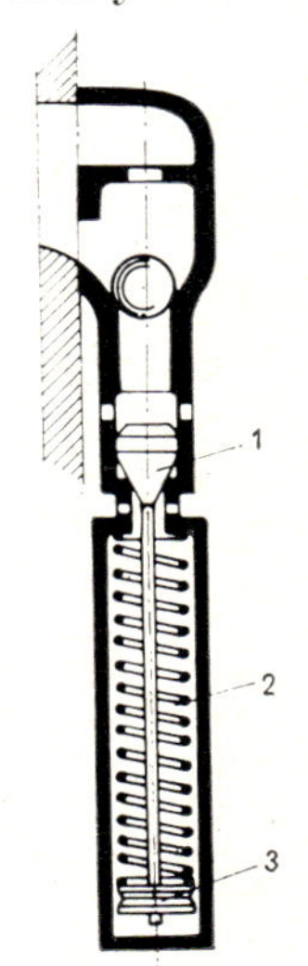

Fig. 2.4—36 Krylov and Issakov's U-1-M differential unloading valve

As an example consider the Krylov—Issakov U-1-M type differential unloading valve (Muraviev and Krylov 1949), shown open in Fig. 2.4—36. Tubing pressure acting on the area A_1 of valve *1* and the pull F_s of spring *2* act to open the valve, whereas casing pressure acting through the valve port upon a smaller area A_2 on the pear-shaped valve acts to close it. The closing condition is

$$A_1 p_T + F_s \leqq A_2 p_C \, ,$$

and the opening condition of the closed valve is

$$A_1 p_T + F_s \geqq A_1 p_C \, ,$$

since in the closed valve the casing and the tubing pressure act on equal surfaces A_1. The nearly constant spring force and the fact that tubing pressure invariably acts on a given surface A_1, whereas casing pressure acts on the smaller A_2 in the open state and on the greater A_1 in the closed state, make the opening pressure differential Δp_{op} small and the closing pressure differential Δp_{cl} great. In the U-1-M valve,

$$p_{\mathrm{op}} = 10^5 + 10^4 \Delta p_{\mathrm{cl}} \, .$$

The maximum closing pressure differential of this unloading valve is 34 bars, and hence its greatest opening pressure differential is 4.4 bars. Values less than these can be set by reducing the spring force by means of nut *3*.

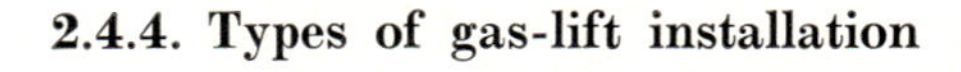

2.4.4. Types of gas-lift installation

(a) Conventional installation

(a.1) *Single completion.* — The simplest completion used in *continuous-flow production* is the so-called *open completion* with a single string of tubing and with no fittings in the well. Injection gas is supplied to the casing annulus and well fluid is produced through the tubing. If the liquid production rate envisaged is quite high, casing flow may be employed, with gas supplied through the tubing and fluid produced through the annulus. A condition for this arrangement is that the well fluid must not erode or corrode the casing and there must be no paraffin deposits. Injection gas is admixed to the well fluid either at the tubing shoe or through a continuous gas lift valve installed in the tubing wall. — Figure 2.4—1 shows a *semi-closed installation*. The annulus is packed off at the well bottom by means of packer *2*. Injection gas passes from the annulus into the tubing through valve *1*. The closed annulus prevents surging of the second type (cf. paragraph 2.3.5b.1). Another advantage is that if the well is shut off or dies, the annulus does not get filled with liquid and has not to be unloaded. — *Intermittent gas lift* normally requires a closed completion. A solution is shown in Fig. 2.4—12. The annulus is packed off by means of packer *3*. In the phase of accumulation, the load on the well bottom is composed of the weight of the liquid column and of the gas column above it, plus the tubing-head pressure. In the production phase, standing valve *4* will close as soon as injection gas raises pressure above the standing valve higher than what prevails below it. Back-pressure retarding the inflow of formation fluid into the well varies as shown by the continuous line in Fig. 2.4—17. The mean flowing *BHP* is seen to be fairly low. The Figure reveals also the influence of the standing valve. The dashed line shows the variation of pressure above the closed standing valve. In the absence of such a valve, the *BHP* would vary according to this line (assuming there is no backflow of fluid into the formation). The increase in mean flowing *BHP* would reduce the daily inflow of liquid into the well.

The gas injection to the tubing may be single-point or multipoint. *Single-point injection* means that gas is fed to the tubing through a single intermitting valve, usually the lowermost one. In *multipoint injection*, the unloading valves above the operating valve open one after another as the rising slug passes them, delivering additional volumes of injection gas to the tubing. Multipoint injection is capable of delivering a copious supply of gas to the tubing section below the slug even if casing pressure is comparatively low. The result is a high slug velocity and reduced fallback. Its advantages predominate in comparatively deep wells with large-size tubing, low injection gas pressure, and in wells where fluctuations in flowing *BHP* preclude designing for optimal-depth single-point injection. The critical requirement of the method is an adequate supply of injection gas from the surface as, in the absence of such, casing pressure will drop soon enough to close the gas lift valve or valves so that the entire liquid slug will fall back and kill production.

The above-outlined differences in well completion entail certain typical features of the two production methods. (i) Injection gas pressure at the tubing shoe is less than or equal to flowing *BHP* in the case of continuous flow and greater than the mean flowing *BHP* in intermittent production. (ii) In continuous flow, all the gas delivered by the formation can be used to lift fluid; in intermittent production, formation gas hardly affects the specific injection gas requirement. (iii) In an intermittent installation, specific injection gas requirement in a given well producing a given fluid remains constant as long as the initial length h of the liquid slug to be lifted remains the same; the specific injection gas requirement does not necessarily increase as formation pressure declines. This will remain the case until the decline of formation pressure or the desire to increase production makes it necessary to reduce initial slug length. In a continuous flow installation, decline of formation pressure will even at a constant formation *GOR* entail a gradual increase in specific injection gas requirement.

(a.2) *Dual completions.* (Largely after Winkler and Smith 1962.) — There are two types of installation, that is (i) both tubing strings are supplied with injection gas from a common conduit or casing annulus; (ii) there are two separate injection-gas supplies in the well. — Both solutions in Fig. 2.4—37 belong to the first group. In solution (a), the well is provided with two concentrically disposed tubing strings. Gas enters into the tubing annulus *1*. The lower zone produces through tubing *2*, the upper one through casing annulus *3*. This solution was popular in early dual-completion practice, as it could be realized with the current types of packer and wellhead equipment then available. Nowadays the completion shown in part (b) of the Figure is preferred, as it can be adapted much more readily to a wide range of production conditions. Valves are usually of the wireline-retrievable type. This permits a fast valve change if design turns out to be wrong or inflow characteristics change. Valve choice may be based on various criteria depending on the inflow characteristics and the production method (continuous or intermittent) to be preferred.

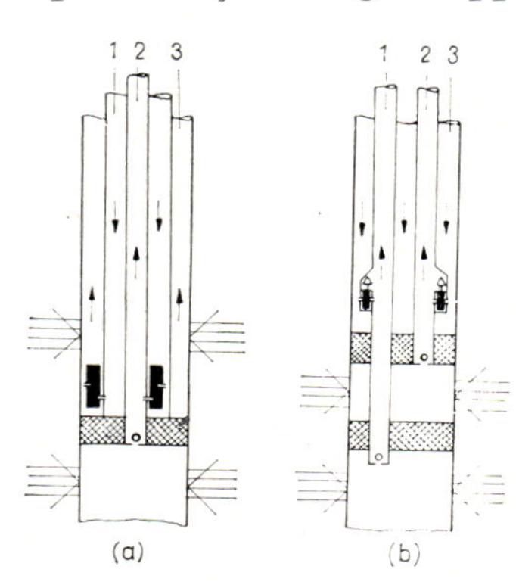

Fig. 2.4—37 Dual gas-lift completion (I)

Both zones continuous-flow. The continuous flow valves of both strings are casing-pressure-operated. The pressure of injection gas entering the casing annulus is to be stabilized at the surface. The lower-zone valve has a great gas throughput with a low flowing pressure drop. The continuous valves of the upper zone are choked so as to give a pressure drop of 7—8 bars. The greater the pressure differential over these valves, the less the change in gas throughput per unit change in tubing pressure (see Eq. 1.5—2). Production is consequently more uniform. Injection gas flow through the lower-zone valve is regulated by means of slight changes in surface pressure. This hardly affects gas flow through the upper-zone valve.

One zone continuous-flow, other intermittent lift. Casing-pressure-operated valves are recommended for both zones. The operating casing pressure of the continuous valve should be less than the opening pressure of the intermitting valve. Also, the continuous valve should be choked so as to bring about a pressure differential of 7—8 bars across it. Gas throughput will thus not be appreciably affected by casing pressure fluctuations due to the intermitting of the other zone. Surface closing pressures of the valve string of the intermittent zone should be equal or slightly different. Valves with a large port (high tubing sensitivity) are recommended. Using this technique, slight errors in intermittent-lift design will cause no appreciable production difficulties, because the annulus pressure is practically independent of the depth of the operating intermittent valve and therefore the casing pressure of the continuous flow zone remains practically constant.

Both zones intermittent lift. Several valve types might be used to economic advantage. (i) Tubing-pressure-operated valves are used to lift the lower-capacity zone. Casing-pressure-operated valves of high gas throughput capacity are used in lifting the higher-capacity zone. Surface closing pressures are set so as to decrease slightly downwards. The cycle frequency of the higher-capacity zone is controlled at the surface. The production of the other zone will be correctly controlled by the tubing-pressure operated valves, provided initial liquid slug length h is sufficient to trigger opening of the valves and the surfacing of the slug does not raise wellhead pressure high enough to make the tubing pressure build-up close the intermittent valve. (ii) Both zones are produced with casing-pressure-operated valves. This solution is justified if the flowing *BHP*s of both zones are so low, or the wellhead pressures built up during production are so high, that tubing-pressure-operated valves are out of question for both zones. The opening pressures of the two operating valves are set at different values. The valve with the lower setting may open and produce one zone (say, zone A) while the other valve remains still closed. When the valve with the higher setting opens to produce the other zone (say, zone B), the surface controller shuts off the flow line of zone A. The string of zone A will fill up with injection gas, which has no outlet, however. As soon as zone B ceases to produce, the flow line of zone A is automatically opened: the gas in the string can now escape without doing any useful work. This gas loss will be less if the valve with the higher opening-pressure setting is used to produce the zone with the lower cycle frequency.

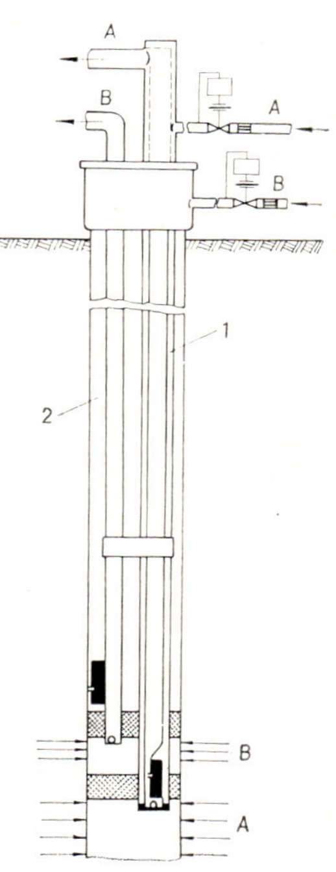

Fig. 2.4—38 Dual gas-lift completion (II)

In the solution shown as Fig. 2.4—38, the two zones may be produced independently, either by continuous flow or intermittent lift. The well contains three strings of tubing. The lower zone receives injection gas through annulus *1*, the upper one through annulus *2*. The larger amount of piping required makes this completion costlier than the foregoing one; also, a given size of production casing will take smaller sizes of tubing. — Dual completions will permit the production of one zone by flowing and another by gas-lifting. The completion required in this case is a simplified version of the above-described ones. Dual completions will prove advantageous also in the case when one zone produces gas and the other produces oil by gas lift.

In favourable cases, gas from the gas zone may be used to gaslift the other zone. Several completions are possible, depending primarily on which zone is deeper and also on whether the gas can be produced through the annulus, or requires a separate tubing string.

(b) Chamber installation

A chamber is essentially a larger-diameter piping attached to the tubing shoe, intended to facilitate intermittent lift. A chamber installation at a given specific injection gas requirement gives rise to a lower flowing *BHP* and hence to a higher rate of production than a conventional one. Its advantages become apparent primarily if the flowing *BHP* is low and casing size is large. A chamber installation with its larger diameter makes the same volume of oil represent less head against the formation than a conventional installation.

Example 2.4—11. In a well characterized by the data given in Example 2.2—8, let the liquid accumulate in chambers of diameters ranging from 50 to 150 mm; the tubing itself is invariably of 2 7/8 in. size. Find the time required for 0.849 m³ of oil to accumulate, as well as the mean flowing *BHP* over this period, and the daily liquid production rate. — The values of h_{la} corresponding to the various chamber diameters can be calculated by means of the relationships $A_{wk} h_{la} = 0.849$. For simplicity, let

$$h_{lr} = \frac{90}{281} h_{la} = 0.320\, h_{la} \quad \text{and} \quad p_g = 0 \text{ bar}$$

in each case; Δp_{w1} and Δp_{w2} can be calculated using Eqs 2.4—24 and 2.4—25 respectively. In the knowledge of these data, we may use Eq. 2.4—21a to find the span of time needed for accumulation, 2.4—22 for mean drawdown and 2.4—23 for daily liquid production. The main data of the calculation are listed in Table 2.4—10. Let us note that, in conformity with the foregoing example, flowing *BHP* is taken to be invariably higher by 2.5 bars than tubing-shoe pressure.

Figure 2.4—39 shows plots v. chamber diameter of the t_2', n_c and q_o columns of Table 2.4—10. The Figure reveals accumulation time to decrease and daily cycle number and production to increase as chamber diameter increases. In the case examined, a chamber of 0.1 m diameter represents

Table 2.4—10

Chamber diameter	h_{la}	Δp_{w1}	Δp_{w2}	$\Delta \bar{p}_w$	t_2'	q_o	n_c
mm	m	bars			s	m³/d	
50	432.3	107.5	83	94.7	1640	30.3	43.9
62	281.0	111.5	95.6	103.5	1503	33.1	47.2
75	192.2	114	107	108	1443	34.6	48.8
100	108.1	116.2	110	113.4	1387	36.3	50.4
150	48.0	117.8	115	115.8	1358	37.1	51.2

an advantage of $(36.3 - 32.8) = 3.5$ m³/d, that is, a round 11 percent, over a conventional installation with 2 7/8 in. tubing ($d_i = 0.062$ m). If the chamber diameter could be increased to 150 mm, this would increase daily production only by a further 0.8 m³, that is, by 2.4 percent of the initial production. If the mean flowing *BHP* is to be reduced in an installation of given chamber diameter, then — other production parameters being equal — it is necessary to decrease the production per cycle, q_{oc}. This, however, entails an increase in specific injection gas requirement.

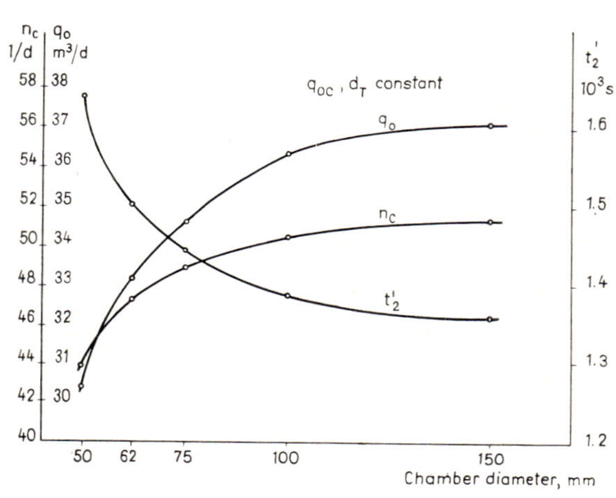

Fig. 2.4—39

Figure 2.4—40 shows five modern versions of chamber installations (largely after Winkler 1956). The common features of all five are that (i) gas injection is controlled by a surface time cycle controller, (ii) at the top of the chamber there is a bleed port or bleed valve to get rid of the formation gas accumulated between production cycles: (iii) the chamber practically reaches down to the well bottom, and (iv) injection gas fed to the chamber top enters the tubing only after having displaced the entire liquid volume.

A surface time cycle controller is required in each case because no liquid will rise opposite valve *1*, so that the pressure there will be no higher at the end of accumulation than at its beginning. Valve *1* is opened by periodical injection of gas from the surface, and remains open until casing pressure drops below its closing pressure. Bleed port *2*, or bleed valve *3*, lets the formation gas accumulated in the top part of the chamber escape into the tubing, thus permitting most of the chamber to be filled by liquid. If the formation gas were not bled off, the volume of liquid producible per cycle would be less and the specific injection gas requirement would be greater. Bleed port sizes at low *GOR* are 2—3 mm. Bleed ports have the drawback that they will bleed also during the liquid production phase. This increases gas requirements somewhat. This drawback is eliminated by the use of bleed valves which can be closed by a pressure differential of about 2 bars, that is, by the pressure buildup of starting production.

Standing valve *4* prevents the pressure rise during production from reacting on the formation. Cases (a), (b) and (e) in Fig. 2.4—40 are packer chamber installations. They can be used if the well is cased to bottom. The so-called bottle-chamber or insert-chamber solutions (c) and (d) are employed when the sand face is uncased. The *ID* of the chamber being

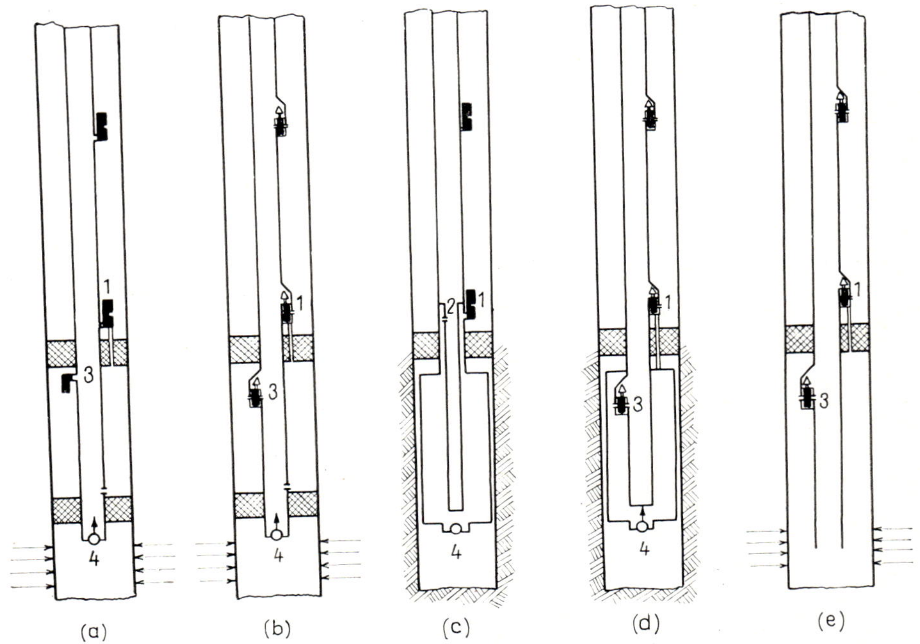

Fig. 2.4—40 Chamber installations, largely after Winkler and Smith 1962 (used with permission of CAMCO, Inc.)

less than the *ID* of the production casing, it does not boost production as much as a packer-chamber installation in a cased-to-bottom completion would. In (e), the well bottom is not sealed off during production. This results in a larger chamber diameter than in an insert chamber, but entails a higher mean *BHP*; also, pressure surges at the end of the production phase may trigger sand inrushes or a cave-in of the sand face.

2.4.5. Injection-gas supply

(a) Surface control of wells

All the control of a continuous-flow well consists in supplying injection gas at a suitable pressure and rate through a suitable choke to the casing annulus. In the knowledge of injection-gas line pressure and prescribed casing pressure and gas injection rate, the choke bore can be calculated using Eq. 1.5—4. Various types of control are used in intermittent lift installations. The most widespread two types are shown in Fig. 2.4—41. In (a), pressure regulator *1* ensures in the line section upstream of a motor valve controlled by a clock-driven time cycle pilot a constant line pressure slightly higher than the maximum opening pressure of the intermitting valve. At pre-set intervals and over pre-set spans of time, time cycle pilot *2* opens motor valve *3* and lets gas pass into the well. Such control does not require instant-action gas lift valves. Sector (a) of the two-pen pressure-recorder chart in Fig. 2.4—42 shows the change v. time of casing- and tubing-head pressures. The scale of tubing-head pressures is greater than that of casing-head pressures. Figure 2.4—43 shows a two-pen pressure chart with traces of $p_{CO} = f(t)$ and $p_{TO} = f'(t)$ recorded by a fictive instrument whose period of rotation precisely equals one production cycle. The time cycle pilot opens at the instant marked *1* and closes at *2*. The top of the liquid slug surfaces at *3*; the top of the lift gas (that is, the bottom of the liquid slug) surfaces at *4*. Between *4* and *5*, the well delivers first a mist, then pure gas to the flow line. — Time cycle pilots of a variety of types are employed. Their common feature is a rotating timing wheel provided with a suitable number of timing pins controlling the opening of the injection-gas line. The closing of the line is controlled either by the pins, or by the tubing pressure build-up caused by the surfacing of the slug. The clock-driven cycle controller for wells produced by intermittent natural flow, described in connection with Fig. 2.3—39, can by a change of assembly be made suitable for controlling gas lift wells also. Figure 2.4—44 shows a comparatively simple time cycle pilot that controls both the opening and the shut-off of the injection gas flow (Wieland 1961b). Motor valve *1* is closed in the state shown in the Figure. The pressure of injection gas flowing through the supply line *2* is reduced in reductors *3* to round 2 bars. Supply gas acts upon the diaphragm through the open valve *4*, depressing it to close the motor valve. If a pin *6* on clock-driven wheel *5* lifts arm *7*, then needle *8* obstructs orifice *9* and opens an annular orifice *11* where the supply gas depressing diaphragm *10* bleeds off. Supply gas now lifts diaphragm *10* making it open the attached valve *12* and close *4*. Supply gas depressing the diaphragm of the motor valve bleeds off through orifice *13*. The push down to close motor valve is now opened by a spring (not shown), opening the line to the passage of injection gas. When timing wheel *5* has rotated far enough to let fall arm *7* back into the position shown in the Figure, it is easy to see that the motor valve will again close. The duration of the motor valve will again close. The duration of the open phase may be adjusted by raising more or fewer timing pins *6* on wheel *5*. — In the control shown as Fig. 2.4—41b, pressure regulator *1* provides a constant gas pressure

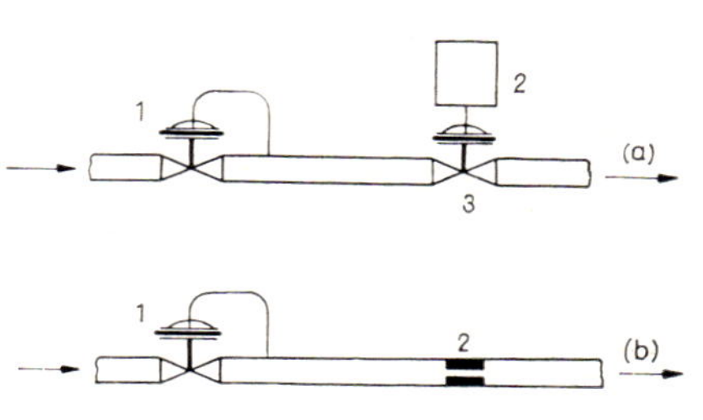

Fig. 2.4—41 Surface control of intermittent-lift installations

p_{gl} upstream of choke 2. According to whether the choke bore is comparatively large or small, there are two different types of control. (i) Comparatively large-bore choke. Valves installed in the well snap-acting, unbalanced. Regulator *1* provides upstream of the choke a pressure corresponding to the opening casing pressure required at the prescribed initial length h_{la}

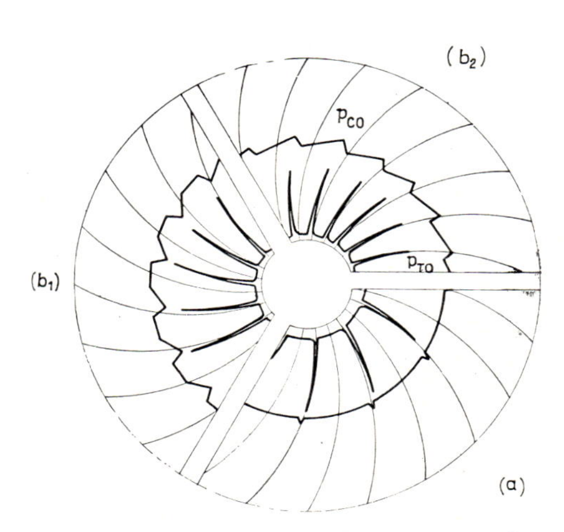

Fig. 2.4—42 Wellhead pressures of intermittent-lift wells under various surface controls

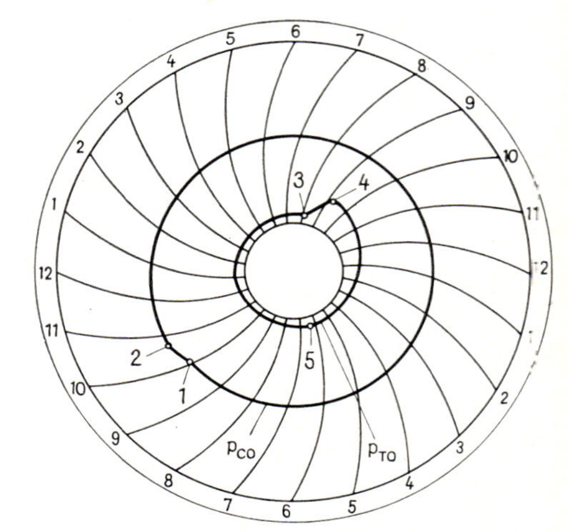

Fig. 2.4—43 Wellhead pressure chart of intermittent-lift well, controlled by clock-driven time cycle pilot, during one production cycle

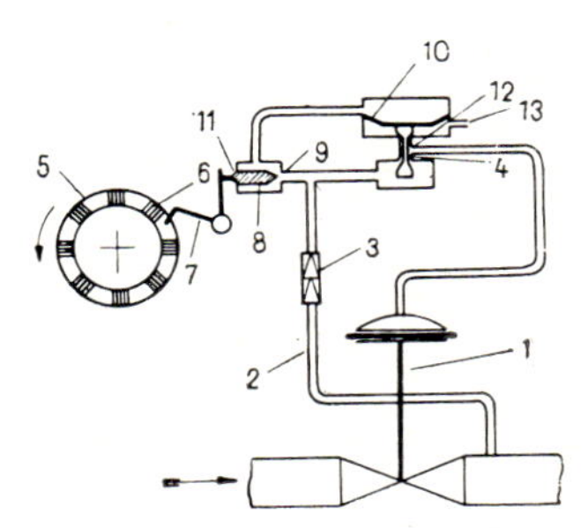

Fig. 2.4—44 Clock-driven time cycle pilot, after Wieland (1961b)

of the liquid slug. The annulus is filled through the choke comparatively fast to the required injection gas pressure, whereupon the regulator shuts off the injection gas line. Production starts when a liquid slug of the required length h_{la} has built up above the operating valve. The diameter of the surface choke is to be chosen so as to be, on the one hand, large enough to deliver to the annulus enough gas of sufficient pressure during the accumulation period and, on the other, not larger than necessary, so as to avoid a slow drop of casing pressure to the closing pressure of the intermitting valve in the production phase. A two-pen pressure-recorder chart of the tubing- and casing-head pressure changes during the cycle thus controlled is shown in part (b_1) of Fig. 2.4—42. (ii) Bore of choke comparatively small. Intermittent valves installed in the well snap-acting unbalanced. The regulator provides upstream of the choke a pressure in excess of the maximum required casing pressure. There is an uninterrupted flow of injection gas into the casing annulus and a corresponding uninterrupted pressure rise during the phase of accumulation. The intermitting valve opens when the resultant of casing pressure and of the likewise rising tubing pressure provides the force necessary to open it. A typical two-pen pressure-recorder chart of this type of operation is shown as part (b_2) of Fig. 2.4—42.

The most positive control permitting to attain a minimum of specific injection gas requirement is provided by the clock-driven time cycle pilot. It is, however, rather substantially costlier than the choke-type controller. It is most expedient to supply wells with injection gas through individual lines from a common constant-pressure source. The total injection gas volume used by the wells and supplied by the source is continually measured; those of the individual wells can be measured periodically or on a spot check basis.

(b) Analysing and trouble-shooting gas-lift installations

Measurements performed to analyse the operation of a gas-lift well are of two kinds: (i) subsurface measurements and (ii) surface measurements. — The first group includes pressure and temperature surveys and liquid level soundings. The most important measurement in the tubing of a continuous flow gas lift well is the pressure bomb survey. It can be performed in any well produced through the tubing rather than the annulus and may be expected to provide highly useful information. The survey is performed by lowering a pressure bomb into the tubing through a lubricator installed on the wellhead. Pressures are measured at a number of points, including a point directly below each valve. In wells with a fast-rising fluid, it may be impossible to lower the bomb against the well flow. This difficulty tends to arise in the top tubing section where low pressure makes the fluid expand and accelerate. The well is then shut off to permit insertion and lowering to a certain depth of the bomb; after reopening, the survey is started at some distance from the surface. Figure 2.4—45 is the record of a pressure bomb survey. The pressure traverse is seen to exhibit two breaks. Injection gas enters the tubing through two valves, 2 and

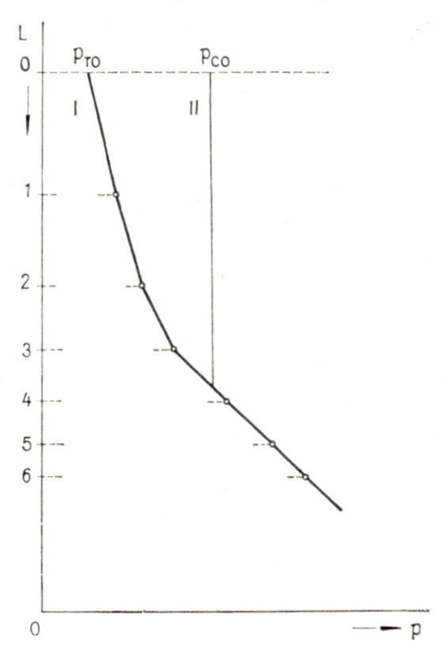

Fig. 2.4—45 Checking gas-lift valve operation by pressure bomb survey (after Brown 1967; by permission of Prentice-Hall, Inc., Englewood Cliffs, New Jersey, USA)

3, contrary to design. This may have several causes, e.g. (i) a dimensioning error (closing pressure of valve *2* less than flowing pressure of gas flowing through valve *3)*, (ii) dome pressure of valve *2* has decreased and cannot close valve, (iii) valve *2* is not suitably packed off in its seat. The Figure further reveals injection gas pressure opposite valve *3* to be higher than necessary. If wellhead pressure p_{TO} can be decreased, or casing pressure p_{CO} can be raised above the opening pressure of valve *4*, or if the valves can be re-spaced so that the depth of valve *4* is less, then the point of gas injection will be at valve *4*. This will reduce flowing *BHP* and increase the rate of production. — In an intermittent well, it is inexpedient or indeed impossible to run a wire-line pressure-bomb survey. The liquid slug may rise fast enough to sweep up the bomb, snarl up or tear the wire line. If a survey cannot be dispensed with, then the instrument should be lowered in the accumulation period and efficient precautions are to be taken to ensure its remaining below the operating valve during the production period. A temperature survey of the tubing string will permit the location of the depth of injection, as expansion will reduce temperature below the ambient value. Undesirable injection of gas may be due to imperfectly sealed tubing joints, a tubing leak, or a valve that has failed to close. Liquid level soundings usually are of a subordinate importance. If required, liquid levels may be sounded by means of an acoustic survey.

Surface measurements to check gas lift wells include measurements of casing and tubing pressures, mainly with recording pressure gauges; metering liquid and gas production; metering injection gas volumes; and pressure measurements.

The continuous recording of casing and tubing pressures is of a particular importance in intermittent wells. The recorder charts will show up correct operation and permit the diagnosis of a variety of malfunctions, such as: (i) A casing pressure drop between production cycles usually indicates a tubing or casing leak or improper valve closure. If the leak is between the tubing and the annulus, then gas will rise in the tubing also in the accumulation period. (ii) If wellhead pressure rises very high during production, then the choking beyond the wellhead has to be reduced. (iii) If casing pressure is normal, and tubing pressure exhibits no periodic increase, then the valve or the tubing is obstructed. (iv) If the tubing pressure build-up during production is small and very short, then cycle frequency is too great, and vice versa. (v) If the opening and closing casing pressures have changed, then the injection gas has started to enter the tubing through a different valve, or dome pressure in the operating valve has changed. (vi) The pressure charts will reveal when the well is able to produce also without injection.

The pressure recorder is installed next to the well but not on the wellhead, because the surfacing of the liquid slug may entail vibrations affecting the record. Recording tubing and casing pressures in continuous-flow gas lift wells will likewise provide useful information. The continuous recording of wellhead pressures is not usually necessary, however. — Figure 2.4—46 shows wellhead pressure recorder charts of three malfunctioning wells (somewhat modified after Brown 1967). In (a), it is primarily the casing pressure diagram that reveals two valves to be in simultaneous operation. The tubing pressure diagram shows the production period to be too long. The cause of the fault is an inadequate supply of injection gas to the tubing through the intended operating valve and too low a cycle frequency. In (b), the well produces at too high a cycle frequency. The rapid drop in both

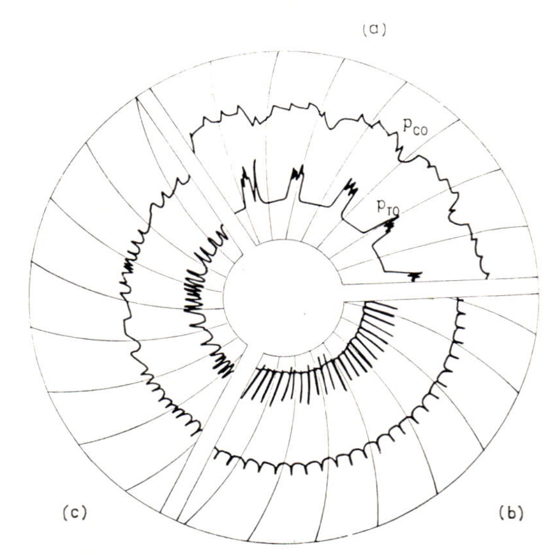

Fig. 2.4—46 Wellhead pressure charts of malfunctioning intermittent-lift wells, after Brown 1967, constructed by using parts of flow diagrams 14—14(12), 14—14(20) and 14—14(23) (by permission of Prentice-Hall, Inc., Englewood Cliffs, New Jersey, USA)

casing and tubing pressure shows liquid production per cycle to be too small. The high cycle frequency entails too high a specific injection gas requirement. In part (c) of the Figure, the time cycle controller is out of kilter.

(c) Gas supply system

Injection and lift gas is supplied by a gas well or by a compressor station. If the gas production rate of the well and/or the injection gas requirements of the gas lift wells tend to fluctuate, then in order to ensure a smooth supply of gas to the compressor station and an adequate covering of the fluctuating well demands the setting up of a suitable surface supply system is required.

Figure 2.4—47 shows a so-called closed rotative gas lift system in which said functions are discharged by various facilities. The Figure is essentially an outline of possible options, and it is not usually necessary to realize all of them. — Oil and gas produced from a number of wells, including a flowing group K_1, a continuous-flow gas lift group K_2, an intermittent lift gas lift group

K_3, and a pumped group K_4 are delivered to test separator *1* and production separator *2*. Oil is collected in stock tank *3*, where it is gauged and treated to a certain extent before removal. Low-pressure gas from the wells is led to compressor station *4*, or into sales line *7*. Gas intended for injection enters the intakes of the compressors through conduit *5*, whereas conduit *6* supplies the compressor engines with fuel. The station compresses gas for repressuring and for gas lifting. The latter is fed through line *8* to the gas lift distribution centers *10*, whereas the former passes through line *9* to the repressuring wells K_{rp}. The well groups K_2 and K_3 are supplied from the

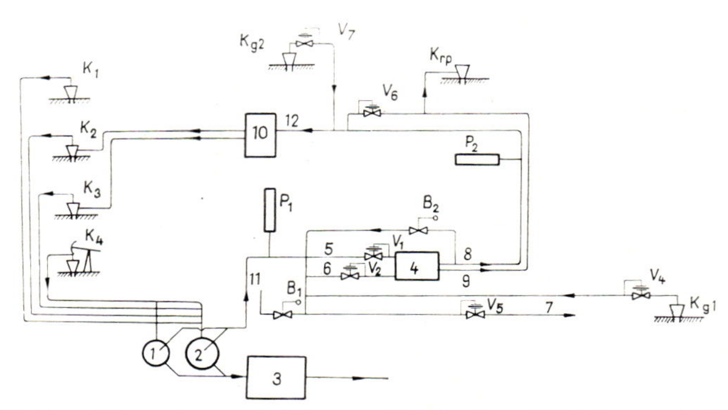

Fig. 2.4—47 Gas lift system, slightly modified after Winkler and Smith (1962; used with permission of CAMCO, Inc.)

distribution centers *10*. — Gas production from a number of flowing, continuous-flow gas lift, and pumped wells is likely to give a fairly smooth gas supply to the compressors. Further smoothing can be achieved by using a high-capacity pipeline or a number of unproductive wells as a buffer gas tank P_1. Regulators V_1 and V_2 regulate the pressure of gas entering the compressors. If pressure drops too low in lines *5* and *6*, valve V_4 automatically supplies make-up gas from the low-pressure gas well K_{g1} to the compressors. If that will not help, valve V_5 throttles or shuts off the sales line. If pressure in the low-pressure system grows too high, then pressure in the sales line is increased first by means of valve V_5; when this has reached a permissible maximum, the excess gas is flared through back-pressure regulator B_1 and vent *11*. The gas storage capacity of high-pressure line *12* is augmented by a buffer tank P_2 similar to P_1. When pressure decreases in this line, valve V_6 automatically connects it with the repressuring line, or alternatively, valve V_7 opens up automatically the high-pressure gas well K_{g2}. If the pressure in the system mounts too high, then by-pass pressure relief regulator B_2 bleeds off the excess pressure into the low-pressure system.

2.4.6. Plunger lift

(a) Operating principles; design features

The plunger lift is a peculiar version of intermittent gas lift. Its main feature is a piston (plunger) inserted in the tubing and separating the rising liquid slug from the gas column lifting it, with the effect of considerably reducing gas break-through and liquid fallback. The arrangement permits the utilization of formation gas pressure to help lift the well fluid. Indeed, plunger lift will permit the unassisted production of some wells that will not produce by continuous natural flow. Gas pressure in the casing annulus acts on the well bottom, wherefore plunger lift does not permit the realization of low *BHP*s. — The two fundamental types of plunger lift employed in production practice are those without and with time cycle control.

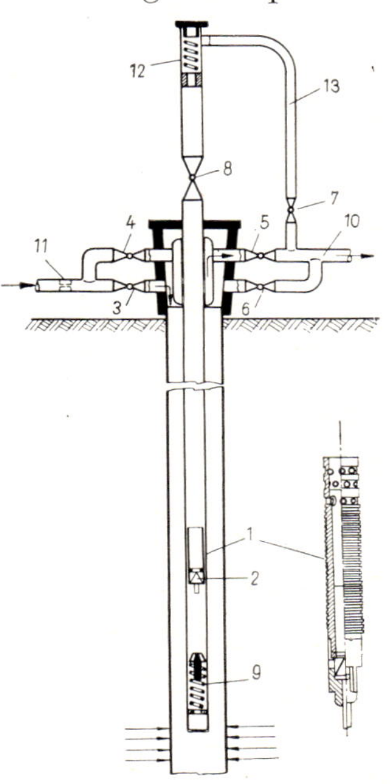

Fig. 2.4—48 Hughes' plunger lift

The operation of the first type essentially agrees with that of the plunger lift patented by Hughes in 1927. It is shown in Fig. 2.4—48. In normal operation, injection gas — if required — enters the annulus through line *11* and open valve *3*. Let us assume that plunger *1* with its valve *2* closed sits on the bottom shock absorber. There is above it a short liquid column in the tubing. When the pressure force of gas accumulating in the annulus and acting upon the plunger exceeds the weight of the plunger plus the weight of the liquid and gas column above it, the plunger rises. Liquid flows through the wellhead perforations of the tubing and the open valve *5* into flow line *10*. Valves *4*, *6*, and *7* are closed. The plunger cannot rise beyond the bumper spring *12*. Pressure drop under the plunger makes valve *2* open and lets the plunger descend. Impact on downhole bumper spring *9* closes valve *2*; this makes the plunger ready to rise again. — This type of plunger lift is uneconomical in low-capacity wells because (i) the plunger starts to rise directly after impact on the downhole bumper spring and to lift such fluid as has accumulated during one full cycle of its travel. Thus if the length of this column is small, only a small portion of gas energy expended will do useful work, since plunger weight is the same irrespective of the weight of liquid above it. (ii) Between plunger and tubing — particularly those of the early rigid-seal type — there may be a substantial gap permitting the fallback of an appreciable fraction of the liquid slug. The relative amount of this fallback is high if the slug is small. Finally, (iii), during the fall of the plunger, gas can escape from the tubing without having done useful work.

The economic benefits of plunger lifting can be extended into the low-capacity range of wells by using a plunger lift controlled by a cycle controller. The well completion itself resembles that in Fig. 2.4—48. The surface equipment is shown in Fig. 2.4—49. Injection gas — if required —

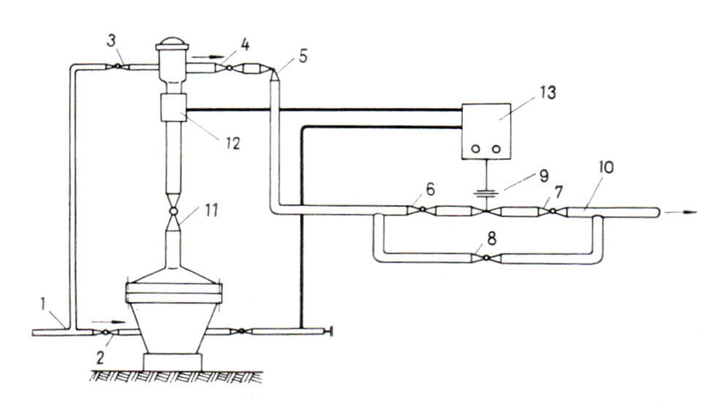

Fig. 2.4—49 Surface equipment of plunger-lift installation controlled by a time cycle controller

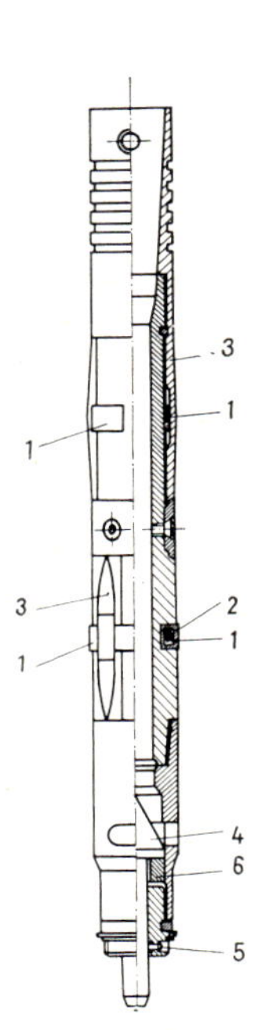

Fig. 2.4—50 National-make elastic-seal plunger

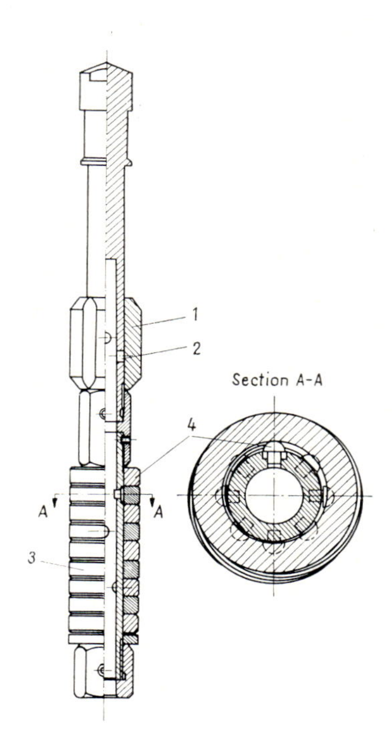

Fig. 2.4—51 Merla's elastic-seal plunger

flows during normal operation into the annulus through line *1* and open valve *2*. Valves *3* and *8* are closed; valves *11*, *4*, *6* and *7*, as well as choked valve *5*, are open. The opening and closing of motor valve *9* is controlled by cycle pilot *13*. There are two widespread types of control, that is (i) opening is initiated by a clockwork mechanism, (ii) opening is initiated by a rise in casing pressure. Closing is controlled in both cases by the surfacing of the plunger. Regardless of the type of control, the result is the same, namely, a decrease of cycle frequency, by letting the plunger rise only when enough liquid has accumulated in the tubing above it. Lubricator *12* contains a mechanical or magnetic sensor detecting the proximity of the plunger; it is equipped to send a pneumatic or hydraulic signal to cycle pilot *13* which thereupon instructs motor valve *9* to shut off the flow line. The volume of liquid lifted per cycle can be varied over a fairly wide range; also, no gas can escape from the well during the plunger's descent.

In order to ensure a better seal between the plunger and the tubing wall, plastic-seal plungers are increasingly employed. The plunger in Fig. 2.4—50 is a construction of the National Co. Split-ring seals *1* are pushed outward by springs *2*. The maximum displacement of the split rings is limited by ribs *3*. Valve *4* stays shut during ascent owing to the pressure differential across the plunger; it is fixed in place by hasp *5*. The valve opens after surfacing and is kept in the open position by magnet *6*. — The sealing element of the Merla-make plunger in Fig. 2.4—51 is the plastic sleeve *1*. Friction against the tubing makes it close opening *2* on ascent and open it on descent. The plastic seal rings *3* can move sidewise independently of each other. During ascent, the small pistons *4*, actuated by the higher pressure within the plunger, push the rings eccentrically against the tubing wall. Section *A-A* shows in an axial view how the aggregate deformation of the rings manages to obstruct the entire aperture of the tubing. Numerous other solutions are known.

Plunger lift represents an advantage when producing waxy oils and those liable to form stable emulsions. Wax deposits in the tubing are scraped off by the plunger as they are formed; mixing leading to the formation of a stable emulsion is limited, because gas and liquid are comparatively well separated during their upward travel. Plunger lift is used also in gas wells producing also water and/or condensate; the latter, settling at the well bottom, result in an increase of *BHP*. Plunger lift with or without cycle control removes the liquid as it forms and keeps the *BHP* at a low value. — In certain cases, gas is produced through the annulus, whereas liquid is produced by plunger lift through the tubing. Plungers can be used to advantage in intermittent gas lift as well as in simple plunger lift installations. In numerous Hungarian oil wells, plungers are employed in completions shown in Fig. 2.4—12, incorporating pressure-operated valves. The presence of the plunger reduces fallback, so that in a favourable case production is improved and specific injection gas requirement reduced. In intermittent wells producing a waxy oil the plunger performs the task of dewaxing, too. The economically attainable flowing *BHP* is in this case generally lower than in plunger lifting in the strict sense, but a higher injection gas pressure is required and formation gas cannot be exploited to lift liquid.

(b) Designing the plunger-lift operation

Evaluating operating data of 145 wells, C. M. Beeson, D. G. Knox, and J. H. Stoddard (1958) have written up relationships describing the operation of cycle-controlled plunger lifts using expanding positive-seal plungers. Table 2.4—11 lists some of the typical data of the wells analysed. The

Table 2.4—11

	Symbol	Unit	d = 2 3/8 in.			d = 2 7/8 in.		
			max	min	mean	max	min	mean
Tubing length	L_T	m	930	3537	2035	1038	3574	2534
Production	q_o	m^3/d	0.7	10.0	5.1	1.6	17.5	7.1
Production per cycle	q_{oc}	m^3	0.02	0.46	0.11	0.03	0.86	0.32
Oil density	ϱ_o	kg/m^3	780	850	835	797	910	857
WOR	R_w	%	0	87	17	0	89	13

fundamental equations derived by the authors using the correlation method are, transposed into SI units, as follows. For 2 3/8 in. tubing:

$$p_{CO\max} - p_{TO\min} = 3.376 \times 10^6 q_{oc} + 187.5 L_T + 2577 q_{oc} L_T + 4.648 \times 10^5; \qquad 2.4\text{—}33$$

$$R_{go} = \frac{10^{-3} L_T}{q_{oc}} (3.018 \times 10^{-3} L_T + 1.043 \times 10^{-5} p_{TO\min} + 25.92) + 117.6; \qquad 2.4\text{—}34$$

$$p_{CO\max} - p_{CO\min} = 3.545 \times 10^5 q_{oc} + 77.61 L_T + 2 \times 10^{-2} p_{TO\min} + 6.827 \times 10^4. \qquad 2.4\text{—}35$$

For 2 7/8 in. tubing:

$$p_{CO\max} - p_{TO\min} = 1.494 \times 10^5 q_{oc} + 130.7 L_T + 1582 q_{oc} L_T + 6.303 \times 10^5 ; \qquad 2.4\text{—}36$$

$$R_{go} = \frac{10^{-3} L_T}{q_{oc}} (1.457 \times 10^{-2} L_T + 9.940 \times 10^{-6} p_{TO\min} + 23.6) + 71.39 ; \qquad 2.4\text{—}37$$

$$p_{CO\max} - p_{CO\min} = 1.647 \times 10^5 q_{oc} + 210.7 L_T + 5.1 \times 10^{-2} p_{TO\min} - 3.014 \times 10^5 . \qquad 2.4\text{—}38$$

In the above equations, $p_{CO\max}$ is the maximum and $p_{CO\min}$ the minimum casing pressure, $p_{TO\min}$ is the least tubing-head pressure during normal production, $q_{o\max}$ is the maximum daily production to be achieved by plunger-lifting a given well; R_{go} is the total specific gas volume required for production. The maximum possible production can be calculated out of mean data concerning the ascent and descent of the plunger. The authors have found that the mean velocity of plunger ascent is 5 m/s until the liquid plug surfaces. Descent velocity in pure gas is about twice this value. Velocities are less during the evacuation of the liquid, on the one hand, and during descent through liquid, on the other. The minimum cycle frequency determined purely by the ascent and descent times of the plunger — that is, assuming that the plunger immediately rebounds from the downhole bumper spring without any rest period — is, for 2 3/8 in. tubing,

$$t_c = 0.295\, L_T + 3023\, q_{oc}\,,$$

and for 2 7/8 in. tubing,

$$t_c = 0.295\, L_T + 2267\, q_{oc}\,.$$

The maximum possible daily production by plunger lift can be calculated by substituting the above expressions into the equation

$$q_{o\max} = n_c\, q_{oc} = \frac{86{,}400\, q_{oc}}{t_c}\ \mathrm{m^3/d}\,,$$

which yields, for 2 3/8 in. tubing,

$$q_{o\max} = \frac{86{,}400\, q_{oc}}{0.295\, L_T + 3023\, q_{oc}}\ \mathrm{m^3/d}\,, \qquad 2.4\text{—}39$$

and for 2 7/8 in. tubing,

$$q_{o\max} = \frac{86{,}400\, q_{oc}}{0.295\, L_T + 2267\, q_{oc}}\ \mathrm{m^3/d}\,. \qquad 2.4\text{—}40$$

Using their own fundamental equations, the authors have prepared nomograms and proposed procedures for operation design. In the following I shall outline a process based on the same fundamental relationships, but somewhat different from the Beeson—Knox—Stoddard method, and in better keeping with our own design principles.

The mean flowing *BHP* is, in a fair approximation,

$$\bar{p}_{wf} = \left[p_{CO\max} - \frac{1}{2}(p_{CO\max} - p_{CO\min})\right](1 + CL_T)$$

where C is the weight correction factor, of dimension $\left[\frac{N}{m^2}\bigg/\frac{Nm}{m^2}\right]$, of a gas column of height of 1 m. Let the daily inflow from the formation be described by the relationship

$$q_o = 86{,}400\, J(p_{ws} - p_{wf})\ \mathrm{m^3/d}\,.$$

Let us substitute, assuming a 2 3/8 in. tubing, the expression of $p_{CO\,max}$ from Eq. 2.4—33 and the expression of $(p_{CO\,max} - p_{CO\,min})$ from Eq. 2.4—35. Rearranging, we obtain the relationship

$$q_o = 86{,}400\, [Jp_{ws} - J(1 + CL_T)\,(0.99\, p_{TO\,min} + 148.7\, L_T + 4.307 \times 10^5) - J(1 + CL_T)\,(2577\, L_T + 3.198 \times 10^6)\, q_{oc}]\ \mathrm{m^3/d}\,. \qquad 2.4-41$$

In the same manner, using Eqs 2.4—36 and 2.4—38, we may derive for 2 7/8 in. tubing

$$q_o = 86{,}400\, [Jp_{ws} - J(1 + CL_T)\,(0.975\, p_{TO\,min} + 25.35\, L_T + 7.81 \times 10^5) - J(1 + CL_T)\,(1582\, L_T + 6.710 \times 10^4)\, q_{co}]\ \mathrm{m^3/d}\,. \qquad 2.4-42$$

Example 2.4—12 serves to elucidate the application of these relationships. Let $d_T = 2\ 7/8$ in.; $d_i = 0.062$ m; $p_{TO\,min} = 2.0$ bars; $L_T = 1440$ m; $J_1 = 7.26 \times 10^{-12}$ m^5/sN; $p_{ws} = 53.9$ bars and $C = 8.64 \times 10^{-5}$ l/m. Let us calculate the operating conditions to be expected at $p_{wf} = 25.5$ bars. The tubing is run to bottom. — In Fig. 2.4—52, line $q_{o1} = f(q_{oc})$ is calculated using Eq. 2.4—42. Let us plot the maximum feasible production v. cycle production using Eq. 2.4—40, and specific injection gas requirement, using Eq. 2.4—37. On the left-hand side of the diagram, the line $q_{o1} = f'(p_{w1})$ characterizing inflow is plotted. It is seen that at a flowing *BHP* of 25.5 bars, daily production will be 1.8 m^3. The cycle production corresponding to point of intersection B_1 is 0.55 m^3; the specific injection gas requirement corresponding to point of intersection C is 200 m^3/m^3; and finally, the value of $q_{o\,max} = 28.6$ m^3/d corresponding to point of intersection D indicates that the designed production is technically feasible. Let $J_2 = 7.26 \times 10^{-11}$ m^5/sN, other well parameters being equal. Inflow into the well is characterized by the line $q_{o2} = f(p_{w2})$. Production, at the same

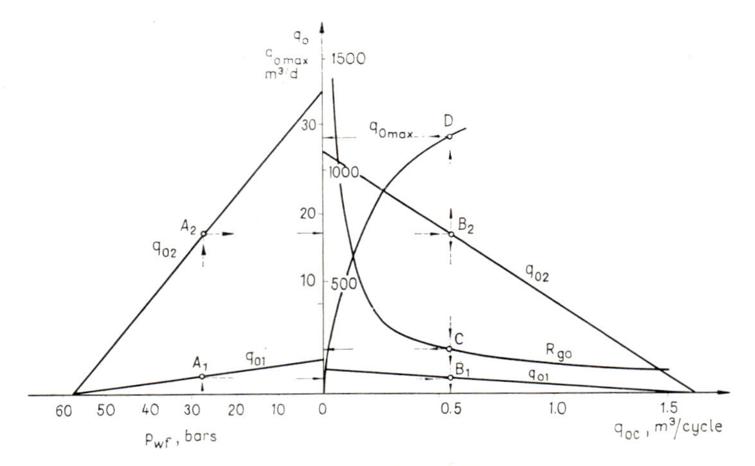

Fig. 2.4—52 Design of plunger-lift operation

flowing BHP of 25.5 bars, is 17.8 m^3; $q_{o\,max}$ and R_{go} remain constant at 28.5 m^3/d and 200 m^3/m^3, respectively. Figure 2.4—52 reveals that the flowing BHP can only be decreased by increasing the specific injection gas requirement. For instance, to establish a flowing BHP of 14.7 bars an increase of gas injection from 200 to 535 m^3/m^3 is required in the low-productivity well, causing only a very slight rise in production, from 1.8 to 2.0 m^3/d. A flowing BHP of 14.7 bars cannot be realized by plunger lift in the higherproductivity well since a daily inflow of 24.5 m^3 is higher than the technically feasible $q_{o\,max} = 16.5$ m^3/d.

The above procedure, as has been shown, permits us to check whether a prescribed flowing BHP can be realized by means of plunger lift, and if so, what specific injection gas requirement is to be expected.

Chapter 3

Producing gas wells

A gas well is a flowing well producing predominantly gaseous hydrocarbons. The gas may contain subordinate amounts of liquid hydrocarbons and water. Condensate, that is, the hydrocarbons produced in gas form but liquid under surface conditions, is a colourless or pale liquid composed of low-molecular-weight hydrocarbons. *GOR* is in the order of ten thousand at least. Gas wells may therefore be regarded also as oil wells with a high (sometimes infinite) *GOR*. Our statements in Section 2.3 hold in many respects also for gas wells. In this chapter we shall aim at presenting those features of gas wells which differ from those of flowing oil wells. The first subject to be tackled will be a productivity analysis of gas wells; the compressibility of gas being much greater than that of most well fluids composed of oil and gas, the flow of gas in the reservoir is governed by relationships other than those discussed in connexion with the performance of oil wells. The often very high pressure, high temperature and possible corrosivity of gas raise the need of completing wells with a view to these features.

3.1. Well testing, inflow performance curves

The characteristic performance relationships of gas wells differ from those of oil wells, because several physical parameters of the gaseous medium in the reservoir are much more dependent on changes in pressure. Assuming flow to be plane radial, steady-state, laminar and single-phase, the solution of Darcy's Eq. 2.1—1 becomes

$$q_{gn} = \frac{\pi hk T_n (p_{ws}^2 - p_{wf}^2)}{\bar{\mu} p_n T \bar{z} \ln \frac{r_e}{r_w}} . \qquad 3.1-1$$

In a given well, the expression

$$\frac{\pi hk T_n}{p_n T \ln \frac{r_e}{r_w}}$$

is practically constant and so is, approximately, also $\overline{\mu z}$. Denoting the constant part by C, we get

$$q_{gn} = C(p_{ws}^2 - p_{wf}^2) .$$

In reality, the mean pressure of gas flowing in the reservoir will vary with *BHP* and this will entail a variation also of the mean compressibility factor $\bar{z}$ and the mean viscosity $\bar{\mu}$. Flow is often turbulent, especially close to the well. For these reasons, Rawlins and Schellhardt have modified the above relationship and derived the following equation describing actual conditions in steady-state flow (Katz 1959):

$$q_{gn} = C(p_{ws}^2 - p_{wf}^2)^n \quad . \qquad 3.1-2$$

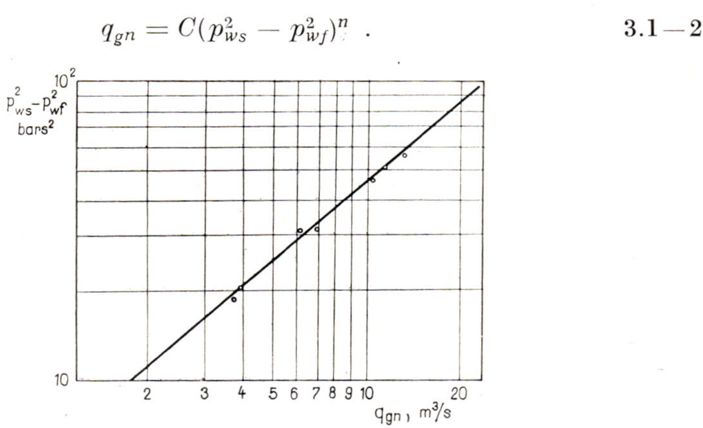

Fig. 3.1—1 Inflow performance curve of gas well ÖK-17, Hungary

This relationship, which is borne out fairly well by actual fact, is usually plotted in a bilogarithmic system of coordinates, in which case the inflow performance curve is a straight line. Figure 3.1—1 shows the inflow performance curve of the Hungarian gas well ÖK—17. The value of n is in the range from 0.5 to 1.0. If it is outside this range, then the well test has been incorrectly run and has to be repeated. Wrong results will be obtained also if liquid accumulates in the well during the test or, if the steady-flow method has been employed and the flow could not stabilize while testing at each individual operating point. The above relationship will be strictly valid for any given gas well if the fluid flowing in the reservoir contains no liquid phase. It is a fair approximation, however, also of conditions in gas and oil wells with high *GOR*s. While testing such wells, special care must be taken to avoid the formation of liquid slugs in the well during the test. In wells producing wet gas it is indicated to produce at a high rate for several hours (up to 24) in order to clean the well. Of the several calculation methods developed to calculate the least gas flow rate that will still prevent the condensation of liquid at the well bottom and the formation of a liquid slug, we shall discuss here the theory and calculation method of Turner, Hubbard and Dukler (1969).

No liquid will settle at the bottom of a well if the velocity $v_{g\min}$ of the gas flow is equal to or greater than the fall velocity of the largest liquid drop (more precisely, than its steady-state or terminal velocity). The fall velocity of smaller droplets is less, so that at the velocity $v_{g\min}$ defined by this hypothesis the entire dispersed phase will be lifted to the surface by the gas flow. The diameter of the largest drop, which is assumed to be spherical for simplicity, is determined by its kinetic energy and surface tension. On the basis of these, the fall velocity of the largest drops has been derived so as to equal, by hypothesis, the minimum gas velocity:

$$v_{g\min} = C \frac{\sigma^{0.25}(\varrho_l - \varrho_g)^{0.25}}{\varrho_g^{0.5}} \qquad 3.1-3$$

where C is a constant whose numerical value is provided by the theoretical considerations referred to. For a hydrocarbon condensate, $\sigma = \sigma_o$ approximately equals 0.02 N/m and $\varrho_l = \varrho_o = 721$ kg/m^3. For water, $\sigma = \sigma_w = 0.06$ N/m and $\varrho_l = \varrho_w = 1007$ kg/m^3. Substituting these into Eq. 3.1—3, and increasing the constant C by 20 percent to be on the safe side, we obtain for condensate

$$v_{gc\min} = \frac{1.71(67 - 4.5 \times 10^{-7}\, p)^{0.25}}{(4.5 \times 10^{-7}\, p)^{0.5}}, \qquad 3.1-4$$

and for water

$$v_{gw\min} = \frac{1.23(45 - 4.5 \times 10^{-7}\, p)^{0.25}}{(4.5 \times 10^{-7}\, p)^{0.5}}. \qquad 3.1-5$$

For gas of temperature T flowing through a section of area A at a velocity v_g, the combined gas law (cf. the derivation of Eq. 1.2—10) gives

$$q_{gn} = \frac{T_n\, p v_g A}{p_n z T}. \qquad 3.1-6$$

Replacing v_g by the expression for $v_{gc\min}$ furnished by Eq. 3.1—4 for wells producing gas plus hydrocarbon condensate, or by the expression for $v_{gw\min}$ furnished by Eq. 3.1—5 for wells also producing water, we get $q_{gn} = q_{gn\min}$ as the least gas flow rate that will still prevent the formation of a liquid slug in the well. — The above relationships hold of course for production through the annulus as well as through the tubing. Gas flow velocity in steady-state flow in a given well is slowest at the tubing shoe, where pressure is greatest; $q_{gn\min}$ is therefore to be determined using the parameters valid there. In exceptional cases involving fast cooling on ascent or slow decrease of pressure, the point of slowest flow may be situated farther up the well.

Example 3.1—1. Find the least rate of production preventing liquid slug formation in a well producing also water, in which the slowest gas flow is at the shoe of the tubing through which the gas is being produced. $d_T = 2\ 7/8$ in. ($d_i = 0.062$ m). $p_{TL} = 100$ bars; $T_{TL} = 330$ K; $p_n = 1.01$ bar; $T_n =$

$= 288.2$ K; $M_g = 21$ kg/kmole ($p_c = 46$ bars and $T_c = 224$ K). — By Eq. 3.1—5,

$$v_{gw\,min} = \frac{1.23(45 - 4.5\times 10^{-7}\times 100\times 10^5)^{0.25}}{(4.5\times 10^{-7}\times 100\times 10^5)^{0.5}} = 1.46 \text{ m/s} ;$$

$$p_r = \frac{p}{p_c} = \frac{100\times 10^5}{46\times 10^5} = 2.18 \quad \text{and} \quad T_r = \frac{T}{T_C} = \frac{330}{224} = 1.47 .$$

Figure 8.1—2 furnishes a $z = 0.83$. By Eq. 3.1—6,

$$q_{gn\,min} = \frac{288.2\times 100\times 10^5\times 1.46\,\dfrac{0.062^2\,\pi}{4}}{1.01\times 10^5\times 0.83\times 330} = 0.459 \text{ m}^3\text{/s} = 3.96\times 10^4 \text{ m}^3\text{/d} .$$

The results of well tests are affected also by the circumstance that the temperature of the flowing gas is modified by the test. If e.g. a flowing gas well is shut in, then the wellhead pressure will first increase, but may subsequently decrease as the well cools off. Testing should therefore be carried out so as to change the temperature of the gas stream little or not at all. This can be achieved by producing the well for a longer period at a comparatively high rate before testing, in order to bring about a comparatively wide warm zone around the well. The temperature of gas rising in a gas well is influenced by a number of factors even if flow is steady. These include the heat released into the surrounding rocks, expansion of the gas, phase transitions during ascent, increase in potential energy and decrease in kinetic energy (Pápay 1970). Little is known to the present author about the accuracy of the various relevant calculation methods published in literature. It is best to determine the *BHP* of the shut-in well by means of a pressure bomb and to calculate the flowing *BHP*s out of the wellhead pressures, because in wells producing wet gas the formation of a liquid slug after shut-off cannot be avoided. The quantity of accumulated liquid is not known. The lengths and consequently the hydrostatic pressures acting on the well bottom of the gas and liquid column in the well are consequently unknown, too. During production, on the other hand, gas velocities prevailing in the tubing may be high enough to sweep up the conventional wireline-operated pressure bomb. If the *ID* of the tubing is large enough as compared with the *OD* of the pressure bomb, and gas velocities are rather low, there can be no objection to subsurface pressure surveys.

Equation 3.1—2 can be established by several well testing methods. Three methods are widely used: the steady-flow test, the isochronal test and the Carter method. — The steady-flow test is used only if reservoir permeability is rather high, as otherwise testing at any operating point may take days and even weeks; flow conditions will not stabilize any sooner. The two other tests are, on the other hand, best suited precisely for the testing of this type of well; flow during these tests is invariably transient. Equation 3.1—2 valid for steady flow is usually established after suitable processing of the data furnished by these one-day or even shorter tests.

3.1.1. The steady-flow test

This test is called, with some ambiguity, also the back-pressure or multipoint test. It fundamentally consists in measuring stabilized open flow of the well with four chokes of different diameter built in in succession, and the flowing *BHP* recorded or calculated for each choke. The static *BHP* is determined out of shut-in data. After the stabilization of flow and *BHP* with a given choke in place, the test can be resumed immediately after changing the choke. This is the feature which gave the test its name. In wells of comparatively low flowing temperatures producing dry gas, successive flow rates should increase as this reduces test duration as compared with the opposite sequence. If the well is in addition of comparatively small capacity, then wellhead pressure is to be reduced by at least 5 percent in the first stage and by at least 25 percent in the fourth. The operating points will thus be far enough apart, which improves the accuracy of establishing the performance curve. If the capacity of the well is comparatively high, one has to be contented with a smaller terminal pressure reduction. — If the well produces liquid, too, or if the flowing temperature is comparatively high, then the flow rate should be highest in the first stage. This results in any liquid accumulated in the well being swept out without the intercalation of a 'purifying interval' in the first case; in the second, the advantage of this measure lies in the faster stabilization of temperatures around the well. It is best to carry out the first test 2—4 hours after opening the well. During subsequent production, pressures and outflow temperatures are recorded at intervals of 30 minutes, until they become stabilized. Then the rate of production is measured. — The line shown in Fig. 3.1—1 has been established by a steady-flow test that has furnished the points plotted in the Figure. The test, performed at a larger-than-usual number of operating points, has resulted in a plot providing a fair fit to a straight line.

3.1.2. The isochronal test

In this test, the well is first produced for a while through a comparatively small-bore choke, and the tubing pressure and gas flow rate are measured at predetermined intervals of time, say, 1/2, 1, 2 and 3 hours. The well is then shut in until the pre-opening wellhead pressure builds up again. Now the well is reopened and produced through a larger-bore choke; tubing pressures and rates of production are measured at the same intervals of time. The procedure is repeated, usually with two larger-bore chokes. During production, flow must not be hampered by any operation. By restarting each phase of the test from the initial wellhead pressure, distortions of the flow pattern about the well by previous stages of testing can be avoided. The piezometric surface visualized above the horizontal plane passing through the well bottom is rather simple, its shape being determined solely by the circumstances of flow in the current phase of the test.

The instantaneous radius of influence of a given gas well has been shown to depend solely on the dimensionless time N_t and not on the rate of production (Cullender 1955). Dimensionless time is

$$N_t = \frac{k_g t}{\Phi \bar{\mu}_g c_g r_w^2} . \qquad 3.1-7$$

Hence, if several tests of equal duration are successively performed at different terminal rates of flow, the radius of influence will be the same for each, provided each test is started from a state of static equilibrium in the formation.

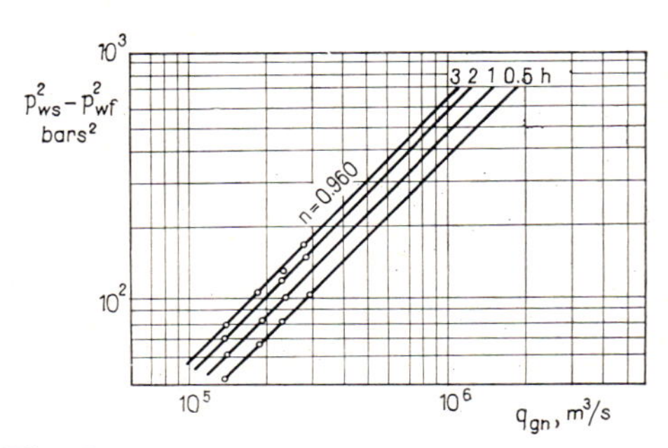

Fig. 3.1—2 Isochronal well performance curves

Each set of points $[(p_{ws}^2 - p_{wf}^2), q_{gn}]$ belonging to a given test duration (isochronal points) defines a well performance curve that can be described by Eq. 3.1—2. The exponent n of the equation is the same for all parameters N_t, and the coefficient, denoted C' to indicate transient flow, decreases with the duration of the test. Figure 3.1—2 shows the performance curves of such a test. In bilogarithmic representation, the curves take the form of parallel lines, shifting towards lower rates of production as time goes by. The equation of the performance curve for stabilized flow can in principle be established in two ways. (i) By producing the well at one of the chokes until the rate of production stabilizes. Substitution of the q_{gn} and p_{wf} values thus obtained into Eq. 3.1—2 permits us to calculate the value of C. (ii) The performance equation for stabilized flow is determined from the isochronal curve. This latter calculation is based on the consideration that the radius of influence r_e' of the well will monotonely increase with time until it attains the radius pertaining to steady-state flow. On the circumference of the circle defined by the radius r_e' reservoir pressure will be $p_e = p_{ws}$, and the *BHP*, p_{wf}, will remain unchanged. The increase of r_e' consequently entails a decrease in the mean pressure gradient of flow within the formation, which in turn reduces the rate of inflow into the well, q_{gn}. When $r_e' = r_e$, flow into the well has stabilized. The factor C' derived from production data over a space of time t is to be multiplied by a reduction factor c; the factor C for stabilized flow is then given as $C = c \times C'$ (Hurst et al. 1963). The reduction factor can be calculated out of the data of *BHP* build-up v. time when the well is shut in after a test phase of duration t, because the rate of pressure build-up is determined by the same parameters as the relationship between the factors C and C'. According to the authors cited,

$$c = \frac{p_{w1} - p_{wf}}{p_{ws} - p_{wf}} , \qquad 3.1-8$$

where p_{w1} is *BHP* after shut-in of the same duration as the preceding test; p_{wf} is flowing *BHP* prior to shut-in; and p_{ws} is the static *BHP*.

Example 3.1—2. Establish a performance equation for stabilized flow if the 8-hour isochronal performance equation is

$$q_{gn} = 3.081 \times 10^{-12}(p_{ws}^2 - p_{wf}^2)^{0.80} .$$

The static reservoir pressure determined from the pressure-build-up curve is $p_{ws} = 173.3$ bars; *BHP* prior to shut-in is $p_{wf} = 149.4$ bars; *BHP* measured 8 hours after shut-in is $p_{w1} = 170.4$ bars. — By Eq. 3.1—8,

$$c = \frac{1.704 \times 10^7 - 1.494 \times 10^7}{1.733 \times 10^7 - 1.494 \times 10^7} = 0.879 ,$$

and hence,

$$C = c \times C' = 0.879 \times 3.081 \times 10^{-12} = 2.708 \times 10^{-12} .$$

Equation 3.1—2 of stabilized performance now becomes

$$q_{gn} = 2.708 \times 10^{-12}(p_{ws}^2 - p_{wf}^2)^{0.80} .$$

In the Carter method, short test runs with just two different bore chokes are sufficient if the rate of production remains unchanged throughout the test (Carter et al. 1963). Description of the method can be dispensed with as it is more closely related by its nature to the subject of reservoir engineering: also, as far as well performance is concerned, it represents no improvement over the isochronal method.

3.1.3. Transformation of the performance equation derived from the steady-flow test into an isochronal performance equation

When performing a steady-flow test, it may happen that production through a given choke does not attain a steady state. This may be due, e.g., to a wellhead pressure change so slow as to be mistaken for zero by the observer, although, given time, it would build up to a significant value. The line connecting operating points thus established is of course wrong, and the exponent n of the performance equation will deviate from the true value. Points incorrectly determined for the above reason can be converted by calculation into the points of an isochronal graph. The significance of the correction resides in the fact that in certain cases it may be an advantage to test the well without intercalated shut-ins. If e.g. the well produces some liquid, a liquid column may accumulate on the well bottom during shut-in and fail to be removed by subsequent low-rate production. The flowing *BHP* can then be measured only by a pressure survey, which may sometimes prove quite difficult. In the following I shall describe Clark's method (Katz et al. 1959) of transforming steady-flow performance equations into isochronal.

The first point $[(p_{ws}^2 - p_{wf1}^2), q_{gn1}]$ determined by the steady-flow test is adopted as the first point of the isochronal graph. The other points of the steady-flow test have to be transformed into isochronal points valid at the instants t_i: t_1 is the duration of test production through the first choke. Transformation is performed by dividing by the correction factor K_i the values $\Delta p_{wi}^2 = (p_{ws}^2 - p_{wfi}^2)$ pertaining to flow rates determined by the steady-flow test. The suffix indicates the serial numbers of the successive phases of the steady-flow test, each belonging to a different choke size. The correction factor is given by the equation

$$K_i = \frac{q_{gni} N_{pti}}{q_{gn1} N_{pt1} + (q_{gn2} - q_{gn1}) N_{pt2} + \ldots + (q_{gni} - q_{gn(i-1)}) N_{pti}} , \qquad 3.1-9$$

where N_{pt} is the dimensionless *BHP* at various instants of dimensionless time, N_t. If $N_t > 100$, then N_{pt} can be calculated using the equation

$$N_{pt} = \frac{1}{2} (\ln N_t + 0.80977) , \qquad 3.1-10$$

where, by Eq. 3.1—7,

$$N_t = \frac{k_g t}{\Phi \bar{\mu}_g c_g r_w^2} .$$

In the case under consideration, the parameter t in Eq. 3.1—7 denotes time passed since the beginning of the steady-flow test.

Example 3.1—3 (after Mihály Megyeri). Data measured on the well Algyő—11 (Hungary), established by steady flow interrupted before complete stabilization, are listed in Table 3.1—1. The following physical parameters were found to remain constant in a good enough approximation throughout the entire test: $k_g = 1.432 \times 10^{-14}$ m²; $\mu_g = 1.895 \times 10^{-5}$ Ns/m²;

Table 3.1—1

Serial number	d_{ch}	q_g	Δp_{wf}^2	t
	mm	m³/s	10² bar²	h
1	10	1.532	262.1	7
2	8	1.322	195.4	14
3	6	0.864	116.7	21
4	4	0.373	53.77	28

$\Phi = 0.223$; $c_g = 4.61 \times 10^{-3}$ 1/bar; $r_w = 0.084$ m. Establish the isochronal performance equation for $t = 7$ h. The equation of the line defined by the plot of the $(q_{gn}, \Delta p_{wf}^2)$ data established by the test is

$$q_{gn} = 3.446 \times 10^{-13}(p_{ws}^2 - p_{wf}^2)^{0.878} .$$

Now by Eq. 3.1—7,

$$N_t = \frac{1.432 \times 10^{-14} \times t}{1.895 \times 10^{-5} \times 0.223 \times 4.61 \times 10^{-8} \times 0.084^2} = 10.42t .$$

Using this equation, find N_t for various values of t, and then, using Eq. 3.1—10, calculate the corresponding values of N_{pt} and K_i using Eq. 3.1—9. Divide the values of Δp_{wf}^2 by the appropriate correction factors. The results are listed in Table 3.1—2. The equation of the line fitted to the points thus established is

$$q_{gn} = 2.054 \times 10^{-11}(p_{ws}^2 - p_{wf}^2)^{0.754} .$$

The isochronal performance graphs established by referring data of a steady flow test to $t = 7$ h are plotted in a bilogarithmic system of coordinates shown as Fig. 3.1—3.

Table 3.1—2

q_g	p_{wf}^2	N_t	N_{pt}	k	p_{wf}^2 corr.
m³/s	$10^{12}\left(\frac{N}{m^2}\right)^2$	10^5			$10^{12}\left(\frac{N}{m^2}\right)^2$
1.532	262.1	2.626	7.787	1	262.1
1.322	195.4	5.252	8.133	1.007	194.0
0.8640	116.7	7.878	8.336	1.052	110.8
0.3730	53.77	10.50	8.479	1.295	41.52

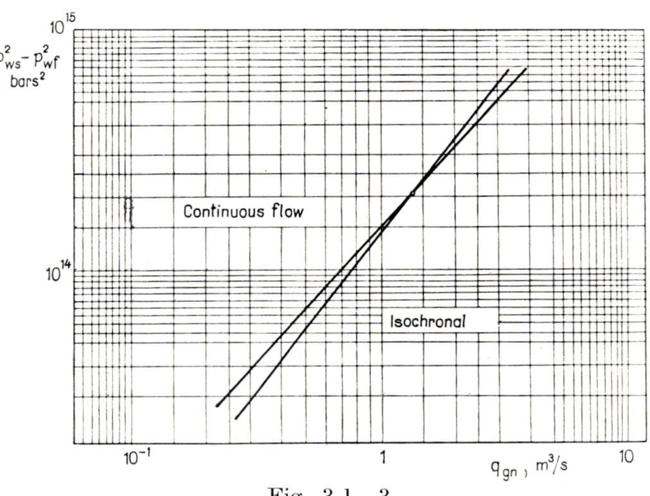

Fig. 3.1—3

3.2. Well completion; dimensioning the tubing

A gas well may be regarded as a flowing gaseous oil well whose well fluid contains little or no liquid. Well completions may thus be identical in principle with those of flowing oil wells. The changed importance of certain production parameters may, however, make it reasonable to change the completion quite considerably. — In dimensioning the tubing it is necessary to see that pressure drop due to flow resistance is comparatively low and the wellhead pressure of the flowing gas is the least permissible value or even less. Pressure drop in a general way is the less, the greater the tubing size. Maximum feasible tubing size is limited by the *ID* of the production casing, together with any other strings of tubing conduits and other equipment in the well. — The pressure drop of gas rising in the tubing will have to be calculated differently according as the gas produced is dry or wet. To a gas comparatively rich in liquid, one may apply Ros' theory (Section 1.4—3e). For dry gas or gas very low in liquid, the considerations in Section 1.2 will apply.

Example 3.2—1. Find the optimum tubing size if $q_{gn} = 500,000$ m³/d; $L_T = 2000$ m; flowing *BHP* declines during production from 190 bars to 90 bars. The least permissible wellhead pressure is $p_{TO\,\mathrm{min}} = 70$ bars. The standard-state density of the gas produced is $\varrho_n = 0.881$ kg/m³. The mean flowing temperature is estimated at 86.1 °C. Wellhead pressures p_{TO} belonging to several *BHP*s and tubing sizes are calculated using Eq. 1.2—4; λ is expressed by means of Eq. 1.2—12.

The results have been plotted in Fig. 3.2—1. The surface (p_{wf}, d_T, p_{TO}) is intersected by a plane parallel to the base plane and passing through $p_{TO\,\mathrm{min}} = 69$ bars in the line $A—B$. Clearly, up to a flowing *BHP* of 140

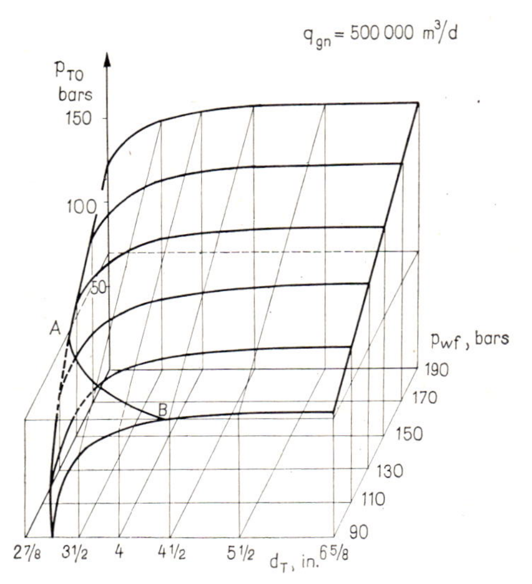

Fig. 3.2—1 Influence of tubing size and bottom-hole pressure upon wellhead pressure of a gas well

bars, the prescribed gas flow rate can be achieved through 2 7/8 in. tubing. At a flowing *BHP* of 90 bars, however, a wellhead pressure of 69 bars will be ensured by 4 1/2 in. tubing only. The pressure energy expended in producing $q_{gn} = 500,000$ m^3 of gas per day is the greater the less the flowing *BHP*. — The threads of the tubing string should provide a perfect seal. This is facilitated by special male and female threads (cf. Fig. 2.3—31), or by the use of plastic seal rings. The sealing of the male-female couplings can be improved e.g. by the use of teflon powder. This will flow under pressure and fill out the minor unevennesses of the thread. — When designing the well completion it is necessary to bear in mind the need for (i)

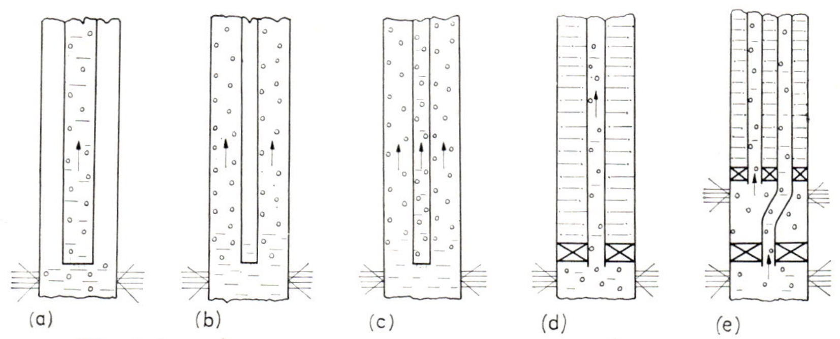

Fig. 3.2—2 Typical gas-well completions, after Speel (1967)

protecting production tubing from damage due to temperature and pressure changes, corrosion and erosion, (ii) an automatic shut-off of the gas flow in case of damage to the wellhead, (iii) the avoidance of the accumulation of a liquid slug on the well bottom during production; (iv) also, temperature and pressure changes during production must not result in a loading of the string to yield or collapse. (v) It should be possible to perform workovers, repairs and shut-offs simply and safely. — The wellhead equipment is similar to that described in Section 2.3.4. A useful review of modern high-pressure gas-well completions has been given by Speel (1967).

Figure 3.2—2 shows sketches of some typical completions. Solutions (a) and (d) are single completions, to be used when the well is produced exclusively through the tubing. The dimensioning of tubing for this type of completion has been discussed early in this section. — A tubing of size exceeding the maximum prescribed by the criterion of total fluid removal can be used if the well is equipped for plunger lifting (Bennett and Auvenshine 1957). During production, the plunger is out of the way in a tubing attachment (lubricator) installed on the Christmas tree (cf. Section 2.4.6). A motor valve under time-cycle control shuts off the flow line 2 to 8 times a day. The plunger then sinks to the bumper spring installed at the tubing bottom. The controller now reopens the flow line and formation gas lifts up the plunger together with the liquid column above it. — The solution shown in Fig. 3.2—2b can be used when the casing is not expected to suffer damage during production (Ledet et al. 1968). The annulus of comparatively large cross-section will produce dry gas because the flow section is greater than the maximum permitted by the criterion of total fluid removal; any liquid will accumulate at the well bottom. Periodic opening of the tubing head will permit gas pressure in the annulus to remove the liquid through the comparatively small-size tubing. The large cross-section of the annulus restricts flowing pressure drop in the gas. The accumulating liquid is 'blown off' rather often, so as to preclude appreciable increases in *BHP*. This completion permits the production of gas at a fast rate. Intermittent production of liquid may be controlled e.g. by the opening and closing of the tubing outlet. The economical removal of liquid accumulated in the tubing, which now plays the role of a 'dewatering string', can be facilitated e.g. by gas lift valves installed close to the tubing shoe, a plunger lift operated in the tubing, sucker-rod pumping or the addition of foam-producing chemicals (Nichols 1968). In the first solution, the gas lift valve opens as soon as a liquid column of sufficient length has accumulated above it. It may be of the differential, or tubing-pressure-operated type, or, in the Baker—Merla system, it may be controlled by a retarder. The second solution is similar to the plunger-lift installation described in the foregoing section. The main difference is that gas is produced through the annulus and the tubing serves for dewatering only. — The sucker-rod installation is quite conventional. The most important thing to be kept in mind is the choice of corrosion-resistant pumps and rods. This solution might be economical in wells producing both gas and water at comparatively high rates but at a comparatively low flowing *BHP*. — Foam-producing chemicals are fed in batches to the well during periodical shut-offs. Their thorough mixing with water and gas is ensured by suitable means. Foaming water can be removed efficiently from the well by gas pressure. This method is most economical in comparatively high *GWR* wells. — In the solution shown in Fig. 3.2—2c, well fluid is produced by continuous open flow through both the annulus and the tubing. Also in this case, the casing must not be damaged by the well fluid. In the large cross-section ensured by the combination of the two conduits, pressure drop due to flow resistance is comparatively small. Gas flow rate in the casing will tend to be below-critical. Periodical shut-offs of the casing head will push the liquid accumulated in the annulus into the tubing whence it is removed by gas pressure. — In the solution shown as Fig. 3.2—2d, the annulus is packed off at the tubing shoe and filled with liquid above the packer. The solution has two purposes: one, to protect the casing string from gas pressures higher than the hydrostatic pressure of the liquid column, as well as from gas corrosion, and two, to permit the fast killing of the well (by opening the valve in the packer, the well bottom can be flooded with the liquid stored in the annulus). The liquid in the annulus is of low-viscosity, non-corrosive to the tubing or casing, and unaffected in its properties by the pressures and temperatures prevailing in the well. Low viscosity results in ease of pumping, fast flooding of the well bottom when the well is to be killed, and easy aeration by inflowing gas. — Density of the liquid is chosen in dependence on formation pressure. Several types of liquid are used. Slightly alkaline fresh water or fresh water with a dissolved inhibitor will often do. Higher-density liquids include $CaCl_2$ or $ZnCl_2$ dissolved in fresh water; densities may range up to 1900 kg/m^3. The pH of these solutions is rather low, however; their corrosive tendencies have to be kept in check by the addition of an inhibitor.

Figure 3.2—2e shows a high-pressure gas well producing two zones. The annulus above the upper packer is filled with liquid. — Figure 3.2—3 shows the wellhead equipment used in the GFR for a well of this type (Werner and Becker 1968). Formation pressure in the strata traversed by the well is 379 bars at a depth of 2500 m. Production casing string *1* is of 13 3/8 in. size. Tubing strings *2* are of 3 1/2 in. size each. The outer annulus of 18 5/8 in. size can be opened to the surface by means of a bleed valve; the annulus between the 9 5/8 in. casing string and the production casing can be opened by means of valve pairs *4*. There is a blow-out preventer *5* closing on the tubings attached to the casing head, surmounted by the tubing head *6*. The Christmas tree assembly *7* is of the monoblock type. Each string of tubing is provided with a pair of main valves *8*, one wing valve *9* and one lubricating valve *10*.

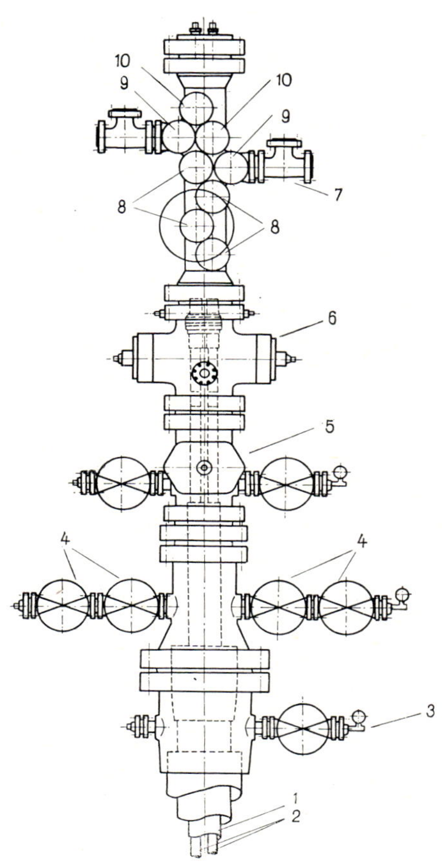

Fig. 3.2—3 Christmas tree of high-pressure dual gas-well completion, after Werner and Becker (1968)

3.3. Corrosion of gas wells; hydrate formation

In gas wells, corrosion hazard usually comes from inside, in the form of CO_2, organic acids, H_2S and corrosive formation waters as the main agents. The effect of CO_2 is described by the following reaction equations:

$$CO_2 + H_2O \rightarrow H_2CO_3 \;;$$

$$Fe + H_2CO_3 \rightarrow FeCO_3 + H_2 \,.$$

CO_2, inactive in itself, becomes corrosive if the well fluid contains water. Dissolved in the water, CO_2 turns into carbonic acid. Significant CO_2 corrosion should be expected if the well is deeper than about 1500 m; bottom-hole temperature is higher than 71 °C; *BHP* is higher than 103 bars; production exceeds 57,000 m^3/d; the partial pressure of CO_2 exceeds 2.1 bars; and the pH at the wellhead is less than 5.4. According to literature, one or several of these limits are exceeded in 90 percent of gas wells producing also condensates (Bilharz and Greenwell 1953).

H_2S likewise causes serious corrosion damage only if the well fluid contains water. Some less dangerous corrosion may take place, however, also in the absence of water:

$$H_2S + Fe \rightarrow FeS + 2H \,.$$

The iron sulphide thus formed is a dark powder or scale, having a higher electrode potential than iron. In the presence of water, a galvanic cell comes to exist; a current starts from the Fe pole towards the FeS pole; the resulting electrolytical corrosion may cause punctures. Hydrogen liberated on the formation of FeS also represents a hazard. It diffuses into the undeteriorated steel and, entering the crystal lattice of the iron, significantly decreases its elasticity. The effect is called hydrogen embrittlement. Hydrogen atoms in steel combine into hydrogen molecules, causing very high local pressures up to 10^6 or 10^8 bars, which may burst or fracture the pipes. Deposits of sulphide powder hamper or prevent the functioning of certain equipment such as storm chokes or gas lift valves.

Internal chemical corrosion can be prevented by the use of inhibitors, the corrosion-resistant coating of pipe surfaces and the choice of corrosion-resistent steels for well completions. Inhibitors can be fed to the well in a variety of ways. In the case of a single open completion, the inhibitor fed to the annulus at the wellhead will mix with the well fluid at the tubing shoe. In a closed installation without liquid in the annulus, a special metering valve in the packer or a small size conduit run in the tubing may serve to feed the inhibitor to the well fluid in the tubing. In some cases, the inhibitor is fed batchwise to the producing formation at times of well shut-off, whence it is gradually transferred to the wells by the fluid. It is, however, usually better to feed the inhibitor direct to the well to be protected. — Coatings such as epoxy resin in layers 0.05—0.08 mm thick may provide adequate protection if temperatures or pressures are not excessive (Speel 1967). Monel metal coatings are highly resistant but rather expensive. — Adequate resistance to highly corrosive gases will be provided by suitably alloyed steels only. Steel containing 9 percent Cr and 0.5—1.0 percent Mo has been found satisfactory in the face of CO_2 corrosion. Recent research has shown that one of the fundamental conditions of H_2S corrosion prevention is that the yield strength of the tubing steel should not exceed 6.2—6.9 N/m², and that the tubing string should be tensioned as little as possible. Figure 3.3—1 represents time up to corrosion-controlled fracture v. rigidity of the pipe (expressed in terms of Rockwell hardness), and tension referred to yield strength (Hudgins 1970). In order to decrease tension in the tubing, Soviet petroleum engineers use completions with the tubing shoe landed on a seat fixed to the casing string (Nomisikov et al. 1970).

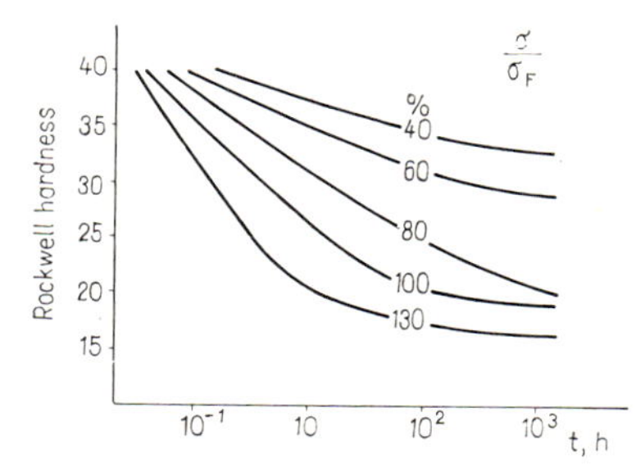

Fig. 3.3—1 Incidence of corrosion failure v. pipe rigidity and tension referred to yield strength, after Hudgins (1970)

Low-strength tubing sufficiently resistant to gas containing H_2S can be made of API H—40 or J—55 steel. A suitable high-strength steel is C—75 (Hudgins 1970). Well equipment faced with the hazard of corrosion should be tested with particular care in every detail. Modern non-destructive testing of tubing is prescribed by API Std 5—A (1968). The methods cited there include magnetic-powder and ultrasonic testing. Equipment should be periodically checked also during production. Caliper surveys are run in the tubing; the wellhead equipment is tested by ultrasonic techniques (Bilharz and Greenwell 1953). The extent of corrosion is checked at short intervals by gamma-ray techniques at numerous points of the Lacq Field collection system (Vermeersch 1968). Since gas containing H_2S is also poisonous, special care must be taken to prevent significant amounts from entering the atmosphere. The well is to be equipped with a storm choke; also, an automatic shut-off device triggered by both too high and too low line pressures is to be installed in the wellhead. It is expedient to transmit to a dispatcher centre the wellhead pressures and temperatures as well as the positions of the more important valves.

Aggressive formation waters may also cause significant corrosion. It was found in wells of the Lacq Field that the decline of formation pressure entails the in-flow of increasing amounts of water, presumably the water films adhering to sand grains (Vermeersch 1968). This causes corrosion, first and foremost in the lower reaches of the tubing string. Farther up the water grows more neutral and, mixing with the water separating from the gas, it loses most of its aggressivity. It was found that corrosion in flow lines may be significant where flow is stratified but much less so in annular-type flow. Hence, to limit flow-line corrosion by formation waters, it is best to have a high gas flow rate, a low flow temperature, and a high flowing pressure. Also, the water produced together with the gas should not be separated at the wellhead but led together with the well fluid to the separator.

Production may be hampered also by hydrocarbon hydrates separating from the gas. At a given pressure, hydration temperature is raised by the presence of H_2S and CO_2 produced together with the gas. Hydrates tend to form first at the wellhead, where gas passing the choke may get cooled significantly by the almost adiabatic expansion. The solid hydrates formed there may completely obstruct the passage of gas, so that their formation is to be prevented. The means of doing so include warming of the gas in a heat exchanger prior to passage through the choke; if the gas is liable to cool in the flow line to a temperature conducive to hydration, some inhibitor, e.g. diethylene glycol is admixed to the well fluid at the wellhead. In favourable cases, it may be simplest to produce the well at a high rate: this will keep the gas warm enough to prevent hydrate formation. A bottom-hole choke installed in the lower reaches of the tubing may also help (cf. point (i) in paragraph 2.3—5b.3). — At too high rates of production, gas flow into the well may sweep in sand grains even from comparatively consolidated sand faces. These sand grains prised off the face by the sudden pressure drop in the formation around the well may cause considerable erosion in the flow string. The maximum rate of production which still does not sweep up sand is to be established by experiment. If this rate is too low to satisfy some other condition, then either a filter is to be installed at the sand face or the sand is to be consolidated by injecting a suitable substance.

Chapter 4

Producing oil wells — 2

4.1. Production by bottom-hole pumps

Production by bottom-hole pumps is a mechanical technique. The fluid entering the well from the formation is lifted to the surface by a pump installed below the producing fluid level. The prime mover of the pump is installed either on the surface, or in the well; in the latter case, it is integral with the pump. The bottom-hole pump unit comprises all the mechanisms and equipment serving the purposes of production. Numerous types of bottom-hole pump have been developed from the mid-nineteenth century onward. According to Coberly, in the decade starting with 1859, deep wells were drilled with wireline rigs whose bit-lifting horsehead was used after well completion also for sucker-rod pumping (API History 1961). The bottom-hole pumps of today can be subdivided as follows.

The sucker-rod pump is a plunger pump performing a reciprocating motion. Its prime mover is installed on the surface. The reciprocating motion of the surface drive is communicated to the pump by a string of sucker rods. The rotating motion of the motor shaft can be transformed into reciprocating motion in various ways. If a crank and a flywheel are used, the installation is called a *crank-type* or *walking-beam-type sucker-rod pump*. In long-stroke hydraulic pumps, a hydraulic means of transformation is adopted; the installation is called a *hydraulic sucker-rod pump*. If the transformation is by wireline and pulley, the installation is called a *derrick-type sucker-rod pump*.

In *rodless bottom-hole pump installations*, the bottom-hole pump may be of plunger or centrifugal or some other type. Hydraulic pumps are driven by a hydraulic engine integral with them, driven in its turn by a power fluid to which pressure is imparted by a prime mover situated on the surface. This type is called a *hydraulic* (*rodless*) *bottom-hole pump*. Centrifugal pumps integral with an electric motor, and lowered to the well bottom, are called *submersible pumps*. Further rodless bottom-hole pumps include electric *membrane pumps* and *sonic pumps*.

Of the sucker-rod type of pump, the walking-beam type is most widespread. According to US data, in 1969, 413,000 wells were produced by this means, making up round 81 percent of the total number of mechanically produced wells (Scott 1969). We shall therefore concentrate in our following discussion on the peculiarities of walking-beam type sucker-rod pumps.

4.1.1. Sucker-rod pumps with walking-beam-type drive

Referring to the sketch of a walking-beam type sucker-rod pumping unit (Fig. 4.1—1), the power of electric motor *1* is transferred by v-belts to a gear reducer *2*. This reduces the rather high rpm of the electric motor to between, say, 3 and 25. This number determines (is equal to) the number of double strokes per minute (spm) of the sucker rod. The stroke of the polished

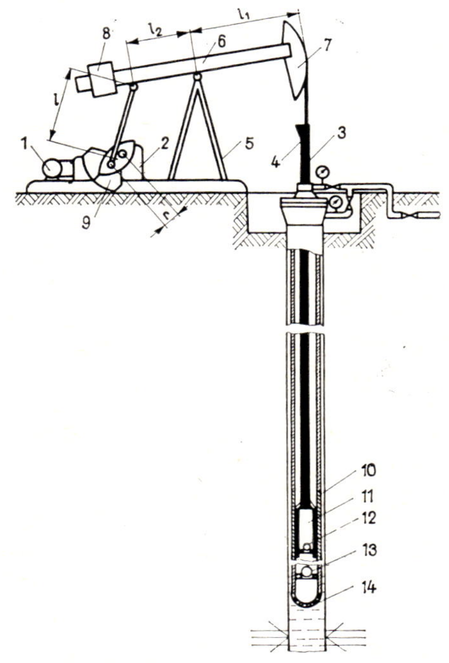

Fig. 4.1—1 Walking-beam-type sucker-rod pump installation

rod *3* is twice the length r of the crank, provided that $l_1 = l_2$. The crank length and hence the stroke are both variable within limits set by the design. The longest stroke that can be realized does not usually exceed 3 m. Power is transferred from the crank to the walking beam by the connecting rod (pitman) of length l. The structure moving the polished rod is composed of a trestle (samson post) *5*, a walking beam *6* and the horsehead *7*. The variation of polished-rod load over the pumping cycle is balanced by one of various means, not to be detailed here. In the case shown in the Figure, this balancing is performed by means of a crank counterweight, *8* and a beam counterweight *9*. The specially made and machined top unit of the rod string, the polished rod, is hung from carrier *4*. Attached to the tubing shoe installed in the well is pump barrel *10*, in which plunger *11* is moved up and down by the rod string. During the upstroke, travelling valve *12* is closed, and the plunger can lift the fluid filling the annular space between tubing and rod. At the same time, standing valve *13* is open, so that fluid may enter the barrel through filter *14*. During the downstroke, the travelling valve is open and the standing valve is closed: the plunger sinks in the fluid filling the barrel.

(a) Loads on the rod string and their effects

Several methods have been developed for calculating the polished-rod load. One of the reasons for this is that a rigorous treatment would be very complicated; it would have to account for a large number of factors, some of which are or but approximately known or totally unknown at the time of designing. The various procedures of calculation are based on various simplifying assumptions. The deviation between calculated and actual data is often quite large with each procedure; this reveals the limits of simulability to be rather narrow, and suggests that comparatively simple relationships may be as satisfactory for design purposes as the most complicated ones. What is to be expected of such a procedure is that, firstly, it should describe to a fair degree of accuracy the variation of polished-rod load v. travel, thus providing insight into operating conditions and their control, and secondly, that it should give results sufficiently accurate to permit the correct choice of the pumping unit to be used on the well under consideration.

One modern method of calculation is that contained in API RP 11L, first published in 1967. Check measurements on 77 wells showed the mean calculated value of $F_{p\max}$ to exceed the mean measured value by as little as 1.41 percent (Griffin 1968). The greatest depth of installation of the bottom-hole pump in the check wells was 3150 m. This method requires the use of auxiliary diagrams given in the standard. In the following we shall discuss a different procedure based on a consideration by Muraviev and Krylov (1949) which, although presumably less accurate, is deemed to give better insight into operating conditions. We shall, however, solve some basic problems also by the new API method.

(a.1) *Rod load for solid-rod strings.* — Rod load is maximum in the top unit of the string, that is, in the polished rod. It is subject to considerable variation during the double-stroke pumping cycle. Instantaneous load is a function of a large number of factors. These can be static and dynamic. The static polished-rod load during the upstroke is

$$F_s = F_l' + F_r + F_{TO} - F_h, \qquad 4.1-1$$

F_{TO} is usually small enough to be negligible. In the case of continuous production, the producing fluid level is very often quite close to the bottom-hole pump, in which case F_h can also be neglected; we shall do so in the sequel. If, however, the producing fluid level is high, F_h may play a significant role. A high producing level is frequently encountered in intermittent-life wells, but sometimes also in continuous-lift ones. It occurs as a transitory phenomenon at the start of pumping in almost every well produced by a bottom-hole pump. During continuous production, then, the static upstroke load in most wells produced by sucker-rod pumps is, in a fair approximation,

$$F_s = F_l' + F_r .$$

Since liquid load equals the weight of the liquid of gravity γ_l above the pump operating at depth L,

$$F_l' = A_p L \gamma_l - A_r L \gamma_l$$

and

$$F_r = L A_r \gamma_r$$

hence

$$F_s = A_p L \gamma_l + L A_r (\gamma_r - \gamma_l) \qquad 4.1\text{—}2$$

If there were no rod string in the tubing, the weight of one metre of liquid column would be $G_l = A_T \gamma_l$; weight per metre of the sucker rod in air $G_r = A_r \gamma_r$, and its wet weight, immersed in the liquid, reduced by buoyancy, is $G_r' = A_r(\gamma_r - \gamma_l)$. Now,

$$G_r' = G_r \frac{\gamma_r - \gamma_l}{\gamma_r} = G_r b \qquad 4.1\text{—}3$$

where b is the weight reduction factor:

$$b = \frac{\gamma_r - \gamma_l}{\gamma_r} \qquad 4.1\text{—}3a$$

and static load can be expressed also as

$$F_s = F_l + F_r b = G_l L + G_r b L .$$

During the downstroke, the static polished-rod load equals the net wet weight of the rod string, that is,

$$F_{s1} = G_r L = G_r' b L . \qquad 4.1\text{—}4$$

The rod string immersed in the liquid is invariably stretched by its own weight. Its stretch is, by Hooke's law,

$$\Delta L_{rp} = \frac{F_r b L}{2 A_r E_r} = \frac{G_r L^2 b}{2 A_r E_r} . \qquad 4.1\text{—}5$$

The ratio G_r/A_r is nearly constant for standard sucker rods (round 8.5×10^4 N/m^3 as an average of the data in Table 4.1—1). In general, $\gamma_r = 7.7 \times 10^4$ N/m^3 and $E = 2.06 \times 10^{11}$ N/m^2. Assuming $\gamma_l = 8826$ N/m^3 and substituting the numerical values into Eq. 4.1—5, we get

$$\Delta L_{rp} = \frac{L^2}{5.5 \times 10^6} . \qquad 4.1\text{—}6$$

The basic stretch of the rod string in a given well fluid thus depends essentially on the length of the rod string alone. By Eqs 4.1—4 and 4.1—6, the string is loaded by its own weight only during the downstroke, and also by the weight of the liquid column acting on the plunger during the upstroke.

Table 4.1—1

Standard sucker rod sizes (largely after Hungarian Std. MSZ 5152)

Nominal rod diameter d_9		A_r[*2]	G_r[*3]		d_3 ± 0.1	d_4 ± 1.0	d_5 ± 0.1	f	l_1	l_3	l_4	k_1	k_2
			$L = 7.62$ m	$L = 9.14$ m					$+1.549$	min	min	± 0.8	$0-0.8$
in.	mm[*1]	cm²	N/m		mm								
1/2[*3]	12.7	1.27	14.2[*4]	13.9[*4]	$25.4 \begin{smallmatrix}+0.13\\-0.25\end{smallmatrix}$	$\leq d_3$	$25.4 \begin{smallmatrix}+0.13\\-0.25\end{smallmatrix}$	—	—	—	69.9	15.9	—
5/8	15.9	1.99	17.4	17.1	34.9	35	38.1	10	28.58	31.8	101.6	22.0	34.9
3/4	19.0	2.84	24.0	23.6	38.1	38	41.3	10	34.92	31.8	101.6	25.5	38.1
7/8	22.2	3.87	34.9	34.1	42.9	43	46.0	10	34.92	31.8	101.6	25.5	41.3
1	25.4	5.07	42.2	41.7	50.8	51	55.6	10	44.45	38.1	101.6	33.3	47.6
1 1/8	28.6	6.42	53.4	52.8	57.2	57	60.3	10	50.80	41.3	114.3	38.1	54.0

*1 ±0.5 mm
*2 Calculated values
*3 After API Std 11 B
*4 Estimated values

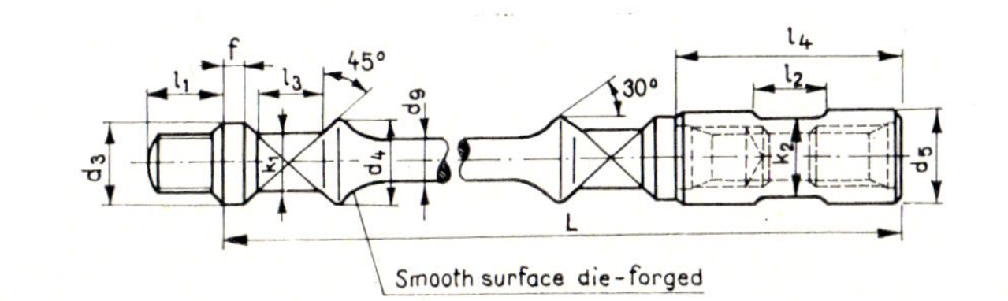

Fig. to Table 4.1—1 Sucker-rod dimensions (full rod)

The change in liquid load entails a change in stretch, which is described, likewise by Hooke's law, as

$$\Delta L_{rv} = \frac{F_l L}{E_r A_r} = \frac{G_l L^2}{E_r A_r}. \qquad 4.1-7$$

The tubing also has a basic and a variable stretch, because it is loaded by its own weight during the upstroke, to which is added the weight of the liquid column during the downstroke. The variable stretch, which is of a primary interest to our present discussion, is

$$\Delta L_{Tv} = \frac{F_l L}{E_T A_T} = \frac{G_l L^2}{E_T A_T}. \qquad 4.1-8$$

We have assumed here a pump barrel diameter equal to the *ID* of the tubing.

At fairly low pumping speeds ($n < 8$ spm) dynamic loads can usually be neglected, and the plunger stroke equals the polished-rod stroke less the stretch of rod string plus tubing. This can be verified, e.g., by the following consideration. Leaving the static load and the basic stretch out of consideration, we shall for the time being identify stretch with the stretch fraction due to load variation. At the top of the polished-rod stroke (point A), the rod string is fully stretched and the plunger is at the top of its stroke (point B). Early in the downstroke of the polished rod, the tensile stress in the rod string gradually decreases to zero which it attains after a travel of ΔL_{rv}. This is when the plunger starts to actually travel downwards. In this phase, polished rod, rod string and plunger move downwards at the same speed. Meanwhile, the weight of the liquid column has been transferred by the closure of the standing valve to the tubing string. This makes the pump barrel fixed to the tubing shoe sink by ΔL_{Tv} against point B. Hence, the plunger will start to move relative to the barrel only when the plunger has 'overtaken' the lowered barrel. By a similar consideration, the stroke reduction for the upstroke is

$$s - s_d = \Delta s = \Delta L_{rv} + \Delta L_{Tv} = \frac{G_l L^2}{E}\left[\frac{1}{A_r} + \frac{1}{A_T}\right] \qquad 4.1-9$$

where we have assumed $E_r = E_T = E$. Introducing the expression $G_l = A_p \gamma_l$, we find that

$$\Delta s = \frac{A_p \gamma_l L^2}{E}\left[\frac{1}{A_r} + \frac{1}{A_T}\right]. \qquad 4.1-10$$

Example 4.1—1. Find the basic stretch of the rod string and the stroke reduction. The dynamic load is negligible. The plunger diameter of the RWT type bottom-hole pump is $d_p = 63.5$ mm; $d_r = 22.2$ mm; $L_w = 1200$ m;

$\gamma_l = 8826$ N/m³; $E = 2.06 \times 10^{11}$ N/m². — The basic stretch of the rod string is, by Eq. 4.1—6,

$$\Delta L_{rp} = \frac{1200^2}{5.5 \times 10^6} = 0.26 \text{ m}$$

$$G_l = A_p \, \gamma_l = 3.17 \times 10^{-3} \times 8826 = 28.0 \text{ N/m}.$$

The values of A_p are listed in Table 4.1—2. By Table 4.1—1, A_r is 3.87 cm². The tubing required for a RWT pump of 63.5 mm diameter is of 3 1/2 in. size, A_T = 16.71 cm². The stroke reduction is by Eq. 4.1—9

$$\Delta s = \frac{28.0 \times 1200^2}{2.06 \times 10^{11}} \left[\frac{1}{3.87 \times 10^{-4}} + \frac{1}{16.71 \times 10^{-4}} \right] = 0.62 \text{ m}.$$

At comparatively high pumping speeds ($n > 8$ spm) and great depths ($L > 1000$ m), the *dynamic factors* cannot be neglected any more when calculating the polished-road load. Dynamic loads may be due to various causes. Some of them can be calculated to a fair approximation (e.g., the transformation of motor-shaft rotation into a vertical alternating motion of the rod string). The play of forces transferred from shaft to polished rod can be given a mathematical formulation valid for a large number of cases. Other loads can be described at least approximately (e.g., those due to the free vibrations of the rod string); finally, there are loads that defy mathematical treatment, such as a crooked hole, highly viscous oil, a gas-rich well fluid passing through the bottom-hole pump, a sandy well fluid, and intermittent flowing of the well. For practical purposes, it will usually do to

Table 4.1—2
Percentage lengths of rod sizes making up tapered string

Plunger			3/4″×5/8″ 19.0×15.9	7/8″×3/4″ 22.2×19.0	7/8″×3/4″×5/8″ 22.2×19.0×15.9			1″×7/8″ 25.4×22.2	1″×7/8″×3/4″ 25.4×22.2×19.0		
d_p		Cross-sect. area A_p									
in.	mm	cm²	Percentage of 3/4 in.	Percentage of 7/8 in.	%			Percentage of 1 in.	%		
3/4	19.0	2.9	27.0	23.8	—	—	—	—	—	—	—
15/16	23.8	4.4	28.9	24.6	—	—	—	—	—	—	—
1	25.4	5.1	29.6	25.0	19.3	23.0	57.7	22.7	16.7	19.0	64.3
1 1/16	27.0	5.7	30.5	25.5	20.3	23.4	56.3	23.1	17.2	19.5	63.3
1 1/4	31.8	7.9	33.2	27.3	22.6	26.1	51.3	24.3	18.7	21.2	60.1
1 1/2	38.1	11.4	37.4	29.6	26.5	30.2	43.3	26.2	21.1	23.9	55.0
1 3/4	44.5	15.5	42.3	32.9	30.6	35.2	34.2	28.4	23.9	27.0	49.1
2	50.8	20.3	48.2	36.5	35.7	41.0	23.3	30.9	27.1	30.7	42.2
2 1/4	57.2	25.7	54.4	40.6	42.0	47.0	11.0	33.8	30.9	34.8	34.3
2 7/16	61.9	30.1	59.8	44.0	—	—	—	36.2	33.0	38.0	29.0
2 1/2	63.5	31.7	63.0	45.0	—	—	—	37.0	35.0	39.5	25.5
2 3/4	69.9	38.7	69.7	50.2	—	—	—	40.6	39.5	44.6	15.9
3 3/4	95.3	71.3	—	75.4	—	—	—	58.2	44.5	50.2	5.3
4	101.6	81.0	—	—	—	—	—	58.5	—	—	—

take into account the force transfer relations of the drive unit, while estimating the other factors or determining them by measurements during production. — In a fair enough approximation valid for many cases, the upper bearing of the pitman (Fig. 4.1—1) reciprocates along a straight vertical line. If this assumption is adopted, then force transfer can be discussed on the analogy of crosshead-type engine drives. The maximum positive acceleration of the upper pitman bearing — or, if the walking-beam arms are of equal length, of the horsehead — takes place at the onset of the polished rod's upstroke; then,

$$a_{max} = r\omega^2\left[1 + \frac{r}{l}\right]. \qquad 4.1-11$$

If the walking-beam arms are of unequal length, then the expression in the brackets is to be multiplied by the ratio of working centres, l_1/l_2. The acceleration at any instant, including the maximal, travels down the rod string at the speed of sound and attains the plunger after a span of time $t = L/v_s$. The plunger will start to lift the liquid column only after that span of time t. Hence, the greatest total dynamic load appears not at the instant when the polished rod starts to rise, but slightly later. In the relationships to be discussed below we shall usually take into consideration the dynamic load on the rod string only because, according to Muraviev, the acceleration of the liquid column can be neglected: the rod string is in the process of stretching when the maximal acceleration is travelling along it, and this fact serves to damp the displacement of the fluid. Acceleration will propagate 4—5 times more slowly in a gaseous fluid than in rod steel: also, the liquid exerts a drag on the tubing wall during its rise. The maximum dynamic load is

$$F_d = \frac{F_r}{g} a_{max} = \delta F_r \qquad 4.1-12$$

where

$$\delta = \frac{a_{max}}{g}$$

is the so-called dynamic factor. Increasing the pumping speed may raise the acceleration of the rod string above the acceleration of gravity, and this may cause operating troubles. In practice, therefore, the maximum allowable dynamic factor is 0.5. Substituting into this formula the values $\omega = (n\pi)30$*, $r = s/2$ and $r/l = 0.25$ (this value may change!) we get

$$\delta = \frac{sn^2}{1440} \qquad 4.1-13$$

and the maximum dynamic load turns out to be

$$F_d = \frac{F_r sn^2}{1440}. \qquad 4.1-14$$

* n is expressed here and elsewhere in Section 4.1 in 1/min (min^{-1}).

The maximum polished-rod load is to be anticipated after the plunge has started to rise; when both static and dynamic loads are maximal, then

$$F_{p\max} = F_s + F_d = F_l + F_r(b + \delta) = A_p L \gamma_l + G_r L \left[b + \frac{sn^2}{1440}\right]. \quad 4.1-15$$

This latter formula is in a fair agreement with actual fact for rod-string lengths of 1000—1200 m. Satisfactory agreement is confined in any well to comparatively low pumping speeds. In practice, numerous other relationships are used to calculate maximum polished-rod load. Let us enumerate some of these. Charny's formula (Muraviev and Krylov 1949):

$$F_{\max} = F_l + F_r \left[b + \frac{sn^2}{1800} \frac{\tan \varphi}{\varphi}\right] \quad 4.1-16$$

where $\varphi = (\omega L)/v_s$ and v_s is the speed of sound (5100 m/s). — The Slonegger or API formula (Eubanks et al. 1958) is

$$F_{\max} = (F_l + F_r)\left(1 + \frac{sn}{137}\right). \quad 4.1-17$$

It gives satisfactory results primarily for low pumping speeds and shallow wells. If the rod string is long, the formula gives a value lower than the actual load. — The relationship most resembling Eq. 4.1—15 is the Mills formula (Eubanks et al. 1958):

$$F_{\max} = F_l' + F_r\left[1 + \frac{sn^2}{1790}\right]. \quad 4.1-18$$

The free vibrations of the rod string may in unfavourable cases — at high pumping speeds in particular — give rise to significant excess dynamic loads. The sudden load changes at the upper and lower ends of the plunger stroke propagate at the speed of sound up the rod string to the point of suspension of the polished rod, and back again after reflection. The frequency of the longitudinal vibration depends solely on the length of the rod string (assuming the speed of sound in the steel to be constant at $v_s = 5100$ m/s):

$$n_O = \frac{5100 \times 60}{4 \times L} = \frac{76{,}500}{L} \text{ min}^{-1}. \quad 4.1-19$$

If the frequency of the free vibration equals, or is a multiple of, the pumping speed, then the free vibrations, damped otherwise, are reinforced by further pulses arriving in phase, and loads may significantly increase. Pumping speeds, giving integer cycle ratios are called synchronous speeds.

The above is rigorously valid only if there is only one sudden load change per stroke. This is the case especially when gaseous oils are being pumped; loading is then sudden, whereas offloading is gradual and comparatively slow. If the oil produced by sucker-rod pump is gasless, no synchronous vibration takes place as a rule, because the longitudinal waves generated by the two sudden load changes per stroke usually attenuate each other. In practice it is usual not to take into consideration the load increment due to synchronous vibration, but if the dynamometer card reveals the presence of such, then the pumping speed is changed sufficiently to displace the frequencies so that the vibrations attenuate each other (n_O, n).

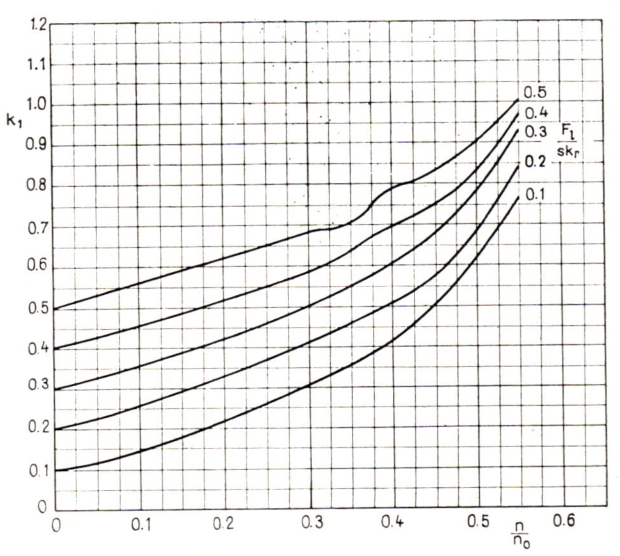

Fig. 4.1—2 After API RP 11L

The calculation procedure published in API RP—11 has been developed by experiments on mechanical and subsequently on electrical analog models. The maximum polished-rod load can be calculated by the slightly modified formula

$$F_{p\max} = F_r b + k_1 s k_r \,. \qquad 4.1\text{—}20$$

The factor k_1 can for various values of F_l/sk_r and n/n_O be read off Fig. 4.1—2; the load bringing about unity stretch in the rod string, is

$$\frac{1}{k_r} = \frac{1}{E}\left[\frac{L_1}{A_1} + \frac{L_2}{A_2} + \ldots\right] \qquad 4.1\text{—}21$$

n_O can be calculated using Eq. 4.1—19.

Example 4.1—2. Find the maximum polished-rod load by means of Eqs 4.1—20 and 4.1—15, if the producing fluid level in the well is at $L_d = 1372$ m; the setting depth of the bottom-hole pump is $L = 1525$ m; $d_p = 1.5$ in.; $n = 16$ min^{-1}; $s = 1.37$ m; the string is tapered, with 30.9 percent of 7/8-in.

and 69.1 percent of 3/4-in. rod; $\gamma_l = 8826$ N/m³; $E = 2.06 \times 10^{11}$ N/m². — Rod string weight is

$$F_r = 0.309 \times L \times G_{r1} + 0.691 \times L \times G_{r2} =$$

$$= 0.309 \times 1525 \times 34.1 + 0.691 \times 1525 \times 23.6 = 4.09 \times 10^4 \text{ N}.$$

By Eq. 4.1—21,

$$\frac{1}{k_r} = \frac{1}{2.06 \times 10^{11}} \left[\frac{1525 \times 0.309}{3.87 \times 10^{-4}} + \frac{1525 \times 0.691}{2.84 \times 10^{-4}} \right] = \frac{1}{4.18 \times 10^4}$$

Liquid load on the total plunger surface is:

$$F_l = A_p L_d \gamma_l = 11.4 \times 10^{-4} \times 1372 \times 8826 = 1.38 \times 10^4 \text{ N}$$

$$\frac{F_l}{sk_r} = \frac{1.38 \times 10^4}{1.37 \times 4.18 \times 10^4} = 0.24 .$$

By Eq. 4.1—19,

$$n_O = \frac{76{,}500}{1525} = 50.2$$

and hence

$$\frac{n}{n_O} = \frac{16}{50.2} = 0.32 .$$

According to Fig. 4.1—2, $k_1 = 0.47$, and hence, by Eq. 4.1—20,

$$F_{p\max} = 4.09 \times 10^4 \times 0.885 + 0.47 \times 1.37 \times 4.18 \times 10^4 = 6.31 \times 10^4 \text{ N}.$$

By Eq. 4.1—15,

$$F_{p\max} = 1.38 \times 10^4 + 4.09 \times 10^4 \left[0.885 + \frac{1.37 \times 16^2}{1140} \right] = 6.26 \times 10^4 \text{ N}.$$

The difference between the $F_{p\max}$ values calculated by the two methods is

$$\frac{6.31 - 6.26}{6.31} 100\% = 0.7\% .$$

The load F_{fr} due to rod and fluid friction can also be regarded as dynamic. Its value may be significant in crooked wells or if the oil is of high-viscosity or tends to freeze at well temperature. This friction cannot be described mathematically, so that it does not figure in our fundamental formulae. Its presence may be detected from the dynamometer cards. It is negligible in most cases.

(a.2) *Rod load for hollow-rod strings.* — In sucker-rod pumping using hollow rods, the bottom-hole pump barrel is usually fixed to the production casing, and the production rises through the hollow rods. The casing annulus is not packed off in most cases; largely after McDannold (1960). On the upstroke, the liquid rises and accelerates together with the rod string.

Maximum polished-rod load is calculated by means of a slightly modified Eq. 4.1—18:

$$F_{max} = (G_l'L + F_r)(1 + \delta) \qquad 4.1\text{—}22$$

where $G_l'L$ is the weight of the liquid held by the hollow rod string; it is, as opposed to the liquid load F_l for the solid-rod string, independent of the plunger diameter. — The minimum liquid load on the downstroke is

$$F_{min} = F_r(1 - \delta) + (A_{ri} - A_p)\gamma_l L(1 \pm \delta) - F_{fr}\,. \qquad 4.1\text{—}23$$

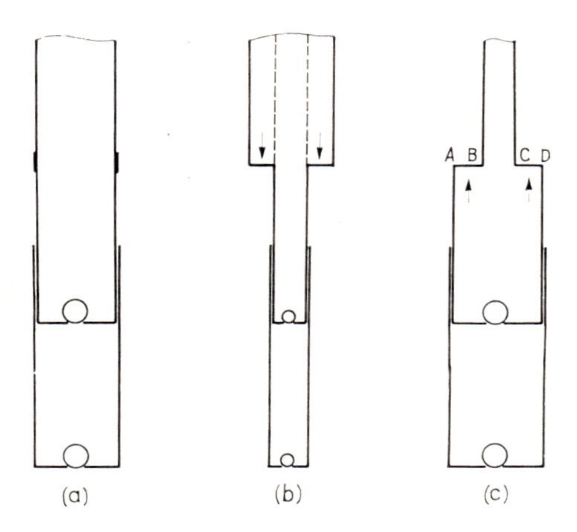

Fig. 4.1—3 Bottom-hole pumps with hollow rods, after McDannold (1960)

In practice, A_{ri} may equal A_p, but it may also be greater or less (Fig. 4.1—3). If $A_{ri} = A_p$, then the second term on the right-hand side of the above equation is zero. If $A_{ri} > A_p$, then the rod string has to carry the additional weight of a 'liquid annulus', that is, F_{min} is greater than in the preceding case. At the same time, δ appears with a negative sign in the third pair of parentheses, because the liquid annulus moving together with the rod string also decelerates together with it. If $A_{ri} < A_p$, then the pressure acting from below on the surface $ABCD$ reduces the rod-string load, and the sign of δ is positive in the third pair of parentheses. F_{fr} is the friction of the unmoving fluid column against the internal surface of the sinking string. Relative displacement between well fluid and rod string takes place during the downstroke only, and hence, so does liquid production. Thus, when calculating the friction loss, the relative rate of flow has to be calculated from twice the daily rate of production. In order to account for variations in crank speed, the velocity thus obtained is further multiplied by1.57.Taking production as a basis, the corrected production used to give the friction loss is

$$q_c = 1.57 \times 2 \times q = 3.14q\,.$$

Putting $\delta = 0$ in Eqs 4.1—22 and 4.1—23, we get the static loads for the up- and downstroke as

$$F_{p\max} = G_l L + F_r$$

$$F_{p\min} = F_r + (A_{ri} - A_p)\gamma_l L \,.$$

The greatest difference is due to the change in liquid load:

$$\Delta F_p = G_l L - A_{ri}\gamma_l L + A_p \gamma_l L \,.$$

But since $A_{ri}\gamma_l L = G_l L$, this simplifies to

$$\Delta F_p = A_p \gamma_l L \,. \qquad 4.1\text{—}24$$

The stroke reduction due to the change in liquid load is

$$\Delta s = \frac{\Delta F_p L}{E A_r} = \frac{A_p \gamma_l L^2}{E A_r} \,. \qquad 4.1\text{—}25$$

(a.3) *Rod string design.*— We shall consider solid-rod strings in what follows below. The maximum stress in the polished rod is obtained by dividing the maximum polished-rod load given by Eq. 4.1—15 by the cross-section of the polished rod. Employing the substitution $F_r = G_r L$, we get

$$\sigma_{\max} = \frac{F_{p\max}}{A_r} = \frac{F_l}{A_r} + \frac{G_r L}{A_r}(b + \delta) \,. \qquad 4.1\text{—}26$$

The maximum stress must be less than the maximum allowable stress σ_{al} given by Eq. 4.1—54. — In practice, rod strings are frequently tapered, that is, composed of standard rod sizes increasing from the plunger up. The reason for this is obvious: the string section directly attached to the plunger, that is, the lowermost rod, is loaded by the liquid column only, whereas the sections farther above are loaded also by the weight of the rods below them. The criterion mentioned in connection with Eq. 4.1—26, i.e. that the maximum stress must be less than the maximum allowable stress must hold separately for any rod of the string. Keeping this in mind, one of two design procedures is employed: (i) Rods of the least standard size are attached to the plunger. The string is made up of this size rod until the maximum stress arising attains the allowable maximum. To this string section, rods of the next greater standard size are attached; the length of this second section is determined by the repeated application of the same criterion. If the two sections do not add up to the required total length, then the string is continued with rods of the next greater standard size. Putting in Eq. 4.1—26 $\sigma_{\max} = \sigma_{al}$, $A_r = A_{r1}$, $G_r = G_{r1}$ and $L = l_1$, we get

$$\sigma_{al} = \frac{F_l}{A_{r1}} + \frac{G_{r1} l_1}{A_{r1}}(b + \delta)$$

and hence

$$l_1 = \frac{\sigma_{al} A_{r1} - F_l}{G_{r1}(b + \delta)} \,. \qquad 4.1\text{—}27$$

The length of the nth section counted from below can be calculated analogously:

$$l_n = \frac{\sigma_{al}(A_{rn} - A_{r(n-1)})}{G_{rn}(b + \delta)}. \qquad 4.1-28$$

The maximum stress in the top end of the last — uppermost — string section designed by this procedure is usually less than the allowable maximum, or the actual stress at the top of any of the string sections farther below. (ii) Another procedure of tapered string design is to ensure that the maximum stress at the top of each string section be equal. In the following, suffix 1 refers to the top section. This principle yields for a two-section tapered string

$$\frac{G_l L}{A_{r1}} + \frac{(G_{r1} l_1 + G_{r2} l_2)(b + \delta)}{A_{r1}} = \frac{G_l L}{A_{r2}} + \frac{G_{r2} l_2 (b + \delta)}{A_{r2}}.$$

Assuming that $A_{r1}/A_{r2} \approx G_{r1}/G_{r2} = C$, and solving for l_2, we get

$$l_2 = L \frac{G_{r1}(b + \delta) - G_l(C - 1)}{G_{r1}(b + \delta)\left(2 - \frac{1}{C}\right)} \qquad 4.1-29$$

and, in the knowledge of l_2,

$$l_1 = L - l_2. \qquad 4.1-30$$

For a three-section tapered string, we may similarly write

$$l_2 = L \frac{[G_l C_3 + G_{r1}(b + \delta)](C_2 - 1)}{G_{r1} C_2 \left(3 - \frac{1}{C_1} - \frac{1}{C_2}\right)(b + \delta)} \qquad 4.1-31$$

$$l_3 = L \frac{G_l(C_1 + C_2 - 2C_3) + G_{r1}(b + \delta)}{G_{r1}\left(3 - \frac{1}{C_1} - \frac{1}{C_2}\right)(b + \delta)} \qquad 4.1-32$$

and

$$l_1 = L - l_2 - l_3. \qquad 4.1-33$$

In the above relationships,

$$C_1 = A_{r1}/A_{r2} \approx G_{r1}/G_{r2},$$
$$C_2 = A_{r2}/A_{r3} \approx G_{r2}/G_{r3},$$
$$C_3 = A_{r1}/A_{r3} \approx G_{r1}/G_{r3}.$$

Equations from 4.1—31 to 4.1—33 permit the compilation of tables indicating for various plunger sizes the length percentages of the individual rod sizes to be used in making up the string. Calculation of the tabulated values requires the assumption of a liquid gravity γ_l and of a dynamic factor δ. In design practice, numerous such tables are used; they are slightly different according to whether one or another fundamental formula has been used to calculate $F_{p\max}$ and according to the adapted values of γ_l and δ. (Table 4.1—2 has been taken out of Jones Bulletin No. 624.)

Example 4.1—3. Design a tapered rod string out of sucker rods of 7.62 m length and 25.4, 22.2 and 19.0 mm dia., respectively. $d_p = 44.5$ mm; $L = 1500$ m; $\gamma_l = 8826$ N/m³; $\delta = 0$. — Equations from 4.1—31 to 4.1—33 shall be employed.

$$C_1 = \frac{5.07}{3.87} = 1.31 \ ;$$

$$C_2 = \frac{3.87}{2.84} = 1.36 \ ;$$

$$C_3 = \frac{5.07}{2.84} = 1.79 \ ;$$

and, by Table 4.1—1, $G_{r1} = 42.2$.

$$G_l = A_p \gamma_l = 15.5 \times 10^{-4} \times 8826 = 13.7$$

$$l_2 = 1500 \frac{[13.7 \times 1.79 + 42.2(0.885 + 0)]\,(1.36 - 1)}{42.2 \times 1.36 \left(3 - \dfrac{1}{1.31} - \dfrac{1}{1.36}\right)(0.885 + 0)} = 438 \text{ m}$$

$$l_3 = 1500 \frac{13.7(1.31 + 1.36 - 2 \times 1.79) + 42.2(0.885 + 0)}{42.2 \left(3 - \dfrac{1}{1.31} - \dfrac{1}{1.36}\right)(0.885 + 0)} = 665 \text{ m}$$

$$l_1 = L - l_2 - l_3 = 1500 - 438 - 665 = 397 \text{ m} .$$

The length percentages of the individual string sections are, from top to bottom, 26.5; 29.2; 44.3; this is rather close to the sequence 23.9; 27.0; 49.1 in Table 4.1—2.

(a.4) *Effective plunger stroke.* — The difference between plunger stroke and polished-rod stroke is correctly given by Eq. 4.1—10 only if dynamic loads can be neglected, the rod string is untapered, and the tubing shoe is not fixed to the casing string.

Influence of dynamic loads. As mentioned above, assuming harmonic motion of the polished rod, the acceleration of the rod string varies at any instant of the cycle; this effect has to be accounted for at comparatively high pumping speeds and great well depths. The magnitude of the acceleration is greatest at the lower stroke end (where its sign is positive) and at the upper stroke end (where its sign is negative). The changes in dynamic load due to this circumstance result in a greater rod-string stretch at the lower stroke end and a smaller one at the upper stroke end than if the basic plus variable static load were only considered. Hence, the plunger or pump shoe passes beyond the end points to be expected under purely static loads: the stroke is somewhat lengthened. This is the phenomenon known as overtravel. — The stretch due to rod-string weight and dynamic loads, at the lower stroke end is, by Eqs 4.1—5, 4.1—11 and 4.1—12,

$$\Delta L_1 = \Delta L_{rp} + \Delta L_{rd1} ;$$

where

$$\Delta L_{rd1} = \frac{F_r L \delta}{2\,A_r E} = \frac{G_r L^2 r\,\omega^2}{2\,A_r Eg}\left(1 + \frac{r}{l}\right) .$$

Stretch at the upper stroke end is

$$\Delta L_2 = \Delta L_{rp} - \Delta L_{rd2}$$

where

$$\Delta L_{rd2} = \frac{G_r L^2 r\omega^2}{2\,A_r Eg}\left(1 - \frac{r}{l}\right) .$$

The difference in stretch between the lower and upper stroke end is

$$\Delta L_1 - \Delta L_2 = \frac{G_r L^2 r\,\omega^2}{A_r\, E g} .$$

Putting $G_r/A_r = 8.5\times 10^4$ N/m³, $r = s/2$ m, $\omega = (n\pi)/30$, $E = 2.06\times 10^{11}$ N/m² and $g = 9.81$ m/s², we get

$$\Delta L_1 - \Delta L_2 = 2.3\times 10^{-10} L^2 n^2 s . \qquad 4.1-34$$

The formula of Coberly (Zaba and Doherty 1956), differing from the above only in the coefficient, was probably derived by a similar consideration. In SI units, it reads

$$\frac{\Delta L_1 - \Delta L_2}{s} = 2.1\times 10^{-10} (Ln)^2 .$$

Hence, taking into account the changes in acceleration due to the motion, assumed to be harmonic, of the polished rod, and the consequent changes in dynamic load, the plunger stroke becomes

$$s_p = s(1 + 2.3\times 10^{-10}\times L^2\, n^2) - \Delta s .$$

The expression in the parentheses, called the Doherty coefficient, is denoted by K; the above formula may accordingly be written as

$$s_p = sK - \Delta s . \qquad 4.1-35$$

The plunger-stroke formula Eq. 4.1—35 is just an approximation in most cases. The main causes of deviation between fact and formula are that, firstly, the angular velocity ω of the crank is not constant; secondly, the upper end of the pitman travels along a circular arc rather than a straight vertical line; and thirdly, there arise complicated vibrations caused by interference, slight shocks and drag.

If the dynamometer card is near ideal (which may well be the case if the pumping speed is low), s_p is easier to determine; it can be directly read off

Fig. 4.1—4. The actual plunger stroke can also be determined from the dynamometer card. Several graphical methods are known, the one to be discussed below is due to Falk (Szilas 1969). The procedure expresses the probable plunger travel in terms of polished-rod travel. The stretch of the rod string under the load can be calculated using Eq. 4.1—7. Plot this relationship to the scale of the dynamometer card in Fig. 4.1—5 (line I). Then draw a parallel to I through the starting point of the chart corresponding to the wet weight $F_r b$ of the rod string (line I′). The intercept a of the line parallel to the axis of abscissae through any point of the chart gives the stretch under the load at that point. Now let us calibrate the ordinate axis

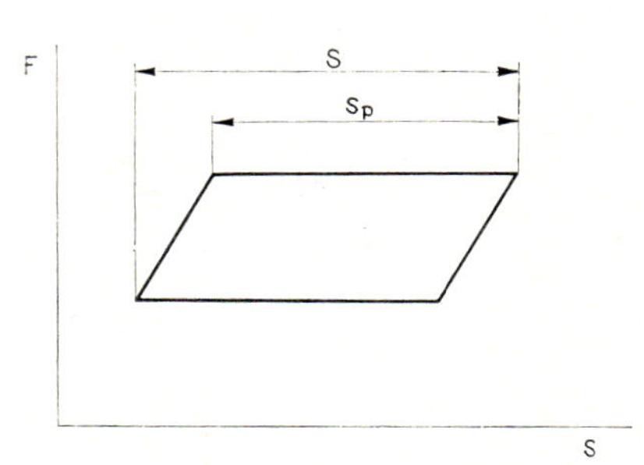

Fig. 4.1—4

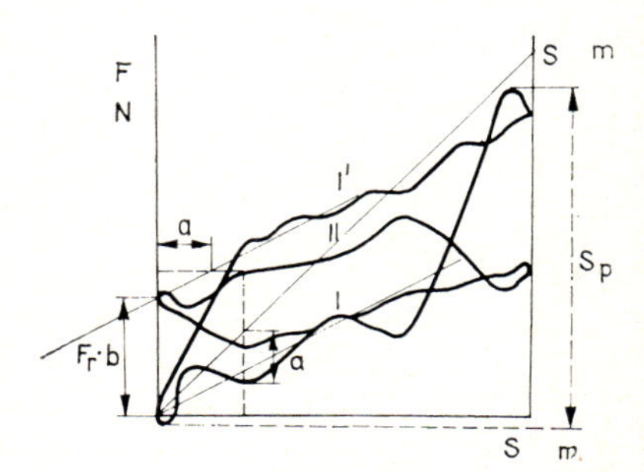

Fig. 4.1—5 Determination of plunger stroke with Falk's method

on the same scale as the abscissa axis, and plot plunger travel v. polished-rod travel. In the absence of stretch, this diagram would be a straight line of unity slope (line II). By adding to each point of this line the corresponding stretch a with the correct sign, we obtain a diagram illustrating the probable plunger travel. The ordinate difference between the lowermost and uppermost point of this diagram gives s_p, the plunger stroke. It is advisable at intervals to check the calculated value by the method just described.

When pumping high-viscosity oil, the oil surrounding the top sections of the rod string may be cold enough to freeze. The oil may then 'grip' the rod string when the polished rod starts on its upstroke: load will then build up steeply for a while before the plunger actually starts moving. It is in particular the top faces of the rod couplings that have to be ploughed through the 'solid' oil above them, which means that a force exceeding the static shear force of the oil is needed to start the string moving. Hence, the load will concentrate for a while in a certain section of the rod string, rather than being distributed over the entire string. The known methods of determining plunger travel will of course fail in this case. On the other hand, friction between fluid and tubing wall does not in itself limit the applicability of these methods.

Influence of the well completion. If the tubing shoe is fixed to the casing then the tubing string will exhibit no variable stretch: the plunger stroke is thus increased. If the rod string is tapered, then the changes in rod size should be taken into account in calculating stretch. For the above reasons, it is advisable to use Eq. 4.1—10 in the following, more general form:

$$\Delta s = \frac{A_p \gamma_l L^2}{E} w \qquad 4.1\text{—}36$$

where w is a factor accounting for the type of well completion. In the general case,

$$w = \frac{a + bC_1 + cC_3}{A_{r1}} + \frac{1}{A_T} \qquad 4.1\text{—}37$$

which holds for a tapered rod string and a non-anchored tubing. If the rod string is tapered and the tubing is anchored, then $1/A_T = 0$, and

$$w = \frac{a + bC_1 + cC_3}{A_{r1}} .$$

If the rod string is non-tapered and the tubing is not anchored, then

$$w = \frac{1}{A_r} + \frac{1}{A_T} .$$

And finally, if the rod string is non-tapered and the tubing is anchored, then

$$w = \frac{1}{A_r} .$$

Here, $a = l_1/L$; $b = l_2/L$; $c = l_3/L$; $C_1 = A_{r1}/A_{r2}$, and $C_3 = A_{r1}/A_{r3}$.

(a.5) *Buckling of the tubing.* — Research in recent years has shown that the variable liquid load causes unanchored tubing not only to stretch, but also to buckle during the upstroke (Fig. 4.1—6a). This may entail several types of trouble. For instance, significant friction may arise between the buckled tubing and the rods tensioned by the liquid load: the rods and tubes may undergo excessive wear and may break or puncture. The interpretation of multiple buckling in the tubing was given by Lubinski and Blenkarn in 1957. According to them, the liquid load acting on the plunger generates an upward force $F = A_p \Delta p$ in the tubing: this is the force giving rise to buckling. The length of the buckled tubing section is determined by finding the depth at which the tubing weight plus the weight of the liquid column equals the buckling force F. This is the critical tubing length (assuming that the fluid level in the annulus is flush with the top of the bottom-hole pump):

$$l_c = \frac{A_p \Delta p}{G_T + G_l'} . \qquad 4.1\text{—}38$$

Fig. 4.1—6 Tubing buckling during pumping, after Lubinski and Blenkarn (1957)

Length l_c measured from the tubing shoe determines the neutral point of the tubing. Up to that point, the tubing will undergo multiple buckling; above it, the tubing will not buckle even during the upstroke. As mentioned above, multiple buckling of the tubing may cause a variety of troubles: (i) friction between tubing and rod string increases the polished-rod load and hence the energy consumption of pumping; (ii) wear of the rod string against the tubing and of the tubing against the casing may cause punctures or breaks in any of these strings; (iii) repeated buckling of the tubing may entail wear or failure of the threaded couplings; (iv) lateral stress on the plunger entails its rapid, uneven wear. — In order to eliminate these harmful effects, it is usual to anchor the lower end of the tubing string to the casing (cf. paragraph 4.1.1d3).

(b) Operating points of sucker rod pumping

(b.1) *Production capacity of pumping.* — The theoretical production capacity of pumping is given by

$$q_t = 1440\, A_p s_p n \ \text{m}^3/\text{d}. \qquad 4.1-39$$

It is assumed that the volumetric efficiency is unity. The analysis of theoretical production capacity is facilitated by considering the volume produced per stroke:

$$V = A_p s_p .$$

Let us replace s_p by its Expression 4.1—35 and, in the latter, let us substitute Δs by its Expression 4.1—36. Then

$$V = A_p\, sK - \frac{A_p^2\, \gamma_l\, L^2}{E} w .$$

Let us find the plunger giving maximum production for a given polished rod stroke s and a given pumping speed n. Differentiating the above equation with respect to A_p, we obtain

$$\mathrm{d}V = sK\mathrm{d}A_p - \frac{\gamma_l L^2 w}{E}\, 2A_p \mathrm{d}A_p .$$

Production is maximum when $\mathrm{d}V/\mathrm{d}A_p = 0$, that is,

$$sK - 2\frac{A_p\, \gamma_l\, L^2}{E} w = 0 . \qquad 4.1-40$$

By Eq. 4.1—36, the second term on the left-hand side of this equation equals $2\Delta s$, and thus the theoretical production capacity is maximum when

$$sK = 2\Delta s \; ; \quad \text{that is,} \quad \Delta s = \frac{sK}{2} .$$

Substituting this into Eq. 4.1—36, we obtain for the cross-sectional area of the plunger providing the maximum theoretical production capacity

$$A_{p\max} = \frac{sKE}{2\gamma_l L^2 w}. \qquad 4.1\text{—}41$$

It is observed that, as opposed to surface reciprocating positive-displacement pumps, the theoretical production capacity of the bottom-hole pump at a given polished-rod stroke is not a linear function of the plunger's cross-sectional area, because increasing the latter increases the liquid load and hence also stroke reduction; that is, the plunger stroke s_p of the pump will be reduced. — Table 4.1—3 lists values of $A_{p\max}$, calculated using Eq. 4.1—41, v. sK and L, for $\gamma_l = 8826$ N/m³ and $d_r = 22.2$ mm.

(b.2) *Volumetric efficiency of pumping.* — The fluid volume actually produced is less than the theoretical capacity furnished by Eq. 4.1—39. The ratio of the effective production q_l to the theoretical q_t, gives the volumetric efficiency of bottom-hole pumping as

$$\eta_v = \frac{q_l}{q_t}. \qquad 4.1\text{—}42$$

Table 4.1—3

Theoretical values of $A_{p\max}$, in cm², for $d_r = 22.2$ mm, and $\gamma_l = 8826$ N/m³

L	sK, m								
m	0.5	0.75	1.0	1.25	1.5	1.75	2.0	2.5	3.0
500	90.2								
750	40.1	60.2	80.3	100.4			Higher than feasible		
1000	22.5	33.9	45.2	56.4	67.7	79.0	90.3		
1250	14.4	21.7	28.9	36.1	43.4	50.6	57.8	72.3	86.7
1500	10.0	15.0	20.1	25.1	30.1	35.1	40.1	50.2	60.2
1750	7.4	11.1	14.7	18.4	22.1	25.8	29.5	36.9	44.2
2000	5.6	8.5	11.3	14.1	16.9	19.7	22.6	28.2	33.9
2500	3.6	5.4	7.2	9.0	10.8	12.6	14.4	18.1	21.7
3000	2.5	3.8	5.0	6.3	7.5	8.8	10.0	12.5	15.0

Volumetric efficiency is a product of the efficiency factor η_{va} characterizing the measure to which the pump barrel is filled with an ideal liquid, and of the efficiency factor η_{vb}, characterizing the measure of leakage in the 'channel' leading the liquid to the flow line, that is,

$$\eta_v = \eta_{va}\eta_{vb}. \qquad 4.1\text{—}43$$

Here,

$$\eta_{va} = \frac{q_1}{q_t} \qquad 4.1\text{—}44$$

and

$$\eta_{vb} = \frac{q_1 - q_2 - q_3 - q_4}{q_1} = \frac{q_l}{q_1} \qquad 4.1\text{—}45$$

where q_1 is the amount of fluid sucked into the barrel; q_2 is slippage past the plunger; q_3 is leakage through the tubing into the casing annulus; and q_4 is slippage past the check valve in the surface conduit connecting the annulus with the tubing, all in m^3/d units at stock-tank conditions. — This interpretation of volumetric efficiency in bottom-hole pump deviates from the one for surface reciprocating positive-displacement pumps. This difference is due to the fact that, in such a surface pump, it is justified to assume that the cylinder is sucked full of liquid, and liquid only, during each stroke. The bottom-hole pump barrel does not get filled up with liquid on each stroke. The slippage loss of a surface pump can easily be determined, whereas in a bottom-hole pump it is not usually possible to separate the slippage loss from the various leakage flows and, moreover, q_1 cannot be measured either.

Filling efficiency η_{va}. The fact that $q_1 < q_t$ may be due to a variety of causes. (i) The theoretical capacity of the pump exceeds the rate of inflow from the formation into the well. The liquid level in the annulus then stabilizes approximately at pump level. (ii) Inflow of oil into the pump barrel is slower than the upward travel of the plunger, so that during the upstroke the liquid 'has not got enough time' to fill the barrel. (iii) Together with the oil, the formation often delivers gas to the well, and if no measures are taken to separate and remove it, it will enter the pump barrel and occupy part of the barrel space. (iv) Even if the well fluid contains no free gas at the pressure p_i and temperature T_i of entry into the pump barrel, the volume of stock-tank oil produced per unit time is less by volume factor B_i than the volume of oil at p_i and T_i.

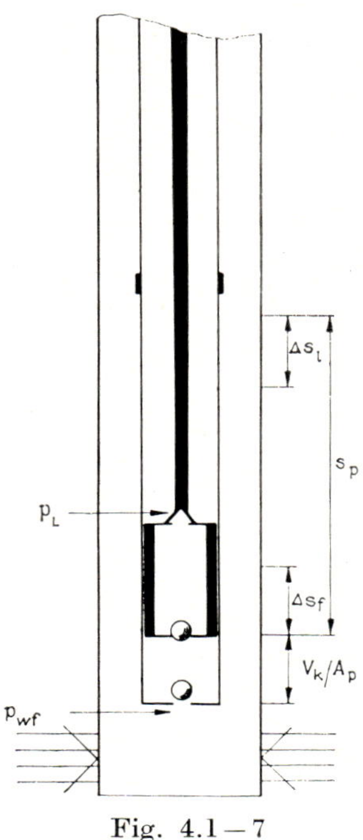

Fig. 4.1—7

ad (i). The filling factor can be determined by dynamometric measurements or level recording in the annulus. The capacity of the pump is to be reduced so as to match it to the inflow. *ad* (ii). Incomplete filling of the pump barrel may be due to the hydraulic resistance of the 'suction channel' being too great (either because it is sanded up or because it was too narrow to start with); or the viscosity of the oil is too high. The latter can be remedied by heating, introducing a solvent, or increasing the depth of immersion of the pump. *ad* (iii). In order to clarify the connexion between gas content and filling efficiency, it is necessary to discuss in some detail the process of pumping a gaseous liquid.

During the upstroke, the barrel is filled with a gas-liquid mixture at a pressure p_i almost equal to the pressure of the liquid column in the annulus (or the production *BHP*) p_{wf} (Fig. 4.1—7). When the plunger is at the upper end of its stroke, the space between the travelling and standing valve si filled with this mixture at this pressure p_{wf}. The pressure above the travelling valve, p_L is comparatively high, nearly equal to the pressure of the liquid column of height L in the tubing. At the onset of the down-stroke, the standing valve shuts off, but the travelling valve opens up only when the sinking plunger has compressed the gas-liquid mixture between the valves sufficiently for its pressure to attain or, indeed, slightly exceed p_L. The plunger then sinks through this high-pressure mix to the lower end of its stroke. If there is no dead space between the valves, the standing valve will open immediately at the onset of the upstroke. If, however, there is such dead space, then the standing valve opens only if the expansion of the gas-liquid mixture, made possible by the rise of the plunger, reduces pressure in the barrel to below the annulus pressure p_{wf}. — Variations of pressure p and polished-rod load F v. length of stroke are illustrated by the dashed lines in parts (a) and (b) of Fig. 4.1—56 (see later), where TV means the travelling valve; SV denotes the standing valve; O and C denote opening and closure, respectively.

By the above considerations, the presence of gas reduces filling efficiency by several causes: during the upstroke, some reduction is due to the opening delay Δs_f of the standing valve, expressed in terms of plunger travel, due in its turn to the significant expansion of fluid in the dead space. This results in a reduction of the effective barrel volume. Moreover, part of the effective barrel volume is occupied by gas rather than liquid. A further reduction in the effective downstroke volume is due to the opening delay Δs_l of the travelling valve expressed in terms of plunger travel. — Filling efficiency is a function primarily of the gas content of the fluid entering the pump, the proportion of dead space to the stroke volume of the pump, and the pressure ratio p_L/p_{wf}. Let us assume that the free gas sucked in at the pressure p_{wf} is uniformly distributed in the oil, and that compressibility of the oil and changes in dissolved-gas content are negligible. With the plunger at the upper end of its stroke, we have

$$V_{o1} + V_{o1}R_{wf} = V_p + V_k$$

where V_{o1} is the volume of oil in the space between the two valves; R_{wf} is the specific gas volume in the same space, at the pressure p_{wf}; V_p is the total stroke volume of the plunger; V_k is the dead-space volume. Solving for V_{o1}, we have

$$V_{o1} = \frac{V_p + V_k}{1 + R_{wf}} . \qquad 4.1\text{—}46$$

With the plunger at the lower end of its stroke, we may write

$$V_{o2} + V_{o2}R_L = V_k$$

where V_{o2} is the volume of oil in the dead space, and R_L is the specific gas volume in the dead space at the pressure p_L. Hence

$$V_{o2} = \frac{V_k}{1 + R_L} . \qquad 4.1\text{—}47$$

Assuming that the filling efficiency is affected by the presence of gas only, introducing $q_1 = (V_{o1} - V_{o2})\,n$ and $q_t = V_p n$ into Eq. 4.1—44 and dividing by n, we obtain the relationship

$$\eta_{va} = \frac{V_{o1} - V_{o2}}{V_p}.$$

Introducing V_{o1} and V_{o2} as expressed by Eqs 4.1—46 and 4.1—47, we get

$$\eta_{va} = \frac{\dfrac{V_p + V_k}{1 + R_{wf}} - \dfrac{V_k}{1 + R_L}}{V_p}.$$

Let $V_k/V_p = k$ and $R_{wf}/R_L = k'$; then,

$$\eta_{va} = \frac{1 + k}{1 + R_{wf}} - \frac{k}{1 + R_{wf}/k'}. \qquad 4.1—48$$

Consequently, the filling efficiency is a function of the effective *GOR* of the liquid sucked into the barrel, of the relative dead-space volume, k and of the pressure-dependent change in *GOR*, k'. Figure 4.1—8 illustrates the relationship 4.1—48 for $R_L = 0.1$. It is apparent that the smaller the relative dead-space volume k, the higher the filling efficiency. The latter is increased also by the decrease of k'. Since $k' = R_{wf}/R_L$, which, in a given well, equals Cp_L/p_{wf}, k' will be small if p_L/p_{wf} is small. At a given p_L, this can be attained by increasing the depth of immersion below the producing liquid level. If the gas content of well fluid contained in the dead-space at pressure p_L is so high that its expansion due to a pressure reduction to $p_{wf} \approx p_i$ is greater than the pump stroke volume, then a so-called gas lock comes to exist, and the sucker-rod pump ceases to produce any liquid. In the formulation of Juch and Watson (1969),

$$\frac{B_i}{B_d} \geq \frac{V_p + V_k}{V_k}.$$

Volume efficiency η_{vb}. Part of the liquid lifted by the plunger may (i) slip back through the clearance between plunger and barrel, between valve balls and seats, and past the seating cone of a rod pump. This leakage q_2 in-

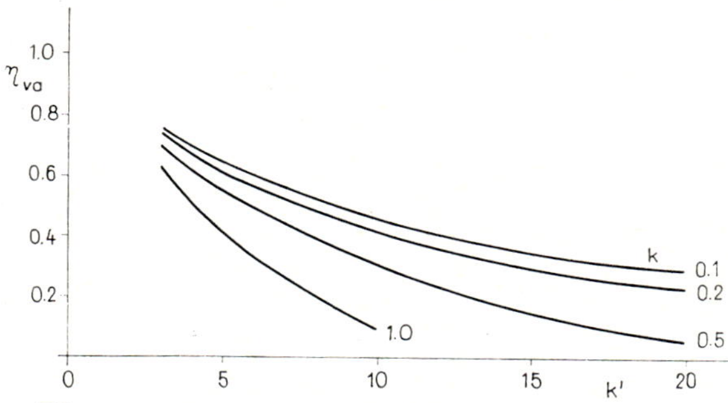

Fig. 4.1—8 η_{va} as a function of k' at $R_L = 0.1$ with k as a parameter

creases as the pump wears, and may attain quite high values. When pumping a sandy well fluid, the fit between barrel and plunger, quite close originally, will deteriorate more rapidly and thus the slippage loss will rapidly increase too. The same clearance will result in a greater slippage loss if oil viscosity is less. (ii) The number and size of leaks in the tubing wall may be significant particularly if the tubing is old. These leaks are due to erosion and to a lesser extent to corrosion. Erosion may be due to contact with the moving rod string, or to repeated stresses at a coupling. The very large number of periodic bucklings, stretchings and contractions will cause wear on the coupling threads and this effect may be enhanced by erosive solid particles or corrosive fluids entering between the threads. A considerable leak may come to exist also if the threads are not cleaned adequately before make-up. These leaks may permit a significant flow q_3 of well fluid into the casing annulus. Leakage through the tubing can be measured rather simply after the running of a new close-fitted sucker-rod pump in which slippage past the plunger may be neglected: the tubing is filled with oil to its open top, and topped up once per minute. If leakage exceeds a certain allowable value, the tubing string must be pulled and pressure-tested length by length. Leakage due to worn threads may be minimized by inverting couplings or by rethreading. Leakage along the threads may be significant even if the tubing pipe is new, if the torque used in make-up is insufficient or if the thread compound is not of the right quality. (iii) The casing annulus of wells pumped by means of bottom-hole pumps is usually connected with the flowline through a conduit incorporating a check valve, so as to permit the gas entering the well to bypass the bottom-hole pump. If the check valve does not close tight, some liquid will leak through it into the annulus. In the arrangement shown as Fig. 4.1—9, pressure gauge *4* will register an increase in pressure after shut-off of the casing valve *3* if the check valve *2* leaks liquid into line *1*.

(b.3) *Operating point of maximum liquid production.* — By our considerations in section (b.1), the plunger size giving maximum production capacity for a given polished-rod stroke and pumping speed can be calculated using Eq. 4.1—41. The maximum feasible production capacity of a given sucker-rod pump can be obtained by operating it with this plunger at the maximum possible stroke and speed settings. However, increasing the polished-rod stroke and the pumping speed entails by Eq. 4.1—14 an

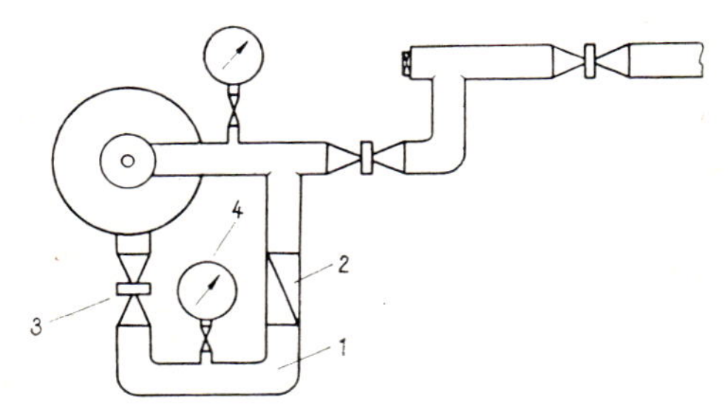

Fig. 4.1—9 Wellhead connections to flowline at a well produced by sucker-rod pump

increase in the dynamic, and hence also in the total, load on the rod string. Increasing n is permissible until the total maximum load attains the allowable maximum, that is, until $\sigma_{max} = \sigma_{al}$. By Eq. 4.1—26, the allowable maximum tension stress in a tapered rod string is

$$\sigma_{al} = \sigma_{max} = \frac{F_l}{A_{r1}} + \frac{\bar{G}_r L}{A_{r1}}(b+\delta) = \frac{A_p L \gamma_l}{A_{r1}} + \frac{\bar{G}_r L}{A_{r1}}\left(b + \frac{sn^2}{1440}\right).$$

This equation may be solved for the maximum plunger cross-section area permitted by rod strength as

$$A_{p\,max} = \frac{1}{\gamma_l}\left[\frac{\sigma_{al} A_{r1}}{L} - \bar{G}_r b - \frac{\bar{G}_r sn^2}{1440}\right].$$

Let us substitute $A_{p\,max}$ by the expression Eq. 4.1—41 of the plunger cross-section area giving the maximum production capacity; then,

$$\frac{s\,KE}{2L^2 w\gamma_l} = \frac{1}{\gamma_l}\left[\frac{\sigma_{al} A_{r1}}{L} - \bar{G}_r b - \frac{\bar{G}_r sn^2}{1440}\right].$$

This equation can be solved for the maximum stroke, $s = s_{max}$, permitted by rod strength. Let us substitute sK by its expression in Eq. 4.1—35: then,

$$s_{max} = \frac{\dfrac{\sigma_{al} A_{r1}}{L} - \bar{G}_r b}{\dfrac{(1 + 2.3\times 10^{-10} L^2 n^2)\,E}{2L^2 w} + \dfrac{\bar{G}_r n^2}{1440}}. \qquad 4.1—49$$

This equation gives s_{max} v. pumping speed n (in spm) for a given rod string, if the plunger used is that one providing the maximum feasible production. The operating point of maximum liquid production may thus be determined by finding the maximum permissible polished-rod stroke for various pumping speeds and also the value of $A_{p\,max}$ for each pair of values (s_{max}, n).

The solution of the problem is complicated by the fact that, in order to calculate s_{max}, one needs the length percentages of the rod sizes making up the tapered string. Now for calculating these, one needs A_p and the dynamic factor δ, which will emerge only somewhat later in the calculation. Solution of the problem thus requires the use of iteration. In view of the reduced number of options, however, the number of steps can be reduced with the help of auxiliary tables to just one or two. Calculation proceeds as follows. A_p is estimated using Tables 4.1—2 and 4.1—3, and, having selected a set of rod sizes, the length percentages of the individual string sections are determined with the aid of Table 4.1—2; s_{max} is then calculated for various feasible pumping speeds. $A_{p\,max}$ is determined for each feasible pair of n and s_{max}. The corresponding values of A_p, s and n yield values of q_t, of which the greatest one is chosen. This is the first step of the approximation. Carrying out the dimensioning of the rod string all over again one gets another triplet A_p, s, n. If this is close enough to the first triplet, no further step is necessary. As a check, σ_{max} is calculated and verified against σ_{al}.

Table 4.1 – 4

Average weight of sucker-rod string in N/m, for the length percentages listed in Table 4.1 – 2

	Plunger diameter, mm													
	19.0	23.8	25.4	27.0	31.8	38.1	44.5	50.8	57,2	61.9	63.5	69.9	95.3	101.6
15.9	17.1	17.1	17.1	17.1	17.1	17.1	17.1	17.1	17.1	17.1	17.1	17.1	17.1	17.1
19.0 × 15.9	18.8	18.9	19.0	19.0	19.2	19.5	19.8	20.2	20.6	21.0	21.2	21.7	—	—
19.0	23.6	23.6	23.6	23.6	23.6	23.6	23.6	23.6	23.6	23.6	23.6	23.6	23.6	23.6
22.2 × 19.0 × 15.9	—	—	21.9	22.1	22.7	23.5	24.6	25.9	27.4	—	—	—	—	—
22.2 × 19.0	26.1	26.2	26.3	26.3	26.5	26.8	27.1	27.5	28.0	28.3	28.4	23.8	31.6	—
22.2	34.1	34.1	34.1	34.1	34.1	34.1	34.1	34.1	34.1	34.1	34.1	34.1	34.1	34.1
25.4 × 22.2 × 19.0	—	—	28.6	28.8	29.2	30.0	30.8	31.8	32.9	33.6	34.1	35.5	37.0	—
25.4 × 22.2	—	—	35.9	35.9	36.0	36.1	36.3	36.5	36.7	36.9	37.0	37.2	38.6	38.6
25.4	41.7	41.7	41.7	41.7	41.7	41.7	41.7	41.7	41.7	41.7	41.7	41.7	41.7	41.7

Example 4.1—4. Find the size of the TL type sucker-rod pump and the characteristics of its operating point providing the maximum theoretical liquid production, if $L = 1550$ m; $\gamma_l = 8826$ N/m³; $s = 1.8$, 1.4, 1.0 m; $n = 20$, 15, 10 min^{-1}; the rod string is to be composed of rods of 22.2, 19.0 and 15.9 mm diameter; $\sigma_{al} = 2.06 \times 10^9$ N/m²; the tubing is anchored. — Let d_p be 50.8 mm. Table 4.1—2 gives the rod-size percentages as 35.7, 41.0 and 23.3. From Table 4.1—4, $G_r = 25.9$ N/m³. — The well-completion factor is 0.354×10^4 by formula 4.1—37. — By Eq. 4.1—49, the longest stroke permitted by rod strength is 1.30 m at $n = 20$, 1.62 m at $n = 15$ and 1.96 m at $n = 10$. The actual feasible stroke settings are 1.00, 1.40 and 1.80 m, respectively. — The cross-section of the plunger providing maximum production capacity is 16.7, 21.5 and 26.0×10^{-4} m² by Eq. 4.1—41. The actual feasible plunger sizes of the TL type pump chosen are 15.5 and 25.7×10^{-4} m². The corresponding and feasible values of n, A_p and s, as well as the values of s_p furnished by Eq. 4.1—35 and the values of q_t furnished by Eq. 4.1—39, are listed in Table 4.1—5. — The operating point chosen is No. 3. The dimensioning of the rod string is now repeated with the parameters of this point, using Eqs from 4.1—31 to 4.1—33: $l_1 = 611$ m, $l_2 = 651$ m and $l_3 = 288$ m.

Table 4.1—5

	n	A_p	s	s_p	q_t
	1/min	10^{-4} m²	m	m	m³/min
1	20	15.5	1.0	0.65	0.02015
2	15	15.5	1.40	1.01	0.02348
3	10	25.7	1.80	0.96	0.02467

Calculating again the values of s_{max} we obtain 1.20, 1.50 and 1.82 m; the corresponding feasible sizes are again 1.00, 1.40 and 1.80 m.

By Eq. 4.1—26, $\sigma_{max} = 2.00 \times 10^8$ N/m². Since this is less than $\sigma_{al} = 2.06 \times 10^8$ N/m², the result is acceptable. Let us add that at high pumping speeds it is sometimes necessary to calculate in addition to the maximum polished-rod stroke permitted by considerations of rod strength also the maximum stroke permitted by dynamic considerations. By Eq. 4.1—13,

$$\delta = \frac{sn^2}{1440} = 0.5 \quad \text{and hence} \quad s'_{max} = \frac{720}{n^2}$$

The values of s_{max} calculated from this relationship are 1.80, 3.20 and 7.20 m, much greater than those obtained above.

(b.4) *Operating point of minimum polished-rod load.* — If the liquid production required is less than the feasible maximum, then the viewpoints of saving on both rod size and power consumption suggest the choice of an operating point that delivers the required production at the least feasible maximum polished-rod load $F_{p\max}$. Several methods for designing such a system are known. The simplest of these proceeds as follows. The theoretical production $q_t = q_l/\eta_v$ is calculated in the knowledge of the volumetric efficiencies of similar wells. The least plunger diameter available for the pump type selected is then chosen, with the longest of the feasible polished-rod strokes. The length percentages of the rod sizes to be used can be taken from Table 4.1—2. With an estimated value for n, s_p is calculated using Eq. 4.1—35; Eq. 4.1—39 is then used to find the actual value of n needed to lift the prescribed production. If n is too great to be realized with the given unit, then the calculation is repeated with the next bigger plunger size. If the pumping speed thus obtained is less than the feasible maximum, then the operating point chosen is feasible. The data thus obtained permit us to calculate the rod-size percentages using Eqs 4.1—31 to 4.1—33; Eq. 4.1—26 then yields the maximum polished-rod stress. The rod material is to be specified accordingly. The procedure is based on the consideration that the less A_p the less the fluid load and the rod-string weight, that is, the less the static load. The greater s, the less the n required to realize the desired q_t. By Eq. 4.1—14, the dynamic load varies as the square of n and as the first power of s. Hence, if the least n is chosen, the dynamic load will also be least.

Example 4.1—5. The pump to be used is an RWA type sucker-rod pump, producing $q_t = 27.8$ m³/d. The operating point should give the least maximum load $F_{p\max}$; $L = 1200$ m; $\gamma_l = 8826$ N/m³; $s = 1.5$, 1.0; $n =$ all integers between 5 and 15; the string is to be composed of rod sizes of 22.2 and 19.0 mm diameter; the tubing is anchored. — The least plunger diameter available is 1 1/2 in. Its cross-sectional area is 11.4 cm². From Table 4.1—2, (a) = 0.296 and (b) = 0.704; the maximum feasible polished-rod stroke s is 1.5 m. — For calculating the plunger stroke let us assume that n is about 10. The well-completion factor is 0.325×10^4 by Eq. 4.1—37. The plunger stroke is given by Eqs 4.1—35 and 4.1—36 as

$$s_p = 1.5(1 + 2.3\times10^{-10}\times1200^2\times10^2) -$$

$$-\frac{11.4\times10^{-4}\times8826\times1200^2}{2.06\times10^{11}}\times0.325\times10^4 = 1.32 \text{ m} .$$

By Eq. 4.1—39,

$$n = \frac{27.8}{1440\times11.4\times10^{-4}\times1.32} = 12.8 \approx 13 \text{ min}^{-1} .$$

The parameters of the operating point ensuring the least polished-rod load are, then, $d_p = 1$ 1/2 in., $s = 1.5$ m and $n = 13$ min⁻¹. — With these data in hand, let us design a two-stage rod string using Eqs 4.1—29 and 4.1—30:

$$l_1 = 354 \text{ m} \quad \text{and} \quad l_2 = 846 \text{ m} .$$

By Eq. 4.1—26, the maximum polished-rod stress is

$$\sigma_{\max} = \frac{11.4 \times 10^{-4} \times 8826 \times 1200}{3.87 \times 10^{-4}} +$$

$$+ \frac{354 \times 34.1 + 846 \times 23.6}{3.87 \times 10^{-4}} \left[0.885 + \frac{1.5 \times 13^2}{1440} \right] = 1.16 \times 10^8 \ \mathrm{N/m^2} \ .$$

The rod material is to be specified so that its allowable stress be greater than this value.

(c) Pumping units and prime movers

In order to select the correct surface unit, one has to know the maximum polished-rod load anticipated, the maximum polished-rod stroke and pumping speed to be used, as well as the maximum driveshaft torque required. A new standard, GOST 5866—66, which was introduced in the Soviet Union in 1967, contains the main parameters of 20 different types of pumping unit. Of these, 9 are so-called basic models and 11 are modified models. The basic models (cf. Figs 4.1—10 to 13) have equal walking-beam arms, whereas all modified models except 7 SK 12—2.5—4000 have different arms, with the arm on the horsehead side longer by 40—50 percent. Hence, the stroke of the modified model is longer than that of the corresponding basic model, and its allowable polished-rod load is less. The basic models have higher depth capabilities whereas the modified models offer higher production capacities. For production rates below 150 m^3/d, the drive required can be chosen by reference to Fig. 4.1—10. Part (a) refers to the basic models, part (b) to the modified ones. Further to be used in selection is Table 4.1—6. Columns 1—7 carry the markings of the individual models. The first number after the letters SK is the maximum allowable polished-rod load in Mp (1 Mp = 9.81 kN in the SI system), the second number is the maximum polished-rod stroke in metres, the third is the maximum torque of the slow shaft of the gear reducer in kp.m (1 kp.m = 9.81 Nm in the SI system). Correlation between the Table and Fig. 4.1—10 is established by means of the Roman numerals in Columns 6 and 9. In constructing the Figure, it has been assumed that $\gamma_l = 8826$ $\mathrm{N/m^3}$, $\eta_v = 0.85$ and $\sigma_{\mathrm{al}} = 1.18 \times 10^8$ $\mathrm{N/m^2}$.

Example 4.1—6. 50 m^3/d of liquid is to be pumped by sucker-rod pump from a depth of 1500 m. Which is the basic model to be selected? — Figure 4.1—10a and Table 4.1—6 reveal the best suited model to be 7 SK—12—2,5-4000, marked VII. Figure 4.1—10 also helps to find the approximate value of the optimum plunger diameter. The fields outlined in bold line and marked with Roman numerals are subdivided by dashed lines into smaller fields marked by Arabic numerals. Each of these corresponds to a plunger diameter, listed in Columns 10 and 11 of Table 4.1—6. In the above example, optimum plunger diameter is 43 mm, corresponding to mark 4.

Table 4.1 – 6

Data of Soviet sucker-rod pumping units (after GOST 5866)

Basic types (a)						Modified types (b)		Sucker-rod pump		
Type code	Speed spm 1/min	Power of electric mover kW	Weight of surface pumping unit kN	Counter balancing	Field of application (a) in the Fig.	Type code	Max. speed spm 1/min	Field of application; (b) in the Fig.	Diam. mm	Code
1	2	3	4	5	6	7	8	9	10	11
1 SK-1,5-0,42-100	4.8—14.7	1.7	9.8	Beam	I	1 SK-1-0,6-100	15	I	28	1
2 SK-2-0,6-250	5.5—15	2.3	15	counter-	II	2 SK-1,25-0,9-250	15	II	32	2
3 SK-3-0,75-400	4.8—15	4.5	25	weight	III	3 SK-2-1,05-400	15	III	38	3
4 SK-3-1,2-700	4.7—15.5	7	39	Combined	IV	4 SK-2-1,8-700	15	IV	43	4
5 SK-6-1,5-1600	5.0—15.5	10	59		V	5 SK-4-2,1-1600	14	V	55	5
6 SK-6-2,1-2500	5.8—15.3	20	88		VI	6 SK-4-3-2500	12	VI	65	6
7 SK-12-2,5-4000	5.2—12.2	28	137	Crank	VII	7 SK-8-3,5-4000	11	VII	68	7
8 SK-12-3,5-8000	5.2—10.0	55	196	counter-	VIII	7 SK-8-3,5-6000	11	VIII	82	8
9 SK-20-4,2-12000	4.8—10.0	55	314	weight	IX	7 SK-12-2,5-6000	13	IX	93	9
						8 SK-8-5-8000	11	VIII		
						9 SK-15-6-12000	8	X		

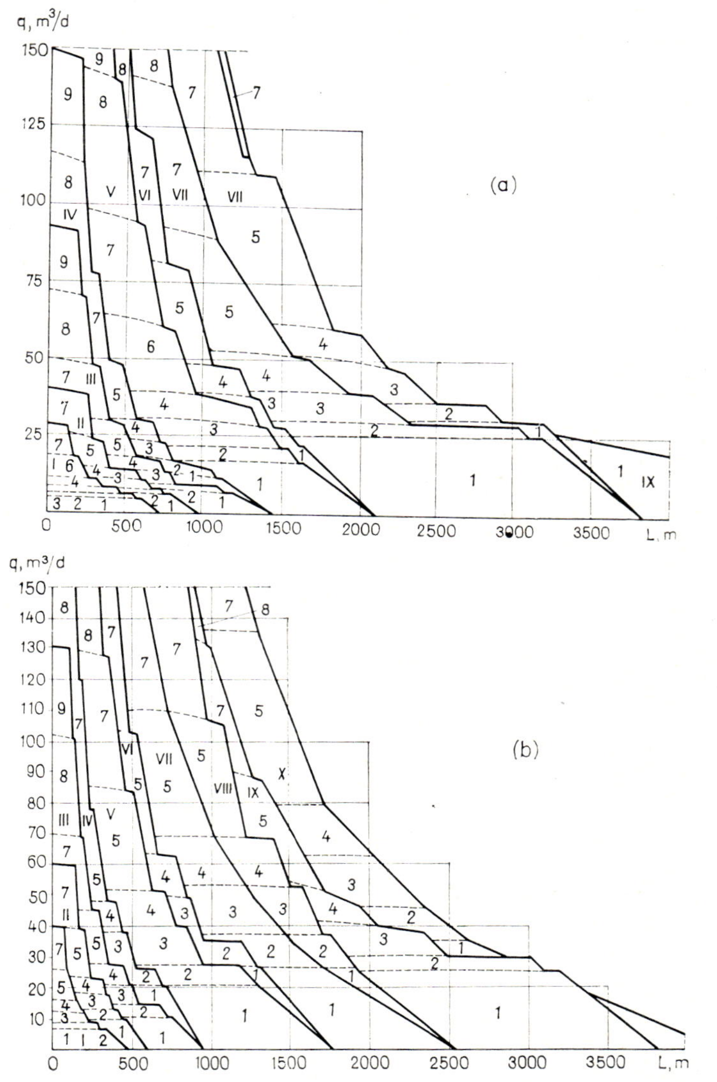

Fig. 4.1—10 Application limits of Soviet GOST 5866-66 bottom-hole pumps, after Adonin (1969)

The peak torque can be calculated by a slightly modified form of a formula in API RP 11L:

$$M_{\max} = \frac{1}{2} s^2 k_2 k_3 k_r \tag{4.1—50}$$

$$k_2 = 1 + k_4 \left[\frac{F_r b}{s k_r} - 0.3 \right] \tag{4.1—51}$$

where the value of k_4 belonging to a given pair of $\frac{F_l}{ks_r}$ and n/n_O can be read off Fig. 4.1—11; k_3 is plotted against these same variables in Fig. 4.1—12; k_r is furnished by Eq. 4.1—21.

In some cases studied by Griffin (1968), the mean peak torque calculated by this method turned out to be greater by 7.62 percent than the mean of the measured values. Hence, the method in question errs on the safe side. A point to be considered is, however, that the formula takes the efficiency

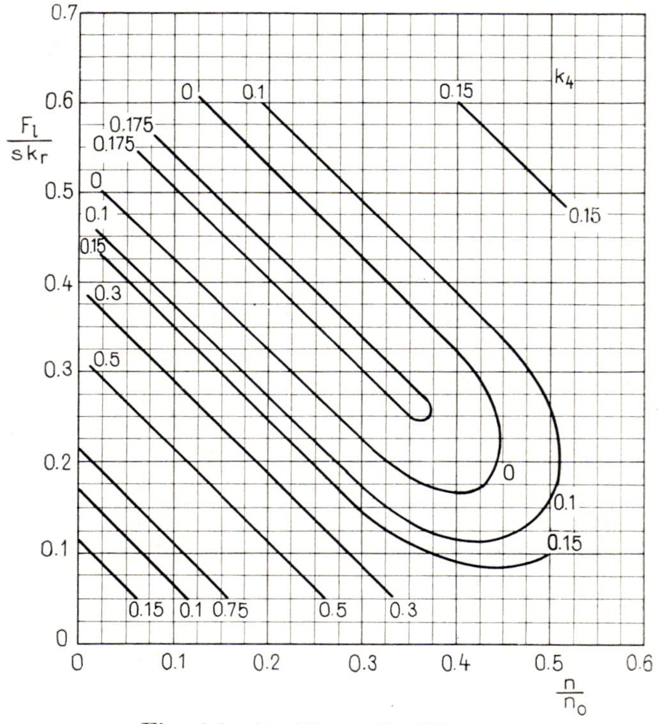

Fig. 4.1—11 After API RP 11L

of the gear reducer to be unity, whereas it is 0.90—0.95 in actual fact. Furthermore, the balancing of the pumping units is not usually perfect, either. It is therefore indicated to multiply the calculated value by a further factor of 1.2.

Example 4.1—7. Find the peak torque of a sucker-rod pump drive if $L = 1200$ m; $s = 1.4$; $n = 9$ min^{-1}; $d_p = 44.5$ mm; $\gamma_l = 8826$ N/m^3; 32.9 percent of the rod string is 7/8-in. size, 67.1 percent is 3/4-in. size; $F_l = A_p L \gamma_l = 15.5 \times 10^{-4} \times 1200 \times 8826 = 1.64 \times 10^4$ N. — By Eq. 4.1—21,

$$\frac{1}{k_r} = \frac{1}{2.06 \times 10^{11}} \left[\frac{1200 \times 0.329}{3.87 \times 10^{-4}} + \frac{1200 \times 0.671}{2.84 \times 10^{-4}} \right] = \frac{1}{6.05 \times 10^4} \text{ m/N} ,$$

$$\frac{F_l}{s\,k_r} = \frac{1.64 \times 10^4}{1.4 \times 6.05 \times 10^4} = 0.19$$

and by Eq. 4.1 – 19,

$$n_O = \frac{76{,}500}{1200} = 63.8 \ \text{min}^{-1}$$

and hence

$$\frac{n}{n_O} = \frac{9}{63.8} = 0.14 \ .$$

Figure 4.1 – 11 gives $k_4 = 0.42$. Furthermore, from Table 4.1 – 4,

$$F_r = 27.1 \times 1200 = 3.25 \times 10^4 \text{ N} .$$

$$b = \frac{\gamma_r - \gamma_l}{\gamma_r} = \frac{7.7 \times 10^4 - 0.8826 \times 10^4}{7.7 \times 10^4} = 0.885$$

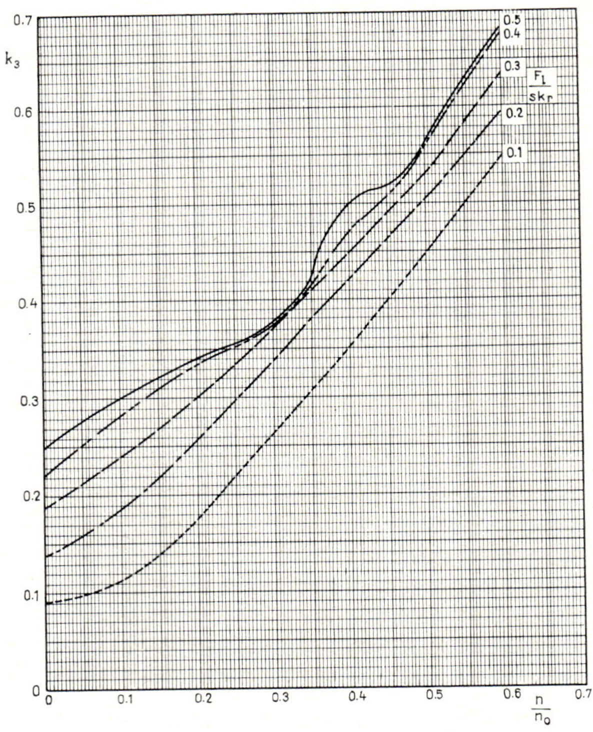

Fig. 4.1 – 12 After API RP 11L

and hence, by Eq. 4.1—51,

$$k_2 = 1 + 0.42\left[\frac{3.25\times10^4\times0.885}{1.4\times6.05\times10^4} - 0.3\right] = 1.02 .$$

Figure 4.1—12 furnishes $k_3 = 0.21$. Now by Eq. 4.1—50,

$$M = \frac{1}{2}\,1.4^2\times1.02\times0.21\times6.05\times10^4 = 1.27\times10^4\ \text{Nm} .$$

Several formulae for determining *the power required of the prime mover* have been published. Of these, we shall discuss one developed by the Soviet institute AZhNII (Kulizade 1960):

$$P = 10^3\,\frac{n}{\eta_m}\,(k_5 + k_6 F_l s)\ \text{kW} \qquad 4.1\text{—}52$$

where k_5 is a factor depending on pumping-unit design. Its values are listed in Table 4.1—7 for some of the earlier Soviet pumping units; η_m is the

Table 4.1—7
Values of k_5

Type	k_5
SKN 2-615*	0.020
SKN 3-915	0.035
SKN 5-1812	0.100
SKN 10-2115	0.160
SKN 10-3012	0.220

* The first number is F_{max} in Mp; the first or first two digits after the hyphen indicate s_{max} in dm; the last two digits indicate n_{max} in min^{-1}

mechanical efficiency of power transfer from prime mover to polished rod. The factor

$$k_6 = 2.14\times10^{-5}\sqrt{\eta_{v'}^2 + 0.28\left(1 + 10^{-5}\,\frac{sn^2}{d_p^2}\right)^2} \qquad 4.1\text{—}53$$

where $\eta_{v'}$ is the capacity factor, the ratio of the effective liquid production to the theoretical production given by the polished-rod stroke, that is,

$$\eta_{v'} = \frac{q_l}{q_t'} = \frac{A_p s_p n \eta_v}{A_p s n} = \frac{s_p\,\eta_v}{s}$$

Example 4.1—8. Find the useful output of the prime mover if $L = 1200$ m; $d_p = 44.5$ mm; the pumping unit is Soviet type SKN 5-1812 with $\eta_m = 0.96$; $s = 1.4$ m; $n = 9\ \text{min}^{-1}$; $\eta_v = 0.42$ and $\gamma_l = 8826\ \text{N/m}^3$. The liquid load is

$$F_l = A_p\,\gamma_l L = 15.5\times10^{-4}\times8826\times1200 = 1.64\times10^4\ \text{N}.$$

From Table 4.1—7, $k_5 = 0.100$. By Eq. 4.1—53,

$$k_6 = 2.14 \times 10^{-5} \sqrt{0.42^2 + 0.28 \left(1 + 10^{-5} \frac{1.4 \times 9^2}{0.0445^2}\right)^2} = 2.00 \times 10^{-5}$$

and hence, by Eq. 4.1—52,

$$P = 10^3 \frac{9}{0.96} (0.100 + 2.00 \times 10^{-5} \times 1.64 \times 10^4 \times 1.4) = 5.3 \text{ kW}.$$

One of the frequently used pumping units is the SKN type unit shown in Fig. 4.1—13. It has a combined crank-and-beam balance. Some pumping units feature either the crank or the beam type balance only. In the air-balanced unit shown in Fig. 4.1—14, on the other hand, the role of the balancing counterweight is assumed by compressed air in a cylinder. Compressed air at 4—5 bars pressure is provided by a compressor driven by the pumping unit. Practice sometimes employs special pumping units, of which we should consider here: (i) the hydraulic drive shown in Fig. 4.1—15; the walking beam is moved by piston *1*, connected to the beam by a bearing; the piston is driven by power liquid provided through line *2* by an electrically driven pump; the piston is controlled by toggle *3*. (ii) Figure 4.1—16 shows the PK-5 type Soviet make pumping unit with an accessory gas compressor; rod *2* of the piston moving in cylinder *1* is connected by a bearing to walking beam *3*; the double-acting compressor sucks gas from

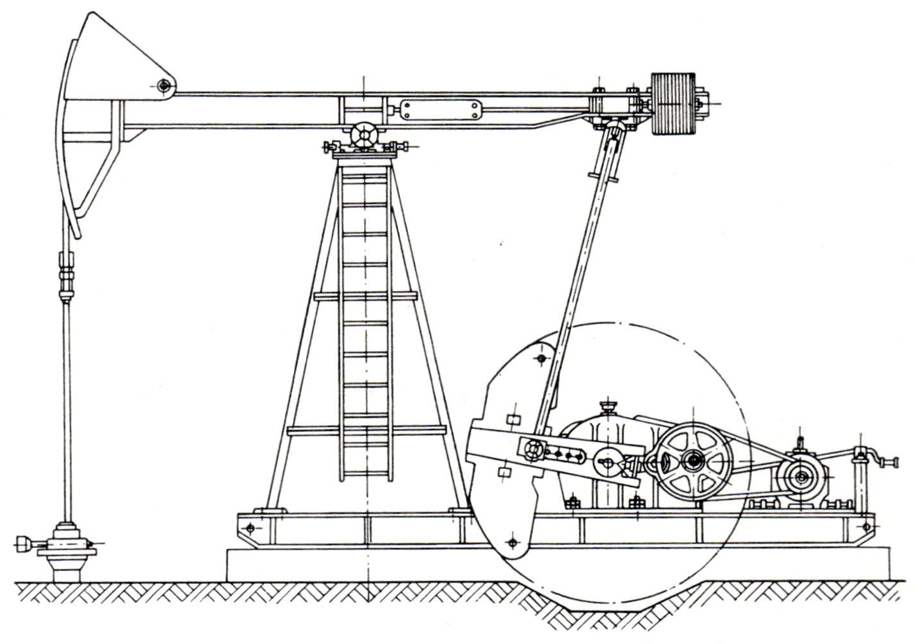

Fig. 4.1—13 SKN type sucker-rod pumping unit

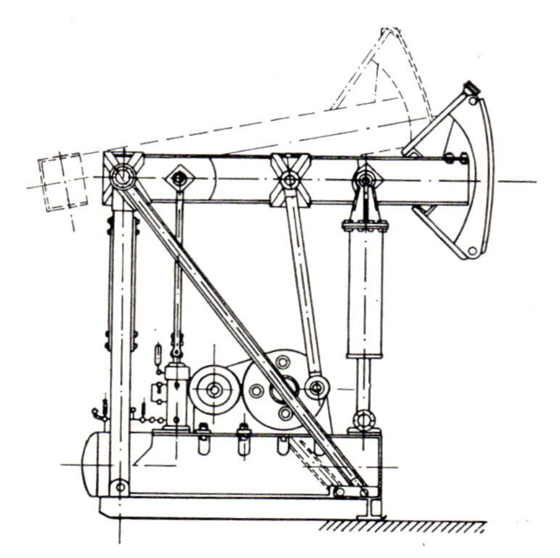

Fig. 4.1 — 14 Pneumatically balanced pumping unit

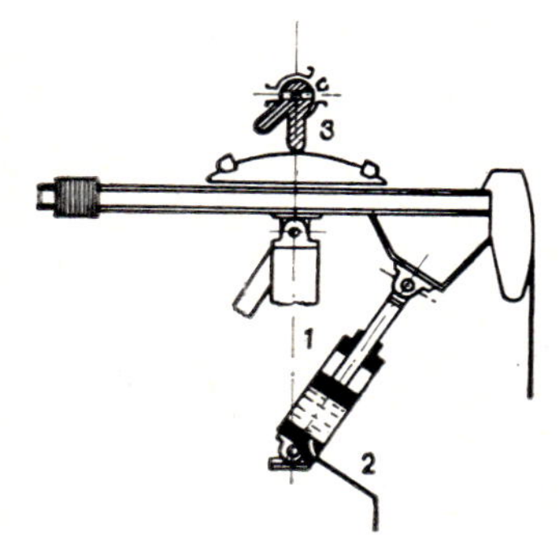

Fig. 4.1 — 15 Hydraulic sucker-rod pump drive

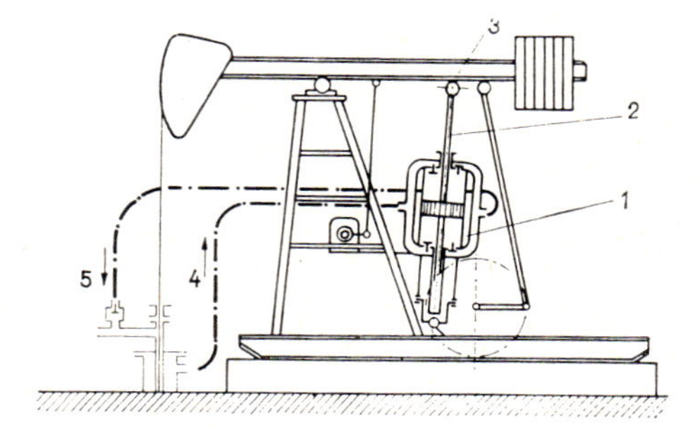

Fig. 4.1 — 16 PK-5 suction-compressor drive

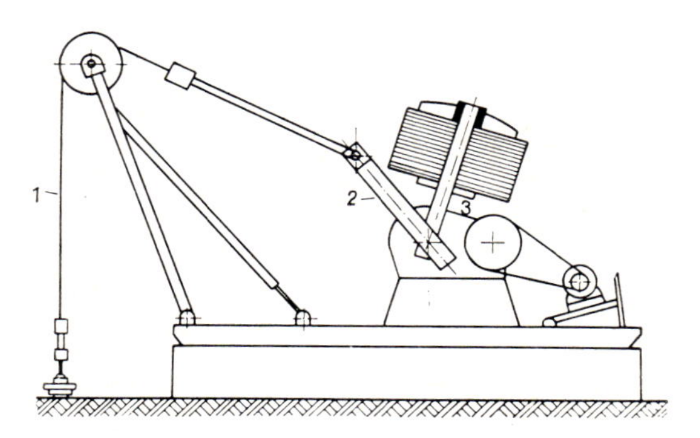

Fig. 4.1 — 17 SBN-5-3015 drive

pipe *4* and pumps it into flow line *5*; this reduces the *BHP*; the unit is therefore suited for attaining especially low *BHP*s; suction pressure is 0.9 bar, maximum discharge pressure is 5 bars. At a pumping speed of 10 spm, the unit marked PK-5-350 moves 350 m^3 per day of gas. (iii) A motion transformer significantly different from the conventional ones is incorporated in the Soviet-make pumping unit SBN 5-3015. The polished rod is suspended from wire rope *1* (Fig. 4.1—17). Drive crank *2* is rigidly fixed to counterweight crank *3*. The structural steel consumption of this solution is much less than that of the conventional ones. The maximum polished-rod stroke is 3.0 m; pumping speeds can be varied from 5 to 15 spm; the maximum allowable torque on the slow shaft of the gear reducer is 2.26×10^4 Nm. The total weight of the pumping unit is 92 kN.

(d) Wellhead and subsurface equipment

Wellhead designs for wells produced by means of sucker-rod pumps differ from those of flowing and gas-lift wells. A frequently adopted arrangement is shown in Fig. 4.1—9 (see earlier). The casing- and tubinghead often agree with those used in other types of wells. On the tubinghead, however, a polished-rod stuffing box is installed. The packoff provided by this device prevents the leakage of liquid from the tubing along the moving polished rod. One possible polished-rod stuffing box design is shown in Fig. 4.1—18. If the oil-resistant rubber packings *1* get worn, and well fluid starts to leak

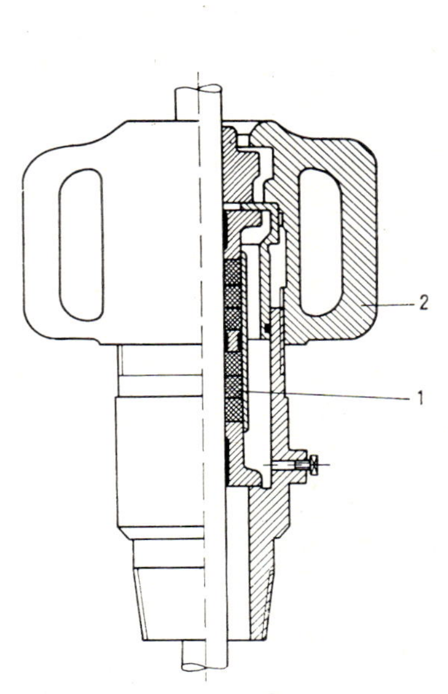

Fig. 4.1—18 Axelson's polished-rod stuffing box

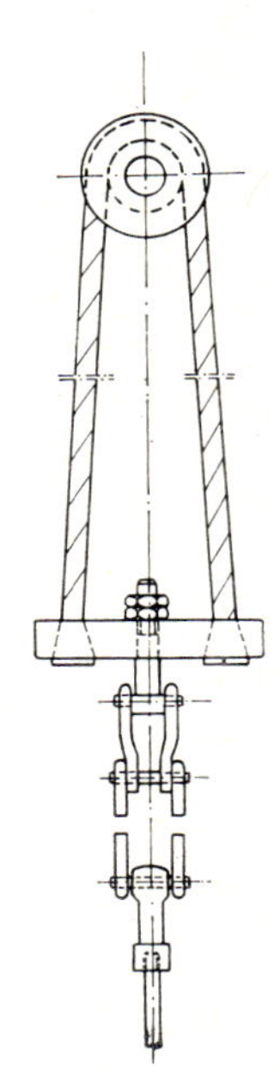

Fig. 4.1—19 Rod string suspension involving Galle chain

out, the packings can be compressed and the seal improved by screwing down ear nut *2*. The top section of the rod string is the so-called polished rod. It is carried by a carrier bar fixed to a hanger cable depending from the horsehead. Its suspension from the carrier bar may follow any one of several designs. The suspension must permit the height of the polished rod relative to the horsehead to be adjusted, in order to correctly adjust the plunger stroke within the pump barrel. In the Soviet Union, suspensions using the Galle chain (Fig. 4.1—19) are most popular. Adjustment is performed by changing the number of chain links. In Hungary, adjustment is usually performed by an Axelson type polished-rod clamp that can, by tightening the bolts, be fixed at any height on the polished rod (Fig. 4.1—20). The polished

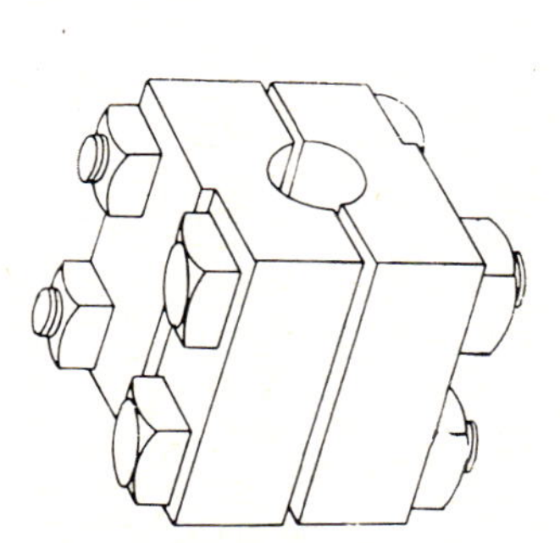

Fig. 4.1—20 Axelson's polished-rod clamp

Fig. 4.1—21 Rod coupling compacted by rolling, after McCurdy and Elkins (1967)

rod is cold-drawn from high-strength alloy steel. Corrosion-resistant alloys are used where well fluids are corrosive. The diameter of the polished rod is usually greater by 10 mm than that of the sucker-rod directly attached to it. Sucker-rod sizes in Hungary are governed by the standard MSz 5154.

(d.1) *Sucker rods.* — Sucker-rod sizes are standardized in numerous countries. The Hungarian standard MSz 5152 derived from API Std 11B prescribes the dimensions given in Table 4.1—1 for solid rods. The various standards of countries differ but slightly from one another, so that formulae, tables and diagrams involving rod sizes are most often directly applicable to any make of rod. The solid sucker rods most often employed wear pin threads on both ends (cf. the Figure belonging to Table diagram 4.1—1). Accordingly, their couplings are of the double-box-thread type. Rods having a box thread at one end and a pin thread at the other with corresponding couplings are also employed. From the late nineteen-fifties on, some rod threads are first machined to an oversize, and then reduced to the correct size by rolling (Fig. 4.1—21). Later on, this idea was adopted to rod couplings. too. It was found that this solution limits corrosion and the build-up of harmful stresses. As a result of both these factors, the frequency of joint failures in rods of this type is much less than in conventional sucker rods (Crosby 1969b). The lengths of sucker rods are likewise standardised. The relevant Hungarian prescriptions are contained in the above-named standard.

By the term sucker rod, a solid rod is most often meant in practice. However, the increasing number of wells producing sandy and heavy crudes, and of small-diameter wells, has led to the development of *hollow sucker rods*. At first, strings were simply made up of standard external-upset tubing of 1 — 1 1/4 in. size. Failures in this type of string were very frequent, however, and they usually took place at the last joint. As a result of some high-pressure development work, however, hollow rod strings made in 1960 could already operate pumps installed up to 2265 m depth. Today hollow rods are made by several manufacturers. These are usually pin-threaded at both ends and joined by appropriate couplings. Figure 4.1 — 22 shows the end and coupling design in longitudinal section of a Varco make hollow sucker rod. Table 4.1 — 8 lists some of the main parameters of Varco make hollow rods conformable to API 11B. In the Soviet Union, successful experiments have been carried out with hollow rods glass-coated on the inside. The glass has the effect of reducing wax deposits (Zotov and Kand 1967).

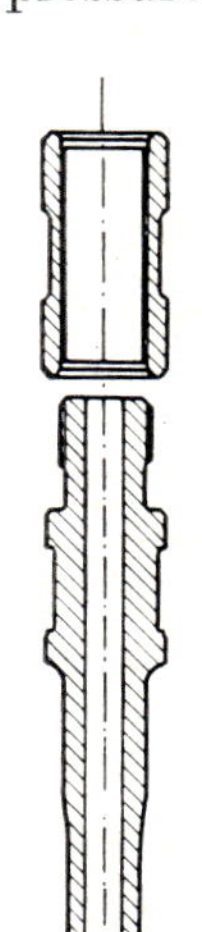

Fig. 4.1 — 22 Varco's hollow rod

The *maximum allowable tensile stress* of the sucker rod is given by

$$\sigma_{al} = \frac{\sigma_B \, \sigma_F}{a \dfrac{\sigma_B}{2} + \sigma_F} \qquad 4.1-54$$

where a is a safety factor whose value is in the range from 1.5 to 2. — Sucker rods are exposed to substantial fatigue due to significant load changes at comparatively short intervals. Even at the rather low pumping speed of 10 spm, the number of annual load changes exceeds five million. In a well 1000 — 1500 m deep, the difference between maximum and minimum load is 1 — 30 kN. The above formula — which, in its original form, is due to Timoshenko — accounts for load changes, and the fatigue limit, respectively, provided no corrosion is to be anticipated. According to more recent designing principles, the complex nature of the recurring stresses forbids us to speak of 'the' fatigue limit of steel, because structural materials have a variety of fatigue limits (Zorkóczy 1968). Designing is facilitated by consideration of the so-called areas of safety shown in diagrams characterizing the individual types of fatigue limit. The type of stress on sucker rods is pulsating tension. This means that the rod is under tension throughout, and that the magnitude of this tension varies more or less periodically. The maximum allowable stress can be determined by means of a modified Goodman diagram. The orthogonal system of coordinates in Fig. 4.1 — 23 is calibrated in minimum stress on its abscissa axis and in maximum allowable stress on its ordinate axis (Jernigan 1971). The plot is constructed as follows. From the origin of coordinates, a line of plus unity slope is drawn. This is the locus of line $\sigma_{max} = \sigma_{min}$. Now the value $\sigma_B/4$ characterizing the

Table 4.1—8
Main data of Varco hollow rods

Data	Symbol	Unit	Nominal size, in.		
			3/4	1	1 1/8
O.D.	d_0	mm	26.7	33.4	28.6
I. D.	d_i	mm	20.9	26.6	15.9
Steel cross-section	A_r	cm²	2.15	3.19	4.43
Capacity per unit length	V_i	l/m	0.344	0.557	0.198
API thread on rod end	—	in.	7/8	1	1
Overall rod length	L_r	m	9.14 ± 0.05	9.14 ± 0.05	9.14 ± 0.05
Rod weight per unit length	G_r	N/m	18.7	27.3	36.5
Maximum allowable load for rod made of N-80 steel	F_{max}	kN	53.0	80.41	107

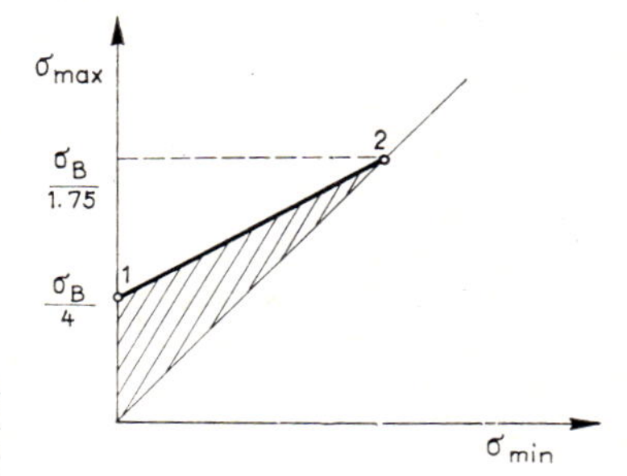

Fig. 4.1—23 Modified Goodman diagram for designing sucker-rod strings, after Jernigan (1971)

rod material to be used is plotted on the ordinate axis. This gives point *1*. After plotting $\sigma_B/1.75$ on the ordinate axis, a line is drawn parallel to the abscissa axis through the point thus obtained. This line intersects the line of minimum stress in point *2*. The line connecting points *1* and *2* is the graph showing the variation of maximum allowable stress v. minimum stress. The shaded area is the area of safety.

Corrosion. The number of wells producing strongly corrosive fluids is comparatively small, but there is almost no well in which corrosion is nil. Corrosion is due primarily to formation water, and to a lesser extent to accessory gases such as hydrogen sulphide, oxygen and sulphur dioxide. Corrosion results in pitting of the rod surface. The pits may, on the one hand, start cracks and, on the other, entail stress concentration. The stress in the section of a deep pit may be ten times as much as in a full, uncorroded cross-section. The harmful concentration of stress and the reduction of the cross-section is further enhanced by the fact that the corrosion pits are deformed by the variable stress on the rods. A greater tensile stress will distend the pits. A pit so distended may catch a particle of metal or a sand grain. In the stress decrease phase, this particle prevents the relaxation of the material around the pit and, serving as a wedge, causes cracking in the surrounding metal. Cracks thus formed tend to propagate until the rod breaks under a stress exceeding the lowered endurance of the material.

The extent of corrosion thus depends in addition to the given rod material and corroding medium also to a significant extent on time and the stress-variation range. This is why it is impossible to successfully simulate in the laboratory conditions affected by a number of secondary factors acting over incomparably longer spans of time. This, however, is not usually necessary, because only rod materials resistant to the kind of corrosion anticipated may be used anyway.

A variety of *steels* are used to make sucker rods. All steels contain Fe in a proportion above 90 percent. To this are added alloying elements increasing the hardness, strength and/or corrosion resistance of the steel. As to composition, rod steels fall into two groups. If the manganese content is less than 0.5 percent, and there are no alloying elements other than Si and C, and traces of P and S as contaminants, the material is called a carbon steel. It is termed an alloy steel if it contains other alloying elements as well, such as Ni, Cr, Cu, Mo, V and B. — The presence of *C* in steel considerably increases strength, hardness and the suitability for tempering. However, it also increases brittleness and lowers corrosion resistance. — *Mn*. This is a deoxidant that reduces brittleness in the presence of sulphur. Otherwise, if added in small amounts, it plays a role similar to that of carbon. — *Si*. A very effective deoxidant. It serves first of all to reduce the grain size of high-strength steels. — *Ni*. A hardener in solid solution in ferrite. It does not form carbides the way some other alloying elements do. It inhibits corrosion brittleness caused by hydrogen sulphide gas in corrosion pits. — *Cr*. This element forms carbides and considerably improves the temperability of steel. It does not provide protection against hydrogen brittleness, but considerably improves resistance to corrosive agents other than hydrogen sulphide. — *Cu*. Added in comparatively small amounts, it improves resistance to atmospheric corrosion. — *Mo*. Enables the steel to be heat-treated to improve its strength. — *V*. Similar to Mo; moreover, it promotes the formation of a fine-grained texture. — *B*. Similar to Mo and V.

Table 4.1—9 (end of book) shows composition and strength parameters of various rod materials. Irrespective of strength criteria, rod steel should be chosen for corrosion resistance according to the following main viewpoints:

Medium surrounding the rod	*The rod is to be made of*
Mildly corrosive	Carbon steel
Contains H_2S	Ni—Mo steel
Strongly corrosive brine	Ni—Cr steel

In order to prevent or limit sucker-rod corrosion, *inhibitors* are sometimes employed. The inhibitor dosed into the annulus flows down to the well bottom where it mixes with the well fluid. The protective action of organic inhibitors is usually due to the fact that their heteropolar molecules, adhering with one end to the metal surface, form an impermeable film that keeps the corrosive medium from direct contact with the steel. Inorganic inhibitors neutralize the corrosive agent by entering into a chemical reaction with it. Inhibitors have the drawback that their application is a never-ending job. Their advantage is that they protect from corrosion not only the sucker rods but all the steel surfaces in contact with the well fluid.

Rods are particularly prone to joint failure if an insufficient *make-up torque* is used. According to Walmsley and Helman, in a rod of 22.2 mm diameter, failures occur predominantly at the joints if make-up torque is less than 206 Nm. In the torque range from 206 to 540 Nm, joint failure is about as probable as failure elsewhere along the rod, whereas at 540 Nm the number of joint failures decreases rather steeply. Thus if joint failures occur fairly often, it is advisable to employ power sucker-rod tongs ensuring correct and uniform make-up torque at all joints.

As a result of cooperation between Bethlehem Steel and DuPont de Nemours, a *flexible rod* built under the trade name *Flexirod* or *Corod* has been applied since 1961 in experimental installations. The flexible rod, whose description was published in 1968, is made up of 37 strands, each of a round 2 mm diameter and of 165 MN/m^2 breaking strength, and encased in nylon 0.25 mm thick (Joy and Coleman 1968). The wirerope thus made up is encased in an outer nylon jacket about 0.6 mm thick. The effective breaking strength of the rope is 186 kN. In 1970, the flexible sucker-rod was introduced into commercial production (Patton 1970). It may be of the same material as the solid rod; flexible rods have been made of C (AISI 1036Md) and K (4651) steels. The wirerope is composed of wires 183 or 366 m

Table 4.1—10

Corod sizes and weights (after Patton 1970)

d		G_r
in.	mm	N/m
11/16	17.5	18.4
3/4	19.1	21.9
13/16	20.6	25.7
7/8	22.2	29.8
15/16	23.8	34.2
1	25.4	39.0

long, but-welded, heat-treated, rolled into an elliptic form, and again heat-treated. The main dimensions are given in Table 4.1—10. The full length of line is then quality-controlled by ultrasonic means, shot-peened, plastic jacketed and wound on a drum of a round 5.5 m diameter for transport to the wellsite. At the well, it is transferred to a special well-completion derrick by means of a sheave-like rodguide. Figure 4.1—24 is the sketch of a pumping unit equipped with a *Flexirod* (Joy and Coleman 1968). Pump *1* is of a special, so-called differential type. It is liquid-loaded also during the down-stroke, so that the *Flexirod* is tensioned throughout (cf. also Fig. 4.1—28).

The portion of the *Flexirod* which emerges to the surface is encased in a hollow polished rod *2*. The role of this latter is restricted to ensuring a satisfactory seal together with the polished-rod stuffing box; it carries no load. The upper end of the *Flexirod* is wound on drum *3* on the samson post. Running and pulling are simple and fast; the pump can be run at speeds up to 1.8 m/s. In 1968, the sucker-rod pump was still run and pulled by means of a well-completion rig suited for the purpose. A pumping unit is being developed, however, that can carry out these operations by itself.

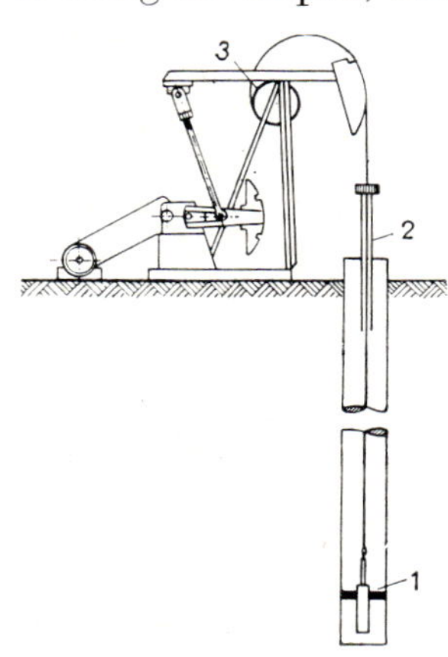

Fig. 4.1—24 Sucker-rod pump with *Flexirods*, after Joy and Coleman (1968)

The *Flexirod* has a number of advantages (Patton 1970). Rod-string weight may be significantly reduced by the fact that standard *Flexirod* sizes differ by 1.6 mm rather than the 3.2 mm for solid rods. Thus e.g. a four-stage *Corod* string may be lighter by 17 percent than the two-stage solid-rod string of the same strength. The smaller rod-string weight entails a smaller polished-rod load and a lower specific power consumption, so that a prime mover of lower rating will do. The probability of failure is greatly reduced because 65—80 percent of all failures in solid rods occur at the joints. The absence of rod couplings permits the selection of smaller-size tubing and hence also of smaller-size production casing. The tendency to wax deposition is considerably reduced because first deposits usually form on the shoulders of the rod couplings. Friction of the rod string against the tubing is likewise reduced. It simplifies the use of plastic-lined tubing, which reduces both tubing corrosion and friction loss in the well fluid. However, an efficient joining of broken *Flexirod* ends at the wellsite is still unsolved today.

(d.2) *Bottom-hole pumps.* Fundamental types. — Sucker-rod pumps may be tubing pumps or rod pumps.

The tubing pump owes its name to the fact that the pump barrel is run with the tubing and cannot be removed without pulling it. The barrel is screwed onto the lowermost length of tubing. In most types of tubing pump, the plunger is run on the rod string, but in some solutions the barrel is run with the plunger in place, and the rod string is fixed to it subsequently. The standing valve can be installed with or without the plunger, but it is invariably removed together with, and often by means of, the plunger. — In the case of the rod pump, both the plunger and the barrel can be run or pulled with the rod string. The barrel is seated on and fixed to a conical seat previously installed at the tubing shoe.

The tubing pump has the advantages over the rod pump that it will accommodate a larger-diameter plunger in a given tubing size, and it is simpler and therefore cheaper. — The advantages of the rod pump are, on the other hand, that it is not necessary to pull the tubing when changing the pump, and so pump changes are cheaper; the plunger is not run 'naked'' so that its surface will not be damaged on running and pulling; dead space is less, which is an advantage when pumping gaseous fluids; certain designs are more trouble-free provided the sand content of the fluid is low.

Several fundamental types of both tubing and rod pumps are known. These may be classified according to various viewpoints. The fundamental types shown in Fig. 4.1—25 have been taken from API Std 11—AX, and slightly modified. Unequivocal specification of a sucker-rod pump includes the nominal size of the tubing, the (basic) plunger diameter, the API standard designation of the pump (found in Table 4.1—11), the lengths of barrel and plunger, and the overall structural length. Plunger diameters of standard sucker-rod pumps are listed in Table 4.1—12. The parts of the Figure featuring pumps of TH and TL type show the plunger and standing valve and barrel. There is an insert between standing valve and barrel. It is needed because the various types of standing-valve puller mounted on the plunger or the standing-valve cage (not shown in the Figure) also require some space. This entails, however, a certain unavoidable dead space. The heavy-walled full-barrel rod pumps shown in parts (c), (d) and (e) of the Figure agree in general design features with the thin-walled full-

Table 4.1—11

API standard designations of sucker-rod pumps

Type of pump	Symbol for full barrel		Liner barrel
	Heavy-walled	Thin-walled	
Tubing type	TH	—	TL
Rod type			
Stationary barrel, top holddown	RHA	RWA	RLA
Stationary barrel, bottom holddown	RHB	RWB	RLB
Travelling barrel, bottom holddown	RHT	RWT	RLT

Table 4.1—12

Sucker-rod pump sizes (after API Std 11—AX)

API code	Nominal tubing size, in.						Corresponding metric sizes	
	1.9	2 3/8		2 7/8		3 1/2		
	Plunger size, in.						in.	mm
RHA	1 1/16	1 1/16	1 1/4	1 1/2	1 3/4	2 1/4	1 1/16	27.0
RHB	1 1/16	1 1/16	1 1/4	1 1/2	1 3/4	2 1/4	1 1/4	31.8
RHT	1 1/16	1 1/16	1 1/4	1 1/2	1 3/4	2 1/4	1 1/2	38.1
RLA	—	1 1/16	1 1/4	1 1/2	1 3/4	2 1/4	1 3/4	44.5
RLB	—	1 1/16	1 1/4	1 1/2	1 3/4	2 1/4	2	50.8
RLT	—	1 1/16	1 1/4	1 1/2	1 3/4	2 1/4	2 1/4	57.2
RWA	—	1 1/4	1 1/2	2		2 1/2	2 1/2	63.5
RWB	1 1/4	1 1/4	1 1/2	2		2 1/2	2 3/4	69.9
RWT	1 1/4	1 1/4	1 1/2	2		2 1/2		
TH	—	1 3/4		2 1/4		2 3/4		
TL	—	1 3/4		2 1/4		2 3/4		

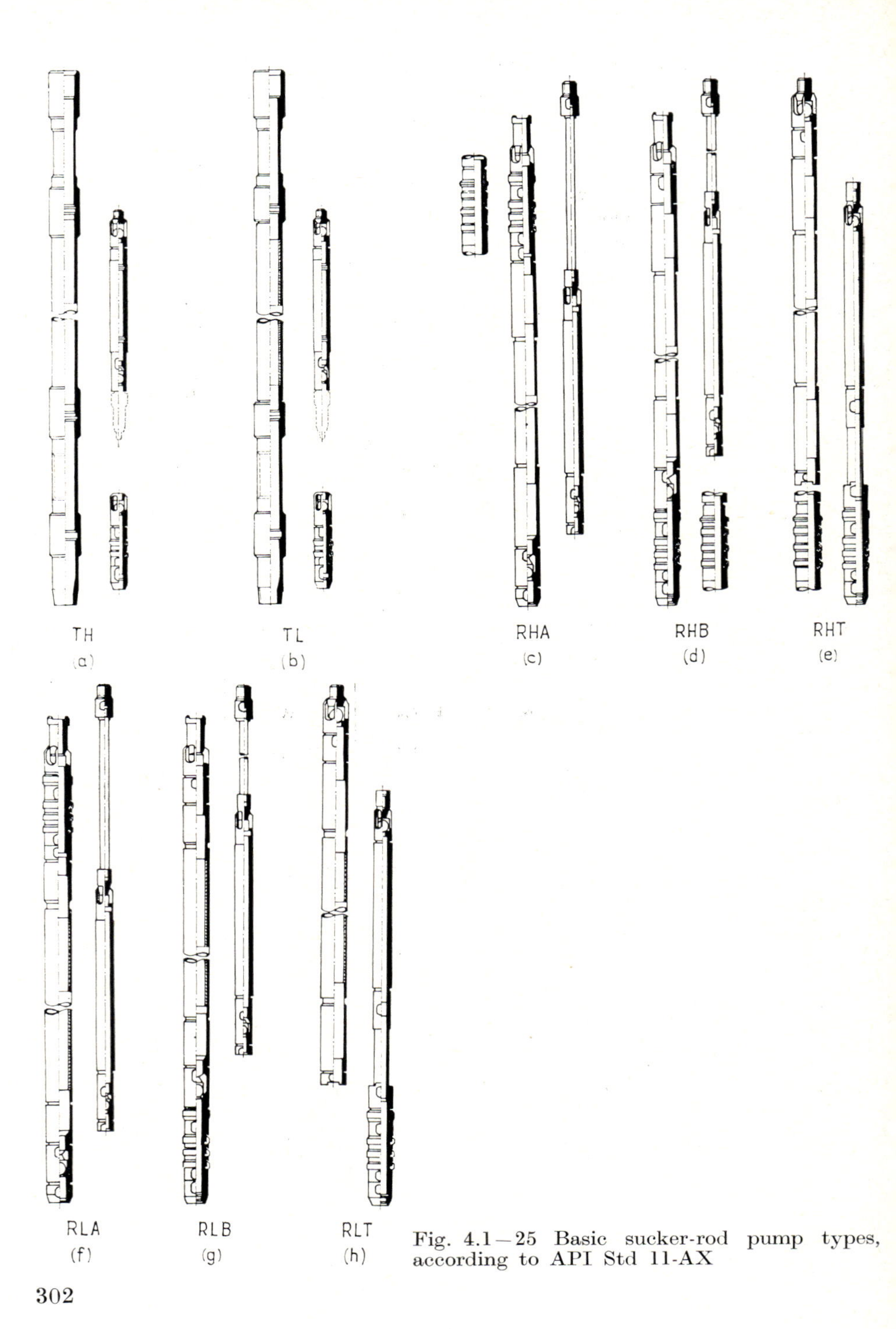

Fig. 4.1—25 Basic sucker-rod pump types, according to API Std 11-AX

barrel pumps of type designation RW. Full barrels are cheaper than barrels with sectional liners. They have, however, the disadvantage that the reworking of a worn barrel is more difficult. The pumps with heavy-walled barrels denoted RH can stand a heavier liquid load without deformation, and can therefore be used at greater depths. The rod pumps shown in parts (c) and (f) of the Figure are provided with top holddown. This is an advantage especially when pumping a sandy crude, for it prevents the settling of sand between the outer barrel wall and the tubing. In the solutions shown as (d) and (g), the bottom holddown permits such settling of sand. It has, however, the advantage that, after pulling the plunger and standing valve of a well previously pumped by means of a tubing pump, it can be installed and operated without letting the producing fluid level rise. In both cases, fixation to the tubing is more elastic, owing to the conic seating surface of the holddown, than it would be in the case of a tubing pump. Also, in crooked wells, the pump has less tendency to seize in the tubing. In the types RHT and RLT, the plunger is fixed to the seating nipple and the pump barrel is travelling together with the rod string. Because of the smaller standing-valve inlet, these types are better suited for lower-viscosity oils. They are less sensitive to sand than the stationary-barrel types with bottom holddown, because turbulency about the barrel limits the settling of sand during operation. They are favourable also when pumping a gaseous fluid. Let us add that it is usual to install above rod pumps a ring-type check valve that prevents the settling of sand risen through the tubing in the event of a stoppage.

For structural details let us consider the TL type tubing pump shown in Fig. 4.1—26 and the RLA type rod pump shown in Fig. 4.1—27, both of USI Axelson make. In Fig. 4.1—26, standing valve *1* is simply dropped into the well prior to installing the pump; falling down the tubing, it finds its own place. It is pulled together with the barrel by latching onto extension *3* of the standing-valve cage the self-latching standing-valve puller *2* on the plunger. In Fig. 4.1—27, the pump barrel is equipped with mandrel *1*, to be seated in a nipple in the tubing string. Holddown under operating conditions is provided by the pressure differential acting on the three seating cup rings marked *2*. The pump can be pulled with a definite jerk; another pump can be installed without damaging the structure.

Besides the fundamental types just described there are other, special pumps. The *casing pump* is a rod pump whose seating nipple is fixed to the casing by means of an anchor packer. The completion involved is of the tubeless type. This solution is restricted to gasless wells where the annulus is not required for producing gas. The absence of the tubing may make this solution highly economical. *Telescopic* or *three-tube sucker-rod pumps* are rod pumps with the middle tube fixed to the tubing, whereas the other two coaxial tubes fitting the stationary one on its inside and outside move together with the rod string. Contrast between the concepts of plunger and barrel is obscured here. Because of the considerable tube lengths usual in this type of pump, a relatively greater operating clearance may be permitted between moving parts than in the more conventional pumps. The three-tube pump is used to advantage in producing well fluids containing fine sand whose grains are smaller than the operating clearances. *The differential sucker-rod pump* (Fig. 4.1—28) is used in conjunction with *Flexirod*-type rod strings (Hood 1968). It has the advantage that, during the downstroke, a downward-directed force acts on the plunger, which permits it to sink at sufficient speed. The differential pump has two plungers. The true

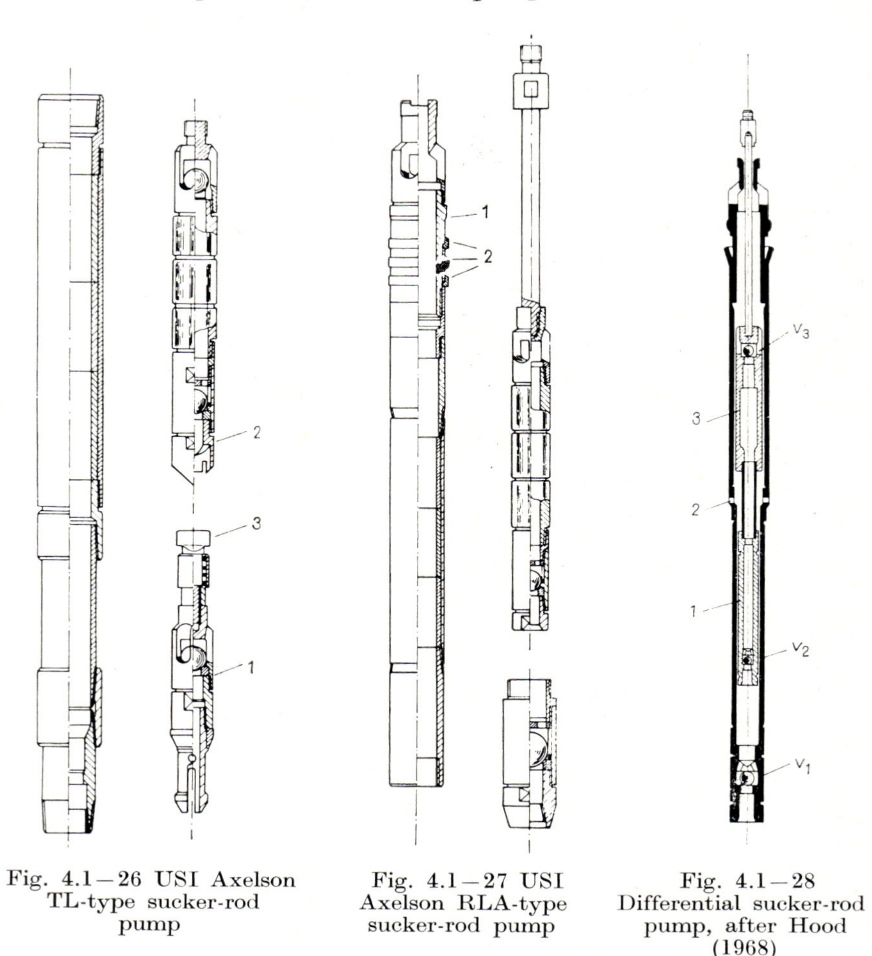

Fig. 4.1—26 USI Axelson TL-type sucker-rod pump

Fig. 4.1—27 USI Axelson RLA-type sucker-rod pump

Fig. 4.1—28 Differential sucker-rod pump, after Hood (1968)

plunger lifting the well fluid is the lower one marked *1*. It operates on the upstroke, in the same way as a conventional sucker-rod pump. On the downstroke, standing valve V_1 closes, whereas travelling valves V_2 and V_3 open. Through orifice *2* the effective cross-sectional area of plunger *3* is subjected to the comparatively small annulus pressure from below, but to the pressure of the liquid column in the tubing from above. The plunger is forced downward by a force proportional to the pressure differential. — Sucker-rod pumps of special design will be discussed in more detail in paragraphs 4.1.1d4—6.

Main structural parts. — *The pump barrel* may be a one-piece barrel made of a cold-drawn steel tube or of cast iron, or a barrel composed of a number of liners, called liner barrel. The liner barrel usually contains several cylindrical liners, each of 1 ft length (or 300 mm according to a Soviet standard), very carefully honed on the inside and at the shoulders. The liners are placed in a close-fitting jacket and held together by two flush collars. Liners are made of wear- and corrosion-resistant alloy steels. The insides of some liners are specially treated, nitrated or provided with a

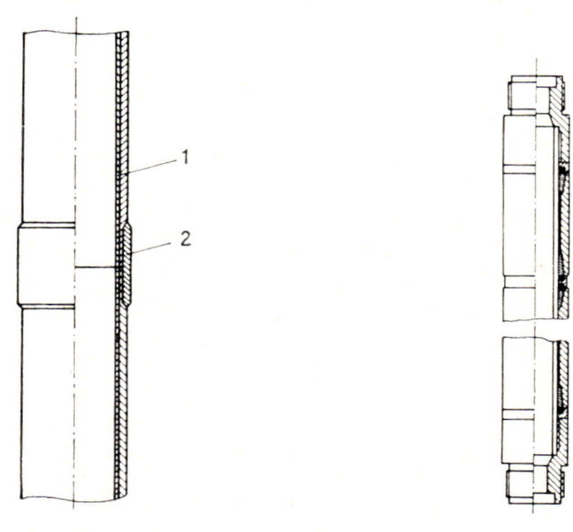

Fig. 4.1—29 Oilwell's Neilsen design pump barrel with steel band

Fig. 4.1—30 Oilwell's Neilsen design O-ring type pump plunger

hard chrome plating. — The advantages of the one-piece barrel are that, for a given nominal size (a given tubing diameter), the plunger may be of greater diameter, and that it is cheaper. The sectional-liner barrel has, on the other hand, the advantages that any length of barrel may be made up of short, precisely honed liners, whereas it is difficult to accurately hone a long one-piece barrel; the short liners enable machining to closer tolerances, which is important especially at the high lift pressures encountered in deep wells; a worn barrel is comparatively cheaper to rehone. Let us point out that high pressures will tend to misalign liners if these are simply placed end to end, and this may cause operating trouble. On the other hand, e.g. in the Neilsen type barrel made by Oilwell (Fig. 4.1—29), liners *1* are locked together by steel bands *2* that prevent their misalignment even at great depths. — In order that a worn barrel may be reworked and reused, it is usual to provide undersize plungers. After a certain amount of wear, the barrel is rehoned and fitted with the standard size plunger. Soviet sucker-rod pumps are furnished with undersize and oversize barrels, both differing in diameter by 1 mm from the standard size. Undersize barrels differing in diameter from the standard API size by 0.04 in. or 1.02 mm are marked '—40'. — Standard pump barrel lengths are usually multiples of 1 ft (0.306 m), or 0.300 m in countries using the metric system. Standard API barrel lengths include 1.52, 2.13, 2.74, 3.35, 3.66 and 4.57 m. Barrels of the biggest sizes (e.g. 12.2 m) may pose a handling problem on the surface and also on running in the well. It is usual to make these big barrels in two halves (each of which may be of the sectional liner type), and join them together on running by means of a special coupling.

Two types of plunger are distinguished: metal plungers and soft-packed ones; the latter are provided with rubber or plastic cups. Metal plungers are made of an alloy steel chosen for strength and resistance to wear and corrosion, and matched to the barrel material. They are case-hardened or provided e.g. with a hard chrome plating. Most plungers are machined in one piece. The low breaking strength of certain alloys necessitates making up of several piece plungers to be exposed to high loads. Figure 4.1—30 shows a Neilson type O-ring plunger made by Oilwell. Plungers may be plain or grooved. The latter may have the advantage that, when pumping a sandy fluid, sand grains will get caught in the grooves rather than scoring the plunger and barrel full length. If, on the other hand, the plunger is operated so as to stroke out of the barrel, its grooves may pick up and carry solid sand particles into the barrel. The advantages of grooved barrels are therefore debatable. — The plunger diameter equals the barrel diameter except for a very narrow clearance. In the Soviet Union, three clearance groups are distinguished (20—70, 70—120 and 120—170 μm). In the US, there are five nominal clearance groups increasing in 0.001-in. steps from 0.001 in. to 0.005 in. (that is, from 25 to 127 μm). The corresponding plungers are termed —1, —2, —3, —4 and —5 fits, respectively. — The correct choice of the plunger-barrel combination best suited for a given well is very essential, in order to minimize slippage past the plunger under the excessive pressure differential building up between plunger ends during the upstroke. Slippage loss can be estimated by the formula

$$q_2 = 0.0164 \frac{d_p \Delta p \Delta d^3}{\mu_l h_p} \text{ m}^3/\text{s} \qquad 4.1\text{—}55$$

(Oil Well Supply, *Bulletin* 1957), where Δp is the pressure differential across the plunger in N/m^2; Δd is the diametral clearance (difference in diameters) in m; and h_p is plunger length, in m.

Example 4.1—9. Find the daily slippage loss past the plunger in oil of 120 and 1.2 cP viscosity, respectively (1 cP$=10^{-3}$ Ns/m^2), if d_p=57.1 mm; Δp = 200 bars; Δd = 0.1 mm; and h_p = 1.22 m. — For the 120-cp oil, Eq. 4.1—55 gives

$$q_2 = 0.0164 \frac{0.0571 \times 2 \times 10^7 (1 \times 10^{-4})^3}{0.120 \times 1.22} = 1.28 \times 10^{-7} \text{ m}^3/\text{s},$$

which equals $1.28 \times 10^{-7} \times 86{,}400 = 0.011$ m^3/d. — For the 1.2-cp oil, the slippage is 100 times this value, that is, 1.1 m^3/d.

The correct choice of plunger fit requires consideration of well-fluid viscosity. As a rule of thumb, the −1 fit is used with oils of low viscosity (1−20 cP), whereas the −5 fit may ensure a satisfactory operation even about 400 cP. Too tight a fit should be avoided, because sand grains suspended in the fluid, which would cause a smaller-clearance plunger to seize, may pass through a larger clearance. Let us point out that oil slipping past the plunger is warmed by friction, so that its viscosity tends to be less than that of the bulk well fluid. If the plunger and barrel are not made of the same

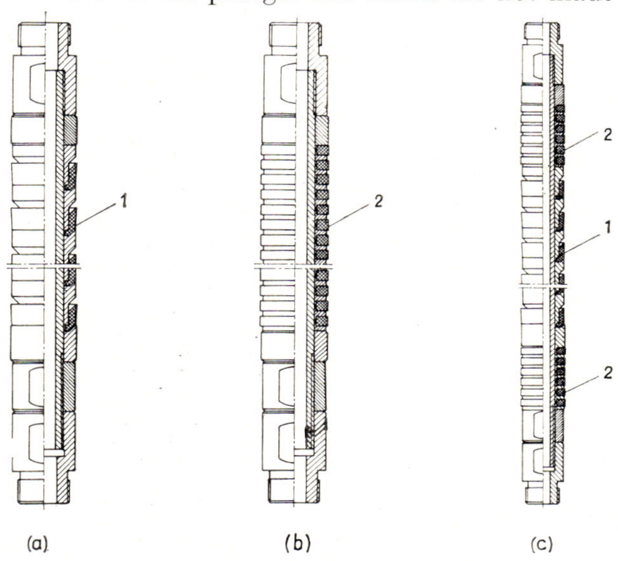

Fig. 4.1−31 Oilwell's soft-packed plunger types

material, differential thermal expansion at the setting depth should be taken into account. The plunger may even seize up if the clearance is too small.

In soft-packed plungers, the diameter of the metal body is significantly less than that of the barrel bore, and packing is provided by valve cups, rings, etc. Such plungers are used at depths less than 1500 m. They have the advantage of longer life when producing a sandy fluid, because the sealing surfaces are hardly worn by the sand. They are usually cheaper than metal-to-metal plungers. In Fig. 4.1−31a, packing is provided by valve cups *1* made of oil-resistant rubber. On the upstroke, the liquid weight on the plunger presses the cups against the barrel, whereas on the downstroke the contracting cups hardly touch the barrel wall. This design is used at comparatively low sand contents. A considerable drawback is that cups will fail all of a sudden, without any preliminary warning, so that repair jobs cannot be scheduled in advance. In (b), packing is provided by valve rings *2* made of oil resistant synthetic rubber. Progressive wear can be detected well enough. Type (c) is a combination of cup and ring-type plungers. It can be used for cleaning up wells following a sand fracturing operation. This type also permits to detect wear. In addition to these types of Oilwell make, other types of soft-packed plungers are also known. — A special type to be mentioned is the combined plunger where the packing is provided by a close metal-to-metal fit along part of the body and by valve cups along the rest. They are used at comparatively great depth.

Valve balls and seats are made of wear- and corrosion-resistant metal, occasionally case-hardened. Figures 4.1—26 and 4.1—27 illustrate some

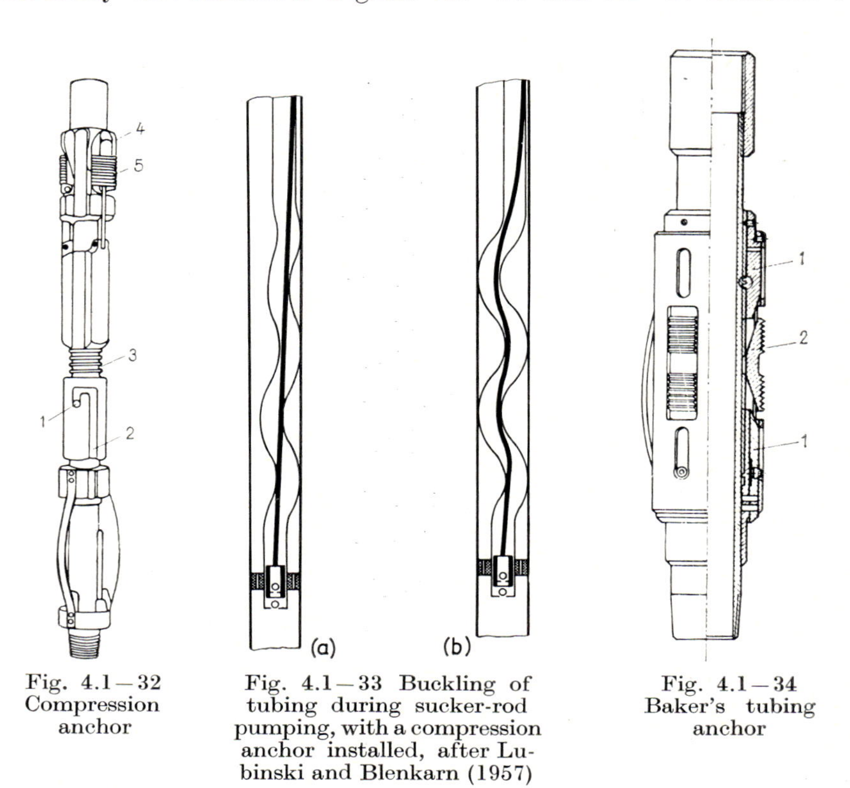

Fig. 4.1—32 Compression anchor

Fig. 4.1—33 Buckling of tubing during sucker-rod pumping, with a compression anchor installed, after Lubinski and Blenkarn (1957)

Fig. 4.1—34 Baker's tubing anchor

popular types. The ball is confined in its motion by a cage of 3 or 4 ribs. The Figure also shows the mode of attaching the seat in the sucker-rod pump. Let us add that, in a tubing pump, the standing valve may simply be fixed by adhesion between mating tapers. In this case, the conical seat is lined with plastic or white metal, which also has favourable adhesive properties.

(d.3) *Tubing anchor.* — In order to increase plunger stroke, a device designed to fix the tubing shoe to the casing used to be employed even in early practice. Since, however, the theory concerning the multiple buckling of the tubing (cf. paragraph 4.1.1a.5) has become widely known, the operation of the anchor was also submitted to a more detailed analysis. The anchor used in earlier practice was usually of the compression type (Fig. 4.1—32): it is a device resembling a hook-wall packer, without the sealing elements.

It can be set at the desired depth by releasing the J-hooks (*1*, *2*). Now spring *3* can press up slips *5* on cone *4*, and these will grip the inside of the casing. Since the slips are arranged so as to slide freely upwards and seize against the casing downwards, this holddown will fix the tubing shoe in the highest position occurring after its release. The tubing may therefore undergo multiple buckling during both the up- and the downstroke (Fig. 4.1—33). Now during the upstroke, the rod string is pulled straight by the load on the plunger. This reduces to some extent the buckling of the tubing (part (a) of the Figure). This is the type of buckling discussed in the section referred to above. During the downstroke, the tubing will stretch, but since the downward movement of the tubing shoe is prevented by the compression anchor, the tubing will buckle again (part (b) of the Figure). The rod string, not loaded by fluid, is not stretched; it can therefore marry the curves of the tubing. The buckling of the latter is, however, limited by the casing. Thus even though the compression anchor prevents the movement of the pump barrel, it does not prevent wear and damage of the rod string, tubing and possibly casing, nor overloads due to friction (Lubinski and Blenkarn 1957).

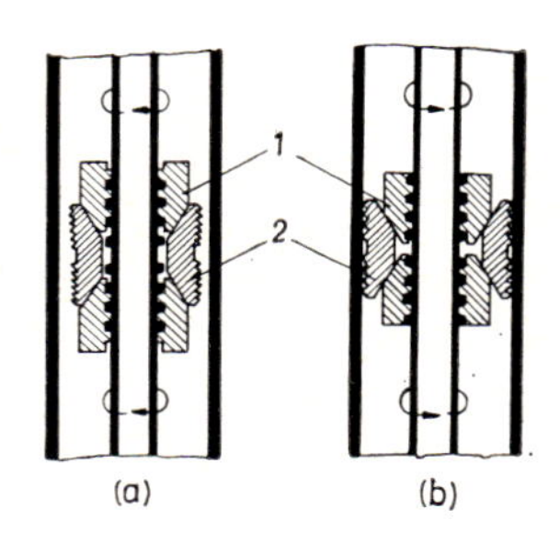

Fig. 4.1—35 Setting and releasing Baker's tubing anchor

The above circumstances have made it desirable to have an anchor which attaches the tubing shoe to the casing in the deepest position occurring, that is, in the fully stretched state. If a compression anchor is installed upside down, a *tension anchor* results. This type of anchor keeps the tubing from both buckling and shortening, and so limits stroke reduction, on the one hand, and eliminates, on the other, the sources of damage associated with the compression anchor. In the tension-anchor types first employed in practice, a short upward travel was needed to make the slips grip the casing wall, and this could result in casing wear or puncture. In order to prevent this, measures were subsequently taken to avoid the anchor's climbing down into the deepest possible position under the gradually increasing loads; notably, the anchor was set in a prestretched state of the tubing. The necessary prestretch is to be determined by calculation. Prestretching increases the tensile stress on the tubing, and this stress will further increase at times of pumping stoppage, when the tubing cools down. If the stress exceeds the allowable value, the tubing may undergo a permanent deformation and even break. In order to prevent this, the tension anchor is provided with a safety device that disengages the slips in overstress situations. This, of course, puts an end to anchor action. — Figure 4.1—34 shows a Baker type *tubing anchor* that can be set at any depth. Once the tubing is run to the desired depth, it is rotated to the left from 3 1/2 to 4 turns. This makes the expander wedges *1* approach each other (Fig. 4.1—35); these then press the slips *2* against the casing wall. These slips are serrated so as to prevent both upward and downward movement. Calculating the necessary amount of prestretch is facilitated by tables and diagrams. — The Guiberson type *hydraulic tubing anchor* (Fig. 4.1—36) is provided with a number of holddown buttons moving in a number of radial cylinders. Whenever tubing pressure exceeds casing pressure, the holddown buttons bear against the inside of the casing. Correct operation requires prestretching also in this case. — No prestretching is required if an *automatic tension anchor* is used. This differs in principle from the basic type (the compression anchor installed upside down) only in that the up-

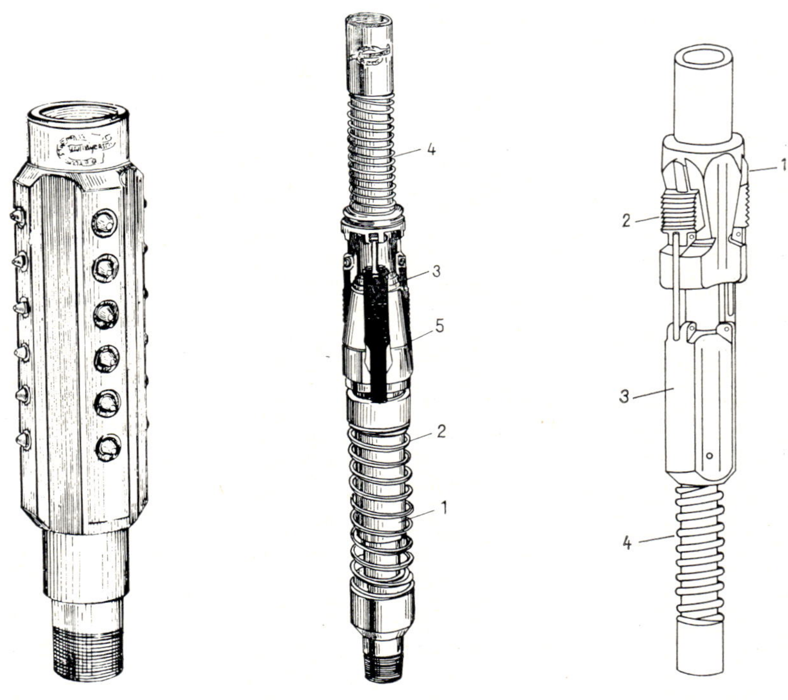

Fig. 4.1—36 Guiberson's hydraulic tubing anchor

Fig. 4.1—37 Guiberson's HM-2 hydromechanical automatic tension anchor

Fig. 4.1—38 Tubing catcher, after Muraviev and Krylov (1949)

ward serrations of the slips immediately grip the casing wall after release; no upward movement at all is required to seat the slips. Hence, this type of anchor automatically sets the tubing in the deepest position. As no prestretching is required, the stretch during operation in the tubing is the least possible. Figure 4.1—37 shows the Guiberson type HM—2 hydromechanical automatic tension anchor. Once tubing pressure exceeds casing pressure by about 14 bars after the onset of pumping, cylinder *1* is moved downward by the pressure differential across it, against the force of spring *2*. This permits spring *4* to press slips *3* downwards, so that, forced outwards by cone *5*, they come to bear against the inside of the casing. Overstress breaks a shear ring, and the slips may then disengage. If prior to retrieval the pressure differential between tubing and annulus is equalized by pulling the standing valve or by any other means, then cylinder *1* is pushed up by spring *2* and the slips disengage as above. — Table 4.1—13 lists some operating data on wells provided at first with no anchor, or an anchor that failed to operate properly, and then reworked to install a correctly operating tension anchor (Taylor 1960). — If the tubing string breaks, part of it will drop to the well bottom, and may get warped sufficiently to become useless. *The tubing*

Table 4.1—13
Effects of correctly functioning tubing anchors (after Taylor 1960)

Tubing		Main operating characteristics	
Diam. in.	Length m	Prior to repair	Subsequent to repair
2 7/8	823	Production 95 m^3/d, substantial wear on rod and tubing strings	Production 111 m^3/d, no wear at all
2 7/8	3473	Many joints had to be changed each quarter	No joint failure
2 3/8	2179	Production 5.6 m^3/d	Production 10 m^3/d
2 7/8	1826	Production 34 m^3/d, 2–3 rod breakages per month	Production 46 m^3/d no rod breakage
2 7/8	1737	Tubing perforated, had to be pulled every 6 weeks	No tubing trouble over 15 months
2 7/8	1890	Rod string had to be pulled every other month	No rod string change necessary, production increased by 20%
2 3/8	2743	Marked wear on rod string in deflected hole	Wear reduced by half

catcher shown in Fig. 4.1—38 will, after a certain length of drop, lock against the tubing and forestall further damage. When the tubing starts dropping, spring *4* presses upwards weight *3*, which is moved upward by inertia anyway, and this makes slips *2* expand over the conic seats *1*. No tubing catcher is required if the tubing is provided with an anchor preventing its downward movement (e.g. one of the types shown in Figs 4.1—34 and 4.1—36).

(d.4) *Equipment for producing high-viscosity and high pour-point crudes.* — Producing oils of several thousand cSt viscosity (waxy crudes) by means of sucker-rod pumps requires special measures, because (i) high friction losses against the tubing and flow line result in an increased polished-rod load during the upstroke; (ii) comparatively narrow travelling-valve inlets may, during the downstroke, represent a hydraulic resistance high enough to prevent the rod string from sinking during the time available: the carrier bar overtakes the polished rod; (iii) flow resistance of the standing valve may prevent the barrel from filling with liquid during the upstroke; (iv) the valve ball moves sluggishly in the narrow cage, and the valve will not open or close on time; (v) during pumping stoppages, the viscosity of the oil in the tubing increases, so that, on restarting the well, the rated polished-rod and prime-mover load have to be exceeded significantly; (vi) considerable friction between barrel and plunger may lift the barrel off its seat.

These same difficulties will arise when pumping crudes of high pour-point at well temperature, but their comparative significance will be different, because high-viscosity crudes tend to be naphthene-based, and their viscosity is then comparatively insensitive to temperature. They will not jell even at 0 °C, but may be pretty viscous even at high temperatures. Typical jelling crudes are paraffin-based, on the other hand. Their apparent viscosity may be very high near the pour point, but it will usually decrease rather steeply under a temperature rise as small as 5 – 10 °C. It is more difficult to start a well producing a crude of high pour point than a high-viscosity one because the starting polished-rod load, proportional to the static shear stress, tends to be very high. Friction in continuous operation may, on the other hand, be less than in the case of a high-viscosity crude. — If the waxy crude enters the well together with some gas, then removing the latter through the annulus may cause a special kind of operating trouble if formation pressure is comparatively high and well temperature is comparatively low. This is due to the following phenomenon. Pumping will inevitably stop at times during continuous production, and regularly during intermittent production. The fluid level will then rise in the annulus. If it attains a height where the temperature is low enough to make the oil jell, then a 'packer' develops in the annulus and stays there even after pumping has been restarted. It will not let the gas produced flow out through the casinghead, so that the gas in question is deflected into the sucker-rod pump together with the oil. This reduces the volumetric efficiency of pumping and, indeed, may result in full gas lock, in which case the pump produces no liquid at all. If production is reduced but not halted, the expansion of gas in the tubing enhances the cooling of the well fluid and thus increases the polished-rod load. All these conditions may considerably augment gas pressure below the jelled oil plug in the annulus, so that the gas may break through the plug, shooting it to the surface or into the flow line with a loud report. This irregular shock load is harmful to both the well and the pumping installation. — In mixed-base oils, the flow properties described are transitional, too. Measures which permit the pumping of high viscosity crudes without the above pitfalls can be subdivided in two groups: installing pumps that operate satisfactorily even if the viscosity of the well fluid is high; and decreasing the viscosity of oil entering the pump.

Valves of conventional sucker-rod pumps will perform better if the clearance between cage ribs and valve ball is at least 2 mm and the cage height is less than usual. Pumps with large valve ports are to be preferred. — Cia. Shell de Venezuela and USI Venezolana have developed a modified sucker-rod pump whose flow resistances are less than those of the API types (Juch and Watson 1969). A sketch of pump design is given in Fig. 4.1—55. A semi-empirical formula was derived to provide the flowing pressure drop of water-cut oil of 800 cP viscosity in pumps of 2 3/8—4 1/2-in. size:

$$\Delta p = 6897 + (A\mu_l + Bq_l)\, q_l\, d_p^{-3.82} \qquad 4.1-56$$

Table 4.1 — 14

Pump type	A	B
Streamlined	1162	6.864×10^4
API	1709	11.28×10^4

where the constants A and B for the new, streamlined, and the conventional, API type pumps are listed in Table 4.1 — 14.

Example 4.1 — 10. Find the flowing pressure drop of water-cut oil, 800 cP viscosity, in a 2 3/4-in. size API sucker-rod pump ($d_p = 69.9$ mm), if $q_l = 100$ m³/day. — By Eq. 4.1 — 56,

$$\Delta p = 6897 + (1709 \times 0.8 + 11.28 \times 10^4 \times 1.16 \times 10^{-3}) \cdot 1.16 \times 10^{-3} \times \times 0.0699^{-3.82} = 5.2 \times 10^4 \text{ N/m}^2 = 0.52 \text{ bar}.$$

The force pressing the sucker-rod pump into its seat in the tubing must exceed the friction force between plunger and barrel during the upstroke. The latter can be estimated (Juch and Watson 1969) by the formula

$$F_{fr} = \frac{3.14\, d_p\, h_p\, \mu_l\, v_p}{\Delta d}. \qquad 4.1\text{—}57$$

Example 4.1 — 11. Find the friction force arising between barrel and plunger if $d_p = 1\ 3/4$ in.; $h_p = 1.22$ m; $\mu_l = 10^4$ cP; $v_p = 1.4$ m/s (to be expected at about $n = 12$ min^{-1} and $s_d = 1.8$ m); $\Delta d = 1.3 \times 10^{-4}$ m. — By the above formula

$$F_{fr} = \frac{3.14 \times 0.0445 \times 1.22 \times 10 \times 1.4}{1.3 \times 10^{-4}} = 1.8 \times 10^4 \text{ N}.$$

The Pleuger type clap-valve pump, designed for the sucker-rod pumping of heavy crudes, can be employed up to 5200 cSt viscosity. In Fig. 4.1 — 39, sleeve *2* performs a reciprocating motion along pump tube *1*. During the upstroke, rod string *3* moves upwards; clap valve *4* covers the seat machined in the plane $A — A$ and closes. Shoulder *5* then presses against crosspiece *6*, lifting sleeve *2* together with the fluid in it. The rising fluid lifts check valve *7* which permits one strokeful of the oil in the tubing to enter the flow line. During the downstroke, check valve *7* closes, clap valve *4* opens, and sleeve *2* re-enters the liquid. This type of pump comes in three sizes: their parameters are given in Table 4.1 — 15. It has been used in the German oilfields

Table 4.1 — 15
Data of Pleuger clap-valve pumps (Brüggemann and de Monyé 1959)

O. D.	Stroke length	Speed	Capacity	Effective valve surface	Max. load
in.	m	1/min	m³/d	cm²	kN
3	1.2	4	14	26.4	25
4	1.2	4	20	38.5	39
5 1/2	1.8	4	64	78.5	118

since 1956 (Brüggemann and de Monyé 1959). — The viscosity of oil entering the well may be decreased by heating or by dilution with low-viscosity oil. Heating may be of one of three types: hot-water, electrical or gas-burner.

Figure 4.1—40 shows a well completion suitable for *hot-water heating* (Walker 1959). Water is heated at the surface, usually in gas-burner boilers, and fed to the well bottom through pipe *1*, usually of 1-in. diameter. This

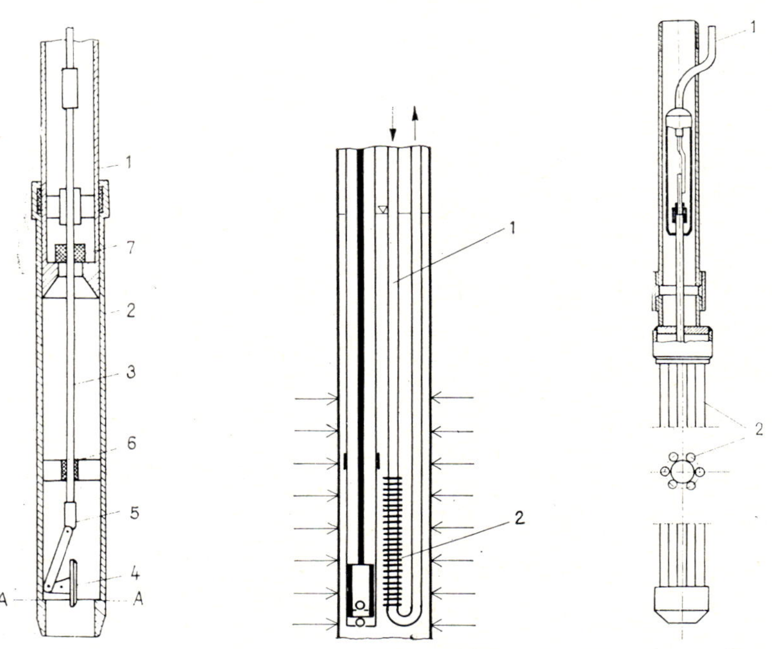

Fig. 4.1—39 Pleuger's clap-valve bottom-hole pump, after Brüggemann and de Monyé (1959)

Fig. 4.1—40 Bottom-hole heating with hot water, in well produced by sucker-rod pump, after Walker (1959)

Fig. 4.1—41 Electric bottom-hole heating in well produced by sucker-rod pump*

pipe is fitted with heat-exchange plates in a height corresponding to from one-half to three-quarters of the perforated well section. In order to limit undesirable heat losses, pipe *1* is usually heat-insulated on the inside or outside. If this is not the case, it is recommended to insert an insulating ring at each coupling, in order to prevert direct contact between hot-water pipe, tubing and casing.

The *electric heater* can be placed either below the pump (bottom-hole heater), or farther up the tubing, so as to envelop the rod string (tubing heater). The first solution is to be preferred if waxy deposits are to be anticipated even at the well bottom, or if the oil is too viscous even at the

* Figure taken from Howel and Hogwood Petroleum Publishing Company, Box 1260 Tulsa, Oklahoma 74101 (1962)

original formation temperature of the well bottom. The tubing heater is usually installed at a height where wax would start to deposit in the absence of heating. Figure 4.1—41 shows an electric bottom-hole heater (Howell and Hogwood 1962). The electric current is led in cable *1* to the six steel-clad heater elements. It may return via another cable strand or in the tubing steel. When designing the heating system one should keep in mind that the oil must not be heated above its coking temperature. The heater elements must therefore invariably be covered with oil. Their surface temperature must not exceed that of the oil by more than 40 °C and must

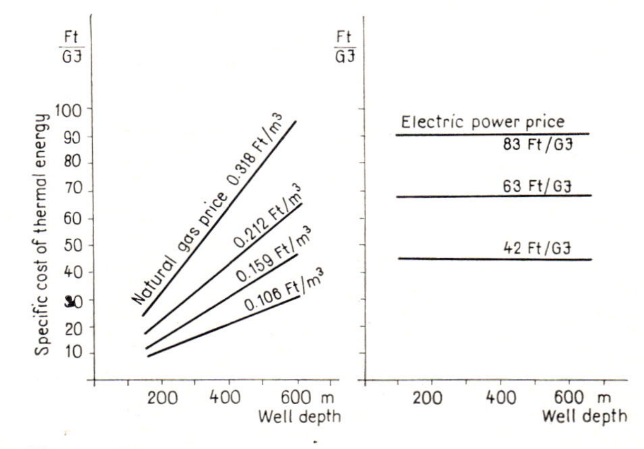

Fig. 4.1—42 Economics of bottom-hole heating methods, after Walker (1959)

not exceed 150 °C in any case. Heating temperature is a function largely of the rate of production, duration of heating and loss of heat to the formation. The heat supply required is given by

$$\Phi = \eta \times q \times \gamma \times c \times \Delta T \qquad 4.1\text{—}58$$

where q is the liquid production rate; ΔT is the temperature increase desired; and η, the efficiency of heating, depends on the heat loss to the surroundings; it equals 0.5 at a rough estimate. The choice of the cable requires special care. It must be protected from both mechanical damage and corrosion. PVC sheets may be used only up to round 80 °C. Up to 93 °C, cables insulated with asbestos and lacquered textile in a lead sheath are employed. At even higher temperatures, copper-insulated cables are recommended. In some well fluids, however, these may be damaged by corrosion. Figure 4.1—42 compares the economics of the above two kinds of heating under given operating conditions (modified after Walker 1959). The cost of delivering one thermal unit at the well bottom increases in the case of hot-water heating with depth and with the price of gas used to heat the water. The cost of electric heating is practically independent of depth and is a function of power cost alone. For instance, if power is bought or produced at 83 Forint per GJ, electric heating at a depth of 600 m roughly equals hot-water heating if the cost of gas is 0.318 Ft per m³. In California, 1959, about 1700 wells were produced with the aid of one type of heating or another.

Bottom-hole heating may also involve direct *gas burning*. Figure 4.1—43 is a sketch of a well completion incorporating a sucker-rod pump provided with a gas burner (Brandt et al. 1965). Burner *1* is installed below the pump. The fuel is a mixture of natural gas and air or propane and air. It is fed to the burner through pipe *2*. Combustion products are led through pipe *3* to

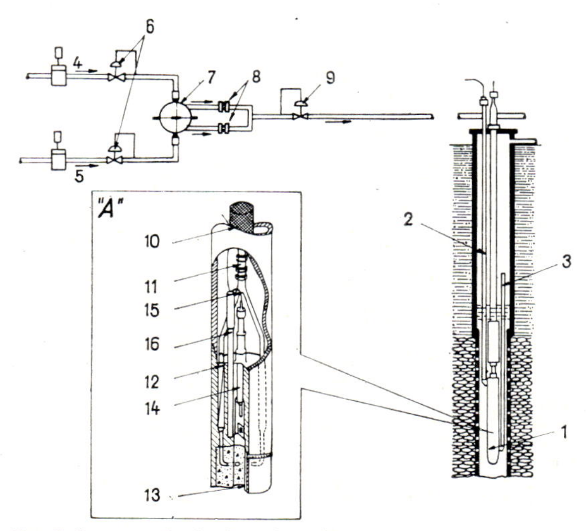

Fig. 4.1—43 Gas-fired bottom-hole heating in well produced by sucker-rod pump after Brandt et al. (1965)

the annulus, which lets them off into the atmosphere. The fuel feed and the gas-to-air ratio are adjusted by a surface control device. Gas arrives through conduit *4*, air through conduit *5* to the pressure-control valves *6*, that feed them to the pressure balancing valve *7* which ensures that gas and air entering the chokes are of equal pressure. The feed rate is controlled by the pressure differential between control valves *6* and the back-pressure valve *9*, interacting with the flow resistance of the system. The correct gas-to-air ratio can be ensured by the proper choice of the chokes *8*. Details of the gas burner are shown in part (A) of the Figure. The fuel attains the two opposed burner mozzles through filter *10* and check valve *11*. The filter is required to hold back solid impurities which could plug the screen installed to prevent flashbacking. The screen is of 0.1 mm mesh size. The combustion chamber is ceramic-lined. This protects the wall of the heater from direct contact with the flame, on the one hand, and prevents coking of the oil, on the other. The length of the stepped burner chamber is round 2.4 m; its minimum diameter is 19 mm. Ignition of the fuel is ensured by glow plug *14*, fed with current through cable *15*. Incorporated in the heater is a thermometer *16*, in series with the glow plug (no temperature measurement is possible during ignition). In the application of this type of heater, it must not be forgotten that some liquid condensate may form during the rise of the combustion product. Flow will thus become two-phase. At comparatively low rates, it may be of the slug type, and the *BHP* required to keep it up is therefore fairly high. This has an adverse effect on production. At higher flow rates, flow is of the annular type, and the pressure gradient required to maintain it is less, and so is, therefore, also the *BHP*. Hence, this type of heating will not usually be applicable in wells producing at rates

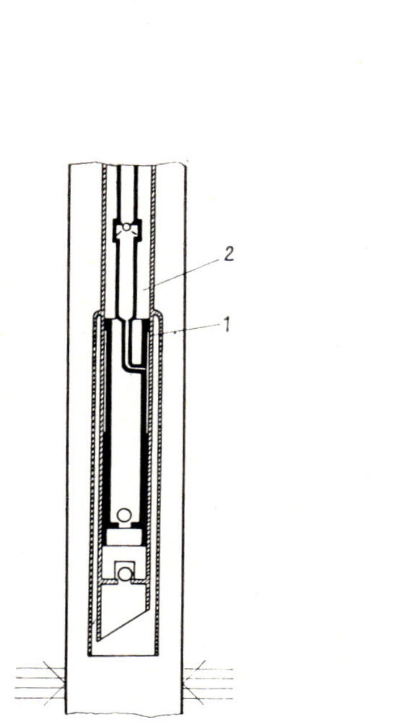

Fig. 4.1—44 Bottom-hole pump suited for solvent injection, after Walker (1959)

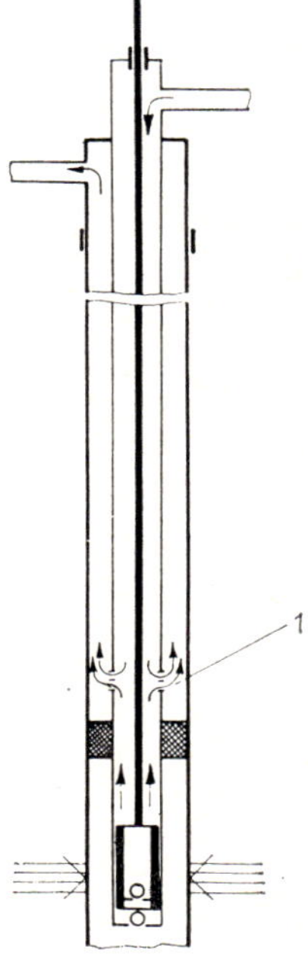

Fig. 4.1—45 Well completion suited for solvent injection, after Walker (1959)

lower than 0.2 m^3/day. The upper limit in wells producing a fluid of low *WOR* is about 32 m^3/day. Applicability is restricted by higher water cont. ents. The method can be used even at depths exceeding 1500 m.

Two possible designs of the submerged pump and of the well completion permitting the addition of a viscosity-reducing solvent are shown in Figs 4.1—44 and 4.1—45 (Walker 1959). In Fig. 4.1—44, a sucker-rod pump is attached to a hollow rod string. Low-viscosity oil is pumped to the well bottom through the hollow rods and through port *1* on the sucker-rod pump. The diluted oil rises to the surface through annulus *2* between the rod string and the tubing. Figure 4.1—45 shows a solution popular in Venezuela. Low-viscosity oil is pumped into the annulus between tubing and rod string. Through perforations *1* it enters the casing annulus where it rises to the surface after mixing with the heavy crude. This method is excluded if the rod string sinks too sluggishly on the downstroke or if the oil is gaseous.

The solutions outlined above are suited in general also for the pumping of high pour-point crudes. If the annulus risks freezing up during pumping stoppages, the well must be filled up with low-viscosity, non-freezing oil directly after stoppage or before restarting. Interrupting such wells may be something of a problem.

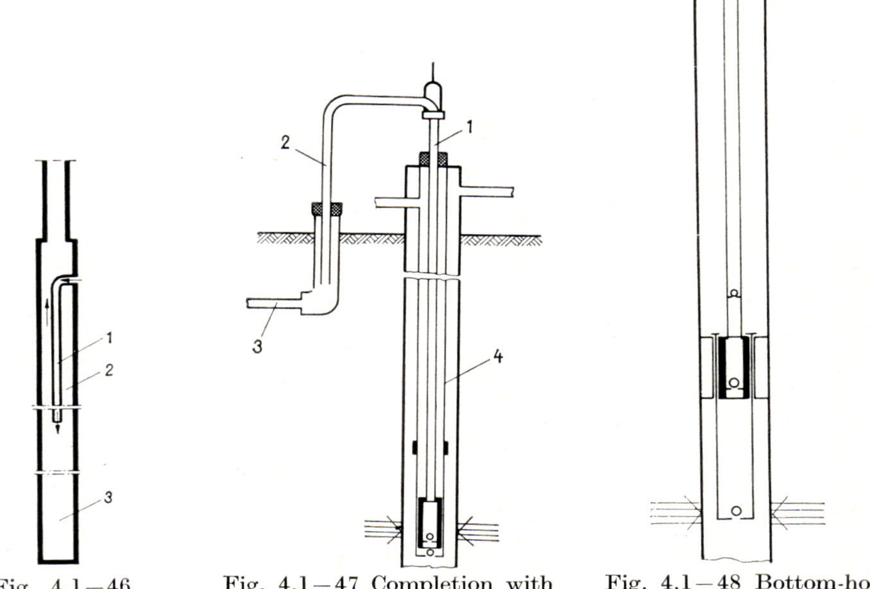

Fig. 4.1—46 Sand anchor

Fig. 4.1—47 Completion with tubing, produced with hollow-rod pump

Fig. 4.1—48 Bottom-hole design with hollow-rod pump seated on an anchor

(d.5) *Production equipment for sandy crudes.* — The purpose of a sand anchor is to separate the sand in the well fluid before it enters the pump barrel. In the sand anchor shown in Fig. 4.1—46, the fluid enters the pump through tube *1* and annular space *2*. The cross-section of annulus *2* is chosen so that the rise velocity of the well fluid in it is less than the settling velocity of the sand. The sand collects in chamber *3*. Since the chamber can be emptied only after pulling it together with the tubing, this solution is uneconomical except if the sand content in the well fluid is rather low. In certain other designs, the sand in the chamber can be dumped onto the well bottom by a sudden jerk on the tubing string. Such equipment may prolong cleaning intervals, but the cleaning operation itself will be more complicated. — The importance of a sand trap is not too great in most cases. The essential thing is to produce the well without damaging the sand face, and if some sand is produced nevertheless, to evacuate it continuously to the surface with the least possible harm to the production equipment. —

The hollow sucker rod has in the first place been developed to facilitate the production of sandy crudes. In the solution shown as Fig. 4.1—47, a tubing pump is fixed to the tubing shoe. Fluid lifted on the upstroke reaches flow line *3* through hollow rod string *1* and branchoff *2*. Hollow rods have the advantage that (i) fluid flows faster in the narrower hollow-rod space than in the rod-to-tubing annulus, which reduces the likelihood of sand grains settling out during continuous production, (ii) if pumping is stopped, the sand in the hollow rods cannot fall between barrel and plunger. The check

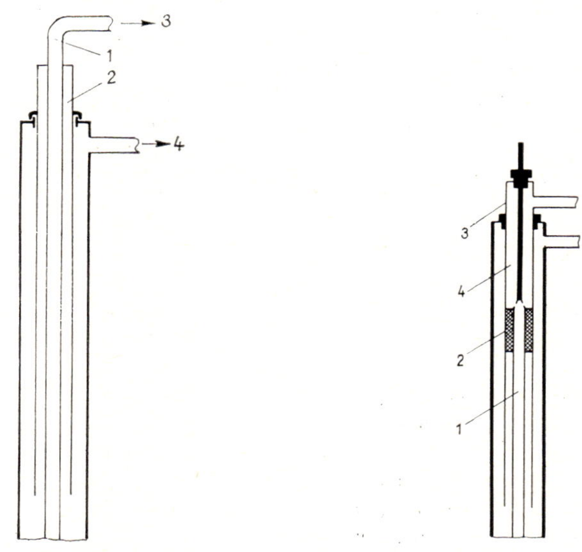

Fig. 4.1—49 Borger-type Christmas tree for hollow-rod pumping

Fig. 4.1—50 Canadian-type Christmas tree for hollow-rod pumping

valve in the hollow rod string further prevents sand settling during stoppages from reaching the pump. — In completions such as these, a tubing used to be run because the plunger used to be of larger diameter than the *ID* of the hollow rods (cf. Fig. 4.1—3c), and the upward force acting during the upstroke caused buckling and early failure in the hollow rods. This hazard was the more grave, the deeper the well. In order to increase the depth of applicability, the annulus between tubing and hollow rod string used to be filled with water (Fig. 4.1—47).

In more recent types of hollow rod (cf. paragraph 4.1.1d.1), the risk of joint failure is rather slight. The depth of installation can be increased significantly even if the pump is not loaded by an outside fluid column. The pump can be seated on an anchor fixed to the casing, as shown in Fig. 4.1—48. — There are more recent solutions also for wellhead assemblies. In the Borger type wellhead equipment (Fig. 4.1—49), hollow-rod string *1* is surrounded by cylinder *2*, polished on the outside, which assumes the role of the polished rod. The rod-string head is connected to the flow line by a flexible conduit. In the Canadian type wellhead completion (Fig. 4.1—50), plunger *2* mounted on hollow rod string *1* moves in cylinder *3*, which is polished on the inside. Hollow-rod string and flow line are connected by conduit *4*. The wellhead equipment much resembles the one for a solid-rod sucker-rod pump completion. Modern hollow sucker-rod strings are to be preferred to solid ones for several reasons in addition to the ease of pumping sandy crudes: (i) Solid-rod string plus tubing is replaced by the cheaper hollow-rod string. (ii) Owing to the greater metallic cross section of the hollow rod string, stretch is less than in a 5/9-in. size solid rod, and the absence of tubing eliminates tubing stretch; hence, stroke reduction due to changes in liquid load tends to be less. (iii) If plunger size is large enough, F_{max} may be less even if rod-string weight is greater (cf. Eqs 4.1—15 and 4.1—22). (iv) Requiring less space, this solution is at a distinct advantage in multiple completions and midi wells. (v) If the well-fluid is non-gaseous, it can be produced through the casing-to-rod annulus, in which case the hollow rod string can be used as a feed line for inhibitors or low-viscosity oil, or to house electric heaters. (vi) The rod string does not have to be pulled if paraffin deposits form. Such deposits can be prevented by a paraphobic lining or coating.

Nevertheless, hollow rods have certain drawbacks as well. (i) In unfavourable cases, the hollow rod string will rub against the inside of the casing. To reduce friction, plastic rod-guides are employed. (ii) At high production rates, fluid friction may be significant in the narrow cross-section of the rod Of course, the well may be produced through the casing annulus if the fluid is non-gaseous, non-corrosive and non-erosive to the casing, and no significant deposits of paraffin are to be anticipated. (iii) If plunger diameter is small, F_{max} may be greater than for a solid rod string.

(d.6) *Production equipment for gaseous and water-cut oil.* — Free gas entering the sucker-rod pump reduces its volumetric efficiency (cf. paragraph 4.1.1b.2). For this reason, it is best to separate as much gas as possible from the liquid entering the well even before it enters the sucker-rod pump and to deflect it into the annulus for delivery to the flow line. The device separating the gas from the fluid at the well bottom is called a gas anchor. The simplest design is the so-called single-body gas anchor (Fig. 4.1—51). Well fluid enters the anchor through ports *1* from where it moves downward in annulus *2*. The annulus is dimensioned so that the rise velocity of the gas bubbles is greater than the rate of downward flow of the fluid. Gas may thus collect and escape through the entry ports and rise up in the casing annulus. According to S. Virnovsky (Muraviev and Krylov 1949), the cross-section area A_a of the gas-anchor annulus (in m^2) should be for high viscosity oil

$$A_a = 130\, A_p\, snv \qquad 4.1\text{—}59$$

for low-viscosity oil

$$A_a = 0.12\, A_p\, sn\sqrt[3]{v} \qquad 4.1\text{—}60$$

for water

$$A_a = 0.00012\, A_p s\,. \qquad 4.1\text{—}61$$

A typical port diameter A_p is 2 mm. The third formula is to be used if the well fluid contains more than 80 percent water. The length of the gas anchor is to be chosen so that the gas bubbles entering its annulus may rise to the top ports during one full stroke cycle of the pump. A typical gas-anchor length equals 20 times the jacket diameter. If the A_a furnished by the above formulae cannot be realized owing to the well being too slim, two- or three-body gas anchors can be used (Fig. 4.1—52). In this case, A_a in the above formula means the sum of the cross-sectional areas of the gas passages in the individual anchor bodies. Installing a gas anchor of sufficient diameter may be difficult or impossible in slim wells. The design shown in Fig 4.1—53 may be used even in such cases (Schmoe 1959). Well fluid and

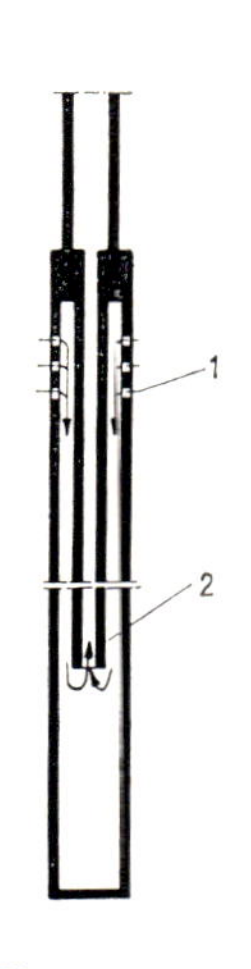

Fig. 4.1—51 Single-body gas anchor

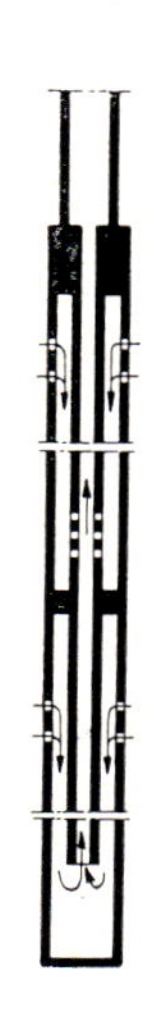

Fig. 4.1—52 Multi-body gas anchor

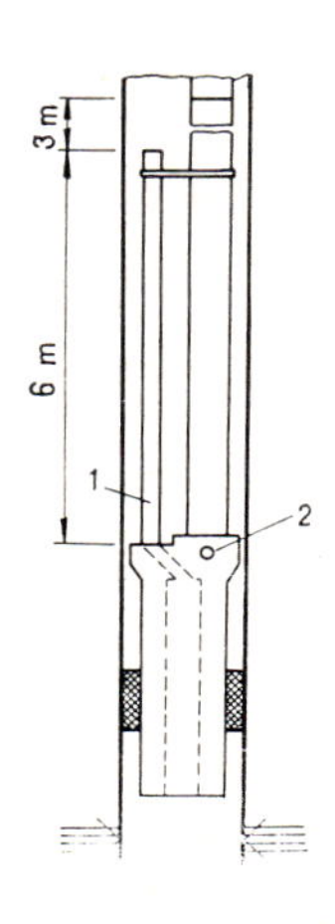

Fig. 4.1—53 Gas anchor for slim holes, after Schmoe (1959)

gas flow past the packer and into the annulus through conduit *1*. Oil enters the sucker-rod pump through port *2*. Experience has shown this device to operate satisfactorily even at *GOR*s up to 2000. Failures may be due, firstly, to the choking of the comparatively slim conduits and, secondly, to packing breakdown. — In light oils, the vibration gas anchor (Fig. 4.1—54) is often used with success. This is a multi-body gas anchor in whose annular space gas separation is promoted by disk baffles mounted on spiral springs. Well-fluid flow keeps the springs and baffles in continuous vibration. Quite often, the gas anchor will not remove all the free gas from the liquid entering the pump. In such cases, a sucker-rod pump design whose volumetric efficiency is not too seriously affected by the gas should possibly be chosen. Of the conventional types, rod pumps are at an advantage because dead space below the lower end of the plunger stroke is comparatively small.

Several special 'gas-insensitive' sucker-rod pumps are also known. Let us consider one of the more advanced designs. It has been mentioned in Section 4.1.1d.4 that novel, modified sucker-rod pumps have been designed for heavy crudes (Fig. 4.1—55). Both the tubing pump marked *T* and the rod pump denoted *R* are provided with a ring valve *1* made of brass, which in effect turns the pump into a two-stage one. Operation is analysed with reference to Fig. 4.1—56 (for which see also p. 277). Part (a) shows various plunger positions; Part (c) shows the variation of polished-rod load v. stroke length in the conventional pump (dashed line)

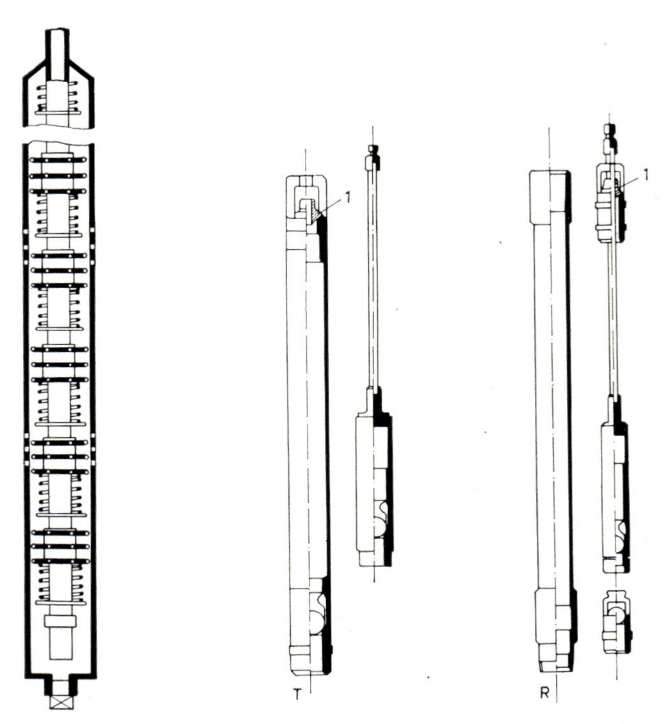

Fig. 4.1—54 Sonic gas anchor

Fig. 4.1—55 Modified bottom-hole pump, after Juch and Watson (1969)

and the novel design (full line). In Part (b), the dashed line shows pressure in the space below the plunger v. stroke length in a conventional pump. The upper part of the full-line diagram shows pressure change in the compression space between plunger and standing valve; the lower part shows the same for the space between plunger and standing valves. The advantage of the modified design when pumping gaseous fluids is that (i) the plunger valve opens earlier and more smoothly during the downstroke; (ii) the rod string is in tension throughout, which reduces the risk of its buckling and reduces or eliminates the liability of fluid pound (cf. paragraph 4.1—1 f. 2); (iii) on the upstroke, the standing valve opens earlier, so that more fluid can enter the barrel. Earlier opening improves the volumetric efficiency of the pump. Gas in high-viscosity oil is particularly deleterious to volumetric efficiency, because the considerable pressure drop on entrance into the barrel permits more gas to separate. Laboratory experiments have shown that even if the gas will separate rather readily, compression during the downstroke will not cause it to redissolve in the crude to any significant extent (Juch and Watson 1969).

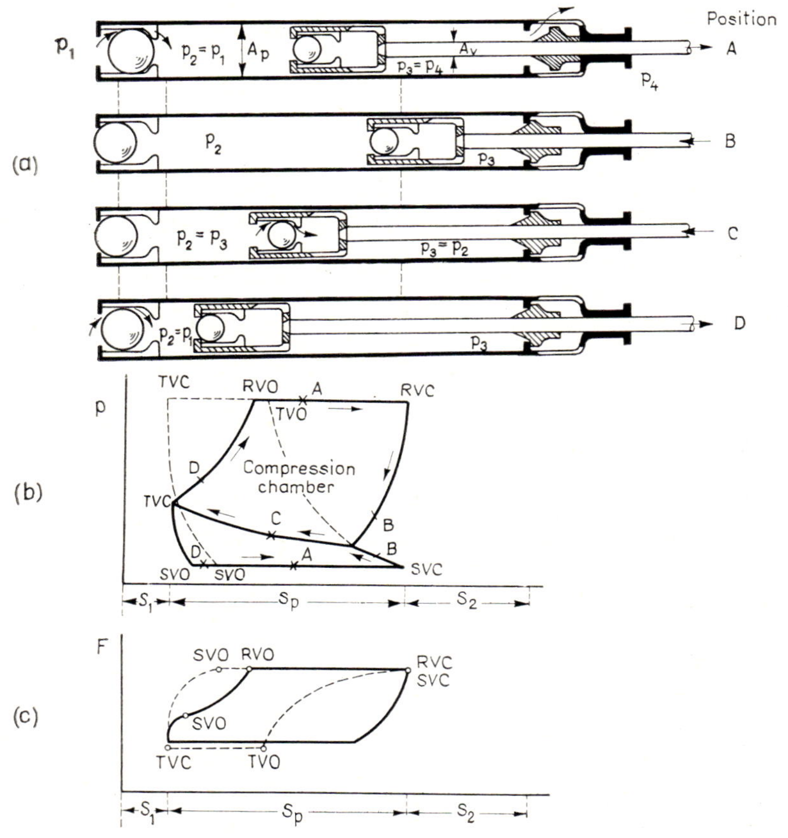

Fig. 4.1—56 Operation of modified bottom-hole pump (Juch and Watson 1969)

The presence of water when producing oil with a sucker-rod pump causes two kinds of harm: (i) it reduces volumetric efficiency, because the lower viscosity of water permits faster slippage past the plunger, and (ii) it reduces lubrication, which makes the barrel and plunger wear faster. These difficulties are partly obviated in the oil-lubricated pump of ARMCO.

During production, liquid may enter both the central bore *1* and the annulus *2* of the pump (Fig. 4.1—57). The upper part of the annulus is filled with oil, whereas its lower part contains water or watery oil. During the downstroke, part of the oil enters space *1* directly; the rest enters space *2*.

Part of the latter also attains space *1* through port *3*. During the upstroke, chamber *4* is under suction pressure, whereas groove *5* is under the discharge pressure of the pump. Pressures being equal between the groove and the space above the plunger, no oil seeps from the former into the latter. There is, however, a sizeable pressure differential between the groove and the space below the plunger; this makes the oil seep downwards and lubricate the plunger with oil throughout. During each downstroke, oil lost from space *2* is made up from the well fluid.

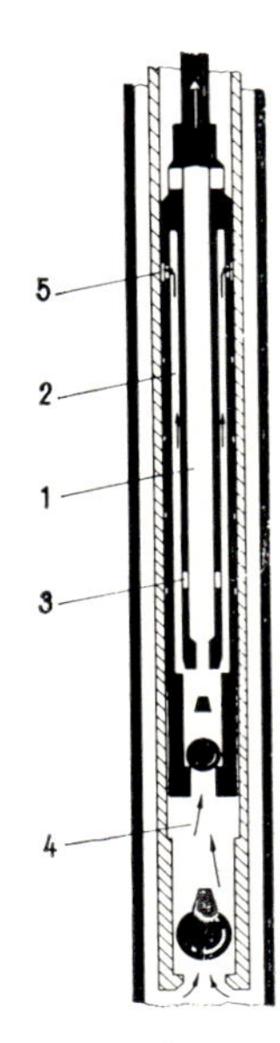

Fig. 4.1—57 ARMCO oil-lubricated bottom-hole pump for wet oil

(e) Well testing

One of the most important instruments of testing wells produced with sucker-rod pumps is the dynamometer, and the most frequent production control procedure is the recording and analysis of dynamometer cards. Essentially a plot of polished-rod load v. stroke, the card permits determination of numerous parameters of operation of the sucker-rod well. We shall not, however, enter into details concerning this instrument and its applications, as these are discussed in sufficient detail in numerous papers and even books dedicated to this single topic (e.g. Zaba and Doherty 1956; Belov 1960, Slonegger 1961, Craft et al. 1962). Let us discuss below the potential testing of wells and the means required to perform it.

Determining performance curves of pumped wells by means of Eqs 2.1—7 or 2.1—8 is in a general way more of a problem than in flowing or gas-lifted wells, because no bottom-hole pressure bomb can be installed in the tubing without pulling the sucker-rod string. For this reason, special methods for measuring *BHP*s have been developed. In wells with a production casing of large enough size, a tubing string of small enough *OD*, and featuring an open completion, the *BHP* can be measured by a pressure bomb run in the annulus. The measurement requires a special wellhead completion (Reneau 1953), and a special running winch. This procedure is fraught with the risk of the wireline carrying the bomb winding itself around the tubing; it must therefore be run and pulled very carefully in order to avoid breaking it. Figure 4.1—58 shows a Halliburton type power-driven measuring assembly and the running of the pressure bomb. The winch is provided with a hydraulic torque convertor and a depth and load recorder. If the bomb gets caught by a coupling during its pulling, or the wireline gets wrapped around the tubing, the hydraulic power transfer will gradually build up the load in the wireline to the allowable limit and no further. During a wait of a few minutes, the bomb will usually disengage itself and pulling can be continued. The assembly permits measurement of *BHP*s in 3 or 4 wells per day, each 1500—2000 m deep. The application of this method is often forbidden by well size and completion type.

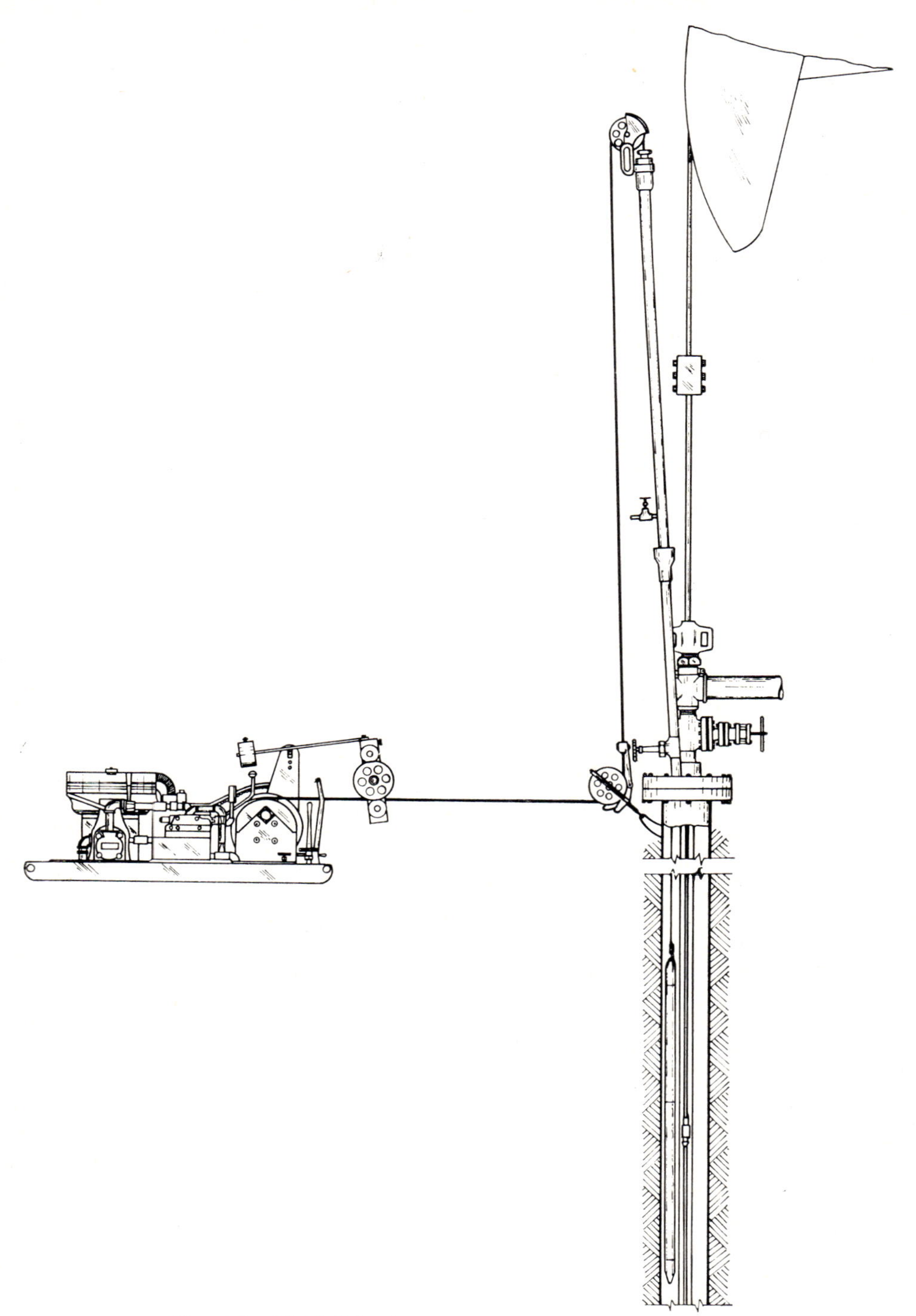

Fig. 4.1–58 Halliburton pressure measuring assembly

Pressure gauges installed in the well. — It may be economical to permanently install pressure gauges below the tubing shoes of key wells. Pressures so measured are transmitted to the surface via an electric cable. In the Maihak type device (one of whose variants is used in process control: cf. paragraph 4.1.1f.1), the measuring element is an elastic wire, one of whose ends is fixed, the other being attached to a membrane whose deformation is proportional to pressure. Tension in the wire, and hence its frequency of vibration, are proportional to membrane deformation. The frequency of vibration triggered in the wire is transmitted to the surface. Another design incorporates a Bourdon tube which turns a disk proportionally to the pressure sensed. The position of the disk is electrically sensed and transmitted to the surface. Such methods have the drawback of being comparatively expensive, which prohibits their use in just any well; also, running and pulling the measuring system requires an excess effort.

Walker's method (Nind 1964). — The basis of the method is to subsequently adjust two different annulus pressures p_{CO} at the casing head while producing the well at a constant rate by means of the sucker-rod pump. (A constant rate of production implies a constant producing *BHP*.) At each operating point, the fluid level in the annulus is measured. We may write the annulus pressures at the two operating points as

$$p_{wf} = p_{CO1} + p_{g1} + (L - L_1)\,\bar{\gamma} \qquad 4.1-62a$$

and

$$p_{wf} = p_{CO2} + p_{g2} + (L - L_2)\,\bar{\gamma}. \qquad 4.1-62b$$

Solving for $\bar{\gamma}$, we obtain

$$\bar{\gamma} = \frac{p_{CO1} + p_{g1} - p_{CO2} - p_{g2}}{L_2 - L_1}. \qquad 4.1-63$$

It is assumed that $\bar{\gamma}$ remains practically constant during the pair of measurements. By measuring fluid levels L_1 and L_2 in the annulus and pressures p_{CO1} and p_{CO2} at the casing head, the average gravity $\bar{\gamma}$ can be derived from the above formula. Now p_{wf} can be calculated using Eq. 4.1—62a or b. The pressure of the gas column can be obtained from Eq. 2.4—3, putting $p_{gi} = p_{Ci} - p_{CO}$:

$$p_{gi} = p_{CO}\left[e^{\frac{0.00118\,L_i M_g}{T\bar{z}}} - 1\right]. \qquad 4.1-64$$

The fluid level in the annulus is usually determined by means of an acoustic survey. The principle of the measurement is illustrated by Fig. 4.1—59 (Muraviev and Krylov 1949). The sound source (ping gun) *1* is mounted on the casing-head. The reflection of the sound wave generated pneumatically or by exploding some cordite is sensed by microphone *2*. This is a tungsten filament bent in the shape of a letter W, to which current is fed by a low-voltage cell. Changes in microphone current due to the incident sound waves are transmitted by amplifier *3* to pen recorder *4*, which traces them on paper strip *5*. This latter is moved at constant speed by an electric motor. The speed of sound is different in different gases at different pressures. For calibration purposes, so-called marker couplings, of larger-than-usual diameter, are installed at various known depths in the tubing string. Reflections from these couplings permit us to calculate the speed of sound and hence the depth of the fluid level.

The gas-breakthrough method (Nind 1964). — A dynamometer installed on the polished rod is used to record a number of diagrams during operation, while annulus pressure is gradually increased by a slow closing of the casing valve. Increasing pressure gradually depresses the fluid level in the annulus. When the level is lowered as far as the pump, the latter is filled with gas and the dynamometer diagram assumes a typical pistol-shaped form (cf. paragraph 4.1.1.f.2). Now annulus pressure is reduced by a few tenths of a bar and production is continued at a steady rate for one or two hours, in order to let operating parameters stabilize with the fluid level at pump height.

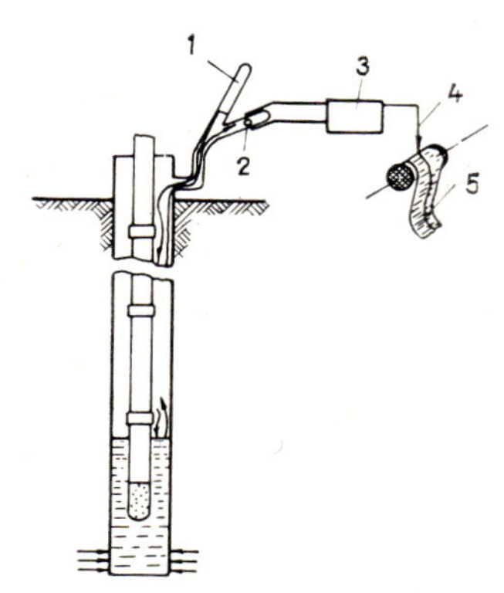

Fig. 4.1—59 Acoustical survey (Muraviev and Krylov 1949)

Knowing the casinghead pressure and the depth of installation of the pump, $p_{wf}=p_{CO}+p_g$ can be calculated.

Agnew's method (1956). — The method is applicable if the sucker-rod pump is operated at a slow enough speed to make dynamic loads and losses negligible. These conditions are ensured by operating the pump at a low speed (if the well is prone to waxing, the measurement is carried out shortly after the installation of a new sucker-rod pump). It is preferable to open the annulus and let the gas escape into the open. By Eq. 4.1—1, the polished-rod load on the upstroke will then be

$$F_s = F_l' + F_r + F_{TO} - F_h .$$

While deducing Eq. 4.1—3 we have observed that

$$F_l' + F_r = F_l + F_r b = A_p L \bar{\gamma}_l - F_r b .$$

The downstroke polished-rod load is, by a slightly modified Eq. 4.1—4,

$$F_{s1} = F_r b.$$

By definition now,

$$F_h = p_{wf} A_p$$

and the above equations resolve to

$$p_{wf} = \frac{A_p L \bar{\gamma}_l + F_{TO} - (F_s - F_{s1})}{A_p} . \qquad 4.1-65$$

We further know that

$$F_r b = F_r \frac{\gamma_r - \gamma_l}{\gamma_r}$$

and hence

$$\bar{\gamma}_l = \frac{F_r - F_r b}{F_r} \gamma_r; \qquad 4.1\text{—}66$$

F_r is rod-string weight in air; it can be calculated from rod-string design and rod parameters. The 'wet' rod-string weight, $F_r b = F_{s1}$ equals the pressure force defined by the lower border of the dynamometer diagram (Fig. 4.1—60). In the knowledge of these values, $\bar{\gamma}_l$ can be calculated using Eq. 4.1—66; F_s can be determined directly from a diagram similar to the one in Fig. 4.1—60. The tubing-head pressure p_{TO} being known, the force loading the plunger, $F_{TO} = A_p p_{TO}$ can be calculated. Knowing these data, p_{wf} can be obtained from Eq. 4.1—65. The accuracy of the result is significantly affected by the accuracy at which forces can be read off the dynamometer card.

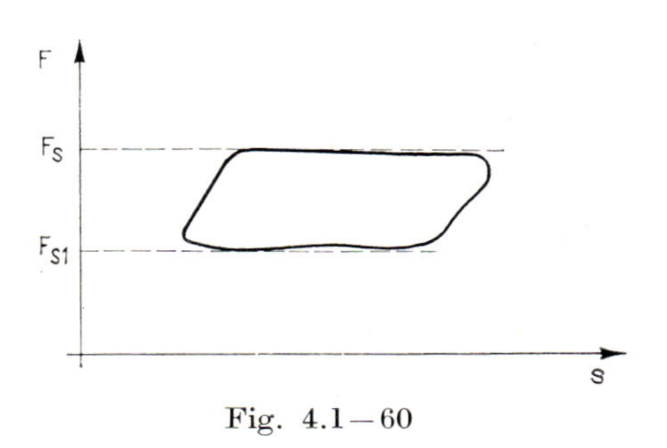

Fig. 4.1—60

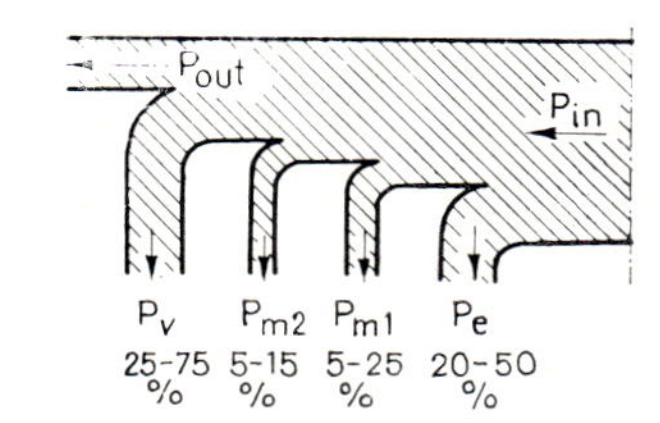

Fig. 4.1—61 Power consumption of sucker-rod pump, after Milinsky et al. (1970)

(f) Operating conditions

Correctly choosing the operating point of a sucker-rod pump installation, and adjusting it to the changes taking place during the pumped life of the well, is a highly important task. Power consumption of 474 electrically driven sucker-rod pump installations was investigated in the Soviet Union (Milinsky et al. 1970). Well depths ranged from 800 to 1100 m, daily production rates from 10 to 80 m^3. Some of the results are shown in Fig. 4.1—61. It is seen that the difference between input power P_{in} and output power P_{out} is spent to cover the electric loss P_e, the surface mechanical loss P_{m1}, the subsurface mechanical loss P_{m2} and the volumetric loss P_v. The volumetric and electric losses are seen to assume a high significance occasionally.

(f.1) *Continuous v. intermittent operation.* — It often happens that, after continuously producing a well of comparatively high rate with a pump of the correct capacity, a decline in well capacity entails a gradual decrease in volumetric efficiency; if the capacity of the sucker-rod pump installation is not adjusted to the gradually growing deficiency of well fluid, a gradual increase in the specific power cost of production will result (Szilas 1964). In order to forestall this, the production rate of the pump is to be reduced so as to match the daily inflow rate. This can be realized in one of two ways: either by maintaining continuous pumping, and reducing pump capacity to the desired value, or by maintaining the capacity and reducing the duration of pumping per day. If both variants are feasible, the more economical one is to be chosen. In order to decide whether continuous or intermittent pumping is more economical, let us assume as a first approximation that (i) the mode of operation does not affect the daily inflow rate of fluid into the well; (ii) the liquid flowing through the sucker-rod pump is gasless; (iii) volumetric efficiency is unity.

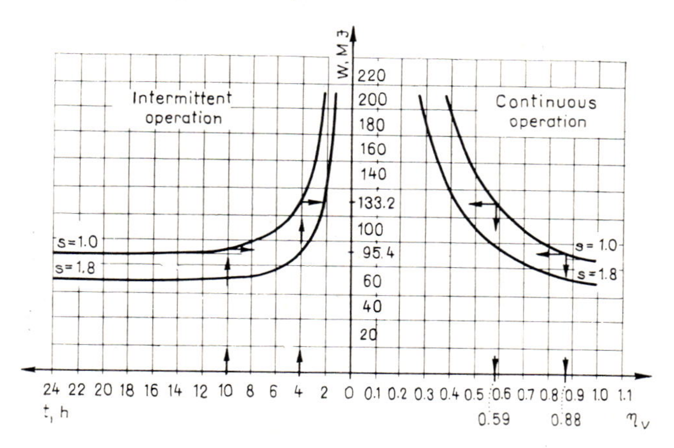

Fig. 4.1—62 Daily power consumption of intermittent and continuous pumping at $q_m = 2500$ kg/d; $L = 1200$ m; $\gamma = 8826$ N/m^3; $d_p = 31.8$ mm; after Szilas (1964)

If the daily production rates of the two modes have been correctly chosen, then they should equal each other as well as the daily inflow rate. The main difference between the two modes is in this case that the same volume of liquid is pumped over 24 h in the continuous mode, and during a number of hours $t < 24$ in the intermittent mode. Power consumption is greater in the intermittent mode, which, involving a higher spm, gives rise to higher dynamic loads. On the left-hand side of Fig. 4.1—62 (assuming a given well and a given rate of production) daily power consumption v. daily pumping time has been plotted for two values of polished-rod stroke. Volumetric efficiency has been assumed to equal unity. Pumping speeds have been determined using the relationship implied by Eqs 4.1—39 and 4.1—42,

$$n = \frac{q_t}{\eta_v A_p s_p} \quad [1/\text{min}]$$

with the substitution

$$q' = 60\, q_t \gamma_l \quad [\text{N/s}]$$

giving

$$n = \frac{q'}{60\, \eta_v\, A_p s_p \gamma_l}. \qquad 4.1\text{—}67$$

Daily power consumption is provided by

$$W = Pt \qquad 4.1-68$$

wherein P has been derived from Eq. 4.1—52; t is daily pumping time in s.

The Figure reveals daily power consumption to be the higher the shorter the daily pumping time. In the case examined, power consumption hardly increases while pumping time decreases from 24 to 12 h. A further reduction in pumping time, however, entails quite a steep rise in power consumption. The figure further shows that, all other parameters being constant, power consumption is higher at shorter polished-rod strokes. Since a daily pumping time of 24 h means continuous operation, the power consumption of a correctly dimensioned continuous pumping operation is less for a given volumetric efficiency than that of the intermittent operation.

The volumetric efficiencies of the two modes of operation are, however, different in the general case, because: (i) the production capacity of the pump decreases in time owing to wear of the moving parts; now, in order to deliver the required production, the initial capacity must exceed the rate of inflow. When wear has reduced capacity to the level of the prescribed production rate, the theoretical capacity must be increased. Repeated capacity increases may be brought about by increasing the pumping speed of continuous operation, or the daily pumping time of intermittent operation. If the prime mover of continuous pumping is an electric motor, changing the pumping speed requires changing of the v-belt sheave. Noticing the decrease of production capacity below a given threshold and changing the sheave requires human intervention; also, except for a short while between changes, mean volumetric efficiency is invariably less than unity. In an automated intermitting operation (see below), the automatic control of daily pumping time permits to continuously adjust production capacity to inflow. (ii) By what has been expounded in Section 4.1.1b.2, the decrease in volumetric efficiency due to the gas content of the well fluid is the greater, the less the *BHP* and the depth of immersion. In continuous operation, immersion is equal to or less than in intermittent operation. The decrease of volumetric efficiency due to the presence of gas may be more costly in continuous operation.

On the basis of the above considerations, we have plotted on the right-hand side of Fig. 4.1—62 the daily power consumption furnished by Eq. 4.1—68 v. volumetric efficiency. The diagram enables us to decide which of the modes is the more economical, provided daily production is equal in the two cases. For instance, we find with reference to the diagram that, in the given well, at a polished-rod stroke of $s = 1.8$ m, the daily power consumption of continuous pumping at a volumetric efficiency of $\eta_v = 0.88$ equals that of intermittent pumping for 10 h/d, at unity efficiency. In other words, if the volumetric efficiency of continuous pumping is 0.88, then intermittent pumping is more economical, as regards specific power consumption, at daily pumping times exceeding 10 h. In reality, the volumetric efficiency of intermittent pumping is less than unity, too; the above consideration applies to ideal intermittent pumping. We have assumed so far that the daily production rate is not affected by the mode of production. In reality, this condition obtains only: if (i) the well bottom is below the sand face (the well ends in a 'sump'): both modes will produce at the same rate if the fluid level cannot rise past the sand face; (ii) if the allowable pumping rate of the well is rather low so that the pump is deep below the fluid level, with the producing level comparatively high. The mean producing fluid level of intermittent pumping may be adjusted to equal that of continuous production. Thus inflow and daily production rates can be equalized.

In cases other than the above, the producing fluid level usually stabilizes near the pump in continuous operation. In the intermittent mode, the producing fluid level is invariably higher, and the daily production rate is that much lower. The efficiency of intermittent production is lowered by the missing production. The necessarily higher pumping speeds of intermittent operation cause more wear and tear in the pumping installation, since lifting a given liquid volume from a given depth requires in a fair approximation the same total number of strokes at given values of polished-rod stroke and plunger diameter. In intermittent pumping, this number of strokes is realized within a shorter span of time. The wearing parts of the pump therefore cover the same aggregate 'friction path', but at a faster average speed in the intermittent case. Now wear is about proportional to the square of the relative speed of the moving parts. Dynamic stresses are also stronger, and failures are more frequent. We may, then, state that if the volumetric efficiency of continuous pumping is comparatively high, it is usually preferable to operate in the continuous mode. Intermittent pumping is justified if it significantly improves volumetric efficiency. If this is the case, automation of the intermittent installation is to be recommended.

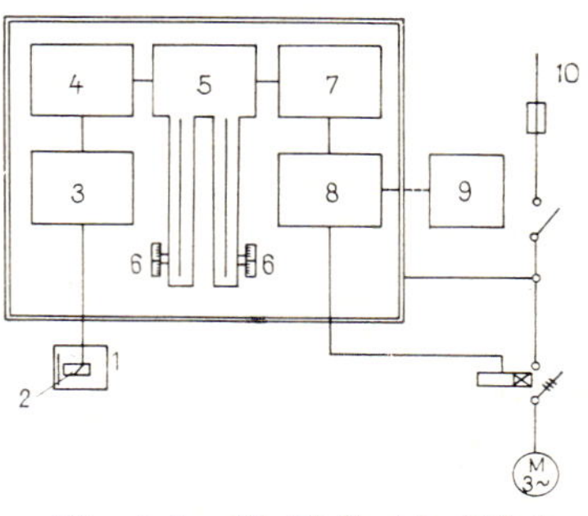

Fig. 4.1—63 Maihak's SR-1 controller, after de Monyé (1959)

Automatic 'formation control' of intermittent pumping. Formation control essentially consists in installing on the well bottom a sensor that signals, on the one hand, when the fluid level has attained a predetermined height during a pumping stop and, on the other, when the well has been pumped off. The electric signals control the start-up and stopping of the pumping unit's prime mover. Figure 4.1—63 is a diagram of a regulating device built by Maihak AG. and denoted the SR-1 (Monyé 1959). Transmitter *1* installed at the tubing shoe incorporates pressure sensor *2* (a steel wire, one of whose ends is attached to a membrane whose position is pressure-dependent). Mounted next to the wire is an electromagnet. At predetermined intervals, automatic trigger *3* installed on the surface feeds low-voltage pulses to the electromagnet coil. The magnet thereupon pulls in and releases the sensor wire, whereupon the wire starts to vibrate at a frequency depending on its constant physical parameters, on the one hand, and on its tensioning by the membrane, on the other. The electric vibrations induced in the coil by the vibrating wire are fed through a cable attached to the tubing to amplifier *4*, and by the latter to discriminator *5*. This latter incorporates two wires whose natural frequencies are adjusted by means of tuning screws *6* to the least and greatest prescribed pressure. When the down-hole transmitter wire is vibrating at the natural frequency of one of the two discriminator wires, the discriminator ignites one of the series-connected thyratrons *7*. The winding of the relay in the anode circuit of the thyratron in question receives a current which opens or closes the

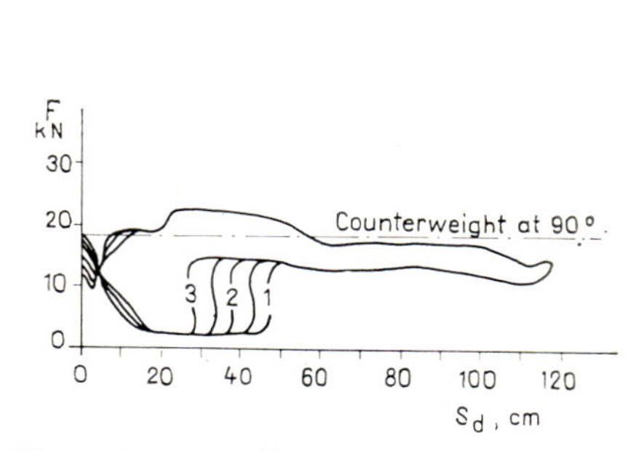

Fig. 4.1—64 Dynamometer cards indicating fluid deficiency, after Martin (1961)

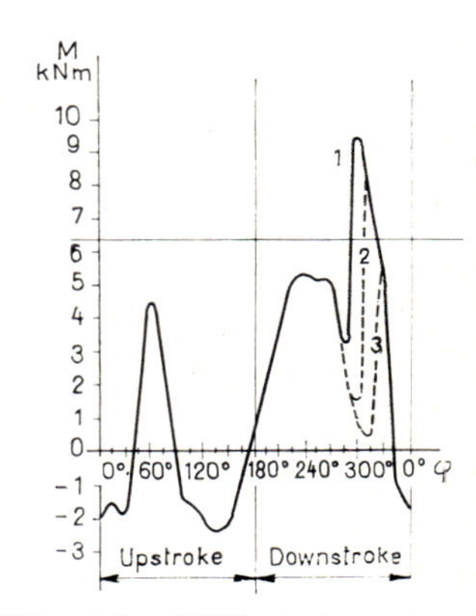

Fig. 4.1—65 Variation of net crankshaft torque over one stroke, after Martin (1961)

main-winding circuit of the prime mover. This control permits us to start pumping when the fluid level has risen to a pre-determined value and to stop it when the well has been pumped off. Of the further units shown in the diagram, *8* is a counter, *9* is a recorder and *10* is an adjusting device.

(f.2) *Fluid pound.* — If the production capacity of a sucker-rod pump exceeds the inflow rate of well fluid into its barrel, then the barrel will not fill up completely with fluid during the plunger stroke. At the beginning of the downstroke, the plunger loaded with the weight of the fluid column in the tubing plus the weight of the rod string suddenly hits the fluid in the barrel after a more or less free drop. This is a dynamic shock as well as a change in static load, and the dynamic forces acting on the plunger are transmitted to the rod string and also to the surface pumping unit. If the forces involved are significant, the phenomenon is called fluid pound. It is particularly harmful to the surface gear reducer and may, even in an apparently well-designed unit, shorten life and lead to early tooth cracking and breakage.

The force of a plunger hitting a gasless well fluid is approximately, according to Juch and Watson (1969),

$$F = 50\, v_p d_p^2 \sqrt{\varrho_l \beta_p}\,. \qquad 4.1-69$$

Gear-reducer load can be readily analysed by examining crankshaft torque.

Figure 4.1—64 shows several dynamometer cards illustrating the 'fluid-deficient' operation of a sucker-rod pump. Curves *1*, *2* and *3* represent possible cases of fluid pound in one and the same well. Determining from these diagrams the peripheral forces on the crankshaft for various shaft positions φ, and calculating torques therefrom, we obtain the diagrams in Fig. 4.1—65. The torque is seen to change significantly and repeatedly during each stroke cycle. Assuming a gear reducer of $M_{al} = 64$ kNm maximum allowable torque, we see this to be exceeded during the downstroke in cases *1* and *2*. No overloading occurs in case *3*. The situation would apparently improve if a larger effective counterbalance were used. Fluid deficiency is not, however, a permanent situation, and thus the balancing required will change from one stroke cycle to the next. — A phenomenon of special interest is the negative torque. Figure 4.1—66a shows a normal case, that is, the gear tooth on the high-speed shaft pushing the gear tooth on the crankshaft. The negative torque (Fig. 4.1—66b) inverts the role of the teeth. Hence, more than once in a stroke cycle, an abrupt snapover of the crankshaft's teeth from the pushed into the pushing position will occur. If the gears are overloaded, this may take the form of a clash cracking or breaking the engaged teeth. Peak loads and load fluctuations may exceed those shown in Fig. 4.1—65 by as much as 50 percent, owing to the dynamic effects left out of consideration here. Hence, in sucker-rod pump installations where the occurrence of fluid pound is anticipated, it is preferable to follow the API procedure of derating the gear reducer by a factor of two. — The phenomenon of fluid pound tends to shorten the life of sucker-rod components, and also of the rod string, primarily because it hastens fatigue. It is liable to cause trouble especially at high pumping speeds.

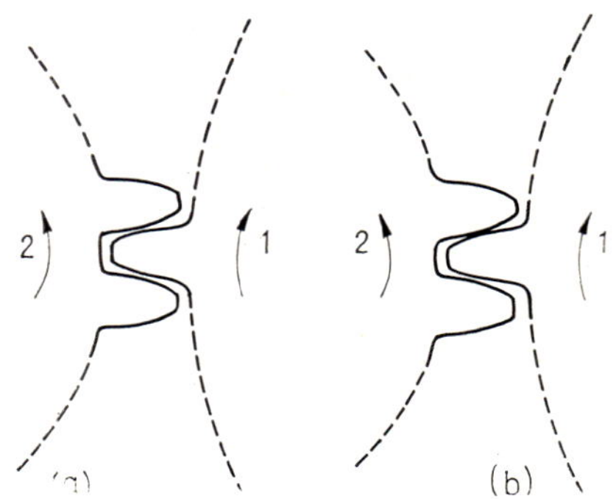

Fig. 4.1—66 Influence of negative torque upon gears, after Martin (1961)

4.1.2. The long-stroke sucker-rod pump

If the production capacity of the sucker-rod pump is to be increased, the options are to increase plunger size, polished-rod stroke and/or pumping speed. A larger plunger size will improve capacity only up to the theoretical maximum, as expounded in paragraph 4.1.1b.1. Increasing pumping speeds is primarily limited by the maximum allowable dynamic load (Eq. 4.1—12). Increasing the polished-rod stroke requires either a significant increase in the dimensions and weight of the surface pumping unit — the weight according to Table 4.1—6 of the Soviet unit 9 SK is 314 kN for a stroke of 4.2 m and a maximum polished-rod load of 196 kN — or a departure from the usual walking-beam drive.

(a) Hydraulic drive

The hydraulic drive permits a significant increase in stroke without too great an increase in structural weight. Figure 4.1—67 shows a hydraulic pumping unit. Its main components are hydraulic cylinder *1*, towering above the wellhead, housing the piston that moves the polished rod; screw pump *2*; drive motor and gear reducer *3*; balance tank *4*, filled with power fluid and gas; a drainage pump; condensate tank *5*, and air accumulator *6*. Operation of the system is explained on the example of an Axelson make hydraulic drive (Fig. 4.1—68). On the upstroke (part a), pump *1* sucks liquid from the balance tank *4* and forces it under the piston in the cylinder. At the upper end of the stroke (part b), pressure above the control valve *2* suddenly increases and forces the valve into the lower position. Reversing valve *3* now rises and changes the direction of fluid flow. On the downstroke, the pump sucks fluid from under the cylinder and forces it into the gas-cushioned balancing tank. The fluid collecting in the condensate tank is likewise delivered to tank *4* by screw pump *6*. When the piston has attained the lower end of the stroke, the control valve rises, the reversing valve sinks and another upstroke begins. Stroke can be adjusted by changing the point where power-fluid conduit *7* enters the cylinder. Pumping speed is determined by the combined duration of the up- and downstroke. The first can be regulated by the rate of flow of power fluid under the piston, the second by correctly choosing the flow resistance in the path of fluid backflow.

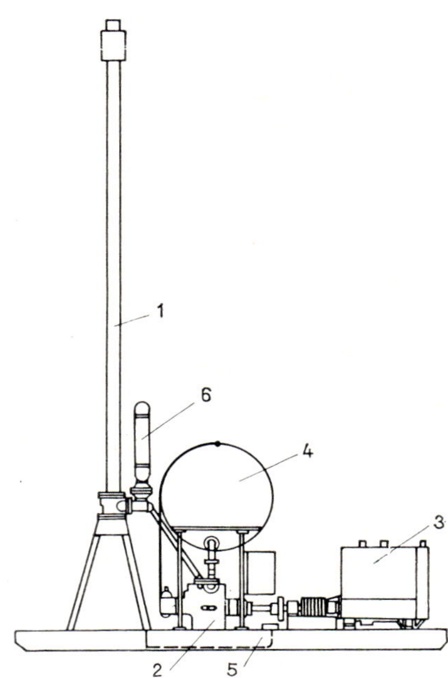

Fig. 4.1—67 Hydraulic long-stroke drive

Table 4.1—16 lists the main parameters of Pelton-made long-stroke hydraulic pumps. It is seen that the weight of even a unit of 9.1 m stroke and 178 kN maximum polished-rod load is only 133 kN, round 42 percent of the walking-beam type unit mentioned above. The longer stroke entails a larger maximum plunger size (cf. Eq. 4.1—41), which further increases production capacity. The longer stroke of the hydraulic drive has extended also the depth range of sucker-rod pumping. Figure 4.1—69 shows the maximum production capacities at various depths of the units listed in Table 4.1—16 ($\varrho_l = 1000$ kg/m^3, $d_r = 25.4$ mm, $\eta_v = 1$). The diagram marked FP shows approximate maximum production rates attainable with walking-beam type pumping units. Comparison of diagrams FP and HP reveals a considerable improvement in production capacity. — The longer stroke limits relative stroke reduction, it ensures a more efficient stroke transfer. In the case of a rod string 2560 m long and a plunger of 1 3/4-in. size, e. g., stroke reduction is 60 percent of polished-rod stroke if the latter equals 1.9 m, and 15 percent if it is 9.1 m. The volumetric-efficiency reduction due to the presence of free gas is also less because the $\Delta s/s$ ratio (cf. paragraph 4.1.1b.2) is more favourable at longer strokes. The better utilization of power is revealed also by the dynamometer card. Figure 4.1—70 shows a typical card of a long-stroke hydraulic unit with $s = 9.1$ m and $n = 2$ min^{-1}. Load is seen to be practically constant during both

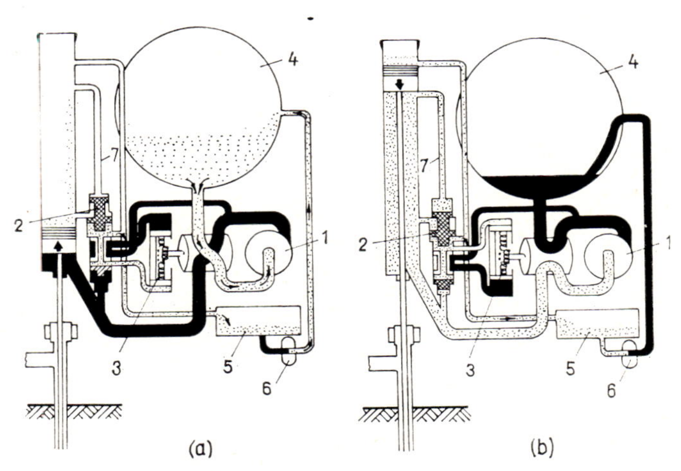

Fig. 4.1—68 Operation of Axelson's hydraulic drive

Table 4.1—16
Typical data of Pelton hydraulic long-stroke sucker-rod pumping units

Type code	F_{max} kN	F_{lmax} kN	F_{emax} kN	s_{max} m	n_{max} 1/min	Weight kN
350-6-10 F-P	67	31	49	3.0	14.4	63
350-7-10 F-P	89	44	70	3.0	12.0	63
350-8-10 F-P	111	53	91	3.0	9.1	63
412-7-20 G-P	89	44	70	6.1	9.1	102
412-8-20 G-P	118	58	91	6.1	7.0	107
412-8.5-20 G-P	133	65	103	6.1	6.3	109
412-9-20 G-P	156	72	116	6.1	5.7	111
512-8-30 H-P	142	107	92	9.1	5.9	130
512-8.5-30 H-P	156	116	105	9.1	5.3	132
512-9-30 H-P	178	130	117	9.1	4.7	133

the up- and the downstroke; the shape of the diagram is near-ideal. The other diagram shown for comparison is that of a walking-beam-type unit (shaded), with $s = 1.9$ m and $n = 10$ min^{-1}. The average upstroke speed is typically the same in both systems. In the walking-beam unit, the load changes significantly over the up- and downstroke; the diagram deviates strongly

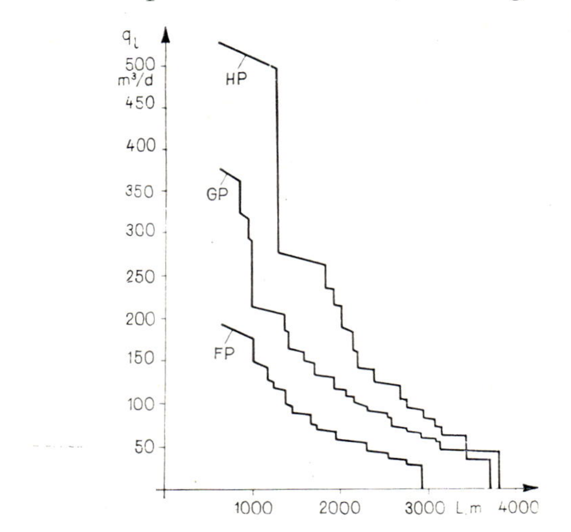

Fig. 4.1—69 Maximum delivery of Pelton's long-stroke hydraulic sucker-rod pumps

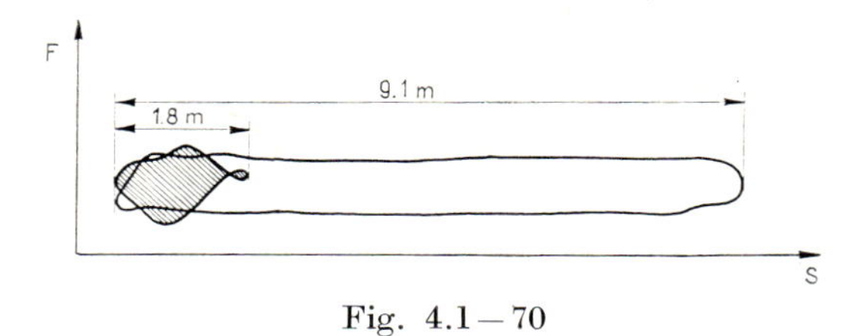

Fig. 4.1—70

from the ideal. In the smooth-running hydraulic drive, dynamic load on the rod string is less; rod life is prolonged; wear of the pump plunger and barrel is reduced; the pump has to be changed less often. Despite the above advantages, long-stroke hydraulic pumps went out of use early in the nineteen-fifties, because the reliability of the drive was found to be unsatisfactory and its maintenance too costly (Metters 1970).

(b) Mechanical drive

The considerable advantages of long-stroke pumping units with hydraulic drive has urged research and design teams to develop a mechanical drive simpler and more reliable than the hydraulic (Metters 1970; Ewing 1970). One of the products of the resulting development work is the Oilwell rig marked 3534. At the top of a derrick, about 16 m high, shown in Fig. 4.1—71, there are two drums turning on the same shaft. The grooves holding the wire-rope windings are of a special design: on the cylindrical part of the drums, groove radii are constant, whereas they decrease along a spiral on the conical drum portion. One drum takes the wire-rope hanging the rod string; the other takes the wire-rope from which depends a counterweight moving up and down in the derrick. The drum shaft is rotated through a gear reducer by an electric motor, likewise installed at the derrick top. This type of drive permits a very economical utilization of motor power. Figure 4.1—72 shows three positions during the upstroke. In part (a), rope *1* connected to the rod string is in contact with rod-guide *2;* its lever arm is zero. The lever arm of counterweight *3* equals the drum radius. Early in the upstroke, an acceleration results purely from the difference in lever arms. As soon as this acceleration peaks out, the motor cuts in and further accelerates the rod string. After having attained a pre-determined top speed, the rod string continues to rise at that speed, while, according to part (b) of the Figure, the lever arms of rod string and counterweight are equal. About 1.8 m before the end of the upstroke, the lever arm of the counterweight begins to decrease, so that the relative torque of the rodstring side increases. The drive motor switches off and the rod string arrives with a smooth deceleration at the upper stroke end, shown in part (c) of the Figure. This solution provides uniform lift speed over 80 percent of the upstroke. Figure 4.1—73 shows the arrangement and suspension of the wire-rope in the spiral groove on the

Fig. 4.1—71 Oilwell's 3534-type mechanical-drive long-stroke sucker-rod pump

Table 4.1—17

Typical data of Oilwell 3534 long-stroke mechanical-drive pumping units

Type code	F_{max}	F_{cmax}	s_{max}	n_{max}	P	Weight
	kN	kN*	m	1/min	kW	kN
3534-75	156	111	10.4	5	56	145
3534-100	156	111	10.4	5	75	156

* Counterbalance effect

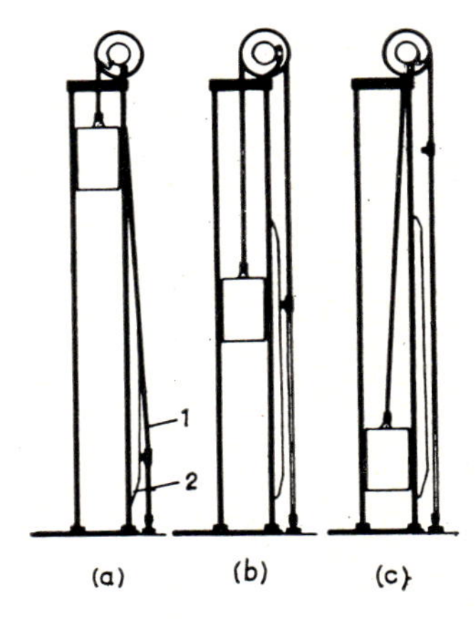

Fig. 4.1—72 Operation of mechanical long-stroke drive, after Ewing 1970 (presented at the 41st Annual California Regional Meeting of the Society of Petroleum Engineers of AIME, in Santa Barbara, California, October 28—30, 1970)

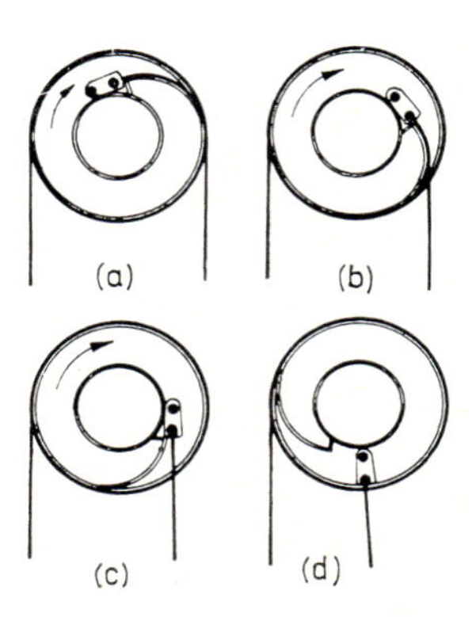

Fig. 4.1—73 Winding and unwinding of suspension rope, after Ewing 1970 (presented at the 41st Annual California Regional Meeting of the Society of Petroleum Engineers of AIME in Santa Barbara, California, October 28—30, 1970)

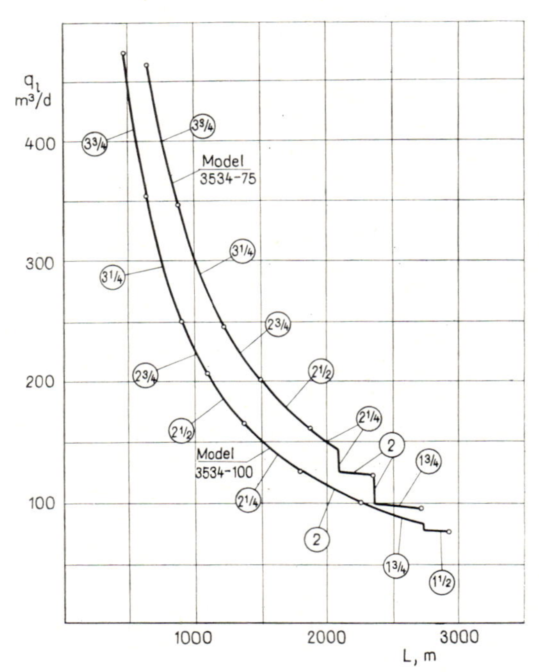

Fig. 4.1—74 Maximum delivery of Oilwell's long-stroke mechanical-drive sucker-rod pumps

drum in various situations. — The data of the drive unit, which has so far been developed in two sizes, are contained in Table 4.1—17. The capacity of the units is given in Fig. 4.1—74 on the assumption of a 90-percent volumetric efficiency. The Figure also shows recommended plunger sizes. — The *Flexirod* has been combined to advantage with the long-stroke mechanical sucker-rod pump drive (Snyder 1970).

4.1.3. Selective sucker-rod pumping of multiple completions

The drive towards reduced production costs has, from the nineteen-fifties on, promoted sucker-rod pumping techniques which permit the selective production of two, more seldom three zones opened in one and the same well. The relevant pumping methods fall into three groups: selective pumping by tandem sucker-rod pumps, by double-horsehead pumps, and by two separate sucker-rod pumps.

(a) Tandem sucker-rod pumps

The tandem pump is a pair of sucker-rod pumps, one below another, driven by a common rod string (see Boyd 1960.). Each pump receives a separate inflow from a separately developed zone, and the produce of each pump is delivered separately to the surface. Dual-zone arrangements fall

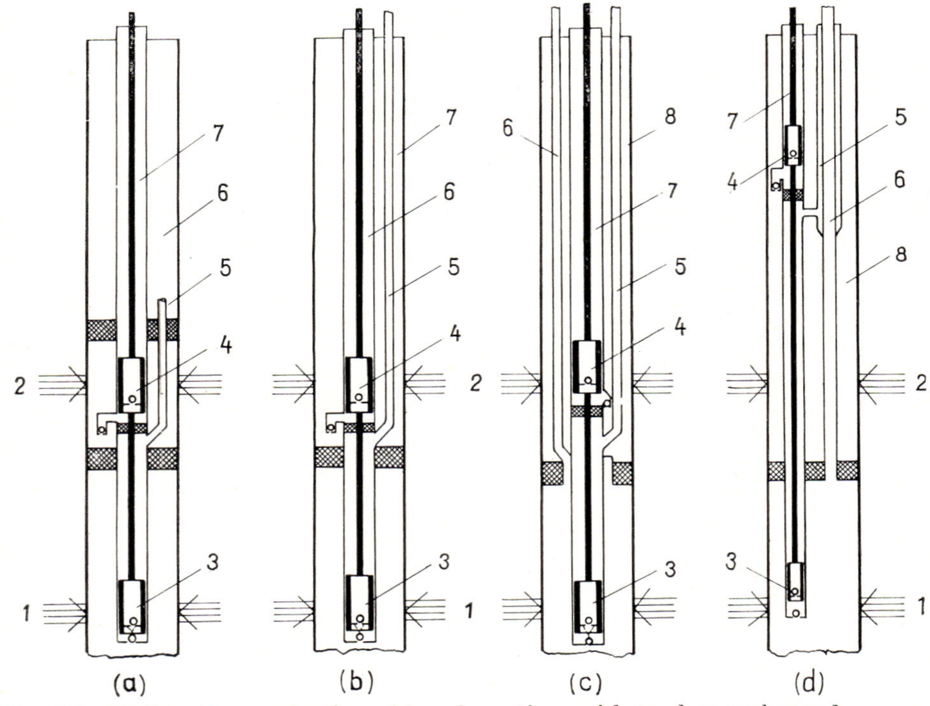

Fig. 4.1—75 Selective production of two formations with tandem sucker-rod pumps, after Boyd (1960)

Table

Operating data of selective production of two

Data of lower pump							
					Barrel		
Max. stroke	Max. torque	Motor power	Setting depth	Nominal size	Length	I. D.	Gross production
m	MNm	kW	m	in.	m	mm	m³/d
1.37	13	11	1365	2	3.7 4.9	31.7	1.4
1.88	18	11	1760	2	3.7 4.9	31.7	13.9
1.88	18	11	1877	2	3.7 4.9	31.7	2.4
1.63	26	29	2614	2	3.7 4.9	31.7	5.7
4.88	73	52	2620	2	7.9 9.1	31.7	5.7
4.88	73	52	3228	2	7.9 9.1	31.7	24.3

into four groups, whose basic principles are outlined in Fig. 4.1—75. In solution (a) pump *3* producing zone *1* delivers its fluid through conduit *5* and annulus *6*. Zone *2* is produced by pump *4* through tubing *7*. This solution is cheaper than the rest of the options. It has the drawback that gas in the individual zone fluids cannot be separated before entering the pumps, and thus pumping at a favourable volumetric efficiency is restricted to gas-less or low-*GOR* wells. Solution (b) differs from the above in that the fluid of the lower zone is led by the conduit *5* up to the surface, which makes the annulus available for removing the gas of the upper zone. Thus the upper zone can be produced at a satisfactory volumetric efficiency even if it is gaseous. This solution is costlier than the one in (a). In solution (c) the fluid of zone *1* is produced through conduit *5* by pump *3*, whereas the gas of this same zone reaches the surface through conduit *6*. Both zones can be produced at a satisfactory volumetric efficiency, even if their fluids are gaseous. This solution is costlier than both foregoing ones. Solution (d) is a variant of solution (c), with the liquid and gas of zone *1* being produced through a pair of coaxial tubing strings, which will fit into a production casing of somewhat smaller size. In all solutions, one of the principal aims is to make running and pulling as simple and troublefree as possible. The combined costs of pump changing, work-overs and paraffin removal are about twice higher than the corresponding costs in a conventional single completion. Design details of solution (b) are shown in Fig. 4.1—76. In wells less than 1800 m deep, completed with a production casing of at least 5 1/2-in. size (125.7 mm *ID*), conventional tubing sizes can be used. Tubing *1* is of 2 3/8-in. size: it takes a rod string made up of 7/8- and 3/4-in. size rods. Tubing *2* is of 1-in. size and can be run separately. It can be attached by means of hold-down *3* to cross-over piece *4*, in whose bore packer *5* ensures

4.1—18

zones with tandem sucker-rod pumps (after Boyd 1960)

	Data of upper pump						
Water content	Setting depth	Nominal size	Barrel		Gross production	Water content	Pump spacing
			Length	I. D.			
%	m	in.	m	mm	m³/d	%	m
47	1344	2	4.9	38.1	10.8	25	21
65	1739	2	4.9	38.1	14.0	25	21
23	1309	2	3.7 4.9	38.1	15.9	16	568
25	2583	2	4.9	38.1	4.9	26	31
—	2590	2	8.5	38.1	3.5	5	30
20	3198	2	8.5	38.1	2.9	19	30

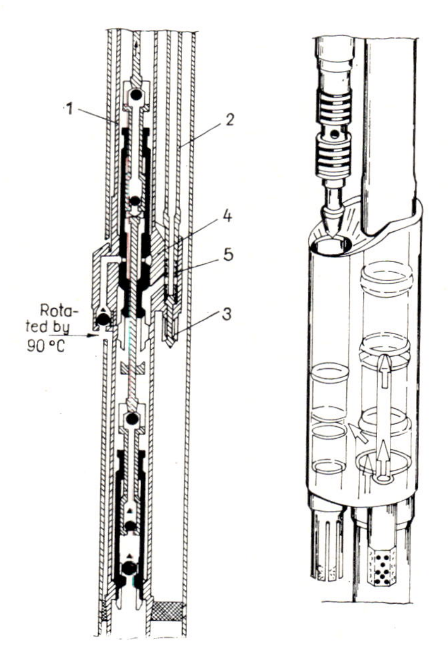

Fig. 4.1—76 'b' arrangement, after Boyd (1960)

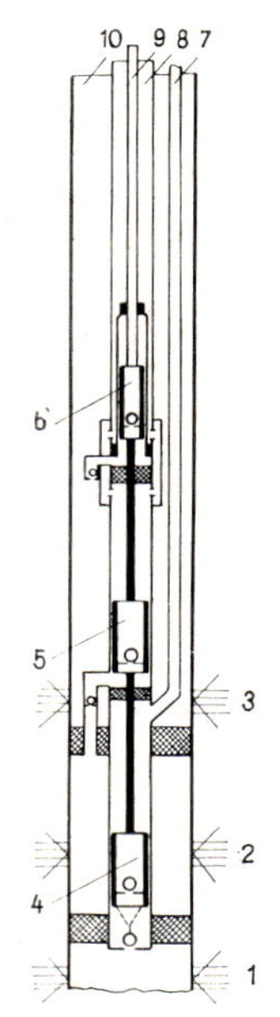

Fig. 4.1—77 Selective production of three formations with tandem sucker-rod pumps, after Boyd (1960)

a leakproof pack-off. In wells deeper than 1800 m, the higher fluid load demands a stronger rod string. The string is usually composed of rods of 1, 7/8 and 3/4-in. size. In the section of 1-in. rods, 2 7/8-in. size tubing is required. Since in deeper wells the casing is thicker-walled, too, the normal 1-in. size tubing will not pass, so that special tubes with slim couplings have to be used. The main operating parameters of wells fitted with tandem sucker-rod pumps of this type are given in Table 4.1—18. The two zones are likely to have different fluid inflow rates. The pumping speeds of the two pumps in tandem are equal, and their strokes differ only inasmuch as stretching of rod strings and tubings is different. The only parameter that can be varied to affect production capacity is the plunger diameter, and even that only within rather narrow limits set by well completion and pump design. The pump producing the lower-capacity zone is therefore often oversized with respect to the rate of inflow. The plunger entering the unfilled barrel may give rise to fluid pound harmful to the entire installation (cf. paragraph 4.1.1f.2). As a prevention, the operating point is sometimes designed with the lower-capacity zone in mind. — It is sometimes necessary to stop pumping one zone while continuing to produce the other. In solution (b), this is achieved for the upper zone by letting the oil produced flow back in the annulus. To solve the same problem for the upper zone, sucker-rod pumps provided with a special travelling valve are used. As long as the tubing head is open, the pump produces in the usual fashion. If, however, the tubing head is shut off, increased pressure keeps the travelling valve permanently open, thus idling the pump. — All the described solutions have the advantage of permitting the selective pumping of two zones at comparatively low cost. Their main drawbacks are that (i) the capacities of the two pumps cannot be varied independently, except within rather narrow limits; (ii) if inflow from one zone is too sparse, or a gas lock develops in the pump barrel for any other reason, fluid pound may occur and result in rod-string failures; (iii) the production of both formations must be stopped for pump changing.

It was in 1959 that a tandem pump was first used to selectively produce three formations. The principle of the solution is shown in Fig. 4.1—77. The production casing is of 5 1/2-in. size; hollow-rod string *9* moves in 2-in. tubing *8*. Tubing *7* is of 1-in. size. The fluid of zone *1* enters pump *4* and reaches the surface through tubing *7*. Zone *2* is produced by pump *5* through tubing annulus *8*. The liquid of top zone *3* is produced by pump *6* through hollow-rod string *9*; and the gas from this zone rises through annulus *10*.

(b) Double-horsehead pumping units

The well completion features two independent strings of tubing, two rod strings and two sucker-rod pumps. The producing zones are separated by a packer. The surface pumping unit is of the walking-beam type with a special horsehead hanging both strings of rods. Both pumps are thus driven by the same prime mover. The main advantage of the solution as compared to the tandem pump is that pumping one zone can be stopped quite independently of the other one. It is, however, much more expensive, as it needs more pipe and more rods.

(c) Two pumping units

The most elastic method of selectively producing two zones is undoubtedly the one involving two entirely independent sucker-rod pumping units. One relevant Salzgitter arrangement is shown in Fig. 4.1—78 (Graf 1957b). The annulus is packed off above zone *1*. The bore of the packer is provided with a check valve. The entire fluid of zone *1* is produced by pump *3* through tubing *5*. The fluid from zone *2* is produced by pump *4* through tubing *6*, whereas the gas from zone *2* is removed through annulus *7*. This arrangement will, of course, work satisfactorily only if the fluid of the lower zone is comparatively gasless.

In practice, 2 3/8-in. tubing is run in a 7-in. production casing. The tubing is of the plain-end type, as external-upset tubes would not go in the casing. In wells close to 2000 m deep, however, the tubing producing the upper zone may under unfavourable conditions (low fluid level in the annulus, high water content, tubing filled with liquid, rod-string load transferred to tubing e.g. during setting) be exposed to a very high tensile stress, so that pipe made of high-strength steel N 80 is employed. In order to facilitate running or pulling, the edges of the couplings are turned down on both tubing strings. One possible wellhead design is shown in Fig. 4.1—79. Both tubings are

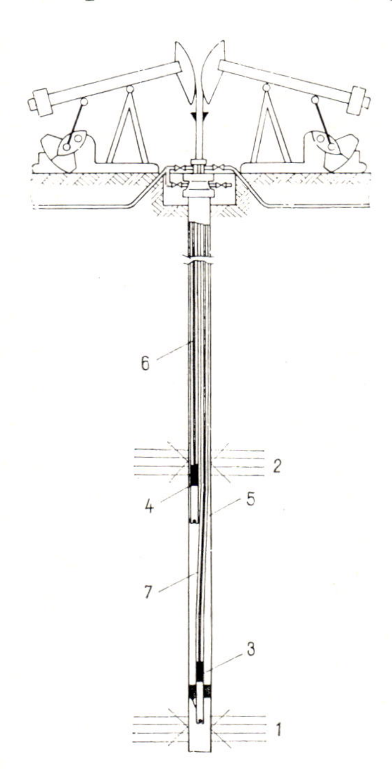

Fig. 4.1—78 Selective sucker-rod pumping with independent pumps

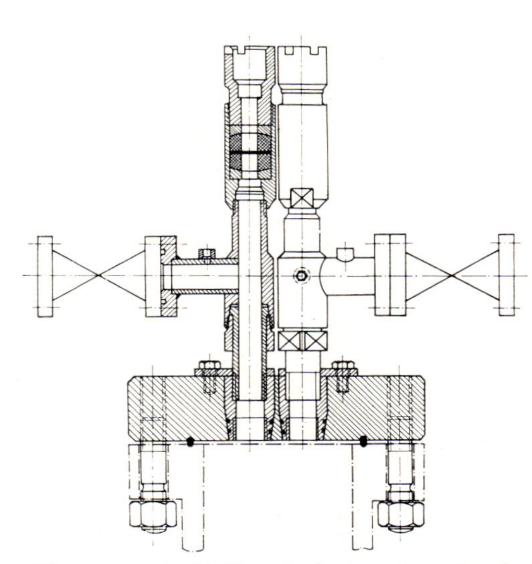

Fig. 4.1—79 Wellhead design for selective pumping with two independent sucker-rod pumps, after Graf (1957b)

hung in the tubing-head by means of mandrel hangers provided with O-ring seals. Tubing axes are deflected far enough to make the least distance between horseheads equal 40 mm. Carrier bars and polished-rod clamps are designed so as to prevent the rod-string hanging equipment from getting entangled. Repair and maintenance costs of producing a single well with two independent pump installations exceed but slightly the cost of producing two separate wells with independent sucker-rod pumps. The saving

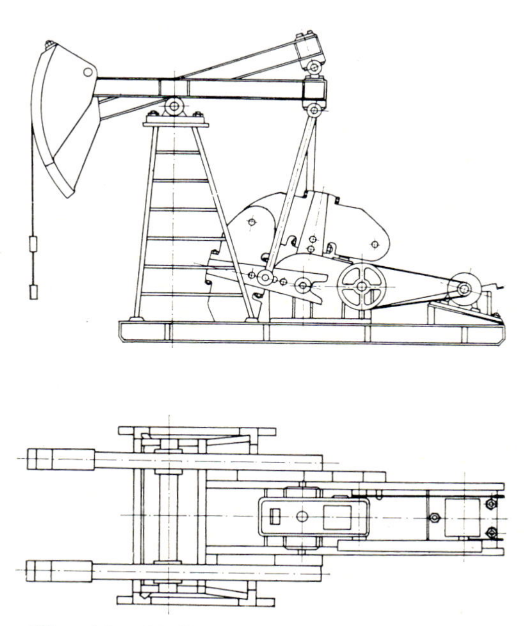

Fig. 4.1—80 Soviet drive unit suited for the production of bunched wells

is thus essentially the investment cost of one well. This well completion has its advantages even if one zone is flowing, and only the other one has to be pumped. Running and pulling is performed under protection of a blowout preventer, especially if the upper zone is flowing.

In the Soviet Union, twin wells drilled side by side are produced in the way illustrated by Fig. 4.1—80 (after Geyman and Gadiev 1960). The pumping unit has a single gear reducer for two separate crank-balanced walking beams. The two cranks are at right angles. This arrangement keeps the torque almost constant, and thus the utilization of prime mover and gear reducer are more efficient than in the conventional set-up. Its drawback is that both zones have to be produced together. Pneumatically balanced double walking beam units based on the same idea are also being built. In principle, this solution can also be employed to selectively produce one and the same well.

(d) Sucker-rod pumping of slim holes

In wells of comparatively small capacity, slim-hole or midi (minimum diameter) completions are often advantageous. Such wells are cased with tubing-size pipes (3 1/2- and preferably even 2 7/8-in. size). This permits to grout into a bore-hole of suitable size several casings of say, 2 7/8 in. size, each of which is to tap one of several zones. The midi completion can thus

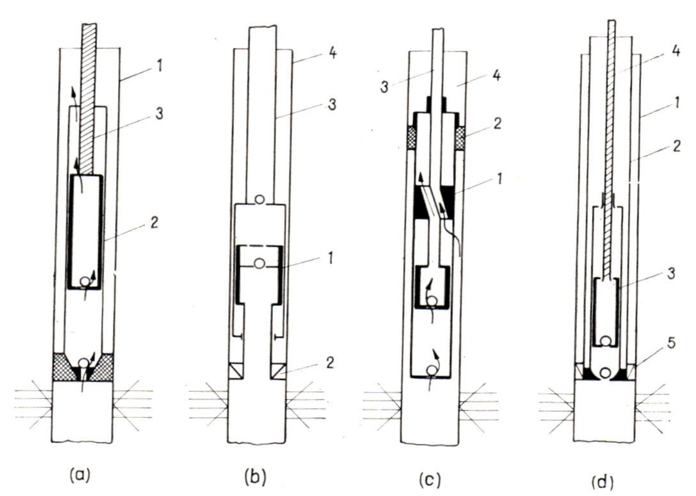

Fig. 4.1—81 Bottom-hole arrangements for the sucker-rod pumping of slim holes, after Corley and Rike (1959) and Crosby (1969b)

be regarded as selective only in the sense that the slim casings, perfectly independent actually, are run in a single common borehole. A survey (Crosby 1969a) found that completion of midi wells produced by mechanical means cost less by 21 percent than that of normal-size wells; 9 percent of the 21 was due to cheaper tubing and wellhead equipment.

Figure 4.1—81 has been compiled after Crosby (1969a) and Corley and Rike (1959). In solution (a), a rod pump fixed to the casing is used. The maximum nominal size of the pump *2* going into the 2 7/8-in. size tubing *1* is 2 in. Size of rod *3* is not restricted. Production capacity is comparatively large. This arrangement has the drawback that it can be employed only if the *GOR* of the well fluid is rather low, and the casing is not exposed to corrosion or erosion. Special rod couplings causing no wear on the casing are to be preferred. Solutions (b) and (c) employ hollow rods, usually of 1 1/4-in. size. Both have the advantage that gas can be produced separately and the volumetric efficiency of pumping is thus better than in (a). In (b), a travelling-barrel pump *1* is used; it is held down by an anchor *2* that does not pack off the annulus. Liquid rises in the hollow rods *3;* gas rises in annulus *4*, between casing and hollow-rod string. Solution (c) employs a special pump *1* fixed to anchor packer *2* that packs off the annulus. It is the gas that rises in the hollow rods and the liquid that rises in the annulus. The advantage of the arrangement is that, here, the liquid flows in the channel of lower flow resistance. Both solutions have the drawback of lower production capacity than (a), also the rod string may cause considerable wear of the casing. Solution (d) is essentially a conventional rod-string arrangement reduced in size. Tubing *2* of 1 1/4- or 1 1/2-in. size is run in casing *1*. Rods *4* are of 1/2- or 5/8-in. size. Maximum pump size is 1 1/4 in. The production capacity of this solution is comparatively small and the completion is comparatively expensive. Its advantage is that the gas can be separated while still in the well, and the rod string will not rub against the casing. The tubing is usually fixed to the casing by means of anchor *5*.

4.1.4. Rodless bottom-hole pumping

All types of sucker-rod pump have the common feature that rod stretch due to the changing fluid load, and hence stroke reduction, increases with depth. As a result, the production capacity of sucker-rod pumps decreases significantly at greater depth. Setting depth is further limited by maximum allowable rod stress. As is shown also in Fig. 4.1 – 74, the maximum setting depth of the sucker-rod pumps used today is about 3000 m, and the maximum feasible production rate out of that depth is about 80 m³/d. — In a rodless bottom-hole pump, not only is the pump at the well bottom, but so is most often also the pump-drive unit. Production capacity in such an arrangement is less depth-dependent than in the sucker-rod pump. The allowable setting depth is greater in some of the types. Lacking a rod string that could rub against the tubing, the rodless pump is at an advantage in inclined wells.

(a) Hydraulic bottom-hole pumps

The common feature of hydraulic bottom-hole pumps is that the pump unit anchored to the tubing shoe incorporates a hydraulic engine whose piston reciprocates in the vertical; to the other end of its piston rod, one or two pump plungers are rigidly attached.* The piston is moved by a power fluid supplied to it from the surface through a separate conduit. The pressure of the power fluid in the surface pumping unit is, in the case of a single plunger,

$$p = L\,\bar{\gamma}_l \frac{A_p}{A_m} + p_s + p_{TO}\,. \qquad 4.1-70$$

A_p is the cross-sectional area of the plunger, A_m is that of the piston; p_s is the fluid friction loss in the entire flow string. The power-fluid supply rate required to drive the pump is

$$q_{lp} = \frac{q_l A_m}{\eta A_p}\,, \qquad 4.1-71$$

* We shall consistently term 'piston' the piston in the engine and 'plunger' the piston in the pump.

where η is the overall efficiency of the bottom-hole pump installation (0.65 on an average), and q_l is the liquid production rate. — If the piston diameter is larger than the plunger diameter, that is, $A_m/A_p > 1$, then, firstly, by Eq. 4.1—70, a lower power-fluid pressure p will do at a given well depth L, and secondly, by Eq. 4.1—71, power-fluid supply rate q_{lp} must exceed liquid production rate q_l. Such arrangements are chosen for deep wells of low productivity. If, on the other hand, $A_m/A_p < 1$, then a higher p is required for the same well depth, and the power-fluid supply rate will be less than the liquid production rate. This arrangement is therefore to be preferred in wells of smaller depth and higher productivity.

The overall efficiency of the pumping installation is significantly reduced by hydraulic losses. These fall into two groups: friction loss and slippage. Friction loss is composed of the sliding-fit friction of the mechanical parts and of the fluid friction of flow. A precondition of operations planning is to know beforehand the friction losses of both the power fluid and the well fluid. Calculation of these is somewhat hampered by the fact that the temperatures of both fluids are affected by well depth as well as by the duration and rate of pumping. At the temperatures prevailing in the well, flow properties of both the power fluid and the well fluid may be regarded as Newtonian. Calculation is usually performed with the aid of simplified formulae, diagrams or nomograms (Coberly 1961).

(a.1) *Double-acting hydraulic pumps.* — The double-acting hydraulic bottom-hole pump is the earliest type of rodless plunger pumps. Kobe make double-acting pumps were first employed commercially in 1932 (Coberly 1961). — Figure 4.1—82 shows a standard Kobe bottom-hole pump. Piston *1* is forced downwards by power fluid fed to it through conduit *2*. At the lower end of the stroke, engine valve *3* rises so that the power fluid can now force the piston upwards. The spent power fluid flows through annulus *4* to port *5* and rises to the surface through casing annulus *6*. Plunger *7*, rigidly attached to the piston, sucks fluid from space *8* and delivers it into annulus *4*. In this set-up, the well fluid mixes with the power fluid in annulus *4* and rises together with it through casing annulus *6*. — Figure 4.1—83 shows a so-called B-type Kobe pump for producing high-capacity wells. This is a more recent design, improved by the use of more modern means of sealing. This permits piston and plunger sizes to be greater for a given overall pump diameter. The pump operates essentially in the same way as the standard pump. Seal rings are marked *1*.

There are various combinations of pump and well completion, including, on the one hand, (i) bottom-hole insert or tubing pumps run on a pipe string, and (ii) free pumps, and, on the other (i) open systems in which the exhaust fluid from the engine is mixed with the discharge from the pump, and (ii) closed systems. — Figure 4.1—84 shows designs of *open-system* completions. The common feature of these is that the spent power fluid and the well fluid get mixed in the well, and rise in one and the same flow channel. In solutions (a) and (b), the pump can be run and pulled on a macaroni tubing string; (a) has the advantage over (b) that it permits the production of also gaseous fluids; its drawback is that it needs more pipe; (c) and (d) are *free-pump types*; (d) is shown in more detail in Fig. 4.1—82. The power fluid attains the pump through a large-diameter tubing, and the spent power fluid plus the well fluid rise through a smaller-size tubing in (c) and through the casing annulus in (d). The pump is installed by placing it into the tubing at the wellhead and letting it sink to its setting depth. For recovery, power-fluid flow is reversed; the fluid underflowing the pump then raises the unit to the surface. In types (c) and (d), the standing valve

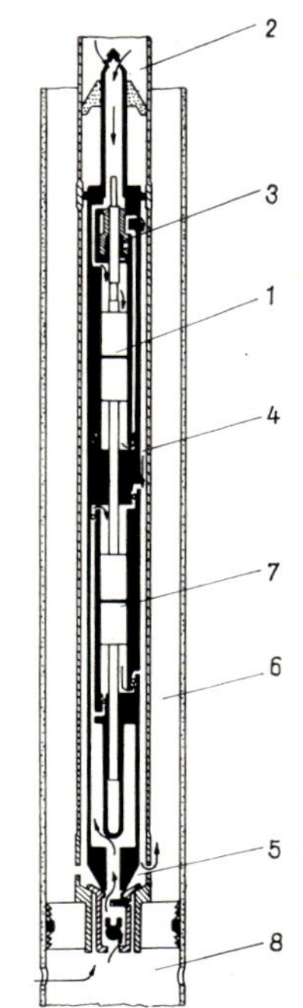

Fig. 4.1—82 Kobe's conventional hydraulic bottom-hole pump

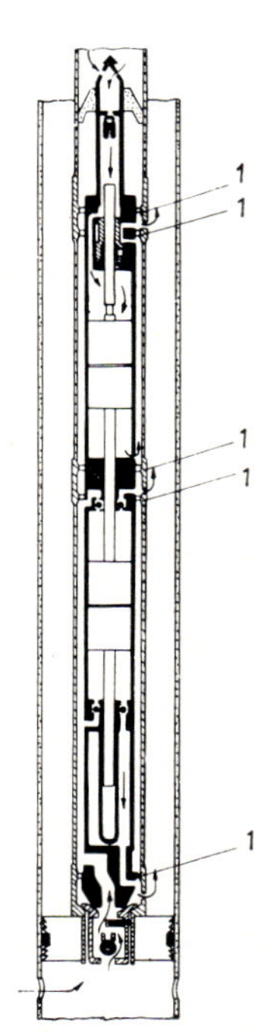

Fig. 4.1—83 Kobe's B-type hydraulic bottom-hole pump

is permanently open in normal operation. It is assisted in staying open by a magnet placed above the ball. The valve closes if the power-fluid circulation is reversed for the purpose of pump recovery. The free pump is thus very easy to change; changing it requires none of the usual work-over operations. The advantage of solution (c) over (d) is that it can also be used to produce gaseous well fluids; its drawback is that it needs more pipe, which makes it more expensive.

The closed system differs from the types just described in that the power fluid and well fluid are kept separate during their upward travel. In the solution shown as Fig. 4.1—85, the spent power fluid rises through annulus *1*, whereas the well fluid rises through tubing *2*. The solution shown is just an example; there is a variety of other solutions including closed systems

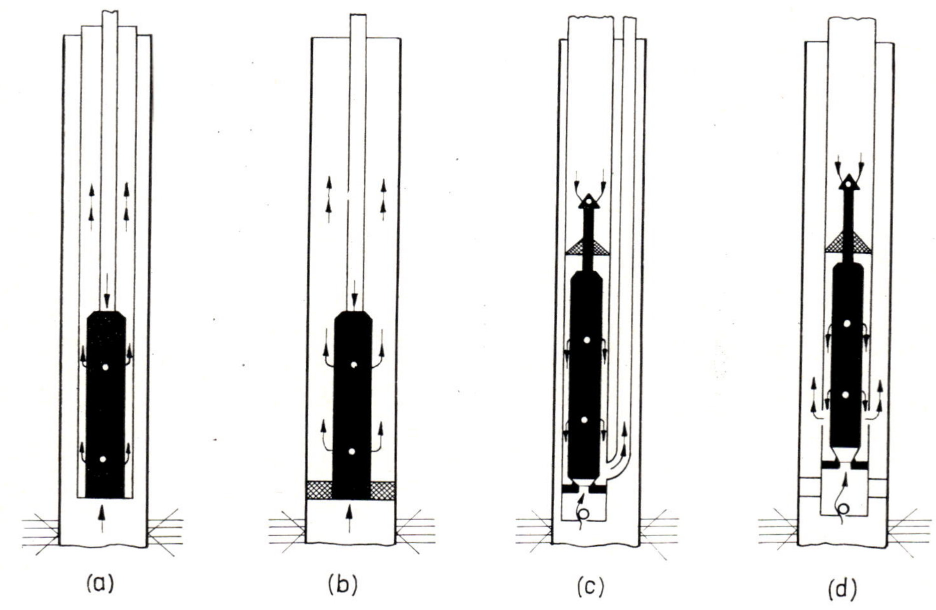

Fig. 4.1—84 Completions involving Kobe's bottom-hole pumps, producing mixed oil

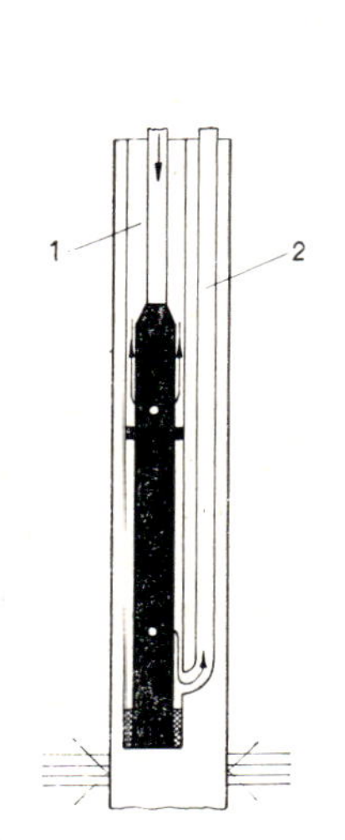

Fig. 4.1—85 Closed type completion involving Kobe's bottom-hole pump

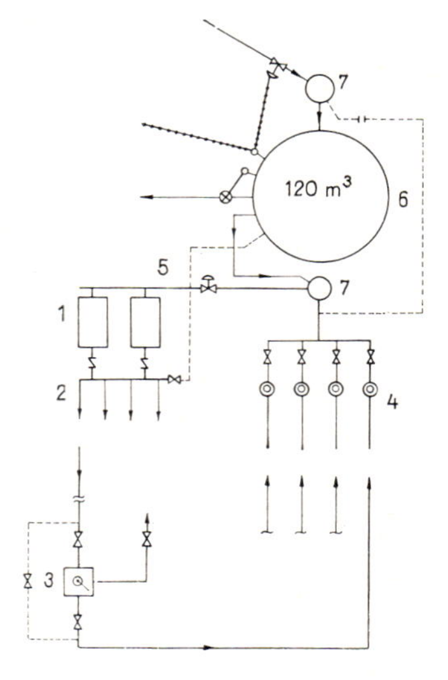

Fig. 4.1—86 Closed power-fluid system, after Bowers (1970)

with free pumps. Table 4.1—19 gives sizes and operating parameters of some typical Kobe pumps after catalogues issued in 1968–69. The production capacity of the pump installation can be further increased by installing two pumps in one well and operating them in tandem. In the Reno field, USA, three 2 1/8×2 1/2 pumps are used to produce 540 m^3/d of oil from a depth of 4604 m. In 1968, this was the rodless hydraulic bottom-hole pump operating at the greatest depth (Hollis 1968). — It is essential that the power fluid be pure; if paraffin deposition is to be anticipated, it should also be warm. The power fluid most often employed is simply a purified and heated well fluid. Earlier, each well used to be equipped with a separate heater-treater. The elasticity and economy of this system was unsatisfactory, however; power-fluid treatment is more and more centralized nowadays, and central pump stations tend to be installed in most fields using this production method. The power-fluid distribution system much resembles the lift-gas distribution systems of gas-lifted wells. Figure 4.1—86 shows a closed power-fluid system after (Bowers 1970). Power fluid is pumped by triplex pumps 1 through distribution line *2* to wells *3*. Spent fluid returns through lines *4* to the intake manifold of the triplexes. Tank *6* stores makeup to the pump intake. The devices marked *7* are gas boots. — In the Wasson field of Western Texas, a closed system using water as the power fluid is employed in order to reduce the fire hazard in this rather densely populated region. The power water returning from the well is filtered on a glass filter: the water thus cleaned of solid contaminations and mixed with an inhibitor is then recycled. Operating costs have turned out to be lower than in systems using power oil, but pump life is shorter. The overall maintenance costs of power-oil and power-water systems are about equal (Bowers 1970).

The tubing strings installed in the well are cleaned by means of a plug of suitable form, made of a special material; the plug is pumped down to the bottom-hole pump, where it dissolves in the hot oil. Scraping the annulus is performed in certain wells by mounting scraper blades on the outside of the tubing, and rotating it occasionally during operating stoppages. If the power oil is hot enough, no paraffin will deposit out of it, nor out of the well fluid. The system actually employed is selected on the basis of economic considerations. No paraffin deposition will occur if the pipes are provided with a paraphobic plastic lining.

(a.2) *Single-acting bottom-hole pumps.* — The operating principle of the single-acting hydraulic pump is the same as that of the sucker-rod pump. The plunger lifts the well-fluid column in the tubing on the upstroke, and sinks in the fluid sucked into the barrel during the downstroke. The only difference is that the plunger is moved by a hydraulic engine installed directly above the pump, rather than by a rod string driven from the surface. Single-acting pumps were called into life in order to improve upon the poor volumetric efficiency of short-stroke double-acting pumps when pumping fluids even of comparatively slight gas content. Single-acting pumps are usually simpler in design than double-acting ones, and are consequently less sensitive to sand, too. The stroke and plunger size of a single-acting pump are usually greater, but its pumping speed is less than that of the double-acting pump. The operating parameters of the two types of pump

Table 4.1 – 19

Sizes and operating parameters of Kobe pumps

Type	Pump size*	Length m	Stroke m	Area ratio A_m/A_p	Rated speed** spm	Theoretical capacity m^3/d	Max. setting depth km	Number of all sizes
Small bore	1 1/4 × 1 × 7/8	2.6	0.38	1.268	82	18	3.8	2
Standard single pump end	2 × 1 × 13/16	2.0	0.30	1.834	91–121	16–22	4.6	22
	2 1/2 × 1 1/4 × 1	2.7	0.46	1.923	75–100	31–41	4.6	
	3 × 1 1/2 × 1 1/4	3.5	0.61	1.686	65–87	58–77	4.6	
	4 × 2 × 1 3/4	4.4	0.76	1.449	58–77	133–176	4.5	
Standard double pump end	2 × 1 3/16 × 1 × 1	3.4	0.30	0.775	91–121	61–81	2.3	15
	2 1/2 × 1 7/16 × 1 1/8 × 1	4.4	0.46	1.129	75–100	74–99	3.4	
	3 × 1 3/4 × 1 1/4 × 1 1/4	5.7	0.61	1.249	65–87	116–155	3.8	
	4 × 2 3/8 × 2 × 1 3/4	7.2	0.76	0.914	58–77	327–433	2.8	
'B' single pump end	2 × 1 3/8 × 1 3/16	2.5	0.30	1.414	91–121	47–63	4.3	8
	2 1/2 × 1 3/4 × 1 1/2	3.2	0.46	1.439	75–100	93–124	4.4	
	3 × 2 1/8 × 1 7/8	4.0	0.61	1.339	65–87	171–228	4.1	
	4 × 2 7/8 × 2 1/2	5.0	0.76	1.383	58–77	339–450	4.2	
'B' double pump end	2 × 1 3/8 × 1 3/16 × 1 3/16	3.3	0.30	0.707	91–121	94–125	2.2	12
	2 1/2 × 1 3/4 × 1 1/2 × 1 1/2	4.7	0.46	0.719	45–100	186–248	2.2	
	3 × 2 1/8 × 1 7/8 × 1 7/8	5.3	0.61	0.669	65–87	342–457	2.0	
	4 × 2 7/8 × 2 1/2 × 2 1/2	6.6	0.76	0.692	58–77	678–900	2.1	

* The figures denote: tubing *ID* × piston *OD* × first pump *ID* × second plunger *OD*
** Can be operated below rated speed

thus exhibit a fairly wide overlap. Plunger sizes are $d_p > 2$ in., stroke range is $s = 0.4-1.7$ m, and pumping speeds range from 6–30 $\min^{-1}$.

A typical single-acting pump is the Byron—Jackson-make Hydralift, shown as Fig. 4.1—87. During the downstroke, travelling valve *2* of plunger *1* is open; standing valve *3* is closed. Reversing valve *5* in piston *4* is open, so that power-fluid pressure may act on the top of the piston also through conduit *6* and bore *7*. Since the top area of the piston is greater than its bottom area, the resultant of the pressure forces acting on the piston is down: the plunger will, of course, copy piston motion. On attaining the lower end of the stroke, reversing valve *5* moves into the lower position (part b of the Figure). Power-fluid pressure now makes the piston rise. The power fluid above the piston enters through bore *8* above the plunger, from where, after mixing with the well fluid, it flows into tubing *9*. During the upstroke, the standing valve is open and the travelling valve of the plunger is closed. The Hydralift, just as the Kobe pump, can be tubing-run or used as a free pump (cf. Fig. 4.1—84c). Nominal pump sizes are from 2 to 5 1/2 in.; the stroke is 1.53 m; production capacity ranges from 24 to 2400 m³/d; maximum setting depth is 4600 m. — The operation of the single-acting hydraulic pump is supervised by means of the hydraulic dynamometer recording pressure changes of the power fluid near the wellhead (Woods 1961). These changes

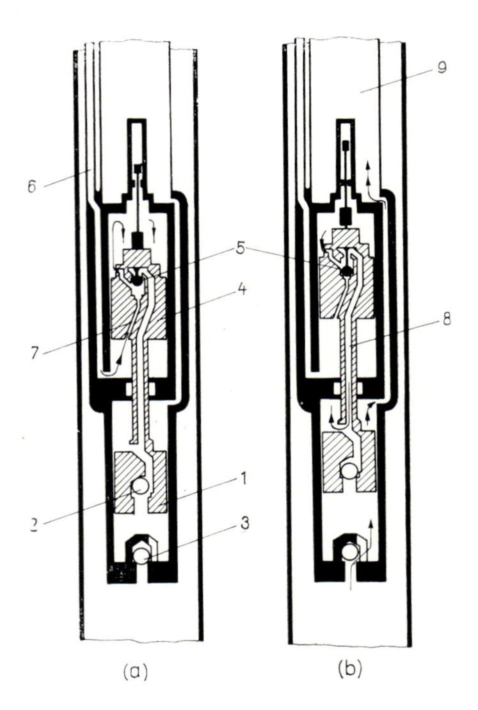

Fig. 4.1—87 Byron—Jackson single-acting hydraulic bottom-hole pump

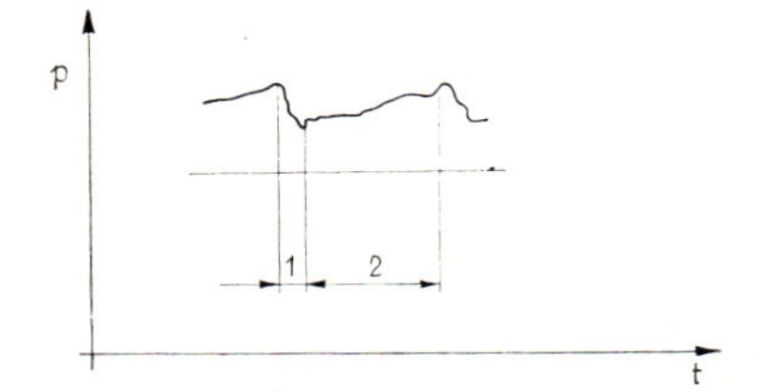

Fig. 4.1—88 Variation of power-fluid pressure at wellhead, after Woods (1961)

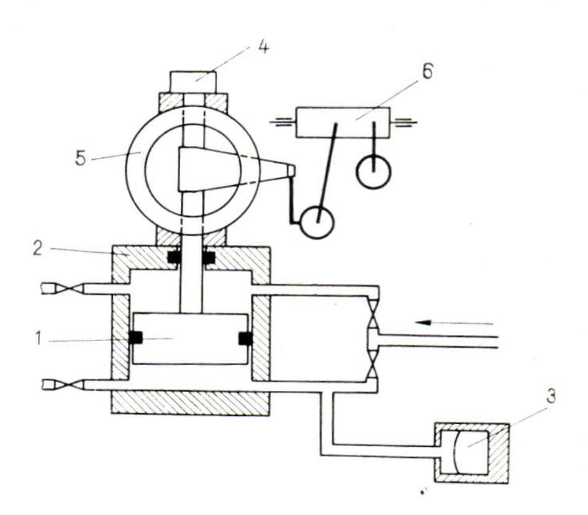

Fig. 4.1—89 Hydraulic dynamometer, after Woods (1961)

permit the determination of first of all the duration of the upstroke (*2* in Fig. 4.1—88) and downstroke *1* and the detection of fluid deficiency. The principle of the hydraulic dynamometer is illustrated in Fig. 4.1—89. The two faces of piston *1* are exposed to equal fluid pressures. That is, the piston can be adjusted to middle height in cylinder *2* at any pressure. Chamber *3* contains nitrogen gas, whose pressure equals that of the power fluid. Pressure changes affect the volume of the nitrogen, which displaces the piston downwards if pressure builds up and upwards if it decreases. The force transmitted by piston rod *4* onto gauge ring *5* varies accordingly. The deformation of the ring, proportional to the force, is amplified and recorded by servo mechanism *6*.

(a.3) *Selective bottom-hole pumping of several zones.* — In the solution shown as Fig. 4.1—90a, pumps are installed in separate tubings run side by side. The power fluid is fed through conduit *2* to pump *1* producing the lower zone and through conduit *4* to pump *3* producing the upper zone. The exhaust of pump *1* rises through annulus *5* and that of pump *3* through annulus *6*. The well fluids are mixed and rise together with the respective exhausts. Gas is removed through tubing *7* from the lower zone and through annulus *8* from the upper zone. In the tandem arrangement shown as Fig. 4.1—90b, the two pumps are mounted in a single body, end to end along a vertical axis. The fluid of the upper zone enters the upper pump through port *1*, and leaves through port *2* into tubing *3*. The engine driving the upper pump receives power fluid through port *4* and exhausts it through port *5* into annulus *6*. The oil from the lower zone is sucked in by the lower pump through port *7* and pumped through port *8* likewise into annulus *6*. The engine driving the lower pump, supplied with power fluid through conduit *9* and port *10*, exhausts it through port *11* into the annulus. This solution has the advantage over the preceding one of requiring less space, but it cannot be used in wells producing a gaseous fluid. — Other solutions are, of course, also possible. For instance, the completion in Fig. 4.1—90a can be modified to be used with a free pump. The slim power-fluid conduits are dispensed with; power fluid reaches the engines through large-size tubings. The liquid from the upper zone is produced separately. The liquid from the lower zone merges with the exhausts of both engines before rising through the annulus. There are two packers; the gas produced from the two zones cannot be separated.

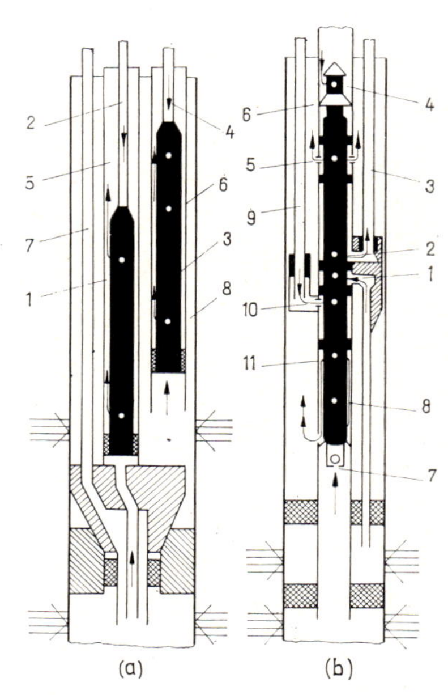

Fig. 4.1—90 Selective production of several formations with hydraulic bottom-hole pumps; well completions

(b) Electric centrifugal submersible pumps

The inventor of the electric centrifugal submersible pump is A. Arutyunoff, whose pumps manufactured by the Reda Company have been in commercial use since 1927 in a variety of oilfields (API History 1961). The importance of the submersible pump is illustrated by the fact that in the Soviet Union, where this means of production had first been introduced in the

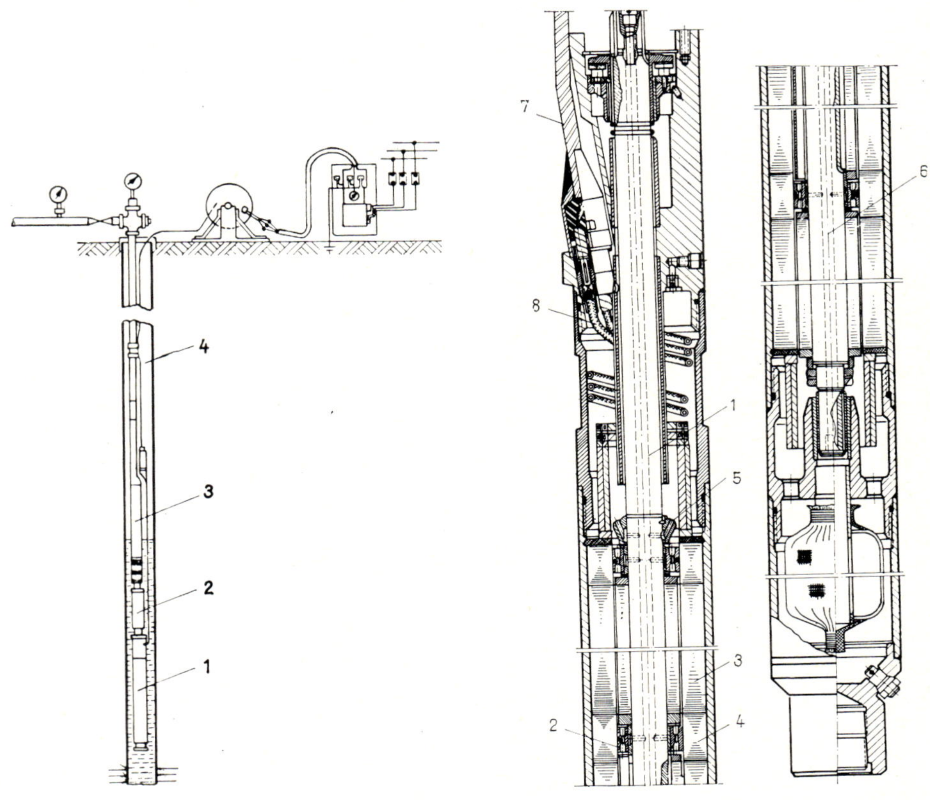

Fig. 4.1—91 Electric centrifugal submersible pump (Muraviev and Krylov 1949)

Fig. 4.1—92 Motor of ECN electric submersible pump, after Bogdanov (1968)

nineteen-fifties, 3400 wells were produced by electric submersible pumps in 1967, which is about 9 percent of all wells produced by mechanical means. The combined output of these pumps was 58 million tons of crude in 1967, about 49 percent of the total amount of crude produced by mechanical means (Zaitsev 1968). — The pumping unit usually includes three main structural units: an electric motor, a so-called protector connected to it, and a multistage centrifugal pump. In pumps producing a gaseous well fluid, a gas separator is inserted between protector and pump. The most typical arrangement is shown in Fig. 4.1—91. The bottom unit is motor *1*;

there are above it protector *2* and pump *3*, the latter attached to the tubing string. The electric motor receives current through a cable *4* fixed to tubing. Figure 4.1—92 illustrates the ECN type dipole, three-phase asynchronous squirrel-cage motor used in the Soviet Union (Bogdanov 1968). The motor built around shaft *1* is split into a number of short sections, separated by bearings to prevent the buckling of the comparatively long shaft. Stator windings *3* are separated by non-magnetic bearing-support stator lamellae *4*.

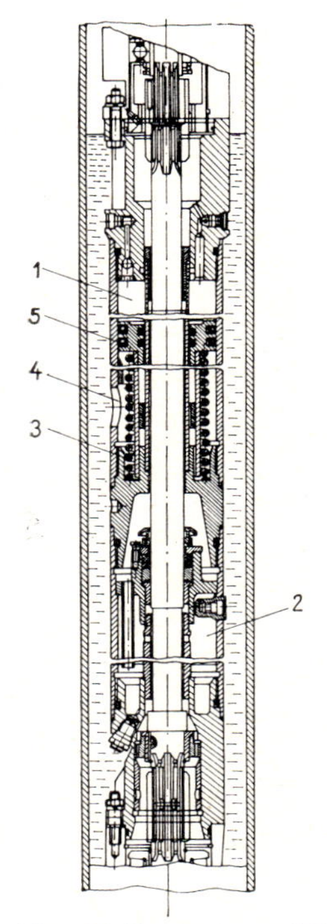

Fig. 4.1—93 Protector of ECN electric submersible pump, after Bogdanov (1968)

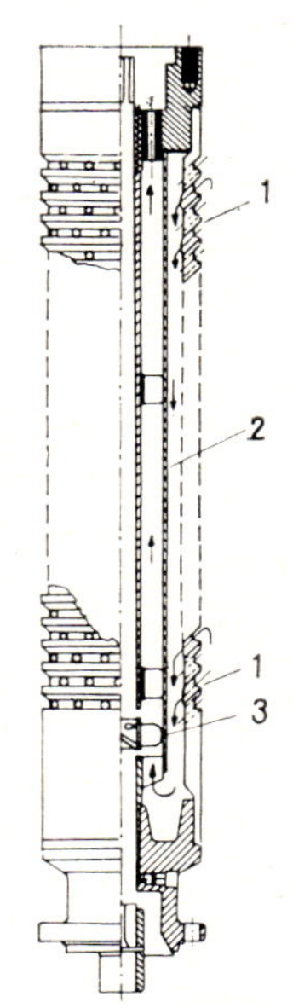

Fig. 4.1—94 Reda's gas separator

Lube oil is circulated by turbine *5* in the motor's interstices and through bore *6*. Power is supplied to the motor through cable *7* and lead *8*. The *OD* of the motor is 103—123 mm; its length is 4.2—8.1 m; its normal operating speed is 3000 rpm. The maximum motor power that can be installed depends primarily on the *ID* of the production casing. Motor power can be increased in Reda pumps by mounting on one shaft two motors in tandem. The maximum power that can thus be installed is 57 kW in 4 1/2-in. casing, 89 kW in 5 1/2-in., 224 kW in 7-in., and 358 kW in 8 5/8-in. casing. — Figure 4.1—93 shows the protector unit of the Soviet ECN pumps (Bogdanov 1968). Its role is to feed lube grease from chamber *1* to the pump and lube oil from chamber *2* to the motor. The pressure required to perform these tasks is supplied by coil spring *3* and the fluid acting on plunger *5* through port *4*. As the grease can through slits along the shaft reach also the bottom of chamber *2*, it keeps under pressure the motor oil, too. The temperature of the motor and hence the volume of the lube oil are variable, depending primarily on the rate of production and on the *WOR* of the well fluid. The protector serves both as a buffer tank and as a makeup tank for lube oil lost by seepage.

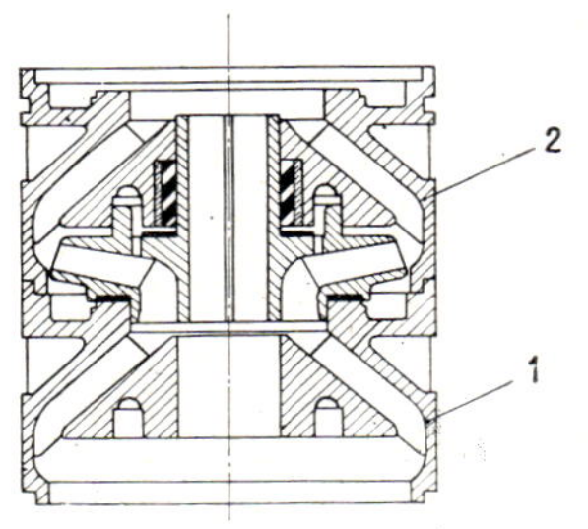

Fig. 4.1—95 One stage of B.J. Centrilift centrifugal pump, after Boley (1967)

Figure 4.1—94 shows a Reda make gas separator. Well fluid enters the separator through ports *1—1*. Gas rises in annulus *2* to escape into the casing annulus. The degassed liquid is fed to the main pump by screw pump *3*. The pressure increase due to screw pump action may make the liquid dissolve some of the separated gas. Centrilift pumps incorporate a centrifugal separator. The well fluid entering the separator body gets into a high-speed single-stage special pump, the centrifugal separator. Its liquid fraction rises along the wall of the chamber above the separator into the main pump. The gas rising along the shaft is led through a port into the annulus. Efficient gas separation is highly important; gas entering the pump does not only reduce its volumetric efficiency, but may also destroy it by cavitation. — Figure 4.1—95 shows one stage of a multi-stage Centrilift pump (Boley 1967). Fluid enters the pump from the gas separator, or, in the absence of such, through some side ports. It flows through diffuser *1* on to impeller *2*. Today's tallest pumps include up to 530 stages. High pump efficiency largely depends on good diffuser and impeller design. Figure 4.1—96 shows efficiency curves of Reda pumps (Arutyunoff 1965). It is seen that pump efficiency may attain 75 percent at high enough rates of production; also, the efficiency of a pump up to production rates of about 1500 m^3/d is the better the higher its capacity. Higher-efficiency pumps could well be built, but this would require increasing axial length even further, and this is not desirable. — Figure 4.1—97 shows characteristics of the G—110 type,

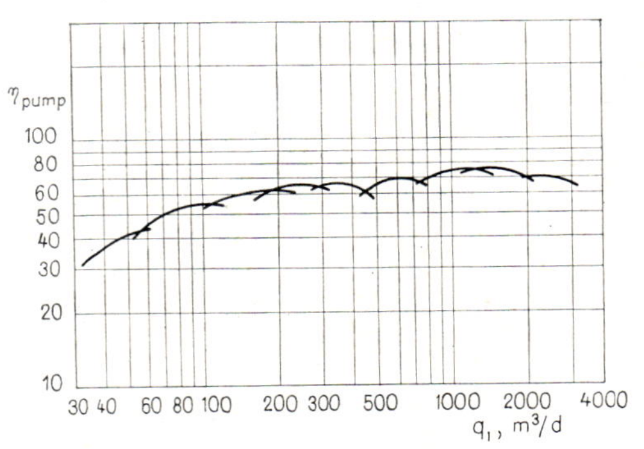

Fig. 4.1—96 Efficiency of Reda's electric submersible pumps

100-stage Reda pump. According to the $q_l - H$ characteristic, head capacity declines gradually from an initial value as output increases. Hence, as opposed to surface centrifugal pumps, the submersible pump is stable at any output. Characteristics are fairly steep about the operating point. The idea behind this is to make the output as independent as possible of changes in lift head due to possible fluctuations in producing fluid level. In the interest of satisfactory pump operation, it is necessary to keep pump output within ± 15 percent of the value belonging to the highest efficiency. Diffusers and impellers are made of a corrosion- and erosion-resistant material. This may be a nickel steel, bronze or plastic. In the Soviet Union,

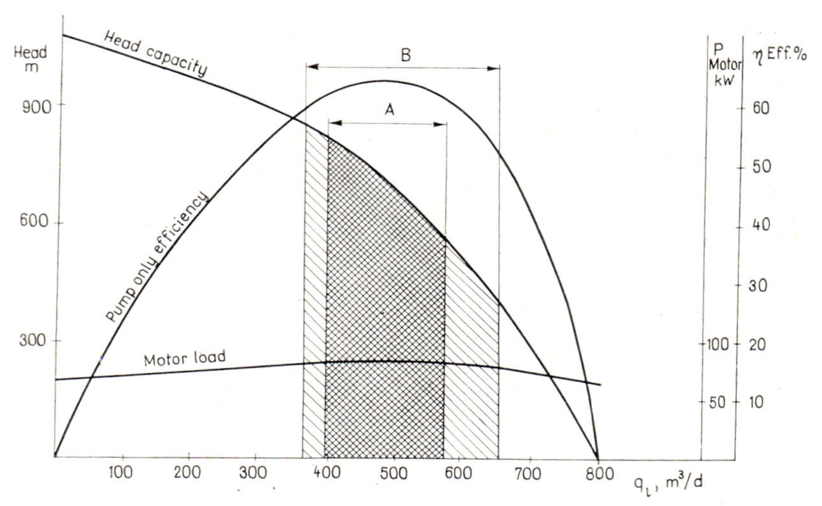

Fig. 4.1–97 Characteristics of the G–110 type Reda submersible pump. *A* optimal range, *B* maximal range

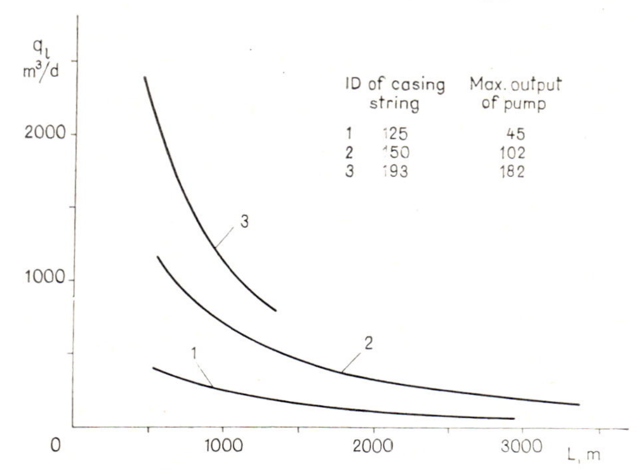

Fig. 4.1–98 Capacity of Reda's electric submersible pumps

favourable results have been attained with impellers made of a plastic denoted P—68. Typically, the weight of an impeller made of this plastic is 9 percent of a bronze impeller (Bogdanov 1968). The capacity of the pump depends primarily on well diameter and lift head. Figure 4.1—98 gives production capacities of single-motor Reda pumps.

The cable feeding power to the pump is three-core, provided with high-grade insulation, chemically inert with respect to the well fluid, corrosion-resistant, impervious to gas, and protected from mechanical damage. Cables used in the GFR have conductors insulated separately with butyl rubber or polyethylene; these are enclosed in a common butadiene rubber sheath. Mechanical protection is provided by a sheath of steel strip, alloyed or provided with a corrosion-resistant coating. The cable is fixed to the tubing at intervals of 4—5 m and run together with it. The maximum allowable ambient temperature of the cable is 93—125 °C (Boley 1967). A special emphasis should be laid on the importance of full impermeability to gas of the insulation. If the high-pressure gas can penetrate pores in the insulator, then drops in pressure may result in blistering and insulation failure.

A ball-type check valve is usually installed above the pump. This plays an important double role. It is essential in any well that the tubing above the pump should not run empty during pumping pauses. If there is no check valve installed or it is leaky, the liquid will flow back through the pump during shut-off. After start-up this may result in burning out the electric motor. In order to prevent this, the pumping unit is provided with an additional protection. A delay switch installed on the surface prevents restarting for a certain interval after stopping. During the interval, the well will partly fill with fluid entering from the formation. The other purpose of the check valve is to prevent sand from settling in the pump when producing a sandy fluid. Sand depositing in the pump may silt up and seize the impellers.

Economical planning of submersible-pump operation requires a fairly accurate idea of the operating point to be anticipated and a pump to match that operating point. A pump thus chosen will produce at a rate corresponding to optimum efficiency. In high-capacity wells, however, production testing often requires the use of submersible pumps, too. It may therefore happen that the first-run pump will be suited for a well test only, and that the definitive pump may only be installed after analysis of the test results. A production test may be performed e.g. as follows.

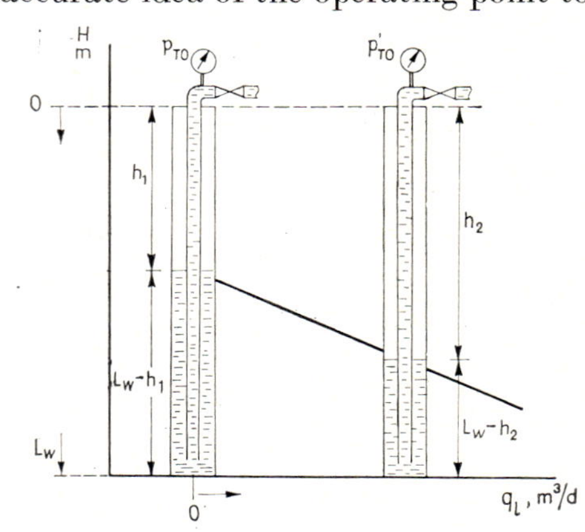

Fig. 4.1—99 Well testing with electric submersible pumps

After a wait long enough to let the static level establish itself in the well, the tubing is filled up with oil taken from the well; the wellhead is shut off and the pump is started up. The pump builds up pressure equal to its maximum head capacity H_0 above the liquid column in the annulus (Fig. 4.1—99), resulting in a pressure p_{TO} on the closed wellhead. We may now write

$$H_o = h_1 + \frac{p_{TO}}{\bar{\gamma}_l}. \qquad 4.1\text{—}72$$

The well is then reopened and produced at a rate q_l. Once this rate is stabilized, the well is abruptly shut off and the wellhead pressure is measured. Assuming that the producing level is not significantly disturbed during the short shut-off time, we may write in a fair approximation

$$H_o = h_2 + \frac{p'_{TO}}{\bar{\gamma}_l}. \qquad 4.1\text{—}73$$

If well depth is L_w and the mean gravity of the liquid column in the well is $\bar{\gamma}_l$, then formation pressure is

$$p_{ws} = (L_w - h_1)\bar{\gamma}_l ,$$

and

$$p_{wf} = (L_w - h_2)\bar{\gamma}_l ,$$

where h_1 and h_2 may be taken from the above equations. In the knowledge of the stabilized rate of production q_l, and using the relationship $q_l = J(p_{ws} - p_{wf})$, the productivity index can be calculated. J and p_{ws} are known, so that the productivity equation characterizing the inflow of liquid into the well may be written up. It is assumed that the exponent n of the productivity equation equals unity. If this is not the case, then the production test must be repeated and n calculated from the results. If the inflow characteristic of the well could not be accurately established beforehand, then, in order to operate the pump at a more economical operating point, production may be slightly corrected in some cases by suitably choking the well output (Muraviev and Krylov 1959).

In operation planning, special attention must be paid to the electric power supply, and to choosing the most economical cable size and voltage drop in the cable. The significance of this problem is revealed by a power diagram (Fig. 4.1—100; Boley 1967) recorded on a well producing 80 m³/d at a lift head of 1850 metres of water column. Cable losses are seen to consume almost one-fifth of total input power. Voltage drop in the cable may be quite significant. In order to supply the pump at the rated voltage, it is

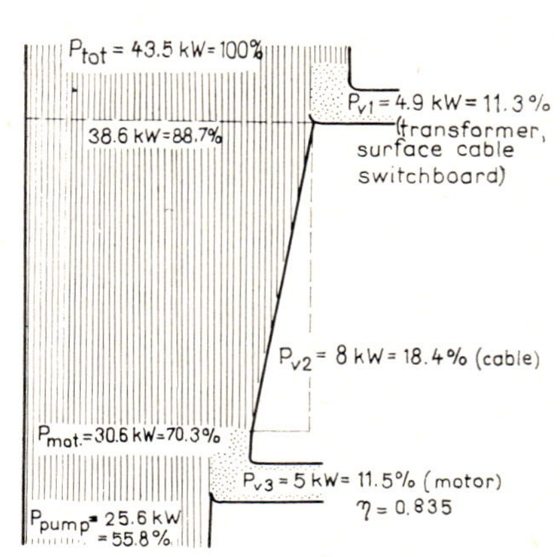

Fig. 4.1—100 Power consumption of electric submersible pump, after Boley (1967)

necessary to carefully calculate the voltage drop to be anticipated and to ensure the required voltage by means of a surface transformer. Rated voltages of submersible pumps range from 400 to 2300 V. Operation planning is considerably hampered if the well fluid liberates gas during its rise. This may result in a significant reduction of flowing pressure gradient, and the lift head and motor power needed may turn out half the value calculated without reckoning with gas separation. A simplified calculation procedure for obtaining the tubing pressure to be expected is given in Smith (1968). He recommends to let gas and liquid rise together in the annulus rather than separately. The lower tubing pressure may significantly reduce the power consumption of pumping.

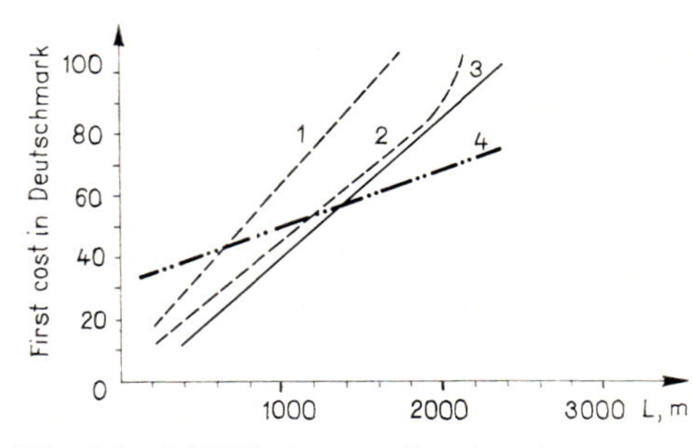

Fig. 4.1 — 101 First cost of various bottom-hole pumping methods, after Boley (1967)

The submersible pump may realize a saving against other means of production especially in medium-depth wells of large or very large capacity. Figure 4.1—101 shows investment costs of production equipment v. depth in a well of 80 m^3/d capacity. Graphs *1* and *2* refer to walking-beam type sucker-rod pumps of various makes, graph *3* to a rodless hydraulic bottom-hole pump, and graph *4* to an electric centrifugal submersible pump. It is seen that at depths exceeding 1400 m the investment cost of the submersible pump is least. Figure 4.1—102 shows the cost of pumping one ton

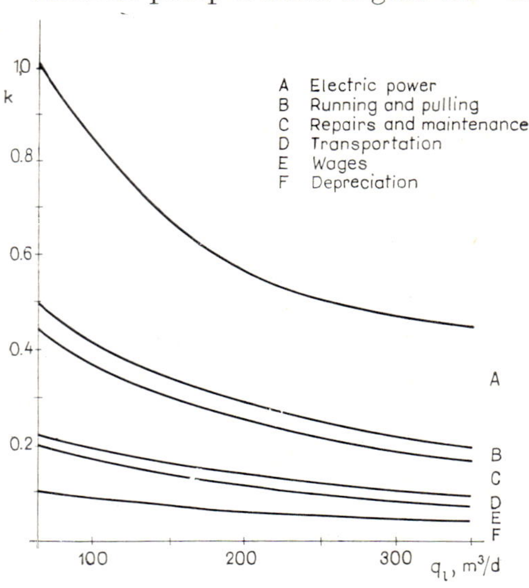

Fig. 4.1 — 102 Cost components of Reda's bottom-hole pumps operation

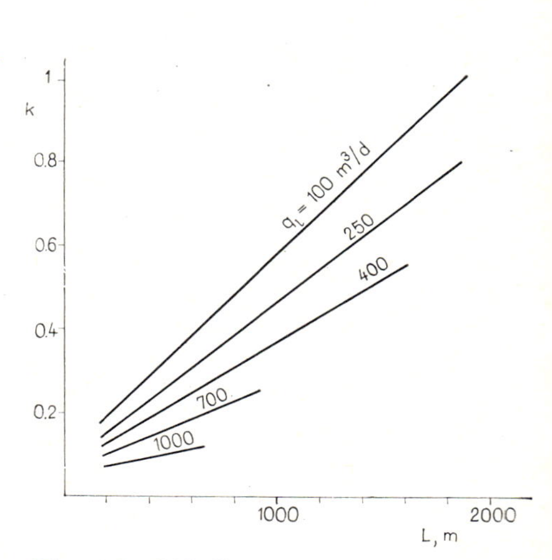

Fig. 4.1 — 103 Comparison of operating costs of Reda's bottom-hole pumps

of liquid with a Reda pump v. the rate of production. Each cost item is seen to decrease with increased rate of production, and so does, consequently, also the overall cost of production. The most expensive single item is power. Figure 4.1—103 shows specific production costs v. well depth for Reda

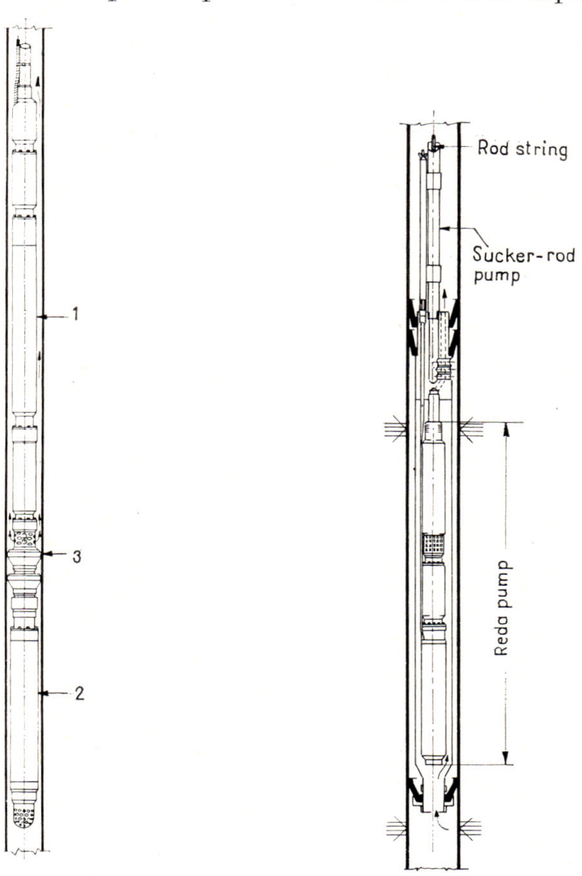

Fig. 4.1—104 Completion for high-capacity wells, after Arutyunoff (1965)

Fig. 4.1—105 Selective production by sucker-rod and electric submersible pumps in one well, after Arutyunoff (1965)

pumps at various rates of production. The diagrams show the specific cost of production to increase almost linearly with depth at any given rate of production. At a given well depth, production cost is the less the higher the production rate.

The most usual single completion for submersible pumps is shown in Fig. 4.1—91. Other arrangements known include the one shown as Fig. 4.1—104, used in wells with 4 1/2-in. production casing if the fluid to be produced is gasless and the rate of production is to be quite high (Arutyunoff 1965). Motor *1* is above pump *2*; between protector and pump there is a packer *3*. This solution permits the utilization of the large casing annulus for producing the well fluid. It has enabled the production of a well at a rate of 430 m^3 of liquid per day in which a conventional arrangement would not have given more than 80 m^3/d. The inverted pumping-unit design has its advantages even at larger casing sizes; e.g. if the sand face is protected by a length of slimmer casing; the pump can then be installed in the larger-size casing with only the intake tube reaching down to sand-face level. Care should be taken that the fluid flow cooling the motor be strong enough. — Even in the selective production of several zones from the same well, there is an application for the submersible pump. Usually only one zone, the one of higher production capacity, is produced by submersible pump; the other is either flowing or produced by some other means. The larger annulus is used to lift the output of the submersible pump. An arrangement of this sort is shown in Fig. 4.1—105 (Arutyunoff 1965), where the lower zone is produced through the annulus by means of a submersible pump and the upper zone through the tubing by sucker-rod pump.

(c) Other types of rodless bottom-hole pumps

Bottom-hole pumps operating on a number of principles and designs different from the above-discussed ones have been patented all over the world. For instance, in the US alone, 165 patents for bottom-hole pumps were applied between 1935 and 1960 (API History 1961). Comparatively few of these have ever found commercial use, however. Let us now describe some of the more important designs. *The sonic pump* of Bodine was first tested in operation in 1953 (API History 1961); its commercial applications are mentioned in literature from the late fifties on (*Petroleum Engineer* 1958). Its principle is illustrated in Fig. 4.1—106. The tubing is hung from hanger plate *2* resting on coil springs *1*. Check valves, usually made of plastic, are installed at each tubing joint, or at least at a large number of these. Mounted on the hanger plate are two eccentric weights *4* suitable for exciting vibrations. Cogwheels *5* keyed onto the excenter shafts ensure the synchronous rotation of the two eccentric weights. One of the shafts is driven by a motor. The opposite rotation of the two shafts generates centrifugal forces whose resultant is a reciprocating vertical force. As a result, the hanger plate starts vibrating at a frequency equal to the rpm of the shafts. The vibration propagates in the tubing at the speed of sound, resulting in periodic stretching and contraction in the individual sections between check valves. On stretching, the open valves submerge in the liquid, whereas on shrinking they close and lift the liquid column. Vibration frequency is 600—1200 min^{-1}; amplitude is 7.6—19 mm. The acceleration of fluid lifted by the check valves is 5—10 times greater than that of gravity, and thus the fluid will go on rising even when the valve is already sinking on the next half-wave. Frequency is chosen so as to equal or to be a multiple of the resonance frequency of the tubing. Tubing size is 2 3/8—4 1/2 in., which permits daily rates of production of 30–160 m^3/d. In experimental wells, daily rates of production up to 400 m^3 have been attained. In order to avoid transverse vibrations resulting in the tubing's rubbing against the casing, centralizers *6*, made of plastic, are mounted at intervals of about 3 m on the outside of the tubing. In the Soviet Union, sonic pumps have been experimented with since the middle fifties. Theoretical relationships

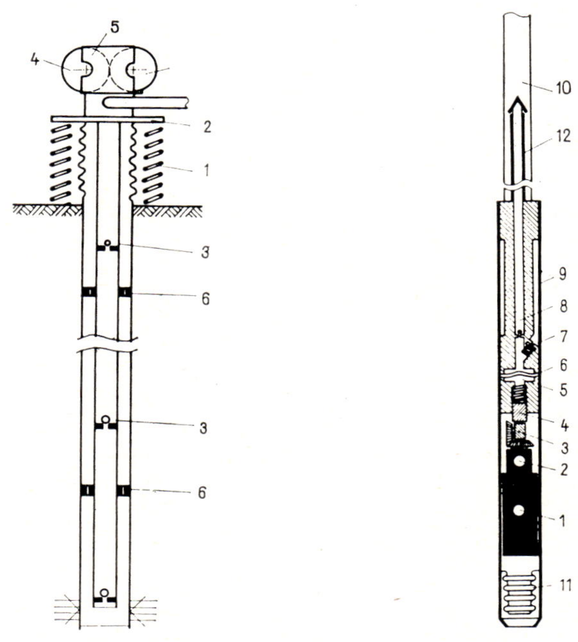

Fig. 4.1–106 Sonic pump (Petroleum Engineer 1958)

Fig. 4.1–107 Pleuger's diaphragm pump (Brüggemann and de Monyé 1959)

describing pump operation have been developed, including a relationship of production capacity v. the excited frequency (Kruman and Geibovich 1970). The overall efficiency of the sonic pump may attain 0.7. It has the considerable advantage of being almost insensitive to sand; it can produce fluids containing up to 80 percent solids. It has the drawback that packing off the casing head is a problem at high pressures, so that producing a gaseous fluid is a rather sensitive job. Neither the tubing nor the annulus is accessible to well testing or dewaxing. Efficiency presumably drops steeply at low rates of production.

In the oil fields of the German Federal Republic, the *Pleuger type membrane pump* (Brüggemann and de Monyé 1959) has been employed since 1953. The principle of the pump is explained by Fig. 4.1–107. A pair of bevel gears mounted on rotor *2* of electric motor *1* drives eccentric disk *3* which generates a reciprocating motion in plunger *4*. The latter, by the intermediary of the liquid in sealed space *5*, drives membrane *6*. The space above the membrane is bounded by intake valve *7* and check valve *8*. The well fluid reaches intake valve through sheath screen *9*. The membrane sucks fluid on the downstroke and lifts it into tubing *10* on the next upstroke, while the intake valve is closed. The motor space ends in bellows *11*. The space between this latter and the pump jacket is filled with oil. The entire space is under the pressure corresponding to the setting depth. Above the check valve, a tube several metres long, fitted with a conical cap on its top, is installed to prevent sand settling out of the pumped fluid from settling in the check valve. — Plunger size in commercially used membrane pumps is 29—38 mm; pumping speed is about 700 spm; a pump of 136 mm *OD* may produce 10—20 m³/d against a head of 30—100 bars. Temperature and viscosity of the pumped fluid must not exceed 70 °C and 300 cSt, respectively. This pump has the advantage that none of its sensitive component parts is in direct contact with the well fluid, so that life is comparatively long even if the fluid is sandy. The tubing section is open, so that testing and dewaxing is easy enough.

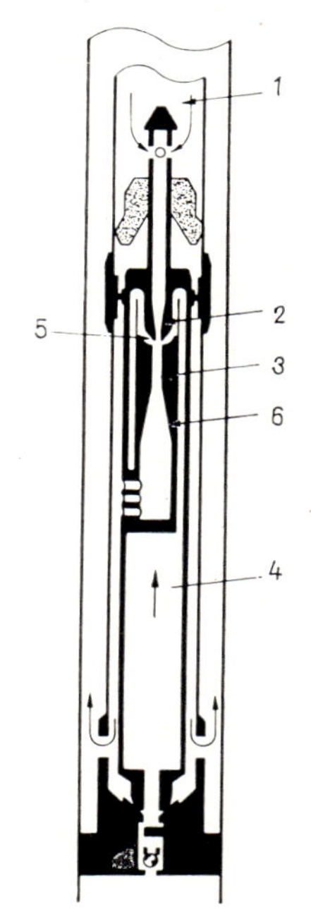

Fig. 4.1—108 Kobe's liquid-jet pump, after Wilson (1973)

The *bottom-hole jet pump* is one type of rodless bottom-hole pump. It has been used in full-scale oil production since 1973. A diagram of the type most preferred in practice is shown as Fig. 4.1—108. A flow q_s of high-pressure liquid is supplied through tubing *1* to nozzle *2*, from where it flows at a decreased pressure and increased velocity into the throat *3*. According to Raabe (1970), this liquid jet entrains the well fluid q entering through the sand face into the direction of arrow *4* essentially as a result of its apparent turbulent shear drag. The well fluid joins the high-pressure liquid after the passage of the annular aperture *5*. The power and well fluids, mixed together, flow from the throat into diffusor *6*. Velocity is reduced there and pressure is increased sufficiently to let the fluid rise to the surface. — The Kobe-make liquid-jet bottom-hole pumps have numerous advantages: the bottom-hole equipment includes no moving parts; the pump has a broad depth range of application; exchange and repairs are simple; the pump is insensitive to the quality of power fluid; it permits the production of corrosive and abrasive fluids as well as fluids of a high *GOR*; it demands little supervision in operation. — The wearing parts are primarily the nozzle and the throat. The first is worn largely by the solids contained in the high-pressure liquid, the latter by cavitation. In order to minimize cavitation it is indicated to install the pump as deep below the dynamic level as possible. The desirable depth of immersion is 0.2—0.25 times well depth. — The efficiency of the pump (the ratio of input work and useful work) depends considerably on the rate of production. In one instance, it was 12—25 percent for a rate of 64 m³/d and 40—57 percent for 159 m³/d. —

Figure 4.1—109 presents as an example the operating chart of a Kobe-make liquid jet pump. The diagram shows the variation of power fluid requirement v. the desired rate of production for the given well dimensions for various dynamic level depths. The line bordering the diagram on the top is where cavitation becomes so intense as to preclude the use of the pump. There is some cavitation also within the upper shaded zone, but

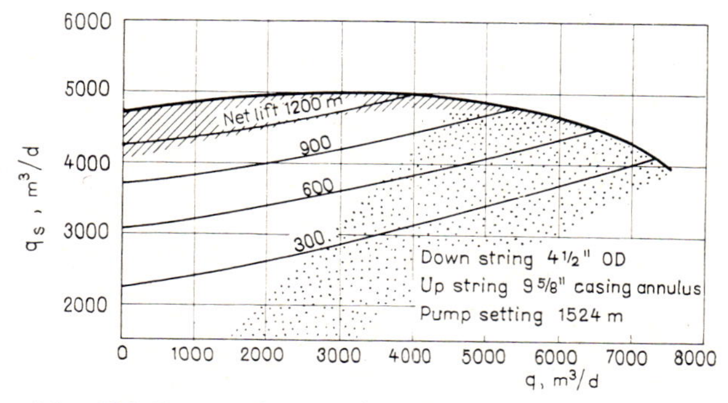

Fig. 4.1—109 Zones of operation of Kobe's liquid-jet pump

still within tolerable limits if a rather significant throat wear can be accepted. Operation is most favourable in the stippled zone on the right-hand side.

A solenoid-driven electromagnetic pump is described by Ioachim (1965). The pump is of the reciprocating-plunger type. Below the single-acting plunger, on a rod fixed to it, there are several iron cores surrounded by a solenoid coil. If the coil is fed current from the surface, then the iron cores will raise the plunger and the fluid column above it. Current is interrupted at the upper end of the stroke; the plunger then sinks back under its own weight. The device is comparatively simple and rather insensitive to sand.

Chapter 5

Choice of most economical production methods

The rational installation of production equipment requires the selection of such means of production and its operation at such operating point or points as make producing the well under consideration at the prescribed rate as cheap as possible. In the above chapters, discussing the various means of production, the reader finds calculation procedures and comparative tables which set the choice of finding the most economical operation techniques and the optimum-size production equipment. Thus e.g. in the case of flowing production, more economical than any mechanical means, the main aim of well design and production planning is to find the tubing size at which flowing production at a prescribed *BHP* can be maintained for the longest possible time. — In the Chapter on gas lift, calculation procedures are aimed at establishing the least specific injection gas requirement in continuous or intermittent operation that will produce the well at the desired rate and at the desired *BHP*. The minimum consumption of injection gas marks the most economical operating point of the production equipment under consideration. — In sucker-rod pumping, the most economical operating point is that resulting in the least fluid load. — The advantages and disadvantages of the various well completions compatible with this or that means of production, to be pondered against economy, have been discussed at the appropriate places.

The above-mentioned procedures permit comparisons of economy only within one and the same group of production equipment. The procedure to be outlined below is suited, on the other hand, for choosing between different types of equipment.

Table 5—1 states the economic advantages of the eleven main types of production methods at various well fluid characteristics, in inclined wells and at low well productivities. It points out, for instance, that the production of gaseous well fluid is an advantage in flowing production, in continuous gas lift and plunger lift, is immaterial in intermittent gas lift and unfavourable to various degrees in the various types of bottom-hole pumping. (The type of pump envisaged in the group of rodless hydraulic bottom-hole pumps is invariably the double-acting hydraulic pump.)

Table 5—2 provides information on the maximum production capacity of the various types of equipment, grouped in the same way as in Table 5—1, at various well depths and producing *BHP*s. In compiling this Table, limits of rational operation have been considered. For instance, continuous gas-lifting a well 1000 m deep at a producing *BHP* of 5 bars may be technically feasible, but the specific injection-gas requirement is so high as to be

Table 5—1
Selection of production methods (I)

Characteristics		flowing	Gas lifting			Pumping						
						Sucker rod pumping			Rodless pumping			
			continuous	intermittent	plunger lift	walking beam type		long stroke unit	hydraulic	submersible centrifugal	sonic	membrane pump
						solid rods	hollow rods					
Well stream	parafinous	A	A	A	—	B	A	B	A	A	C	A
	viscous	B	C	C	C	B	B	B	B	C	C	C
	emulsifying	B	B	A	A	A	A	—	A	A	B	A
	gaseous	+	+	—	+	B	B	A	B	A	C	B
	sandy	A	A	B	B	B	A	B	C	—	—	A
Inclined well		—	—	—	—	A	A	A	—	—	A	—
Stripper well		—	B	A	A	B	B	C	B	C	A	A

Profitable +
Indifferent —
Disadvantageous A
Detrimental B
Inapplicable C

prohibitive in practice. The selection table indicates at the same time that $q_{l\max}$ in gas-lift type methods of production varies significantly with producing *BHP* and well depth, whereas in bottom-hole pumping the *BHP* is almost irrelevant. Although the selection tables give but a rough orientation, they usually permit the reduction of the number of methods to be given a closer look to just one or two.

Example 5.1. Find the most economical means of production, if $L_w = 1800$ m; $q_l = 10$ m³/d; $R_{go} = 10$ m³/m³; $p_{TO} = 15$ bars; the well fluid contains 2—3 percent sand and paraffin deposits are liable to occur. — Table 5—1 states that options are restricted by the sand content to continuous gas-lifting, and further to hollow-rod, electric submersible, sonic and membrane type bottom-hole pumps. Paraffin deposits exclude the sonic pump; the low production rate excludes electric submersibles, and continuous gas lift is liable to be uneconomical. Table 5—2 states membrane-type bottom-hole pumps to be unsuited for a depth of 1800 m. Hence, sucker-rod pumping with a hollow-rod string is the obvious choice.

Table 5—2 also gives some indication as to how the economy of the individual production means depends on the rate of production. The consideration to be given below is aimed at proving that, in low-productivity wells, optimum economy will lie with the means of production having low depreciation and repair and maintenance costs even though specific cost of power is comparatively high.

Table 5—2

Selection of production methods (II)

			Gas lifting			Pumping						
						Sucker rod pumping			Rodless pumping			
						walking beam type		long stroke unit				
P_{wf} bars	L_w m	flowing	continuous	intermittent	plunger lift	solid rods	hollow rods		hydraulic	submersible centrifugal	sonic	membrane pump
	1000	1	0	1	1	3	3	6	6	6	3	1
5	2000	1	0	1	0	1	1	4	6	6	?	0
	3000	0	0	0	0	0	0	2	6	5	?	0
	1000	5	5	2	3	3	3	6	6	6	3	1
20	2000	3	3	1	3	2	2	4	6	6	?	0
	3000	2	2	1	2	0	0	2	6	5	?	0
	1000	6	6	2	3	3	3	6	6	6	3	1
50	2000	6	6	2	3	2	2	4	6	6	?	0
	3000	4	4	1	2	0	0	3	6	5	?	0

	$q_{l\max}$, m³/d (within reasonable economic limits)
0	0
1	0— 50
2	50—100
3	100—200
4	200—300
5	300—400
6	>400

Economic comparison of production equipment should preferably be based on the so-called direct specific cost (k, in Ft/Mg).* The items figuring in annual direct cost are the depreciation of the production equipment A, repair and maintenance costs B, and the cost of power c. Cost items independent of the type of production equipment (cost of drilling, overheads, wages of well-maintenance and work-over personnel, etc.) are left out of consideration. Of the items in the direct cost, A and B are almost independent of the rate of production, whereas power cost c is an approximately linear function of that variable. Let the specific cost of power be c Ft/Mg; the yearly power cost, at an annual production rate of $q_{an} = 365\ q_l$, is cq_{an}. The annual direct cost is, then,

$$K_{an} = (A + B) + cq_{an}\ .$$

Dividing both sides of the equation by q_{an}, we get

$$k = \frac{K_{an}}{q_{an}} = \frac{A + B}{q_{an}} + c \quad \text{Ft/Mg}. \tag{5.1}$$

* Ft is the symbol for Forint, the Hungarian currency unit.

Figure 5.1 is a plot of Eq. 5.1 for two different types of production equipment. The one is characterized by $A + B = 45\,000$ Ft per year and $c = 6$ Ft/Mg, and the other by $A + B = 7500$ Ft/y and $c = 40$ Ft/Mg. Equipment *1* is seen to be more economical at rates of production above 3 Mg/d, and vice versa. The relationships imply that, at low rates of production, gas lift for instance is more economical than bottom-hole pumping.

Equation 5.1 merely serves to provide a comparison in principle of two types of equipment. In real-life situations, choice between two options is

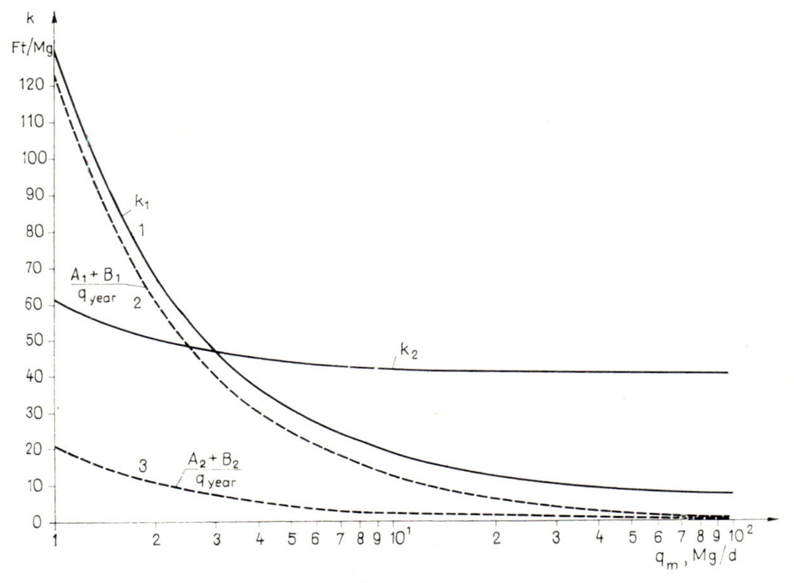

Fig. 5.1

made possible by the procedure illustrated by Fig. 5.2. In part (a), Graphs *1*, *2* and *3* of Fig. 5.1 are plotted in a bilogarithmic system of coordinates. Graphs *1* and *2* refer to walking-beam type sucker-rod pumps, Graph *3* to an intermittent gas lift. The ordinate difference between Graphs *1* and *3* defines a certain cost fraction. If this fraction is spent on compressing injection gas, and the amount of gas thus compressed is just sufficient to produce the well, then the specific cost of the crude thus produced is precisely equal to that produced by the sucker-rod pump. Dividing the ordinate differences *1*—*3* belonging to various rates of production by the corresponding specific costs of compression (corrected for losses), we obtain the specific injection-gas consumption R_{gl} that will result in a cost equal to that of sucker-rod pumping. Part (b) of the Figure is a plot of R_{gl} v. q_l in the range from 1—10 Mg/d (Graph *4*). Let us plot in this system of coordinates the actual or anticipated specific injection-gas consumption of production v. daily production rate q_l (Graph *5*). The ordinate difference between graphs *4* and *5* at various rates of production indicates the injection-gas saving in m³ per Mg of crude against the hypothetical gas-lift operation equal in cost to sucker-rod pumping. Multiplying this hypothetical saving by the corresponding annual production rate q_{an} and the specific cost of compression, and plotting the resulting values, we obtain a graph of annual cost saving Δk_{an} v. daily rate of production (Graph *6*). In the case assumed

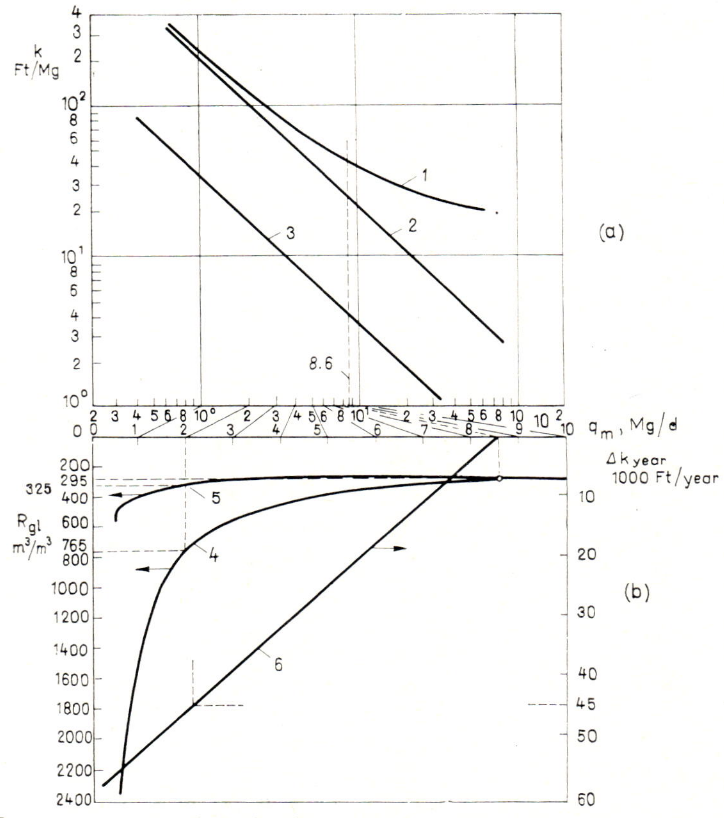

Fig. 5.2 Economy v. rate of production in sucker-rod pumping and intermittent gas-lift, after Szilas (1957)

in the Figure, intermittent gas-lift production of a well is more economical at rates below 8.6 Mg/d, and sucker-rod pumping is more economical above it. In a well producing 2 Mg/d, 45 000 Ft/y is saved by changing over from sucker-rod pumping to intermittent gas lift. — In operations planning, Fig. 5.2 is to be prepared for various well depths or well groups of various depths.

Chapter 6

Gathering and separation of oil and gas

6.1. Line pipes

6.1.1. Steel pipes

Steel pipes used in transporting oil and gas are predominantly either hot-rolled seamless or so-called spiral-welded pipes. Axially welded pipes made by electric fusion-arc or resistance welding or submerged-arc welding are used to a lesser extent. Table 6.1 — 1 gives some characteristic sizes of plain-end pipes larger than 60.3 mm after API Stds 5L — 1971 and 5LX — 1971. Nominal size in inches equals the *OD* of the pipe. The *OD* tolerance varies with the diameter and mode of fabrication of the pipe. The maximum admissible tolerance is ± 1 percent. The tolerance on wall thickness likewise depends on pipe size and mode of fabrication. The maximum admissible tolerances are $+20$ and -12.5 percent. The Table lists the least and greatest standard wall thicknesses for any given size. Within these extremes there are of course several other standard wall thicknesses. In Column 6 of the Table are the standard wall thicknesses (and hence standard *ID*s) belonging to each nominal size. Pipe ends are bevelled to facilitate butt welding. Unless there is an agreement to the contrary, the bevel angle is $30°$ $(+5, -0°)$ as measured from a plane perpendicular to the pipe axis. The height of the unbevelled pipe face is to be 1.59 mm (± 0.79 mm).

Some characteristic data of pipe materials and their strength are listed in Table 6.1 — 2. Threaded-end pipes for joining with couplings are also made within the 2 3/8 in. — 20 in. nominal size range. These pipes are made, however, invariably of A25, A and B grade steel. As the tensile and yield strengths of steels increase, the difficulties of failureproof welding tend to increase, too. Characteristically, it was as late as the early sixties that a niobium-alloyed steel of 41.4 bars yield strength as weldable as the lower-strength pipe steels was produced. According to information published in 1965, gas pipelines of X100 steel have since been built in the USA. The yield strength of that steel is $\sigma_F = 75.1$ bars. Improving steel quality has a considerable economic importance. A pipeline of a given *ID* and pressure rating will be the cheaper the less the wall thickness of the pipe. This, of course, requires higher-strength steels. For instance, the unit-weight prices of steel pipes current in Hungary are all within the range of 1 — 1.2 times the price of the cheapest steel, whereas their yield-strength range is 1 — 1.9 times that of the weakest steel. Since permissible operating pressure varies by Eq. 6.1 — 1 directly as both wall thickness (and hence, weight) and yield strength, the economy of using higher-strength steel is obvious.

Specific costs of transporting both oil and gas will decrease as throughput increases, provided the oil or gas is transported in sufficiently large-size

Table 6.1—1

API Standard dimensions of seamless and axially welded line pipe above 2 3/8 in. size (after API Stds 5L—1971 and 5LX—1971)

Nominal size	OD d_o	Nominal weight G	Wall thickness s	ID d_i	Number of sizes	
in.	mm	N/m	mm	mm	Std 5LX	Std 5L
1	2	3	4	5	6	7
2 3/8	60.3	29.62	2.11	56.1	11	11
2 3/8	60.3	131.9	11.07	38.2	12	12
2 7/8	73.0	36.1	2.11	68.8		
2 7/8	73.0	200.0	14.02	45.0	12	12
3 1/2	88.9	44.2	2.11	84.7		
3 1/2	88.9	271.4	15.24	58.4		
4	101.6	50.7	2.11	97.4	12	12
4	101.6	182.7	8.08	85.4	11	11
4 1/2	114.3	57.2	2.11	110.1	16	16
4 1/2	114.3	402.4	17.12	80.1		
6 5/8	168.3	84.8	2.11	164.1	19	20
6 5/8	168.3	662.7	18.26	131.8		
8 5/8	219.1	165.9	3.18	212.7	16	16
8 5/8	219.1	887.1	18.26	182.6		
10 3/4	273.0	257.9	3.96	265.1	14	14
10 3/4	273.0	1126.0	18.26	236.5		
12 3/4	323.8	337.7	4.37	315.1	19	17
12 3/4	323.8	1621.0	22.22	279.4		
14	355.6	405.2	4.78	346.0	19	16
14	355.6	1912.0	23.28	307.9		
16	406.4	463.9	4.78	396.8	22	20
16	406.4	2611.0	28.58	349.2		
18	457.2	522.5	4.78	447.6	23	21
18	457.2	3267.0	31.75	393.7		
20	508.0	676.1	5.56	496.9	24	22
20	508.0	3996.0	34.92	438.2		
22	558.8	744.4	5.56	547.7	26	24
22	558.8	4798.0	38.10	482.6		
24	609.6	926.6	6.35	596.9	26	24
24	609.6	5470.0	39.67	530.3		
26	660.4	1004.0	6.35	647.7	17	15
26	660.4	3901.0	25.40	609.6		
28	711.2	1083.0	6.35	698.5	17	14
28	711.2	4213.0	25.40	660.4		
30	762.0	1161.0	6.35	749.3	17	14
30	762.0	4526.0	25.40	711.2		
32	812.8	1239.0	6.35	800.1	17	14
32	812.8	4838.0	25.40	762.0		
34	863.6	1317.0	6.35	850.9	17	14
34	863.6	5150.0	25.40	812.8		
36	914.4	1395.0	6.35	901.7	17	14
36	914.4	5462.0	25.40	863.6		
38	965.2	1835.9	7.92	949.4	15	15
38	965.2	5774.0	25.40	914.4		
40	1016.0	1932.0	7.92	1000.2	15	15
40	1016.0	6086.0	25.40	965.2		

(Cont. on the next page)

Table 6.1—1 (Cont.)

Nominal size	OD d_o	Nominal weight G	Wall thickness s	ID d_i	Number of sizes	
in.	mm	N/m	mm	mm	Std 5LX	Std 5L
1	2	3	4	5	6	7
42	1066.8	2236.0	8.74	1049.3	14	14
42	1066.8	6398.0	25.40	1016.0		
44	1117.6	2344.0	8.74	1100.1	14	14
44	1117.6	6710.0	25.40	1066.8		
46	1168.4	2451.0	8.74	1150.9	14	14
46	1168.4	7022.0	25.40	1117.6		
48	1219.2	2558.0	8.74	1201.7	14	14
48	1219.2	7335.0	25.40	1168.4		

Table 6.1—2

Strength and composition of steel line pipe (after API Stds 5L—1971 and 5LX—1971)

Grade	API Std	σ_F MN/m²	σ_B MN/m²	Alloying elements and impurities	Note
A-25	5L-1971 (normal)	173	310	C, Mn, Ph, S	Not used for spiral-weld line pipes
A	,,	207	331	C, Mn, Ph, S	
B	,,	241	413	C, Mn, Ph, S	
X 42	5LX-1971 (high-test)	289	413	C, Mn, Ph, S	
X 46	,,	317	434	C, Mn, Ph, S	
X 52	,,	359	455[1] 496[2]	C, Mn, Ph, S	
X 56	,,	385	490[1] 517[2]	C, Mn, Ph, S Nb, V, Ti	
X 60	,,	414	517[1] 538[2]	Nb, V, Ti	
X 65	,,	448	531[1] 551[2]	C, Mn, Ph, S Nb, V, Ti or by agreement	

[1] For pipe diameter less than d_0 = 508.0 mm, and larger, with wall thickness greater than 9.52 mm
[2] For pipe diameter d_0 = 508.0 mm and larger, with wall thickness 9.52 mm and less

pipes optimal at the given throughput. Big-diameter pipelines are usually cheapest to build using spiral-welded steel pipe. This technology permits giving a pipe a smaller wall thickness than that of hot-drawn steel pipes of equal size. Also, the operating pressure ratings of spirally-welded pipes are higher than those of axially-welded pipes of the same size and wall thickness (see farther below). Spirally-welded pipes can be made of any steel grade listed in Table 6.1—2, with the exception of A25. Their sizes conformable to API Std 5LS agree with those of the high-test pipes of API Std 5LX, so that Table 6.1—1 can be used to find their standard sizes, too.

The range of requirements facing pipe steels has lately been extended rather considerably by the fact that more and more oil and gas wells have been drilled under arctic climates. The low temperatures occurring there will substantially reduce the ductility of pipe steel. A parameter permitting the assessment of steel strength from this viewpoint is, first of all, the critical or brittle-transition temperature established by the notch impact-bending

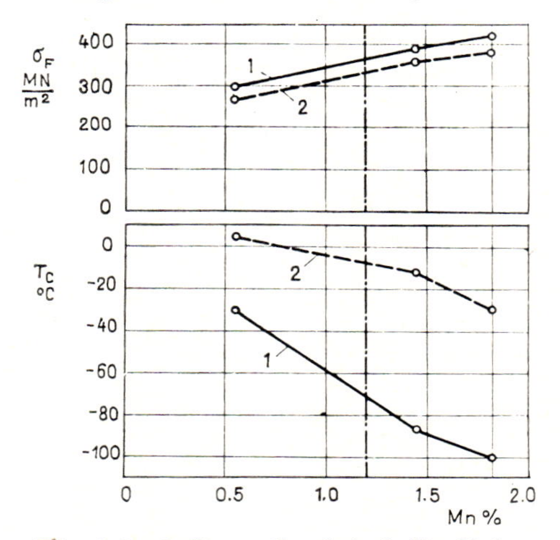

Fig. 6.1—1 Strength of steel St 52.3 as affected by Mn content at low temperatures, at Al content of 0.05 percent (*1*) and 0.01 percent (*2*); after Haarmann (1970)

test. The addition of Mn up to 2 percent will raise the yield strength of the steel and lower its brittle-transition temperature. A comparatively slight addition of Al (0.05 percent) will, however, raise the yield strength and substantially lower the brittle-transition temperature at any Mn content (Fig. 6.1–1). Pipes used under cold climates will therefore usually contain slight amounts of Al (Haarmann 1970).

6.1.2. Aluminium pipes

It was in the sixties that the advantages of alloyed aluminium pipe in transporting oil and gas came to be recognized. The advantages include a lower unit weight entailing easier handling and transportation; better resistance to impacts, rough handling, and temperature fluctuations; comparative ease of welding; easier laying over hilly terrain or along curved traces, owing to greater flexibility; the oxide layer forming on the surface of aluminum is resistant to most types of corrosion to be expected in the ground; it is not corroded by hydrogen sulphide, which makes it preferable when oil or gas containing this gas is to be conveyed; it is not attacked by salt water, which makes it the natural choice for the undersea flowlines of offshore wells. Table 6.1—3 lists the quality and strength parameters of alloyed aluminum pipe after API Std 5LA. Table 6.1—4 gives the main sizes of aluminum pipe after the same standard. Aluminum pipes are not very widely used as yet, mainly because the tensile and yield strengths of the aluminum alloys available for the purpose are lower than those of the pipe steels, and perhaps also because their advantages are not known widely enough.

Table 6.1—3
Strength and composition of aluminium alloy line pipe
(after API Tentative Std. 5LA—1962)

Alloying elements	Max. weight percent of alloying elements in materials designated			
	6063	6151	6051	6062
Si	0.20—0.60	0.70—1.3	0.40—0.80	0.40—0.80
Fe	0.35	0.50	0.70	0.70 0
Cu	0.10	0.10	0.15—0.40	0.15—0.4
Mn	0.10	0.40—0.80	0.15	0.15
Mg	0.45—0.90	0.40—0.80	0.80—1.2	0.80—1.2
Cr	0.10	—	0.15—0.35	0.04—0.14
Zn	0.10	—	0.25	0.25
Ti	0.10	0.20	0.15	0.15
Others, each	0.05	0.05	0.05	0.05
Others, total	0.15	0.15	0.15	0.15
Al	Rest			
σ_B tensile strength MN/m^2	207	289	262	262
σ_F yield strength MN/m^2	173	255	241	241
Least elongation on 2-in. length, percent	8	10	10	10

Table 6.1—4
Dimensions and weights of aluminium alloy line pipe
(after API Tentative Std. 5LA—1962)

Nominal size in.	*OD* mm	*ID* mm	Wall thickness mm	Tolerance in		Nominal weight N/m
				OD mm	Wall thickness mm	
2	60.3	57.0	1.6	±0.79	±0.30	3.65
2 1/2	73.0	68.8	2.1	±0.79	±0.30	5.64
3	88.9	84.7	2.1	±0.79	±0.30	6.90
3 1/2	101.6	97.4	2.1	±0.79	±0.30	7.91
4	114.3	110.1	2.1	±0.79	±0.30	8.92
5	141.3	135.8	2.8	+1.57 −0.79	±0.36	14.5
6	168.3	160.2	2.8	+1.57 −0.79	±0.36	17.3

6.1.3. Plastic pipes; plastic-lined steel pipes

The oil and gas industry have been using increasing amounts of plastic pipes since the forties. In 1948, the value of plastic pipes used in the USA was a mere $ 0.5 million; in 1960 it attained 100 million, and prediction for 1975 is 250 million. Plastic pipes have the advantage of lighter weight, which facilitates handling, storage, transportation and laying; they are resistant to both external and internal corrosion; the relative roughness of their inner surface is low enough for such pipes to be regarded as hydraulically smooth. Hence, friction losses at any given throughput are lower than in steel pipes of the same size; also certain plastics have paraphobic properties preventing wax deposition, which considerably reduces the cost of scraping

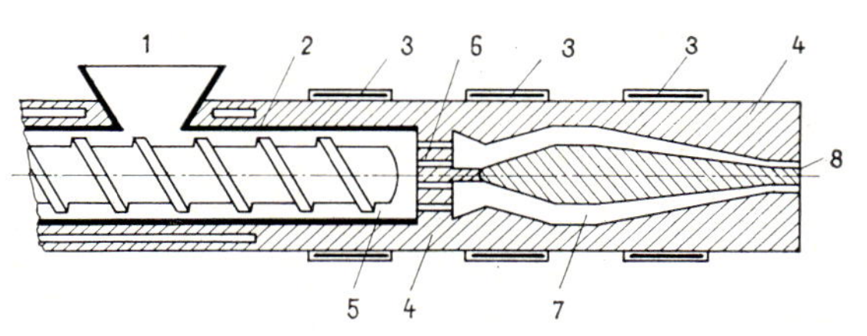

Fig. 6.1—2 Extrusion press for plastic pipe

when transporting waxy crudes; their thermal and electrical conductivities are low; the maintenance costs of the pipeline are lower. — The drawbacks of plastic pipes are as follows. Strength, especially in the case of heat-softened (thermoplastic) substances is low and is further lowered, rather substantially, by any increase in temperature. Their application is therefore restricted to rather low pressures and a rather narrow temperature range. Thermal expansion is considerable, up to 15 times that of steel; constancy of size and shape is rather poor; resistance to physical influences including fire is slight. All these drawbacks are much reduced in pipes made of heat-hardened (thermosetting) resins. Plastics used to make line pipes fall into two big groups, (i) heat-softened (thermoplastic), and (ii) heat-hardened (thermosetting).

Thermoplastic materials are usually synthetic polymers. Polymerisation essentially consists in the hookup of short monomeric molecules into polymeric molecules of considerable length. These plastics are made into pipes by means of the extrusion press shown in Fig. 6.1—2. Such pipes are accordingly called extruded plastic pipes. A powder of the plastic to be used is poured into funnel *1*, from where it is pressed by screw *2* through channel *5* in electrically heated (*3*) barrel *4* through perforated plate *6*. Its passage through the perforations serves to eliminate all 'memory of torsion' and to arrange the long molecules of the plastic more or less in the direction of the future pipe axis. The plastic mass is then extruded through channel *7*. Its final size and shape are imparted by the ring aperture in extruder head *8* (Oates 1962). Thermoplastic materials have the advantage that scrap and reject pipes can be reused to make pipe no worse than that which is made of primary material.

Thermosetting plastics or molding resins are usually formed by polycondensation of monomers. The uniting of monomers in this fashion gives rise to some byproduct such as water, CO_2 or ammonia. Pipes are made similarly to cast-iron pipes, by static or centrifugal casting. The resin is cast when still liquid, and then solidified and aged by chemical means or by heating. After aging and cooling, the pipe will keep its size and shape and will not change them even on further heating. The material of the ready pipe cannot be re-used. Pipes made of such plastics are, however, rather heavy and brittle. Resins are therefore usually combined with fiberglass textiles; either resinsoaked fiberglass textile is wound onto a pipe core (in which case the pipe wall grows from inside outward), or by centrifugal casting (by centrifugally spraying the resin onto a fiberglass-textile shell held in place by a casting cylinder, in which case the pipe wall grows from outside inward). The technology used must not permit air bubbles to form, as these will act as local foci of weakness.

Thermoplastic materials most often employed are as follows. *PVC* was first produced in the middle thirties in Germany, by the polymerisation of vinyl chloride, a derivative of acetylene. Several means of improving its quality have been devised since. Pipes made of pure PVC (PVC—1) have comparatively high tensile strength but are brittle and liable to breakage. Brittleness is reduced by adding 3 percent synthetic rubber to pure PVC (PVC—2). This product is resistant to both mechanical and corrosive action. Both types of PVC are paraphobic vis-à-vis certain paraffin compounds. According to Crumback (1964), increasing the ductility of PVC—1 by other means resulted in a cheap high-quality pipe material: *Cralastic* or *ABS*. This is a variant of the synthetic rubber Buna first used in Germany during World War II. It is essentially a copolymer of acrylonitrile-butadiene and butadiene-styrene. Its strength depends comparatively little on temperature. It has no paraphobic property, though. *CAB* is a mixed esther of cellulose with butyric and acetic acids. Its strength is least among all extrudable plastics, and is even further reduced rather substantially by acidic or alkaline solutions. The drop in strength is quite fast, however, and any further loss of strength is insignificant. CAB is paraphobic vis-à-vis certain paraffins. *Polyethylene* results from the polymerisation of ethylene at high pressure, in the presence of benzene peroxide catalyst. Its specific weight is 0.92, least among all types of plastic. It is especially resistant to chemical influences, and is therefore popular as a corrosion-resistant lining.

Of the thermosetting plastics, those most suitable for making pipes are polyesther and epoxy resins. Polyesthers are polycondensation products of polyvalent alcohols (glycol, glycerine etc.) with polyvalent acids (phtalic, maleic acid, etc.). Epoxy resins are produced by condensing diene with epichlorhydrine in the presence of caustic soda. Both resins will harden cold in the presence of a suitable catalyst, but also on heating without a catalyst. Hardening results from the formation of a three-dimensional lattice. The product is a resistant, high-strength plastic. Polyesther resins have considerable shrinkage; that of epoxy resins is less pronounced. Resins combined with glass fibers (fibercast) are of even higher strength

Table

Typical parameters of plastic pipe materials (condensed

	$d^{24}_{15,5}$	Tensile strength σ_B $\frac{MN}{m^2}$	Long-range tensile strength σ_{eff} $\frac{MN}{m^2}$	Impact strength $\frac{kNm}{m^2}$	Elastic modulus $\frac{GN}{m^2}$
	1	2	3	4	5
PVC-I	1.38	48.3	17.9	1.7	0.34
PVC-II	1.35	41.4	13.8	31	1.7
Cralastic	1.06	32.4	13.8	21—31	1.6
CAB	1.20	31.0	6.9	3.8	1.0
Polyethylene	0.92—1.00	6.9	2.6	34	0.1
Polyesther resin	1.55	276	51.7	31	11
Epoxy resin	?	276	103	?	—
Steel	7.90	386	—	73	207

than pure resins; also, their resistance to heat and to water absorption is more favourable.

Tensile strengths of plastics v. temperature are shown in Fig. 6.1—3. The ordinate is calibrated in percent of the tensile strength at room temperature. Decrease with increasing temperature is seen to be linear in the case of PVC, whereas in some fiberglass-filled resins strength is highest at 80 °C.

Figure 6.1—4 (Greathouse and McGlasson 1958) refers to PVC pipes, but the graphs of other plastics are of a similar nature. The graphs are plots of increase in diameter v. time for various tangential or hoop stresses σ_t. Clearly, as opposed to steel pipes, the increase in diameter is a slow process (creep) in plastics. There is a critical tangential or hoop stress

$$\sigma_t = \frac{p(r_i^2 + r_o^2)}{r_o^2 - r_i^2}$$

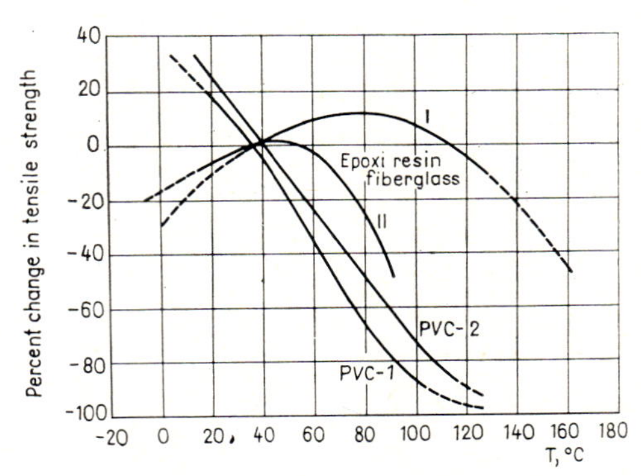

Fig. 6.1—3 Tensile strength of plastics v. temperature (Greathouse 1958)

6.1—5
and modified from Greathouse and McGlasson 1958)

Temperature range of application C°	Water adsorption in 24 h %	Decrease in (3) in weak acids and bases in 30 d %	Decrease in (3) in oil over 18 months %	Maintenance cost index; steel = 1	Wax deposition
6	7	8	9	10	11
— 6 + 60	0.06	?	?	16	±
—12 + 60	0.07	0	?	0	±
—29 + 60	0.3	0	15—19	5.4	—
—23 + 49	2.1	12—23	0—24	8.9	±
—46 + 49	0.01	?	?	?	?
—68 + 66	0.3	highly	favourable	?	?
? +126	?	highly	favourable	?	?
—46 +316	0.0	—	—	1	—

below which the increase in diameter will cease after a certain time. If, however, this critical hoop stress is exceeded, the pipe will go on swelling until it ruptures. Tensile strength in Column 2 of Table 6.1—5 is strength under a 'short-duration' internal overpressure. *Sustained tensile strength* is, on the other hand, the value belonging to the critical hoop stress at a given temperature.

Paraphobic behaviour means that the paraffin crystals forming in the fluid will not adhere to the pipe wall and that, consequently, no wax deposit will form in the pipe. Since a given plastic will usually be paraphobic to certain paraffin compounds but not to all, the degree of paraphobia is to be determined experimentally in any particular case of application. The plastic will not, of course, inhibit the formation of paraffin crystals, and thus the

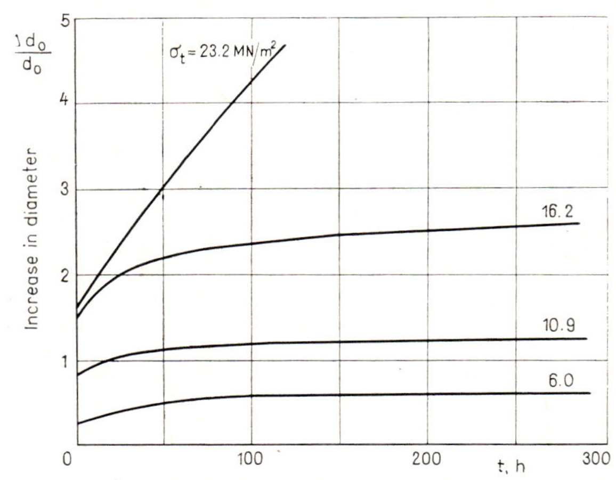

Fig. 6.1—4 Increase of diameter v. time of PVC-1 pipe (Greathouse and McGlasson 1958)

precipitation of more or less loose aggregates of such crystals at slow rates of flow is to be reckoned with in any case, and means for their removal are to be provided.

According to data published in 1961, 98 percent of the plastic pipelines then in use were made of thermoplastic pipe (Pylant 1961). The percentages of various kinds of thermoplastics used in pipes serving various purposes in oil and gas fields are given by Table 6.1—6. Prescriptions concerning the materials, manufacture, sizes and testing of plastic pipes are contained in API Tent. Std. 5LP—1968 (for thermoplastics) and API Tent. Std. 5LR—1971 (for resins).

Table 6.1—6
Plastic pipe in oil and gas field service
Percent of total footage in ground (after Pylant 1961)

Use	ABS	PE	PVC	CAB	Others	Total
Crude gathering	58	8	18	12	4	100
Gas gathering	1	97	1	1	0	100
Salt water disposal	52	6	15	4	23	100
Miscellaneous	1	58	28	4	9	100

The fact that plastics combine some advantageous properties with a strength that is lower than that of steel and is influenced, moreover, by various external circumstances, gave birth to the idea of combining the two materials by lining steel pipes with plastic. The first experiments of thus lining pipe intended for transporting corrosive oil and gas date back to 1947. Commercial-scale application was initiated in 1959. New pipe can be plastic-lined at the pipe yard, but pipelines already in use can also be lined on-site. Experience has shown the plastic best suited for this kind of lining to be epoxy resin. The term, however, covers quite a large group of substances; experiments have shown only one of some 60 epoxy compounds to be suitable in all respects (Kut 1969). Prior to preparing the lining of 37—50 μm thickness, the pipe wall is cleaned first mechanically, e.g. with a wire brush designed for this purpose, and then chemically, by an acid treatment, finally rinsed with water containing phosphates in a low concentration, and dried. Pipelines in use are lined with epoxy resin in the following way. First, a special plug is inserted into the pipe. This consists of rubber cones placed between two rubber disks and held together by rubber couplings. The liquid plastic is filled in between two such plugs. Both ends of the pipeline are then filled with compressed air, and pressures are adjusted so as to drive the plugs slowly along the pipe. A film of plastic will be left behind on the pipe wall; its thickness depends on the flow properties of the plastic, the clearance between the plugs and the pipe wall, and the rate of advance of the plugs. A good lining is smooth and durable and is not peeled off by a brush scraper. Its application reduces the hydraulic roughness of the pipe, and thereby improves its throughput capacity; it reduces or prevents the deposition of paraffin on the wall, and eliminates the hazard of corrosion; the pipe will not introduce rust scales into the fluid, and the wear on pipe fittings and meters will thus be reduced. A special advantage in low-pressure pipelines is that the plastic will plug small punctures or cracks, and thereby reduce or stop leakages. According to Soviet data, the price of a high-quality lining is 15—20 percent of the pipeline's first cost. — Recommendations concerning materials, preparation and testing of the internal linings of gas pipelines are contained in API RP 5L2—1968.

6.1.4. Wall thickness of pipes

The wall thickness of pipes used in transporting oil and gas can be calculated by a formula for thin-walled pipes,

$$s = \frac{d_0 \Delta p}{2\, e \sigma_{al}} \qquad 6.1\text{—}1$$

Δp is the pressure differential across the pipe wall. In steel pipes, e is a figure of merit of weld seams. Its value is 1 for seamless pipe, and 0.7—0.9 for axially welded pipe. In spiral-welded pipe, one should according to Siebel divide e by the ratio $\sigma_\alpha/\sigma_\varphi$, where σ_α is the tangential stress perpendicular to the seam and σ_φ is the tangential stress in a cross-section of the pipe perpendicular to its axis. In spiral-welded pipe used in practice, the angle α, the pitch of the seam, is greater than 40°, and thus $\sigma_\alpha/\sigma_\varphi = 0.8—0.6$. The greater α, the less is the ratio $\sigma_\alpha/\sigma_\varphi$. If, then, $\alpha \leq 40°$ and $\frac{e}{\sigma_\alpha/\sigma_\varphi} \geq 1$, this means that the locus of least strength is in the steel sheet itself rather than in the weld seam, and it is permitted to put $e = 1$ in Eq. 6.1—1 (Stradtmann 1961). According to DIN 2413,

$$\sigma_{al} = \frac{\sigma_F}{k} \qquad 6.1\text{—}2$$

if temperature is below 120 °C and the internal pressure is non-fluctuating. Here, σ_F is yield strength and k is a safety factor whose value ranges between 1.6 and 2.0. (Incidentally, yield strength in the strict sense is the stress at which permanent deformation sets in under a constant, non-fluctuating stress. This phenomenon cannot be observed on every metal and alloy. It is usual in such cases to consider a permanent deformation of 0.2 percent as yield and the corresponding stress as yield strength, and to denote it $\sigma_{0.2}$ rather than σ_F.) — According to the Hungarian prescriptions in force, the safety factors given in the Table on p. 382 apply to pipelines transporting crudes and oil products (1) and gas and liquefied gas (2).

Equation 6.1—1 essentially agrees with the Barlow outside-diameter formula used in the USA. Safety factors for gas pipelines are subdivided in four groups also in the USA, on the basis of considerations similar to those outlined above, and the values of k to be used are also close to those given under (2) in the Table (Tube Turns Division 1969). When using

	(1)	(2)
(a) Open country, ploughland, woods etc.	1.3	1.4
(b) At a distance of 100—200 m from inhabited areas, railroads, arterial roads, etc.	1.5	1.7
(c) Within 100 m from same	1.7	2.0
(d) In industrial and densely inhabited areas, under railroads, arterial roads and water-courses	2.0	2.5

plastic pipe, σ_{al} in Eq. 6.1—1 is to be replaced by σ_B and the righthand side is to be multiplied by a temperature factor k_t, that is,

$$s = \frac{d_o \Delta p}{2\sigma_B} k_t \qquad 6.1-3$$

where σ_B is the effective or sustained tensile strength described in connection with Fig. 6.1—4. For the plastics discussed there, its value can be read off graphs similar to those in Fig. 6.1—3 for various operating temperatures.

Example 6.1—1. Find the required wall thickness of spiral-welded pipe of $d = 10\ 3/4$ in. nominal size, made of X52 grade steel, to be used in a gas pipeline laid across ploughland and exposed to an operating pressure of 60 bars. — From Table 6.1—2, $\sigma_F = 359$ MN/m². By the above considerations, $k = 1.4$. Hence, by Eq. 6.1—1,

$$s = \frac{0.273 \times 60 \times 10^5}{2 \times \frac{359 \times 10^6}{1.4}} = 3.04 \times 10^{-3} \text{m} = 3.04 \text{ mm}$$

Table 6.1—1 shows the least standard wall thickness at 10 3/4 in. nominal size to be 3.96 mm. This thickness may, then, be chosen.

6.2. Valves; pressure regulators

6.2.1. Valves

Valves used in oil and gas pipelines may be of gate, plug or ball, or globe type. All of these have various subtypes.

(a) Gate valves

The gate valve is a type of valve whose element of obstruction, the valve gate, moves in a direction perpendicular to both inflow and outflow when the valve is being opened or closed. Its numerous types can be classified according to several viewpoints. In the following we shall consider them according to their mode of closure. Typical examples are shown as Figs 6.2—1—6.2—4. Of these, 6.2—2 and 6.2—3 are conformable to API Std.

6D—1971. The main parts of a gate valve are termed as follows: handwheel *1*, stem *2*, gland *3*, stem packing *4*, bonnet *5*, bonnet bolts *6*, body *7*, wedge or disc *8*, and seat *9*. The element of obstruction in the valve shown as Fig. 6.2—1 is wedge *8*. This valve is of a simple design and consequently rather low-priced, but its closure is not leakproof. First of all, it is difficult to grind the wedge to an accurate fit, and secondly, the obstructing element will glide on metal on closure; deposits of solid particles will scratch the surfaces and

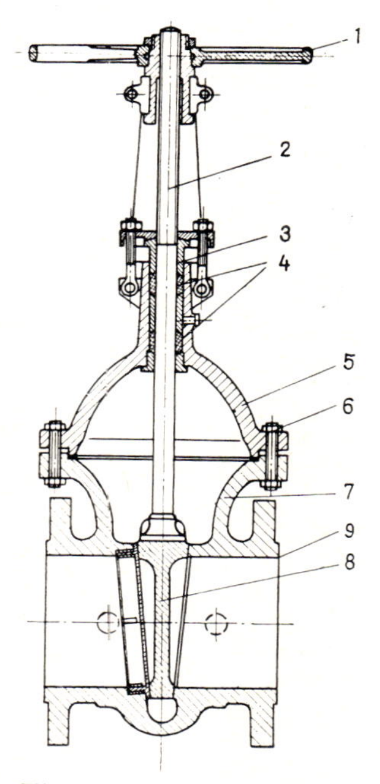

Fig. 6.2—1 Wedge-type rising-stem gate valve

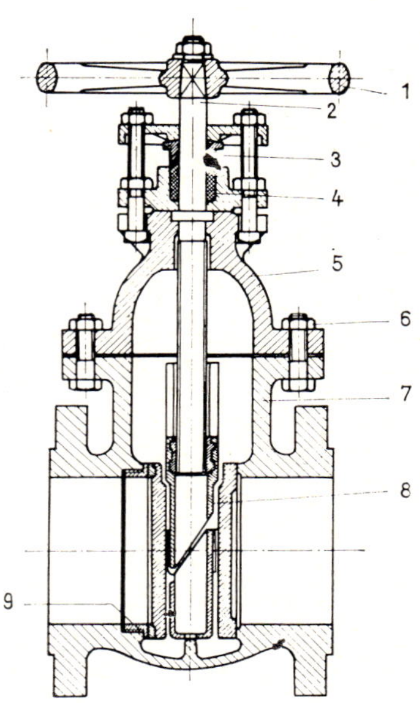

Fig. 6.2—2 Disk-type non-rising-stem gate valve, after API Std 6D-1971

thus produce leaks. Fairly widespread earlier, this type of valve has lost much of its popularity nowadays. — The valve in Fig. 6.2—2 is of the double-disk non-rising-stem type. Its tightness is much better than that of the foregoing type, because the two disks do not glide on the seats immediately before closure or after opening; rather, they move nearly at right angles to the same. The two elements of obstruction, machined separately, are more easily ground to a tight fit. The valve in Fig. 6.2—3 is of the rising-stem, through conduit type with a metallic seal. It has stem indicator *1* as a typical structural feature. One of the advantages of this design is that the inner wall of the valve is exposed to no erosion even if the fluid contains solids. The elements of obstruction cannot wedge or stick. Overpressure in the valve is automatically released by safety valve *2*. Its drawback is that, owing to the metal-to-metal seal, tightness may not always be sufficient for the purposes envisaged in the oil and gas industry. Furthermore, the fit of the disks may be damaged by scratches due to solid particles. The biggest valves of this type now in use are made for 1000 mm nominal size and 66 bars operating pressure rating (Laabs 1969). — Figure 6.2—4 shows a through conduit-type, non-rising-stem floating-seat valve. The so-

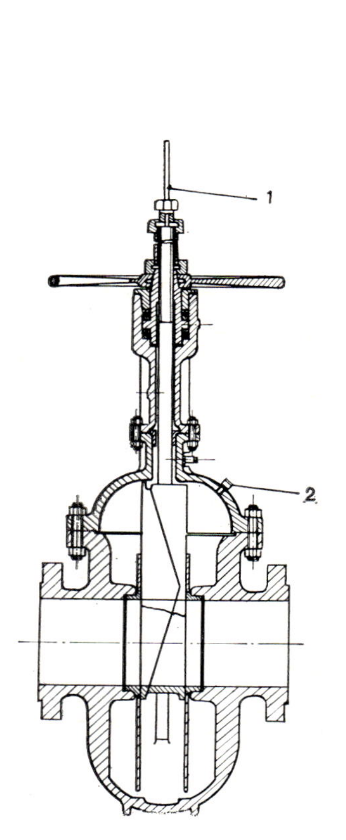

Fig. 6.2—3 Full-opening rising-stem gate valve, after API Std 6D-1971

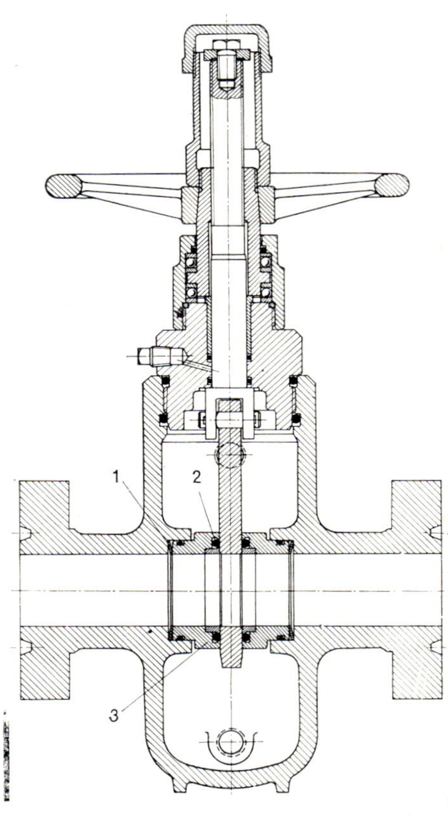

Fig. 6.2—4 Grove's G-2 full-opening floating-seat gate valve

called floating seats *3*, capable of axial movement, and provided with 0-ring seals *2*, are pressed against the valve gate by spring *1*. The pressure differential developing on closure generates a force which improves the seal. (Concerning this mechanism cf. the passage on ball valves in Section 6.2.1b.) Small solid particles will not cause appreciable damage to the sealing surfaces, but coarse metal fragments may. For high-pressure large-diameter applications, variants of this type of gate valve with the valve body welded out of steel plate are gaining in popularity.

The valves described above have either rising or non-rising stems. Rising stems require more space vertically. In a rising stem, the thread is on the upper part, likely to be exposed to the outer environment. In a non-rising stem, the thread is on the lower part, in contact with the well fluid. It is a matter of correctly assessing the local circumstances to predict whether this or that solution will more likely prevent damage to the threads. In rising-stem valves the handwheel may or may not rise together with the stem. In modern equipment, the second solution is current (cf. e.g. Fig.

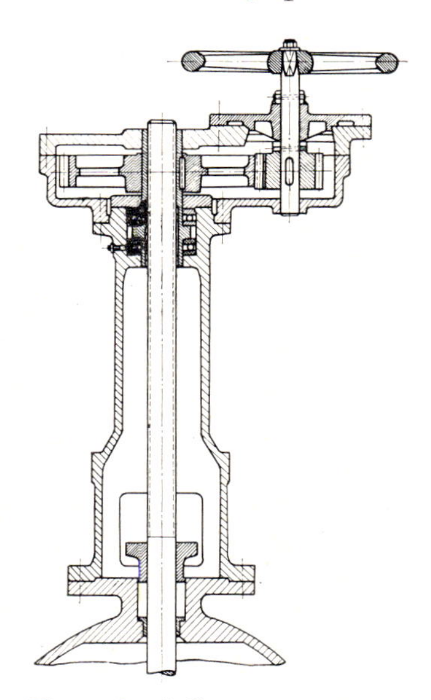

Fig. 6.2—5 Spur-gear gate-valve stem drive

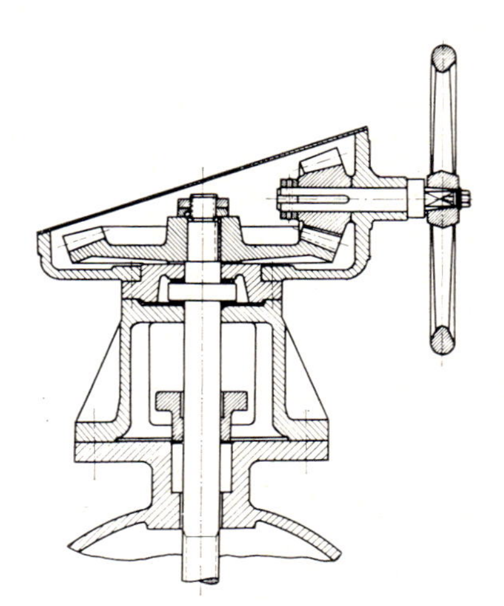

Fig. 6.2—6 Bevel-gear gate-valve stem drive

6.2—3). Turning the stem may be effected directly, by means of a handwheel mounted on the stem (as in all types figured above), but also indirectly by transferring torque to the stem by a spur or bevel gear (Figs 6.2—5 and 6.2—6).

The valve may, of course, be power-operated, in which case the drive may be pneumatic, hydraulic or electric. Figure 6.2—7 shows a rising-stem valve with electric motor and a bevel gear drive. — Gate valves used in oil and gas pipelines usually have flange couplings, but in some cases plain-end valve bodies are used with flanges attached by welding, or threaded-end ones with couplings. At low operating pressures, the outer face of the flange, to lie up against the opposite flange, is either smooth or roughened. Ring gaskets used are non-metallic, e.g. klingerite. At higher pressures, it is usual to machine a coaxial groove into the flange, and to insert into it a metallic gasket, e.g. a soft-iron ring. — Operating-pressure ratings of gate valves range from 55 to 138 bars according to API Std 6D—1971. The standard enumerates those ASTM grades of steel of which these valves may be made. — Gate valves are to be kept either fully open or fully closed during normal operation. When partly open, solids in the fluid will damage the seal surfaces in contact with the fluid stream.

(b) Plug and ball valves

In these valves, the rotation by at most 90° of an element of obstruction permits the full opening or full closing of a conduit. The obstructing element is either in the shape of a truncated cone (plug valve) or of a sphere (ball valve). Figure 6.2—8 shows a plug valve where gland *1* can be tightened to the desired extent by means of bolts. There are also simpler designs. The contacting surfaces of the plug and seat must be kept lubricated in most plug valves. In the design shown in the Figure, lubricating grease is injected at certain intervals between the surfaces in contact by the tightening of screw *2*. Check valves *3* permit lubrication also if the valve is under pressure. Surfaces in contact are sometimes provided with a molybdene sulphide or teflon-type plastic lining. In full-opening plug valves, the orifice in the plug is of the same diameter as the conduit in which the valve is inserted. This design has the same advantages as a through conduit-type gate valve.

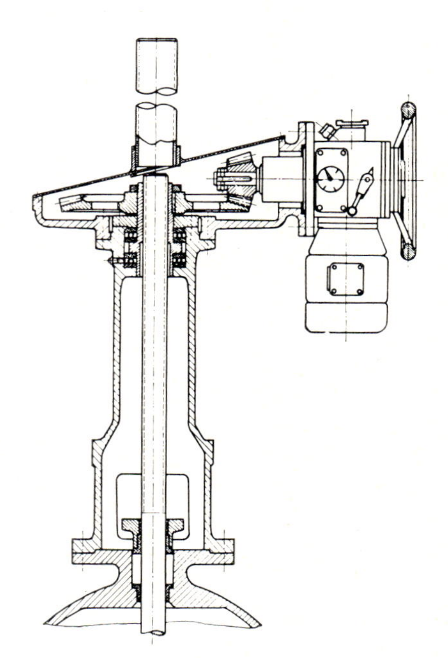

Fig. 6.2—7 Bevel-gear gate-valve stem drive, operated by electric motor

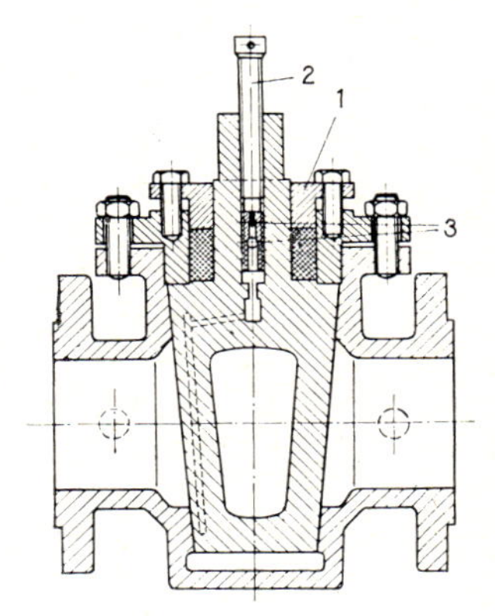

Fig. 6.2—8 Plug valve

In the oil and gas industry, ball valves have been gaining popularity since the early sixties. They have the substantial advantage of being less high and much lighter than gate valves of the same nominal size, of providing a tight shutoff and of being comparatively insensitive to small-size solid particles (Belházy 1970). There are two types of ball valve now in use

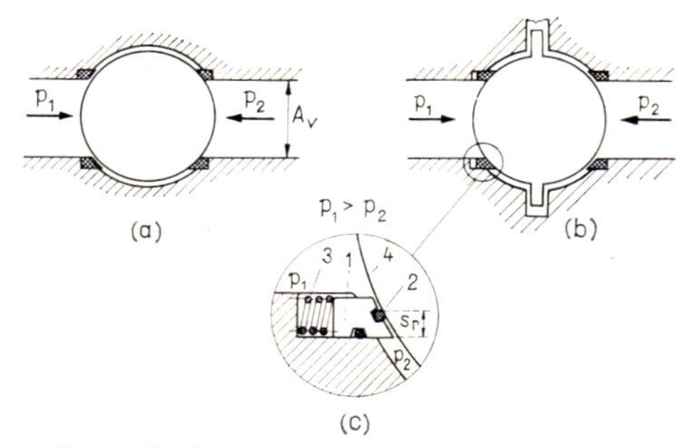

Fig. 6.2—9 Ball valve types, after Perry (1964)

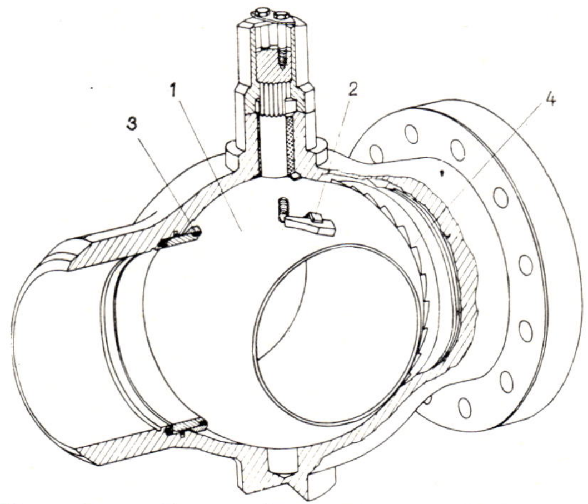

Fig. 6.2—10 Cameron's rotating-seat ball valve

(Perry 1964) shown in outline in Fig. 6.2—9. Solution (a) is provided with a floating ball, which means that the ball may be moved axially by the pressure differential and may be pressed against the valve seat on the low-pressure side. This latter seat is usually made of some sort of plastic, e.g. teflon. The resultant force pressing the ball against the seat is $F = (p_1 - p_2)A_v$, that is, it equals the differential in pressure forces acting upon a valve cross-section equal to the pipe cross-section A_v. If F is great, then the pressure on the plastic seal ring is liable to exceed the permissible maximum. It is in such cases that solution (b) will help. An upper and a lower pin integral with the ball will prevent pipe-axial movements of the ball. Sealing is provided by a so-called floating seat, exposed to two kinds of force. Referring to part (c) of the Figure, spring *3* presses metal ring *1* and the plastic O-ring seal *2* placed in a groove of the ring with a slight constant force against ball *4*. At low pressures, this in itself ensures sufficient shutoff. Moreover, it is easy to see that one side of surface $s_r d_v \pi$ is under a pressure p_1, whereas its other side is under pressure p_2. A suitable choice of ring width s_r will give a resulting pressure force $(F_1 - F_2)$ just sufficient to ensure a tight shutoff at higher pressures but not exceeding the pressure rating of the plastic O-ring. — Figure 6.2—10 shows a Cameron-make

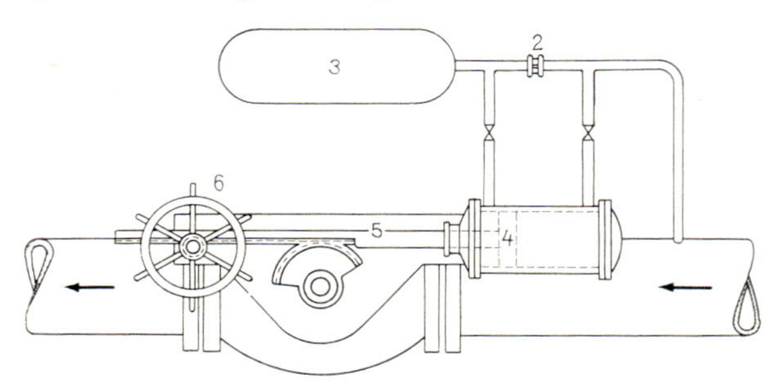

Fig. 6.2—11 Rockwell's automatic gas-line shutoff

rotating- and floating-seat ball valve. On each opening of ball *1*, seat *3* is rotated by a small angle against the pipe axis, by key *2* engaging the teeth *4* on the seat. This solution tends to uniformize wear and thereby to prolong valve life.

Ball and plug valves should be either fully open or fully closed in normal operation, just as gate valves should. The ball or plug can, of course, be power-actuated. Valve materials are prescribed by API Std 6D—1971. Of the power-driven valves available, let us mention the pneumatically-operated Rockwell-make valve shown in Fig. 6.1—11, used to automatically shut off gas pipelines. The valve will close automatically whenever some operating trouble causes pressure in the line to drop below a prescribed minimum. Valve *1* is fully open in normal operaton. Tank *3* is filled through choke *2*, with the gas transported in the pipe at a pressure equal to the normal line pressure. When a drop in line pressure occurs, the pressure differential across piston *4* makes the piston and the rack *5* attached to it move to the right and thus to close the valve. The venting of gas into the atmosphere from the damaged pipe section is thus prevented. The valve can be reopened by means of handwheel *6* (Palm 1962).

(c) Globe valves

The wide range of globe valve types has its main use in regulating liquid and gas flow rates. A common trait of these valves is that the plane of the valve seat is either parallel to the vector of inflow into the valve body, or the two include an angle of 90° at most. The element of obstruction, the so-called inner valve, is moved perpendicularly to the valve-seat plane by means of a valve stem. In the fully open position, the stem is raised at least as high as to allow a cylindrical gas passage area equal to the cross-sectional area of the pipe connected to the valve. A valve type in common use is shown in Fig. 6.1–12. The design of this type may be either single- or double-seat (the latter design is shown in the Figure). Single-seat valves have the advantage that inner valve and seat may be ground together more accurately to give a tighter shutoff. A high-quality valve should not, when shut off, pass more than 0.01 percent of its maximum (full-open) throughput. Double-seat valves tend to provide much lower-grade shutoff. The permissible maximum leakage is 50 times the above value, that is, 0.5 percent (Sanders 1969). The pressure force acting upon the inner valve of a double-seat valve is almost fully balanced; the forces acting from below and from above upon the double inner-valve body are almost equal. The force required to operate the valve is thus comparatively slight. The inner valve in a single-seat design valve is, on the other hand, exposed to a substantial pressure differential when closed, and the force required to open it is therefore quite great. According to its design and opening characteristic, the inner valve may be quick-opening, linear or equal-percentage. Part (a) of Fig. 6.2–13 shows a so-called quick-opening disk valve. It is used when the aim is to either shut off the fluid stream entirely or let it pass at full capacity ('snap action'). The characteristic of such a valve is graph (a) in Fig. 6.2–14. The graph is a plot of actual throughput related to maximal throughput v. actual valve travel related to total valve travel. Graph

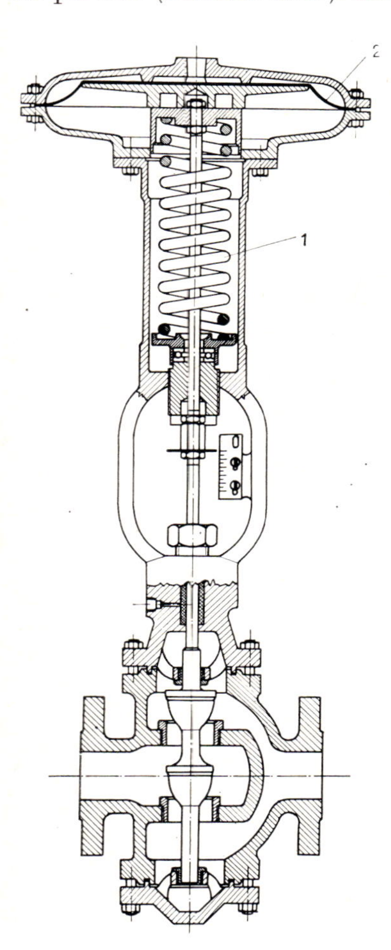

Fig. 6.2–12 Fisher's balanced diaphragm motor valve

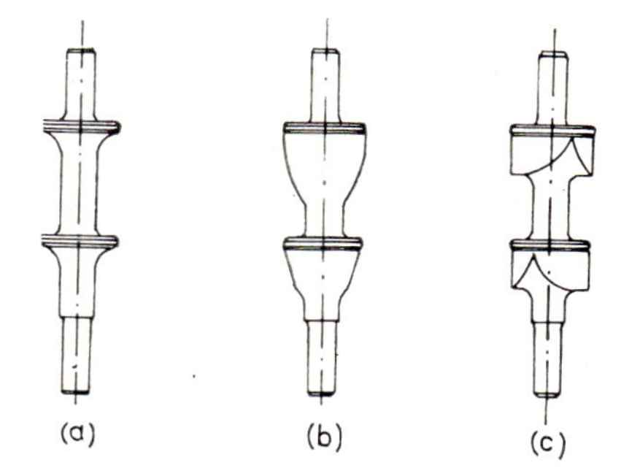

Fig. 6.2–13 Inner-valve designs

(a) shows that, in a quick-opening valve, throughput attains 0.8 times the full value at a relative valve travel of 0.5. The inner-valve design shown in part (b) of Fig. 6.2—13 provides the linear characteristic of graph (b) in Fig. 6.1—14. In such a valve, a given through-put invariably entails a given increment in throughput. This type of regulation is required e.g. in fluid-level control where, in an upright cylindrical tank, a given difference in fluid level invariably corresponds to a given difference in fluid content. The inner-valve design shown in part (c) of Fig. 6.2—13 gives rise to a so-called equal-percentage characteristic (part c, in Fig. 6.2—14).

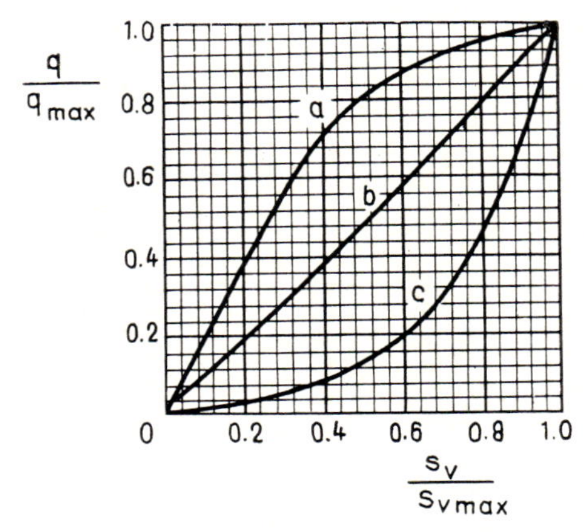

Fig. 6.2—14 Response characteristics of inner-valvedesigns

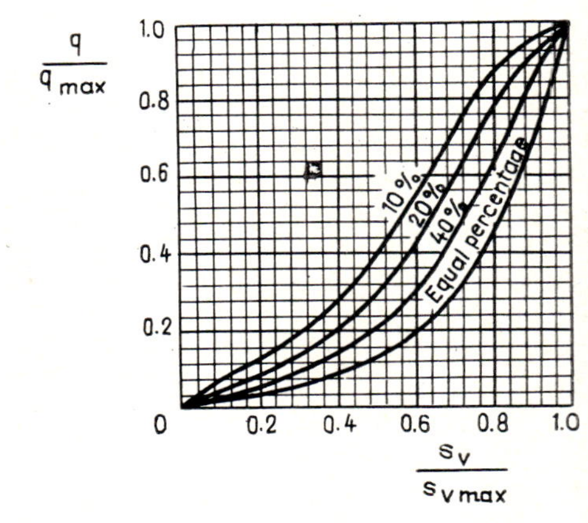

Fig. 6.2—15 Modified equal-percentage valve characteristics, after Reid (1969)

In a valve of this design, a given valve travel will invariably change the throughput in a given proportion. Let e.g. $q/q_{\max} = 0.09$, in which case the diagram shows $s_v/s_{v\max}$ to equal 0.40. If this latter is increased by 0.20 to 0.60, then $q/q_{\max}$ becomes 0.2, that is, about 2.2 times the preceding value. A further increase in relative valve travel by 0.2 (that is, to 0.8) results in a relative throughput of 0.45, which is again round 2.2 times the foregoing value. This is the type of regulation required e.g. in controlling the temperature of a system. To bring about a unit change of temperature in the system requires, say, 9 thermal units when the system is loaded at 90 percent but only 5 units if it is loaded at 50 percent (Reid 1969).

The above characteristics will strictly hold only if the head loss caused by the valve alone is considered; if, however, the flow resistance of the pipes and fittings attached to the valve is comparatively high, then increasing the throughput will entail a higher head loss in these. Figure 6.2—15 (after Reid 1969) shows how an equal-percentage characteristic is displaced if the head loss in the valve makes up 40, 20 and 10 percent, respectively, of the total head loss in the system. The less the percentage due to the valve itself, the more the characteristic is distorted, and the less can the desired equal-percentage regulation be achieved. Throughput of a valve is usually characterized by the formulae

$$q_l = K_v \sqrt{\frac{\Delta p}{\varrho_l}} \qquad 6.2\text{—}1$$

for liquids and

$$q_g = K_v' \sqrt{\frac{p_1^2 - p_2^2}{\varrho_g T}} \qquad 6.2\text{—}2$$

for gas. K_v and K_v' are factors characteristic of the valve's throughput, equalling throughput under unity pressure differential (respectively, unity pressure-square differential in the case of gas). The K factor will, of course, change as the valve travel is changed. In a good regulating valve, it is required that the K_v (or K_v') value obtaining at maximum throughput (at full opening) should be at least 50 times the value at the minimum uniform throughput permitted by the valve design (Reid 1969).

A widely used type of control valve is the Fisher-make double-seat spring-loaded motor valve of the normal-open type shown in Fig. 6.2—12. (This latter term means that the valve is kept open by spring *1* when no actuating gas pressure acts upon diaphragm *2*.) The valve is two-way two-position, because it has two apertures (one inlet and one outlet), and the inner valve may occupy two extreme positions (one open and one closed). — Other diaphragm-type motor valves include the three-way two-position motor valve shown as Fig. 6.2—16. At zero actuating-gas pressure, the fluid entering the valve through inlet *1* emerges through outlet *2*. If gas pressure is applied, then the upper inner valve will close and the lower one will open, diverting fluid flow to outlet 3. Figure 6.2—17 shows a three-way three-position motor valve. The diaphragm may be displaced by actuating-gas

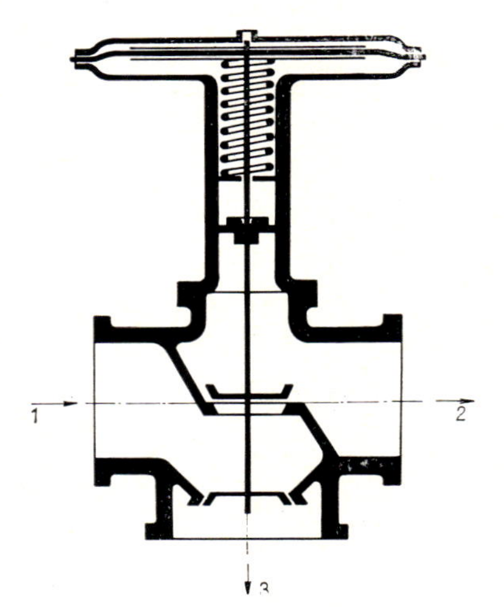

Fig. 6.2—16 Three-way two-position motor valve

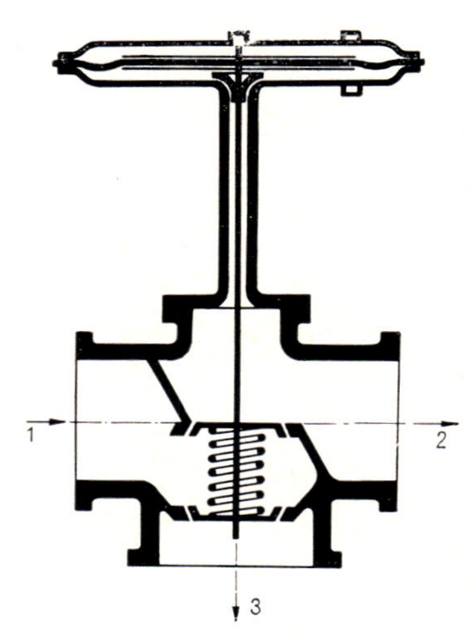

Fig. 6.2—17 Three-way three-position motor valve

pressure acting from either above or below. At zero actuating-gas pressure, the inner valves are pressed against their respective seats by the spring between them, and both outlets are closed. If gas pressure depresses the diaphragm, the upper inner valve is opened and the fluid may emerge through outlet *2*. If gas pressure lifts the diaphragm, the lower inner valve is raised while the upper one is closed, and the fluid stream is deflected to pass through outlet *3*.

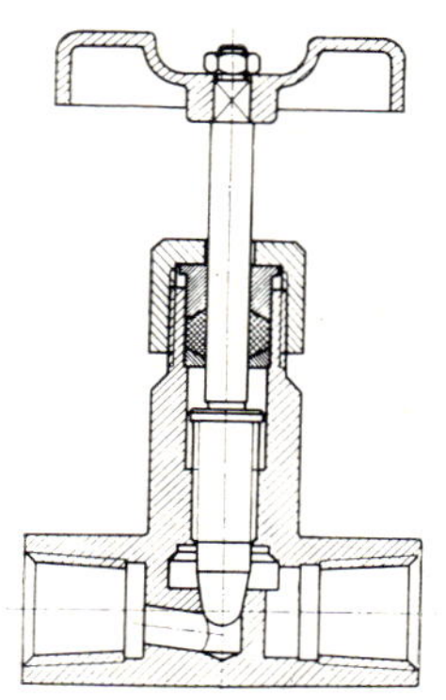
Fig. 6.2—18 Needle valve

Of the host of special-purpose valves, let us specially mention the needle valve, one design of which, shown as Fig. 6.2—18, can be used to control high pressures in small-size piping of low throughput, e.g. in a laboratory. It is of a simple construction easy to service and repair. — Check valves permit flow in one direction only. Figure 6.2—19 shows a Grove make 'Chexflo' type check valve. The tapering end of elastic rubber sleeve *1* tightly embraces the cylindrical part of central core *2* if there is no flow in the direction shown in part (a) of the Figure. Even a slight pressure differential from left to right will, however, lift the sleeve from the core and press it against the valve barrel. The valve is thus opened. A pressure differential in the opposite sense will prevent flow by pressing the sleeve even more firmly against the core. — Actuating gas for diaphragm-controlled motor valves is often controlled in its turn by a pilot valve. Figure 6.2—20 shows an electromagnetic pilot valve as an example. Current fed into solenoid *1* will raise soft-iron core *2* and the inner valve attached to it. — Valve choice is governed by a large number of factors, some of which have been discussed above. Table 6.2—1 is intended to facilitate the choice of the most suitable design.

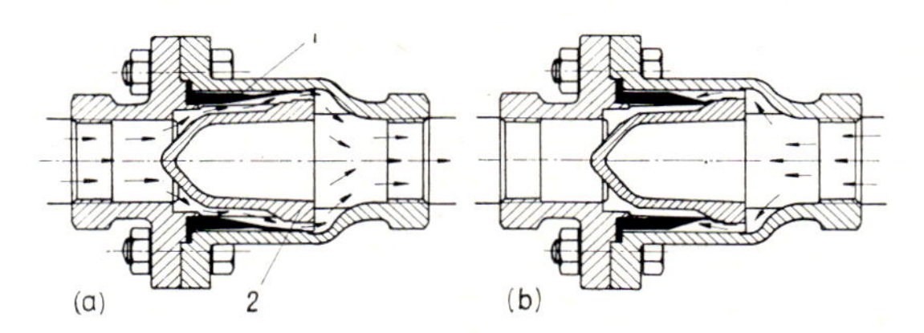

Fig. 6.2—19 Grove's 'Chexflo' check valve

6.2.2. Pressure regulators

Gas and gas-liquid streams are usually controlled by means of pressure regulators without or with a pilot. A rough regulation of comparatively low pressures (up to about 30 bars) may be effected by means of automatic pressure regulators without pilot. These may be loaded by a spring, pressure

Table 6.2—1

Advantages of various valve designs (personal communication by J. Bognár)

Design features	Low head loss	Pipe scraping is possible	Long life	Tight shutcff	Ease of operation	Speed of operation	Applicability			Small space demand	Quick access
							High pressure	Corrosive medium	Corrosive surroundings		
Full opening	×	×									
Seal of Element of obstruction — metallic			×				×				
Seal of Element of obstruction — elastic				×							
Seal of Element of obstruction — mixed			×	×			×				
Exchangeable seat								×			
Element of obstruction balanced					×		×				
Stem turning on bearing					×		×				
Inside-thread stem									×		
Outside-thread stem								×			
Bolted bonnet										×	
Ring-held bonnet											×
Bolted stuffing box										×	
Thread stuffing box											×
Lever-operated						×					
Power-operated					×						

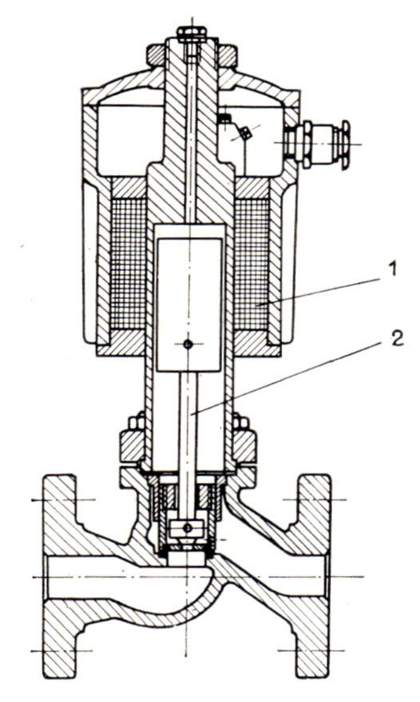

Fig. 6.2—20 MAW magnetic pilot valve

or deadweight, or any combination of these. Figure 6.2—21 shows an automatic regulator loaded by a deadweight plus gas pressure, serving to regulate the pressure of outflowing gas. Weight *1* serves to adjust the degree of pressure reduction. If downstream pressure is higher than desired, the increased pressure attaining membrane *3* through conduit *2* will depress both the diaphragm and the inner valve *4*, thereby reducing the flow cross-section and the pressure of the outflowing gas. If downstream pressure is lower than required, the opposite process will take place. Automatic

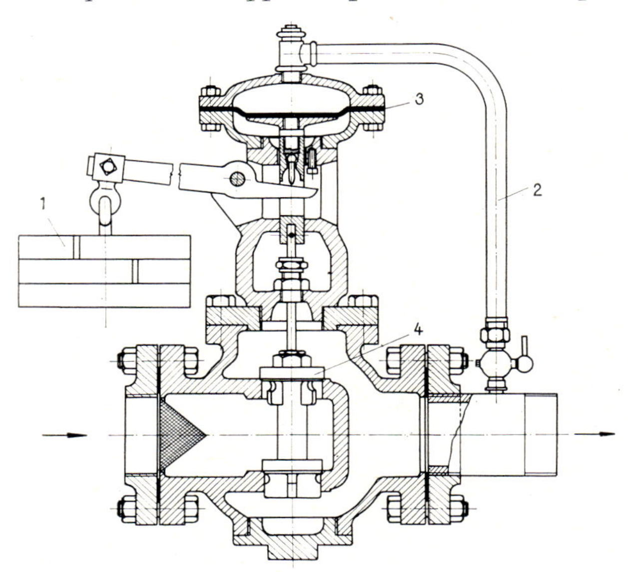

Fig. 6.2—21 Deadweight-loaded, automatic pressure regulator

pressure regulators have the advantage of simple design and a consequent low failure rate and low price, but also the drawback that, especially at higher pressures, they will tend to provide a rather rough regulation only. Control by the regulated pressure is the finer the larger the diaphragm surface upon which said pressure is permitted to act. A large diaphragm, however, entails a considerable shear stress on the free diaphragm perimeter. A thicker diaphragm is thus required, which is another factor hampering fine regulations.

A pilot-controlled pressure regulator is shown in outline in Fig. 6.2—22 (Petr. Ext. Service 1956). The regulated pressure acts here on a pilot device rather than directly upon the diaphragm of the motor valve. Pressure of gas supplied from a separate source is reduced to a low constant value by automatic regulator *1*. This supply gas attains space *3* through orifice *2* of capillary size. If vent nozzle *4* is open, then supply gas is bled off into the atmosphere. If downstream pressure increases, Bourdon tube *5* strives to straighten, thereby forcing the lower end of flapper *6* towards nozzle *4*. Gas pressure will therefore increase in space *3* above diaphragm *7* and the inner valve will travel downwards. Control sensitivity may be adjusted by displacing pivot *8*, whereas screw *9* serves to adjust the degree of pressure reduction.

In pressure regulation, normal-open motor valves are often replaced by normal-closed ones. These latter may be of the type e.g. of the motor valve shown in Fig. 6.2—12, with the inner valve inserted upside down, that is, closing when pressed against the seat from below. Choice between the two options is to be based on a consideration of whether an open or a closed valve will cause less havoc, should the actuating-gas pressure fail. If the pressure conduit of the motor valve is hooked up to the upstream side, then the device will, without any further change, regulate the upstream pressure. If vent nozzle *4* is installed on the opposite side of flapper *6*, then control will be inverted, that is, pressure above the diaphragm will decrease as the controlled pressure increases. Figure 6.2—23 presents the main modes of hookup of pressure regulators involving the usual types of pressure-loaded motor valve (after Petr. Ext. Service 1956; cf. also Table 6.2—2).

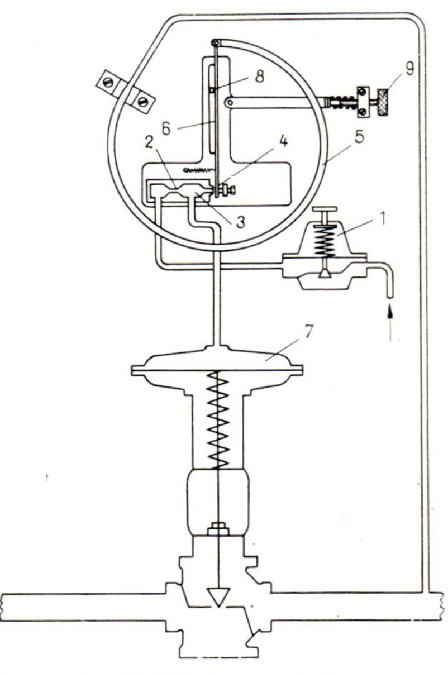

Fig. 6.2—22 Pilot-operated pressure regulator (Petr. Ext. Service 1956)

Figure 6.2—24 is a sketch of a Grove-make gas-dome-type pressure regulator. On installation of the valve, dome *1* is filled through conduits not shown in the Figure with gas at a pressure sufficient to exert upon diaphragm *2* a force bringing about the pressure reduction desired. This gas is trapped in the dome by means of needle valves. The lower surface of the diaphragm is exposed to the reduced pressure. It is primarily the

Table 6.2—2
Behaviour of main gas pressure regulator types

	Controlled gas flow is					
	downstream			upstream		
Control	direct	reverse	reverse	reverse	direct	direct
Downward valve travel will	close	open	close	close	open	close
On actuating-gas failure, valve will	open	close	close	open	close	close
On increase of controlled pressure, valve will	close	close	close	open	open	open
Type code	Ia	Ib	Ic	IIa	IIb	IIc

pressure differential across the diaphragm that determines the position of inner valve *3*. The regulating action is improved by the gas space of the dome being divided in two by capillary *4*. If the controlled pressure rises above a certain maximum, a sudden pressure increase in the space below the capillary will occur and generate a considerable back-pressure. The great advantage of this type of reducing valve is its simplicity and reliability

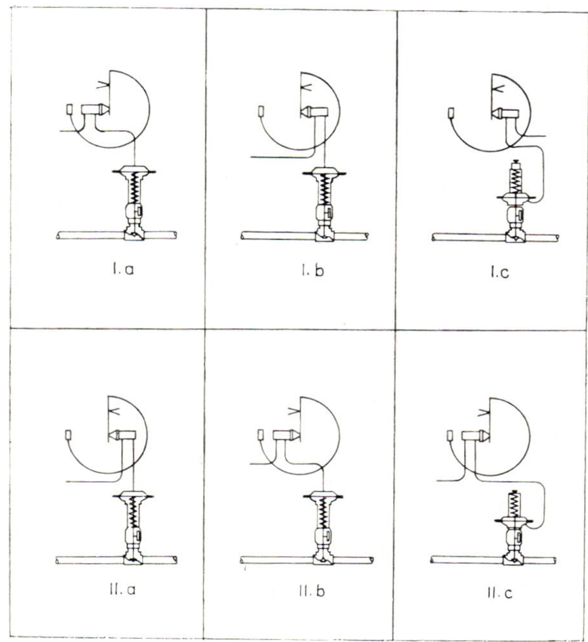

Fig. 6.2—23 Pressure-regulator hookups (Petr. Ext. Service 1956)

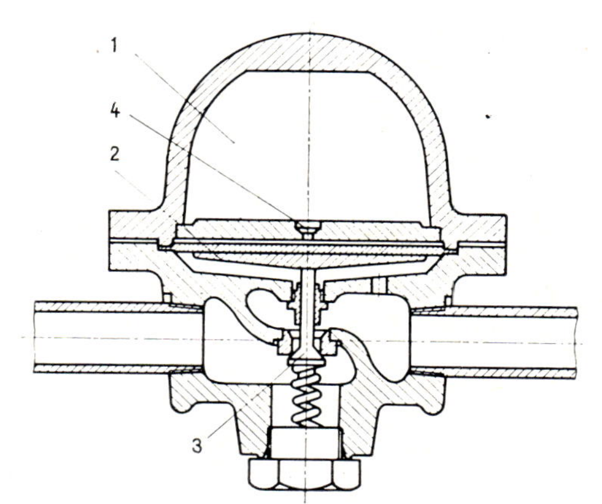

Fig. 6.2—24 Grove's gas-dome pressure regulator

of operation. Its disadvantage is that the regulating pressure in the dome depends, of course, on ambient temperature. This design is therefore best suited for controlling low-rate gas flows in places of constant temperature. — The Grove Model 16 pressure regulator is likewise suited for controlling small gas throughputs (Fig. 6.2—25), e.g. to provide pressure-loaded motor valves with supply gas. The various sizes of this model can reduce the input pressure of round 70 bars to practically any pressure not less

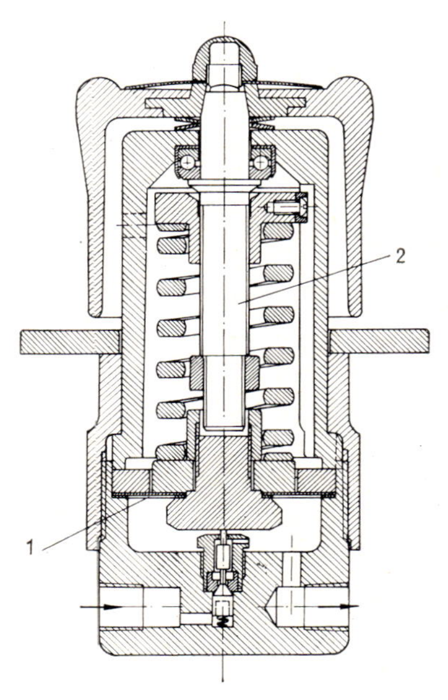

Fig. 6.2—25 Grove's M-16 type pressure regulator

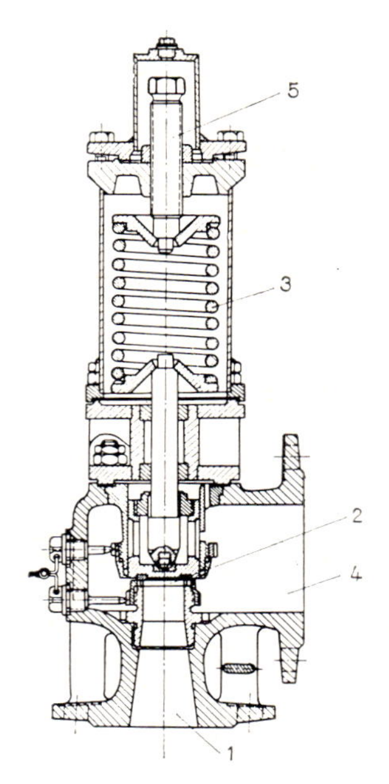

Fig. 6.2—26 Giproneftmash's SPPK-16 type spring-loaded safety valve

than 0.07 bar. Its operation resembles that of the gas-dome pressure regulator, the main difference being that diaphragm *1* is loaded by a spring adjusted by screw *2* rather than by the temperature-dependent pressure of gas in a dome. — Safety valves will blow off fluid or gas whenever pressure in the vessel containing them exceeds a given maximum. Figure 6.2—26 shows a SPPK—16 type Giproneftmash-make spring-loaded safety valve. If the force acting on valve disk *2* in the space connected with port *1* exceeds the downward force of spring *3*, the valve opens and permits the fluid to vent through outlet *4*. Opening pressure may be adjusted by modifying the spring force by means of screw *5*.

Most of the numerous other known types of pressure regulators and safety valves embody operating principles similar to those outlined above.

6.3. Internal maintenance of pipelines

On the walls of oil and sometimes also of gas pipelines, wax deposits will tend to form. Wax is composed of paraffin compounds, ceresin, asphalt and sand. The deposit reduces the cross-section open to flow, and its removal must therefore be provided for. Liquid hydrocarbons and water may condense in pipelines transporting gas. This liquid, which may contain solids,

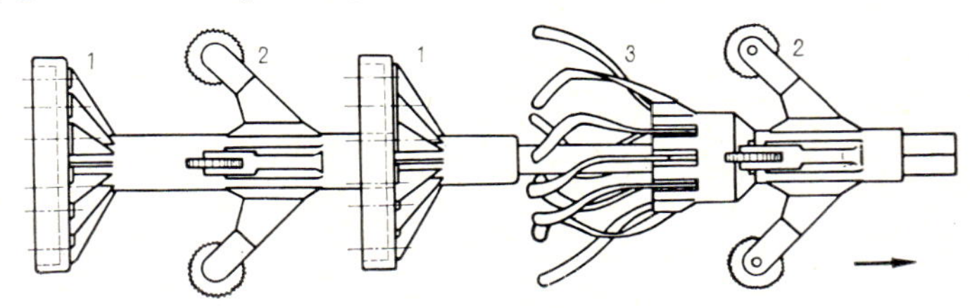

Fig. 6.3—1 Go-devil

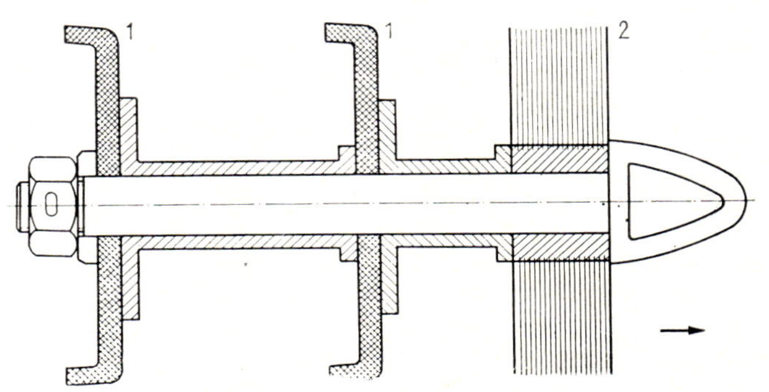

Fig. 6.3—2 General Descaling scraper for small-size pipelines

too, will reduce on the one hand the throughput capacity of the pipeline; on the other, liquid entering gas burners together with the gas may cause serious trouble. The accumulation of liquid in gas pipelines must therefore be prevented. Deposits are usually removed by means of scrapers (go-devils) or cleaning pigs inserted in the pipeline, at intervals depending on the rate of deposition, and moved along by the fluid or gas stream.

The go-devil (Fig. 6.3—1) is the type in longest use for cleaning the inner walls of oil pipelines. It is used up to nominal diameters of 12 in. Disks of leather or oil-resistant synthetic rubber mounted on plates *1* ensure 'piston action', the moving along of the scraper in the oil stream. Spring-loaded sawtooth rollers *2* ensure coaxial advance. The deposits are scraped off the pipe wall by spring-loaded blades *3*. Certain scrapers are articulated, to help them negotiate sharp bends. This type of scraper has the drawback of having a large number of parts liable to break off in the pipe and cause all manner of trouble. In the scraper used for small-size pipes shown as Fig. 6.3—2, piston action is ensured by rubber cups *1;* scraping is performed by round wire brush *2*. In cleaning large-size pipelines, revolving pigs are used (Fig. 6.3—3). The front disk *3* of the scraper carrying piston cups *1* and round brushes *2* is provided with tangential jet nozzles through which a small amount of fluid may flow from behind the scraper. The jets shooting from these nozzles make the pig rotate slowly, and help to loosen the deposit piled up by the advance of the scraper. Both effects reduce the hazard of the pig's seizing up in the pipeline.

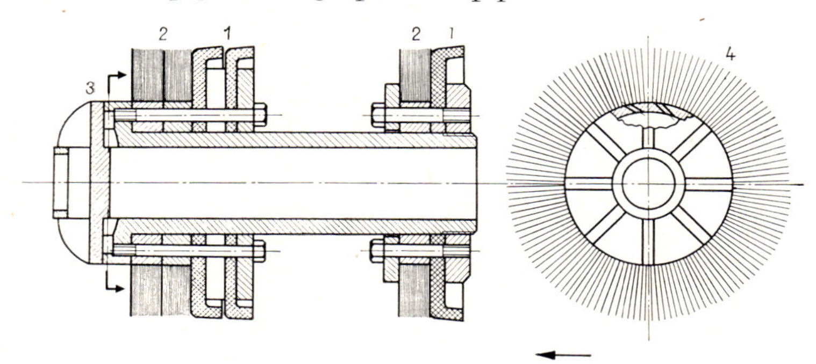

Fig. 6.3—3 General Descaling rotating scraper for large-size pipes

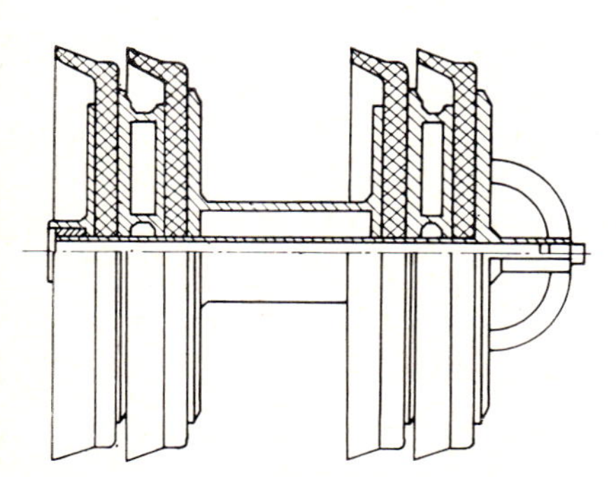

Fig. 6.3—4 General Descaling GDSS 90-4 pipeline pig

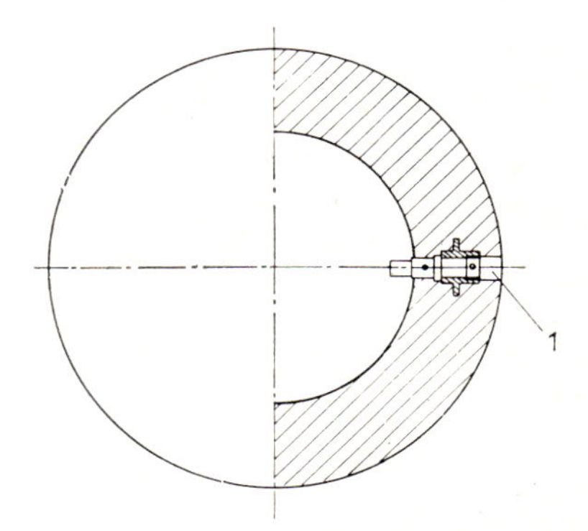

Fig. 6.3—5 General Descaling 'Piball'

Scrapers may be used also to separate plugs of different liquids or plugs of liquid and gas ('batching'). If there is no wax deposit in the pipe, it is simpler, cheaper and less liable to cause trouble to insert a batching piston (pipeline pig) rather than a scraper. A pig of General Descaling make, GDSS 90—4 type is shown as Fig. 6.3—4. Its swab cups are made of neoprene. This pig will pass through any bend whose radius is at least half as great again as the pipe radius. Branch-offs of the same diameter as the mainline are no obstacle to its application. A pipeline batching ball serving also for the removal of liquids from gas lines is the Piball, likewise of GD make (Fig. 6.3—5). The ball of synthetic rubber can be inflated with a liquid through valve 1 so that its diameter exceeds by 2 percent the *ID* of the pipe. Experiments have shown that the Piball removes all but 0.02 – 0.04 percent of the liquid in the pipe. Its advantages include longer life owing to its spherical shape and inflatability. It is compatible with lease automation and will readily pass through pipe fittings. Scrapers fitted with polyurethane blades, inflatable polyurethane pipeline balls and polyurethane-foam balls are also used; practical experience has shown these to be rather

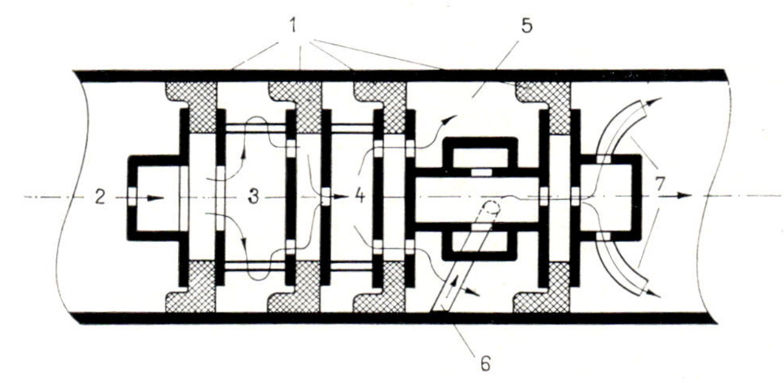

Fig. 6.3—6 Sypho-pig, after 'PLI Staff' (1970)

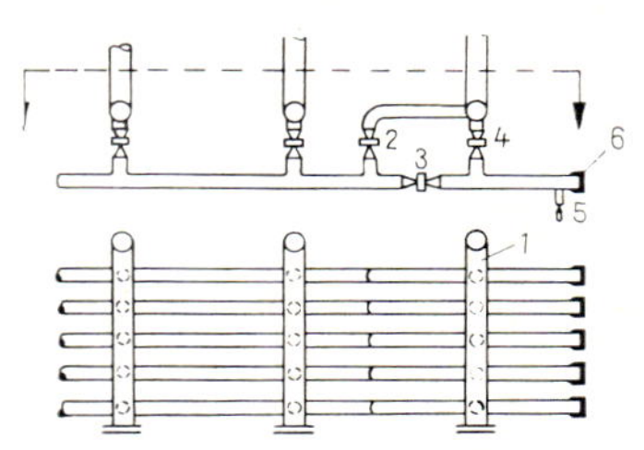

Fig. 6.3—7 Scraper trap in front of separator station

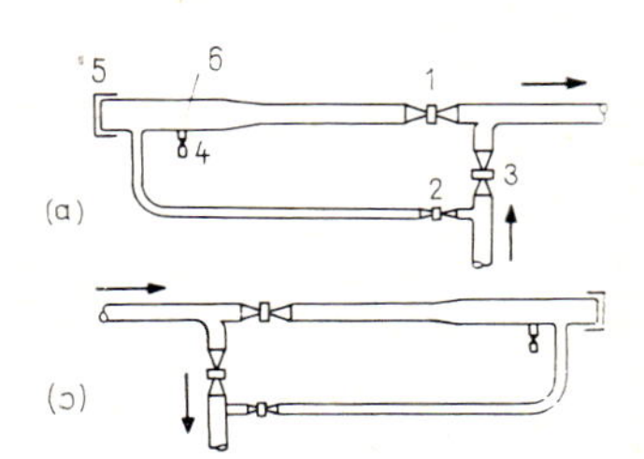

Fig. 6.3—8 Scraper traps at head and tail end of pipeline

long-lived (Zongker 1969). — A fairly complete removal of liquid may be achieved by means of the so-called sypho-pig, shown schematically in Fig. 6.3—6. The device, fitted with four swab cups *1*, is pushed forward by dry gas in the pipeline. Part of the dry gas, entering through orifice *2*, passes through the tortuous paths indicated by arrows into chambers *3* and *4*, where it dries the pipe wall by evaporating the liquid film adhering to it, and carries the vapour into chamber *5*. Liquid in chamber *5* is expelled by the gas through tube *6* and jetted through injectors *7* in front of the pig. The jets, causing turbulency in the liquid pushed forward by the pig, prevent solids from settling out of it and from seizing the pig (PLI Staff 1970). In some cases, balls or cylinders made of paraffin and asphalt kneaded together or of a substance slowly soluble in oil are also used for paraffin-removal.

Scrapers are introduced into the flowlines of wells at the wellhead, and retrieved at scraper traps installed next to the separator station (Fig. 6.3—7). During the scraping operation, the well fluid is directed through gathering line *1* into the first separator, with valve *2* closed and valves *3* and *4* open. After the arrival of the scraper, valve *2* is opened; valves *3* and *4* are closed. The cut-off pipe section containing the scraper is bled off through relief valve *5*; the scraper can be removed after opening cover *6*. — Figure 6.3—8 shows launching and receiving scraper traps installed at the beginning and at the end, respectively, of an oil pipeline. During normal operation, valves *1* and *2* of the launching trap [part (a) of the Figure] are closed, while valve *3* is open. After depressurizing through valve *4* the section cut off by valves *1* and *2*, cover *5* is removed and the scraper is inserted into oversize barrel *6*. After replacing cover *5*, valves *1* and *2* are opened and valve *3* is closed. Instead of gate valves, handwheel-operated plug valves are often employed in these applications. The operation of the receiving scraper trap [part (b) of the Figure] is much similar to that of the launching trap. — Long pipelines are often provided with intermediate scraper stations (Fig. 6.3—9), where the worn scraper arriving on side *1*

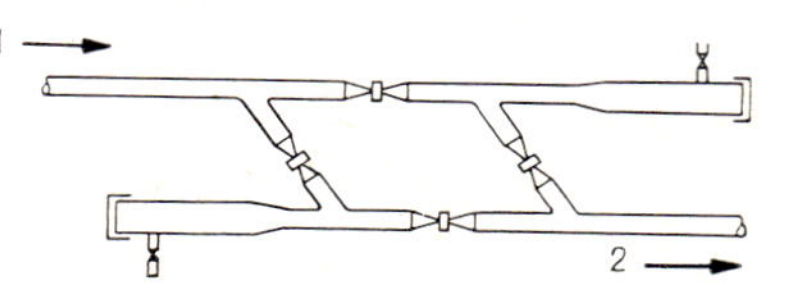

Fig. 6.3—9 Intermediate scraper trap on a pipeline

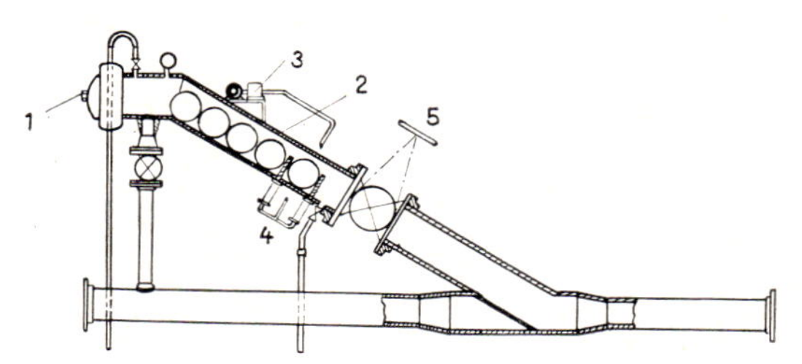

Fig. 6.3—10 General Descaling pipe ball feeder

is retrieved and a new one is inserted on side *2*. — Pipeline balls may be fed into the line by remote control. Figure 6.3—10 shows the GD hydraulic ball feeder. After removal of cover *1*, several balls can be inserted into pipe section *2*. At the desired instant, a hydraulic control unit *3* (not shown in detail in the Figure) actuates release device *4* which lets a ball roll into the pipeline through open valve *5*. It is possible to monitor the passage of the ball through suitably chosen pipe sections. The sections in question are perforated and a fitting is installed in the perforation which includes a small feeler reaching radially into the flow section. The passing ball will touch the feeler; a device attached to this latter will mechanically or electrically signal the passage of the ball.

6.4. Separation of oil and gas

6.4.1. Equilibrium calculations*

The oil and gas composing the well fluid are usually separated in separators. A separator is essentially a vessel whose interior is kept at the prescribed separation pressure and temperature. Separator temperature is either determined by the temperature of the inflowing well stream and the ambient temperature, or adjusted by means of heating or cooling equipment. The prescribed separator pressure is usually maintained by a pressure regulator installed in the gas line of the separator. Oil and gas emerge from the separator through separate outlets. The composition and relative abundance of the separation products can be predicted either by laboratory tests, or — in a fair approximation — by theoretical considerations and/or using diagrams based on practical experience. In calculations based on auxiliary diagrams it is assumed that the well stream and the discharged oil and gas do not vary in composition with time; that separation is of the flash type; and that the system is in thermodynamic equilibrium at the given pressure and temperature. The calculations are based on the following three equations:

$$n = n_L + n_V \qquad 6.4-1$$

$$z_i n = n_L x_i + n_V y_i \qquad 6.4-2$$

$$K_i = \frac{y_i}{x_i} \qquad 6.4-3$$

K_i is the equilibrium ratio of the ith component in the liquid-gas system. By Eq. 6.4—1, then, the total number of moles in the system equals the number of moles in the liquid and gas phases taken separately. By Eq. 6.4—2, the total number of moles of any component in the system equals the number of moles of that component taken separately in the liquid and gas phases. The above considerations imply

$$\sum_{i=1}^{m} x_i = \sum_{i=1}^{m} y_i = \sum_{i=1}^{m} z_i = 1 \qquad 6.4-4$$

Dividing both sides of Eq. 6.4—2 by n, we get

$$z_i = \frac{n_L}{n} x_i + \frac{n_V}{n} y_i .$$

Let $n_L/n = z_L$ and $n_V/n = z_V$; then

$$z_i = z_L x_i + z_V y_i . \qquad 6.4-5$$

By Eq. 6.4—4, the sum of the mole fractions of the components is unity, separately in the liquid and gas as well as in the entire liquid-gas system. By Eq. 6.4—5, the mole fraction of a given component in the system equals

* Largely after Amyx (1960)

the sum of its mole fractions in the gas and liquid phase, taken separately. Introducing y_i and then x_i from Eq. 6.4—3, we get

$$x_i = \frac{z_i}{z_L + K_i z_V} \qquad 6.4\text{—}6$$

and

$$y_i = \frac{z_i}{z_L/K_i + z_V}\,. \qquad 6.4\text{—}7$$

These are the fundamental equations of equilibrium in separation.

In practice, separation is usually performed in several stages. Stages are series-connected, the last stage being the stock tank. Three-stage separation, for instance, takes place in two separators and one stock tank (Fig. 6.4—1).

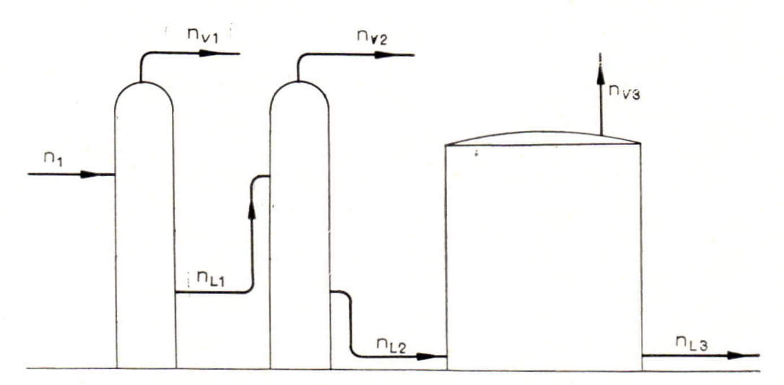

Fig. 6.4—1 Schematic diagram of three-stage separation

The numerical suffixes attached to the symbols introduced above refer to the respective separator stages. The liquid discharged from the first separator enters the second separator. The mole number of this stream is

$$n_{L1} = z_{L1} n_1.$$

The mole number of the liquid entering the third separator (which is the stock tank in the case considered) is

$$n_{L2} = z_{L2} n_{L1} = z_{L1} z_{L2} n_1\,.$$

The mole number of the liquid obtained from the third separator is

$$n_{L3} = z_{L3}\, n_{L2} = z_{L1}\, z_{L2}\, z_{L3}\, n_1\,.$$

The mole number of the gas discharged from the first separator is

$$n_{V1} = z_{V1} n_1.$$

The mole number of gas discharged from the second separator is

$$n_{V2} = z_{V2} n_{L1} = z_{V2} z_{L1} n_1\,.$$

The mole number of the hydrocarbon vapour discharged from the third separator (or simply evaporating from it, if the tank is open) is

$$n_{V3} = z_{V3}\, n_{L2} = z_{V3}\, z_{L1}\, z_{L2}\, n_1 \,.$$

In three-stage separation, the gas-oil ratio is generally the volume ratio of the gas discharged from the first two stages to the liquid collecting in the third stage (stock-tank oil), that is,

$$R = \frac{(n_{V1} + n_{V2})\, V_{\mathrm{mol}}}{n_{L3} \dfrac{M_{L3}}{\varrho_{L3}}}$$

where V_{mol} is the molar volume of the gas molecules in the standard state, in m^3/kmoles; M_{L3} is the molar mass of the liquid collecting in the third stage, in kg/kmoles; and ϱ_{L3} is the density of the liquid collecting in the third stage, in kg/m^3. Dividing both the numerator and denominator of the above formula by n_1, we get

$$R = \frac{\dfrac{n_{V1} + n_{V2}}{n_1}\, V_{\mathrm{mol}}\, \varrho_{L3}}{\dfrac{n_{L3}}{n_1}\, M_{L3}} = \frac{(z_{V1} + z_{V2}\, z_{L1})\, V_{\mathrm{mol}}\, \varrho_{L3}}{z_{L1}\, z_{L2}\, z_{L3}\, M_{L3}} \qquad 6.4\text{—}8$$

and for two-stage separation,

$$R = \frac{z_{V1}\, V_{\mathrm{mol}}\, \varrho_{L2}}{z_{L1}\, z_{L2}\, M_{L2}} \,. \qquad 6.4\text{—}9$$

The practical usefulness of equilibrium calculations depends first and foremost on the accuracy of the equilibrium ratios K_i. The equilibrium ratio of the ith component depends, in addition to separator temperature and pressure, also on the composition of the well fluid, and accurate values may only be expected from the laboratory testing of the fluid to be treated. The values of K_i may, however, be determined in a fair approximation also by various methods based on auxiliary diagrams. The processes in current use relate the composition of the system to convergence pressure. It is assumed that if two systems of hydrocarbons agree as to convergence pressure, then the equilibrium ratios of their components will likewise be equal at the given pressure and temperature. Hence, in order to find K_i, it is necessary to know the convergence pressure of the system and the equilibrium ratios belonging to various convergence pressures. Determining in the laboratory the equilibrium ratios of the components of a given hydrocarbon system at various pressures and at a given temperature, one obtains families of curves similar to the two families shown in Fig. 6.4—2. These curves all end in one point, lying on the line $K = 1$ and characterized by a certain pressure. The pressure thus determined is the apparent convergence pressure. It is 69 bars for the first family shown in the Figure, and 345 bars for the second (Amyx 1960). If the laboratory experiment is performed at the critical temperature of the system, then the apparent convergence pressure agrees with the critical pressure. At any other temperature, convergence is merely apparent, because the system has its bubble point at a pressure lower than the apparent convergence pressure and there is just one phase instead of two at the point of convergence. In the zone between the critical pressure and the apparent convergence pressure, then, the equilibrium constant lacks a physical meaning: the corresponding curve sections and the point of

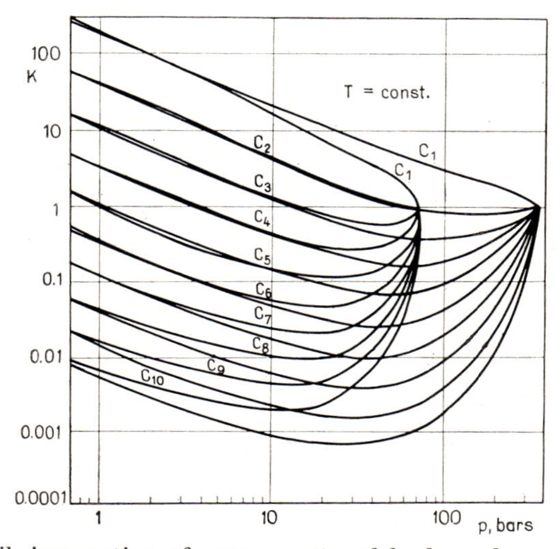

Fig. 6.4—2 Equilibrium ratios of components of hydrocarbon systems of 69 and 345 bars, apparent convergence pressure; from Natural Gasoline Association of America's *Equilibrium Ratio Data Book* (1957; used with permission of Natural Gas Processors Association)

apparent convergence itself are merely the results of extrapolation. The apparent convergence pressure of a given system of hydrocarbons can be determined by various means. Its importance is substantial especially at high pressures. At pressures below cca 7 bars, however, the system's composition, and hence, knowledge of its exact convergence pressure, lose much of their importance (Amyx 1960). Experiment has shown that in establishing the K values required for calculating equilibria in low-pressure separators, a satisfactory accuracy will be achieved if convergence pressure is simply assumed to equal 345 bars (5000 psia). K-isotherms may be found in the NGAA *Equilibrium Ratio Data Book* (1957). Figure 6.4—3 shows the K-isotherms of isobutane for an example.

In the usual laboratory-analysis report, it is usual to state the content or mole fraction of all components heavier than hexane by a single figure denoted C_{7+} (heptane-plus). The composition of this mixture of hydrocarbons may vary widely, depending on the composition of the crude in hand. Since, however, the pressure curves and critical parameters of components heavier than hexane differ very little, the system may be satisfactorily characterized by a median K-value. The equilibrium ratios of this fraction can be estimated using the graphs in Fig. 6.4—4 (after Katz and Roland; in Amyx 1960). Let us finally point out that, even though both the density and molar mass of the C_{7+} fraction are required for equi-

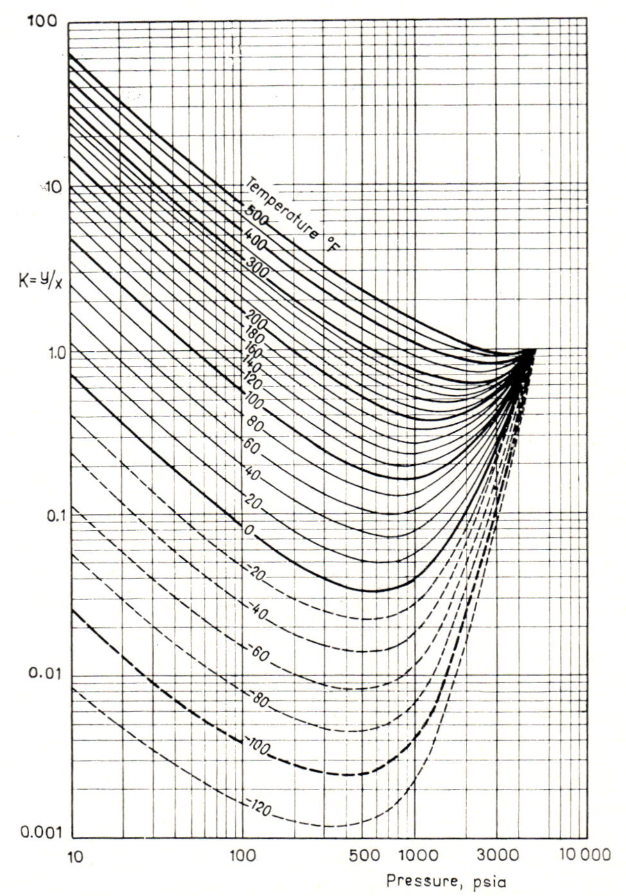

Fig. 6.4—3 Equilibrium ratios of isobutane at 5000 psia apparent convergence pressure; from Natural Gasoline Association of America's *Equilibrium Ratio Data Book* (1957; used with permission of Natural Gas Processors Association)

librium calculations, only one of the two data need be established in the laboratory, as the other one can be calculated using Cragoe's formula:

$$M = \frac{44.29\varrho}{1030 - \varrho}. \qquad 6.4\text{—}10$$

Equilibrium calculation in the case of low-pressure separation proceeds as follows: (i) The K_i values for the prevailing temperature and pressure are read off the NGAA graphs for 345 bars convergence pressure, (ii) the equilibrium ratio of the C_{7+} fraction is read off Fig. 6.4—4, (iii) a value for z_V is estimated, and x_i is calculated for each component using Eq 6.4—6. (iv) If $\sum_{i=1}^{m} x_i$ is unity, then the assumed z_V is correct, and so is, of course, $(1 - z_V) = z_L$. If $\sum_{i=1}^{m} x_i$ differs from unity, the calculation has to be repeated with a different z_V. Equilibrium calculations are usually performed by computer nowadays, thus bypassing the time-consuming successive approximation by pencil and paper. — Table 6.4—1 contains physical constants published by CNGA and NGAA for equilibrium calculations.

Example 6.4—1. With reference to the hydrocarbon system characterized by Columns 1 and 2 of Table 6.4—2, and assuming three-stage separation, (i) what are the compositions of the gas effluents of the individual separator stages? (ii) what will be the composition of the stock-tank oil collecting in the atmospheric-pressure stock-tank? (iii) what will be the 'useful' *GOR*, that is, the volume of gas discharged from the first two stages of separation per m³ of stock-tank oil? Let separation temperature be 4 °C in all three stages, and let pressure be 6 bars in the first stage, 2.1 bars in the second, and 1.01 bar in the third. The standard state is at $p_n = 1.01$ bar and $T_n = 15.6$ °C.

The main results of the equilibrium calculation for the three separator stages are presented in Tables 6.4—2—6.4—4. The K_i values were taken from the diagrams for 5000 psia convergence pressure of the NGAA Equilibrium Data Book; the values for C_{7+} were read off Fig. 6.4—4. The composition of gas discharged from the three separator stages is given in Columns 5 of said Tables, whereas the composition of the stock-tank oil is given in

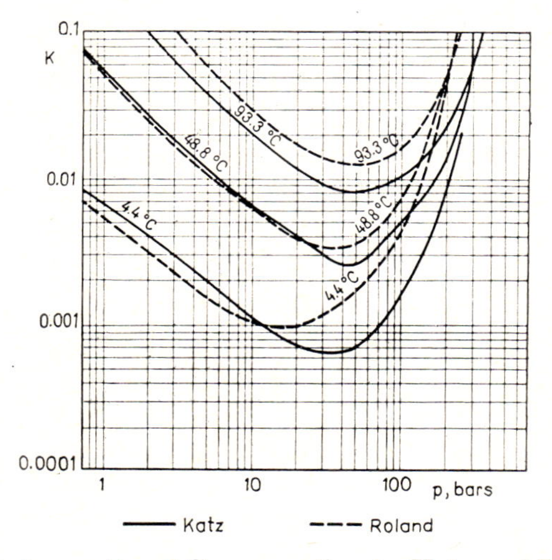

Fig. 6.4—4 Equilibrium ratio of C_{7+}, according to Katz and Roland (Amyx 1960, p. 341; used with permission of McGraw-Hill Book Company)

Table

Physical parameters of hydrocarbons

	Symbol	Unit	CH_4	C_2H_6	C_3H_8
Molar mass	M	$\frac{kg}{kmole}$	16.042	30.068	44.094
Critical temperature	T_c	K	190.6	305.5	370.0
Critical pressure	P_c	bar	46.6	48.8	42.6
Critical density	ϱ_c	kg/m³	161.9	203.1	226.2
Critical molar vol.	V_{mc}	m³/kmol	0.09901	0.1480	0.1950
Gas density (std.)	ϱ_{gn}	kg/m³	0.6750	1.265	1.855
Gas density (rel.)	ϱ_{gr}	—	0.554	1.038	1.522
Liquid density	ϱ_l	kg/m³	299.5*	372.6	506.8

* Pseudo-gravity

Standard state is at $T_n = 15.6°C$, $p_n = 1.01$ bar

Column 4 of Table 6.4—4. The useful *GOR* can be calculated using Eq. 6.4—8. In that relationship, the molar gas volume at the given parameters of state is

$$V_{mol} = \frac{R'T_n}{p_n} = \frac{8.314 \times 288.8}{1.01 \times 10^5} = 0.02377 \text{ m}^3/\text{mole} = 23.77 \text{ m}^3/\text{kmole}.$$

Let us point out that the exact value of the molar gas constant in the SI system is $R = 8.31433 \pm 0.00044$ J/mole K. We have used the value 8.314, of sufficient accuracy for the task in hand. We have neglected z_n, the compressibility factor in the standard state, or have put it equal to

Table 6.4—2

First separator stage; $p_1 = 6$ bars, $T_1 = 4$ °C

Component	z_{v1}	K_{i1}	$z_{L1} = 0.0742$, $z_{v1} = 0.9258$	
			x_{i1}	y_{i1}
1	2	3	4	5
C_1	0.7354	28.0	0.0283	0.7920
C_2	0.1503	3.80	0.0418	0.1590
C_3	0.0120	0.94	0.0127	0.0119
iC_4	0.0087	0.325	0.0232	0.0075
nC_4	0.0273	0.222	0.0976	0.0217
iC_5	0.0036	0.080	0.0243	0.0019
nC_5	0.0045	0.065	0.0335	0.0022
C_6	0.0104	0.026	0.1059	0.0028
C_{7+}	0.0478	0.0015	0.6327	0.0010
Total:	1.0000		1.0000	1.0000

6.4 – 1
(modified after CNGA and NGAA data)

$i-C_4H_{10}$	$n-C_4H_{10}$	$i-C_5H_{12}$	$n-C_5H_{12}$	$n-C_6H_{14}$	$n-C_7H_{16}$	C_7+
58.120	58.120	72.146	72.146	86.172	100.198	
407.2	425.0	461.1	470.6	507.8	540.0	
36.6	38.0	33.2	33.5	29.9	27.4	
233.0	226.6	234.2	231.9	234.6	234.5	
0.2491	0.2578	0.3078	0.3109	0.3671	0.4270	
2.446	3.036	3.036	3.036	3.626	4.216	
2.006	2.006	2.491	2.491	2.975	3.459	M/28.96
561.9	582.3	623.1	629.0	662.5	686.5	

Table 6.4 – 3
Second separator stage; $p_2 = 2.1$ bars, $T_2 = 4$ °C

Component	$z_{v_2} = x_{i_1}$	K_{i_2}	$z_{L_2} = 0.9612$, $z_{V_2} = 0.0388$	
			x_{i2}	y_{i2}
1	2	3	4	5
C_1	0.0283	78.0	0.0071	0.5533
C_2	0.0418	10.5	0.0306	0.3210
C_3	0.0127	2.60	0.0120	0.0311
iC_4	0.0232	0.840	0.0233	0.0196
nC_4	0.0976	0.560	0.0993	0.0556
iC_5	0.0243	0.200	0.0251	0.0050
nC_5	0.0335	0.150	0.0346	0.0052
C_6	0.1059	0.060	0.1099	0.0066
$C_7{}^+$	0.6327	0.004	0.6581	0.0026
Total:	1.0000		1.0000	1.0000

Table 6.4 – 4
Third separator stage; $p_3 = 1.01$ bar, $T_3 = 4$ °C

Component	$z_{v_3} = x_{i_2}$	K_{i_3}	$z_{L_3} = 0.9798$, $z_{V_3} = 0.0202$	
			x_{i3}	y_{i3}
1	2	3	4	5
C_1	0.0071	165.0	0.0016	0.2718
C_2	0.0306	21.8	0.0215	0.4695
C_3	0.0120	5.40	0.0110	0.0593
iC_4	0.0233	1.80	0.0230	0.0413
nC_4	0.0993	1.20	0.0989	0.1187
iC_5	0.0251	0.430	0.0254	0.0109
nC_5	0.0346	0.315	0.0351	0.0111
C_6	0.1099	0.115	0.1119	0.0129
$C_7{}^+$	0.6581	0.0067	0.6716	0.0045
Total:	1.0000		1.0000	1.0000

unity. — For finding liquid density ϱ_{L3} we have used Fig. 6.4—5, which is Standing's nomogram (1952) transposed into the SI system. It furnishes density of a liquid hydrocarbon containing C_1 and C_2. The starting data are listed in Table 6.4—5. The density of the liquid containing C_{3+} is, then,

$$\varrho_{LC3+} = \frac{\sum_{i=3}^{7+} M_i x_{i3}}{\sum_{i=3}^{7+} \frac{M_i x_{i3}}{\varrho_{Li}}} = 688 \text{ kg/m}^3 .$$

The content of C_2 in the liquid C_{2+} is 0.03 mass percent, and that of C_1 in the liquid C_{1+} is 0.66 mass percent. Owing to the low values of C_{1+} and C_{2+}, Fig. 6.4—5 gives no deviation; that is, we may retain $\varrho_{L3} = 688$ kg/m³. — From Table 6.4—5, M_{L3} is 99.48 kg/kmole. — Introduced into Eq. 6.4—8, the data thus obtained give a *GOR* of

$$R = \frac{(0.9258 + 0.0388 \times 0.0742) \times 23.77 \times 688}{0.0742 \times 0.9612 \times 0.9798 \times 99.48} = 2185 \text{ m}^3/\text{m}^3 .$$

If separator pressure exceeds 7 bars, or separator temperature is below −5 °C, then it will in general be necessary to determine the apparent convergence pressure and to read the K_is off the corresponding NGAA curve

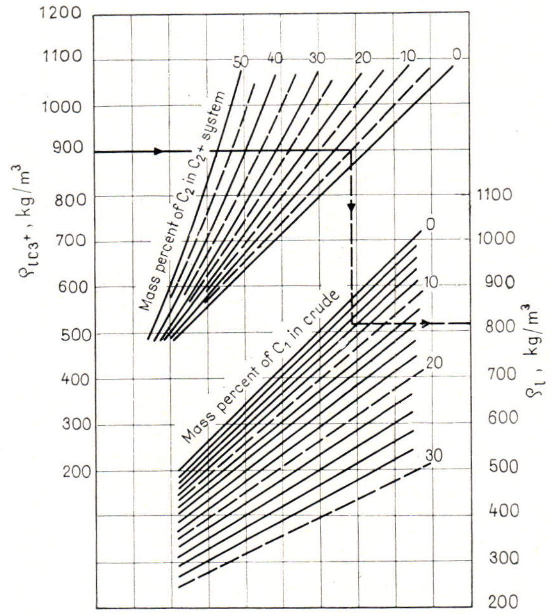

Fig. 6.4—5 Apparent density of crude containing methane and ethane, at $T = 15.6$ °C and $p = 1.01$ bar, after Standing 1952 (reproduced by permission of the copyright owner—copyright © Chevron Research Company 1951; all rights reserved under the International Copyright Convention)

Table 6.4—5

Component	x_{i3}	M_i	$x_{i3}M_i$	ϱ_{Li}	$\frac{x_{i3}M_i}{\varrho_{Li}}$
1	2	3	4	5	6
C_1	0.0016	16.04	0.03		
C_2	0.0215	30.07	0.65		
C_3	0.0110	44.09	0.48	506.8	0.0010
iC_4	0.0230	58.12	1.34	561.9	0.0024
nC_4	0.0989	58.12	5.75	582.3	0.0099
iC_5	0.0254	72.15	1.83	623.1	0.0029
nC_5	0.0351	72.15	2.53	629.0	0.0040
C_6	0.1119	86.17	9.64	662.5	0.0146
C_{7+}	0.6716	115.0	77.23	710.0	0.1088
Total:	1.0000		99.48		0.1436

family. Since equilibrium calculations are rather time-consuming, it is recommended to find out beforehand whether two separate phases do exist at all at the temperature and pressure in question. If the well fluid is composed purely of hydrocarbons, this may be found out as follows. On reducing the temperature of the mix at a given pressure, the dew point (E) is reached first and the bubble point (H) is reached thereafter (Fig. 6.4—6). If therefore, e.g. in so-called low-temperature separation, well fluid is introduced into the separator after cooling, it is necessary to examine whether the temperature before cooling is below the dew point and the temperature after cooling is above the bubble point. If both answers are yes, then the mix is two-phase in both cases and it is worth while to perform the equilibrium calculation. At the *bubble point*, one mole (say) of vapour keeps equilibrium with a liquid drop of evanescent size. Hence, in a very good approximation, $z_L = 0$ and $z_V = 1$, and by Eqs 6.4—4 and 6.4—6,

$$\sum_{i=1}^{m} x_i = \sum_{i=1}^{m} z_i/K_i = 1. \qquad 6.4\text{—}11$$

If the value obtained is greater than unity, the mix is two-phase, because in order to make said value approach unity it is necessary to increase the K_is and, by Fig. 6.4—3, this requires raising the temperature if the pressure is fixed. Now by Fig. 6.4—6, this is possible only if the mix is in the two-phase domain. -- At the *bubble point*, one mole (say) of liquid keeps equilibrium with a gas bubble of evanescent size. Hence, in a very good approximation, $z_L = 1$ and $z_V = 0$, and Eqs 6.4—4 and 6.4—7 yield

$$\sum_{i=1}^{m} y_i = \sum_{i=1}^{m} z_i K_i = 1. \qquad 6.4\text{—}12$$

If the value of the equation is greater than unity, this means once more that the mix is two-phase, because in order to make said value approach unity it is necessary to reduce the K_is; now, by Fig. 6.4—3, this requires lowering the temperature if the pressure is fixed, and by Fig. 6.4—6 this is possible only if the mix is in the two-phase domain.

The well fluid often contains some water, too. It is necessary to keep in mind when performing calculations concerning such a fluid that *the common vapour pressure of immiscible liquids* is entirely independent of the proportion of the components in the liquid phase, being equal to the sum

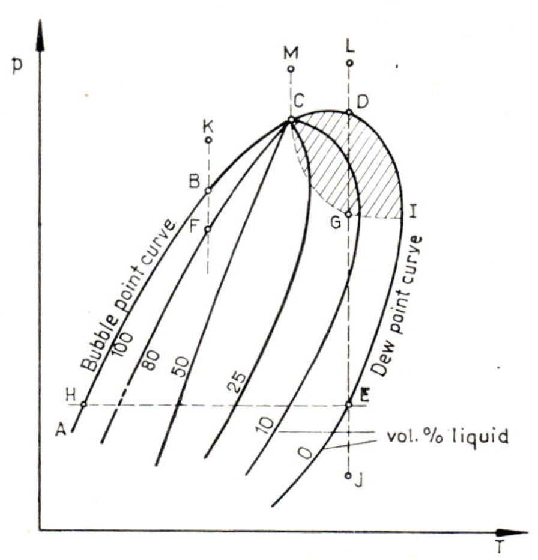

Fig. 6.4—6 Phase diagram of natural gas containing a significant amount of gasoline

of the vapour pressures of the components, taken separately at the temperature considered.

Example 6.4—2. Find the vapour pressure of a water-pentane mix at 50 °C.

Component	Vapour pressure bars
H_2O	0.12
C_5H_{12}	1.60
	1.72

On heating this mix, the bubble point will occur at that temperature where the sum of the two vapour pressures equals the external pressure. The bubble-point temperature of the mix is, then, lower than that of any individual component (water-vapour distillation). It is often necessary to determine the temperature at which water starts to condense out of the hydrocarbon-water system (that is, to establish the dew point of the mixture.) Since each component generates its partial vapour pressure independently of the other components, the task is essentially to find the temperature at which the sum of partial pressures equals the vapour pressure

Example 6.4—3 (after Maddox 1963). What is the dew point at 1.72 bar pressure of vapour composed of one kg H_2O and 10 kg n-pentane; which component will condense first? The calculation results are given in Table 6.4—6. On cooling from a higher temperature, the system will first attain 80.9 °C, and hence, water will condense at the dew point.

Table 6.4—6

Component	m	M	n	z	p_{part} 1.72z	T_p part.
	kg	kg/mole			MN/m²	°C
1	2	3	4	5	6	7
H_2O	1.0	18	0.0555	0.285	0.49	80.9
n-C_5	10.0	72	0.139	0.715	1.23	41.7
			0.1945			

6.4.2. Factors affecting recovery in the separator*

(a) Separator pressure

Figure 6.4—7 shows liquid recovery from a wellstream characterized in Table 6.4—7. It reveals that, e.g. at 70 bars pressure, one million m³ of well fluid will release 100 m³ liquid C_{7+}. The curves show that, on increasing separator pressure, all components will at first condense to an increasing extent. The increment in condensate due to unity increase in pressure is highest in the case of methane and grows less as the molecular weight

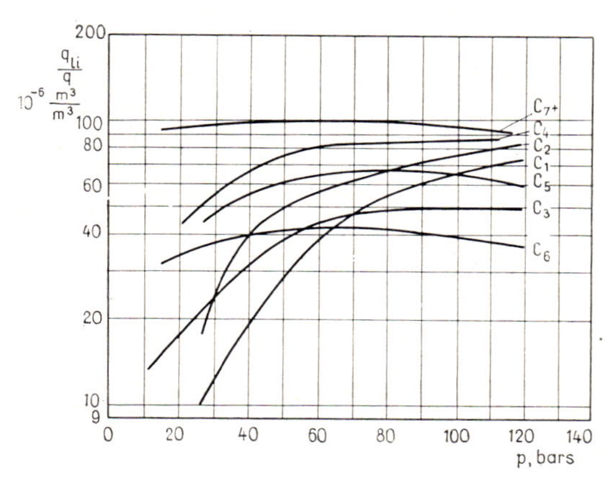

Fig. 6.4—7 Influence of pressure upon liquid recovery of components at T = 27 °C after Campbell (1955)

* After Campbell (1955) and Whinery and Campbell (1958).

Table 6.4—7

Component	z_i
C_1	0.7118
C_2	0.1503
C_3	0.0478
i-C_4	0.0104
n-C_4	0.0343
i-C_5	0.0096
n-C_5	0.0114
C_6	0.0092
C_{7+}	0.0152
Total:	1.0000

increases. The curves of components heavier than butane exhibit, however, peaks at which the liquid recovery is maximum for a given temperature, being reduced on further pressure increase by retrograde vaporization. This phenomenon is of considerable interest as regards the production of a liquid stable at atmospheric pressure. — Graph *1* in Fig. 6.4—8 is a plot v. separator pressure of liquid yield at a separator temperature of 27 °C; Graph *2* represents the volume of stock-tank oil stable at atmospheric pressure and 38 °C temperature, likewise v. separator pressure. The volume of stock-tank oil, composed largely of pentane and heavier components, will increase up to a pressure about of 41 bars to decrease again at higher pressures. A volume of liquid equal to the ordinate difference between the two graphs will evaporate in the tank at the given atmospheric temperature. Other well fluids separated at other temperatures will give curves displaced against those given in the Figure, but their essential features will be the same.

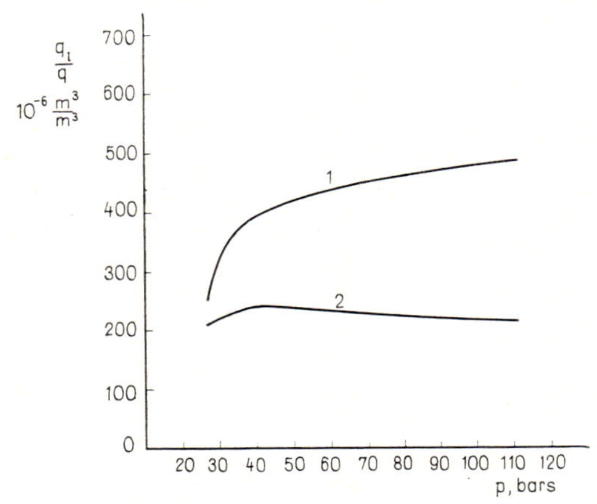

Fig. 6.4—8 Influence of pressure upon liquid recovery of separator at $T = 27$ °C, after Campbell (1955)

(b) Separator temperature

Figure 6.4—9 gives yield curves for various hydrocarbons out of a given wellstream, for a pressure of 28 bars and various temperatures. The lower the temperature, the less will be the increment in C_5 and C_6 yield due to further decrease of temperature. The maximum recovery of C_{7+} is at about 10 °C. — Graph *1* in Fig. 6.4—10 represents the total relative liquid yield of the

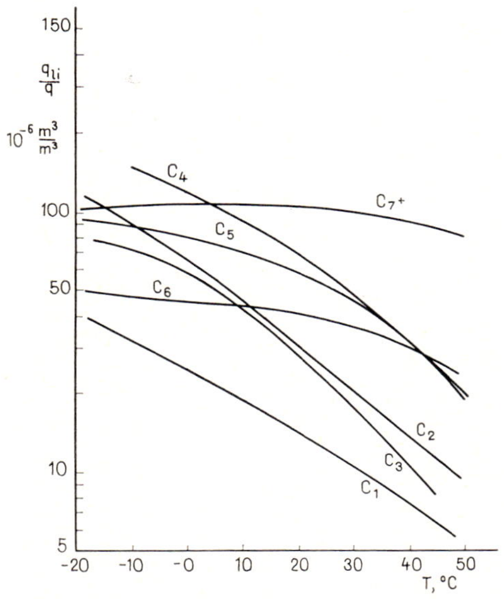

Fig. 6.4—9 Influence of temperature upon liquid recovery of components at $p = 28$ bars, after Campbell (1955)

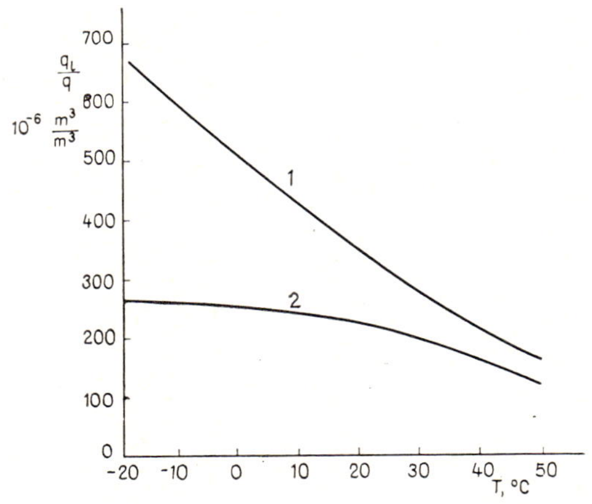

Fig. 6.4—10 Influence of temperature upon liquid recovery, at $p = 28$ bars, after Campbell (1955)

separator; Graph *2* represents the relative yield of stable liquid, capable of storage at atmospheric pressure; both v. separator temperature. Stable-liquid recovery is seen not to increase below about −10 °C. If the primary aim is to recover a maximum amount of stable liquid, temperatures between −1 and −10 °C will be preferred. Let us point out that at temperatures below +10 °C, the pentane and hexane content of the stable product will increase and this will result in a decrease of its gravity. If the primary aim is to increase the total liquid yield, e.g. in order to recover from the gas as much propane and butane as possible, then it may be preferable to lower the temperature even below −26 °C. Low-temperature separation, however, necessitates some special equipment that requires a considerable throughput to make it pay out. Separator temperatures higher than 30 °C are not to be recommended as a rule.

(c) Composition of the wellstream

Figure 6.4−11 is a plot of stable stock-tank oil yield v. temperature, at two different separator pressures, for each of the three wellstreams characterized in Table 6.4−8. All three curves are seen to flatten out at temperatures below about −8 °C, and to become practically independent of temperature and pressure. Recovery of stable stock-tank oil is governed below temperatures of −7 °C largely by wellstream composition. The factor having the most profound influence on stable-liquid yield is the molar ratio of C_{5+}. Table 6.4−8 lists, in addition to the compositions of the three wellstreams marked *A*, *B* and *C*, also the absolute values of z_{3+} and z_{5+}, and their values related to the yield for wellstream *A*. The relative stable-liquid yields of the three wellstreams are plotted in Fig. 6.4−11. The values at −7 °C, marked in the Figure by dashed lines, have also been listed in the Table. The yields related to the yield of fluid *A* are seen to agree pretty well with the relative values of z_{5+}. It is further

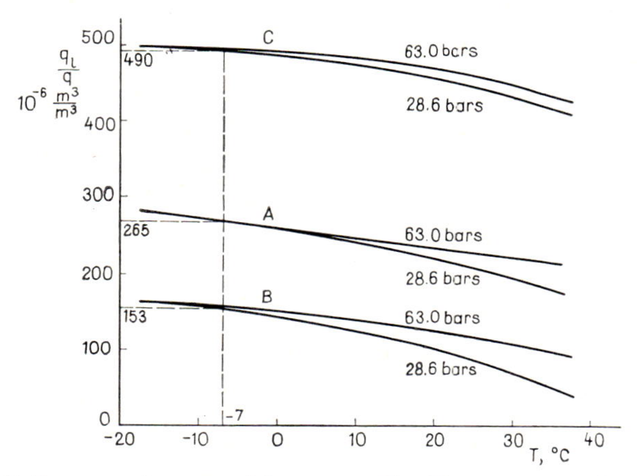

Fig. 6.4−11 Influence of composition upon liquid recovery, after Campbell (1955)

Table 6.4—8

Component	z_i		
	A	B	C
C_1	0.7118	0.7001	0.7354
C_2	0.1503	0.1503	0.1503
C_3	0.0478	0.0742	0.0120
i-C_4	0.0104	0.0104	0.0087
n-C_4	0.0343	0.0343	0.0273
i-C_5	0.0096	0.0096	0.0036
n-C_5	0.0114	0.0114	0.0045
C_6	0.0092	0.0063	0.0104
C_{7+}	0.0152	0.0034	0.0478
Total:	1.0000	1.0000	1.0000
z_{3+}	0.1379	0.1496	0.1143
$z_{3+\mathrm{rel}}$	1.0	1.1	0.82
z_{5+}	0.0358	0.0211	0.0627
$z_{5+\mathrm{rel}}$	1.0	0.59	1.8
$\frac{q_L}{q}$ read off $10^{-6}\ \frac{m^3}{m^3}$	265	153	490
$\frac{q_L}{q}$ read off-rel.	1.0	0.58	1.8
$\frac{q_L}{q}$ calc. $10^{-6}\ \frac{m^3}{m^3}$	272	152	492

apparent that C_{3+} is not representative at all of stable-liquid yield. Stable-liquid yield to be expected on separation below —7 °C can be estimated by means of the formula

$$\frac{q_l}{q} = 8160 \times z_{5+} - 20. \qquad 6.4-13$$

The last row of Table 6.4—8 lists the values furnished by this formula for the three wellstreams A, B and C. There is a fair agreement with the q_l/q values read off Diagram 6.4—11. Relationships characterizing recovery above —7 °C are more complicated. Figure 6.4—12 is the plot of an empirical relationship furnishing the separator pressure ensuring maximum total liquid yield v. well-fluid gravity, for the temperature range between 21 and 27 °C.

(d) Stage separation

In Sections (a), (b) and (c) we have analysed the effects of temperature, pressure and wellstream composition, tacitly assuming two-stage separation with one separator and one stock-tank. If the number of separation stages is more than two, this will improve the recovery of stable stock-tank oil but diminish the total-liquid yield. Hence, evaporation loss in open tanks will be less after stage separation. Increasing the number of stages from two to three brings a considerable improvement. According to an investigation on 13 wells, the recovery of stable stock-tank oil was improved by 8 percent on an average (with a range from 3 to 22 percent). On increasing stage number from three to four, the improvement in re-

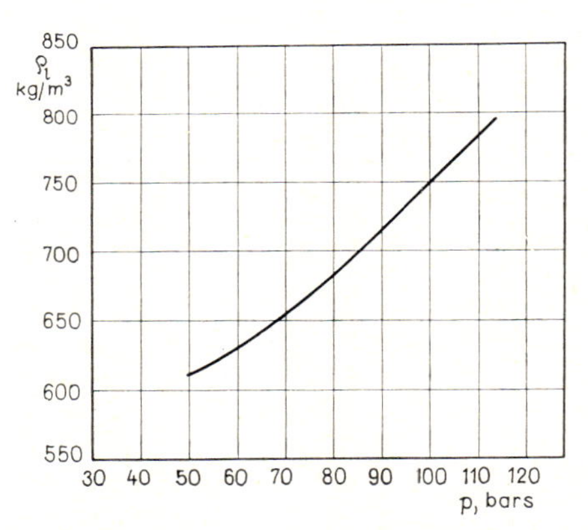

Fig. 6.4—12 Separator pressure ensuring maximum liquid recovery in the 21—27 °C temperature interval, after Campbell (1955)

covery tends to be much less, and four-stage separation is not usually economical. Figure 6.4—13 shows what percentage of the liquid q_{l1} discharged from the first separator stage (composition characterized in Table 6.4—9) will collect as stable oil (q_{l3}) in the stock tank as a function

Table 6.4—9

Component	z_i (A)	z_i (B)	z_i (C)
C_1	0.1191	0.0826	0.1018
C_2	0.2129	0.1060	0.0272
C_3	0.1686	0.0986	0.0348
i-C_4	0.0498	0.0403	0.0216
n-C_4	0.1779	0.1634	0.0276
i-C_5	0.0544	0.0713	0.0177
n-C_5	0.0661	0.0924	0.0220
C_6	0.0564	0.1130	0.7473*
C_{7+}	0.0948	0.2324	
Total:	1.0000	1.0000	1.0000
p_1, bar	27.6	27.6	27.6
t_1, °C	−17.8	26.7	26.7
q_{L1}, $10^{-6}\ \frac{m^3}{m^3}$	658	293	131

* Hexane-plus

of second-stage pressure, provided first-stage pressure is 27.6 bars and third-stage pressure is 1.01 bar. Liquids *A* and *B* passing from the first-stage into the second derive from one and the same wellstream, but first-stage temperature was −17.8 °C for liquid *A* and 26.7 °C for liquid *B*, resulting in a lower C_{5+} content for *A*. Figure 6.4—13 and Table 6.4—9 once more reveal stable-liquid recovery to be the better the higher the C_{5+} content of the liquid entering stage two. The Figure reveals further that stable stock-tank oil yield is maximal in all three cases at the same, second-stage pressure of about 3.5 bars. Pressure in the first separator stage is limited by the maximum possible wellhead pressure or, more precisely, by the maximum incoming flowline pressure. Third-stage pressure is either atmospheric or is determined by the design features of the gas-gathering system. The only pressure that may be freely chosen within limits in order to achieve maximum stable stock-tank oil yield is, then, the second-stage pressure.

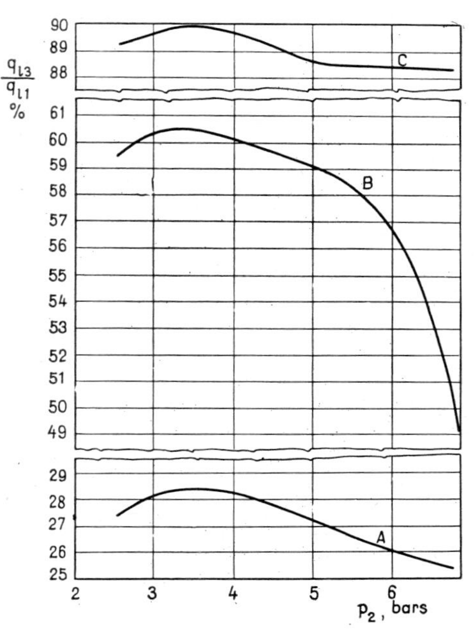

Fig. 6.4—13 Stable stock-tank oil recovery in three-stage separation, after Campbell (1955)

The most favourable second-stage pressure can be determined either by in-plant experiment, time-consuming equilibrium calculations, or approximative calculations. Equilibrium calculations are to be carried out for numerous values of pressure, in the manner outlined in the foregoing section, which is a tedious procedure. The methods of approximation are more rapid but less accurate. Several such methods are known. In a rough approximation, second-stage pressure can be calculated from the relationship

$$p_2 = \sqrt{p_1 p_3} \; . \qquad 6.4\text{—}14$$

The p_2 furnished by this formula is usually higher than optimal.

A more accurate result for a low-molecular-weight gas-liquid mix can be obtained by a procedure which expresses second-stage pressure by one of two equations (Whinery and Campbell 1958). The first one is used when the relative molecular weight M_r of the well fluid, referred to air, is higher than 1, and the second one when it is lower than 1. The equations are for $M_r > 1$,

$$p_2 = 16\,A p_1^{0.686} + 2.96 \times 10^5 (A + 0.057) \qquad 6.4\text{—}15$$

and for $M_r < 1$,

$$p_2 = 8\,Ap_1^{0.765} + 5.75 \times 10^5(A + 0.028). \qquad 6.4\text{—}16$$

The A factor is plotted v. M_r in Fig. 6.4—14. The procedure may be used if the third stage is atmospheric, in which case the result is accurate to within ±5 percent.

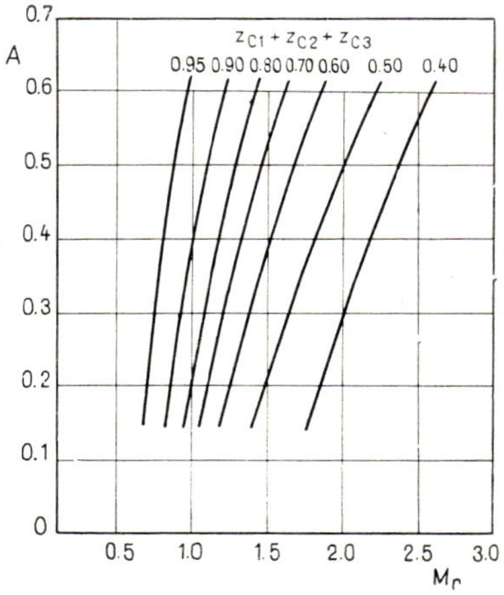

Fig. 6.4—14 Determining the A factor, after Whinery and Campbell (1958)

Example 6.4—4 (after Whinery and Campbell 1958). Let the well fluid be characterized by Table 6.4—10. Let first-stage pressure be 34.5 bars, and let the third stage be atmospheric. What is the most favourable second-stage pressure?

$$M_r = \frac{39.92}{28.96} = 1.38$$

$z_{C1} + z_{C2} + z_{C3} = 0.7$, and Fig. 6.4—14 yields $A = 0.421$. Since M_r is greater than 1, we take Eq. 6.4—15;

$$p_2 = 16 \times 0.421\,(34.5 \times 10^5)^{0.686} +$$
$$+ 2.96 \times 10^5(0.421 + 0.057) =$$
$$= 3.47 \times 10^5\ \mathrm{N/m^2} = 3.47\ \text{bars}.$$

Table 6.4—10

Component	z_i	M	1×2
	1	2	3
C_1	0.40	16.01	6.40
C_2	0.20	30.07	6.01
C_3	0.10	44.09	4.41
C_4	0.10	58.12	5.81
C_5	0.10	72.15	7.22
C_6	0.05	86.17	4.31
C_{7+}	0.05	115.22	5.76
Total:	1.00		39.92

6.4.3. Basic separator types

The compositions of the liquid and gas discharged from a separator differ somewhat from those predicted by calculation. In addition to the fact that calculation procedures based on auxiliary diagrams do not model real-life conditions quite accurately, this is due to the circumstance that the liquid will contain gas bubbles and the gas will contain mist. Of these, it is the liquid content of the gas that may cause trouble (in the transportation system). Modern separators are designed so that the mist content of the discharged gas be less than 0.1 g/m³.

In most gas-liquid separators, the elements ensuring separation may be divided in three groups. (i) Means of rough separation. These serve to effect a first separation of liquid and gas, mainly by centrifugal force and the force of gravity. They include the inlet separating element, baffles, liquid passages, and that part of the separator between liquid surface and inlet (in vertical separators) or above the liquid level (in horizontal and spherical separators). (ii) Mist extractor. In order to separate mist not settled out by centrifugal force or gravity, a mist extractor is placed in the way of the gas stream. The mist extractor may be of the vane or coalescing-pack type, or it might be a hydrocyclone, often installed outside the separator body. (iii) Oil collector. The liquids separated from the wellstream and condensed in the separator collect in the bottom part of the separator vessel. Liquid level is maintained within given limits by an LLC (liquid-level-control) device, in order to prevent gas from entering the oil outlet, and liquid from rising up to the elements mentioned above, and to ensure sufficient retention time for the gas bubbles to break out of the liquid.

Shapewise, separators may be vertical cylindrical, monotube or dual-tube horizontal, or spherical. In the sections below I shall describe typical examples for each of these.

(a) Vertical separators

The wellstream enters the vertical or upright cylindrical separator through an inlet installed at about two-thirds height. The inlet may be radial or tangential [(a) and (b) in Fig. 6.4—15]. Tangential inflow has the advantage that it exploits for separating liquid and gas a centrifugal force which may exceed the force of gravity by up to two orders of magnitude: furthermore, the liquid will move in the separator in a spiral along the shell, while gas will rise in the central space along the separator axis. This considerably reduces the likelihood of contamination with mist as against the case of radial inflow. The inlet separating element shown in Fig. 6.4—16, a development on the radial-inflow idea, has lately been adopted by numerous makers of separators. Its use substantially promotes the separation of liquid and gas directly after inflow. Figure 6.4—17 is a schematic diagram of a modern vertical separator. The wellstream enters separator *2* through inlet *1* and the separating element mentioned above. Liquid discharge is controlled by motor valve *5*, operated by float *3* through pneumatic pilot *4*.

Bobbing of the float is prevented by float shield *6*. Mist in the rising gas is extracted by vane-type device *7* from which it drips back into the separator space. Separator pressure is controlled by regulator *9*, installed in gas outlet *8*. Solids settling on the separator bottom can be removed through drain *10*. Overpressure protection is provided by spring- or deadweight-loaded relief valve *11*. Numerous relief-valve designs are known, including the spring-loaded one in Fig. 6.2 – 26.

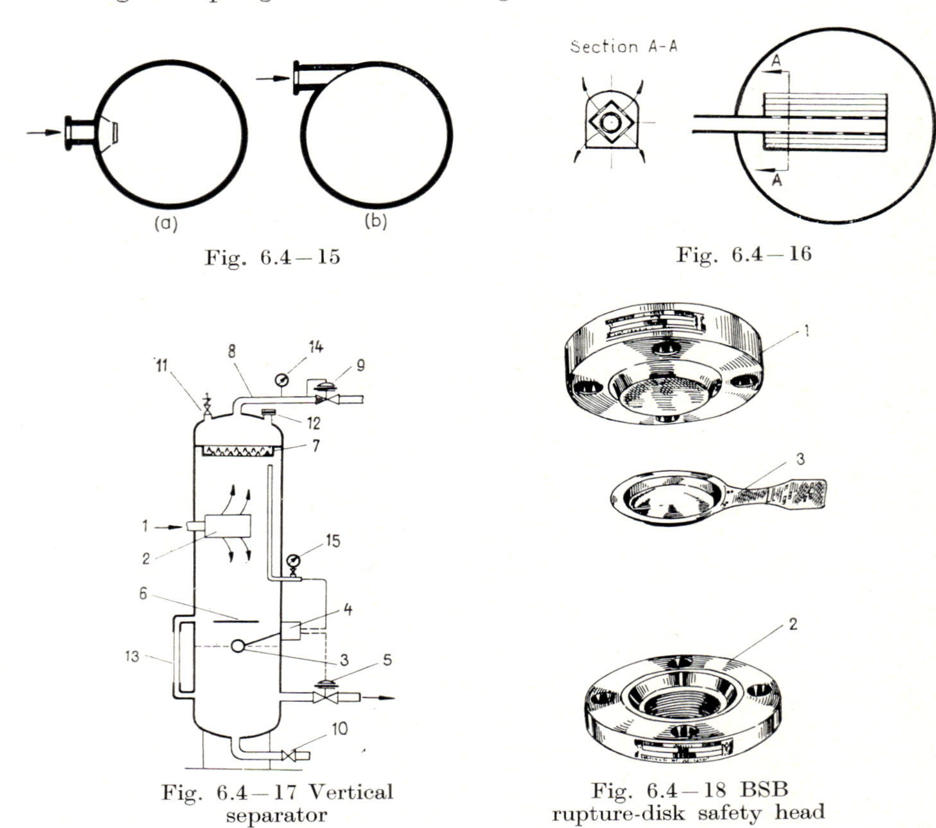

Fig. 6.4 – 15

Fig. 6.4 – 16

Fig. 6.4 – 17 Vertical separator

Fig. 6.4 – 18 BSB rupture-disk safety head

Valves providing full-capacity relief are usually rather large and therefore comparatively expensive. Safety relief of separators is therefore usually ensured by two devices: (i) a relief valve of comparatively small capacity, and (ii) a so-called rupture-disk safety head (*12*). If the overpressure in the separator is too great to blow off through the relief valve, then it will rupture a suitably dimensioned disk which opens an outlet of large enough size. A BSB make rupture-disk safety head is shown in Fig. 6.4 – 18. Disk *3* is sandwiched between flanges *1* and *2*. This double protection has the characteristic that the more frequent slight overpressures are blown off through the relief valve which closes again automatically as soon as separator pressure has returned to normal. Substantial overpressures cause disk rupture, after which it is necessary to close down the separator and to change the disk. — Each separator is equipped with a gauge glass (*13* in Fig. 6.4—17), and a manhole not shown in the Figure. Gas-outlet pressure is measured by gauge *14*, pressure below the mist extractor by pressure gauge *15* installed on the actuating-gas line.

The float-type level sensor of the separator may actuate a regulator either mechanically or pneumatically. Figure 6.4—19 shows a Fisher-make lever-type float-operated oil discharge valve (the float to be attached to the lever is not shown). Pneumatic discharge-control valves tend to provide

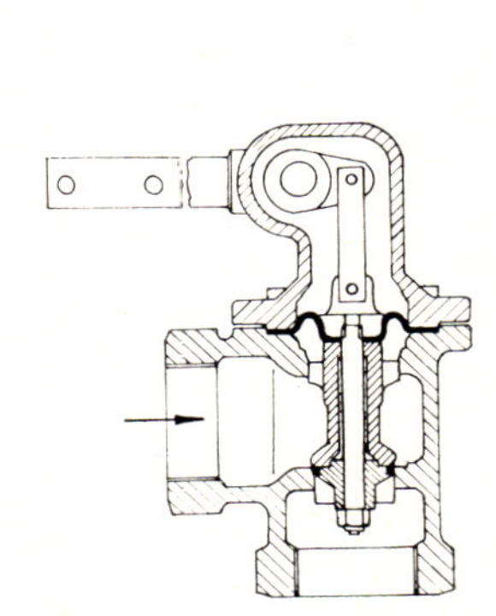

Fig. 6.4—19 Fisher's mechanical liquid level control

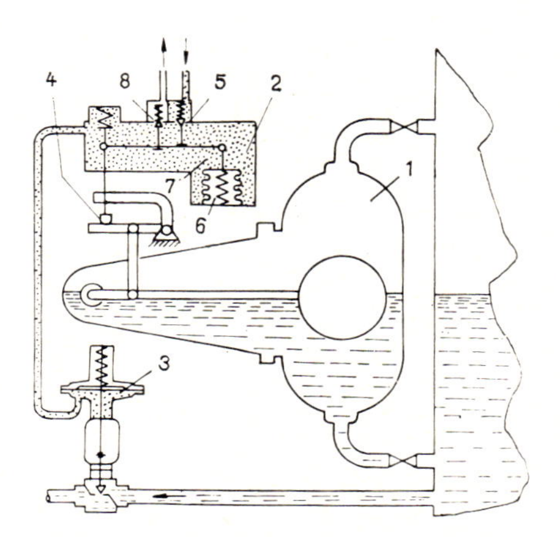

Fig. 6.4—20 Fisher's pilot-operated pneumatic float-chamber type liquid-level control

finer control, but they are more costly than the mechanical variety. The principle of a popular type of pneumatic control is illustrated by Fig. 6.4—17. In another design, the float is installed in a separate chamber connected with the separator space. This has the advantage that, in case of float failure, the unit can be replaced without interrupting separator operation. A float-chamber type liquid-level controller of Fisher-make equipped with a pilot-operated pneumatic motor valve is shown in Fig. 6.4—20. If the liquid level is at middle height in float chamber *1*, pilot *2* feeds supply gas of about 1.6 bar pressure under the diaphragm of motor valve *3*. Liquid inflow and outflow rates of the separator are equal in this position. If inflow exceeds outflow, liquid level will rise in the float chamber. Lever *4* will rise and open valve *5*. Gas pressure will increase in the pilot and hence also under the motor-valve diaphragm, increasing the opening of the inner valve in valve *3*. At the same time, bellows *6*, compressed by the increased outer pressure, depresses the right-hand hinge of lever *7* and through it closes supply-gas valve *5*. This stops the pressure increase under the motor-valve diaphragm. Valve opening will remain constant as long as liquid level remains at the same height. If it starts to decrease, then some of the gas is let out of the motor valve through bleed valve *8*, and the above-outlined process will take place in reverse.

(b) Horizontal separators

Monotube horizontal separator (Fig. 6.4—21). The stream enters the vessel through inlet *1*, to impinge tangentially upon the vessel wall. Separation of oil and gas is promoted by baffle plates *2*. Liquid level is controlled by discharge valve *4*, operated by float *3*. Gas is discharged through outlet *5*, liquid through outlet *6*. Pressure regulation, safety and other equipments are much the same as in vertical separators.

Dual-tube horizontal separator (Fig. 6.4—22). The wellstream enters the upper separator tube through inlet *1* and screen plates *2*. Part of the fluid flows through the first vertical conduit into lower tube *3*. The gas, still laden with mist, flows towards perforated vanes *4*. These promote the coagulation of the mist particles. The liquid phase thus formed may flow down the vertical conduits into tube *3*, from where it is discharged into the outlet through diaphragm motor valve *6* controlled by float-operated pneumatic regulator *5*. Gas is discharged through outlet *7*.

Fig. 6.4—21 Monotube horizontal separator

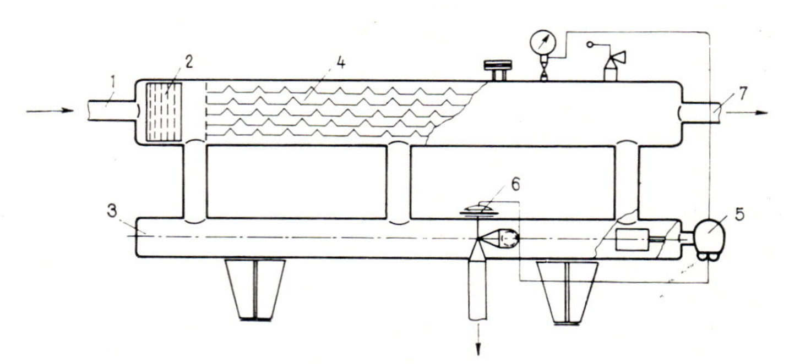

Fig. 6.4—22 Dual-tube horizontal separator

(c) Spherical separators

The separator shown in Fig. 6.4—23 is provided with a bottom gas outlet. The wellstream enters through inlet *1*, to flow downward in the space bounded by spherical cap *2*. Oil collects in the bottom space. Its level is sensed by float *3* and its discharge through valve *5* is controlled by pneumatic pilot *4*. Mist is extracted in coalescing pack *6*. Gas is discharged through outlet *8* provided with pressure regulator *7*.

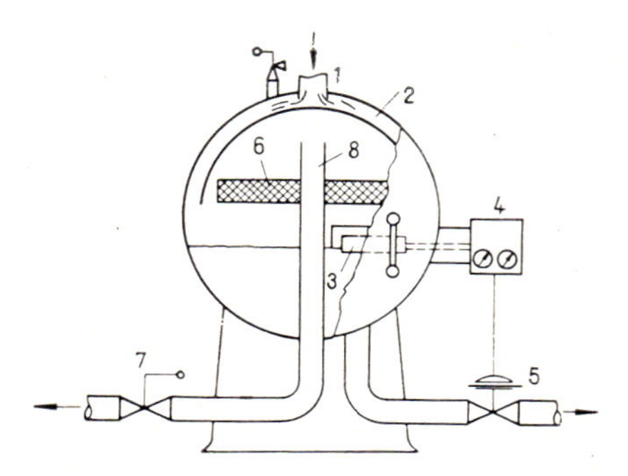

Fig. 6.4—23 Spherical separator

6.4.4. Separator selection

(a) Choice of separator type

Vertical separators have the advantage of being insensitive to level fluctuations, and therefore permitting level control by simple means because, owing to the upright cylindrical shape, the liquid volume change per unit level change is slight, and, on the other hand, level changes will not affect gas-flow cross-section, nor, consequently, the mist content of the gas. They are therefore best suited for separating rapidly fluctuating ('heading') wellstreams. Floor space required per unit gas capacity is less than in the other types. Cleaning is relatively simple, which makes this type of separator suitable for handling sandy crudes. In the horizontal separator, gas capacity (that is, the number of standard volume units of gas that can be freed of liquid to the desired degree per unit of time) is greater for a given cylinder size than in the vertical separator. This is due to several circumstances. In the vertical separator, diameter must be sufficient to make the gas rise slower than the settling velocity of mist particles. Height is primarily determined by the space requirement of the elements described in Section 6.4.3a. Shell height of the current designs is, accordingly, about 3 m. Further increase in height will not improve gas capacity. — The capacity of horizontal separators increases with length. The wellstream entering the vessel will first lose the larger mist drops near the inlet. Mist particle size will gradually decrease from the inlet towards the gas outlet. The longer the separator the less will be the final mist-particle size, and hence, the mist content, of the gas effluent. (Particles below 0.01 mm size cannot be removed by settling alone.) The gas discharge does not meet the inflow, so that mist contamination is unlikely, which permits a higher flow velocity to be adjusted. The gas–liquid interface is comparatively extensive, so that the breakout of gas bubbles from the liquid takes rather a short time. The monotube horizontal separator is therefore smaller and cheaper than the vertical separator of the same gas capacity. The difference in cost is further increased if high-pressure separation is necessary, because a vertical separator of 3 m height, whose wall must be fairly thick to withstand high pressures, can be substituted by a longer horizontal vessel of less diameter and smaller wall thickness. — Dual-tube horizontal separators are more expensive than monotube ones, but the well-separated liquid and gas space prevent any subsequent mixing of the two phases, and the liquid level is always smooth. This is an advantage especially when handling light liquids. Heading of the well hardly affects separation efficiency. Both horizontal separator types are easier to install, handle and service than the vertical ones.

Table 6.4—11
Advantages and drawbacks of separator types

Separator type		I/q_g*	Economy at high gas capacity, q_g	Economy at high gas pressure	Mud, sand	Foaming oil	High-viscosity, freezing crude	Heading	Liquid level control	Portability	Installation, maintenance	A/q_g**
Vertical		3	2	3	2	3	2	2	1	3	2	1
Horizontal	monotube	2	1	1	3	1	1	3	4	2	1	4
	dualtube	2	1	1	3	1	3	1	2	2	1	3
	sphere	1	3	2	1	3	4	4	3	1	1	2

* First cost per unit gas capacity
** Floor space required per unit gas capacity

The main advantage of the spherical separator is that its first cost per unit gas capacity is least. It is to be preferred where well yields are comparatively low and uniform. It is easy to displace, install and clean. It requires little space especially in two-stage separation, because one sphere may be mounted on top of the other. Table 6.4—11 serves to facilitate comparison of separator types' advantages and drawbacks. A lower rating means a greater advantage. (For choice primarily among horizontal separators see the work of Maher and Coggins, 1969.)

(b) Separator sizing

Separator *ID* is determined in keeping with the above considerations by the desired gas capacity. This term means the maximum gas throughput that can be handled by the separator under the prescribed pressure and temperature with mist content kept below 0.1 g/m³. Since internal design and differences in wellstream composition will both affect capacity, it is recommended to perform sizing on the basis of empirical formulae or procedures employing such formulae, as, e. g., Ch. K. Gravis' method (1960), adapted below to the SI system of units. Gas-capacity can be read off Fig. 6.4—25 in the knowledge of the capacity factor

$$f = f_1 f_l f_g f_h .$$

The gas-capacity base factor f_1 may be read off Fig. 6.4—24 as a function of *ID* for separators of various type and length. f_l is a liquid-density correction factor; its value is unity if $\varrho_l = 848$ kg/m³; f_g is a gas-density correction factor, whose value is unity if $\varrho_{gr} = 0.7$; f_h is a vessel-length

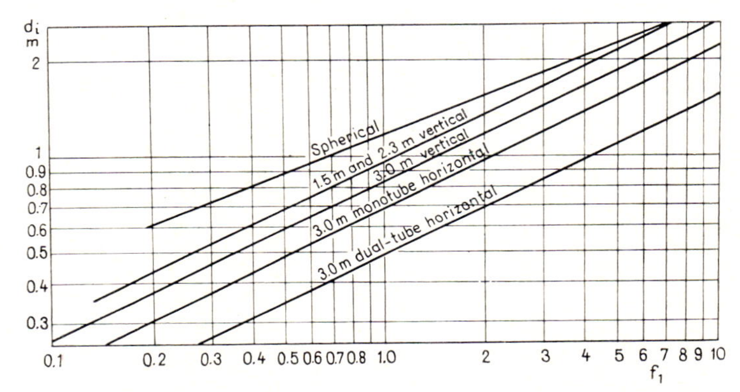

Fig. 6.4—24 Gas capacity factor of separators, after Gravis (1960)

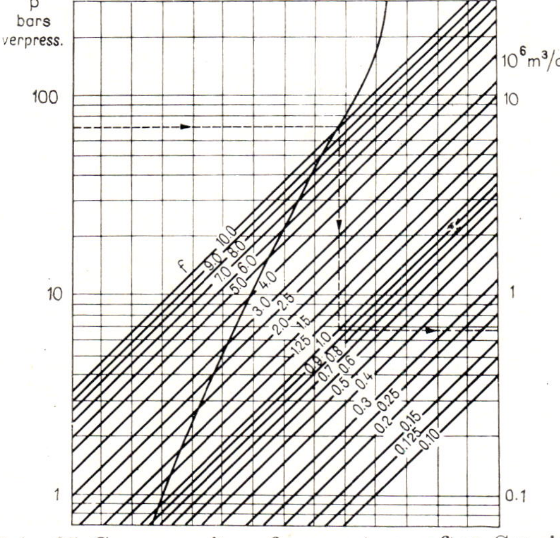

Fig. 6.4—25 Gas capacity of separators, after Gravis (1960)

correction factor whose value is unity for any vertical or spherical separator and for horizontal separators of 3 m length. The values of these factors for conditions other than those stated above can be read off Figs from 6.4—26 to 6.4—28. In the knowledge of f and the required separator pressure p, the daily gas capacity can be read off Fig. 6.4—25.

Strictly speaking, the procedure assumes a separator temperature of 15.6 °C. Capacity is slightly less at higher and slightly more at lower temperatures. It is assumed that the pour point of the oil and the hydrate-

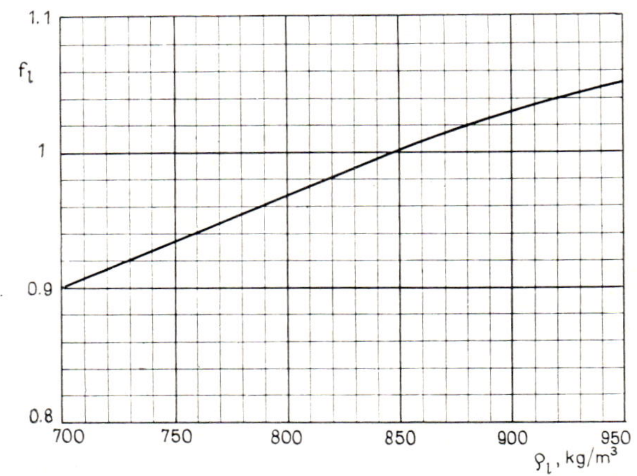

Fig. 6.4—26 Correction factor of liquid density, after Gravis (1960)

formation temperature of the gas are both below separation temperature, foaming of the oil is average, and the in-flow of wellstream is stabilized. For horizontal monotube separators, it is further assumed that liquid takes up one-half of the inner space. In dual-tube separators, liquid is in the lower tube, the upper one being reserved entirely for gas. The gas capacity of a dual-tube horizontal separator is therefore approximately twice that of a monotube one of the same diameter.

Example 6.4—5. Find the gas capacity of a vertical separator if $d_i = 0.9$ m, $p = 50$ bars, $\varrho_l = 760$ kg/m^3, $\varrho_{gr} = 0.650$, and $h = 3$ m; from Fig. 6.4—24, $f_1 = 1.18$; from Fig. 6.4—26, $f_l = 0.94$; from Fig. 6.4—27, $f_g = 1.05$; from Fig. 6.4—28, $f_h = 1$. The capacity factor is, then,

$$f = 1.18 \times 0.94 \times 1.05 = 1.16$$

From Fig. 6.4—25, $q_g = 650 \times 10^3$ m^3/d.

Retention time of oil is to be chosen so that the gas bubbles have sufficient time to break out of the oil. In non-foaming oil, this time is 1 to 3 min. Liquid space is the space between the oil outlet and the level maintained by the regulator. If liquid-space volume is e.g. 0.16 m^3, and retention time is 1 min, then liquid capacity is $1440 \times 0.16 = 230$ m^3/d. Liquid capacity as a function of ID can be estimated using Fig. 6.4—29. If the oil has a tendency to foam, retention time must be increased (even up to 20 min), and capacity is reduced accordingly. Foaming can be reduced considerably by adding to the wellstream small quantities of silicon (0.0017—

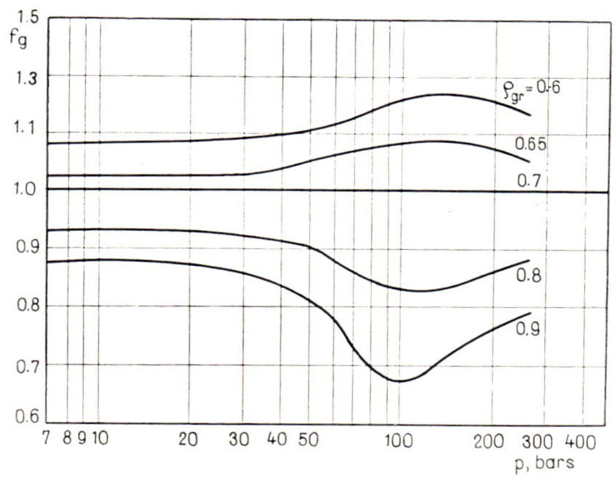

Fig. 6.4—27 Correction factor of gas density, after Gravis (1960)

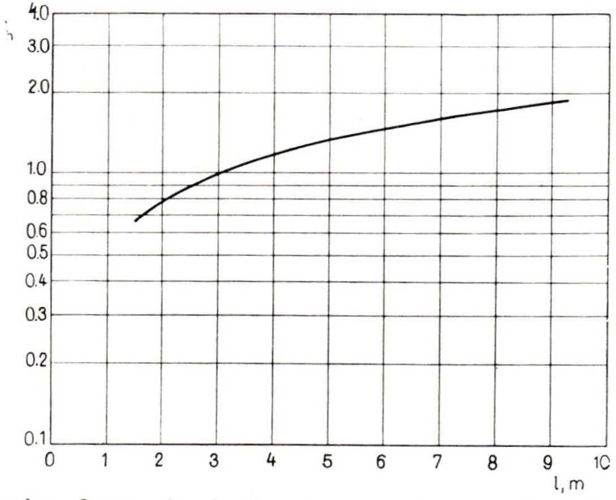

Fig. 6.4—28 Correction factor for horizontal-separator length, after Gravis (1960)

0.025 ppm). Capacity increases up to 40 percent have been recorded (Nematizadeh 1969). This procedure will pay for itself especially if the aim to be achieved is a comparatively short-range temporary increase of capacity.

Example 6.4—6. Find, assuming an average tendency to foam, the liquid capacity of the separator described in the foregoing example, for a retention time of 1 min. — Figure 6.4—29 gives $q_l = 650$ m³/d.

The liquid discharge rate from the separator is determined jointly by the pressure differential between separator and tank and the flow resistance of the discharge line. When designing and installing the separator, it is to be ascertained that the maximum discharge rate determined by these factors be at least as high as the liquid capacity of the separator. The size of the lower tube of a dual-tube horizontal separator is determined by the prescribed retention time of the liquid: in practice, it is at least as large as the upper tube. The liquid level in spherical separators is usually about 13 cm below the horizontal median plane of the sphere.

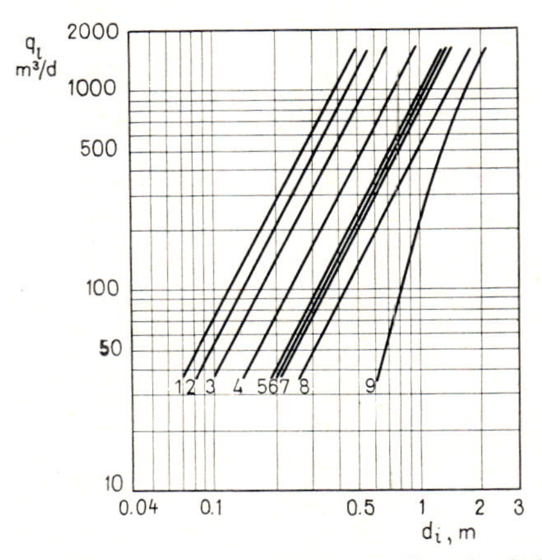

Fig. 6.4—29 Liquid capacity of separators, after Gravis (1960); *1* monotube horizontal, h=6.1 m; *2* dual horizontal, h=4.6 m; *3* dual horizontal, h=3.0 m, and monotube horizontal, h=6.1 m; *4* dual horizontal, h=1.5 m and monotube horizontal, h=3.0 m; *5* vertical, h=4.6 m; *6* monotube horizontal, h=1.5 m; *7* vertical, h=3.0 m; *8* vertical, h=1.5—2.3 m; *9* spherical

If the wellstream is not stabilized (the well produces by heads or slugs or its production is intermittent), separators should be sized for the maximum instantaneous flow rate to be expected, rather than for the daily mean production rate. This is especially important when sizing spherical separators.

6.4.5. Special separators

(a) Cyclone separators

This is a type of two-phase separator which ensures high-efficiency liquid separation without a mechanical mist separator, merely by a suitable flow-path design. Figure 6.4—30 is the schematic diagram of a modern cyclone separator, after (Barrett 1970). Tangentially entering through conduit *1*, and deflected into an approximately spiral flow path, the fluid descends between the vessel shell and bell baffle *2*; most of the liquid flows along the vessel wall into liquid space *3*. The gas vortex still containing mist rises through tube *5* above baffle *4*. The least-pressure point of the vortex is above the central orifice in plate *4*. Some of the mist particles in the gas impinge upon the lower surface of bell *2* and, coalescing, drip down into the liquid space. Most of the mist rising in tube *5* is thrown by vortex motion against the inner tube wall. Suction from the low-pressure centre generated above plate *4* sucks this liquid through apertures *6* and conduit *7* to orifice *8*. The vortex rotating there blows the liquid towards the separator wall, where it may sink into the liquid space. Gas discharged through outlet *9* has 99.94 percent of its liquid content removed in this type of separator. There are also horizontal separators operating on this same principle. The DEMCO centrifugal separator (sand trap) shown schematically in Fig. 6.4—31, serves to remove solid contaminants such as fine sand from liquids. Inflow is through conduit *1*. Solid particles, thrown against the conical shell, sink down along it to be drained through outlet *2*, whereas the pure liquid is discharged through outlet *3*. It is usual to connect several such units in parallel to an inlet

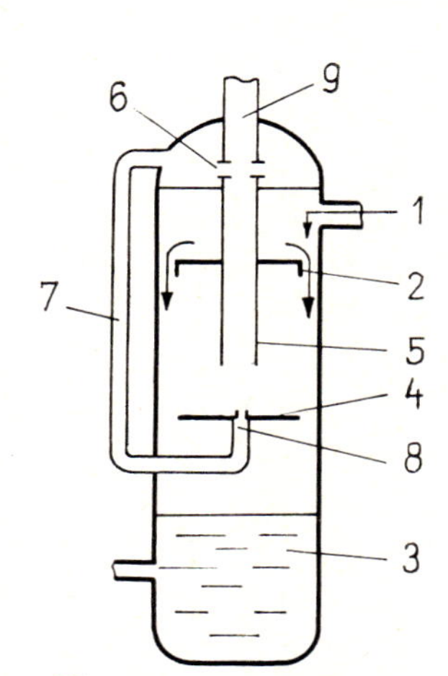

Fig. 6.4—30 Cyclone separator, after Barrett (1970)

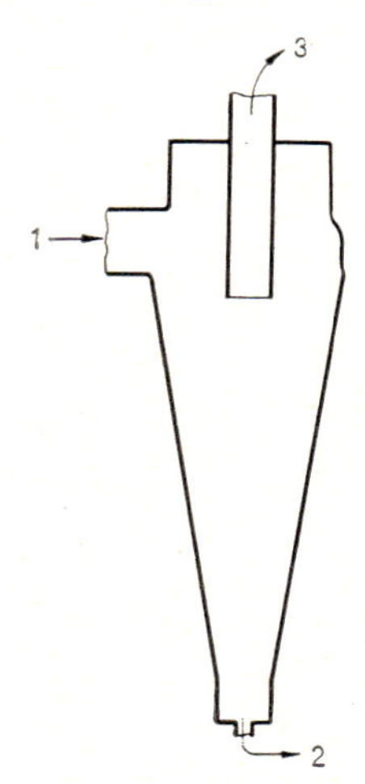

Fig. 6.4—31 DEMCO centrifugal separator

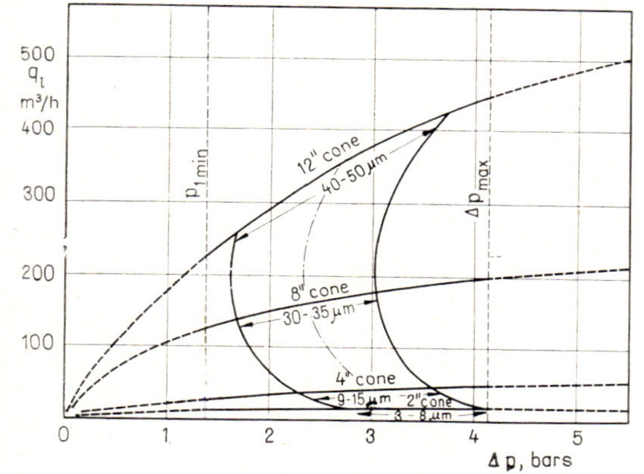

Fig. 6.4—32 Capacity of DEMCO centrifugal separators

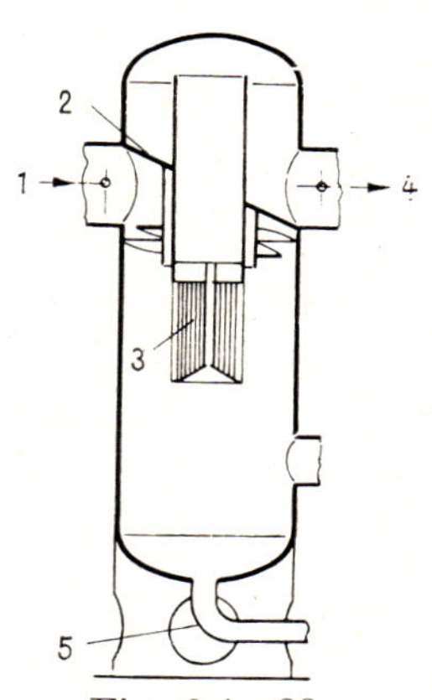

Fig. 6.4—33 MOKVELD separator

and an outlet manifold. The separation capacity of a centrifugal separator of given size depends on the pressure differential between inlet and outlet and on the particle size of the solids to be separated. Figure 6.4—32 is a capacity diagram of DEMCO separators. Separation efficiency is highest in the shaded domain. The separator will remove 90 percent of the solids of particle size indicated in the Diagram. In that domain p_{1min} is the least recommended outflow pressure; Δp_{max} is the maximum pressure difference to be applied. — Figure 6.4—33 shows the MOKVELD separator (scrubber) serving to remove solid and liquid particles above 10 μm size. The primary fluid enters through inlet *1*. Gas flows downward below dividing plate *2*, and reaches outflow aperture *4* through baffles *3*. Liquid drops and solids removed from the turbulent gas are drained through conduit *5*. The scrubber will remove from 98 to 100 percent of the particles above 10 μm size.

(b) Three-phase (oil-water-gas) separators

In addition to oil and gas, the wellstream often contains substantial amounts of water. Water occurs either as a separate phase or in emulsion with oil. In the former case, so-called three-phase separators which separate oil from water as well as liquids from gas can be used to advantage.

The three basic types of three-phase separator are shown in Figs from 6.4—34 to 6.4—36. Water is drained through a pilot-controlled valve operated by a float weighted so as to float at the water–oil interface. In type (a), oil and water levels are maintained by floats *1* and

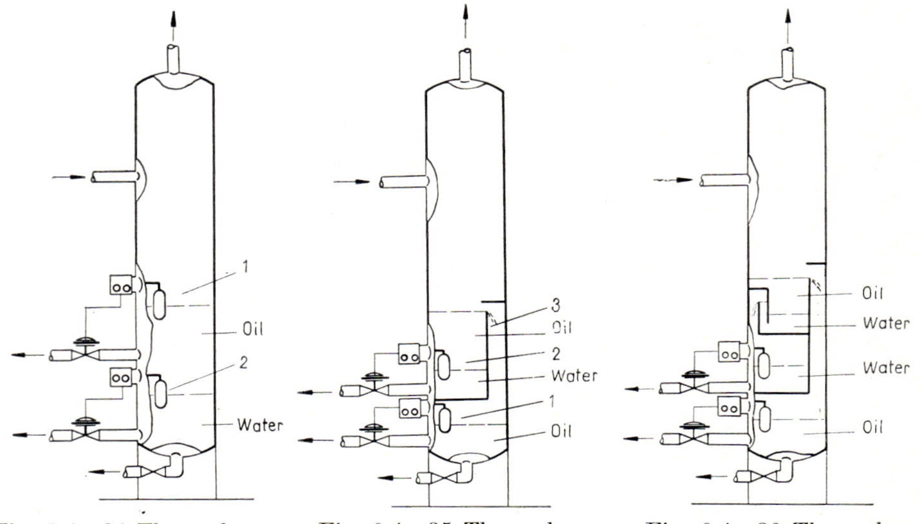

Fig. 6.4—34 Three-phase vertical separator, type (a); after Broussard and Gravis (1960)

Fig. 6.4—35 Three-phase vertical separator, type (b); after Broussard and Gravis (1960)

Fig. 6.4—36 Three-phase vertical separator, type (c); after Broussard and Gravis (1960)

2 alone. The gravity difference between water and oil is usually rather slight, and so is the differential buoyancy maintaining the weighted float at the water-oil interface. Sudden slugs of incoming liquid may make the floats bob up and down, which is detrimental to accurate level maintenance. This type is comparatively cheap and simple; the entire bottom space is available for settling; sand and mud deposits are comparatively easy to remove. In type (b) floats *1* and *2* are assisted in maintaining liquid levels by weir *3*. Settling space is, however, less than in type (a), and the internal fittings tend to hamper cleaning. In design (c), two separate liquid levels are maintained by weirs and floats, and there is no interface float. Water and oil collect in separate compartments. This has the advantage that, should any one of the float-operated level controls fail, water and oil will discharge separately, even though both will be mixed with gas. Its drawback is that the settling space is even smaller than in type (b), cleaning is cumbersome, and the entire device is rather expensive. Designs (b) and (c) above, as applied to spherical separators, are shown in Figs 6.4—37 and 6.4—38. In addition to the typical features of spherical separators, these have the advantage that, despite of the weirs in them, their cleaning is simpler than that of the corresponding vertical separators.

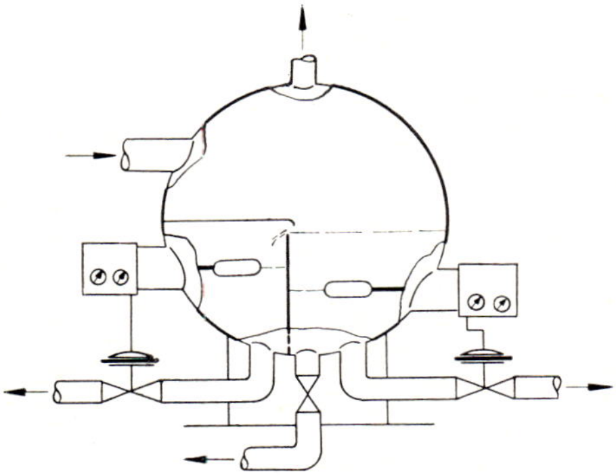

Fig. 6.4—37 Three-phase spherical separator, type (b); after Broussard and Gravis (1960)

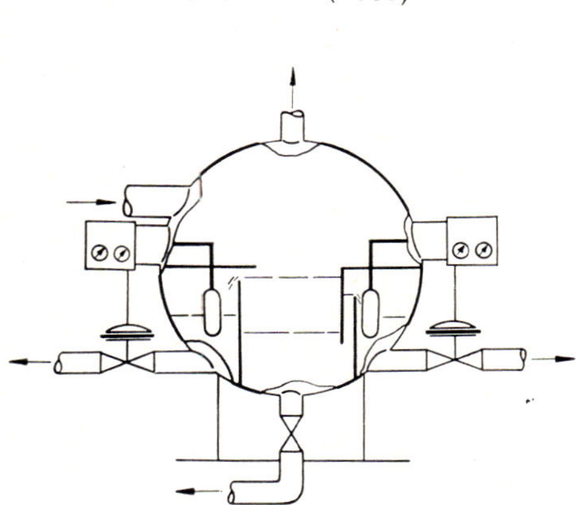

Fig. 6.4—38 Three-phase spherical separator, type (c); after Broussard and Gravis (1960)

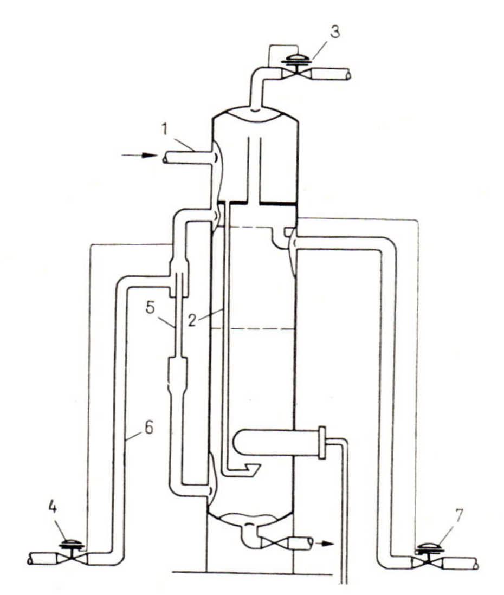

Fig. 6.4—39 Heated three-phase vertical separator, after Campbell (1956)

If the wellstream emulsion can readily be broken by heating, the three-phase separator shown in Fig. 6.4—39 (Campbell 1956) will prove economical. The wellstream enters through inlet *1* and attains the heated space through conduit *2*. Dewatered oil rises to the surface. Diaphragm motor valve *4* regulates the discharge of water, DMV *7* regulates that of oil. Valve *4* opens whenever conduit *6* is entirely filled with water, as the pressure acting upon its diagram from below will exceed in that case the supply-gas pressure acting upon it from above. Valve *7* operates in a similar way.

(c) Automatic metering separators

Metering separators meter the weight or volume of the liquid discharge. Depending on whether the metered output is total liquids or oil and water separately, metering separators may be two- or three-phase. Different designs will have to be adopted according to whether the liquid and gas are easy to separate, or tend to form considerable foam, or the crude is of high-viscosity. Most metering separators measure the volume rather than the weight of the liquid discharge. — Figure 6.4—40 after V. Smith shows a

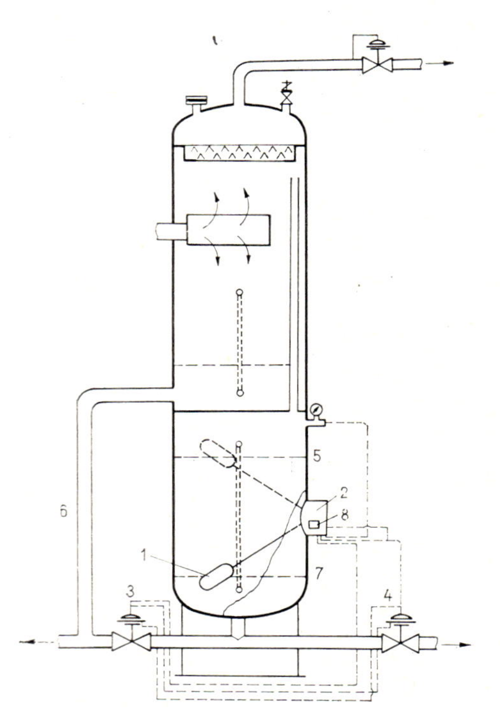

Fig. 6.4—40 Automatic metering separator, according to V. Smith; after Frick 1962, Vol. I, pp. 11—20 (used with permission of McGraw-Hill Book Company)

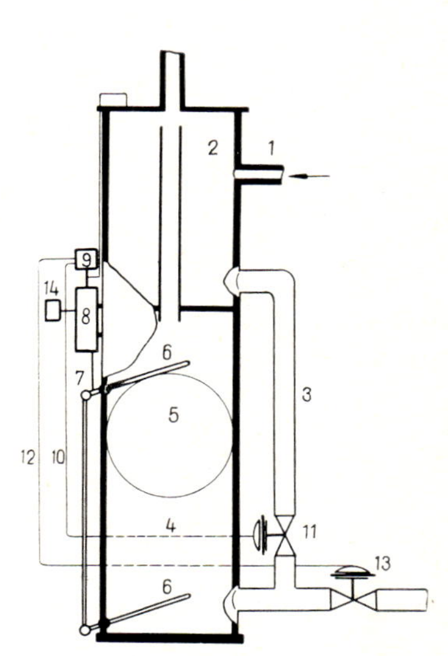

Fig. 6.4—41 Automatic metering separator for low-capacity wells, after McGhee (1957)

metering separator suitable for the two-phase separation of non-foaming oil (Frick 1962). The separator vessel is divided horizontally into two compartments, the upper one of which is the separator proper. Its design is essentially that of a conventional vertical separator (e.g. Fig. 6.4—17). The extreme positions of float *1* in the lower compartment (metering chamber) control by the intermediary of pilot *2* the valves *3* and *4*. Whenever the liquid level in the metering chamber is below upper float level *5*, valve

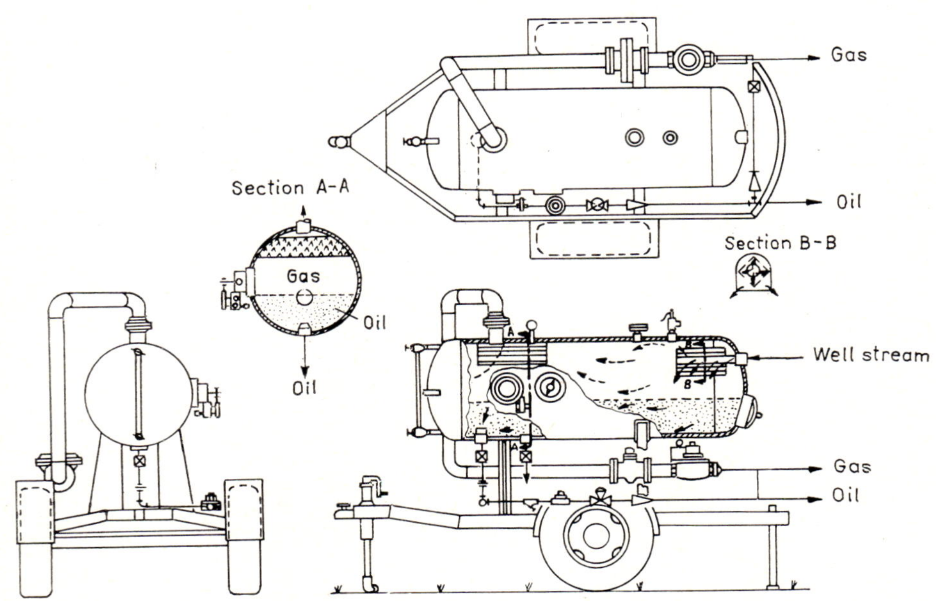

Fig. 6.4—42 Two-phase portable separator, according to J. D. Kimmel; after Frick 1962, Vol. II, pp. 29—10 (used with permission of McGraw-Hill Book Company)

3 is open and valve *4* is closed, and liquid may flow through conduit *6* from the upper compartment into the metering chamber. At the upper float level (position shown in dashed line), pilot *2* closes valve *3* and opens valve *4*. As soon as liquid level has sunk to the lower float level, valve *3* is re-opened and valve *4* is closed. The number of drainages of the metering chamber is recorded by counter *8*. This latter may also be equipped for a digital print-out of the metered liquid volume, by multiplication of the count with metering-chamber volume.

The separator in Fig. 6.4—41 permits the automatic metering of low-capacity wells (McGhee 1957). The wellstream enters separator compartment *2* through inlet *1*. — When discharge is inoperative, degassed oil drains through conduit *3* into metering chamber *4*. The rising oil level lifts steel-ball float into the position shown in the Figure. The ball in turn lifts lever pair *6—6* and pulls down stem *7*. This actuates pilot *8*, which makes supply-gas distributor feed gas through line *10* in order to close motor valve *11*. At the same time, pressure is reduced in line *12*, so that valve *13* will open. On discharge, the ball, sinking to the bottom of the chamber, depresses the lower lever *6*, and the pilot commands the gas distributor to close valve *13* and to open valve *11*. Every downstroke of stem *7* is counted by counter *14*. The gas capacity of a separator of 0.3 m *ID* and 2 m height is about 6000 m^3/d.

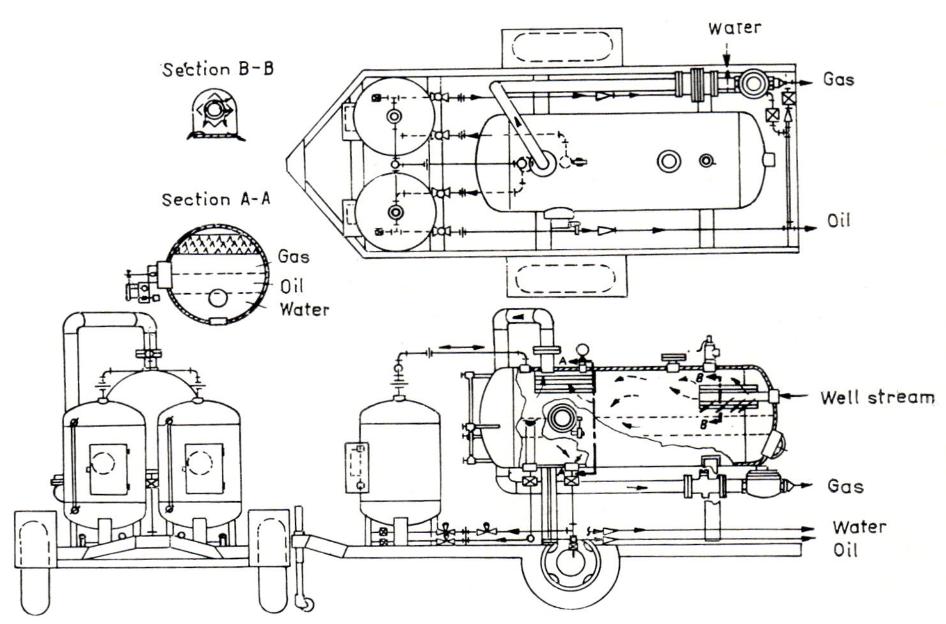

Fig. 6.4—43 Three-phase portable separator, according to J. D. Kimmel; after Frick 1962, Vol. II, pp. 29—12 (used with permission of McGraw-Hill Book Company)

Portable separators are used primarily around exploration wells and in single-well gathering and separating systems. These, too, may be two- or three-phase. They usually meter liquid by a positive-displacement or a dump type meter, and gas by an orifice meter. In three-phase separators, the separation of water from oil may be performed without heating (if the phases will separate readily) or with the aid of heating. In the latter case, a demulsifying chemical agent may or may not be added. Heating may be direct gas-firing, or electrical. Separators may be upright or horizontal. In the following we shall restrict discussion to unheated horizontal portable separators, assuming that, if three phases are to be separated, water is entirely removed from oil by gravity alone during the liquid's stay in the separator. In the two-phase separator shown after J. D. Kimmel (Fig. 6.4—42), liquid passes through a positive-displacement meter; the portable three-phase separator shown in Fig. 6.4—43 (Frick 1962) is equipped for the dump metering of both oil and water. Separator tubes are 1.8—2.1 m long; diameters are 0.30—1.2 m. Maximum operating pressure is of 166 bars; maximum standard-state gas capacity is a round 500×10^3 m^3/d for two-phase, and a round 340×10^3 m^3/d for three-phase separation. Maximum liquid capacity is a round 750 m^3/d of oil for two-phase separation, and 750 m^3/d of oil plus 380 m^3/d of water for three-phase separation. The devices weigh 0.5—2 tons.—When using a dump meter, the meter factor has to be determined in each new application, since owing to the gas content of the liquid and to the fact of metering being carried out at near-separator pressure and temperature, standard-state volume will invariably differ from the metered volume. Moreover, a certain instrument error and some contamination of the liquid must also be reckoned with. The meter constant is always less than unity; that is, the volume reading is invariably greater than the standard-state volume of the liquid or the tank volume. Dump meters have the advantages that: (i) The correct operation of the meter is easy to verify; the meter is in good working order if it contains no deposits and is not deformed, and its valves open and close accurately. (ii) They are less sensitive to sand and other solids than positive-displacement meters. (iii) Metering accuracy is independent of throughput in the range from 0 to $q_{l\max}$. (iv) With gravity checks at intervals, they are suited also for the metering of foaming oil. (v) They can be serviced and repaired even with the separator onstream. (vi) Even if the control equipment fails, they will not meter gas as if it were liquid. Their drawbacks are: (i) First cost and installation cost are higher than in a positive-displacement meter. (ii) Metering is not continuous. (iii) Making the oil pass through the dump meter requires a non-negligible excess pressure in the separator. (iv) Deposits of wax will cause inaccurate metering. (v) Dump meters require more space and are heavier than positive-displacement meters. (vi) Especially if the oil is high-viscosity, the head loss due to metering is higher. — Metering capacity depends on dump-vessel volume.

Of the positive-displacement meters, nutation-disk type meters are most often employed for their simplicity, ruggedness, comparative width of measuring range and cheapness. They are, however, less accurate than oval-gear meters. Positive-displacement meters have the following advantages: (i) Liquid output and metering are continuous. (ii) These meters lend themselves also for the metering of high-viscosity oils. (iii) Pressure drop due to passage through the meter is negligible. (iv) Temperature compensation, that is, reduction of the metered volume to the standard-state volume is cheaper in most types than in a dump meter. Drawbacks include the following: (i) The liquid to be metered must be gasless. The meter will measure gas as if it were oil: indeed, fast changes in flow velocity due to gas slugs may damage or wreck the meter. (ii) Sand, mud, salt or any other solids are harmful to the meter and tend to falsify results. (iii) Accuracy is to be checked at intervals by means of a proving tank or a master meter. (iv) Accurate metering is restricted to a rather narrow output range.

6.4.6. Low-temperature separation

If the wellstream is cooled to a low temperature before entry into the separator, separation will produce dry gas and a liquid rich in low-boiling-point hydrocarbons (condensate). The dew point of dry gas is lower, which makes it better suited for transportation in pipelines. If the liquid in the wellstream is of low boiling point in its entirety, then the liquid recovered will be raw gasoline rich in propane and butane. Low-temperature separation, usually below −5 °C, is a comparatively recent procedure. Still,

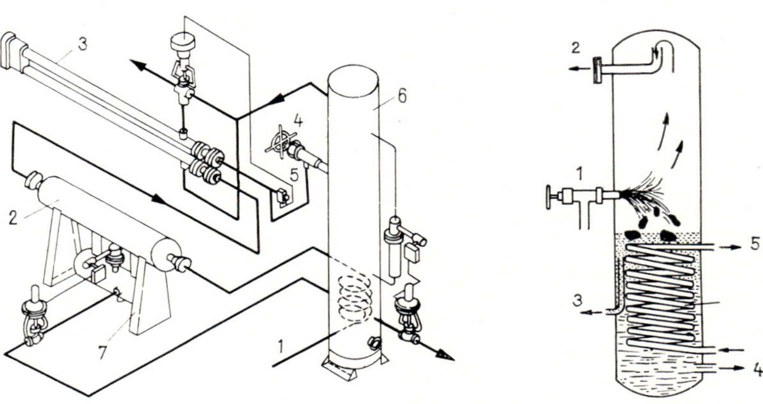

Fig. 6.4—44 Low-temperature separator, type a.1.1, after Frick (1962; used with permission of McGraw-Hill Book Company)

Fig. 6.4—45 Low-temperature separator, after Mapes (1960)

numerous variants have been devised. These may be divided in two large groups depending on whether the wellstream is cooled by its own near-adiabatic expansion, or by external means. The main technical difficulty to be overcome in low-temperature separation is that at the low temperature involved, combined with the pressure prevailing in the separator, gas hydrate will tend to form and, if nothing were done to forestall it, the separator would 'freeze up' pretty fast. In order to ensure continuous operation, various measures are taken to prevent freezing. In the following we shall further subdivide the two main types of low-temperature separation according to these measures.

According to E. C. Young, the main methods of low-temperature separation are as follows (Frick 1962). (a) Cooling is by expansion of the wellstream; (a.1) without hydrate inhibitor, (a.1.1) no external heating applied, (a.1.2) external heating applied; (a.2) with hydrate inhibitor, (a.2.1) without stabilization, (a.2.2) with stabilization; (b) cooling is by a refrigerant in a heat exchanger; (b.1) refrigerated by absorption; (b.2) refrigerated by compression.

Subtype (a.1.1). No inhibitor is added to the wellstream and no heating is employed to melt the hydrate. Figure 6.4—44 shows equipment used for this purpose after Young. Its main parts include inlet liquid separator *2*, gas-to-gas heat exchanger *3*, and low-temperature separator *6*. The low-temperature separator itself is shown in more detail in Fig. 6.4—45, after Mapes (1960). Wellstream enters through choke *1*. The solid hydrates freezing out of it collect on the vessel bottom. Outlets are: *2* for gas, *3* for liquid hydrocarbons, *4* for water. Solid hydrates are heated above decomposition temperature by warm wellstream flowing through coil *5*. Pressures and temperatures may e.g. be as follows: gas temperature upstream of choke, 30 °C; downstream of choke, 0 °C. Inflow temperature

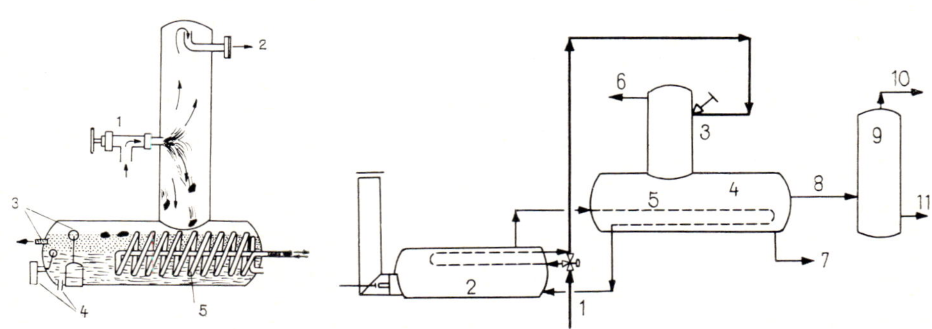

Fig. 6.4—46 Combined upright-horizontal low-temperature separator, after Mapes (1960)

Fig. 6.4—47 Low-temperature separator, type a.1.2, after Mapes (1960)

of heating wellstream, 50 °C; outflow temperature, 30 °C. Such cooling of the well fluid may be ensured according to experience by letting the wellstream expand from 140 bars to, say, 5 or 10 bars. Low-temperature separators are not necessarily vertical: they may be horizontal, combined (Fig. 6.4—46), or spherical as well. — Referring once more to Fig. 6.4—44, the inflowing wellstream *1* is seen to be used in this setup to decompose hydrates collected at the separator bottom. The liquid is separated from the cooled well fluid in inlet-liquid separator *2*, for two reasons; one, to reduce the liability of the choke's freezing up; and two, gas is easier to cool if separated from the liquid of greater heat capacity, from which it could absorb more heat. The gas separated from the liquid then enters into heat exchanger *3* where it is cooled by the cold dry gas discharged from the separator to a temperature just above the hydrate-formation point upstream of the choke. This is achieved by regulating the quantity of dry gas passing through the heat exchanger by means of thermostat-controlled three-way motor valve *3*. This setup will function satisfactorily only if the heat delivered by the inflowing wellstream is sufficient to melt the gas hydrates at the separator bottom, and, on the other hand; if the temperature of the wellstream upstream of the choke is above the hydrate-formation point at the given pressure.

Subtype (a.1.2). Heat required to warm the wellstream and to decompose the gas hydrates collecting in the separator is taken from an external source. In this case, the gas-to-gas heat exchanger is replaced by a line heater. Gas is heated with either steam or hot water in a heat exchanger, or part of the wellstream is passed through a coil in the liquid space of a boiler; the proportion of heated to unheated wellstream is controlled by the temperature upstream of the choke. The second solution is the more up-to-date. Gas hydrates are likewise decomposed by either water or steam taken from the boiler. This latter setup is illustrated by Fig. 6.4—47. The wellstream enters through conduit *1* to be divided by a three-way valve; part of it enters boiler *2*, to pass from there through choke *3* into combined separator *4*. Dry gas is discharged through outlet *6*. Gas hydrates are heated to the required temperature by hot water circulating in conduit *5*. Water is drained through outlet *7*. Hydrocarbons are led through conduit *8* into second-stage separator *9*, where some gas can still be recovered from the liquid.

6.5. On-lease oil storage

6.5.1. Storage losses

Storage losses are comprised of leakage and evaporation loss. Leakage can be avoided if tank plates and fittings are tight and tanks are filled and emptied with proper care. Evaporation losses are due to the fact that the tank space above the stored crude is saturated with hydrocarbon vapours that will escape into the atmosphere. Evaporation losses fall into filling and breathing losses. Breathing is due to the daily fluctuation of tank temperature. Warming of the tank makes vapours expand, to escape partly into the atmosphere through the tank vents. On cooling, the tank will suck in air from the atmosphere. This process is further complicated by rain and other precipitations. Breathing loss is the higher the higher the vapour pressure of the product stored, the greater the temperature fluctuation, and the larger the vapour space. Filling losses will occur as the rising liquid level in a tank being filled expels the vapours contained in the vapour space. Evaporation has the following unwelcome consequences. (i) Hydrocarbon vapours escaping into the atmosphere are lost for good, (ii) the light-fraction content, and hence, the value per ton, of the stored product decreases, (iii) the environment is polluted by the escaping vapours; moreover; (iv) an explosion hazard is created, and (v) the air entering the tank accelerates the corrosion of the inside tank wall.

Several ways to reduce or altogether stop evaporation losses are known. In order to assess the economy of otherwise of the solution proposed, it is necessary to determine the losses, e.g. by closing all apertures of the tank into the atmosphere, save one, and by measuring the quantity and composition of the gas escaping through the remaining aperture over a longer span of time. For estimating evaporation losses of light hydrocarbons in big tanks, a procedure requiring no separate measurement has been devised (O'Brien 1951). The method is based on Figs 6.5—1 — 6.5—3. Figure 6.5—1 shows Reid vapour pressures v. temperature for various hydrocarbons. As regards evaporation loss, it is the surface temperature of the oil stored in the tank that is relevant. A statistic of data

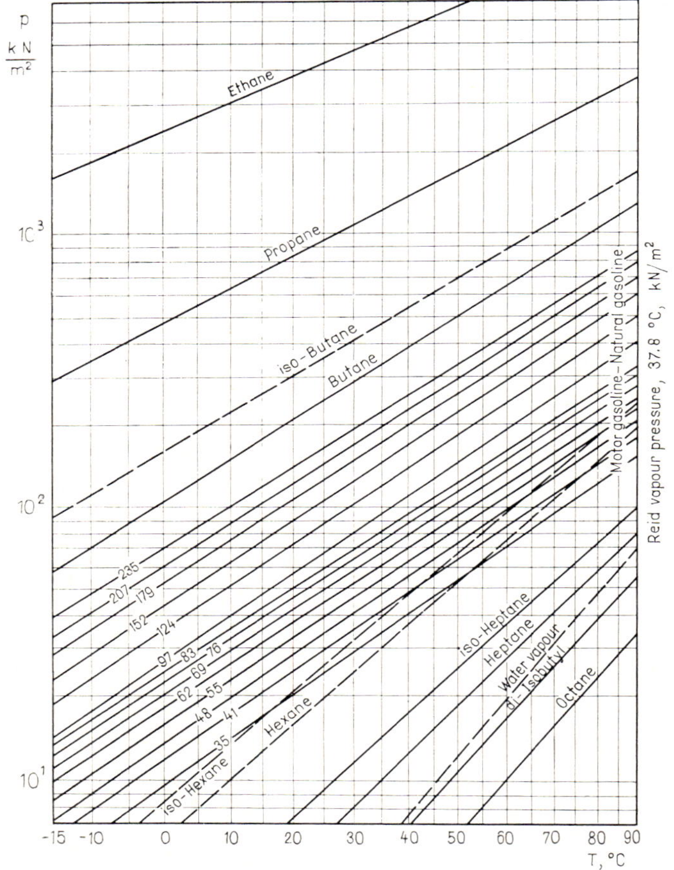

Fig. 6.5—1 Vapour pressures, after O'Brien (1951)

collected all over the USA has revealed surface temperatures to exceed mean annual temperatures by 5.5 °C on an average. Figure 6.5—2 is a plot of breathing losses over one year v. tank volume, for various vapour pressures. In compiling this diagram tanks have been assumed to be painted a silvery colour and to be half full; the vapour space has been assumed to expand by 20 percent. Figure 6.5—3 is a plot of filling loss v. tank volume for various vapour pressures. It has been assumed that the vapour escaping during filling has the same gasoline content as the vapour space. The tank is painted a silvery colour and is half-full on a long-term average.

Example 6.5—1. Find by O'Brien's method the breathing loss over one year and the filling loss during the filling of 10^6 m^3 of liquid in a tank of 8700 m^3 capacity, if the liquid in the tank is gasoline of 90 kN/m^2 Reid vapour pressure and its mean surface temperature is 25 °C. — By Fig. 6.5—1, vapour-space pressure is $p_v = 63$ kN/m^2. Accordingly, breathing

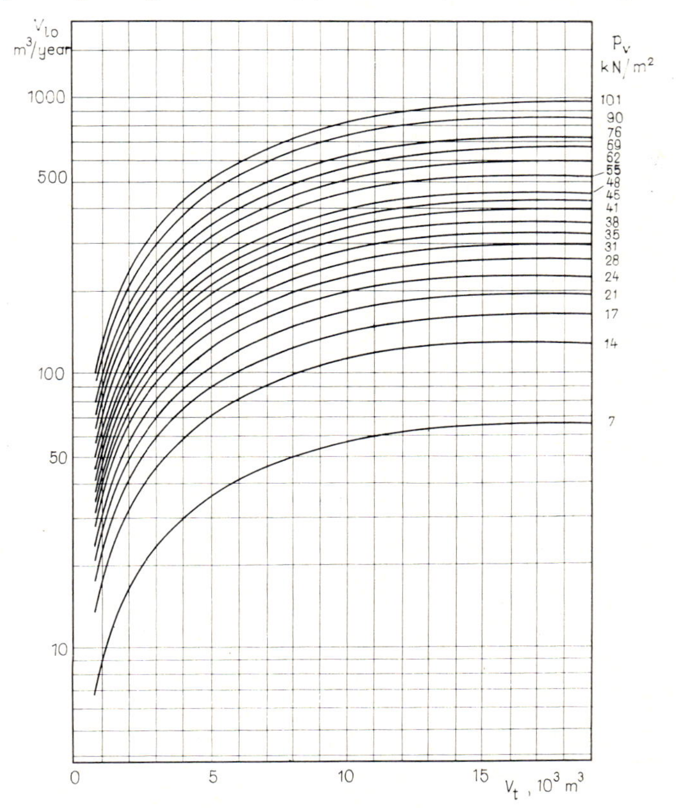

Fig. 6.5—2 Breathing loss, after O'Brien (1951)

loss over one year is 485 m^3/year (Fig. 6.5—2). By Fig. 6.5—3, the loss factor q_{lo}/q_l is 0.00162, that is, 1620 m^3 for a liquid throughput of 10^6 m^3.

Losses in an upright cylindrical tank may be reduced: (a) By keeping temperature fluctuations within limits. (i) The tank is painted a silvery or aluminium colour. Evaporation losses of a crude of 0.837 kg/m^3 gravity for tanks of different colours are listed in Table 6.5—1. (ii) The tank is partly or entirely covered with a heat-insulating layer. If covering is partial, it is applied to the tank roof. (iii) The tank is cooled by the evaporation of a sheet of water on the tank roof, or by sprinkling with water. A so-called water roof may reduce evaporation losses by 25 to 40 percent, sprinkling by more than 50 percent. (iv) The tank is built underground.

Table 6.5—1
Influence of colour to which tank is painted on breathing loss, for oil of 837 kg/m³ gravity

Colour of paint	Evaporation loss percent per year	Gravity after one year of storage
Black	1.24	840.3
Red	1.14	839.7
Grey	1.03	838.6
Aluminium	0.83	838.1

An earth cover of 30 to 40 cm above the tank may prevent all breathing losses. (b) By closing the tank to the atmosphere, that is, by choosing a tank whose permissible internal pressure is higher than the vapour pressures that may arise. Evaporation losses will cease if pressure at the liquid surface is higher than the vapour pressure of the liquid. (c) By special tank-roof designs. (d) By means of vapour-recovery installations.

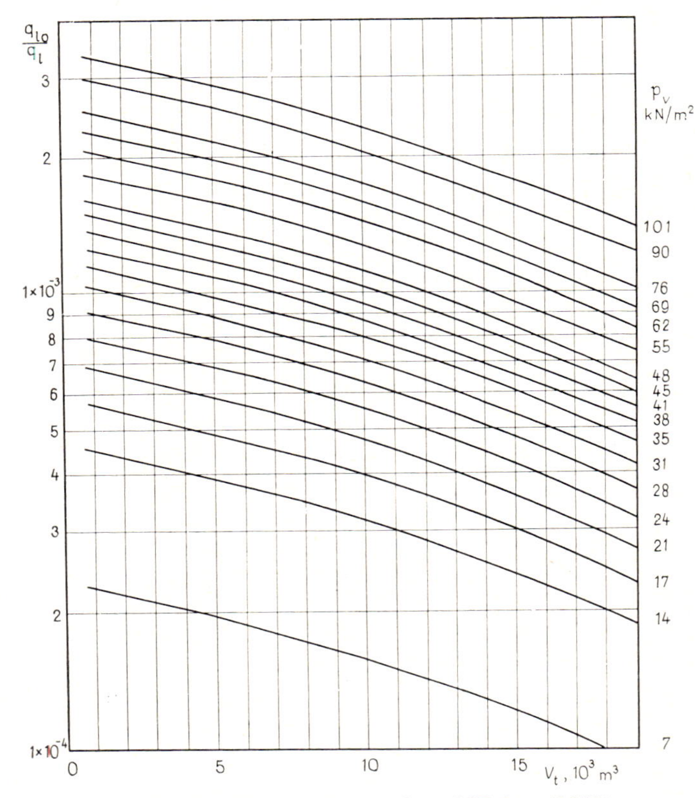

Fig. 6.5—3 Filling loss, after O'Brien (1951)

6.5.2. Oil storage tanks

Oil tanks on a lease are installed partly at well testing stations, and partly at the central oil gathering station of the lease. They are usually made of steel sheet or concrete. Most of them are above-ground, but some are buried underground. — Pits dug in the soil, as the simplest means of liquid storage, are used temporarily, if at all, next to exploration and after well blowouts. Oil-storage pits are to be dug in level ground, if possible. The pit walls and bottom are lined with stamped clay (cf. Fig. 6.5—4). In order to minimize seepage losses, a layer of water 10—15 cm

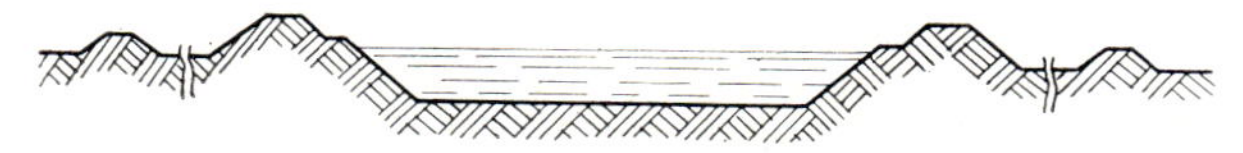

Fig. 6.5—4 Storage pit

high is poured in under the oil. Pits serving as slush pits or for the storage of fracturing fluid are first injected with a silicon compound and then with a polymer expanding in water. In such pits, seepage loss is less than 0.1 percent of the liquid volume stored (*Oil and Gas Journal*, 1969). A similar treatment will reduce oil seepage, too.

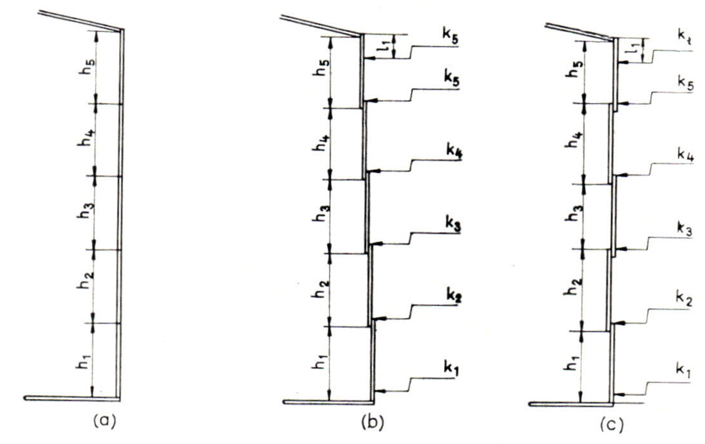

Fig. 6.5—5 Hoop welding patterns of tanks (Petr. Ext. Serv. 1955)

The most widespread type of on-lease oil-storage tank is upright cylindrical, made of sheet steel. The sheet used is usually smooth, seldom corrugated. Smooth sheet is joined by bolting, riveting or welding. Bolted and riveted tanks can be taken apart without loss of sheet, and re-erected elsewhere. Welded tanks can be installed faster; they require less sheet for a given volume. However, their dismantling and re-erection entails a considerable loss of sheet. In welded tanks, rings are either butt- or lap-welded. (cf. Fig. 6.5—5; butt welds (a), telescopic lap-welds (b) and alternating-ring lap-welds (c)). Bottoms are flat; roofs are usually conical. The necessary wall thickness of upright cylindrical tank shells can be calculated in view of the maximum liquid pressure to be expected as

$$s = \frac{rp}{\sigma_{al}} = \frac{rh\varrho_l g}{\sigma_{al}\, k} \qquad 6.5\text{—}1$$

where r is radius, h is the maximum height of liquid column in the tank, ϱ_l is the maximum liquid gravity to be expected, σ_{al} is maximum permissible stress, and k is a weld-seam factor.

Example 6.5—2. Let $h = 3$ m, $d = 5$ m, $\varrho_l = 1000$ kg/m^3, $\sigma_{al} = 140$ MN/m^2. What wall thickness is to be chosen? — By Eq. 6.5—1,

$$s = \frac{2.5 \times 3 \times 1000 \times 9.81}{140 \times 10^6} = 5.3 \times 10^{-4}\ \text{m} = 0.53\ \text{mm}\,.$$

Providing the tank with the necessary rigidity and properly joining the sheets demand a wall thickness of 3—4 mm at least, depending on tank size and steel grade.

Up to about 600 m^3 capacity, tanks are made of sheet of constant thickness, determined by the above considerations, whereas larger tanks are telescoped (higher rings are made of thinner sheet). The quantity of metal needed to build a tank of given volume V_t will be different for various height-to-diameter ratios. Metal volume is

$$V_{me} = d_i \pi h s_1 + \frac{d_i^2 \pi}{4}(s_2 + s_3)$$

and since

$$V_t = \frac{d_i^2 \pi}{4} h$$

we have

$$V_{me} = \pi\, s_1 h \sqrt{\frac{4V_t}{h\pi}} + \frac{V_t(s_2 + s_3)}{h}\,.$$

The least metal volume for a given tank volume may be determined by calculus. The derivative of V_t with respect to h is

$$\frac{dV_{me}}{dh} = \frac{\pi\, s_1}{2}\sqrt{\frac{4V_t}{h\pi}} - \frac{V_t(s_2 + s_3)}{h^2}\,.$$

which implies that metal consumption will be a minimum if

$$\frac{h}{d_i} = \frac{(s_2 + s_3)}{2s_1}\,. \qquad 6.5\text{—}2$$

Introducing this result into the above volume formula we get the most favourable tank height:

$$h_{opt} = \sqrt[3]{\frac{V_t(s_2 + s_3)^2}{\pi s_1^2 k'}} \qquad 6.5\text{—}3$$

where k' is the overlap factor. — It can be shown in similar way for telescoped tanks (Shishchenko and Apriesov 1952) that

$$h_{opt} = \sqrt{\frac{\sigma'_{al}(s_2 + s_3)}{\varrho g\, k'}} \qquad 6.5-4$$

where $\sigma'_{al} = \sigma_{al} k''$. The value of k'' ranges from 0.72 to 0.77. It accounts for strength reduction due to welding and riveting.

The optimal tank shapes given by the above formulae may be slightly modified by standard steel sheet sizes. On the other hand, as regards gauging the contents of a tank, it is better if the tank is taller, because a given error in level determination will cause the less error in volume, the slenderer the tank. In sizing tanks it should be kept in mind that tank weight per unit volume is the less the larger the tank (Fig. 6.5—6). It is therefore more economical, in so far as this is a prime consideration, to use a few big tanks than numerous small ones. Prescriptions concerning tanks are laid down in API Stds 12B, 12C, 12D and 12F. Aluminium tanks are treated in API Std 12G. Let us point out that tanks suitable for dismantling are not usually made in sizes larger than 1600 m^3. Larger tanks are of welded construction. Still, there is no objection to the making of small tanks by welding.

In most upright cylindrical steel tanks, vapour pressure above the liquid must not exceed 0.4—0.5 kN/m^2. For radially ribbed decks, this value may attain 0.2—0.5 bar, but such designs need substantially more steel. — If a tank is to withstand a higher pressure — a few bars, say — then a horizontal cylindrical shape is adopted. The cylindrical shell is closed on both ends by an ellipsoidal or spherical dome. Such tanks should be made in a shop and not on-site. For oils of particularly high vapour pressure, the spherical and spheroidal tanks usual in the storage of light distillation products should be employed. — The fittings of upright cylindrical tanks

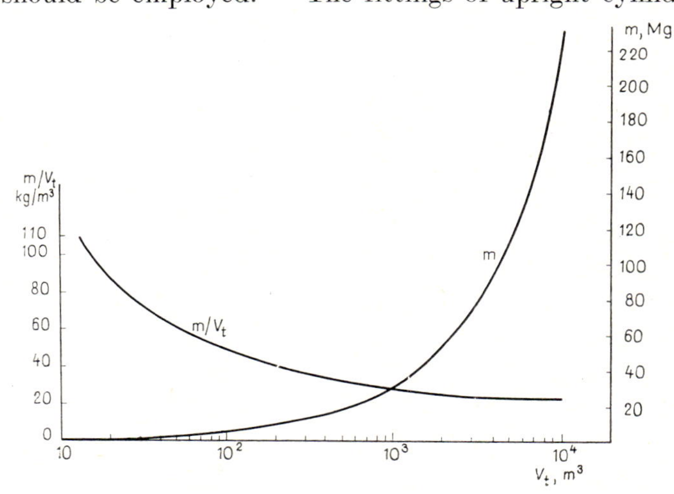

Fig. 6.5—6 Typical data of Soviet tanks, after Shishchenko and Apriesov (1952)

are shown in Fig. 6.5—7 as manhole on deck *1*, clean-out port on shell *2*, inlet studs *3*, breathing valve *4*, gauge hatch *5*, water drain *6*, outside ladder or stairs *7*, inside ladder *8*, drain stud, with swivelling suction pipe *9*, level gauge *10*, hookup to foam-type fire extinguisher *11*.

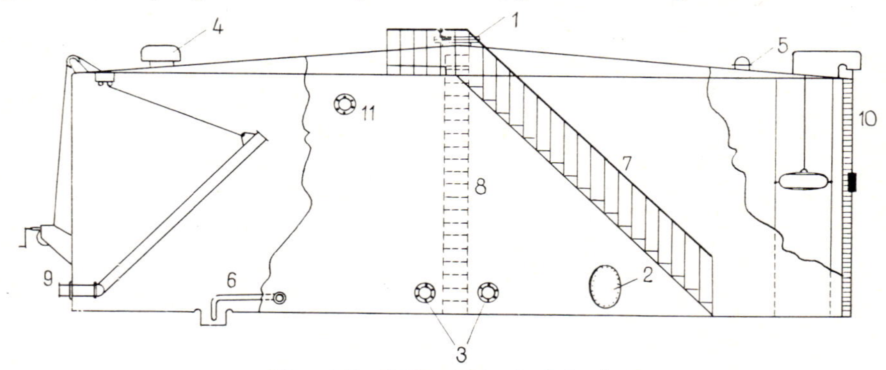

Fig. 6.5—7 Upright steel tank

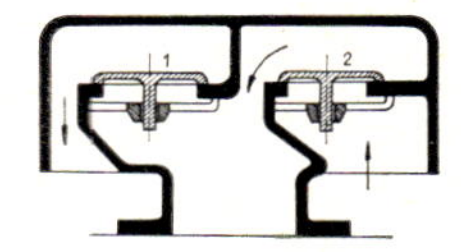

Fig. 6.5—8 Mechanical breathing valve after Shishchenko and Apriesov (1952)

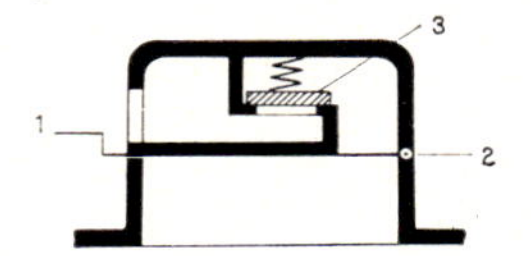

Fig. 6.5—9 Gauge hatch closed by mechanical breathing valve

Table 6.5—2
Diameters of mechanical breathing valves (after Shishchenko and Apriesov 1952)

Pumping rate	Valve diameter
m^3/h	mm
0— 50	50
50—100	100
100—200	150
200—400	200

Breathing valves, installed on the roof apertures of tanks, open at 0.15—0.30 kN/m^2 pressure differential, both positive and negative. In the mechanical breathing valve shown as Fig. 6.5—8, valve *1* opens to overpressure in the tank, valve *2* opens to vacuum. Valve capacity depends on the maximum inflow/outflow rate of liquid into/from the tank (cf. Table 6.5—2, after Shishchenko and Apriesov 1952). In some instances, the gauge hatch is designed as a mechanical-type breathing valve (Fig. 6.5—9); *1* is the datum level of liquid level gauging. Overpressure is released by the entire hatch cover pivoting about *2*; vacuum is compensated through disk valve *3*. In order to eliminate the hazard of the mechanical valves' sticking or seizing, hydraulic valves are used in some instances (Fig. 6.5—10). Antifreeze liquid poured in cup *1* will be raised in annulus *2* by overpressure and in annulus *3* by vacuum. This type of valve is designed so that the liquid will not be spilled out of annulus *2* or *3* until the maximum operating pressure of the mechanical valve has been exceeded. Breathing valves are often provided with flame arresters, in order to help localize tank fires. Breathing valves will reduce evaporation losses by 15 to 20 percent, depending on the nature of the crude stored.

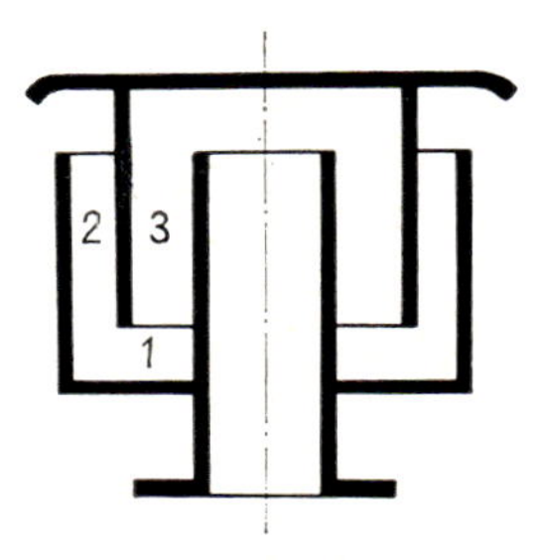

Fig. 6.5—10 Hydraulic safety valve

Figure 6.5—11 shows schematically two types of liquid-level gauge. The device in part (a) permits an approximate determination of liquid level. Line *2*, hanging float *1*, emerges through liquid-filled airlock *3* to pass in front of calibrated tape *4*. Index *5* will give an approximate reading of liquid level in the tank. Its purpose is to provide first-glance information to the operator rather than to permit accurate liquid-level measurement. In design (b), the bobbing of float *1* is limited by guide wires *2*, which makes for a higher accuracy in measurement. The perforated tape attached to the float line is wound on meter drum *3*, whose number of revolutions will be proportional to the liquid level in the tank. Instrument dial *4* is calibrated directly in volume units. Line *2* and the perforated tape are tensioned by weight *5*. — An automatic device permitting level gauging to better than 1 mm accuracy and compatible with a telemetering system is described in Wafelman (1969). — By means of the swing pipe, oil can be extracted from any height within the tank. This permits us to

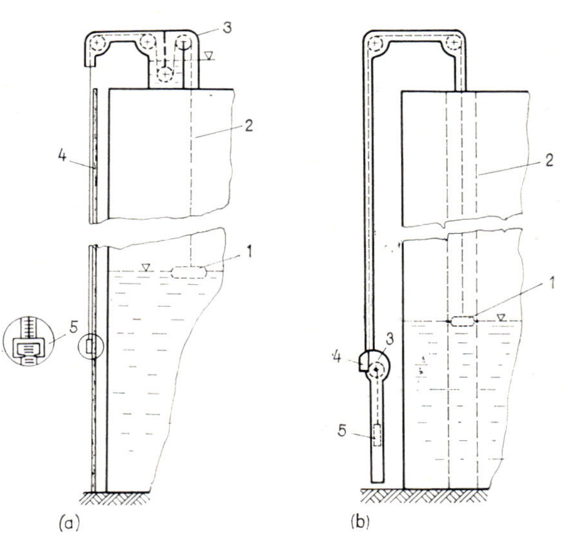

Fig. 6.5—11 Liquid level indicators, after Chilingar and Beason (1969)

separate oil layers of various degrees of contamination, or oil containing water, or other impurities. — The inlet pipes of some tanks are fitted with clap valves, operating as back-pressure valves, that can be actuated from the outside (Fig. 6.5—12). This has the advantage that broken-down valves in the fill line may be replaced without having to drain the tank. — Tanks may be provided with internal heating if the crude stored in them requires warming. Steam is generally used for heating. Radiators may be plane or spiral, installed either near the bottom of the tank, or so as to heat the oil drained from the tank only (Pektyemirov 1951).

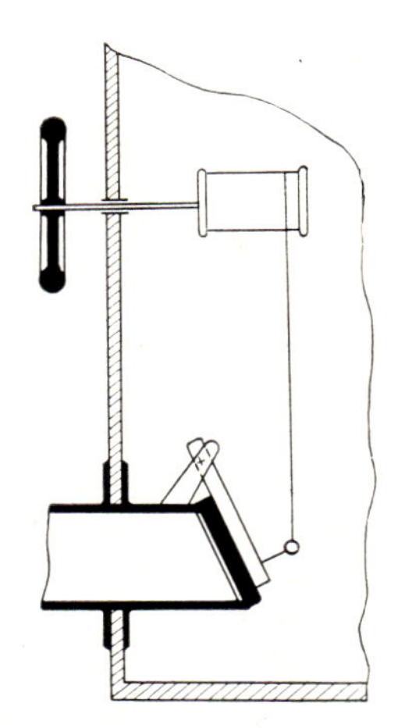

Fig. 6.5—12 Clap valve on fill conduit

Figure 6.5—13 shows the layout of a three-tank battery. Tanks *3* are filled through fill lines *1* and *2*. By manipulating valves *4* and *5*, any tank may be opened to the fill line. Oil is discharged from the tank through drain line *6*, which usually leads to the pump station. Water can be drawn off through line *7*. If the tank battery is manually operated, a means for the operator to see when oil starts to drain through the water drawoff must be provided. The vapour spaces of tanks are connected by vent line *9* hooked up to vertical pipes *8*, which permits the vapour expelled from any tank being filled to enter the vapour spaces of the other two tanks, or a tank to suck in wet gas from the two other tanks when being drained. If gas pressure in the tanks nevertheless exceeds a certain value set on pressure regulator *10*, then the excess wet gas is bled off into a gathering line through conduit *11*. If on the other hand pressure in the vapour spaces drops below a value set on pressure regulator *12*, then this regulator admits dry gas from line *13* through vent line *9* into the vapour spaces of the tanks. If the gas stream required to adjust pressure in the tanks cannot be handled by regulators *10* and *12*, breather valves *14* enter into action. A further measure of safety is introduced by emergency relief valves *15*, which may be hydraulic or of some other design. — It is essential that pressure regulators be matched to each other and to the permissible pressure ranges and gas flows to be expected. Figure 6.5—14 (modified from Chilingar and Beason 1969) represents the pressure ranges covered by the

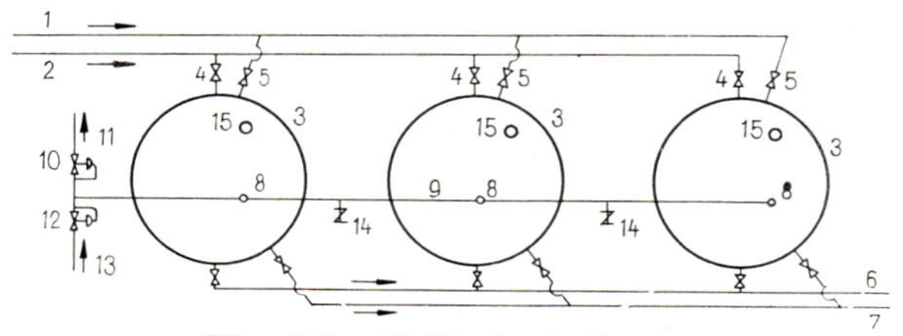

Fig. 6.5—13 Tank station

various regulators and valves of the above-outlined system. Shaded width in each column is proportional to valve travel. — In designing emergency relief valves it should be kept in mind that the volume of a hydrocarbon increases about 15 times on burning. It is this expanded gas stream that has to be released by the valve.

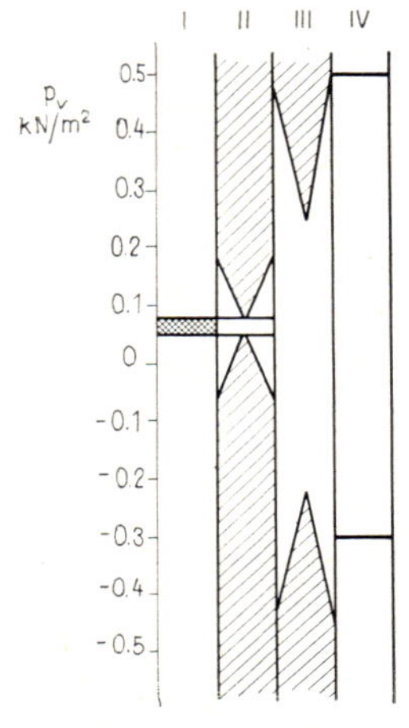

Fig. 6.5—14 Operation ranges of tank pressure regulating devices, modified after Chilingar and Beason (1969, p. 76; by permission of the authors and Elsevier Publishing Company)

Little (1963) describes a vapour recovery system where the vapour is drawn off the tanks at the central gathering station of an oil lease. Wet gas is passed through a scrubber and centrifugal compressor driven by a 22 kW electric motor. The system paid itself out in 76 days. — Tanks with special evaporation-reducing roofs are used primarily for the storage of light petroleum products, but their use may be justified also for comparatively light crudes. Several designs are known, of which the floating-roof tank is the one of particular interest for crude storage. The roof is not fixed to the tank side but floats on top of the liquid. There is practically no vapour space, and hence evaporation loss is either very low or nil, depending on the design. There are several solutions even within the group of floating-roof tanks. Figure 6.5—15a shows a circular-pontoon design. Rain falling on the roof is drained through pipe *1*. Several ways have been devised to seal the roof against the upright cylindrical shell. Figure 6.5—15b shows one of these (McKenna make). The essential thing is that slide *3* mounted on roof *2* fit tank wall *4* as snugly as possible.

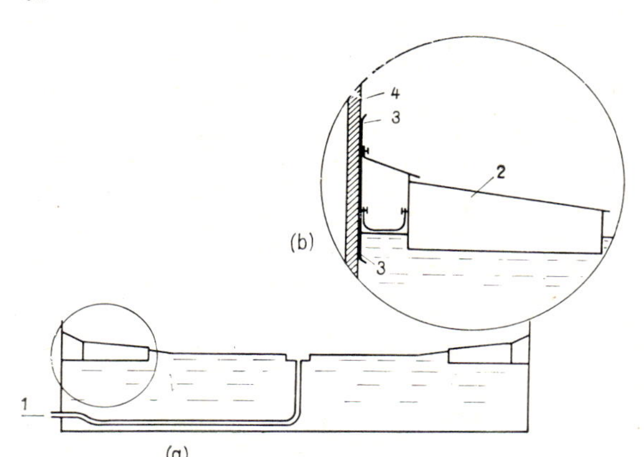

Fig. 6.5—15 McKenna's floating-deck tank

6.5.3. Oil tank gauging

Tanks whose liquid content is determined by liquid-level gauging must be calibrated before use. Calibration may be performed either by calculation based on measurements of circumference and diameter, or directly by filling with known volumes of liquid. Of the calculation methods, the one based on circumference measurement is the more accurate. In this method the circumference of every hoop is measured with a steel tape. In welded tanks, measurements are performed at one-fifth of plate height below each horizontal weld seam, irrespective of the type of welding. The heights

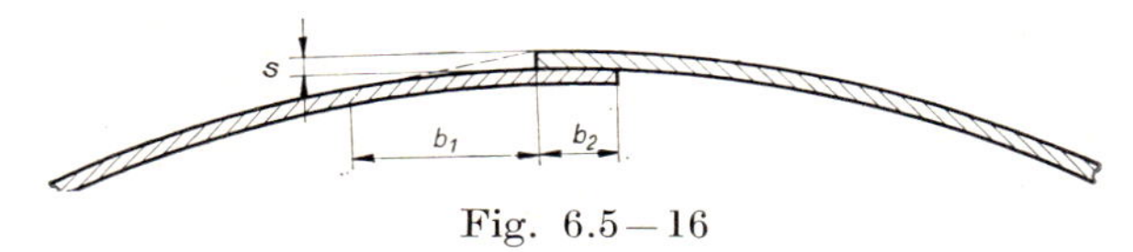

Fig. 6.5—16

where measurements are to be carried out on riveted tanks are shown by the arrows in Fig. 6.5—5. Hoop height is measured for each constant-diameter hoop. Sheet thickness s must also be measured (the value adopted is the average of several measurements). If the measured circumference is s_c, and the number of overlaps is n, then approximate tank volume per unit height is

$$V = \frac{s_c^2}{4\pi} - \left(s_c \times s + b_1 sn + \frac{1}{2} b_2 sn \right) \qquad 6.5-5$$

with b_1 and b_2 identified in Fig. 6.5—16.

Accurate knowledge of tank volume demands the measurement also of the tank fittings that reduce or increase tank volume, or the establishment of their volume by calculation, with their elevation within the tank taken in due account. Tank volume is reduced by the inside ladder, the profiles joining wall to bottom, the ties between roof and floor, heating tubes, etc. The swivelling portion of the suction tube is usually left out of account. Tank volume is augmented by manhole domes, pipe studs, etc. Roughness of the tank bottom is usually accounted for by pouring in a sheet of water of sufficient height to cover all unevennesses. — Tanks of less than 80 m^3 nominal capacity are measured empty, but after complete filling up with liquid at least once. Tanks above 80 m^3 nominal capacity should be at least two thirds full during measurement. — When calibrating a tank with liquid, the tank is either gradually filled with, or gradually drained of, accurately known amounts of water. The results obtained are interpolated to yield capacity differences for each centimetre of level difference. This is a rather lengthy procedure if an accurate result is desired, and therefore it is used primarily to calibrate deformed tanks of irregular shape.

Tanks are calibrated (i) after installation or displacement, (ii) if tank fittings are changed, or (iii) if the tank gets deformed. — The calibration results are transformed into a table which states cumulative tank volume for each centimetre from the bottom upward. Prescriptions in force concerning calibration are contained in API Std 2551. — Deformation of the shell of an upright cylindrical tank under the load of the liquid in it has been described by mathematical formulae (Withers 1970).

The liquid content of a tank may be gauged in several ways. In earlier practice, gauging with steel tape was largely used. Today a small fraction only of all gauging is carried out in this way in the USA. The LACT method that has superseded direct gauging will be described in Section 6.6.

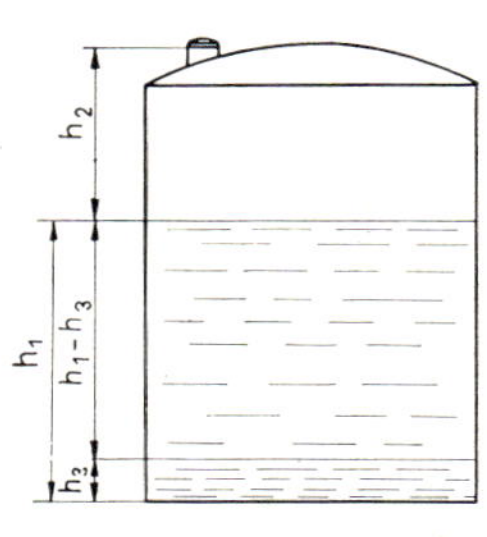

Fig. 6.5—17

Two widespread methods of tape gauging are outage and innage measurement. Innage is measured by lowering a tape weighted with a bob to the tank bottom and, after having withdrawn it, recording the height to which it is wetted with oil (h_1: Fig. 6.5—17). The tank volume corresponding to h_1 can be read off the calibration table. In outage measurement, the height corresponding to $h_1 + h_2$ in the Figure is measured directly after calibration, and the datum level of measurement is marked on the gauge hatch. All subsequent measurements are referred to this datum. The tape with the bob is now used to establish the height h_2, and the height of the oil column is calculated from the relationship $(h_1 + h_2) - h_2 = h_1$. (Of course, the calibration table may be constructed so as to furnish volumes directly as a function of outage height.) The outage method is to be preferred if sediment is settled on the tank bottom. The quantity of water separated at the tank bottom can be established by smearing the bottom end of the tape with a special paste or by hanging from it a bob calibrated in centimetres, with a strip of special paper stuck to it. The paste or paper strip will change its colour when coming into contact with water. The water volume corresponding to water level h_3 can be read off the calibration table.

The so-called actual oil volume established by gauging is to be corrected for two factors. First of all, the oil may contain dispersed, unseparated water, and, second, the volume of pure oil must, in the knowledge of actual temperature, be transformed to standard volume, or, in the knowledge of actual gravity, into mass. In order to check the purity of the oil, a sample must be taken in a manner prescribed in the standards, and the water content of the sample and the gravity of the pure oil must be determined in the laboratory. The average temperature of the liquid in the tank must likewise be determined in a prescribed manner. The relevant prescriptions are contained in API Stds 2543—2548.

6.6. Oil and gas gathering and separation systems

The wellstreams of wells producing crude oil and natural gas are conveyed to centres located on the oilfield. A system comprising piping, pipe fittings and central facilities permits us to separate the liquid from the gas, to measure the quantity of both, to adjust their properties so as to fall within sales-contract and/or other specifications, and to transport them to the consumers or refineries. In the present chapter we shall follow the journey of gas up to where it is discharged from the separator, and that of oil to the intake of the pipeline's driver pump station. It will be assumed throughout that the water produced together with the crude can be removed by simple settling and that no special measures have to be taken to protect equipment from corrosion by the wellstream. Gathering and separating systems fall into three great groups. All three types of system

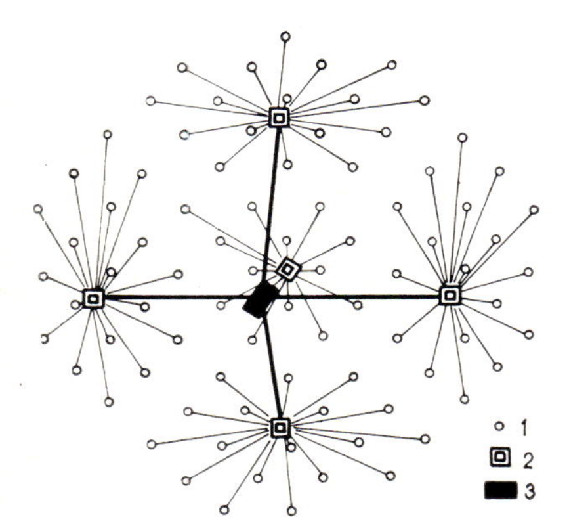

Fig. 6.6—1 Well-centre gathering system, after Graf (1957a)

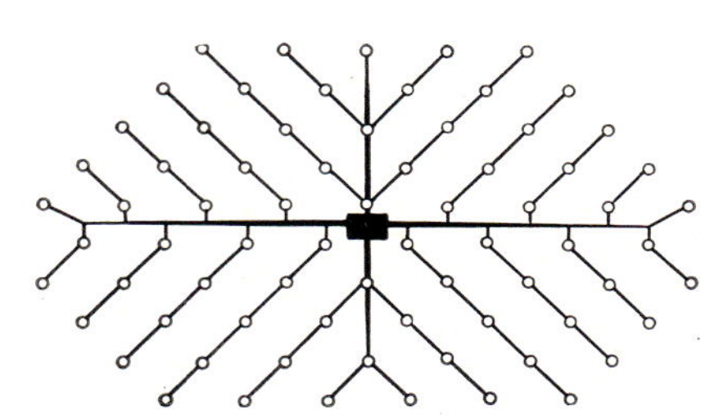

Fig. 6.6—2 Common-line gathering system, after Graf (1957a)

start at the wells and end at the storage tanks of the oil pipeline or in the intake of the pipeline driver pump. The first group includes production systems of extremely high-capacity wells. Each well has its own facilities for separation and metering, possibly also for treatment. This setup is seldom economical. A more frequent type of system involves gathering and separating facilities permitting the common handling of several wellstreams. Figure 6.6—1 shows a so-called well-centre system after Graf (1957a). Individual wells *1* on the lease are connected to well centres *2*. Each wellstream is transported to the well centre in an individual flowline. At the well centres, the wellstreams at least of the wells selected for individual testing and metering are kept separate; their oil, gas and water rates are metered; then either the united wellstreams are transported as they are to the central gathering station *3* of the lease, or the gas separated at each well centre is introduced into the gas gathering line, whereas the liquid is forwarded to the central gathering station. In the third group (likewise after Graf 1957a) several wells produce into a common flowline (Fig. 6.6—2). Oil, gas and water production of individual wells are metered at intervals by means of portable well testers installed at the well sites. All other treatment takes place at the central station. — Of the above-named three groups, the well-centre system is the most widespread and we shall restrict our following discussion to that system.

6.6.1. Viewpoints for designing gathering systems with well-testing centres

In outlining design principles for finding the most favourable gathering and separating system for any oil lease, we shall for the time being disregard whether the system in question is operated under manual control, by means of local automation, or by process control. Discussion shall be production-centred, and the decision as to which means of control is most efficient and most economical will be regarded as a separate subsequent task. The principles to be observed in designing are as follows. (a) The wellhead pressures of the wells should be as low as feasible. The resulting main advantages are: longer flowing life; lower specific injection-gas consumption in gas-lift wells; higher yield of wells produced with bottom-hole pumps in the last phase of production. (b) The hydrocarbon loss of the system should be a minimum. (c) The system should be easy to oversee as far as control and checking are concerned. This makes for disciplined production, and permits fast intervention in times of operating troubles. (d) Metering the individual and common oil, gas and water production of the wells and testing for impurities in the liquid produced should be ensured to the necessary accuracy. (e) When determining oil storage volume, the offtake rate to be expected is to be taken into consideration, together with interruptions in offtake due to breakdowns to be expected, as well as the settling time possibly required for the removal of water and impurities. (f) Expansion of facilities made necessary by the bringing in of further wells should require the least possible modifications to the existing installations; said modifications should be possible to achieve without disturbing the wells already in production. (g) Specific cost, referred to the volume unit of oil and/or gas produced, of installing and operating the system should be as low as possible. (h) Safety prescriptions should be meticulously observed.

The realization of these principles may raise the following, partly contradictory, viewpoints. — *ad* (a). (i) Head loss in the flowline between wellhead and separator should be as low as possible. Hence, sudden breaks in flowline trace and sudden changes in cross-section should be avoided; flowline size, length and trace should be chosen with a view to attaining minimum head loss; if the crude is waxy and/or sandy, measures are to be taken to prevent the formation of bothersome deposits in the flowline and its fittings; in the case of high-viscosity or waxy crudes, reducing head loss may be achieved by heating the crude at the wellhead, heat-insulating the flowline, or injecting a friction-reducing chemical; if water is readily separated from the wellstream, it is recommended to install a water knockout next to each well. (ii) Separator pressure should be as low as possible. In order to keep it so, it is usually recommended to install separators higher than storage tank level at the well centres so as to make the oil flow by gravity from the separators into the tanks. The less the pressure needed to convey gas from the separator through the gathering line to the compressor station, the better. This can be achieved primarily by a gathering-line network of low flow resistance (big size, small aggregate length, efficient liquid knockout or scrubbing) and by using low-intake-pressure compressors. — *ad* (b). (i) Wellstreams from flowing wells of high wellhead pressure should be directed into a high-pressure separator; stage separation is recommended. (ii) The tank system should be closed if possible. (iii) Evaporation losses of open storage tanks, and (iv) oil and gas leaks should be kept at a minimum. — *ad* (c). Particularly in manually operated systems and those with local automation, care should be taken to concentrate all the gathering and separation facilities at the well centre and the central gathering station. In remote-controlled systems, designs ensuring the fast supply of meaningful information should be strived at. Information should be supplied both to the men working on the lease and to the remote-control centre. — *ad* (d). (i) The number of separators enabling individual wells to be tested at the well centre should be sufficient to permit measuring the oil, gas and water production of each well at intervals of 4 to 7 days. (ii) The number of common separators handling the wellstreams that are not separately metered should be composed of uniform units for each stage. The number of separators required is then determined by unit capacity. (iii) In on-lease gas metering, the accuracy of $\pm 1-2$ percent of the orifice meter is adequate. The quantity of gas fed into a sales line or a transmission pipeline should be determined more accurately, if possible. (iv) If the liquid output is measured in the storage tanks at the well centres, then the number of so-called test tanks metering the liquid production of individual wells should agree with the number of test separators. Common tanks metering the production of the remaining wells should number at least two per centre. (v) For accounting within the lease it is usually sufficient to meter liquid at an accuracy of about 0.5 percent. It is desirable to meter oil delivered outside the lease at an accuracy of at least 0.2 percent. — *ad* (e). Optimum storage-tank volume may differ widely depending on the nature of the gathering, separation and transmission facilities. It is recommended in the usual case to have storage capacity for two-three days' production. — *ad* (f). Well centres are to be designed so that bringing wells in or off, and changing separator, tank or metering capacities might be achieved without affecting the equipment in operation. All the equipment and fittings of well centres within a lease should be uniformized; fittings should be transportable to the well centre ready for installation, and it should be possible to install them without welding. — *ad* (g). (i) The mechanical production equipment of wells should be chosen, and phased during the productive life of the wells, so as to minimize specific production cost over the entire life of the lease. (ii) Gas from wells having a high wellhead pressure should be led at the lowest possible pressure loss into the compressor station. One and possibly two compressor stages may thus be saved temporarily. (iii) The number and location of well centres, as well as the location of the central gathering station, should be chosen so as to minimize total cost. (iv) Temporary piping should be joined by couplings. (v) Well testing centres, usually hand-operated early in the life of the lease, should be devised so as to require the least possible modification when converting to automation. — It is usually preferable to install an LACT (Lease Automatic Custody Transfer) system rather than a central gathering station early in lease life.

6.6.2. Hand-operated well testing centres

When choosing the most suitable well centre design, several types may enter into consideration depending (i) on the nature of the wellstream to be handled, (ii) on the role to be assigned to the central gathering station, (iii) on the climatic conditions at the selected location. Figure 6.6—3a shows a type of tank station or well centre suitable for the handling of low-viscosity gaseous crudes containing no sand. Oil and gas separated from the crude are conveyed through flow lines *1* and fill lines *2* into test separators *3* or production separator *4*. Oil flows through line *5* into meter-

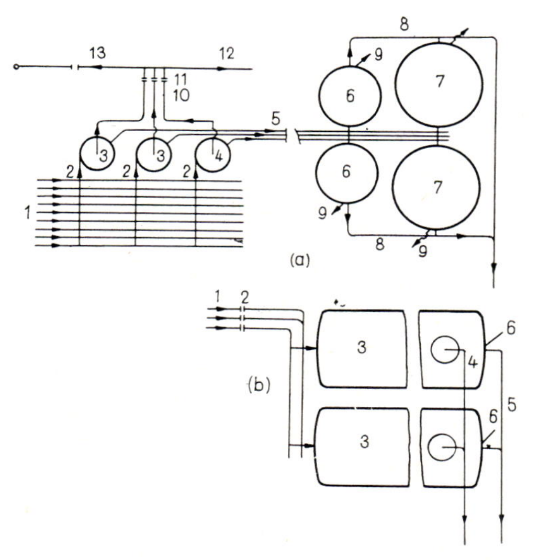

Fig. 6.6—3 Well centre for handling low-viscosity gaseous sand-free crude

ing tanks *6* or storage tanks *7*, which may be drained through lines *8*. Lines *9* serve for water drain. Tanks are upright cylindrical, so-called open sheet steel tanks. Separators discharge gas through lines *10* and orifice meter run *11* into gas gathering line *12*. Gas may also be vented or flared through line *13*. — Figure 6.6—3b shows a schematic diagram of a 'closed' tank station or well centre. The separator-side layout is the same as above. Liquid discharged from the separators flows through lines *1* and orifice liquid meter run *2* into horizontal cylindrical closed tanks *3*. Hydrocarbon vapours enter through line *4* into the gas gathering line. Tank outlets are *4* for gas, *5* for oil and *6* for water. — Well fluid entering the separator is often heated in order to promote easier separation of gas and water and to prevent the formation of gas hydrate plugs in the separators and at the meters. Heating is by hot water or steam through a heat exchanger. — Figure 6.6—4 is a detail of a well centre upstream of the tanks, used for the handling of low-viscosity sandless crudes, with two-stage separation and the evaporation loss to be expected in the open tanks reduced by vacuum stabilization (Scott 1965). The wellstreams enter the well centre through flowlines *1*. One wellstream is introduced into second-stage test separator *2*, of about 70 kN/m^2 pressure, whereas the remaining wellstreams enter production separator *3*. Test separator *2* discharges gas through line *4* into the gas gathering line *5*, and liquid through line *6* into stabilizer *7*, which is under a vacuum of about 4 kN/m^2. The first-stage separator *3*

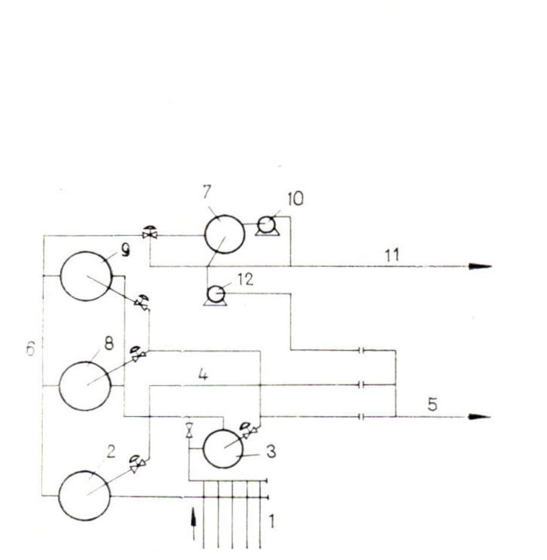

Fig. 6.6—4 Vacuum stabilization at well centre, after Scott (1965)

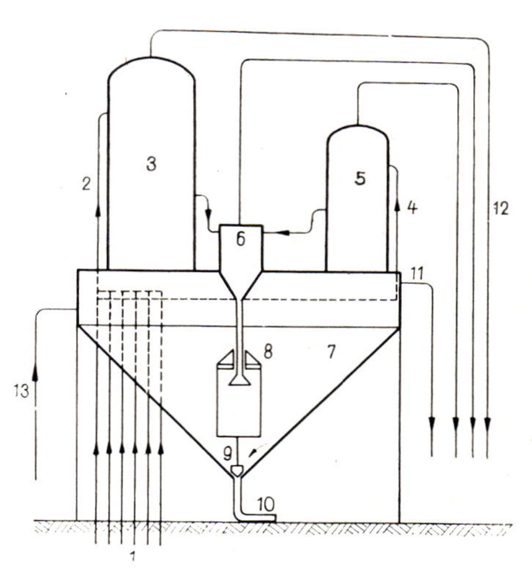

Fig. 6.6—5 Soviet well centre design for handling sandy crude

discharges the common liquid production into second-stage separators *8* and *9*. These discharge oil into the stabilizer, and gas into line *5*. Gas separating in the stabilizer is fed by compressor *12* at the required pressure into gas line *5*. Pump *10* transports oil at an excess pressure of about 70 kN/m^2 through line *11* into the storage tank. Gas takeoff from the stabilizer column is continuous, whereas oil offtake is batchwise, because pump *10* (in a manner not shown in the Figure) circulates oil in the stabilizer until the liquid level in the vessel attains a predetermined height. Reflux is then stopped and discharge into line *11* is initiated. This goes on until the liquid level drops to a predetermined height. The economy of this installation is revealed by the fact that, in the case discussed by Scott, the stabilizer paid itself out in 160 days. Economy is impaired by higher gas gathering line pressure, as this increases the power cost of gas compression.

In a Soviet type tank station used to handle gaseous, cut and sandy crudes (Fig. 6.6—5), wellstreams enter through flowlines *1* and fill lines *2* and *4* into production separator *3* or test separator *5*. Separators discharge liquid through vacuum separator *6* into cone-bottom storage tank *7*. Weighted float *8* at the oil-water interface operates valve *9* which serves for the automatic drain of water and sand from the tank through line *10*.

Pure oil is discharged through line *11*. Separators *3*, *5* and *6* deliver gas through line *12* and meter runs to the gas gathering line. The settled muddy sand can be slurried up with water introduced through pipe *13*. — The schematic layout of a tank station for handling waxy crudes low in gas and sand is shown as Fig. 6.6—6. Wellstreams enter through heated flowlines and distributors *2* into the test and production separators *3* and *4*. Oil is discharged through line *5*, the low gas production through *6*. The water drain is not shown in the Figure. Flowlines are usually heated with steam conveyed in internal or external, coaxial piping. Lines among tanks are conducted in covered concrete ducts sunk into the ground (dashed lines) that can be heated by the piping steam. — This design is suited also for handling high-viscosity oil low in gas and sand, whose pour point is lower than the ambient temperature. Heating the lines in these cases may not always be necessary. — Figure 6.6—7 shows a gathering and separating system (Sostoyaniye) designed by Gidrovostokneft Institute of the USSR and employed in the Soviet Union. This system has the peculiarity that well centres perform first-stage separation only. The comparatively dry gas furnished by the separators is discharged into a gas gathering line; the liquid, which still contains significant quantities of gas, is conveyed to the central gathering station which is some 100 km away. This is where another two or, if the stock tank is also counted, three stages of separation take place. The main advantage of this system is that it permits centralization of a significant portion of the gathering and separating equipment for production spread over a vast area. In the solution shown in the Figure,

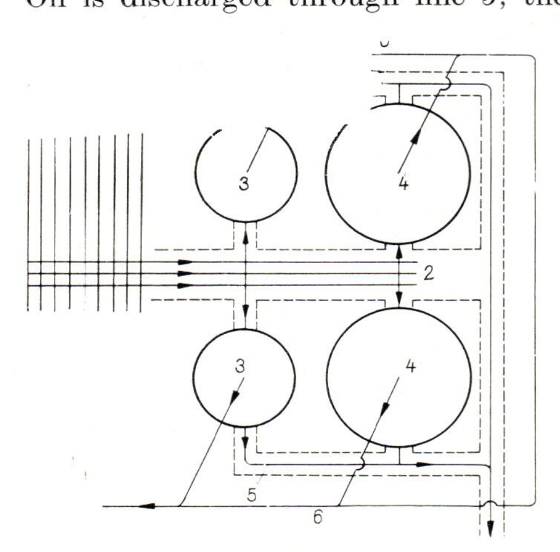

Fig. 6.6—6 Well centre for handling low-gas high-viscosity crude

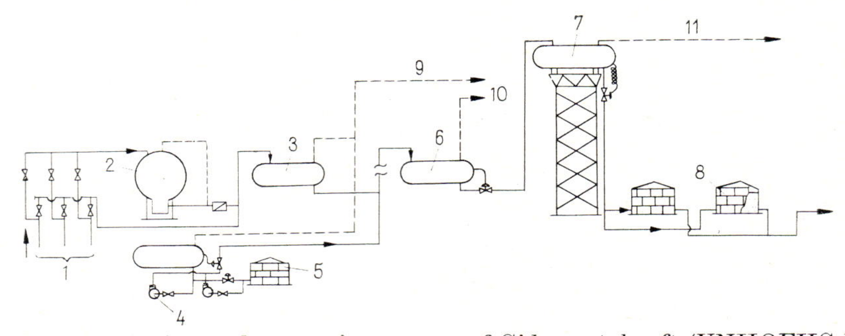

Fig. 6.6—7 Gathering and separating system of Gidrovostokneft (VNIIOEHG 1969)

flow lines *1* deliver wellstreams through metering units *2* into first-stage separator *3*. In the upper part of the Figure, separator pressure is sufficient to convey the liquid effluent to the faraway second-stage separator *6*. In the lower variant (with metering equipment not shown), pressure of separator *3* is insufficient to deliver the liquid to separator *6*, so that pump *4* and tank *5* have been added at the well centre. Separator *6* delivers liquid to elevated third-stage separator *7* which in turn discharges into storage tanks *8*. Gas from each separator stage enters the gathering line of the gasoline recovery station through lines *9*, *10* and *11*.

Graf (1957a) describes a simplified system in which no gas line emerges from well centres. Only the wellstream of the well being tested is being separated; the gas thus produced is vented or flared. The full output of the producing wells is conveyed by means of a screw pump to the central station. Any further separation and metering takes place at that station.

6.6.3. The automated system

Oilfield automation is a process that started a long time ago. The elementary tasks of oil production, gathering and separation have been handled for many a decade by equipment demanding no human intervention (liquid level controls in separators; gas pressure regulators, etc.) Early in the nineteen-fifties, a new wave of automation brought about the relatively rapid development of an automatic system termed LACT (Lease Automatic Custody Transfer), taking care of central oil gathering, treatment, metering and transmission (Resen 1957), as well as the development of automatic well testers (Saye 1958). The devices composing these systems were at first controlled by local automatisms, but connected systems were soon developed where metering results and the main situation parameters of peripheral units could be read off and recorded at a central control station. Even this system permitted substantial savings to be achieved. According to Wiess (*Oil and Gas Journal* 1964) the Gulf Coast Division of Sun Oil Company introduced centralized automation at this level on 19 leases, some of which were offshore. Some of the leases were already in production at the time, while the rest awaited development. The total cost of staggered automation was estimated at USA $ 500,000. This expenditure was balanced by three factors: (i) savings on investment (as against conventional equipment), (ii) increased sales income, and (iii) reduction of maintenance cost (Table 6.6—1). The Table shows that the increased income and the reduced maintenance cost together totalled USA $ 375,000 per year. Adding to this the difference in annual depreciation attributable to the reduction in investment cost we see that the investment of $ 500,000 paid itself out in slightly more than one year. By the late sixties, LACT had become so widespread that 60—70 percent of oil produced in the USA was treated and transferred to transmission lines by this method. This phase of automation was followed early in the sixties by a second wave, represented by Computer Production Control.

Table 6.6 – 1

Savings due to automation of an oilfield

Reduction in first cost	1000 $	%
Tanks	303	50
Piping and fittings	36	6
Other surface equipment	137	23
Offshore platform	128	21
Total	604	100
Annual increase in sales income		
Reduction of oil gravity	51	41
More efficient use of equipment	41	33
Sale of tank vapours	19	15
Saving in gas power	13	11
Total	124	100
Annual reduction in maintenance cost		
Payroll	177	41
Maintenance items	48	19
Transportation	16	6
Miscellaneous	10	4
Total	251	100

The central computer acquires, records and processes all production, situation and safety data concerning the lease. In this off-line system, it is the dispatcher who, in the possession of the information presented by the computer, will intervene if necessary. Intervention will usually be by push-button-actuated remote control. As a stage of further development, on-line computer control was developed, where even intervention by remote control is initiated by the computer. Late in the sixties, after the advent of third-generation computers employing integrated circuits, there were as many as 40 CPC systems operative on the various oil and gas fields of the world (Graf 1970). Production control by computer is considered a highly significant advance, "the biggest thing since rotary drilling in hydrocarbon production" (Pearson 1969). Its main advantages are reduced operating costs, reduced production downtime due to operating trouble, a more elastic and versatile optimization of production, and faster intervention in critical situations. All tasks that can be uniquely defined in the possession of all relevant information, and can be performed by the intermediary of remote-control-effectors, are set to CPC. This liberates most of the lease operators for tasks inaccessible to CPC; e.g. the operator opening and closing wellhead valves or handling scraping operations, or the gauger gauging liquid levels in tanks will become redundant, but the tasks of mechanics performing repairs and maintenance on production equipment will become more sophisticated and possibly more rewarding. No clerks recording production data and writing periodic reports will be required, but production experts and systems analysts analysing the production processes recorded and giving instructions for new production targets will. Automatic operation will become much more flexible under the direction of computer programs than of local automatisms, as programs permit flexible adaptation to changes in requirements facing production without changing peripheral units. The Humble Oil and Refining Co. possesses, for instance, 500 process-control programs that are routinely used in all production units (Scott and Crosby 1970). Reduction of production downtime by fast intervention and more efficient trouble prevention may increase oilfield production by several percent (Harrison 1970).

In the last few years, several descriptions of oil fields operated by CPC have been published. It appears from these that the system will prove economical for both big fields with some thousands of producing wells and smaller operators. A special analysis of where to employ automation, to what extent, and what degree of centralization to aim at is, however, indispensable in any particular case. Reports on automatic oil production control published in 1970 include in addition to the above-cited papers Bleakley 1970a, b; Burrell et al. 1970; Michie et al. 1970; Cox and Underrinner 1970; Chapin and Woodhall 1970; *World Oil* 1970; Martin 1970.

In the following we shall be concerned with the peripheral production equipment of automated leases only. This equipment can be adapted to either human control or local automation or computer process control. The consideration on which this choice of subject is based is as follows. Development of a system of automatic production requires familiarity with three fields of endeavour. The first is oil and gas production. Specialists in this field will formulate production requirements. The second field is automation. Its specialists will supply and develop the hardware required for automatic operation. The third field is systems analysis and software preparation. Software is prepared by mathematicians on the basis of tasks formulated by production specialists, and complemented and updated to adapt it to changing conditions and advances in hardware. In the present book, we restrict discussion to the first field.

Let us point out that, in the first phases of developing (and, indeed, producing) an oilfield, it will rarely be possible to divine the optimal way of exploiting the field. A well-founded prediction of this will not be possible until the development is almost complete and the analysis of at least a shortish period of production has been performed. Specialists designing surface production equipment will have to gear their work to the production plans based on information acquired during these phases. This is when the definitive system may be designed, and the automation system can be chosen and realized. The final, most thoroughly automated system will thus be the product of step-by-step evolution from lesser to greater automation, governed throughout by the obvious aim that production equipment and automation elements once installed might be fitted at least possible excess cost into the definitive system (Terris 1965).

(a) Automated well centres

The purpose of an automated well centre is to carry out without direct human intervention the following tasks. (1) Cycling according to plan of intermitting wells. (2) Hookup according to plan of individual wells to the test or production separator. (3) Shutoff of wells in the case of metering or transmission system malfunction. (4) Separation of oil and gas composing the wellstreams. (5) Metering and recording of oil, gas and water production of wells being tested (possibly also of the entire set of wells hooked up to the production separator), and possibly the reporting of information acquired to a processing centre. (6) Transmission of liquid and gas.

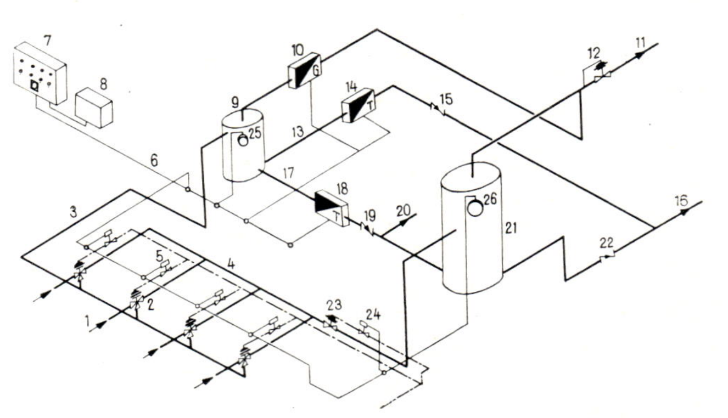

Fig. 6.6—8 Automatic well centre, after Saye (1958)

Carrying out tasks (1)–(3) requires a choice of valves and a suitable hookup that will open or close under instructions from the program control or from malfunction warning equipment. Further required at each well is a flow-line shutoff actuated by pressure changes in the flowline, that will open or close the line, and possibly start up or shut down the installed production equipment. — Figure 6.6—8 shows one of the possible layouts of an automatic tank station (Saye 1958). From flowlines *1*, wellstreams pass through motor valves *2* to be deflected either into line *3* and three-phase test separator *9*, or into line *4* and three-phase production separator *21*.

The diaphragm motor valves actuated by solenoid-operated pilots can be hooked up in different ways. In the design shown as Fig. 6.6—9, each flowline is branched, and each branch carries a two-position two-way motor valve actuated by a solenoid-operated pilot. The valves are double seat, that is, pressure-compensated, and normal-closed as a rule. This design is used in connection with high-pressure wells, where opening and closing a non-compensated valve would require too much power. Figure 6.6—10 illustrates the hookup of three-way, two-position motor valves. The closing of pilot *1* opens motor valve *2* and directs the wellstream into the production separator. If the pilot feeds supply gas over the diaphragm of the motor valve, the wellstream is deflected towards the test separator. In this design, primarily suitable for comparatively low-pressure wellstreams, wells cannot be shut in one by one. In the case of system malfunction, all pilots close and all wellstreams are directed into the production separator. The inlet of this latter will automatically be closed by pilot *3* and motor valve *4*. Figure 6.6—11 shows the hookup of three-way three-

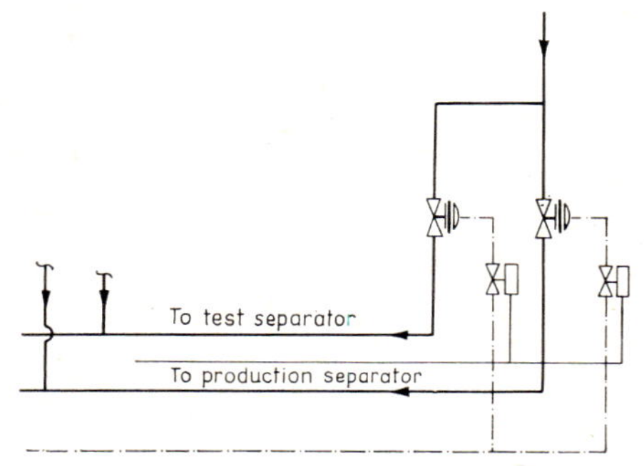

Fig. 6.6—9 Hookup of two-way two-position valves, after Saye (1958)

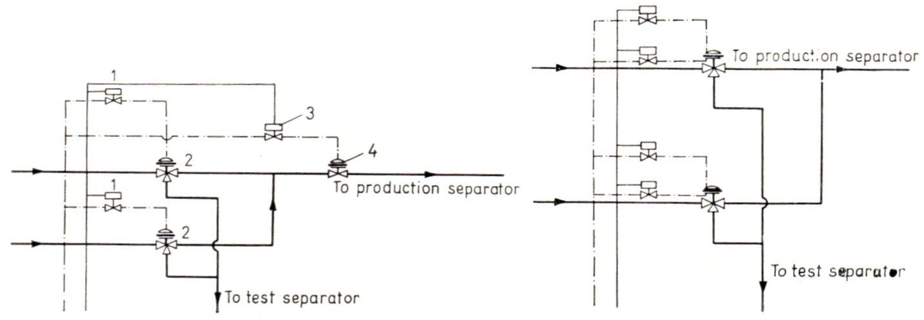

Fig. 6.6—10 Hookup of three-way two-position valves, after Saye (1958)

Fig. 6.6—11 Hookup of three-way three-position valves, after Saye (1958)

position valves. Deflection from the test into the production separator and conversely is controlled in each flowline by two solenoid-operated pilot valves feeding power gas above or below the motor valve diaphragms. If both pilots are closed, the motor valve will shut the well.

Wellhead shut-in. If the maximum allowable operating pressure of the flowline is higher than the pressure in the shut-in well, then opening and shutting in the well can be performed at the tank station. In flowlines of low pressure rating, excessive pressure buildup can be forestalled by installing an electrical remote-control-valve at the wellhead. This solution is fairly costly, however, especially if the distances involved are great. It is more to the point to install on the wellhead a valve that will shut in the well if pressure passes an upper or lower (line rupture!) limit (cf. Fig. 2.3—26). In flowing and gas-lift wells, the safety device may be a pneumatically controlled motor valve, whereas in pumped wells pressure switches are employed. The pressure switch will cut the power supply to the prime mover (or cut the ignition circuit, if the prime mover is an internal-combustion engine) if pressure is too high or too low. Moreover, the pressure switch may control the opening or closing of a pneumatically or electrically operated wellhead shut-in valve.

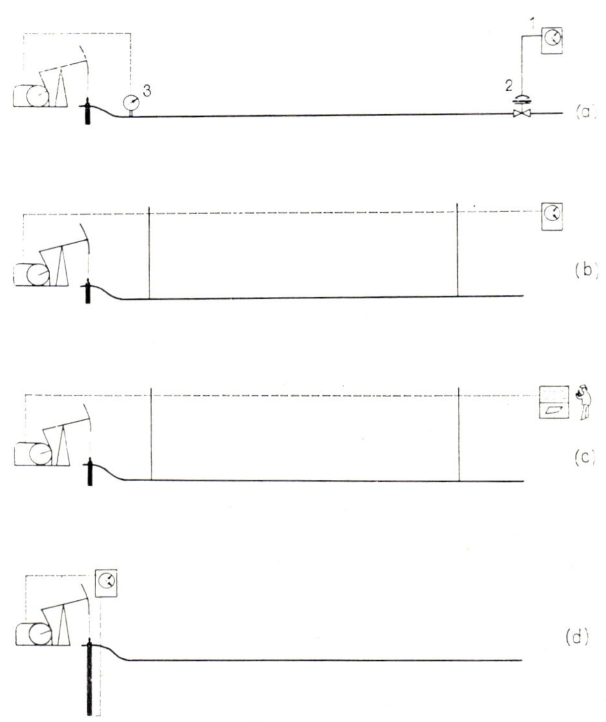

Fig. 6.6—12 Options of automating sucker-rod pumping

Figure 6.6—12 shows as an example four ways of controlling the intermittent production of a well by sucker-rod pump. In (a), a centrally installed clockwork-type cycle controller *1* opens and closes valve *2*, likewise installed at the well centre. Whenever flowline pressure exceeds a certain limit, a pressure switch installed next to the well cuts out the prime mover. In (b), the centrally installed cycle controller clockwork starts the prime mover by electric remote control. In (c), starting and stopping is initiated by an operator, who can also check pump operation by a dynamometer diagram transmitted to him on his call (this is a solution widespread in the Soviet Union). In (d), the pump is started by a clockwork cycle controller installed next to the well and stopped by a *BHP* sensing device. Setup (d) has the advantage that the duty cycle of the pump is controlled solely by the quantity of fluid accumulated in the well; also, the sensor, easy to install next to the well, will permit the recording of *BHP* and its changes, and hence supply the most relevant information concerning the correct adjustment of operation.

In the automatic tank station shown schematically as Fig. 6.6—8, the effluents of test separator *9* are metered by gas meter *10*, oil meter *14* and water meter *18*. Gas and oil discharged by production separator *21* are metered at the central station only. Pressure switches *25* and *26* send signals to process control *7* whenever separator pressures exceed a certain limit. If this happens in the test separator, then the control automatically re-

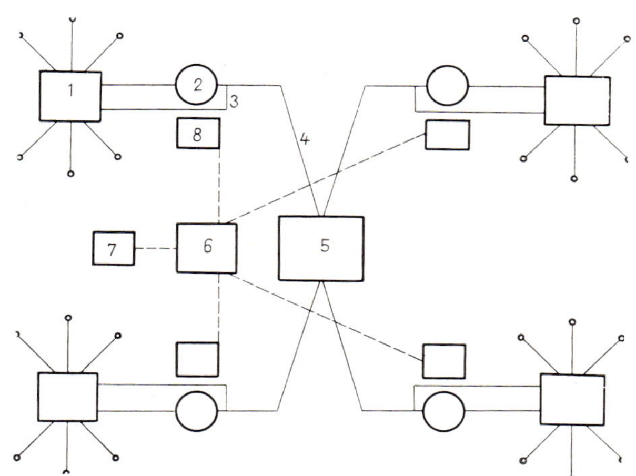

Fig. 6.6—13 Gathering system, after Saye (1958)

routes the wellstream of the tested well into the production separator. If overpressure occurs in the production separator, then inflow is shut off by pilot *24* and motor valve *23*. This entails a pressure rise in the flowlines, as a result of which wellhead valves will shut in wells automatically (if flowlines are low-pressure-rated) and stop the prime movers of pumping units. — Figure 6.6—13 illustrates a gathering system including one central metering and transmission station and four tank stations (Saye 1958). Production lines *3* bypassing metering equipment *2* at well centre *1* discharge all oil into gathering line *4* that delivers it to central station *5*. Central control *6* through substations *8* performs the cycling according to program of all wells and their alternative hookup to the test or production separator. Substations report metering results and other data to central recording facility *7*. — Oil metering by instrument usually obviates the need for the conventional tank batteries of well centres.

The dominant aim nowadays is for automatic well testers to be factory-made skid-mounted portable units whose installation should require just the time for connecting up the flowlines of wells and the oil, gas and water outlets, rather than 'do-it-yourself' rigs built in weeks and possibly months by the pipe-fitter gang on the lease. Portable well centres can be expanded by any number of new units as and when the increasing number of pro-

Fig. 6.6 — 14 Portable well tester of National Tank Co.

ducing wells requires. Today's well testers may need no more horizontal space than a few m^2. Figure 6.6 — 14 shows as an example the National Tank Co.'s portable automatic multi-well test unit.

(b) Automatic custody transfer

The prime purpose of a central gathering station in an oilfield is to separate most of the water, if any, from the crude flowing in from the well centres, to meter at a high accuracy the separated oil whose water content should not exceed a few tenths of a percent, and to deliver it to the inlet manifold of the transmission-line driver pump, or possibly into the storage tank of the pipeline 'in' pump. In certain cases, the crude is also stabilized at the central gathering station. Up to the mid-fifties, central gathering stations were conspicuous by a number of large tanks serving to store crude prior to transmission, and to gauge it by one of the procedures outlined in Section 6.5—3. A modern tank station of those times resembled Fig. 6.5—13.

Automatic gauging and transmission of crudes by the LACT system was first realized in 1954. As a result of this revolutionary step ahead, 60—70 percent of the crude produced in the USA was metered and transmitted by this means in 1960 (Scott 1967). In an LACT set-up, the watery crude enters through a dehydrator into the storage tank for pure oil. When the tank is filled to a certain level, and the transmission equipment is ready to receive oil, the crude stored in the tank is automatically transferred to the transmission equipment. After discharge from the tank, the BS&W (basic sediment and water) content of the crude is measured.

If this is below the limit contracted for, then the oil is continuously metered into the receiving equipment of the transmission line. If BS&W exceeds the limit contracted for, then the crude is redirected from the storage tank into the dehydrator. In the first phases of evolution of LACT, some of the old tanks were retained and used as metering tanks. Crude was permitted to rise to a certain high level, after which it was pumped out until it sank to a certain low level. The tank volume between the two levels was accurately calibrated. The actual crude volume thus determined had to be corrected for water content and reduced to the standard state. In order to do so, it was necessary to establish the mean temperature of liquid in the tank, its actual gravity and its water content on outflow. The first LACT equipment built to perform all these tasks was installed by Gulf Oil Co. in 1954 (*Oil and Gas Journal* 1956). — Weir-type LACT equipment has the advantage that it can be installed in existing tanks that used to be controlled and gauged by hand; its correct operation can be checked at any time by manual gauging; also, the accuracy of measurement is not impaired by the presence of gas. Its disadvantage is that accurate metering depends on the correct functioning of a number of valves and floats; crude retention time may be long; paraffin deposits on the tank wall, and sediment settling out on the tank bottom, will reduce metering accuracy; liquid temperatures in the tank have to be established prior to every draining, and actual oil volumes have to be reduced to standard state by calculation; the system is open, so that significant evaporation losses may arise.

Today's modern LACT equipment is the result of fast, efficient evolution. The essential difference from early types is that oil is metered by positive-displacement meters rather than by draining known tank volumes, and meters will directly indicate or print out oil volumes reduced to standard state. Figure 6.6—15 shows the layout of such a modern LACT system

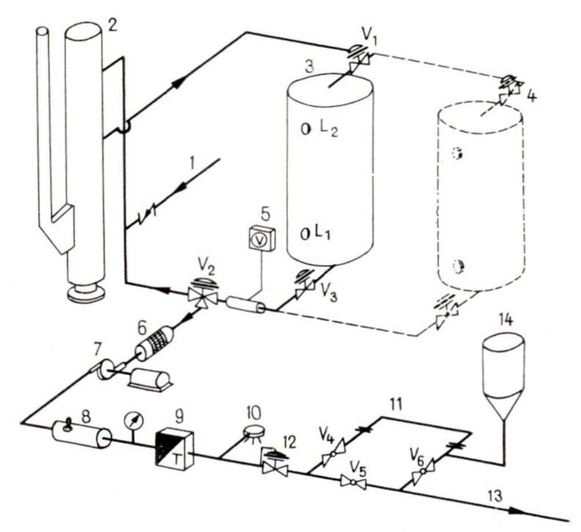

Fig. 6.6—15 Automatic central oil handling station with positive-displacement meter, after Resen (1957)

after Resen (1957). Crude arriving through line *1* from the automatic well tester is fed to dehydrator-demulsifiers. Pure crude is transferred from there into stock tank *3*. If level in *3* attains float L_2, valve V_3 opens and crude can pass through BS&W monitor *5* to valve V_2. If water content exceeds the limit contracted for, valve V_2 returns the oil into the demulsifier. If purity is satisfactory, supplementary transfer pump *7* is started; it will deliver oil sucked through strainer *6* into gas eliminator *8* and through it into the suction line of the main pump *13*. Oil is metered by positive-displacement meter *9* which, correcting the result for flow temperature, either shows on a dial the metered oil volume reduced to the steady state or transmits it to the control centre. Sampler *10* automatically samples the outflowing oil, thus permitting to check BS&W content at intervals in the laboratory. Instead of the sampler, a BS&W check instrument may also be installed downstream of the positive-displacement meter. Bypass *11* (the calibration loop) that can be cut off by means of valves V_4 and V_6 can be replaced by a positive-displacement meter prover in series with meter *9*. This latter can be calibrated also by means of prover tank *14*; *4* is a safety-reserve tank whose presence is optional. — The high accuracy (not less than 0.1 percent) of the system is due to the following circumstances. (i) Pressure regulator valve *12* ensures constant pressure downstream of the meter; the pump delivers at a nearly constant rate, and thus the leakage error of the positive-displacement meter may be kept at a minimum (cf. Section 6.6—3c). (ii) Strainer *6* removes solid impurities still present in the oil, thus preventing wear and malfunction of the meter. (iii) Oil pressure is raised slightly above stock-tank pressure by the transfer pump, and the oil may thus dissolve gas bubbles still present in it, if any. Degassing is facilitated also by gas eliminator *8*. — A schematic diagram of the strainer is shown as Fig. 6.6—16. The wire screen retains solid impurities. These are either sand grains produced together with the oil, or small, often sharp grains of tramp iron and metal that fall into the oil during its treatment. The schematic design of the gas eliminator is shown

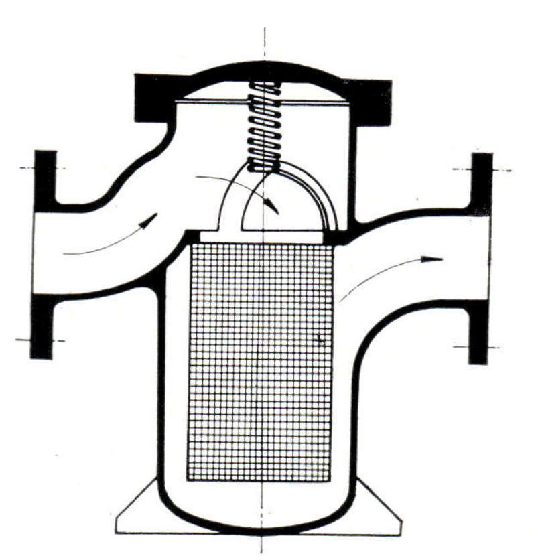

Fig. 6.6—16 Oil screen (Petr. Ext. Serv. 1955)

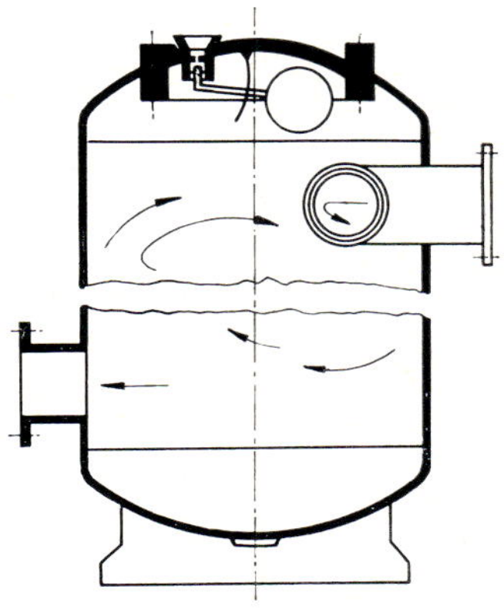

Fig. 6.6—17 Gas extractor (Petr. Ext. Serv. 1955)

in Fig. 6.6—17. In this unit, the flow velocity of the crude drops to a value slightly lower than the critical rise velocity of gas bubbles, so that these may break out and accumulate in the dome of the vessel, depressing the oil level. When the float has sunk far enough, it permits the gas to escape into the atmosphere.

LACT systems with positive-displacement metering have the advantage over weir systems that they may be used in connection with closed, horizontal tanks, which permits to avoid evaporation losses. Another substantial advantage is that all equipment downstream of the stock tank can be portable, skid-mounted, permitting fast installation where required. —

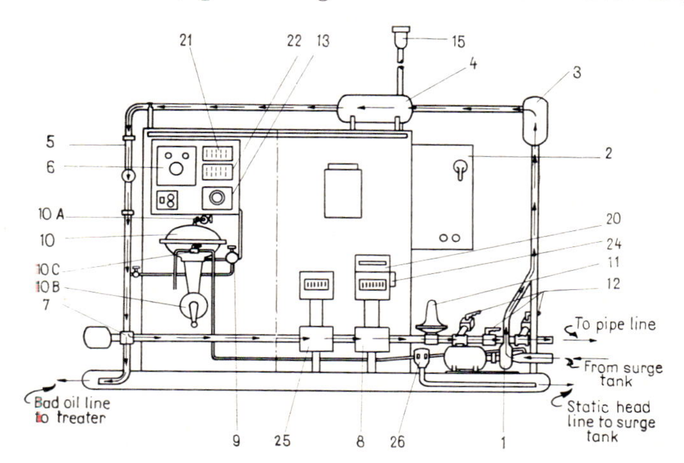

Fig. 6.6—18 Jones and Laughlin portable LACT equipment

Figure 6.6—18 is a diagram of one of the Jones and Laughlin S. D.'s portable LACT units. Once oil in the preparation tank has attained a certain high level, switch *26* trips on and starts through switchgear *2* the electric motor driving pump *1*. When oil pressure has risen high enough, valve *11* opens and discharges oil into the pipeline. In its journey from the pump to the pipeline, the crude passes through strainer *3*, gas eliminator *4*, BS&W monitor *5*, three-way valve *7*, meters *25* and *8*, valve *11*, and valve *12* closing off the calibration loop. If BS&W monitor *5* indicates on control panel *6* that BS&W surpasses the allowable maximum, then valve *7* diverts oil flow towards a treater. During discharge into the pipeline, electrical sampler *9* removes a small quantity of liquid and fills it into pressure-proof vessel *10*. The sample is agitated at intervals by means of manual mixer *10B*. Quantities for laboratory analysis may be drained through four-way valve *10C*. The rest of the sample is discharged, by suitably turning valve *10C*, from tank *10* into the delivery pump, where it joins the main liquid stream. The unit is provided with auxiliary equipment that will shut off oil flow in the case of malfunctions, if the oil stream passing through the unit deviates from prescriptions, or if the daily or monthly limit of oil to be delivered has been attained. Transfer capacity of these portable units is 32—1300 m^3/d; their weight is 4.2—21 kN.

If the central LACT station receives crude from several leases, each sort of crude is fed to a separate LACT unit. Some units, however, are designed so as to be able to handle several inflows. Designing, installing and operation of LACT systems as well as requirements concerning accuracy of measurement and safety are treated in API Std. 2502.

(c) Metering procedures

One of the fundamental conditions for realizing automatic well testers and, *a fortiori*, LACT systems, was the availability of metering instruments furnishing more accurate data than before on the quantity of oil, its BS&W content, and the volume of gas produced; instruments whose metering results could be transmitted in digital form without the necessity of human intervention. The instruments in question include temperature-compensated positive-displacement meters for oil, capacitance type BS&W monitors for water-content determination, and mass flow meters of gas volume.

(c.1) *Positive-displacement metering.* In these meters, the liquid or gas enters a metering space of known capacity, confined by pistons or plungers of various designs and performing motions of various types, and is then expelled therefrom. These meters may be considered as hydraulic motors of high volumetric efficiency that absorb only just as much energy from the passing fluid as is required to keep in motion the moving parts of the instrument. The volume of fluid metered is determined by the number of strokes if the piston is reciprocating, or of revolutions if it is of the rotating type. These meters have been used for some three decades for liquid and gas metering, and have been realized in some 700 designs. Basic types include nutating-disc, oscillating-piston, rotary-vane, oval-gear, reciprocating-piston and birotor meters.

For the metering of liquid hydrocarbons, the oval-gear and nutating-disc meters are the most popular. Figure 6.6—19 is a schematic illustration of the operating principle of the oval-gear meter (Reppisch 1958). Every

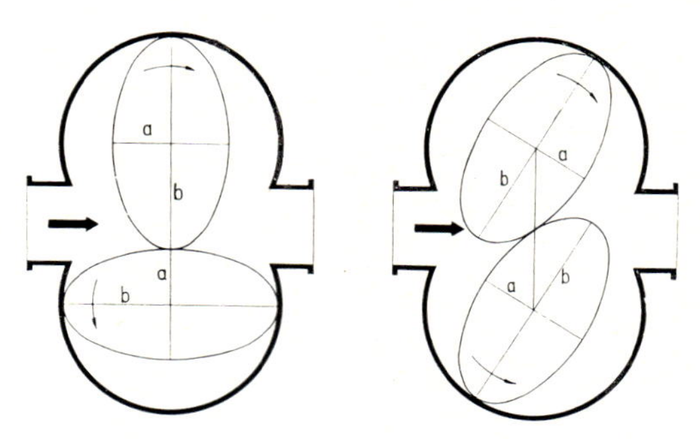

Fig. 6.6—19 Principle of oval-gear positive-displacement meter, after Reppisch (1958)

revolution of the pair of oval wheels provided with spur gears lets pass a theoretically constant amount of liquid. Between the engaging gear-teeth of the two wheels, on the one hand, and the gearbox and the wheel sweeping by it, on the other hand, there are gaps of a few hundredths of a millimeter width. Some of the liquid will pass unmetered through these

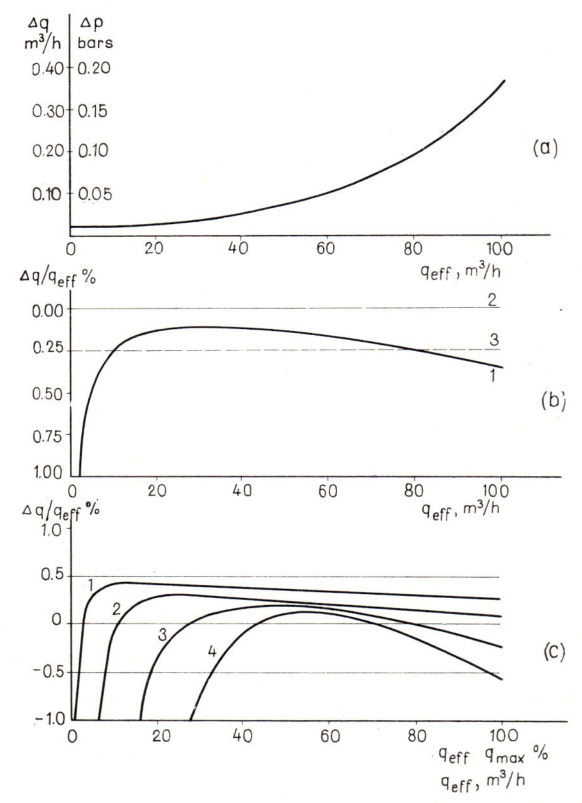

Fig. 6.6—20 Characteristics of oval-gear positive-displacement meter, after Reppisch (1958)

gaps. At higher flow rates, pressure differential due to friction and flow resistance across the meter will increase, and so will leakage through the gaps, as an almost linear function of the pressure differential. Accurate metering requires knowledge of the relative leakage loss. — The upper part (a) of Fig. 6.6—20 shows pressure differential Δp across an oval-gear meter, and leakage Δq, assumed to be proportional to Δp, v. the effective flow rate q_{eff} through the instrument. Part (b) of the Figure shows the relative leakage $\Delta q/q_{eff}$ (or in other words, the percentage error of metering) likewise v. q_{eff} (curve *1*). In the ideal case of no leakage, the instrument will indicate the actual throughput (line *2*). In effect, however, actual throughput invariably exceeds by the leakage loss the volume indicated by the meter, if no measures to the contrary are taken. By suitably adjusting the variable gear ratio between oval gears and counter, however, displacing line *2* into the position of line *3*, so as to make the counter consistently indicate more oil by 0.25 percent than it would in the ideal case, the error is made zero where line *3* and curve *1* intersect, positive between the two points of intersection (that is, the indicated throughput will exceed the actual), and negative outside the two points of intersection.

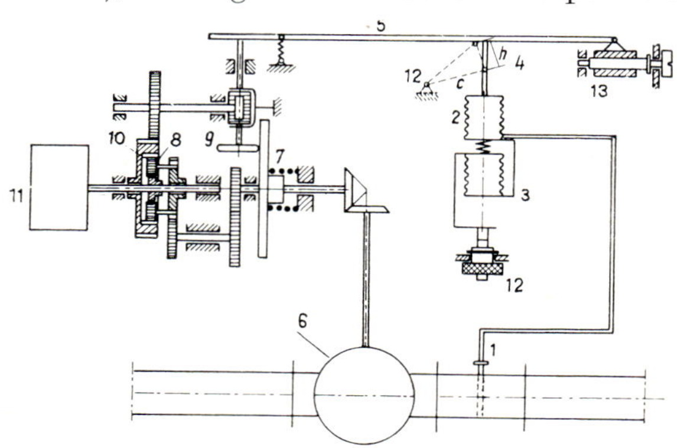

Fig. 6.6—21 Bopp—Reuther temperature-compensated positive-displacement meter, after Ebert and Kalkhof (1961)

In the case shown, at throughputs of 6—100 m^3/d, the maximum error may be ±0.15 percent. The error will be greater at low and high throughputs, and particularly heavy at very low throughputs. This leakage curve will, however, hold only for liquids of a given viscosity as, obviously, leakage loss depends not only on gap width but also on the viscosity of the liquid handled. Part (c) of Fig. 6.6—20 shows after Reppisch (1958) the percentage metering error for 32-cP lubricating oil (*1*), 4.4-cP diesel fuel (*2*), 1.9-cP kerosene (*3*) and 0.48-cP motor gasoline (*4*). Clearly, in order to set a given error level when metering liquids of various viscosity, it is necessary to adjust the gear ratio between oval gear and counter. It further emerges from the Figure that a given error level will hold over a broader range of effective throughput for higher-viscosity liquids, and vice versa.

By the above consideration, the gear ratio has to be changed when changing over to a different crude, but also if the temperature of a given crude is appreciably changed. Metering will be most accurate if the crude is of near-constant viscosity (that is, grade and temperature), and flow rate is even. Using the auxiliary equipment described in connection with the LACT system in Section 6.6.3b, accuracy may attain ±0.1 percent. Modern oval-gear meters are complemented with a thermocouple-controlled regulator that permits the meter to indicate or record throughput in terms of standard volume rather than the effective volume at the actual flow temperature. Figure 6.6—21 represents a Bopp–Reuther-make temperature compensating unit (Ebert and Kalkhof 1961). Bellows *2* is displaced proportionally to the temperature sensed by thermometer *1*. Bellows *3* compensates the travel of bellows *2* for changes in ambient temperature. The travel is transmitted by stem *4* to level *5*. Disk *7* rotates at a speed proportional to the speed of the oval gears in meter *6*. On the one hand, it drives the planetary gears of the bicyclic drive *8* and, on the other, it drives disk *9* by friction; as a result, the sun wheel of the bicyclic drive *10* will rotate at a speed proportional to that of disk *9*. Now this latter speed is proportional to temperature. The instrument has been so designed that counter *11* driven by the axle of the planetary shows the reduced volume. Temperature correction is based on the relationship

$$V_n = V[1 - \beta_T \Delta T + (\beta_T \Delta T)^2]$$

At not too high flow temperatures, the second-degree term is small enough to be negligible, and the formula assumes a simpler, linear form more widely used in practice. If, however, flow temperature is comparatively high, the second-degree term must also be taken into account. In the Bopp—Reuther meter this has been achieved by introducing pivoting arm *12* shown dashed. Screw *13* permits to adjust β_T, depending on crude grade; screw *14* serves to set the standard temperature. — The instrument is furnished with several optional accessories. A built-in printer will record the volume of liquid metered. This is an advantage especially if the printout is required to serve as a ticket to be used for billing consumers or a different accounting unit of the same company. The instrument may further be provided with an attachment that will remote-signal the throughput rate proportional to instantaneous speed of rotation, or total throughput proportional to the total revolutions counted.

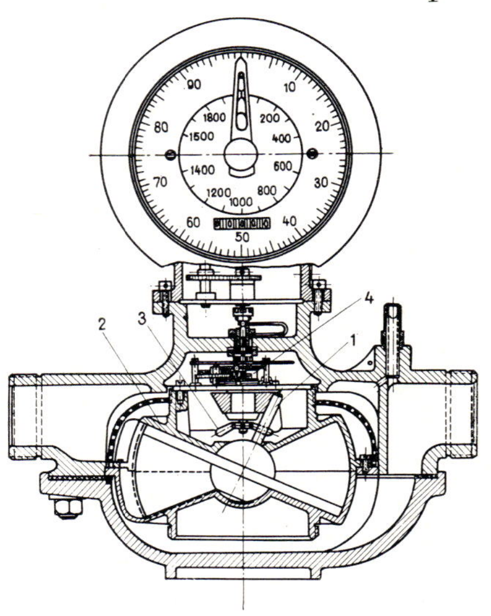

Fig. 6.6—22 Soviet DE-4U type nutating-disk meter

For a comparatively lower-accuracy metering of crude and products, the nutating-disk meter determining actual liquid volume is used. Figure 6.6—22 shows the DE-40 type ‘benzoshchetchik’ used in the Soviet Union. In the metering chamber, pin *1* describes a conical trajectory. Attached to the pin and perpendicular to it there is nutating disk *2* whose perimeter is in gliding contact with the metering-chamber wall. Liquid flowing through the chamber forces the disk to nutate. Each revolution of the disk lets pass a volume of liquid equal to the volume of the metering chamber. The revolutions of pin *1* are transmitted by arm *3* to counter *4* which indicates both the actual throughput rate and the total volume passed.

The threshold of sensitivity of the meter that can be attached to 40-mm nominal-size piping is 0.1 l/h, and its range of measurement is 0.6 — 10 m^3/h. The admissible error is ± 1.0 percent from 0.6 to 1.2 m^3/h, and ± 0.5 percent from 1.2 to 10 m^3/h.

Meter calibration. Calibrating a meter may be performed in one of several ways: (i) using a prover tank, (ii) using a ball-type prover, or (iii) by means of a series-connected calibrating meter. This latter is invariably calibrated by means of a prover tank (cf. API Std 1101). — The idea of the *ball prover* is to connect in series with the instrument to be calibrated a

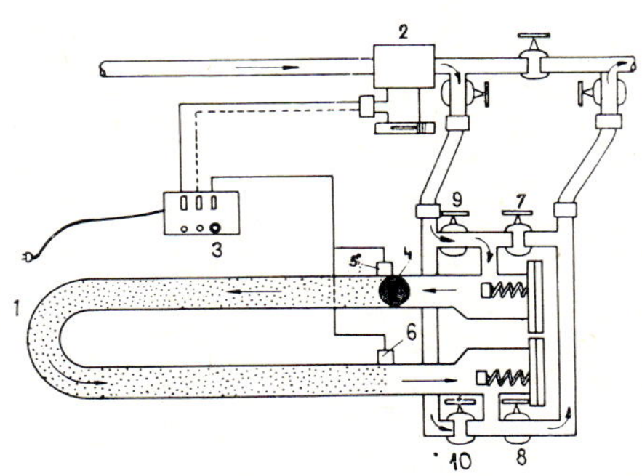

Fig. 6.6—23 Shell Pipe Line Co.'s ball-type meter prover

comparatively small-volume instrument which permits to meter throughput at a very high accuracy. One such instrument that will meter both ways, shown schematically as Fig. 6.6—23, has been developed by the Shell Pipeline Corp. In U-tube *1*, lined with epoxy resin, the liquid to be metered flows as shown by the arrows. Electronic pulse counter *3* of metering instrument *2* is turned on when ball *4* passes sensor *5* and turned off when it attains sensor *6*. Accurate metering is made possible by the fact that the volume of the U-tube can be determined to a high accuracy, and the passage of the balls is sensed by the sensors at a high precision. In the design shown in the Figure, the passage of one USA gallon (3.79 dm^3) will trigger 1000 pulses. The maximum error is 3.79 cm^3, corresponding to one pulse. The U-tube volume must be at least 10,000 times the least volume that can be metered, that is, at least 10 gal. (37.9 dm^3) in this case; on the other hand, it must equal at least 0.5 percent of the maximum hourly output of the meter to be calibrated. In the setup shown as Fig. 6.6—23, flow in tube *1* can be reversed by opening valves *7* and *8*, closing valves *9* and *10*, and the ball will then trigger sensor *6* first and sensor *5* second. Reverse flow can also be used for calibration. — The procedure of calibrating positive-displacement meters is described in Byers (1962—1963, Parts 1, 2, 4).

(c.2) *Water-content monitoring.* — There is a substantial difference in dielectric behaviour between water and oil. (The relative dielectric constant of the latter is about 80, that of the former is about 2—2.5.) This is the circumstance exploited in the capacitance-type BS&W monitor. In the monitoring loop shown in Fig. 6.6—24, a closed pipe *2* of length s is installed coaxially in oil pipe *1*. The arrangement functions as a capacitor of length s, with wet oil as the dielectric. The capacitor (whose capacitance depends, besides the constant geometry of the arrangement, on the dielectric constant of the liquid) is hooked up to a resonant circuit. The eigenfrequency f_n of the circuit depends on the water content of the liquid in the capacitor. This frequency is fed together with the constant frequency output f_1 of a generator into a mixing stage, and the beat frequency $(f_n - f_1)$ of the mixer is fed to a servo mechanism. The servo mechanism is designed so as to readjust the original frequency f_n of waterless oil by turning a variable capacitor. This turning, suitably amplified, serves as a measure of water content against an experimentally calibrated dial. The instrument

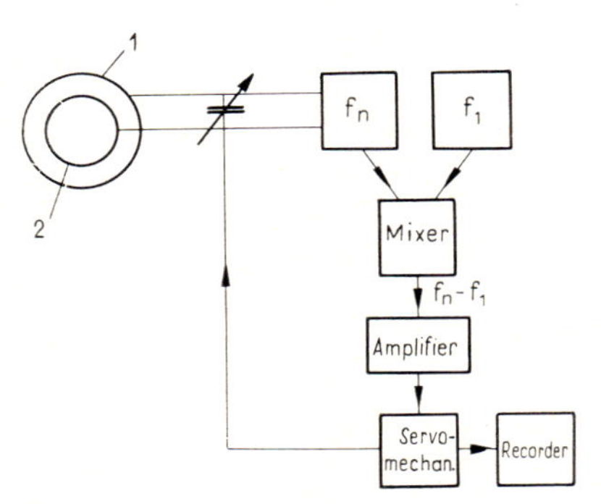

Fig. 6.6—24 Capacitance-type BS&W monitoring

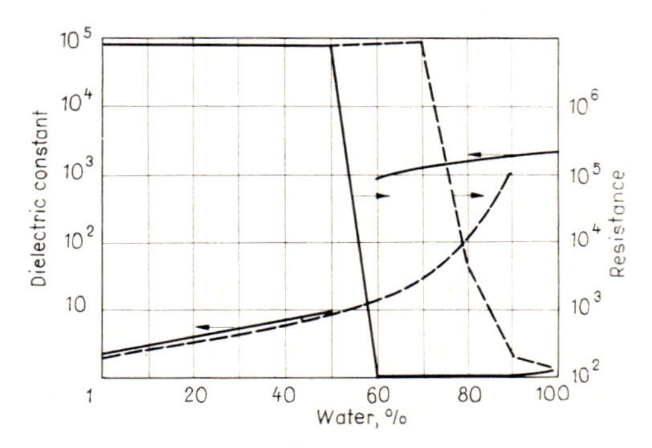

Fig. 6.6—25 Dielectric constant and resistance of a water–oil mix v. the water content, after Wood (1958)

is suited also for remote signalling. — Experience has shown this instrument to operate up to about 50 percent water only. Figure 6.6—25 (after Wood 1958) shows the apparent electric resistivity and dielectric constant v. water content of a given grade of oil mixed with an artificial brine of 9 weight percent salt content. The experiments were performed with O/W emulsions prepared respectively by fast and slow mixing (full and smooth dashed curves, respectively). Clearly, at about 50—60 percent water content the dielectric constant of fast-mixed emulsions takes a sudden jump with a concomitant drop in resistivity. In slow-mixed emulsions, the dielectric constant rises gradually, but resistivity takes a sudden drop between 70 and 80 percent. Similar phenomena are observed in mixes of other oils with brines of different concentration. The reason for this is that, at comparatively high water contents, the original W/O emulsion inverts into an O/W emulsion—earlier in fast mixes, later in slow ones—and salt water short-circuits the capacitor cylinders.

Recent investigations have shown the error of capacitance-type BS&W meters to be significant, up to 7 percent. This has two main causes: the dielectric constant of water will be affected, one, by its content of clay minerals, especially montmorillonite, and, two, by the frequency used in the measurement. Scatter can be reduced considerably by employing a high frequency—10 mcps seems best (Thompson and Nicksic 1970).—The BS&W monitor is not indifferent to installation and hookup. Figure 6.6—26 shows one of the most favourable hookups, immediately downstream of the separator. Liquid is discharged through line *2* from float-controlled separator *1*. Once the liquid in the separator has risen to the high level, control *3* supplies power gas taken from the gas-outlet reducer *4* through line *5* above the diaphragm of motor valve *6*, which opens the liquid discharge line. Water content is monitored by capacitance-type monitor *7*; liquid volume is metered by meter *8*. Bypass line *10* around the metering systems can be remotely opened through remote-controlled

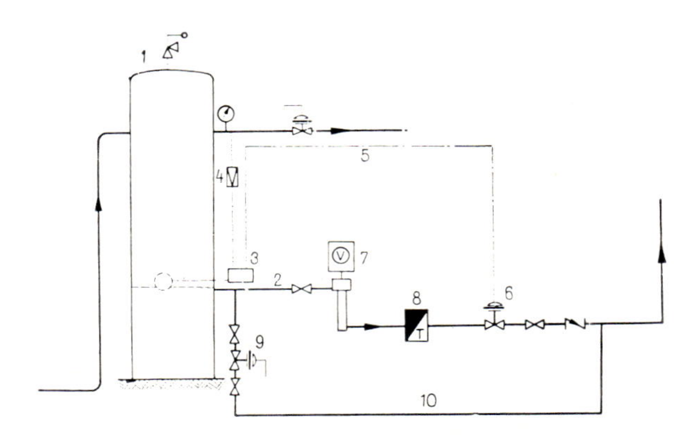

Fig. 6.6—26 Hookup of BS&W monitor downstream of separator, after Wood (1958)

motor valve *9*. The chart of the BS&W monitor is moved by the clockwork only when the separator is delivering liquid. The advantage of this hookup is that it provides continuous monitoring; pressure during discharge from the separator is constant; and the liquid is mixed rapidly and uniformly.—According to Wood, capacitance-type monitoring provides results close to those of sampling from the stock tank, performed by the usual procedures but with the care befitting a calibration experiment. Samples taken from the flowline at the wellhead give widely divergent results, owing presumably to the frequent slug-type flow of the two phases, which tends to discredit this sampling procedure. — Calibration of capacitance-type BS&W monitors is outlined in Byers (1962—1963, Part 3).

(c.3) *Gas metering*. An exhaustive discussion of the subject would require a separate book. In the present work, I shall merely cite the main relevant publications (Stearns et al. 1951, Spink 1958, Mokhanov 1962, Herning 1950, American Gas Association 1955) and give a brief outline of gas metering problems in automated production equipment.

The most widespread means of gas metering is the orifice meter. The conventional meter has several drawbacks. Gas volume referred to the standard state T_n, p_n, passed by the meter between the instants t_1 and t_2, is given by the formula

$$q_{gn} = cd^2 K_1 K_2 K_3 \frac{T_n z_n}{p_n} \cdot \sqrt{\frac{\Delta pp}{TMz}}\,, \qquad 6.6-1$$

where c is a constant; K_1, the basic orifice flow factor, and K_2, the expansion factor depend on the pressure p and temperature T of flow; K_3, the orifice plate expansion factor, depends on flow temperature. This means that Δp, p and T are to be continuously monitored, and so is M if gas quality is variable — which adds up to a total of four physical quantities. Finally, compressibility factor z is a function of p, T and M. Clearly, even if Δp and p are constants throughout the metering period, calculating the standard-state gas throughput takes quite some time. An additional difficulty arises if Δp and p are not constant, as that requires finding the rms value of the instantaneous values' product between t_1 and t_2 that is,

$$\overline{\sqrt{\Delta pp}} = \frac{\int_{t_1}^{t_2} \sqrt{\Delta p_t\, p_t}\,\mathrm{d}t}{t_1 - t_2}.$$

This value used to be estimated by manual calculation. Later on, planimeters performing integration under the square-root sign were employed, but these still tied down people. Next, meters automatically recording $\sqrt{\Delta pp}$ by mechanical, electric, pneumatic or electronic means were devised. Even these meters required, however, the use of Eq. 6.6—1 to calculate q_{gn} by hand or by computer. — The metering accuracy provided by all this complicated equipment and calculation procedures was not better than ± 1.5 percent. It was primarily this state of facts that urged designers to develop meters based on new principles.

It can be shown that if gas is metered in terms of mass rather than standard-state volume units, then the relationship $q_{gn} = K\sqrt{\Delta p \varrho_g}$ will hold, where the factor K is independent of the absolute flow pressure p as well as of T, M, and z. The formula shows, however, that employing this principle requires the continuous monitoring, in addition to the pressure drop across the orifice, also of the effective density ϱ_g of the gas being metered. Nevertheless, the number of parameters to be monitored is reduced to just two. Lately, a number of advanced methods for density monitoring have been developed (Upp 1969, O'Donnell 1969, Sens 1969). The data furnished by these monitors can be fed to a computer which then performs the rms operation (Hennig 1970, Jacobsen 1970). — In Holland, standard gas volume is usually metered by means of turbine velocity meters. The error of these is significant, too, amounting to ± 2 percent in one day, ± 1 percent over a month. Design of mass flow meters conformable to the prescriptions of ISO 541 is underway; the error of these is to be below 0.1 percent. The designs of various mass flow meters and the principles of choosing the instrument best fitting any given application are discussed by Puscoiu (1970).

Chapter 7

Pipeline transportation of oil

7.1. Isothermal flow of a Newtonian oil

Temperature of crude oil flowing in a pipeline will usually vary along the line. If, however, oil viscosity is low and the inflow temperature is close to the soil temperature, then flow may be regarded as isothermal for practical purposes. (See, in general, Gyulay 1942.)

7.1.1. Pressure traverse and maximum capacity of pipelines

Let the specific energy content of flowing oil be W'_1 J/N at the head end and W'_x J/N at a distance of l_x m from the head end. The decrease in specific energy content is

$$W'_1 - W'_x = \frac{p_1 - p_x}{\varrho g} + \frac{v_1^2 - v_x^2}{2g} + z_1 - z_x = h_{sx} . \qquad 7.1-1$$

The decrease in specific energy content is equivalent to friction loss; that is, by Eq. 1.1—1,

$$p_s = \lambda \frac{v^2 l \varrho}{2d_i} .$$

We shall examine below the variation of pressure loss with distance along the line. Let therefore $p_s = p_{sx}$ and $l = l_x$. Let us further divide both sides of the equation by ϱg and let us put $p_{sx}/\varrho g = h_{sx}$; then,

$$h_{sx} = \lambda \frac{v^2 l_x}{2gd_i} = \xi_s l_x \qquad 7.1-2$$

where

$$\xi_s = \frac{\lambda v^2}{2gd_i} \qquad 7.1-3$$

is the hydraulic gradient.

Each term of Eq. 7.1—1 permits of a twofold physical interpretation. They may be considered, on the one hand, as energy contents of a liquid body of unit weight, in J/N units, and on the other, as liquid column heights in m, the density of the liquid being ϱ.

In isothermal flow, ξ_s is constant, and therefore specific energy content decreases linearly with distance along the pipeline. The specific external (mechanical) potential energy $p_x/\varrho g$ is then equivalent to pressure head h_x, specific kinetic energy $v_x^2/2g$ to velocity head, specific internal potential energy z_x to the geodetic head, and specific energy loss by friction, h_{sx}, to friction head; $v_1 = v_2$, and the velocity-head difference is zero. Eq. 7.1—1 may therefore be rewritten as

$$h_{sx} = h_1 - h_x + z_1 - z_x .$$

The way the components of the liquid's specific energy content vary in a pipeline laid over undulating terrain is shown in Fig. 7.1—1.

The pressure head h_1 at head end K required to move oil at a flow rate q through pipe of *ID* d_i laid over a given terrain can be determined as follows.

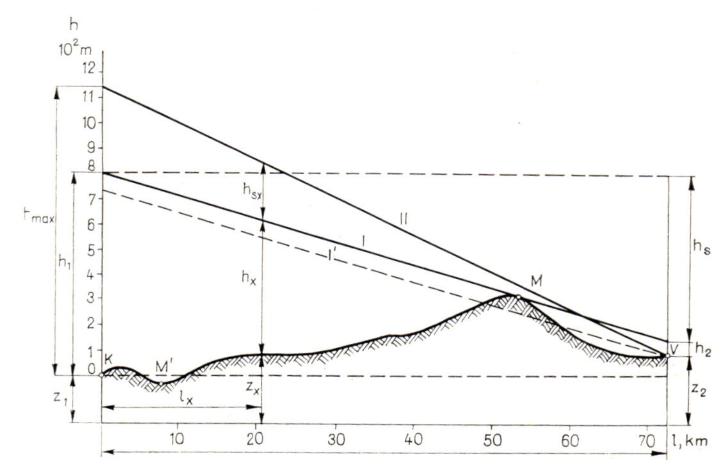

Fig. 7.1—1 Pressure traverse of pipeline laid over an undulating terrain

Using Eq. 7.1—3, the gradient ξ_s is calculated. A line with a slope corresponding to this gradient is drawn from tail end V to head end K. This is the pressure or piezometric traverse I'. If this line intersects the ground profile, then the pressure traverse must be displaced parallel to itself until it will merely touch the ground profile (line I). The initial pressure head h_1 is the above-ground section of the ordinate at the head end K. It is recommended for safety to augment h_1 by 30—50 m, with due attention to the operating conditions to be expected. The point M where the pressure traverse is tangent to the ground profile is called critical. This is where the pressure head of the flowing liquid is least. Hydraulic gradient is greater between points M and V than between K and M. Consequently, if there is no throttling at the tail end, flow is free beyond the critical point and oil will arrive at atmospheric pressure at the tail-end tank. If throttling is applied at the tail end, the oil will have a pressure head of h_2 at the end point; V is usually above ground level, at the maximum possible liquid level of the tail-end tanks. Another critical point as far as pipe strength is concerned may be valley point M'. It is necessary to calculate the pressure head h

above M and find out whether pressure $p = h\varrho g$ does not exceed the maximum allowable operating pressure of the pipeline.

Example 7.1—1. Let a pipeline of *OD* $d_o = 219.1$ mm and *ID* $d_i = 209.5$ mm ($s = 4.78$ mm) be laid over the terrain shown in Fig. 7.1—1. The throughput at the temperature of transportation is $q = 150$ m³/h; $\varrho = 830$ kg/m³; $\nu = 7.5 \times 10^{-6}$ m²/s; $l = 72.4$ km, and the equivalent roughness of the pipe is $k = 0.2$ mm. Determine the pressure traverse and find the head-end pressure head h_1. — Flow velocity is

$$v = \frac{q}{\frac{d_i^2 \pi}{4}} = \frac{0.04167}{0.2095^2 \times 0.7854} = 1.21 \text{ m/s}$$

$$N_{\mathrm{Re}} = \frac{v d_i}{\nu} = \frac{1.21 \times 0.2095}{7.5 \times 10^{-6}} = 3.38 \times 10^4$$

$$\frac{k}{d_i} = \frac{2 \times 10^{-4}}{0.2095} = 9.5 \times 10^{-4} .$$

From Fig. 1.1—1, $\lambda = 0.0253$. — By Eq. 7.1—3, the friction-head gradient is

$$\xi_s = 0.0253 \frac{1.21^2}{2 \times 9.81 \times 0.2095} = 0.00901 \text{ m/m} .$$

The ordinate at K corresponding to the intercept of the pressure traverse traced backwards from V is

$$h_1' = 72{,}400 \times 0.00901 + 83 = 652 + 83 = 735 \text{ m}$$

where 83 m is the geodetic-head difference $(z_1 - z_2)$ between point K and V. In the Figure, the pressure traverse in question is shown by dashed line I'. As this enters the ground, the head-end pressure head actually required is higher: it is obtained by displacing the pressure traverse parallel to itself, making it tangent to the ground profile at the point M. Accordingly $h_1'' = 805$ m. For safety, let us augment this by 40 m, giving $h_1 = 805 + 40 = 845$ m. The pressure head of the pump to be installed at point K should, then, be

$$p_1 = h_1 \varrho g = 845 \times 830 \times 9.81 = 6.88 \times 10^6 \text{ N/m}^2 .$$

Assuming that pumping is concentrated at K, *the greatest throughput of the pipeline* can be determined by the following procedure. Let us plot above the ground profile at K the pressure head h_{max} corresponding to the maximum allowable operating pressure of the pipe (Fig. 7.1—1). Joining this point with V (or with critical point M, if there is one) we obtain a pressure traverse whose slope corresponds to the maximum feasible pressure gradient ξ_{sOmax}. In the knowledge of this latter, the maximum oil throughput, q_{Omax} can be determined as follows. Let the two constants in Eq. 1.1—10 be $a = 0.194$ and $b = 0.189$. Then

$$\lambda = 0.194 N_{\mathrm{Re}}^{-0.189} . \qquad 7.1\text{—}4$$

Introducing this into Eq. 7.1—3, wherein $g = 9.81$ we obtain

$$\xi_s = 9.89 \times 10^{-3} \frac{\nu^{0.189} v^{1.81}}{d_i^{1.19}}$$

and since

$$v = \frac{q}{d_i^2 \frac{\pi}{4}}$$

substitution into the above equation yields

$$\xi_s = 1.53 \times 10^{-2} \frac{\nu^{0.189} q^{1.81}}{d_i^{4.81}}, \qquad 7.1\text{—}5$$

whence

$$q = 10.1 \frac{\xi_s^{0.552} \times d^{2.66}}{\nu^{0.104}}. \qquad 7.1\text{—}6$$

Introducing into Eq. 7.1—6 the graphically determined value of $\xi_{sO\max} = \xi_s$ we may calculate $q_{O\max} = q$, the maximum liquid throughput of the pipeline if the only pump station is at the head end.

Example 7.1—2. Let us find the maximum throughput of the pipeline characterized in Example 7.1—1, if API standard pipe made of X—42 steel is used, and terrain belongs to category (a) characterized in Section 6.1—4, that is, the safety factor $k = 1.3$; let $e = 1$. — The pressure rating is calculated using Eq. 7.1—2. From Table 6.1—2, $\sigma_F = 289$ MN/m², and hence,

$$\sigma_{al} = \frac{289}{1.3} = 222 \text{ MN/m}^2.$$

The maximum allowable internal overpressure by Eq. 6.1—1 is

$$\Delta p = \frac{2e\sigma_{al} s}{d_o} = \frac{2 \times 1 \times 222 \times 10^6 \times 4.78 \times 10^{-3}}{0.2191} = 9.69 \times 10^6 \text{ N/m}^2$$

and

$$h'_{\max} = \frac{\Delta p}{\varrho g} = \frac{9.69 \times 10^6}{830 \times 9.81} = 1190 \text{ m}.$$

Let us reduce this value by 40 m for safety. Hence, $h_{\max} = 1190 - 40 = 1150$ m. Plotting $h_{\max}$ at the point K and joining it with V we get pressure traverse *II*. Clearly, the pressure traverse passes above the critical ground-profile section, and the safety reserve of pressure is sufficient also from this point of view. — The maximum pressure gradient is

$$\xi_{s\,O\max} = \frac{1150 - 83}{72{,}400} = 0.0147 \text{ m/m}.$$

The maximum liquid throughput by Eq. 7.1—6 is

$$q_{O\max} = 10.1 \frac{0.0147^{0.552} \times 0.2095^{2.66}}{(7.5 \times 10^{-6})^{0,104}} = 0.0525 \text{ m}^3/\text{s} = 189 \text{ m}^3/\text{h}.$$

7.1.2. Increasing the capacity of pipelines

Even the maximum throughput $q_{0\max}$ of an existing pipeline may be insufficient to transmit the yield q_1 to be removed from a field. Increasing the throughput of an existing pipeline should be performed with a least-cost consideration in mind. Often this can be achieved by letting the existing pipeline stay as it is and by laying a second one alongside, or by installing one or several intermediate pumping (booster) stations along the existing pipeline.

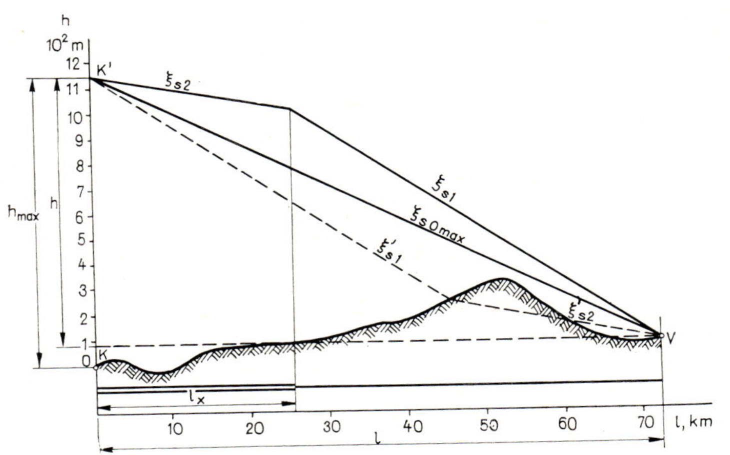

Fig. 7.1—2 Pressure traverse of twin (looped) pipeline

A double line is called 'looped', it may be of two types: (i) new line of same length as old line but usually of different diameter (complete loop), (ii) new line shorter than old line but of the same *ID* (partial loop).

Solution (i) may be regarded as two independent pipelines with the size of the new line chosen so as to deliver a flow of $q_2 = q_1 - q_{0\max}$ under the original input pressure $h_{\max}$. Owing to the identity of trace and of input and output pressures the pressure gradient of the new line is the same as the maximum feasible gradient $\xi_{s0\max}$ of the existing line. The task in hand is, then, to determine the size of pipe required for a line as long as the existing one to deliver fluid precisely at the rate q_2 under a pressure gradient $\xi_{s0\max}$. From Eq. 7.1—5 we obtain

$$d_i = 0.419 \frac{\nu^{0.0393} \times q^{0.376}}{\xi_s^{0.208}}. \qquad 7.1\text{—}7$$

Putting $\xi_s = \xi_{s0\max}$ and $q = q_2$, one may calculate the *ID* of the new line d_i.

(ii) If the new line is to have the same *ID* as the existing one, and $q_1 < 2q_{0\max}$, the length of the new line may be less than that of the existing one. The new line is usually laid beginning at either the tail or the head end of the old one (Fig. 7.1—2). Terrain permitting, the new line should be started at the tail end, as this will give a lower average pressure in the entire system. A pressure traverse for the loop can be determined by the following consideration. When pumping at the prescribed rate $q_1 > q_{O\max}$, the pressure gradient in the single line section is $\xi_{s1} > \xi_{sO\max}$. By Eq. 7.1—5

$$\xi_{s1} = 1.53 \times 10^{-2} \frac{\nu^{0.189} \times q_1^{1.81}}{d_i^{4.81}} .$$

In the double section, the two pipes, of equal size, and subjected to the same head-end end pressure, will both deliver oil at the rate $q_1/2$. The pressure gradient in this section is

$$\xi_{s2} = 1.53 \times 10^{-2} \frac{\nu^{0.189} \left(\frac{q_1}{2}\right)^{1.81}}{d_i^{4.81}} .$$

Dividing the first equation by the second and solving for ξ_{s2}, we get

$$\xi_{s2} = \xi_{s1} \left(\frac{1}{2}\right)^{1.81} = 0.285\, \xi_{s1} . \qquad 7.1—8$$

In the knowledge of the two pressure gradients, the loop pressure traverse and the length of the new line can be determined graphically or by calculation. Graphically it is done by tracing a pressure traverse of slope ξ_{s2} forward from head end K' and another pressure traverse of slope ξ_{s1} backward from tail end V, if construction is to be started at the head end. The new line is to extend to the point where the two traverses intersect. If it is desired to lay the new line starting from the tail end, then the two gradients in the construction should be interchanged. Calculation of the new line's length is performed according to the following geometrical consideration (Fig. 7.1—2):

$$h = \xi_{sO\max}\, l = \xi_{s1}(l - l_x) + \xi_{s2}\, l_x$$

that is,

$$l_x(\xi_{s1} - \xi_{s2}) = l(\xi_{s1} - \xi_{sO\max})$$

and

$$l_x = l \frac{\xi_{s1} - \xi_{sO\max}}{\xi_{s1} - \xi_{s2}} . \qquad 7.1—9$$

Example 7.1—3. Under the conditions stated in the previous example, existing pipeline capacity is to be increased to 222 m³/h by building a new line. The design of said new line is as follows. (i) For a complete loop,

$$q_2 = 222 - 189 = 33 \text{ m}^3/\text{h} = 0.00917 \text{ m}^3/\text{s}.$$

By Eq. 7.1—7,

$$d_i = 0.419 \frac{(7.5 \times 10^{-6})^{0.0393} \times (9.17 \times 10^{-3})^{0.376}}{(1.47 \times 10^{-2})^{0.208}} = 0.108 \text{ m}.$$

The next standard API (high-test) pipe has $d_o = 114.3$ (= 4 1/2 in.). For X—42 grade steel, with reference to Example 7.1—2,

$$\sigma_{al} = 222 \times 10^6 \text{ N/m}^2.$$

By Eq. 6.1—1,

$$s = \frac{\Delta p d_o}{2e\sigma_{al}} = \frac{9.69 \times 10^6 \times 0.1143}{2 \times 1 \times 222 \times 10^6} = 0.0025 \text{ m} = 2.5 \text{ mm}.$$

The next standard pipe-wall thickness is 3.18 mm, and hence, $d_i = 107.9$ mm. (ii) In the case of a partial loop, by Eq. 7.1—5,

$$\xi_{s1} = 1.53 \times 10^{-2} \frac{(7.5 \times 10^{-6})^{0.189} \times 0.06167^{1.81}}{0.2095^{4.81}} = 0.0196 \text{ m/m}$$

and by Eq. 7.1—8,

$$\xi_{s2} = 0.285 \times 0.0196 = 0.00559 \text{ m/m}.$$

The length of the new line is, by Eq. 7.1—9,

$$l_x = 72{,}400 \frac{0.0196 - 0.0147}{0.0196 - 0.0056} = 25340 \text{ m}.$$

The pressure traverse of the loop is shown in Fig. 7.1—2.

Increasing throughput by means of booster pump stations. Pump stations inject oil into the pipeline at the maximum allowable pressure of the pipe used; input pressure is the same at each station (except, usually, the last one). The slope of the pressure traverse — which is likewise equal in all sections — depends on the throughput q to be pumped through the given pipeline. The first pump station is installed at head end K. A pressure traverse of slope equal to the pressure gradient is to be drawn starting from h_{max} plotted at K, until it intersects the ground profile. This is where the first booster station is to be installed. Now h_{max} is plotted again at this point and the construction is repeated until the last booster station will deliver the oil to or beyond the delivery end V (Fig. 7.1—3).

Example 7.1—4. The capacity of the existing pipeline characterized in the previous example is to be increased to 340 m^3/h by installing booster stations. Find the locations of said booster stations along the pipeline. — By Eq. 7.1—5, the pressure gradient to be expected at a throughput of 340 m^3/h = 0.0945 m^3/s is

$$\xi_s = 1.53 \times 10^{-2} \frac{(7.5 \times 10^{-6})^{0.189} \times 0.0945^{1.81}}{0.2095^{4.81}} = 0.0470 \text{ m/m}.$$

The maximum allowable discharge head is 1150 m according to Example 7.1—2. Performing the construction in the knowledge of ξ_s and h_{max} it becomes apparent that at station III, a discharge head h_{max} would be excessive. The discharge head required can be found by tracing a pressure traverse backward from V, in which case the head in question is the height difference at point III between the ground profile and the pressure traverse in that vertical. The discharge head required is 870 + 40 = 910 m.

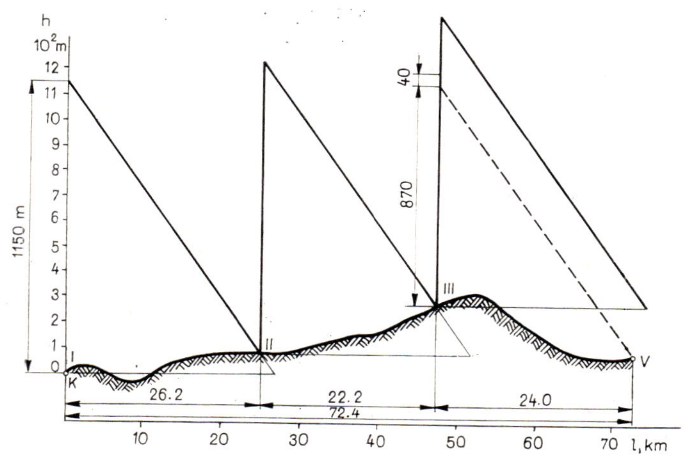

Fig. 7.1—3 Pressure traverse with booster stations

7.1.3. Capacity of branching pipelines

A pipeline may deliver oil to more than one refinery. In Fig. 7.1—4, for instance, three lines emerge from point B on a 8-in. line. With a throughput q flowing in the main line, it is possible to determine the unthrottled partial throughputs of the three lines. By Eq. 1.1—10, approximately,

$$\lambda = a\left(\frac{d_i v}{\nu}\right)^{-b}.$$

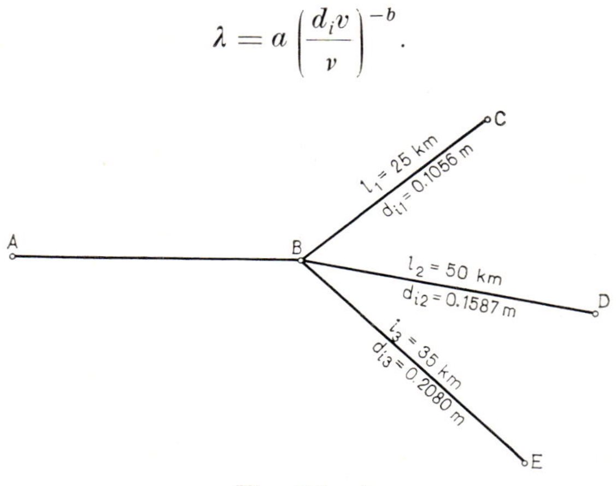

Fig. 7.1—4

Putting this into Eq. 7.1—12 yields

$$h_s = a \frac{v^{(2-b)} \nu^b}{2gd_i^{(b+1)}}.$$

Since by Eq. 7.1—1 the head loss h_s over horizontal terrain equals the pressure or discharge head $\Delta p/\varrho g = h$, it is true for any branch line that

$$h = a' v^{(2-b)} = a'' q^{(2-b)}. \qquad 7.1\text{—}10$$

Since

$$\lg h = \lg a'' + (2 - b) \lg q$$

the $q - h$ relationships of the branch lines, plotted on log-log paper, are mutually parallel straight lines. Summing the capacities q_1, q_2, q_3 of the branch lines at any discharge head h, one finds a point (h, q_4) which defines a so-called equivalent-line curve parallel to the lines $h - q$; this curve will, at any given discharge head, define the sum of the throughputs of the branch lines. If therefore the throughput q of the main line is known, then the point (q, h') on the equivalent-line curve will define the pressure head h' required at the branch point B, and the respective throughputs q_1', q_2' and q_3' of the three branch lines at this head.

Example 7.1—5. Oil of 9.5 cSt viscosity and 830 kg/m³ gravity is being transported between points A and B in Fig. 7.1—4 at the rate of 200 m³/h. Find the delivery rates at the end points C, D, E and the pressure head required at the branch point B. Let the equivalent roughness of the pipes be $k = 0.1$ mm. — At a throughput of 30 m³/h, the friction losses of the individual branch lines are, by Eq. 7.1—2,

$$h_1' = 0.0318 \frac{0.952^2 \times 25{,}000}{2 \times 9.81 \times 0.1056} = 348 \text{ m}$$

$$h_2' = 99.0 \text{ m}$$

$$h_3' = 19.0 \text{ m}.$$

In order to find the slopes of the curves of the branch lines let us calculate the friction head at a throughput of, say, 100 m³/h e.g. for branch line *3*:

$$h_3'' = 0.0275 \frac{0.817^2 \times 35{,}000}{2 \times 9.81 \times 0.208} = 158 \text{ m}$$

Plotting the pairs of q and h on log-log paper, one can trace the characteristic of branch *3*; the lines for branches *1* and *2* will be parallel to this line and pass through the points h_2^1 and h_3^1 (Fig. 7.1—5). Let us find the throughputs of each branch line if the pressure head at point B is, say, $h = 100$ m. From the Figure, $q_1 = 14.7$, $q_2 = 30.2$ and $q_3 = 77.1$ m³/h. The throughput of the equivalent line is the sum of the throughputs of the branch lines, that is, 122 m³/h. The equivalent-line characteristic passes through the point (100 m; 122.0 m³/h) and is parallel to the three other lines. For the prescribed throughput of 200 m³/h, pressure head at node *B* will have to be 238 m. The Figure indicates for that case throughputs of 24.1 m³/h for branch *1*, 49.5 m³/h for branch *2* and 126.4 m³/h for branch *3*.

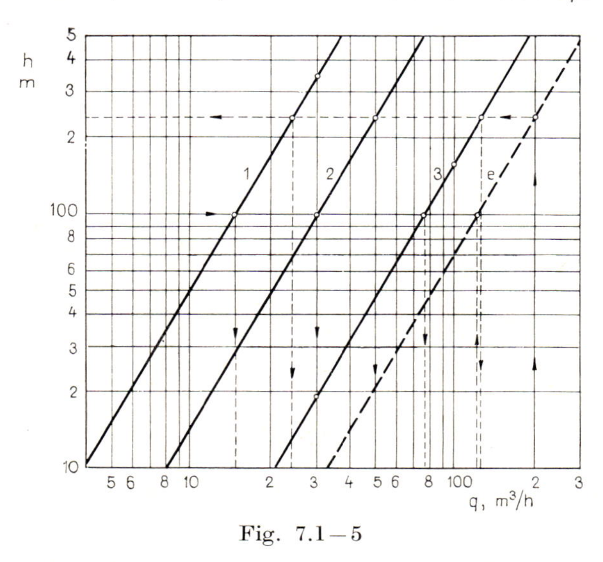

Fig. 7.1–5

7.1.4. Optimum trace and size of pipelines

In the absence of arguments to the contrary (which will be discussed next), the economically optimal trace for a pipeline is the straight line joining the two end points. Arguments to the contrary include the following. (i) The straight trace would traverse country where laying the pipeline would be too costly (swamp, lake, rock, etc.). (ii) It would penetrate the safety zones of objects designated by the competent authorities (buildings, cemeteries, defense establishments), or intersect main roads, railroads, (iii) Deviation from the straight trace would entail a substantial saving by avoiding damage to crops and the tying down of valuable land. (iv) It is to be preferred to lay a line along a road or railroad, as this will reduce laying costs and facilitate and accelerate repairs. (v) Oil losses resulting from a pipe puncture will be less if the line is laid over undulating terrain.

Optimum pipe size is that size permitting transport of oil at lowest cost. Finding optimum pipe size may be divided into two basic tasks according to whether desired annual throughput during the life of the line is to be considered constant or variable. In the following we shall restrict discussion to the first task. The second task is much more complicated. In reality, a pipeline will be operated over several decades. If it transports throughout its life the output of one, relatively well-known, production area the desirable annual throughput can be determined on the basis of the production plan. Starting from said throughput, the pipe size ensuring the least cost of transportation over the entire life of the lease or of the pipeline may be chosen. In so doing, however, one has to keep in mind that the economics of such a system may be changed fundamentally by installing booster stations for a limited space of time (a few years, say), notably during the peak delivery period of the field. The pumps may then be moved elsewhere as and when required. The situation is further complicated if the

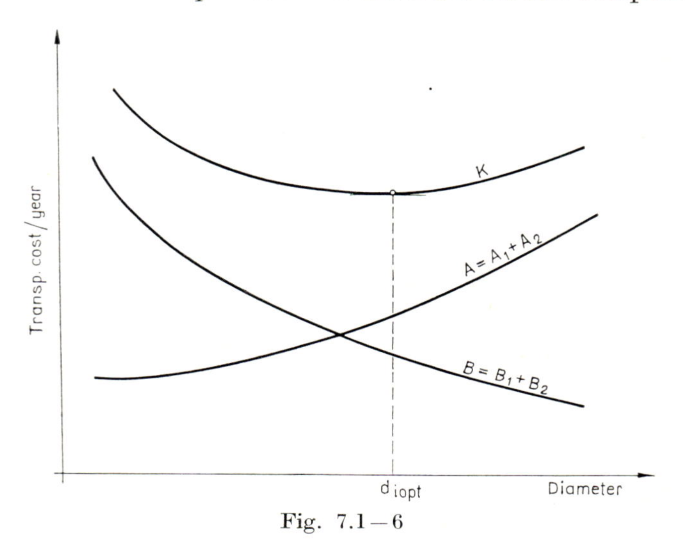

Fig. 7.1—6

transportation plan is upset by some unexpected factor such as the discovery of a new field whose output is to be accommodated by the existing pipeline.

If the annual throughput may be considered constant, then the optimum pipe size for the line can be determined by the following consideration. Cost of transportation is a sum of two components, depreciation plus interest on the investment (to be called depreciation for short), and operating plus maintenance cost. Increasing pipe size will increase first cost and decrease operating cost. Figure 7.1—6 shows plots of depreciation A, operating cost B and total cost K v. pipe size for a given throughput $q =$ $=$ const. Optimal pipe size $d_{i\,\mathrm{opt}}$ is seen to belong to the minimum of the curve $K = \mathrm{f}(d_i)$.

$d_{i\mathrm{opt}}$ may be determined in one of two ways. (i) Estimating costs for various pipe sizes, a plot similar to the one in Fig. 7.1—6 is prepared and $d_{i\mathrm{opt}}$ is determined graphically. (ii) Optimum pipe size is calculated using a suitable mathematical formula, e.g. the one introduced by Smith—Brady—Donnell.

The total annual cost of transportation is

$$K = A_1 + A_2 + B_1 + B_2 \,. \qquad 7.1\text{—}11$$

Here, depreciation A_1 is defined as

$$A_1 = (a_1 + a_2 d_i) l z_1$$

where a_1 is a cost component independent of pipe size, e.g. ditch cutting and backfilling within limits, and supervision in Ft/m*; $a_2 d_1$ is the cost component depending on pipe size (pipe price, transportation to the site, welding, painting, insulation, pressure testing), a_2 is in Ft/m^2 units; z_1 is the annual depreciation rate of the ready-to-operate pipeline, and l is the length of the pipeline in m.

The annual depreciation of a pump station is

$$A_2 = P b z_2 = \frac{q \varrho g h b z_1}{\eta} \qquad 7.1\text{—}12$$

where pump-station output $P = \frac{q \varrho g h}{\eta}$, W; b is unit first cost of the station, in Ft/W; z_2 is the annual depreciation rate of the pump station; η is the overall efficiency of the installation.

Using Eqs 7.1—2 and 7.1—5, we get

$$h = 1.53 \times 10^{-2} \frac{\nu^{0.189} q^{1.81} l}{d_i^{4.81}} .$$

Putting this into Eq. 7.1—12, we find

$$A_2 = 1.53 \times 10^{-2} \frac{q^{2.81} \varrho g \nu^{0.189} b z_2 l}{d_i^{4.81}} . \qquad 7.1\text{—}13$$

The annual operating cost is

$$B_1 = P c e f ,$$

where c is operating time over the year, in s; e is specific power consumption (J/J for electric motors and kg/J for internal-combustion engines); f is the price of power or fuel (in Ft/J or Ft/kg). — Using the symbols introduced when deriving Eq. 7.1—13,

$$B_1 = 1.53 \times 10^{-2} \frac{q^{2.81} \varrho g \nu^{0.189} c e f l}{d_i^{4.81}} \qquad 7.1\text{—}14$$

The annual cost of maintenance, B_2, may be considered constant. — Let us replace the terms on the right-hand side of Eq. 7.1—11 by the above-defined expressions, and determine the derivative $\mathrm{d}K/\mathrm{d}(d_i)$. Putting the results equal to zero, and solving for the diameter, we get by the rules of calculus the pipe size that ensures the least cost of transportation,

$$d_i = \sqrt{3.18 \times 10^{-3} \frac{q^{2.81} \varrho g \nu^{0.189} (b z_2 + c e f)}{a_2 z_1 \eta}} . \qquad 7.1\text{—}15$$

* Ft is the abbreviation for Forint, the Hungarian monetary unit.

In calculating optimal pipe size we have so far tacitly assumed the pipeline to operate without interruption, that is, at 100 percent exploitation. In reality, transportation is most often intermittent to a greater or lesser degree. The pauses in transportation provide the flexibility needed to handle excess output due to non-uniform production, downtimes, surges, etc. on the input side. A lower exploitation efficiency will, however, augment the unit cost of transportation. — Transportation costs may be analysed for various (constant) pipe sizes and variable throughput. Let us replace by the

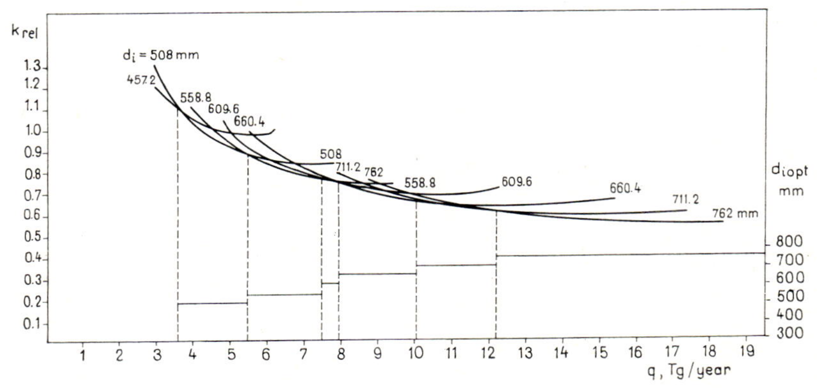

Fig. 7.1—7 Optimum pipe size v. throughput, after Cabet (1966)

above-defined expressions the terms on the right-hand side of Eq. 7.1—11. Let us contract all elements except d_i and q into constants denoted C; then,

$$K = C_1 + C_2 d_i + C_3 d_i^{-4.81} q^{2.81}.$$

Let the length of the line be one km, and let us divide both sides of the equation by the annual oil throughput, $C_4 q \varrho$. Let $K/C_4 q \varrho = k$ denote the unit cost of transportation in Ft/kg km: then,

$$k = C_4 q^{-1} + C_5 d_i q^{-1} + C_6 d_i^{-4.81} q^{1.81}. \qquad 7.1\text{—}16$$

This relationship defines a family of curves $k = f(q)_{di}$. The family in Fig. 7.1—7 has been plotted using a similar relationship. The ordinate is calibrated in terms of a relative specific transportation cost k_{rel} referred to some base value (derived from a diagram in Cabet 1966). In the Figure, the optimal sizes in the individual throughput ranges are indicated by separate lines. Clearly, the greater the desired throughput, the larger the optimum pipe size, and the less the unit transportation cost. This family of curves cannot, of course, be used directly in design work, as the constants in Eq. 7.1—16 will differ from one case to the next.

Table 7.1—1 lists after R. Cabet (1966) the main cost components of the Le Havre—Paris pipeline built in 1964. Depreciation and debt servicing are seen to be substantial. Depreciation is in principle the annual decrease in the value of an installation. Book depreciation and actual depreciation often do not agree. Operating companies prefer book depreciation to be as high as possible. In France for instance, the period of amortization prescribed by law is 30 years for buildings, 20—25 years for pipe, 10 years for pumps, motors and valves, and 5 years for automation and control equipment. In reality, however, since the introduction of cathodic protection, pipes have practically unlimited life. Even so, although the actual life of equipment and fittings may be longer than the amortization periods mentioned above, these periods may nevertheless be technologically justified, as the progress of technology may make obsolete the equipment in use by putting on the market equipment and fittings more advanced both technically and economically. The relative depreciation (moral wear) of the existing installations will thus be accelerated.

In the above calculations we have used what is called the technical cost, which is suitable for comparing various solutions in order to find the most favourable one. Actual cost is invariably higher owing to extraneous cost components such as overheads, debt servicing, etc. that are superimposed on all variants. The taking into account of these is justified only if pipeline transportation is compared to some different means of transport such as tank barges on a river. The importance of debt servicing is revealed by Table 7.1—1. — Cost components generated by such a project in the USA are usefully summed up in White (1969).

Table 7.1—1
Transportation cost components of the Le Havre—Paris oil pipeline
(after Cabet 1966)

Cost components	10^6 Francs	%
Depreciation	6,534	30.7
Debt servicing	2,830	13.3
Operating cost	11,931	56.0
Payroll	4,035	18.9
Power	2,850	13.4
Maintenance	1,688	7.9
Telecommunications	275	1.3
Others	3,083	14.5
Total	21,295	100

7.2. Non-isothermal flow

If oil viscosity is comparatively high and the temperature of the flowing oil differs appreciably from that of the line's environment, flow can no longer be regarded as isothermal. Pipelines transporting oil are most often buried in the ground. The thermal behaviour of the soil will accordingly affect the temperature of the flowing oil.

7.2.1. Thermal properties of soils

The temperature of soil of a given nature, undisturbed by human hand, is determined on the one hand by insolation, and on the other, by heat flowing from the interior of the Earth towards its surface. Insolation has a double period, a daily and a yearly one. The internal heat flow, due primarily to radioactive decay, is constant in a fair approximation at any geographical location. Its magnitude is less by about two orders of magnitude, in the depth range of 1—2 metres we are concerned with, than the effects of insolation, and we shall accordingly disregard it in the sequel. Assuming the soil to be thermally homogeneous, temperature change per unit time is described by Fourier's differential equation

$$\frac{\partial T}{\partial t} = a_s \nabla^2 T \qquad 7.2-1$$

where a_s is the temperature-distribution or diffusion factor defined by the relationship

$$\partial_s = \frac{\lambda_s}{c_s \varrho_s}. \qquad 7.2-2$$

Assuming the surface to be level in an approximation sufficient for our purposes, the above equation describing temperature variations may be used satisfactorily also in its one-dimensional form,

$$-\frac{\partial T}{\partial t} = a_s \frac{\partial^2 T}{\partial h^2}. \qquad 7.2-3$$

Let us characterize the temperature change on the ground surface, that is, at $h = O$ by the relationship

$$\Delta T_O = \Delta T_{Oa} \sin \omega t$$

where $\omega = 2\pi/t_p$. The other symbols are identified in Fig. 7.2—1. Under physically meaningful initial conditions, Eq. 7.2—3 has the solution:

$$\Delta T_h = \Delta T_{Oa} \exp\left[-h\sqrt{\frac{\pi}{a_s t_p}}\right] \sin\left[\frac{2\pi t}{t_p} - h\sqrt{\frac{\pi}{a_s t_p}}\right]. \qquad 7.2-4$$

The extreme temperature fluctuation ΔT_{ha} at depth h arises when the sine term equals unity, that is,

$$\Delta T_{ha} = \Delta T_{Oa} \exp\left[-\sqrt{\frac{h^2 \pi}{a_s t_p}}\right]. \qquad 7.2-5$$

This equation and Fig. 7.2—1 reveal how the amplitude decreases with depth. Daily fluctuation is almost imperceptible at 1 m depth, being less than 0.1 °C in most cases. The yearly wave will cause measurable temperature changes to depths of 25—30 m, depending on the thermal properties of the ground. Beneath a depth of 30 m, temperature will be controlled solely by the terrestrial heat flow. Figure 7.2—1 further reveals that the temperature wave *II* at depth h is displaced in phase against the ground-surface wave. Phase shift is described by

$$\Delta t_p = \sqrt{\frac{t_p h^2}{4\pi a_s}}. \qquad 7.2\text{—}6$$

Example 7.2—1. Find the extremes of the daily temperature fluctuation at 0.3 and 1.0 m depth, and the phase shift at 0.3 and 1.0 m depth if $a =$

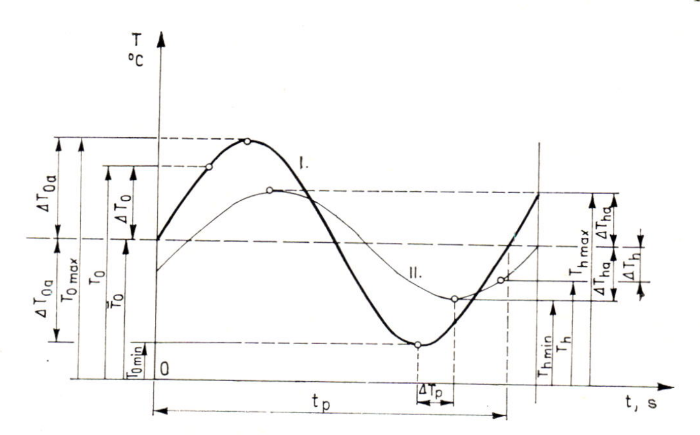

Fig. 7.2—1 Fluctuation in time of soil temperature

4.9×10^{-7} m²/s, and surface temperature varies from +22 to +2 °C. — In the notation of Fig. 7.2—1, surface amplitude is

$$\Delta T_{Oa} = \frac{T_{O\max} - T_{O\min}}{2} = \frac{22 - 2}{2} = 10 \text{ °C}.$$

By Eq .7.2—5, the amplitude at depth 0.3 m is

$$T_{ha} = 10 \exp\left[-\sqrt{\frac{0.3^2 \pi}{4.9 \times 10^{-7} \times 86{,}400}}\right] = \pm 0.7 \text{ °C}$$

whereas at depth 1.0 m it is

$$T_{ha} = \pm 0.0016 \text{ °C}.$$

The mean temperature is, in a fair approximation,

$$\bar{T}_h = \bar{T}_O = T_{O\min} + \Delta T_{Oa} = 2 + 10 = 12 \text{ °C}.$$

Temperature extremes at 0.3 m depth are

$$T_{h\max} = \bar{T}_h + \Delta T_{ha} = 12 + 0.7 = 12.7 \text{ °C}$$

$$T_{h\min} = \bar{T}_h - \Delta T_{ha} = 12 - 0.7 = 11.3 \text{ °C}$$

at 1.0 m depth, they are

$$T_{h\max} = 12.002\ °C \quad \text{and} \quad T_{h\min} = 11.988\ °C$$

that is, the fluctuation range at 1.0 m depth is evanescent. Phase shift at 0.3 depth is stated by Eq. 7.2—6 to be

$$\Delta t_p = \sqrt{\frac{86{,}400 \times 0.3^2}{4 \times \pi \times 4.9 \times 10^{-7}}} = 35.5 \times 10^3\ \text{s} = 9.9\ \text{h}.$$

Thermal conductivity of the soil depends on the conductivity of the soil matrix, on grain size distribution, on the bulk density of the dry soil, and on humidity. Makowski and Mochlinski, using an equation of Gemant, have developed a relationship describing the thermal conductivity of humid soil (Davenport and Conti 1971). They considered the dry soil substance to be made up of two grain size fractions: sand (0.002 — 2 mm particle size) and clay (less than 0.002 mm particle size). Let the weight percent of clay be S_c and the weight percent of humidity referred to the total dry substance be S_w. The thermal conductivity of the humid soil is, then,

$$\lambda_w = (A \lg S_w + B)\, 10^C \tag{7.2—7}$$

where $A = 0.1424 - 0.000465\, S_c$; $B = 0.0419 - 0.000313\, S_c$; $C = 6.24 \times \times 10^{-4}\, \varrho_{sd}$.

The determination of the temperature distribution factor requires knowledge of the bulk density and specific heat of the soil. In humid soil,

$$\varrho_{sw} = \varrho_{sd}(1 + 0.01\, S_w) \tag{7.2—8}$$

and

$$c_{sw} = \frac{c_{sd} + 0.01\, S_w c_w}{1 + 0.01\, S_w} = \frac{c_{sd} + 41.9\, S_w}{1 + 0.01\, S_w}. \tag{7.2—9}$$

The simplification is based on the fact that the specific heat of water, $c_w = 1$ kcal/kg K $= 4187$ J/kg K. — Introducing the factors defined by the above equations into Eq. 7.2—2, we get for humid soil

$$a_{sw} = \frac{\lambda_{sw}}{\varrho_{sw} c_{sw}} = \frac{A(\lg S_w + B)}{\varrho_{sd}(c_{sd} + 41.9\, S_w)}\, 10^C. \tag{7.2—10}$$

Figure 7.2—2 is the diagram of Makowski and Mochlinski (in Davenport and Conti 1971) based on Eq. 7.2—7, and transposed into the SI system of units. It gives the thermal conductivity of humid soil v. percent humidity for sands and clays of different bulk density. Figure 7.2—3 (from the same source with modification), shows heat diffusivities v. percent humidity, likewise for sand and clays of different bulk density.

Various methods have been devised for determining the in-situ thermal conductivity of undisturbed soil. Of these, we shall outline the transient needle-probe method after Makowski and Mochlinski (1956). The transient needle probe is an electrically heated device which can in effect be regarded as a linear source of heat. It is inserted into a hole drilled in the soil at the place where the measurement is to be performed, and heated electrically so as to generate a heat flow of constant magnitude. Two thermocouples are attached to the outer, metallic sheath of the probe. They are used to establish its surface temperature at frequent intervals. It can be shown that if heat flow is constant, the temperature of the soil (on the outside of the metal sheath) at a distance equalling probe diameter r from the linear heat source is given after the elapse of time t by

$$T = \frac{\Phi^*}{4\pi \lambda_s}\left[-\mathrm{EI}\left(\frac{r^2}{4a_s t}\right)\right].$$

If $r^2/4a_s t$ is low enough, that is, if t is large as referred to r^2/a_s, then

$$\frac{1}{\lambda_s} = \frac{4\pi}{\Phi^*} \frac{T_2 - T_1}{\lg \frac{t_2}{t_1}} \qquad 7.2-11$$

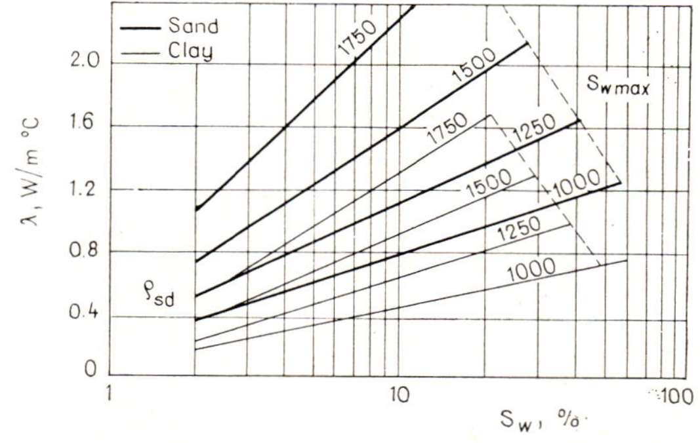

Fig. 7.2-2 Influence of water content upon thermal conductivity of sand and clay, after Davenport and Conti (1971)

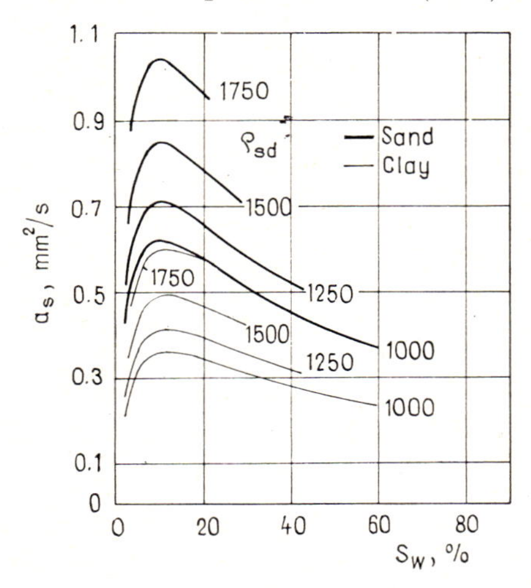

Fig. 7.2-3 Influence of water content upon temperature distribution factors of sand and clay, after Davenport and Conti (1971)

that is, if heat flow Φ^* is constant, temperature varies as the logarithm of time. Plotting this relationship on log paper one obtains a straight line of slope $\Phi^*/4\pi\lambda_s$, whence thermal conductivity can be calculated if Φ^* is known. In reality, steady heat flow will take some time to set in and the relationship will become linear after some time only. This time is, however, short enough (of the order of 10 s). In pipeline-industry practice, the equation is used in a modified form which takes into account that the thermal properties of the probe differ from those of the soil, and contains the electric current generating the heat flow — a directly measurable quantity — rather than the heat flow itself. The depth of penetration of the heat flow into the soil is a mere 20—30 mm, and so the measurement is to be repeated a number of times at any location in order to provide a meaningful mean value.

Thermal conductivity in the undisturbed soil is a function of instantaneous humidity, which is affected by the weather. Accordingly, it is necessary to take soil samples simultaneously with the probe measurements, and to determine on these the humidity of the soil, the bulk density of the dry substance, the clay-to-sand ratio and the specific heat. In the knowledge of these data and using Eqs 7.2—7 and 7.1—11, the thermal conductivity and diffusivity of the soil may be extrapolated to any humidity (Davenport and Conti 1971). A longer-term average diffusivity may be determined recording a full temperature wave on the surface and simultaneously at a depth of h m below ground. Both Eqs 7.2—5 and 7.2—6 are suitable for calculating a_s in the knowledge of temperature period t_p and amplitude ratio $\Delta T_{ha}/\Delta T_{0a}$ and phase shift Δt_p. — The temperature wave may be recorded over a full year, or over the duration of a shorter 'harmonic' wave.

7.2.2. Temperature of oil in steady-state flow, in buried pipelines

The flow temperature of oil injected into a pipeline will usually differ from soil temperature, the soil being mostly cooler than the oil. Assuming the heat flow pattern to be steady-state in and around the pipeline, the variation of axial oil temperature in the pipe can be determined along the pipeline as follows. Part of the potential energy of oil flowing in the pipeline, transformed into heat, will warm the oil. A relative increase in temperature will result also from the solid components' separating out of the oil, which is an exothermic process. Oil temperature is reduced, on the other hand, by the transfer of heat from the pipeline into the lower-temperature environment.

In calculating the heat generated by friction let us assume the friction gradient along the pipeline to be constant on an average. A pressure-force differential $\Delta p d_i^2 \pi/4$ along a length l of pipe generates heat in the amount $Q = \Delta p d_i^2 \pi l/4$. Neglecting the reduction in flow rate due to the increasing density of the cooling liquid, we have $l = vt$ and

$$v = \frac{q}{\frac{d_i^2 \pi}{4}}.$$

Using this relationship we may write up the heat generated by friction per unit of time over unit distance of flow:

$$\Phi^* = \frac{Q}{lt} = \frac{\Delta p}{l} q \text{ [W/m]}. \qquad 7.2\text{—}12$$

Let us assume that a temperature drop of 1 °C = 1 K causes the solidification of ε kg of solids (to be called paraffin for simplicity) per kg of oil. Let the heat liberated by the solidification of 1 kg of paraffin be $\varkappa$, in which case a temperature drop of 1 °C will liberate heat in the amount of $q\varrho\varepsilon\varkappa$, W/K in a liquid of gravity ϱ flowing at a rate q per second; ε and $\varkappa$ are temperature-dependent, but we shall assume their mean values to be constant over the temperature range we are concerned with.

Let k denote the heat flow per unit of time into the soil from a unit length of pipe per unit temperature differential, if the whole differential is $T_f - T_s$ (where T_f is the axial temperature of the oil in the pipeline, and T_s is the original soil temperature at the same depth). k is the heat transfer coefficient per unit length of pipe, which we assume to be constant all along the line. The temperature of flowing oil will decrease by $\mathrm{d}T_f$ over a length $\mathrm{d}l$. The change in heat content of liquid flowing at the rate q under the influence of this temperature difference equals the algebraic sum of the heat generated, on the one hand, and the heat lost to the environment, on the other, over an infinitesimal length of pipe $\mathrm{d}l$, that is,

$$q\varrho c\, \mathrm{d}T_f = \Phi^* \mathrm{d}l + \varepsilon q \varrho \varkappa\, \mathrm{d}T_f - k(T_f - T_s)\mathrm{d}l \qquad 7.2\text{—}13$$

or, in a different formulation,

$$q\varrho c\, \mathrm{d}T_f - \varepsilon q\varrho\varkappa\, \mathrm{d}T_f = \Phi^* \mathrm{d}l - kT_f \mathrm{d}l + kT_s \mathrm{d}l.$$

Let

$$q\varrho c - \varepsilon q\varrho\varkappa = A \quad \text{and} \quad \Phi^* + kT_s = B.$$

Rearranging and writing up the integration, we have

$$\int \frac{A}{B - kT_f} \mathrm{d}T_f = \int \mathrm{d}l.$$

Solving this equation under the initial conditions $l = 0$, $T = T_1$, and resubstituting the values of A and B, the flow temperature at the end of a pipeline of length l turns out to be

$$T_{f2} = T_s + \left(T_{f1} - T_s - \frac{\Phi^*}{k}\right) \exp - \left[\frac{kl}{q\varrho(c + \varepsilon\varkappa)}\right] + \frac{\Phi^*}{k}. \qquad 7.2\text{—}14$$

Let us point out that oil temperature will vary also radially in any cross-section of the pipeline. In turbulent flow, radial decrease is slight, but it may be significant in laminar flow. Figure 7.2—4 shows one of the possible combinations of flow-velocity and temperature-distribution profiles after Chernikin (1958). Clearly, in this case the axial temperature T_f of the oil is somewhat higher than the average temperature.

Example 7.2—2. Let the *ID* of the pipeline be $d_i = 0.2$ m, and its length $l = 20$ km. Oil of gravity $\varrho = 850$ kg/m³ flows at a steady rate of $q = 100$ m³/h through the pipeline. Injection temperature, $T_{f1} = 50$ °C; soil temperature, $T_s = 2$ °C; $\Phi^* = 7$ W/m; $\varepsilon = 3.0 \times 10^{-3}$ 1/°C; $\bar{c} = 1900$ J/(kg °C); $\bar{\varkappa} = 2.3 \times 10^5$ J/kg; $k = 2$ W/(m °C). Find the tail-end temperature T_{f2} of the oil (i) assuming that paraffin deposition takes place over the entire

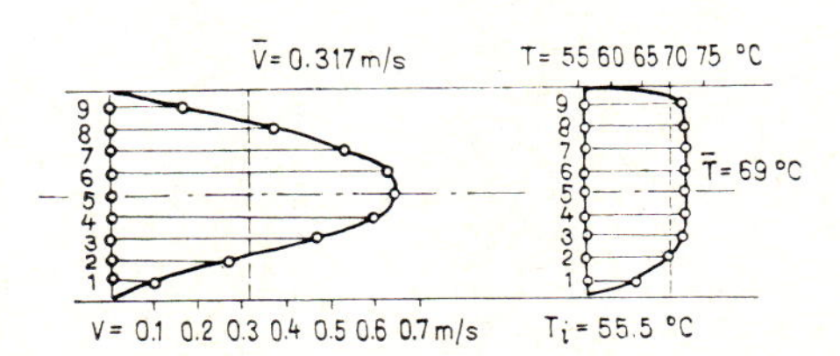

Fig. 7.2—4 Rate-of-flow and temperature traverses of oil flowing in a pipeline, after Chernikin (1958)

temperature range of flow, (ii) assuming that paraffin deposition is negligible, and (iii) neglecting both the effects of paraffin deposition and friction loss.

(i) By Eq. 7.2—14,

$$T_{f2} = 2 + \left(50 - 2 - \frac{7}{2}\right) \times$$

$$\times \exp\left[-\frac{2 \times 2 \times 10^4}{0.0278 \times 850(1900 + 3.0 \times 10^{-3} \times 2.3 \times 10^5)}\right] + \frac{7}{2} = 28.6 \text{ °C} .$$

(ii) Assuming that $\varepsilon = 0$,

$$T_{f2} = 2 + \left(50 - 2 - \frac{7}{2}\right) \exp\left[-\frac{2 \times 2 \times 10^4}{0.0278 \times 850 \times 1900}\right] + \frac{7}{2} = 23.7 \text{ °C}.$$

(iii) Assuming that $\varepsilon = 0$ and $\Phi^* = 0$,

$$T_{f2} = 2 + (50 - 2) \exp\left[-\frac{2 \times 2 \times 10^4}{0.0278 \times 850 \times 1900}\right] = 21.7 \text{ °C}.$$

The heat of solidification of paraffin (Chernikin 1958) is rarely taken into account in practical calculations. It may, however, be of considerable importance, as revealed by Example 7.2—2. Figure 7.2—5 is a diagram of heats of solidification v. melting temperature (after Grosse 1951) for various paraffins and other high-molecular-weight hydrocarbons. The latent heat of the components forming deposits in oil may, according to the data in the Figure, vary rather considerably depending on solid-phase composition. This enhances the importance of knowing the composition and/or thermal behaviour of the components solidifying within the pipeline's temperature range.

As a preliminary indication whether or not the heat generated by friction should be taken into account, the following rule of thumb may be helpful:

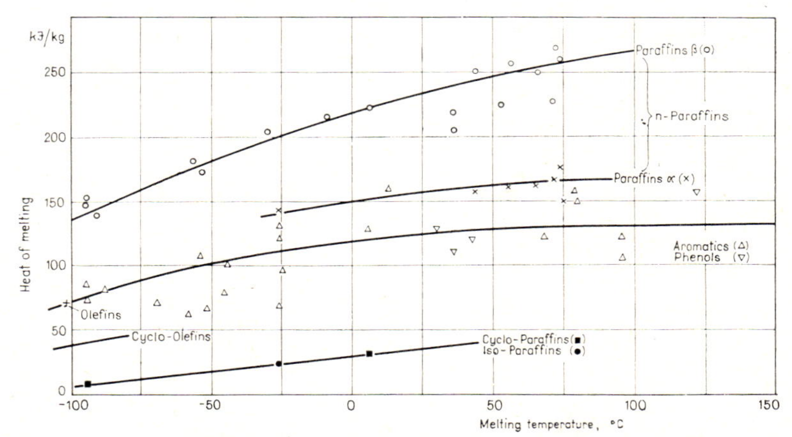

Fig. 7.2—5 Melting heats of solid hydrocarbons, after Grosse 1951 (used with permission of VDI-Verlag GmbH, Düsseldorf)

heat generated by 100 bars of friction loss will raise the tail-end temperature of oil flowing in the pipeline by round 4 °C, irrespective of pipe size and throughput (Haddenhorst 1962).

Equation 7.2—14 is most often used in the simplified form called Sukhov's equation in Soviet literature; it assumes ε and Φ^* to be both zero:

$$T_{f2} = T_s + (T_{f1} - T_s) \exp\left(\frac{kl}{q \varrho c}\right). \qquad 7.2\text{—}15$$

7.2.3. The heat-transfer coefficient

In practice, the heat-transfer coefficient is used in two forms. The factor k^* expressed in W/(m^2.K) is heat flow from unit surface of pipe into the ground for unity temperature differential; k, in W/(m.K) units, is the same for unit length rather than unit surface of pipe. The two factors are related by

$$k = \pi d k^*. \tag{7.2—16}$$

In pipelining practice, pipes are usually buried in the ground, laid in a ditch dug for the purpose, and covered with backfill. Assuming the soil to be homogeneous as to heat conductivity, the heat flow pattern for steady-state flow is described by the stream-lines and the orthogonal set of isotherms shown in Fig. 7.2—6. Next to the pipe, isotherms are circular in a fair approximation, with their centres the lower down on the vertical axis,

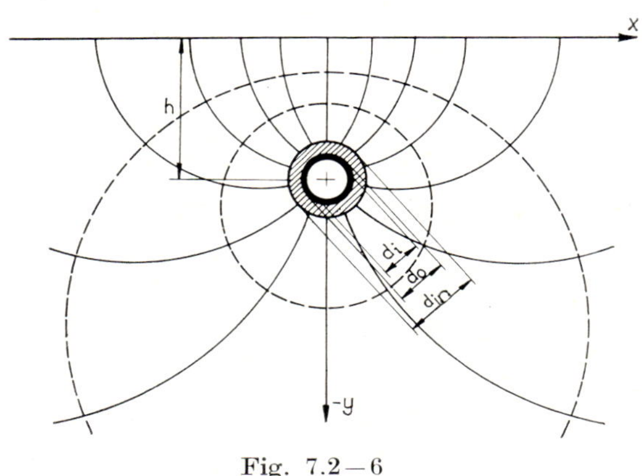

Fig. 7.2—6

the lower the temperature that they represent. In the case illustrated, and assuming that heat retention by the steel wall of the pipe is negligible, the heat transfer factor is

$$k = \frac{\pi}{\dfrac{1}{\alpha_1 d_i} + \dfrac{1}{2\lambda_{\text{in}}} \ln \dfrac{d_{\text{in}}}{d_o} + \dfrac{1}{\alpha_2 d_{\text{in}}}}. \tag{7.2—17}$$

The first term in the denominator describes internal convective thermal resistance, the second the effect of the thermal insulation applied to the pipe, and the third the thermal insulation by the soil itself.

The first term in the denominator of Eq. 7.2—17, the internal-convection factor α_1, may be determined by one of several relationships. It equals the heat flow per unit wall surface, per unit temperature differential, from the pipe axis at temperature T_i to the pipe wall at temperature T_f. The relationships used are found to have the common property of starting from the Nusselt factor N_{Nu}:

$$\alpha_1 = N_{\text{Nu}} \frac{\lambda_f}{d_i}. \tag{7.2—18}$$

Heat transfer by convection has three typical cases, each of which corresponds to a zone of flow (Dobrinescu and Bulau 1969). These are: the turbulent zone at N_{Re} above 10^4, the transition zone for N_{Re} between 2300 and 10^4, and the laminar zone for N_{Re} below 2300. In turbulent flow, according to Sieder and Tate,

$$N_{\mathrm{Nu}} = 0.027\, N_{\mathrm{Re}}^{0.8}\, N_{Pr}^{1/3} \left(\frac{\mu_f}{\mu_i}\right)^{0.14} \qquad 7.2\text{—}19$$

where μ_f is the dynamic viscosity of oil at the pipe axis and μ_i is the same at the pipe wall.

For the zone of transition, Ramm proposes the following, extended formula:

$$N_{\mathrm{Nu}} = 0.027\, N_{\mathrm{Re}}^{0.8}\, N_{\mathrm{Pr}}^{1/3} \left(\frac{\mu_f}{\mu_i}\right)^{0.14} \left(1 - \frac{6\times 10^5}{N_{\mathrm{Re}}^{1.8}}\right). \qquad 7.2\text{—}20$$

In the turbulent and transition zones, heat transfer by convection is rather significant, and so temperature differences between the pipe's axis and wall are not usually taken into account in these two zones. Heat transfer by convection is significantly less in laminar flow; moreover, in liquid undergoing a forced motion, free convection currents will also form, which will modify the velocity profile. This modification is more significant in vertical than in horizontal flow strings, though. It is most expediently characterized in horizontal pipelines by the product of the Prandtl and Grashof numbers (Ford 1955):

$$N_{\mathrm{Gr}}\, N_{\mathrm{Pr}} = \frac{d_i^3 c \varrho g \beta_T (T_f - T_i)}{\nu \lambda_f}. \qquad 7.2\text{—}21$$

For laminar flow, N_{Nu} v. the product $N_{\mathrm{Gr}} N_{\mathrm{Pr}}$ can be determined using the Gill—Russell relationship illustrated by Fig. 7.2—7 (Ford 1955). If the abscissa is greater than 5×10^4, then

$$N_{\mathrm{Nu}} = 0.184 (N_{\mathrm{Gr}} N_{\mathrm{Pr}})^{0.32}. \qquad 7.2\text{—}22$$

Of the factors figuring in Eqs 7.2—21 and 7.2—22, specific heat c and

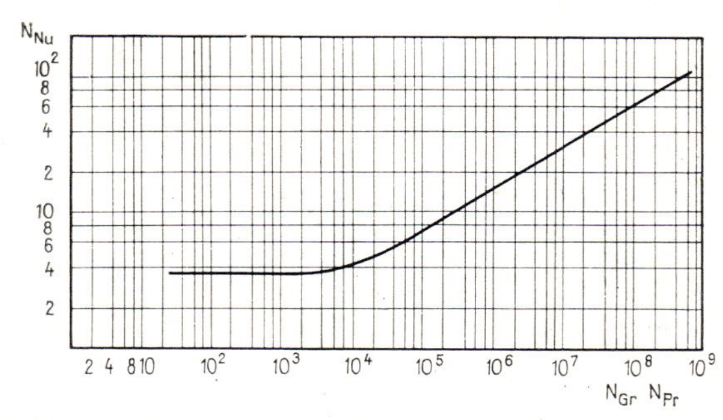

Fig. 7.2—7 Relationship $N_{\mathrm{Nu}} = f(N_{\mathrm{Gr}} N_{\mathrm{Pr}})$ for finding α_1, according to Gill and Russell; after Ford (1955)

thermal conductivity λ_f can be calculated using Cragoe's formula transformed into the SI system, by Dobrinescu and Bulau (1969):

$$c = \frac{762.5 + 3.38\,T}{\sqrt{\varrho_4^{20}}} \qquad 7.2-23$$

$$\beta_T = \frac{1}{2583 - 6340\,\varrho_4^{20} + 5965(\varrho_4^{20})^2 - T} \qquad 7.2-24$$

$$\lambda_f = \frac{0.134 - 6.31\times 10^{-5}T}{\varrho_4^{20}}\,. \qquad 7.2-25$$

Furthermore,

$$\varrho_T = \varrho_n - \alpha_T(T - T_n) + \alpha_p\,p. \qquad 7.2-26$$

The physical state of the oil would, at a first glance, be described in terms of the average temperature between the head and tail ends of the pipeline section. According to Orlicek and Pöll (1955), however, heat transfer depends primarily on the nature and state of the liquid film adhering to the pipe wall. The correct procedure would consequently be to determine the physical parameters of the oil at the mean film temperature. No relationship suitable for calculating this temperature has so far been established, however. For practice it seems best to establish parameters for the mean wall temperature of the pipeline section examined. — Viscosity depends markedly on temperature, less so on pressure in the range of oil transmission in pipelines. The $\nu = \mathrm{f}(p, T)$ graphs state results of laboratory measurements. The temperature-dependence of viscosity in Newtonian liquids is best described by Walther's equation

$$\lg\lg(10^6\nu + a) = b + c\lg T$$

a, b and c are constants depending on the nature of the oil; a ranges from 0.70 to 0.95: putting it equal to 0.8 gives an accuracy sufficient for practical purposes. Hence,

$$\lg\lg(10^6\nu + 0.8) = b + c\lg T. \qquad 7.2-27$$

This equation implies that oil viscosity v. temperature will be represented by a straight line in an orthogonal system of coordinates whose abscissa (T) axis is logarithmic and whose ordinate (viscosity) axis is doubly logarithmic. This is the Walther—Ubbelohde diagram, whose variant conformable to Hungarian standard MNOSz—3258—53 is shown as Fig. 7.2—8. In principle, two viscosity values determined at two different temperatures, plotted in this diagram, define a straight line representing the viscosity v. temperature function. In practice, it is indicated to measure viscosity at three temperatures at least, and to use the straight lines joining the three points plotted merely as an aid in interpolating between plots. It should be ascertained whether or not the oil exhibits anomalous flow behaviour at the lowest temperature of measurement, or evaporation losses at the highest. — Pressure tends to increase viscosity somewhat. According to Dobrinescu and Bulau (1969), dynamic viscosity at pressure p is described by

$$\mu_p = \mu_a(0.9789 + 0.0261\ \varrho_4^{20})^{1.02p}. \qquad 7.2-28$$

This relationship gives a 13-percent increase in the viscosity of a 900 kg/m³ density oil between atmospheric pressure and $p = 50$ bars.

The second term in the denominator of Eq. 7.2—17 is proportional to the thermal resistance of the heat insulation about the pipe; in addition, to the nature of the insulator matrix, the thermal conductivity of the material as

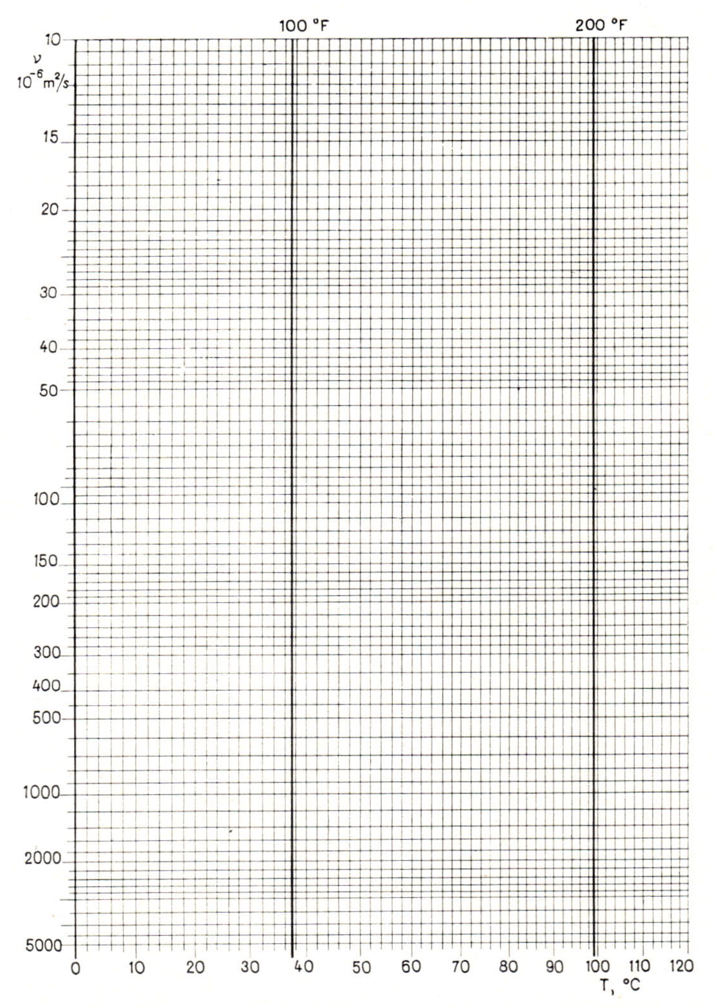

Fig. 7.2—8 Walther—Ubbelohde grid according to Hungarian Standard MNOSZ-3258—53

it is dependent primarily on pore volume and pore distribution. It is primarily the air in the pores that insulates. The thermal conductivity coefficient of air at room temperature (λ_a) is as low as 0.023 W/(m. K). If temperature differences between different points of the hole wall make the air in the pores form conduction currents, then heat will flow through the pores by free convection in addition to conduction. The less the mean pore size, the less significant is convection. The resultant thermal conductivity of 'series-connected' small pores and the pore walls separating them is less than it would be for the same overall pore volume if the pores were larger. On the other hand, the smaller the pores in a given matrix, the less the compressive strength of the insulator. Low-strength insulators will be crushed or pulverized by comparatively weak forces. Hence, reducing pore size in a given matrix is limited by the minimum requirement as to compressive strength. A material is a good thermal insulator if its conductivity is less than 0.08 W/(m. K); if it contains no substance that corrodes the pipe; if it is chemically inert vis-à-vis air and water; if its compressive strength exceeds a prescribed minimum; if it keeps its properties over long periods of time; and if it can be fitted snugly against the pipe wall. — Let us point out that water infiltrated into the pores will considerably impair insulation, since the thermal conductivity of water at room temperature is about 0.58 W/(m. K), that is, about 25 times that of air. The exclusion of water from the insulator pores must therefore be ensured. This is achieved partly by choosing insulators with closed, unconnected pores, and partly by pro-

Table 7.2—1

Parameters of pipeline insulators (after Balcke 1949)

Insulator	ϱ_{in} kg/m³	λ_{in}; W/mK					σ_z bars
		0	20	50	100	150	
		at mean temperature °C					
Slag wool	225	0.042	0.043	0.047	0.053	0.060	—
covered with galvanized sheet	300	0.045	0.047	0.051	0.057	0.064	6
Glass wool	100	0.034	0.036	0.039	0.049	0.058	—
covered with galvanized sheet	375	0.035	0.037	0.041	0.050	0.059	6
Burnt diatomaceous-earth bricks							
light	370	0.070	0.072	0.075	0.081	0.088	3—4
medium	480	0.087	0.089	0.093	0.100	0.107	4—7
heavy	650	0.116	0.120	0.127	0.139	0.145	12—20
Burnt diatomaceous-earth paste*							
light	350	0.063	0.064	0.066	0.070	0.073	2—3
medium	500	0.075	0.078	0.081	0.085	0.088	2—3
heavy	700	0.104	0.108	0.116	0.125	0.129	3—4

* To be applied to hot pipeline only

viding a waterproof coating. — The main parameters of insulators most popular in earlier practice are listed in Table 7.2—1. The materials figuring in the Table are open-pored without exception, and must therefore be provided with a waterproof coating in every application. In the last few years, the use of polyurethane foam is on the increase. Shell Australia Ltd. laid a 8-in. pipeline 56 km long, insulated with 51-mm thick polyurethane foam wound with polyethylene foil (Thomas 1965). Polyurethane foam is extremely porous (about 90 percent porosity) and light (about 34 kg/m^3). Pores are unconnected and filled with freon. Its heat conductivity is 0.016—0.021 W/(m. K). It can be used up to a temperature of 107° C. Insulation is usually applied in the pipe mill, but ends are left bare so as not to hamper welding. After welding, joins are insulated on-site with polyurethane foam, over which a prefabricated, pre-warmed polyethylene sleeve is slipped. On cooling, the sleeve contracts and provides a watertight seal. — Soviet pipeliners use a device that permits the on-site insulation with polyurethane foam of welded-together pipeline sections. The pipe, warmed to about 60 °C, is drawn through a foam-spraying chamber. Correct size is ensured by a template fixed on the chamber wall (Zeinalov et al. 1968). High-grade heat insulation is expensive; it costs approximately as much again as the pipeline itself (Gautier 1970).

The third term in the denominator of Eq. 7.2—17 describes the thermal resistance of soil surrounding the pipeline. Assuming the original undisturbed temperature of the soil at pipeline depth to be T_s, and the ground-surface temperature to be equal to this value, it can be shown that

$$\alpha_2 = \frac{2\lambda_s}{d_{in}\,\mathrm{ch}^{-1}\left(\frac{2h}{d_{in}}\right)}. \qquad 7.2\text{—}29$$

The approximation

$$\alpha_2 = \frac{2\lambda_s}{d_{in}\ln\frac{4h}{d_{in}}} \qquad 7.2\text{—}30$$

will be the better, the greater is h/d_{in}. In most practical cases, that is, at about $h/d_{in} = 2$, the two formulae will furnish closely agreeing results; d_{in} denotes the *OD* of the pipeline as installed, together with insulation and coatings. In the absence of such, $d_{in} = d_o$, the *OD* of the pipe. — Determining the heat conductivity of the soil was discussed in Section 7.2—1. — In determining the heat transfer factor k valid when the pipeline is operative, one should keep in mind that the porosity of the backfill is greater than that of the undisturbed soil, and that its thermal conductivity is consequently lower. A procedure accounting for this circumstance has been devised by Ford, Ells and Russell (Davenport and Conti 1971). The backfill will compact with time, and the effective heat-transfer factor will slightly increase in consequence. The heat transfer factor is affected by the wind, the plant cover, snow if any, soil moisture and the state of aggregation of water. Under a temperate climate, soil frost will rarely enter into consideration, but in permafrost regions it may play an important role. The way wind reduces k, or the plant or snow cover increase it, are too little known as yet to be evaluated in quantitative terms. The influence of soil moisture is known, however (it was discussed in Section 7.2—1). It is accordingly to be expected that the heat transfer factor will vary over the year as a function of soil moisture, influenced in its turn by the weather.

Figure 7.2—9 is a plot of k^* factors measured over three years along three sections of the 'Friendship' pipeline. It is seen that the factor is lowest in June, July and August, the hot dry months of a Northern-

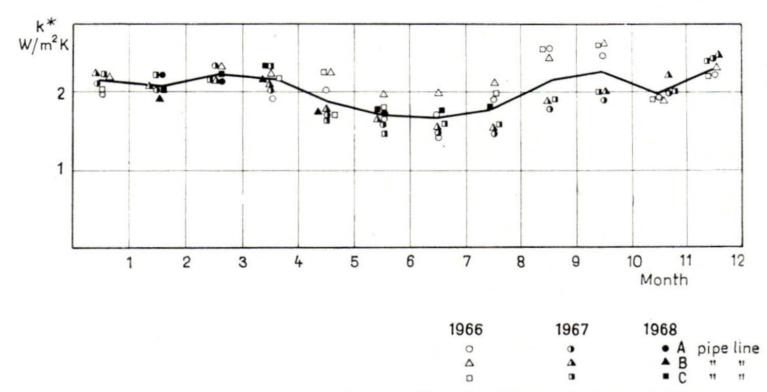

Fig. 7.2—9 Heat transfer factor v. time, after Rabinovitch and Kuznetsov (1970)

hemisphere continental climate. — If the heat transfer factor k is to be determined by calculation, and flow is laminar, then the computation of α_1 requires the knowledge of the inside wall temperature T_i, too (cf. Eq. 7.2—21). Under steady conditions, heat flow from a one-metre pipeline section into the soil is

$$\Phi^* = k(T_f - T_s) = \alpha_1 d_i \pi(T_f - T_i) = \frac{2\pi\lambda_{in}(T_o - T_{in})}{\ln\dfrac{d_{in}}{d_o}} = \alpha_2 d_i \pi(T_{in} - T_s). \qquad 7.2\text{—}31$$

The inside and outside wall temperatures of steel pipe are taken to be equal, that is, $T_i = T_o$. The multiple equation No. 7.2—31 may be decomposed into three independent equations:

$$k(T_f - T_s) = d_i \pi \alpha_1 (T_f - T_i) \qquad 7.2\text{—}32$$

$$d_i \pi \alpha_1 (T_f - T_i) = \frac{2\lambda_n(T_i - T_{in})}{\ln\dfrac{d_{in}}{d_o}} \qquad 7.2\text{—}33$$

$$d_i \pi \alpha_1 (T_f - T_i) = d_{in} \pi \alpha_2 (T_{in} - T_s). \qquad 7.2\text{—}34$$

The relationship needed to determine T_i is Eq. 7.2—32. The calculation can be performed by successive approximations. A value for T_i is assumed first, and, using it, the physical parameters of the oil required to calculate

$N_{Gr} N_{Pr}$ and N_{Nu}, are established. As a result, Eq. 7.2—18 will furnish a value for α_1. Introducing this into Eq. 7.2—17, a k-value is calculated. Now Eq. 7.2—32 is used to produce another k-value, likewise using the previously determined α_1. If the ks furnished by the two procedures agree, then the value assumed for T_i is correct. If this is not the case, calculation has to be repeated with a different value for T_i. — Equation 7.2—33 or 7.2—34 may

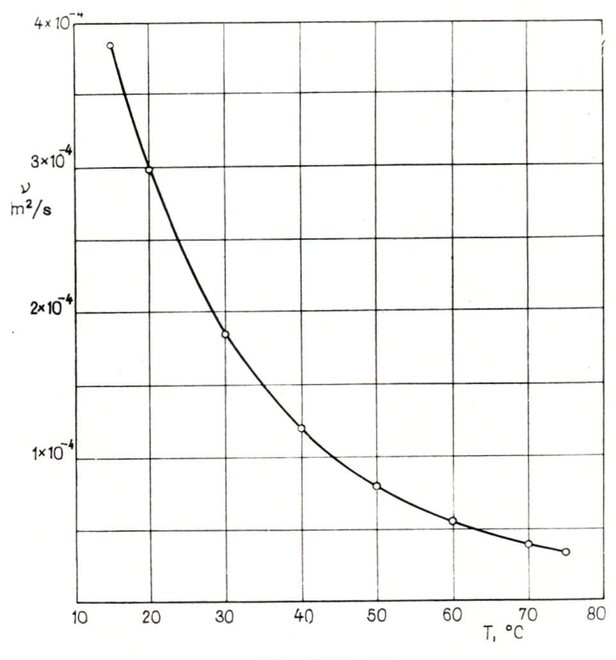

Fig. 7.2—10

be employed, e.g. in possession of measured T_{in} and T_i values for a given pipeline, to calculate the thermal conductivity λ_{in} of the insulation under operating conditions. This procedure permits us to verify whether the insulation provided has the effect aimed at, or whether it has deteriorated since its installation (Szilas 1966).

Example 7.2—3. Find the heat-transfer factor of a pipeline with $d_o = 0.1143$ m, $d_i = 0.1023$ m, $h = 1$ m. Variation of viscosity v. temperature for the oil of density $\varrho_{20} = 887$ kg/m³, flowing laminarly in the pipeline, is given in Fig. 7.2—10; $\lambda_s = 1.76$ W/(m. K), $T_f = 318.4$ K, $T_s = 273$ K; the temperature coefficient of density, $\alpha_T = 0.65$ kg/(m². K). — Let us assume that wall temperature T_f is less by 9.5 K than oil temperature T_f (that is, $T_i = 308.9$ K). Accordingly, the physical parameters of the oil are furnished by Eqs 7.2—23 — 7.2—26 as

$$\varrho_T = 887 - 0.65(308.9 - 293) = 876.6 \text{ kg/m}^3$$

$$c = \frac{762.5 + 3.38 \times 308.9}{\sqrt{0.887}} = 1918 \text{ J/(kg. K)}$$

$$\beta_T = \frac{1}{2583 - 6340 \times 0.887 + 5965 \times 0.887^2 - 308.9} = 7.442 \times 10^{-4} \text{ K}^{-1}$$

$$\lambda_f = \frac{0.134 - 6.31 \times 10^{-5} \times 308.9}{0.887} = 0.129 \text{ W/(m. K)}.$$

From Fig. 7.2—10, the viscosity of the oil at 308.9 K (= 35.9 °C) is 1.423×10^{-4} m^2/s. By Eq. 7.2—21,

$$N_{Gr} N_{Pr} = \frac{0.1023^3 \times 1918 \times 876.6 \times 9.81 \times 7.442 \times 10^{-4}(318.4 - 308.9)}{1.423 \times 10^{-4} \times 0.129} = \\ = 6.798 \times 10^6 .$$

Since $6.798 \times 10^6 > 5 \times 10^4$, N_{Nu} will be furnished by Eq. 7.2—22 as

$$N_{Nu} = 0.184(6.798 \times 10^6)^{0.32} = 28.26 .$$

By Eq. 7.2—18,

$$\alpha_1 = \frac{0.129 \times 28.26}{0.1023} = 35.66 \text{ W/(m}^2\text{. K)}.$$

By Eq. 7.2—30,

$$\alpha_2 = \frac{2 \times 1.76}{0.1143 \times \ln \dfrac{4 \times 1}{0.1143}} = 8.67 \text{ W/(m}^2\text{. K)}.$$

Now the heat-transfer factor is, by Eq. 7.2—17,

$$k = \frac{\pi}{\dfrac{1}{35.66 \times 0.1023} + \dfrac{1}{8.67 \times 0.1143}} = 2.448 \text{ W/(m. K)}$$

whereas Eq. 7.2—32 yields

$$k = \frac{0.1023 \times \pi \times 35.66(318.4 - 308.9)}{318.4 - 273} = 2.400 \text{ W/(m. K)}.$$

As a further approximation, let $T_i = 308.7$ K, in which case Eq. 7.2—17 yields 2.449 W/(m. K), and Eq. 7.2—32 yields 2.450, so that $k = 2.45$ W/(m. K) is acceptable.

7.2.4. Calculating the head loss for the steady-state flow of a Newtonian oil

(a) Chernikin's theory

Equation 7.1—2 will hold for an infinitesimal length dl of pipeline only, as the temperature of the oil flowing through the pipe may be considered constant, and so may, in consequence, also λ and v:

$$\mathrm{d}h_s = \frac{\lambda v^2}{2\,g\,d_i}\,\mathrm{d}l. \qquad 7.2\text{—}35$$

Equation 1.1—10 may be used to express λ also for laminar flow:

$$\lambda = a\left(\frac{1}{N_{\mathrm{Re}}}\right)^b = a\left(\frac{\nu}{d_i v}\right)^b \qquad 7.2\text{—}36$$

where a and b are constants that, in laminar flow, depend on the flow type only. Substituting this expression of λ and the expression

$$v = \frac{q}{\dfrac{d_i^2\pi}{4}}$$

into Eq. 7.2—35, we get

$$\mathrm{d}h_s = \beta\frac{q^{(2-b)}\nu^b}{d^{(5-b)}}\,\mathrm{d}l \qquad 7.2\text{—}37$$

where the constant coefficient is

$$\beta = \frac{4^{(2-b)}\,a}{2\,g\,\pi^{(2-b)}}.$$

Viscosity v. temperature in a Newtonian oil may be described over a limited temperature range also by Filonov's explicit formula

$$\nu = \nu_O e^{-nT} \qquad 7.2\text{—}38$$

where ν_O is kinematic viscosity at 0 °C and

$$n = \frac{\ln\dfrac{\nu_{\mathrm{I}}}{\nu_{\mathrm{II}}}}{T_{\mathrm{I}} - T_{\mathrm{II}}}$$

where ν_{I} and ν_{II} are experimentally determined viscosities at the temperatures T_{I} and T_{II}. Let us point out that T is to be written in °C. ν_O may be considered as a fictive, extrapolated value. It has no physical meaning unless the oil is Newtonian at the temperature considered. At a distance l

from the head end of the pipeline, the temperature of the flowing oil is expressed by Eq. 7.2—15 as

$$T_f = T_s + (T_{f1} - T_s)e^{-ml}$$

where

$$m = \frac{k}{q\varrho c}.$$

Introducing this into Eq. 7.2—38, we obtain

$$\nu = \nu_0 e^{-n[T_s + (T_{f_1} - T_s).e^{-ml}]}.$$

Substituting this latter equation in Eq. 7.2—37, we get

$$\mathrm{d}h = \beta \frac{q^{(2-m)} \nu_O^b}{d^{(5-b)}} e^{-bnT_s} e^{-bn(T_{f_1} - T_s)e^{-ml}} \mathrm{d}l. \qquad 7.2—39$$

Let $bn(T_{f1} - T_s) = B$, and

$$\beta \frac{q^{(2-b)} \nu_o^b}{d^{(5-b)}} e^{-bn\, T_s} = A$$

then

$$h_s = \int_0^l A\, e^{-B_e{}^{-ml}}\, \mathrm{d}l.$$

Let $Be^{-ml} = u$; then, $e^{-ml} = u/B$. Hence,

$$l = -\frac{1}{m} \ln \frac{u}{B} \quad \text{and} \quad \mathrm{d}l = -\frac{1}{mu} \mathrm{d}u.$$

Substituting the constants into this equation yields

$$h_s = -\frac{A}{m} \int_{u=B}^{B'} \frac{e^{-u}}{u} \mathrm{d}u.$$

Since

$$\int_{u=B}^{B'} \frac{e^{-u}}{u} du = \int_{u=B}^{\infty} \frac{e^{-u}}{u} \mathrm{d}u - \int_{u=B'}^{\infty} \frac{e^{-u}}{u} \mathrm{d}u = \mathrm{Ei}(-B') - \mathrm{Ei}(-B),$$

we have, in the original notation,

$$h = \beta \frac{q^{(2-b)} \nu_O^b}{d^{(5-b)}} e^{-bnT_s} \frac{1}{m} \times$$

$$\times \{\mathrm{Ei}[-bn(T_{f1} - T)] - \mathrm{Ei}[-bn(T_{f1} - T_s)e^{-ml}]\}.$$

Introducing the $\nu_O = \nu_1 e^{nT_{f_1}}$ implied by Eq. 7.2—38, and the

$$(T_{f1} - T_s)\, e^{-ml} = T_{f2} - T_s$$

implied by Eq. 7.2—15, we arrive at

$$h_s = \beta \frac{q^{(2-b)} \nu_1^b}{d^{(5-b)}} \, l \, \frac{e^{bn(T_{f1}-T_s)}}{ml} \times \qquad 7.2\text{—}40$$

$$\times \{\text{Ei}\,[-bn(T_{f1} - T_s)] - \text{Ei}\,[-bn(T_{f2} - T_s)]\}.$$

This equation may be contracted to read

$$h_s = \beta \frac{q^{(2-b)} \nu_1^b}{d^{(5-b)}} \, l \, \Delta_l \, \Delta_r \qquad 7.2\text{—}41$$

in which case, with reference to Eq. 7.2—37, it may be interpreted as the product of the pressure head loss of a liquid in isothermal flow whose viscosity is ν_1 at the head end of the pipeline, of the correction accounting for cooling along the pipeline section and of the radial temperature correction Δ_r. In laminar flow, $\beta = 128/\pi g$, and $b = 1$; in turbulent flow, the corresponding values are $0.201/g$ and 0.25, if friction factor λ is determined using Eq. 1.1—4 (Chernikin 1958). In Eq. 7.2—41,

$$\Delta_l = \frac{e^{bn(T_{f1}-T_s)}}{ml} \{\text{Ei}\,[-bn(T_{f1} - T_s)] - \text{Ei}\,[-bn(T_{f2} - T_s)]\}$$

Δ_l is, then, the correction factor accounting for axial cooling. It shows how many times the head loss is increased owing to the fact that oil temperature gradually decreases from T_{f1} to T_{f2} during flow. — Δ_r is the correction factor accounting for radial cooling. Its value would be unity if oil temperature were constant over any pipeline cross section. However, temperature in effect decreases from the pipe axis towards the wall (cf. e.g. Example 7.2—4), and hence mean viscosity in the pipe is higher than the axial viscosity. Δ_r accounts for this effect. Chernikin makes use of the Sieder—Tate relationship

$$\Delta_r = \varepsilon \left(\frac{\nu_i}{\nu_f} \right)^{\omega} \qquad 7.2\text{—}42$$

ν_i and ν_f are average viscosities of oil at the pipe wall and along the axis, respectively, at the arithmetic average of the head-end and tail-end temperatures of the pipeline section considered; ε and ω are constants (ε is 0.9 in laminar and 1 in turbulent flow; ω ranges from 1/3 to 1/4 in laminar and from 1/3 to 1/7 in turbulent flow). In pipelines transporting viscous crudes, the transition from laminar to turbulent flow may be expected at about $N_{\text{Re}} = 2000$.

Equation 7.2—41, suitable for comparatively fast calculation, has the following limitations. (i) It assumes heat transfer factor k to be constant all along the pipeline, whereas actually it may vary substantially especially along non-insulated lines. (ii) In calculating the friction factor of turbulent flow, it employs the Blasius equation which refers to smooth pipe. Accuracy is easy to improve by introducing different constants a and b (cf. e.g. Eq. 7.1—4). (iii) Equation 7.2—38 is an approximate expression of the temperature-dependence of oil viscosity, valid over a rather limited temperature range only. (iv) Constants ε and ω have rather broad ranges and therefore Δ_r cannot be determined to a desirable degree of accuracy.

(b) Ford's theory (with modification)

This procedure, more lengthy than the foregoing one, establishes the head loss of oil flowing in a pipeline by a finite-difference procedure. It permits us to account for the variation in k, and to use accurate friction-factor and viscosity values (Ford 1955). Below we shall present this method in a slightly modified form.

Starting from the initial axial oil temperature T_{f1}, we assume an axial temperature drop ΔT_f of a few °C to take over a place comparatively short, as yet unspecified length l_1 of pipe. It is assumed that the average temperature over this length of pipe is, in a fair approximation,

$$\bar{T}_{f1} = T_{f1} - 0.5\,\Delta T_f. \qquad 7.2-43$$

Using Eqs 7.2—15, 7.2—17 and 7.2—30, it is possible to calculate the length l_1 over which the axial temperature drop is precisely ΔT_f. In order to perform the calculation it is necessary to determine the heat transfer factor as indicated in Example 7.2—3 and, in the process, to find also the average wall temperature T_{i1}. Continuing the process, one obtains related values of T_f, T_i, T_{in} and l, which permits plotting graphs of temperature v. length.

Viscosity values corresponding to the Reynolds number $N_{\mathrm{Re}} = 2000$, characterizing the transition from laminar to turbulent flow, are calculated using Eq. 1.1—2; oil temperature T_f corresponding to the 'critical' viscosity of the oil to be conveyed is then read off the corresponding $T - \nu$ graph. The point along the line where this temperature sets in will be where turbulent flow passes into laminar. In turbulent flow, as far as head loss calculations are concerned, pipeline temperature may be equated with T_f. In laminar flow, however, the temperature differential between pipe axis and pipe wall is so significant that the average of the two, which will in fact determine the behaviour of the crude (its effective viscosity first of all), has to be determined separately. It was found in practice that this is done most simply by reading said temperature off the experimentally established diagram presented as Fig. 7.2—11, where it is plotted against the product of the Grashof and Prandtl numbers. In the knowledge of the diagram it is possible to establish the variation of corrected oil temperature T_{fk} v. position along the pipe.

By the considerations outlined in Section 1.1, it is possible to find the density and friction factor corresponding to the temperatures T_f (or T_{fk}) at the key points of both the turbulent and laminar flow sections along the pipe string. Planimetering will furnish the mean values $\bar{\varrho}$ and $\bar{\lambda}$ for both zones of flow. Substituting the relationship

$$v = \frac{q_m}{\dfrac{d^2\pi}{4}\bar{\varrho}}$$

into Eq. 1.1—1, we obtain the friction loss as

$$p_s = \frac{8\bar{\lambda} q_m^2 l}{\pi^2 \bar{\varrho} d_i^5} . \qquad 7.2-44$$

Introducing the values of $\bar{\lambda}$ and $\bar{\varrho}$, we obtain the values p_{sT} and p_{sL}, valid respectively for the turbulent and laminar zone. The full head loss is

$$p_s = p_{sT} + p_{sL}. \qquad 7.2-45$$

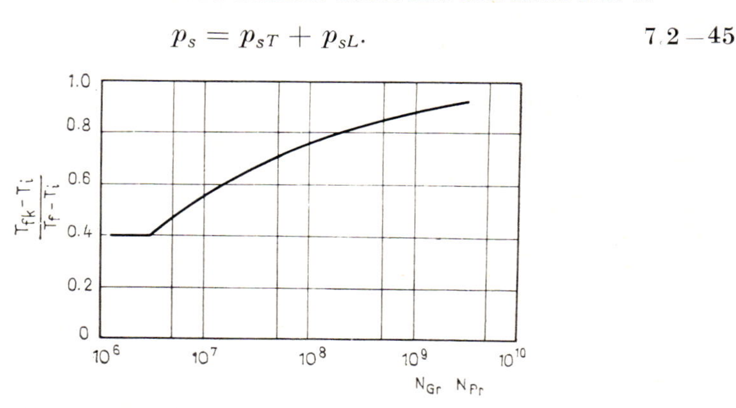

Fig. 7.2—11 Relationship for determining T_{fk}, after Ford (1955)

Example 7.2—4. Find the friction loss in an uninsulated horizontal pipeline buried in the ground, if $d_i = 102.3$ mm; $d_o = 114.3$ mm; $k/d_i = 4 \times 10^{-4}$; $l = 9000$ m, $T_{f1} = 74$ °C; $T_s = 0$ °C; $h = 1$ m; $\lambda_s = 1.76$ W/(m. K); $\varrho_{20} = 887$ kg/m³; $q = 50$ m³/h; $\alpha_T = 0.65$ kg/(m³. K), and crude viscosity varies with temperature as shown in Fig. 7.2—10. — Let $\Delta T_f = 4$ °C; the mean axial temperature $\bar{T}_{f1}$ of the first pipeline section is, then, 74 — 4/2 = = 72 °C. Let us find N_{Re} at this temperature. If it exceeds 2000, then the flow is turbulent. In this case, in an approximation sufficient for our purposes, $\bar{T}_{fk1} = \bar{T}_{f1} = 72$ °C. Let us find c for this temperature using Eq. 7.2—23, then, using Eq. 7.2—15, the length l_1 over which the crude's temperature drops to 70 °C from the initial 74 °C; k is established using Eq. 7.2—17, under the assumption that $1/\alpha_1 = 0$. The procedure is continued until N_{Re} attains the critical value of 2000. In the laminar zone of flow, this procedure is modified as follows. Starting from an initial axial temperature T_{fn}, one establishes the length of pipeline l_n along which the axial temperature drops 4 °C. Here, the mean axial temperature equals T_{fn} — 4/2. To this value, an average wall temperature is arbitrarily assigned, then k_n is calculated in the manner illustrated by Example 7.2—3, and l_n is determined in the knowledge of k_n from Eq. 7.2—15. In a similar way, related values of T_f, T_i and l are calculated until Σl attains or exceeds 9000 m, the full length of the pipeline.

Making use of Fig. 7.2—11, T_{fk} is determined for each pair of T_f and T_i calculated for the laminar zone. Figure 7.2—12 is a plot of temperatures T_f, T_i and T_{fk} v. distance from the pipeline's head end, and of oil gravity ϱ

at the temperatures T_f and T_{fk}. The mean density found by planimetering is $\bar{\varrho}_T = 861.1$ kg/m³ in the turbulent section and $\bar{\varrho}_L = 879.4$ kg/m³ in the laminar section. The N_{Re} and λ values at the key points, at temperatures T_f and T_{fk}, are likewise plotted in Fig. 7.2–12. The mean values of λ have also been planimetered; they are $\bar{\lambda}_T = 0.0432$ in turbulent and $\bar{\lambda}_L = 0.0654$ in laminar flow.

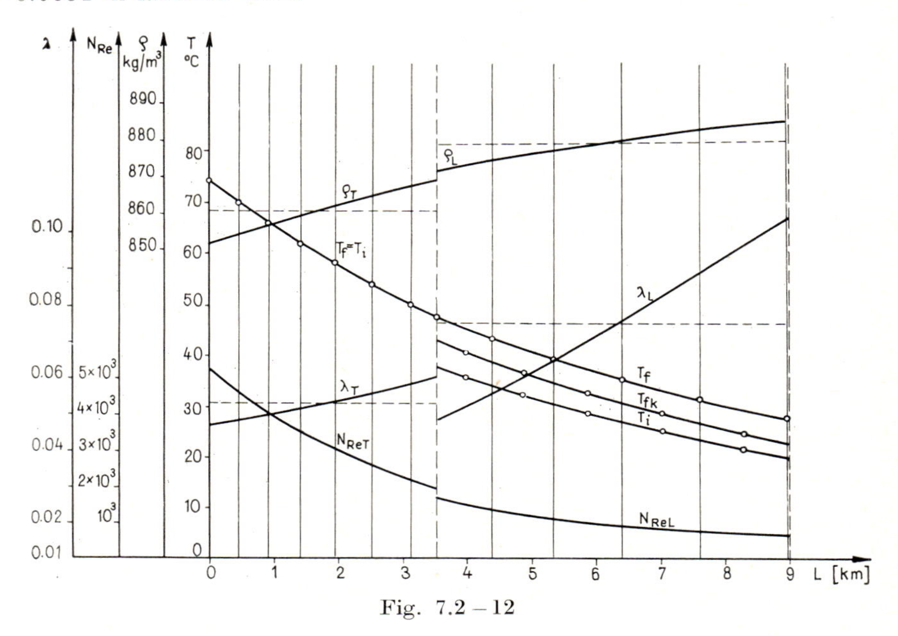

Fig. 7.2–12

Friction loss in the turbulent section is

$$p_{sT} = 1.95 \times 10^6 \text{ N/m}^2$$

and in the laminar section

$$p_{sL} = 4.46 \times 10^6 \text{ N/m}^2.$$

The full friction loss is, then,

$$p_s = 6.41 \times 10^6 \text{ N/m}^2.$$

All the other relevant intermediate results of the calculation are listed in Table 7.2–2.

7.2.5. Temperature of oil in transient flow, in buried pipelines

In the foregoing sections we have assumed the heat flow about the buried pipeline to be steady. This is the case if the pipeline carries a crude of inflow temperature, flow rate and physical parameters constant over a com-

Table 7.2—2

Serial number	$T_{fn} - T_{fn+1}$	$\bar{T}_{fn}$	T_{fkn}	T_{in}	ϱ_n	ν_n
	K	K	K	K	kg/m³	10^{-4} m²/s
1	347 — 343	345.0	—	345.0	853.2	0.361
2	343 — 339	341.0	—	341.0	855.8	0.413
3	339 — 335	337.0	—	337.0	858.4	0.474
4	335 — 331	333.0	—	333.0	861.0	0.546
5	331 — 327	329.0	—	329.0	863.6	0.633
6	327 — 323	325.0	—	325.0	866.2	0.736
7	323 — 320.4	321.7	—	321.7	868.4	0.837
8	320.4—316.4	318.4	313.6	308.7	873.6	1.16
9	316.4—312.4	314.4	309.6	305.1	876.2	1.38
10	312.4—308.4	310.4	305.6	301.6	878.8	1.65
11	308.4—304.4	306.4	301.6	298.1	881.4	1.97
12	304.4—300.4	302.4	297.7	294.6	883.9	2.37
13	300.4—296.4	298.4	294.1	291.3	886.3	2.82

$N_{Gr}\,N_{Prn}$	N_{Ren}	λ_n	α_1	k_n	c_n	Δl_n	$\Sigma\Delta l_n$
10^6	—	—	W/(m² · K)	W/(m · K)	J/(kg · K)	m	m
—	4972	0.0378	—	3.11	2047	450	450
—	4341	0.0392	—	3.11	2033	473	923
—	3769	0.0410	—	3.11	2019	500	1423
—	3259	0.0428	—	3.11	2005	529	1952
—	2806	0.0448	—	3.11	1990	563	2515
—	2405	0.0469	—	3.11	1976	602	3117
—	2061	0.0493	—	3.11	1964	420	3537
6.866	1512	0.0423	35.76	2.45	1935	858	4395
5.567	1268	0.0505	33.52	2.42	1920	948	5343
4.467	1060	0.0604	31.32	2.38	1906	1059	6402
3.545	883	0.0725	29.13	2.34	1829	1197	7599
2.773	733	0.0873	27.00	2.30	1878	1378	8977
2.135	614	0.1042	24.86	2.24	1865	23	9000

paratively long span of time, and the temperature of the soil outside the warmed-up envelope about the pipeline is subject to no change. Strictly speaking, this condition is never satisfied. If, however, departure from the steady state is comparatively slight, procedures outlined in Sections 7.2—2 and 7.2—4 remain applicable.

In numerous cases of considerable practical interest, however, deviation from the steady state may be sufficient to make the steady-state relationships unsuited for even an approximate description of the situation. Transient heat-flow situations of this nature will tend to arise primarily in the following cases: (i) when starting up a hot pipeline: the hot oil flowing into the pipeline that, initially, is as cold as the ground, will gradually warm the environment; (ii) flow in the pipeline is stopped and then started again;

(iii) flow rates in the pipeline are subject to fluctuations, (iv) the nature and hence the thermal properties of the crude transported are subject to fluctuations. — Any meaningful attempt at solving these problems requires the knowledge of the relationships characterizing the processes of warming and cooling. These will permit the calculation of pressures arising on start-up and restarting, and the variation of said pressures with time.

Numerous models and calculation methods have been devised to describe the transient heat-flow and liquid-flow situations and to solve partial problems all subsumed under the above-stated general problems. To mitigate the complications caused by the large number of variables involved, simplifying assumptions have been introduced, and, as a result of these, individual models may rather substantially differ from one another. The first attempts at a synthesis have just recently been published, but the development of theories readily applicable in practice has not by far been accomplished as yet. One publication to be given special mention is Tugunov's academic thesis (Tugunov 1968), which summarizes 39 papers written on the subject by the author and his co-workers. Similarly interesting syntheses were presented at the Symposium on 'Waxy crudes in relation to pipeline operations' (London, November 1970). Some of the lectures presented were concerned with transient-flow problems (Davenport and Conti 1971). It seems, however, too early yet to attempt a synthesis of the results achieved so far in the scope of the present book. In the following I propose to present a transient model which describes the temperature changes of flow v. length and time in a pipeline shut down after steady-state flow (Szilas 1968).

It is assumed that (i) the pipeline is embedded in an infinite half-space filled with soil homogeneous as to thermal behaviour, (ii) the temperature of the soil in contact with the pipeline can be described by the Chernikin model (1958) based on the linear-heat-source theory of Carslaw and Jaeger (1947). — The solution referred to has been used by Chernikin to characterize oil temperatures in non-insulated pipelines, neglecting the difference between the thermal behaviour of oil and soil. In the following I shall derive the 'cooling model' for an insulated pipeline. Let us assume that, at a given instant t, oil temperature is constant all over a certain cross-section of the pipeline, including also the steel pipe cross-section; that is, $T_f = T_o$, where suffix o refers to the OD.

Heat flowing through the wall of unity-length pipe into the cooler ground over an infinitesimal length of time $\mathrm{d}t$ reduces the temperature of oil and steel by $\mathrm{d}T$:

$$\left[\frac{d_i^2 \pi}{4} \varrho_f c_f + \frac{(d_o^2 - d_i^2)\pi}{4} \varrho_{st} c_{st}\right] \mathrm{d}T_o = -k(T'_o - T'_{in})\,\mathrm{d}t$$

where the symbol T' refers to the transient nature of the process. The heat transfer factor is, after the necessary changes to Eq. 7.2—17:

$$k = \frac{\pi}{\dfrac{1}{2\lambda_{in}} \ln \dfrac{d_{in}}{d_o}}.$$

Let

$$\frac{\pi}{4k}\left[d_i^2 \varrho_f c_f + (d_o^2 - d_i^2) \varrho_{st} c_{st}\right] = A \, .$$

Then, after rearranging,

$$A\left[\frac{\mathrm{d}T_o}{T'_o - T'_{\text{in}}}\right] = -\mathrm{d}t. \qquad 7.2-46$$

Cooling will affect not only oil temperature T'_o, but also the outer surface temperature T_{in} of the insulation. It is to describe this change that we shall make use of the Carslaw—Jaeger—Chernikin relationship. Let P_1 be the

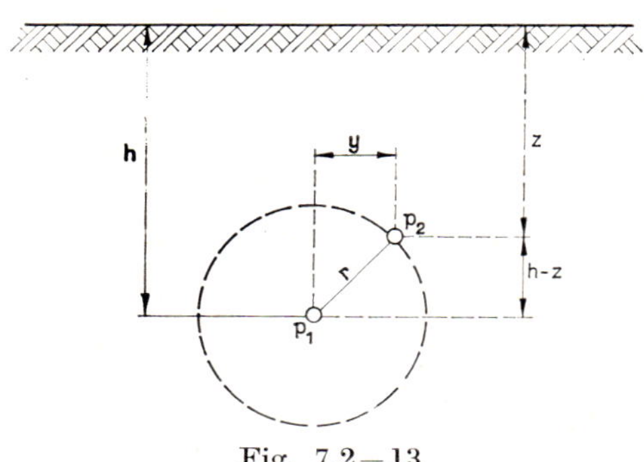

Fig. 7.2—13

image of the projection of the linear heat source on a plane perpendicular to it (Fig. 7.2—13); then the difference in temperature between point P_2 lying in the plane of projection and defined by the coordinates y and z and the undisturbed soil will be

$$T'_o - T_s = \frac{\Phi^*}{4\pi \lambda_s}\left\{\mathrm{Ei}\left[-\frac{h^2}{r^2}\frac{1}{N_{\mathrm{Fo}}}\right] - \mathrm{Ei}\left[-\frac{1}{4N_{\mathrm{Fo}}}\right]\right\}. \qquad 7.2-47$$

If $t = \infty$, then the Equation reduces to the following form valid for steady-state flow:

$$T_o - T_s = \frac{\Phi^*}{2\pi \lambda_s} \ln \frac{2h}{r}. \qquad 7.2-48$$

Now making use of Eqs 7.2—47 and 7.2—48,

$$\frac{T'_o - T_s}{T_o - T_s} = \frac{1}{2 \ln \frac{2h}{r}}\left\{\mathrm{Ei}\left[-\frac{h^2}{r^2}\frac{1}{N_{\mathrm{Fo}}}\right] - \mathrm{Ei}\left[-\frac{1}{4N_{\mathrm{Fo}}}\right]\right\}$$

where

$$N_{\mathrm{Fo}} = \frac{a_s t}{r^2}$$

is the Fourier factor. In a given pipeline, that is, if h, r, λ_s, ϱ_s and c_s are given, this reduces to the simpler form

$$\frac{T'_o - T_s}{T_o - T_s} = \varkappa$$

where

$$\varkappa = f(t) = \frac{1}{2 \ln \frac{2h}{r}} \left\{ \mathrm{Ei}\left[-\frac{h^2}{r^2} \frac{1}{N_{\mathrm{Fo}}} \right] - \mathrm{Ei}\left[-\frac{1}{4N_{\mathrm{Fo}}} \right] \right\}. \qquad 7.2-49$$

This relationship, in the form given here, describes the process of warming up. In the case of cooling down after steady-state flow, on the other hand,

$$\frac{T'_o - T_s}{T_o - T_s} = 1 - \varkappa$$

that is,

$$T'_o = (1 - \varkappa)(T_o - T_s) + T_s \,. \qquad 7.2-50$$

Chernikin used this relationship to determine the surface temperature of an 'oil cylinder' of radius r. He assumed the difference in thermal properties between oil in a non-insulated pipeline and the surrounding soil to be negligible. In an insulated pipeline, this relationship seems suited to determine the external temperature of the insulation, that is,

$$T'_{\mathrm{in}} = (1 - \varkappa)(T_{\mathrm{in}} - T_s) + T_s. \qquad 7.2-51$$

If this expression of the external, variable temperature T'_{in} of the pipeline is introduced into Eq. 7.2—46, then this latter furnishes the time variation of the temperature of oil enclosed in an insulated pipeline. The complicated nature of this relationship made us adopt the following simplified method for solving it. — The relationship $\varkappa = f(t)$ for a given case can be plotted graphically using Eq. 7.2—50. The individual sections of the curve thus obtained are approximated well enough by a relationship of the form

$$\varkappa = a + b \ln t. \qquad 7.2-52$$

Introducing this relationship into Eq. 7.2—51, and the result into Eq. 7.2—46, we get

$$A\left[\frac{dT_o}{T'_o - B + C \ln t}\right] = -\,dt \qquad 7.2-53$$

where

$$B = (1 - a)(T_{\mathrm{in}} - T_s) + T_s$$

and

$$C = b(T_{\mathrm{in}} - T_s).$$

The general solution of Eq. 7.2—53 is

$$T'_o = B - C \ln t + \left\{ C\, \mathrm{Ei}\left[\frac{t}{A}\right] c \right\} e^{-\frac{t}{A}} \qquad 7.2-54$$

where c is the constant of integration. The initial condition of $T'_o = T_o$ if $t = 0$ gives the particular solution

$$T'_o = B - C \ln t + \left\{ C \, \mathrm{Ei}\left[\frac{t}{A}\right] T_o - B - C(0.5772 - \ln A) \right\} e^{-\frac{t}{A}} . \qquad 7.2-55$$

This relationship permit us to calculate the variation of temperature $T'_o = T'_f$ v. the time t elapsed after shutdown of the pipeline, at any pipeline section situated at distance l from the head end of the line.

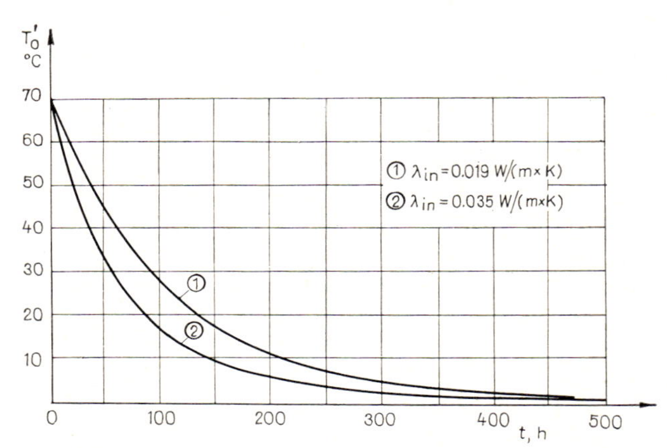

Fig. 7.2—14 Temperature v. time of oil shut-in in an insulated pipeline (Szilas 1968)

Figure 7.2—14 shows the time variation of the temperature of oil enclosed in an insulated pipeline. In plotting these curves using Eq. 7.2—55 we have assumed the temperature prevailing during steady-state flow to be 70 °C at the instant of shutdown. Furthermore, $\varrho_o = 968$ kg/m^3; $c_o = 0.55$ kg/(m^3. K); $T_s = 0$ °C; $d_i = 0.273$ m; $d_o = 0.292$ m; $d_{in} = 0.392$ m; $h = 1.1$ m; $\lambda_{in1} = 0.035$ W/(m. K) (glasswool); $\lambda_{in2} = 0.019$ W/(m. K) (polyurethane foam).

Using Eq. 7.2—55 to establish cooling curves at various pipeline sections and plotting the oil temperatures belonging to given cooling times t v. position along the pipeline one obtains a family of $T'_o = \mathrm{f}(l)_t$ curves. The family shown in Fig. 7.2—15 includes the curve for steady flow at $t = 0$, calculated using Eq. 7.2—15. Oil has been assumed to flow in the pipeline characterized above, except that steel is merely wrapped in tar paper against corrosion. The initial inflow temperature is $T_{o1} = 70$ °C and further, $q_m = 12.5$ kg/s, and $l = 11{,}200$ m. Cooling is seen to be rather rapid in the first few hours, and to slow down rather considerably thereafter. The original cooling rate is reduced to 44 percent after the first day, but the undisturbed soil temperature is not attained even in 30 days.

Warming of a cold pipeline by hot oil flowing in it will be discussed largely after Davenport and Conti (1971). Hot oil introduced into a cold line cools much faster, of course, than it would in steady-state flow in a warmed-up pipe. The heat-transfer factor valid for such situations may be several times the factor for the steady state. Pumping time until steady heat flow sets in may be found in a fair approximation for an uninsulated pipeline buried in the ground by comparing steady-state and transient Nusselt numbers. If the thermal insulation provided by the soil alone is taken into account, then, with Eq. 7.2—30 in mind, the heat-transfer factor to be expected in the steady state is

$$k = \alpha_2 d_o \pi = \frac{2\pi\lambda_s}{\ln \frac{4h}{d_o}} . \qquad 7.2\text{—}56$$

Eq. 7.2—18 may be rewritten in the form

$$N_{\mathrm{Nu}} = \frac{k^* d_o}{\lambda_s} . \qquad 7.2\text{—}57$$

Referring to Eqs 7.2—16 and 7.2—56,

$$N_{\mathrm{Nu}} = \frac{k}{\lambda_s \pi} = \frac{2}{\ln \frac{4h}{d_o}} . \qquad 7.2\text{—}58$$

Assuming the pipeline to be embedded in an infinite half-space of soil, and the heat transfer between ground surface and air to be inifinite, it is possible to show that the transient Nusselt number is a function of the Fourier number:

$$N_{\mathrm{Nu}} = 0.362 + \frac{0.953}{N_{\mathrm{Fo}}^{0.33}} . \qquad 7.2\text{—}59$$

The Fourier number is defined as

$$N_{\mathrm{Fo}} = \frac{a_s t}{d_o^2} . \qquad 7.2\text{—}60$$

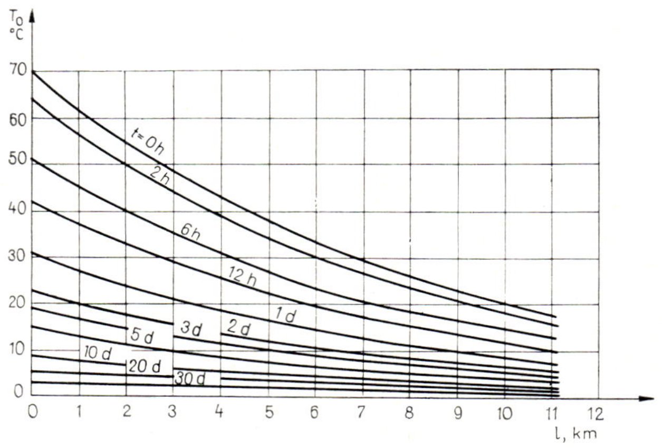

Fig. 7.2—15 Temperature traverses of oil shut-in in a pipeline, at various instants after shut-in (Szilas 1968)

Equation 7.2—59 may be used to calculate the time it takes for oil flowing in an uninsulated pipeline of *OD* d_o laid in soil of thermal diffusivity a_s to attain the steady temperature, that is, for the Nusselt number to attain the value furnished by Eq. 7.2—58. Since in this equation, owing to the approximation employed, N_{Nu} is a function of h and d_o only, substituting it into Eq. 7.2—59 yields

$$N_{Fo} = \left[\frac{0.953}{\dfrac{2}{\ln \dfrac{4h}{d_o}} - 0.362} \right]^3 \qquad 7.2\text{—}61$$

that is, the time required for steady heat flow to set in is a function of the h/d_o ratio only.

In reality, the hypothesis of the infinite half-space will not model the actual heat flow too accurately after the heat flow issuing from the pipeline has attained the ground surface. Heat loss will exceed the calculated value by 10 to 15 percent thereafter. The heat flow pattern will change, however, very slowly, and the time it takes to attain the steady state is infinite in principle. The accuracy of λ_s is not as a rule better than 10 percent. Accuracy is not, then, impaired any further if the above relationship is considered to hold also for the steady state.

Flow in pipelines is not usually interrupted long enough for the oil in the pipe to cool down to the undisturbed soil temperature. On restarting, then, the initial temperature is higher than the undisturbed soil temperature, and it takes less time to attain the steady or a near-steady state. Finding this time for an uninsulated pipeline may be performed by the following consideration. We prepare a family of curves similar to that in Fig. 7.2—15 for the case under consideration. Each curve of the family will be represented in a fair approximation by a 7.2—15-type equation. These equations will differ only as to the value of k, but even k will be constant all along a given curve. Putting $q_Q = q_m$, and $c = c_o$, one may write up for any curve

$$k = \frac{\ln T_1 - \ln T_2}{l} q_m c_o . \qquad 7.2\text{—}62$$

Knowing the duration of cooling one may pick out that curve from Fig. 7.2—15 which will hold when flow is restarted. It is necessary to calculate the time it would take for the oil flowing in the pipe to attain this state if pumping hot oil were started through an entirely cold environment. In order to find it, we use Eq. 7.2—62 to determine the k value corresponding to the actual cooling curve and then Eq. 7.2—58 to determine the transient N_{Nu}; Eq. 7.2—59 is then used to calculate N_{Fo} and Eq. 7.2—60 to find the equivalent pumping time. This is the reduction in the time needed to attain steady heat flow against the case when pumping is started at the undisturbed-soil temperature.

7.2.6. Startup pressures of Newtonian oils and their reduction

A pipeline may in principle be warmed up simply by pumping into it the high-viscosity crude to be transported. This would, however, require long pumping, often of unjustified duration. Accordingly, the cooled-down pipe is usually flushed first with hot light oil and the high-viscosity crude is not introduced until the line has sufficiently warmed up. Pressure conditions in the warming-up pipeline can be described by Ford's method, which is based on relationships referring to the steady state (Ford 1955).

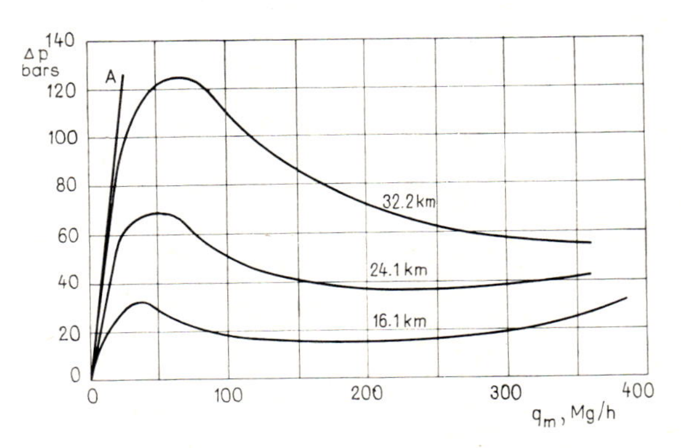

Fig. 7.2—16 Friction loss v. throughput at various lengths of 12-in. pipeline, after Ford (1955)

Figure 7.2—16 shows pressure v. throughput graphs of high-viscosity crudes flowing in 12-in. pipelines of different length. Any point on the curves states the pressure drop of steady flow at the rate q_m chosen in the pipeline of the given length. Inflow temperature of the oil is round 66 °C in each line, and soil temperature is 4 °C. All three curves exhibit peak pressure drops at comparatively low throughputs. After this peak, the driving-pressure requirement decreases first, to increase again at higher throughputs. Assuming initial temperature to be constant, these curve shapes are due to the following: (i) greater throughputs entail higher friction losses in isothermal flow (line $O-A$), (ii) the increase in mean cil temperature reduces viscosity which in turn reduces friction loss. At low throughputs, oil temperature hardly differs from soil temperature, and flow is laminar as a rule. The friction loss can be derived from Eq. 1.1—1 as

$$p_s = k''_L q_m \qquad 7.2-63$$

that is, a linear function of throughput. This is the relationship illustrated by line $O-A$, tangent to the 32.2-km curve. — At infinite throughput, the average oil temperature equals the inflow temperature. Flow may in that case be assumed turbulent, and can, by Eq. 7.1—10, be characterized by the relationship

$$p_s = k''_T q^{(2-b)} \qquad 7.2-64$$

where $b < 1$. The right-hand ends of the curves in Fig. 7.2—16 approach the line defined by this equation as an asymptote. In plotting any point of these curves we have assumed flow to be steady state. The steady state is not attained instantly, however. Let us assume that a throughput of 100 Mg/h is conveyed through the 24.1-km line. Let us now increase this throughput to 150 Mg/h. At first, the increase in throughput will hardly raise the average temperature of the oil above the preceding steady-state value. Assuming flow to be laminar, a throughput increased by half requires a driving pressure increased by half, that is, pressure drop will suddenly increase from 51 to 76 bars. As time goes on, the average temperature increases gradually, and as a result, p_s decreases to and stabilizes at about 40 bars.

It follows from the above considerations that pressure drop on startup is invariably higher than that shown in the Figure, the deviation being the higher, the greater the ratio between the throughput to be attained in one step and the steady throughput shown in the Figure. It is when the step q is infinitesimal that the values in the Figure prevail. Even in this — unrealistic — case it is apparent, however, that startup requires more pressure than steady-state flow. For instance, conveying 350 Mg/h of oil through the 32.2-km line entails a pressure drop of 55 bars, but startup will require a peak pressure of at least 124 bars. If for a delivery pressure of 1 bar the pump can exert 56 bars pressure at most, then this equipment will deliver 14 Mg/h of oil throughput at most. It is therefore fundamental that, if the pipeline is to be heated with the oil to be conveyed, the pump must be chosen so as to be able to deliver the peak pressure. In order, furthermore, to make startup fast enough, it is desirable that the pump should exceed said peak by 20—25 percent at least.

Figure 7.2—16 further reveals the pressure change to be expected when steady flow is reduced. Let e.g. the steady throughput through the 24.1-km line be 150 Mg/h, and let us reduce this flow to 100 Mg/h. In the first instants of reduction, the temperature of oil flowing in the pipeline will not change; hence, by Eq. 7.2—63, pressure drop will suddenly decrease by one-third, from 39 to round 26 bars. Owing to the lower flow rate, however, the mean oil temperature will gradually decrease and stabilize at a lower value, while the pressure loss climbs to 52 bars. — Reducing the startup or peak pressure may be achieved by several means. At a point somewhere along the pipeline, the company installs a booster or a heater station or both. Let us assume that the operating conditions characterized by Fig. 7.2—16 prevail and that a booster-heater station is installed at the middle of the 32.2-km line, that is, at a distance of 16.1 km from both ends. If the intermediate station is used to heat the oil to be conveyed to the inflow temperature of the head end (66 °C), then the line may be considered as comprising two 16.1-km segments, whose behaviour is characterized by the lowermost curve in Fig. 7.2—16. Peak pressure drop will, then, be 31 bars as opposed to the preceding 124 bars. Even if pumping at the intermediate station is avoided, peak pressure will not exceed $2 \times 31 = 62$ bars. Once the line is warmed up, heating at the intermediate station may be ceased (and so may pumping, if any). The peak pressure drop to be attained by heating but no pumping at the intermediate station in the 32.2-km line does not significantly exceed the pressure drop of conveying 350 Mg/h of oil in steady flow. If intermediate heating is unfeasible, peak pressure may be reduced by intermediate pumping alone. In the 32.2-km line, this requires a booster pump of round 70 bars pressure. The booster station will, in order to halve the pressure differential, be installed not halfway along the line, but at a distance of about 23 km from the head end.

We have tacitly assumed so far that startup is performed by pumping the same quality crude as is to be conveyed in the steady state. This would, however, often render startup too slow; other, lower-viscosity liquids are used for startup in these cases, such as a light oil or water. Lower viscosity reduces the pressure drop against what is to be expected with the high-

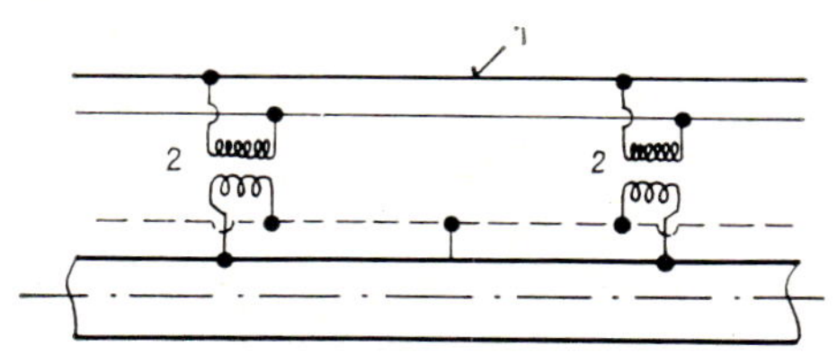

Fig. 7.2—17 Electric pipeline heating, after Pasqualini (1950)

viscosity crude; that is, at the maximum allowable operating pressure of the pipe, throughput will be higher and heating will accordingly be faster. If the startup liquid is water, heating is accelerated also because the specific heat of water is more than twice that of oil. The drawback of this method is that, on closing down the line, the pipe has to be flushed with the low-viscosity liquid, and the transmission of this latter requires a substantial amount of power. Economy is improved if the water can be put to some use at the tail end, or if the line must at intervals carry light oil anyway. — In short pipelines startup pressure can be reduced also by electrically heating the cooled line. There are several methods for this. In the one shown as Figure 7.2—17 (Pasqualini 1950), voltage taken from electric transmission line *1* is reduced to 50—60 V by transformers *2*. The secondaries of the transformers are hooked up to the pipeline in the way shown in the Figure. The oil is heated by the current flowing in the pipe. The use of a.c. is indicated as d.c. will tend to result in electrolytic corrosion. Gautier (1970) describes a solution where the heating cables are fixed to the outer pipe wall in an axial cement-graphite coat, and the entire arrangement is insulated with polyurethane foam. This solution is expensive and uneconomical except in special situations. The heat insulation will cost as much again as the steel pipeline. The heating equipment will cost again the price of the pipeline, thus adding up to three times the original cost. This solution is used, if at all, to maintain the temperature in a shutdown line rather than to heat an already cooled down one.

7.2.7. Startup pressure and flow of pseudoplastic oils

Pressure drop of pseudoplastic oil in isothermal flow in a pipeline can be calculated on the basis of the considerations in Section 1.3—5. — If flow is non-isothermal, the line of thought in Section 7.2—2 is modified inasmuch as establishing the convection-heat-transfer factor α_1 will differ somewhat from the procedure used for Newtonian oil. The error committed in substituting actual viscosity in Eq. 7.2—21 by the apparent viscosity valid at the shear rate considered is likely to be negligible when calculating the N_{Gr} N_{Pr} product. With this modification, the pressure drop can be determined as indicated in Sections 7.2—4b and 1.3—5.

According to an earlier opinion, the pressure required to break out a crude gelled into a pipeline is given by the relationship

$$\Delta p = \frac{4l\tau_e}{d_i} \qquad 7.2-65$$

derived from Eq. 1.3—14 by putting $\tau = \tau_e$. By this formula, assuming a near-atmospheric delivery pressure p_2, the pressure $p_1 = p_2 + \Delta p$ required to break out a crude gelled in a pipeline of given size is directly proportional to the crude's yield stress τ_e and the length l of the pipeline. Research in the last few years has shown that the breakout pressure in commercial-size pipelines may be significantly less (Verschuur et al. 1971, Perkins and Turner 1971), and that its magnitude is influenced also by a number of other factors. Equation 7.2—65 of breakout pressure would hold if the gelled oil in the pipeline filled the entire pipe volume and if it were incompressible, too. It would then be necessary to apply at the head end of the pipeline a pressure sufficient to generate a shear stress equal to the yield stress along the entire length of the pipeline. This would give rise to a liquid oil film next to the wall, and the gelled crude would 'flow' out of the pipe as a solid plug. This situation may be approximated fairly well in the laboratory and in short in-plant pipelines. In commercial-size pipelines, however, the gelled crude will not fill the pipeline entirely, and moreover, the elasticity of both the oil and the pipe must be taken into account. As a result, breakout pressure will usually be significantly less than that furnished by Eq. 7.2—65. In the following, we shall discuss the factors influencing the breakout pressure of oils of structural viscosity largely after Verschuur et al. (1971). Let us point out, however, that said factors have so far been investigated at constant axial temperatures only, and that the relationships derived are based on this same assumption.

If transmission of a warm waxy crude is stopped, the crude will gradually cool, and a tri-dimensional paraffin lattice of appreciable mechanical strength may form, constituting a gel with the liquid oil with which it is combined. The paraffin lattice has a well-defined permeability. The liquid phase will, under a comparatively low pressure gradient, seep through the lattice without breaking it. This seepage is governed by Darcy's law: the flow rate varies linearly as the pressure gradient. At very low flow rates, the lattice will undergo some slight elastic deformation, but its structure will hardly change, and it will essentially stay in place. If the lattice is broken up by some influence, a heterodisperse suspension comes to exist, whose flow behaviour is much more favourable than that of the gelled oil with the lattice in it, and a comparatively low pressure gradient will be sufficient to start flowing the entire system filling out the pipeline. In order to reduce the breakout pressure it is, then, the destruction of the gel, of the tri-dimensional paraffin lattice, that is to be aimed at. In addition to the ratio of shear stress to yield stress, the destruction of the gel is affected by two further important circumstances: (i) the reduction in the specific volume of the crude by cooling, and (ii) the elasticity of the oil and pipe.

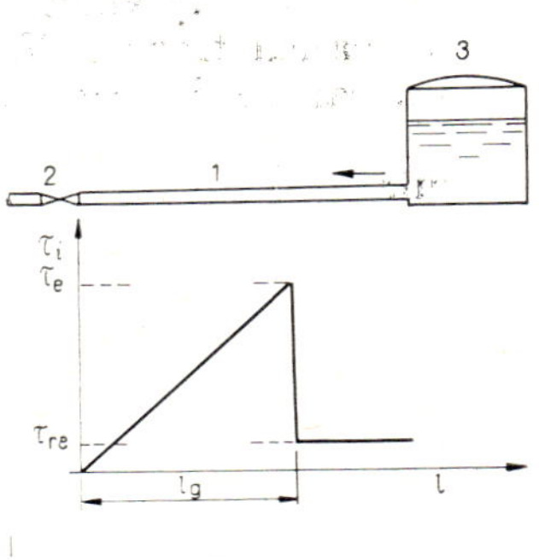

Fig. 7.2—18 After Verschuur et al. (1971)

Shrinkage will tear into pieces the originally continuous oil cylinder enclosed in the pipeline. The gel will in places separate from the pipe wall, and the ends of the plugs forming on contraction will draw apart. Shrinkage may destroy the paraffin network. The fissures thus formed and the measure of discontinuity may be influenced significantly by the circumstances of cooling: notably, cooling starts at the pipe wall, and oil will start to solidify next to the pipe. Near the pipe axis, oil will keep flowing virtually unhampered, and may, if there is enough supply, fill up the contraction gaps due to cooling. — Consider the model shown as Fig. 7.2—18, where flow has stopped in line *1* and oil is cooling down in it. Valve *2* has been closed since the beginning of cooling, but the valve towards tank *3* is open, so that, during cooling, oil may flow into the pipeline from the tank. The rate of longitudinal displacement due to contraction on cooling is

$$v = \beta_T \, l \frac{dT}{dt} \qquad 7.2\text{—}66$$

where β_T is the coefficient of thermal contraction, l is the distance from the closed valve, and dT/dt is the rate of cooling of the oil. — Assuming liquid oil to seep through the paraffin lattice at a rate governed by Darcy's law, the pressure gradient of flow is

$$\frac{dp}{dl} = v \frac{\mu}{k} \qquad 7.2\text{—}67$$

where k is the permeability of the paraffin network and μ is the dynamic viscosity of the liquid phase. — The instantaneous pressure gradient and the shear stress along the pipe wall are, by Eq. 1.3—14, related as

$$\frac{dp}{dl} = \frac{4\tau_i}{d_i} \qquad 7.2\text{—}68$$

which implies

$$\tau_i = \frac{ld_i}{4} \frac{dT}{dt} \beta_T \frac{\mu}{k}.$$

The shear stress along the wall will, then, increase from the closed valve backwards until it attains the yield stress $\tau_i = \tau_e$. Once it attains that value, the gel structure breaks up, and the rest of the pipeline will be filled with a heterodisperse suspension characterized by a lower, remanent shear stress τ_{re}. The equation can be solved for the length $l = l_g$ of the oil plug that remains solid:

$$l_g = \frac{4}{d_i \frac{dT}{dt}} \frac{k \tau_e}{\beta_T \mu}. \qquad 7.2-69$$

Analysing laboratory data, k can be determined using Eq. 7.2—67 and τ_e using Eq. 7.2—68. In laboratory-size piping, the gel structure will be readily destroyed by fast cooling; establishing the constants of the equations consequently requires a great deal of circumspection. In commercial-size pipelines, the likelihood of gel breakdown is reduced if, on gradual cooling, the cavities of contraction are filled with liquid oil. 'Makeup' is not necessarily required to fill these cavities. If a high-pressure pipeline is shut down, cooling will reduce oil volume and hence pressure. This, in turn, will bring about the (relative) expansion of the oil and the contraction of the pipe. All this will reduce and possibly forestall the formation of gaps in the gel filling the pipeline. As far as reducing the breakout pressure is concerned, the shorter the solid plug in the pipeline the better. In the interest of this, the filling-up of the cavities with 'makeup' oil is to be reduced or prevented.

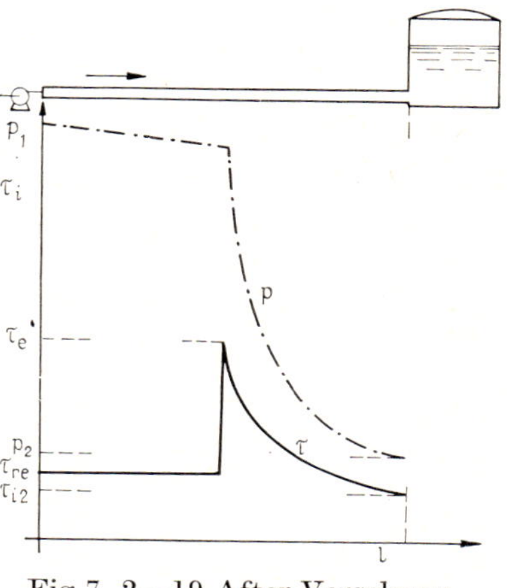

Fig.7. 2—19 After Verschuur et al. (1971)

Another factor affecting breakout pressure is the elasticity of the pipe and of the oil filling it. As a result of elasticity, the displacement and deformation of the gel will increase with pressure. Assuming the model shown as Fig. 7.2—19, where the tail end of the pipeline is under the hydrostatic pressure prevailing at the tank bottom, shear stress along the pipe wall and pressure v. distance along the pipeline are as shown in the Figure. The gel plug filling the pipeline will first undergo elastic deformation on the pumping of further oil into the pipe. Let displacement thus caused be

$$z_g = \frac{\tau_i d_i}{E_g} \qquad 7.2-70$$

where E_g is the modulus of elasticity of the gel. This relationship holds until shear stress along the wall attains the yield stress τ_e. The corresponding displacement is the critical deformation z_{ge}. In that situation,

$$\frac{1}{E_g} = \frac{z_{ge}}{\tau_e d_i} = \frac{\varepsilon_{ge}}{\tau_e} \qquad 7.2-71$$

where ε_{ge} is the critical deformation referred to the ID of the pipe, $\varepsilon_{ge} = z_{ge}/d_i$.

By the above consideration, if $\tau_i/\tau_e \leqq 1$, then

$$z_g = \frac{\varepsilon_{ge}}{\tau_e} d_i \tau_i \,. \qquad 7.2-72$$

The values of z_{ge} and τ_e are determined in U-tube tests at the laboratory. Factor c represents the effective compressibility of the system, including the elasticity of the pipeline, and the compressibility of the gel. The relative deformation due to a pressure p is, then,

$$\frac{dz_g}{dl} = pc \,. \qquad 7.2-73$$

The local pressure gradient is determined by Eq. 7.2–68. By Eqs 7.2—68, –72 and –73,

$$\frac{d^2 p}{dl^2} - \frac{4\tau_e c}{\varepsilon_{ge} d_i^2} p = 0 \,. \qquad 7.2-74$$

Assuming that $p = p_2$, $\tau_i = \tau_{i2}$ and $l = 0$ at the end of the pipe, this equation can be solved to yield the length $l = l_g$ of the pipe section in which the oil stays gelled:

$$l_g = \frac{1}{2} d_i \sqrt{\frac{\varepsilon_{ge}}{c\,\tau_e}} \ln \left[\frac{\tau_e + \sqrt{\tau_e^2 - \tau_{i2}^2 + \left(\frac{p_2}{2}\sqrt{\frac{\tau_e c}{\varepsilon_g}}\right)^2}}{\frac{p_2}{2}\sqrt{\frac{\tau_e c}{\varepsilon_g}} + \tau_{i2}} \right] . \qquad 7.2-75$$

This equation reveals the length of pipeline filled with gelled oil to be maximal if τ_{i2} is least. Pressure at the end of the pipe is then close to atmospheric. It is therefore to be recommended to apply pressure on breakout also at the tail end, in order to reduce l_g, either by shutting down, or by injecting oil from, said tail end. Equation 7.2 – 75 will correctly furnish the length of pipeline filled with gelled oil if the oil in the pipe is cavityless. In that case, c, the compressibility factor of the oil is approximately 7.5×10^{-10} m^2/N. If, however, there are cavities filled with vapour escaping from the oil, effective compressibility is very greatly increased. A cavity volume of 2 percent has been found to increase compressibility 250 times. This may substantially reduce the length of the gelled oil plug. This is another argument in favour of reducing makeup during the cooling of oil.

In addition to the factors outlined above, the gel parameters affecting startup are appreciably influenced by the mechanical and thermal prehistory of the gel, and by the fact that gel temperature will not as a rule be constant along the pipe (it will decrease as outlined in the previous section). Little is known as yet about the quantitative consequences of these factors.

7.2.8. Pipelines transporting hot oil

Heating oil to reduce its flowing pressure drop may be performed either by spot or line heating. In spot heating, oil is heated either before introduction into a pipeline section, or at booster stations. In this type of transmission, temperature will decrease downstream of the heater even if heat flow is steady-state. In line heating, the oil is heated all along the pipeline. This solution is usually restricted to shorter pipelines. In long-distance oil pipelines, spot heating is usually resorted to. One of the possible arrangements of a spot-heating system is shown schematically as Fig. 7.2—20. Oil stored in tank *1* is heated by steam circulating in heat-exchange coils, to a temperature T_{ta} lower than the start-up temperature T_1 of the pipeline, but sufficient to permit the oil to flow to pump *3*. This latter drives the oil through heat exchanger *4* into pipeline *6*. Heat absorbed in *4* and power dissipated in the pump will heat the oil from T_{ta} to T_1. Steam heating the coils in the tanks and in heat exchanger *4* is produced in boiler *5*. At the delivery end, oil enters tank *7*. At the end of the delivery period, pump *3* sucks light oil from tank *2* to flush out the high-viscosity crude; during shutdown, the line stays filled with the non-freezing liquid. On re-starting transmission, the cool pipe and its environment are heated with light oil, taken from tank *2* and heated before injection. During the break in the transmission of the high-viscosity crude, the light oil is returned by pump *9* through line *6* into tank *2*.

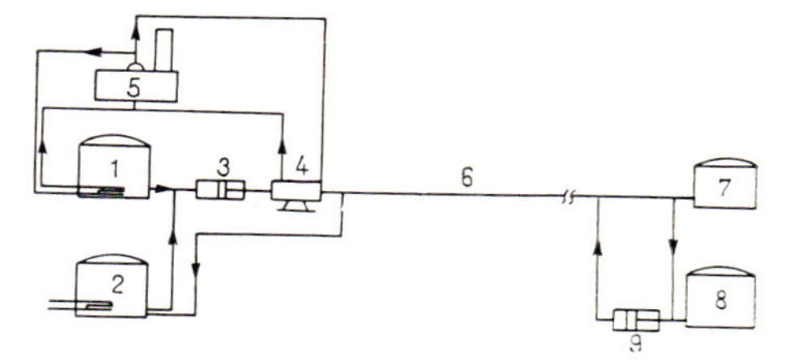

Fig. 7.2—20 Oil transmission system with spot heating

The economics of a pipeline transporting hot oil will be affected rather considerably by breaks in transmission. These breaks may be due to several causes. In planning pipeline operation it is necessary to reckon with variations in the throughput expected of the line. A transmission line is often started up even before the oilfield it is to serve has been fully developed; production is then less than the expected peak. Referring again to Diagram 7.2—16, it is necessary to consider the way the lowering of throughput affects the pressure drop in the pipeline. For instance, considering the diagram of the 24.1-km line, one sees that in the throughput range from 150 to 350 Mg/h, the pump discharge pressure required will hardly change. In the 32.2-km line, on the other hand, transporting 150 Mg/h of oil requires half as high a pump pressure again as transporting 350 m Mg/h. If the maximum allowable working pressure of the pipeline is less than the 88 bars to be expected at the throughput of 150 Mg/h, then continuous transmission is out of question. The higher pumping rate resulting in a lower pressure drop can be achieved only by intermittent transmission. — Further to be provided are means to permit the fast and possibly low-cost restarting of transmission after unforeseen breaks and malfunctions. With reference to the family of cooling curves in Fig. 7.2—15, it is easy to see that, after any interruption of transmission, the pump pressure needed for restarting is the less the sooner the high-viscosity crude in the pipeline is displaced by light oil. Figure 7.2—21 shows as an example (Ells and Brown 1971) how in a given case pump pressure varies in time if the displacement of crude in the pipe with light oil is started (i) immediately after interruption (Curve *1*) and (ii) after a delay of six hours (Curve *2*).

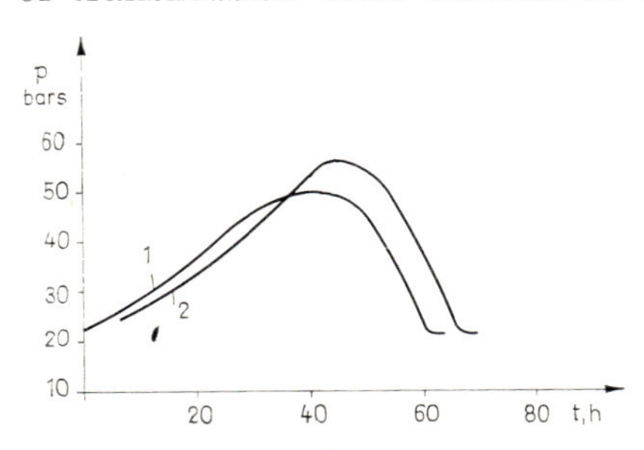

Fig. 7.2—21 Starting pressure v. time, after Ells and Brown (1971)

The equipment and operation of spot heating are to be designed so as to keep the cost of transporting the given quantity of oil over the given pipeline as low as possible. The number of technically feasible variants is so high that no theory taking into account all factors influencing the choice of the optimum solution has so far been put forward. There are, however, methods that permit optimization of individual parameters. For instance, Tugunov (1968) published a procedure for finding the optimum thermal insulation of a pipeline. Tugunov assumes the head-end temperatures of the flow, and the length and diameter of the pipeline, to be known, and the oil to be Newtonian. His variables are the heat insulation to be provided and, accordingly, also the number of intermediate heater stations. — Westphal (1952) compares the economics of operation of insulated and uninsulated pipelines of a given size at various rates of throughput. He finds that, in the case considered, uninsulated pipe will be more economical at comparatively low throughputs and vice versa. — Jablonsky devised a procedure for finding the optimal start-up temperature of oil. T_{opt} is that temperature at which the aggregate cost of heating and pumping is lowest. In his method of calculation, Jablonsky assumed heat transfer factor k to be constant along the pipeline. Abramzon (1968) solved the same problem for variable k, whereas Muradov and Mametklychev (1970), accounted also for the heat liberated by the solidification of paraffin.

As an example for a modern spot-heated hot-oil pipeline let us cite the one built by Getty Oil Co., one of the largest of its kind (O'Donnell 1968, Griffith 1962). The uninsulated 20-in. underground line of 280 km length transports round 26,000 m³ of crude per day. Tankage is restricted to the outgoing station. At the booster stations, there are heaters, pumps and accessories only. The total pump power is 9.3 MW; the aggregate heating power is 59 MW. The line is operated by a Motorola type process control which acquires and processes 250 measurement and position data every 15 s. The control ensures the correct delivery temperature; it governs according to program and records the transmission of various batches of oil; controls the opening and closure of the right valves when scrapers pass booster facilities, the programwise operation of screw pumps on startup, the cut-in

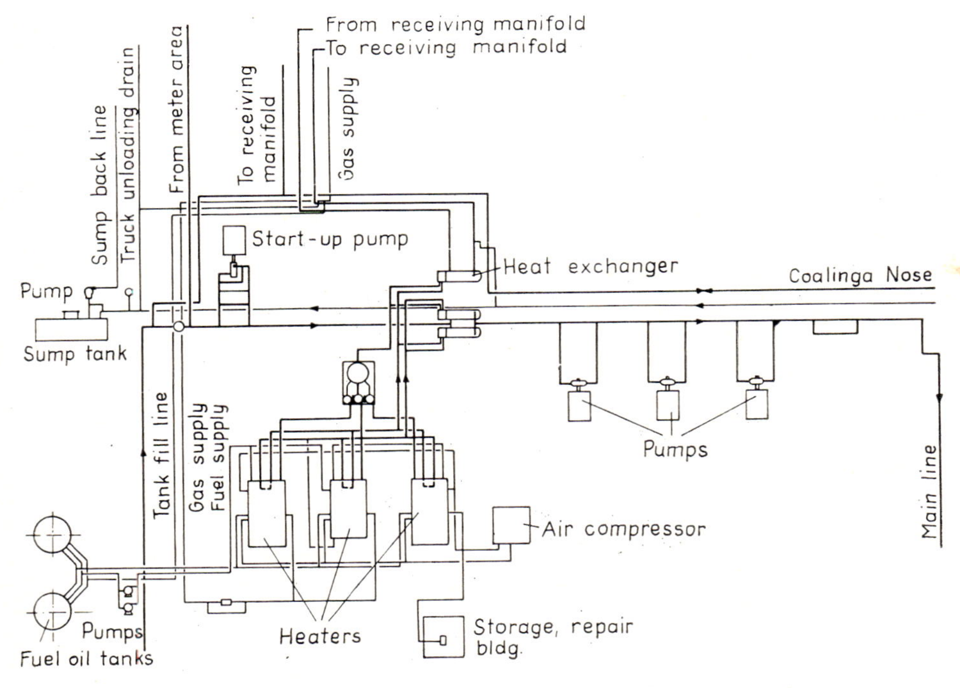

Fig. 7.2—22 Coalinga station, outgoing facilities, after O'Donnell (1968)

of safety equipment, and a number of other operations. Figure 7.2—22 shows outgoing facilities at the Coalinga station; Fig. 7.2—23 shows the layout of a booster station (O'Donnell 1968).

In line heating, heat may be supplied by hot water, steam or electricity. In the two former cases, the heating medium will flow in pipes parallel to the oil pipeline. The heating pipe may be either coaxial within or without the oil pipe (Fig. 7.2—24a, b), or parallel to it on the outside (c, d). In (a) the temperature of the outer pipe wall is lower and heat loss is consequently reduced, but the oil line cannot be scraped. If the heater line is non-coaxial (c, d), heat transfer from heater line to oil line is worse, but the size of the oil pipe may be less for a given pressure drop. The use of heater lines is restricted as a rule to short pipelegs (less than one km in length). Table 7.2—3 (after Pektyemirov 1951) helps in choosing the number and size(s) of heater lines. A modern way of insulating the parallel-heating arrangement is dis-

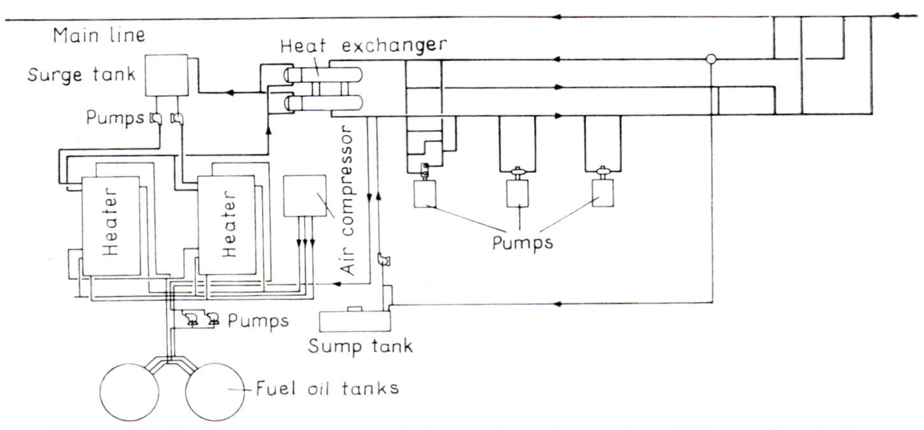

Fig. 7.2—23 Booster pump and heating station, after O'Donnell (1968)

cussed by Eichberg (1970). The 3-in. pipe is provided with two parallel heater lines of 1 1/2 in. size, conveying hot water. The common insulation is a polyurethane wrapping under a common PVC-foil sheath. Directly after installation on the pipe ensemble, this latter can be fastened for good by a sort of zip fastener. — A novel way of electrically heating high-viscosity crudes and fuel oil piped overland has been developed in Japan. In the system, called SECT, one or more pipes of 1/2—1 1/2 in. size are fastened to the outside of the pipeline to be heated, and electrical cables are installed in them. The current passes partly in the wall of the pipe to be heated, and thus warms it directly, ensuring a heat flow of 15—160 W/m. This insulated line appears economical at flowing temperatures below 50 °C and daily repeated warming.

Means of removing gelled oil from spot-heated pipelines must be provided to take care also of cases of protracted breakdown when none of the means outlined above suffices to break out the gel. If breakdowns of this sort are to be expected, the pipeline is to be divided up into shorter lengths and the gel is to be broken segment by segment by means of comparatively low-power portable pumps (Hungarian Patent SI—IIII. 1971. T/1891). As a last-resort means of starting up small-size pipelines, the method shown in

Table 7.2—3

Compatible sizes of oil pipe and of steam pipe used to heat it (after Pektyemirov 1951)

Oil pipe size in.	Steam pipe	
	number	size, in.
2—3	1	1/2
4—5	1	3/4
6	1	3/4—1
8	2	3/4

Fig. 7.2—25 may be used. Pipeline *1* is provided with blowoff pipes *2*. The length of line between two blowoff pipes equals the length of the longest gelled crude plug whose breakout is feasible at the maximum allowable pressure of the pipe. The gelled oil is expelled section by section through the blowoff pipes by injecting low-viscosity oil, water or gas (Szilas 1966).

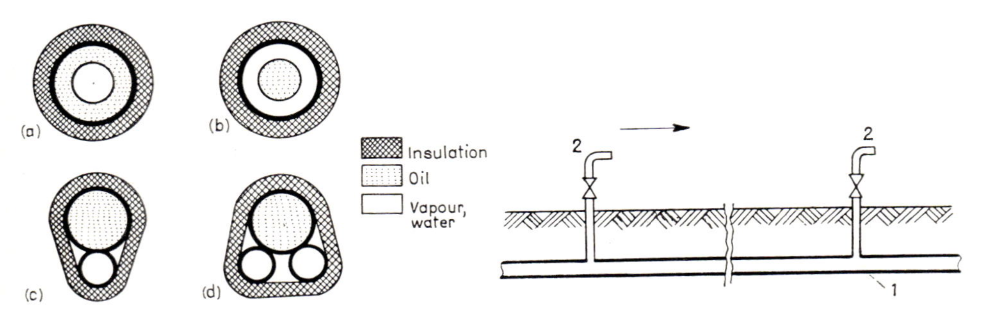

Fig. 7.2—24 Line heating designs

Fig. 7.2—25 Arrangement for the breakout of gelled oil from a pipeline, after Szilas (1966)

7.3. Means of improving flow characteristics

7.3.1. Heat treatment

As has been discussed in Section 1.3.1, which describes flow behaviour of thixotropic pseudoplastic crudes and the causes underlying it, the macrostructure of the solidified paraffins, as a function of the rate of temperature decrease, may vary rather widely. If the crude is heated to a temperature high enough to dissolve all the paraffins and is then suddenly chilled, then a multitude of minute independent paraffin crystals will form, which do not aggregate into a connected three-dimensional lattice even after longer storage at the terminal temperature. The flow behaviour of this dispersed system may be rather more favourable than that of a crude in which paraffin, solidified on slow cooling, constitutes a connected three-dimensional lattice. Heat treatment exploits these properties of paraffin crystallization in improving flow behaviour. Prior to designing the heat treatment system it is necessary to establish by laboratory testing the temperature at which all paraffins dissolve, on the one hand and, on the other, the slowest rate of cooling that will still ensure the desired improvement in flow behaviour and, finally, the lowest terminal temperature, below which chilling will practically not affect flow behaviour any more.

The diagram in Fig. 7.3—1, compiled at the Petroleum Engineering Department of the University of Heavy Industries Miskolc, (Hungary), points out the importance of correctly choosing the heating temperature. A crude from the Algyő field in South Hungary, of 20 °C temperature originally, was treated by first heating to various temperatures and then cooling back to 20 °C at a constant rate. In the case investigated, the liquid with the worst flow properties came to exist after heating to about 50 °C. On further increasing the heating temperature, flow properties improve somewhat, but do not attain those of the original untreated oil, until about 70 °C. Optimum warming is attained between 80 and 85 °C. The

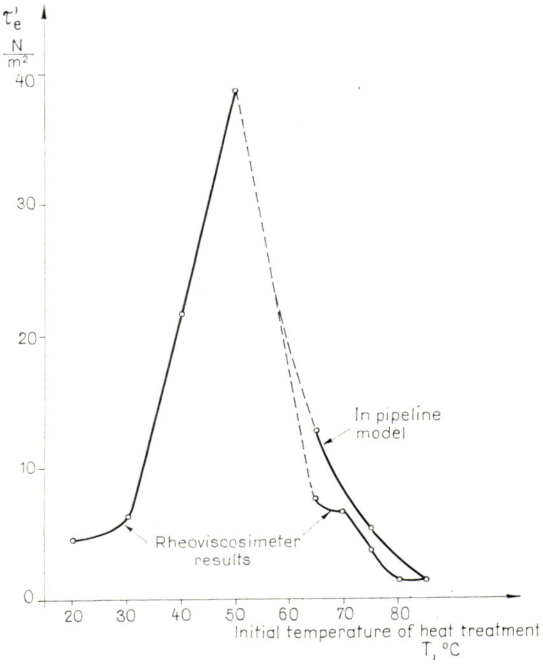

Fig. 7.3—1 Influence of starting temperature of heat treatment upon apparent yield stress, after Szilas (1971)

explication for this situation is that warming will dissolve part of the previously deposited paraffin. If at the onset of cooling there are some vestiges of a paraffin lattice, the separating paraffin will deposit on these, cementing them together. On return to the original temperature, a new lattice will form, differing from the original one in strength and shear resistance (Szilas 1971). — Rapid cooling from a high enough initial temperature will lead to the formation of small separate crystal grains, because the depositing paraffins do not have enough time to find the nuclei of crystallization, previously formed or present in the liquid to start with, and thus to constitute mixed crystal aggregates of ever-growing size. The nuclei will adsorb the resinous components first of all, which reduce the affinity of association among paraffin crystals. The minute crystal grains thus developing may stabilize to a considerable extent. Cooling rate should typically be 0.3—1.0 °C/s. Increasing the rate will not improve flow properties beyond a certain limit. This limit is to be established by experiment. Too fast a rate will increase treatment cost without any resulting advantage. According to Milley, optimum cooling is comparatively fast down to a temperature a few °C above the cloud point, and may be slower below it. It is to be recommended to perfect the optimum furnished by the laboratory tests in a pilot experiment. In heater-treaters of commercial size, the temperature gradient between the crudes next to the cooling tubes may be quite significant. — The terminal temperature of cooling is to be chosen so as to solidify most of the paraffin in the crude, because if a substantial percentage of paraffin remains in solution after treatment, it will deposit on existing nuclei in the course of slow cooling in the storage tank or in the pipeline and thereby worsen the flow properties of the crude. — Changes in flow properties as a result of heat treatment may be widely different in different quality crudes. A substantial improvement is to be expected if the crude is high in paraffin and comparatively low in light fractions. Typically, for instance, heat treatment of the Algyő crude had the same effect as the addition of 10 percent gasoline. On the other hand, after the addition of 10 percent gasoline, heat treatment does not improve flow properties any further.

Table 7.3—1 presents the variation in flow properties of various Soviet crudes v. various initial temperatures of heat treatment. — The first commercial application of heat treatment took place at Nahorkatiya and Moran in India (*Oil and Gas Journal* 1963). Even today these are the largest heater-treater stations described in literature; they are modernized gradually as time goes by (Chandrasekharan and Sikdar 1970). At Nahorkatiya, the crude whose pour point ranges from 29 to 34 °C and whose paraffin content is 15.4 percent, is transported through a 16-in. pipeline of 402 km length to a refinery where part of it is refined; the rest is transported through a 14-in. line to another refinery at a distance of 765 km from the first one. The first line segment was designed for a throughput of 2.75 Tg, the second for a throughput of 2.0 Tg per year. The lowest soil temperature at pipe depth 1.2—1.8 m is 18 °C. Heat treatment is required between October and April. It entails a significant improvement in flow behaviour. Typically, at 18 °C temperature and a shear rate of 13 s^{-1}, the apparent viscosity is 0.6 Ns/m^2 in untreated oil and 0.1 Ns/m^2 in treated oil. — The Nahorkatiya crude heat treater station is shown as Fig. 7.3—2 after Chandrasekharan and Sikdar (1970). The crude is heated in gas-fired heat exchangers to between 90 and 95 °C. Subsequent cooling to 65 °C is effected in a heat exchanger, using crude to be preheated as the coolant. In the next step, the static crude is chilled with water to 18 °C. Each of the 36 treater tanks is an upright cylindrical tank 6.1 m high and of 9.2 m diameter; each tank contains 127 one-inch chilling tubes. It is the cooling water that flows in the tubes and the crude that occupies the space between them. Stationary cooling takes place in the following steps. The empty tank is filled with crude, which is chilled by the circulation of cold water at a rate of 0.6 °C/s. After cooling to 18 °C, the crude is pumped to a storage tank, and the tubes are used to circulate

Table 7.3 – 1
Heat treatment efficiency parameters of Soviet crudes
(after Mikhalkov and Vosrikova 1967)

Oilfield	Warm-up temperature °C	Viscosity, 10^3 Ns/m², at			Pour point	Temperature where τ_e' appears
		0	10	20		
		°C				
Caspian	X	—	124	34	19	>20
	70	—	178	46	>23	>20
	80	—	180	39	23	>20
	90	—	58	15	12	>20
	98	—	31	14	0	>20
Tengutinsk	X	45	21	7	15	20
	40	48	227	8	18	>23
	50	35	17	6	14	21
	60	25	11	5	2	14
	70=75	26	9	?	0	14
	85	18	11	5	0	14
Zhirnovsk	X	53	28	16	−1	17
	25	70	32	16	?	?
	30	51	31	16	?	?
	35	41	25	16	<−17	<0
	40	40	24	16	<−17	<0
	90	40	26	17	<−17	<0
Korobkovsk	X	11	6	?	<−21	0
	40	12	6	?	<−15	?
	50	9	5	?	<−21	<0
	70	8	5	?	<−21	<0
	80	7	5	?	<−21	<0
Slyachovsk	X	12	11	?	+1	+17
	40	15	9	?	−6	+14
	45	?	8	?	−11	+7
	55	14	?	?	−14	+4
	70	12	7	?	?	?
	80	11	6	?	−21	<0
	95	12	6	?	−21	<0

X without heating
>23 higher than 23 °C
<−17 lower than −17 °C
— unmeasurable
? not available

hot water, which again raises tank temperature to 65 °C. This melts the deposited paraffin and prepares the tank for the reception of another batch of crude. These heating and cooling operations are performed by means of five water circuits, each separate from the others. (i) High-pressure hot water heated with gas to 145 °C and held under sufficient pressure to keep it below its bubble point is used to heat low-pressure hot water and to regenerate the concentrated lithium bromide coolant of the absorption cooler. (ii) Low-pressure hot water is used to heat the

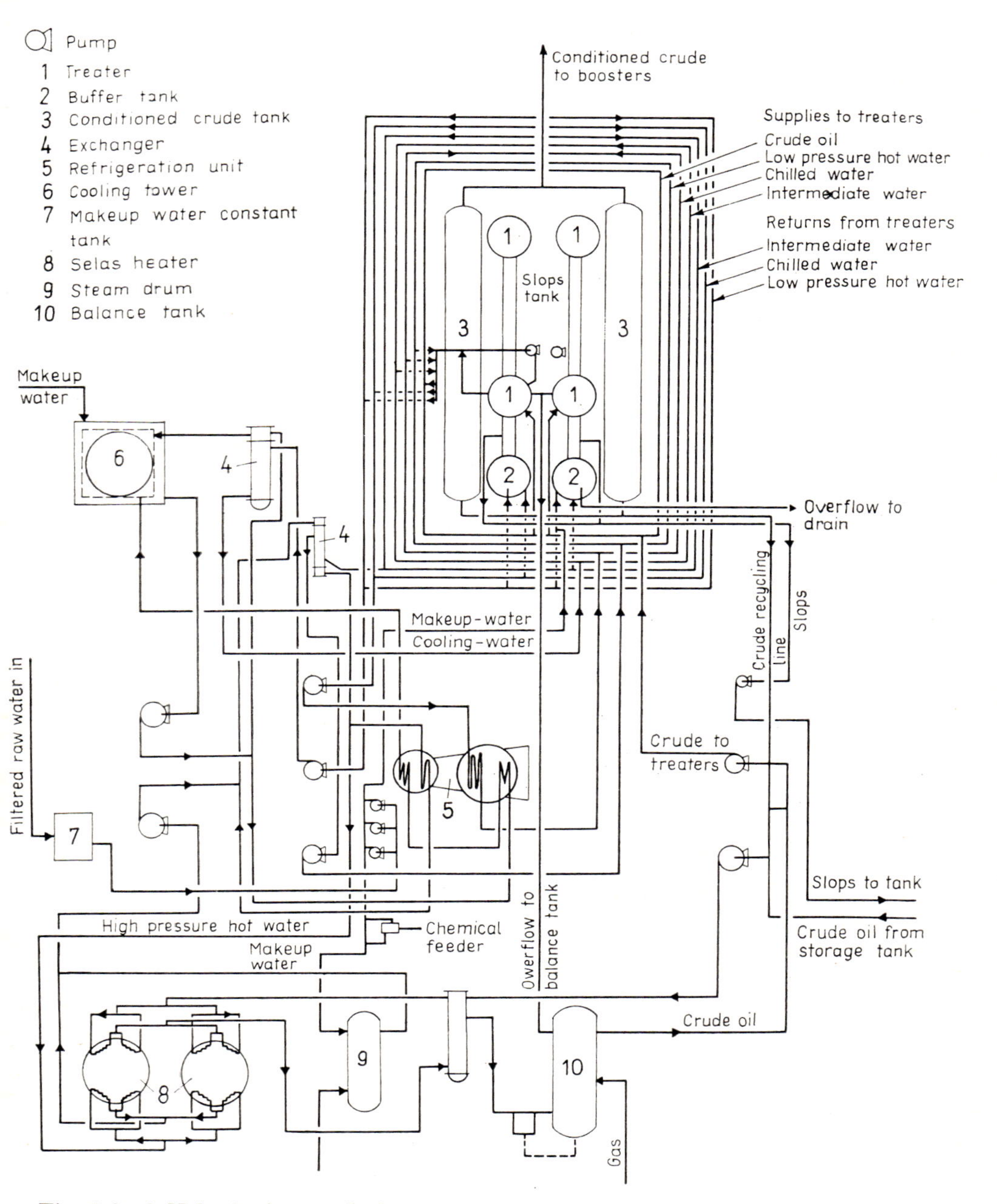

Fig. 7.3—2 Nahorkatiya crude heat treater station (Chandrasekharan and Sikdar 1970)

treaters to 65 °C after the chilling of each batch. (iii) 'Intermediate' water is used to cool oil in the first phase of cooling. It is kept at a temperature of 30 °C by 'cooling' water cooled in a cooling tower. (iv) Cold water is used to chill the crude to 18 °C in the treater tanks. This water is cooled to 14 °C in an absorption refrigerator. (v) Cooling water, circulated in an open circuit including a cooling tower, is used to cool the intermediate water. — The 324 valves of the conditioning plant are controlled by an electric Flex-o-Timer. The water circuits are controlled by a program controller which adjusts the required temperatures.

7.3.2. Solvent addition

The mixing together of two crudes or products comparatively low in malthenes, differing from each other in flow behaviour, will result in a mixture whose flow properties will usually fall between the properties of the two components. It is possible to improve by this means the flow behaviour of high-paraffin, high-viscosity gelling crudes, e.g. by the addition of a lower-viscosity crude, or a liquid product such as gasoline, kerosene or diesel fuel. Attempts have been made at predicting the behaviour of mixtures starting from those of the components. No laws of general scope have been derived as yet, but important research was carried out to determine empirical relationships (Aliev et al. 1969).

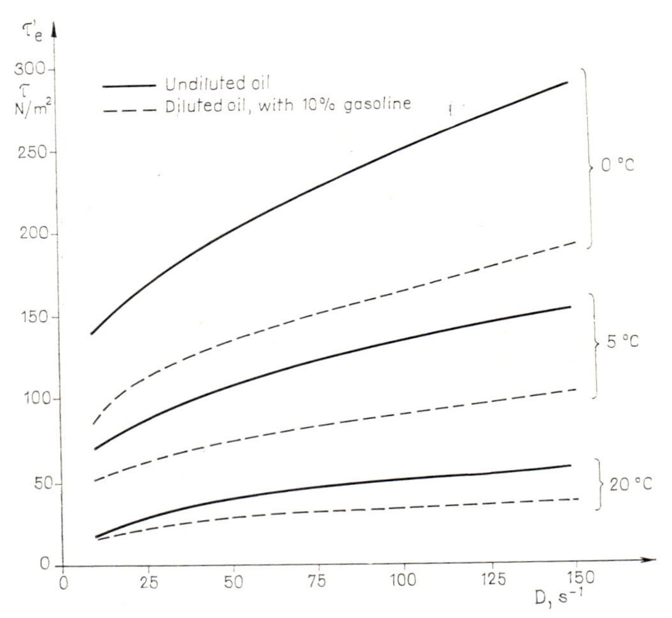

Fig. 7.3—3 Steady-state flow curves of Algyő crude, untreated and diluted with 10 percent gasoline

The exact manner of determining the influence of a solvent on steady state transmission is to record flow curves of the crude and of the mixture at various temperatures, and to calculate, in the manner explained in Section 1.3.5, pressure gradients v. throughput at the various temperatures to be expected during pipeline operation. Figure 7.3—3 shows steady flow curves for three temperatures of an Algyő crude determined by means of a Haake-type rotation viscosimeter at the Petroleum Engineering Department of the University of Heavy Industries, Miskolc (Hungary).

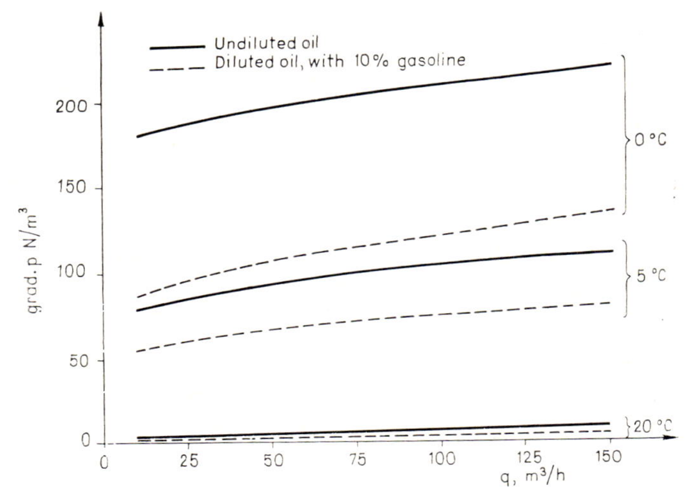

Fig. 7.3—4 Steady-state gradient curves of Algyő crude, untreated and diluted with 10 percent gasoline

Using the method outlined in Section 1.3.5, we have established from these flow curves the pressure gradients to be expected in a 12-in. transmission pipeline. The gradient v. rate of flow curve is plotted in Fig. 7.3—4. Clearly, both the crude alone and the mixture exhibit pseudoplastic behaviour. The addition of gasoline reduces the pressure drop of flow. The reduction is the greater, the cooler the flowing crude. Observations show the improvement due to the solvent to be the greater the greater the structural viscosity of the crude.

The solvent will reduce apparent yield stress, τ_e' too. Figure 7.3—5 is a plot of yield stress v. temperature, as determined in a pipe model, for the same Algyő crude as before, untreated, and with 10 percent gasoline added. The drag reduction due to the addition of gasoline is obvious. Owing to the uncertainties of the measurement method, no conclusions of a more general validity can be drawn from these two graphs.

The economics of diluted transport flow is considerably improved if the solvent is an unrefined product (raw gasoline, low-viscosity crude) produced in the vicinity of the waxy crude, which can be refined together

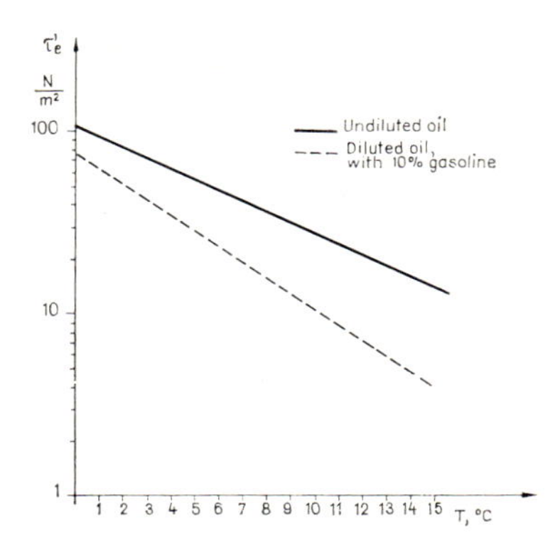

Fig. 7.3—5 Yield stress v. temperature of Algyő crude, untreated and diluted with 10 percent gasoline

with the crude after arrival at the refinery. Costs are higher if no natural solvent is available, and the solvent used has to be recovered by reboiling after delivery and returned to the head end of the pipeline. The schematic diagram of a transmission system of this sort is shown as Fig. 7.3—6 (Szilas 1966). Pump *1* sucks high-viscosity crude from storage tank *2*.

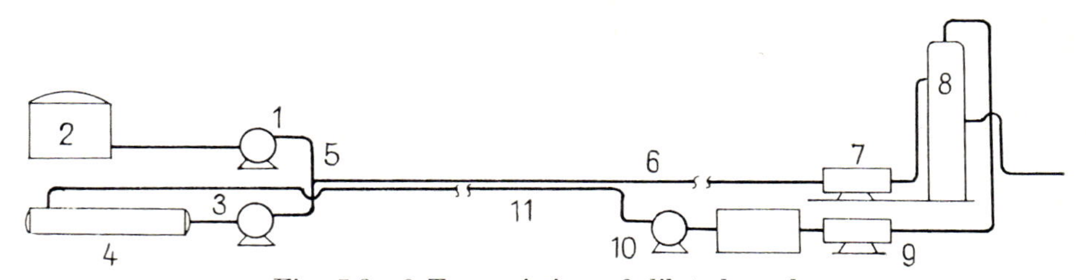

Fig. 7.3—6 Transmission of diluted crude

Pump *3* sucks solvent from horizontal tank *4*. The two pumps inject the two liquids in the prescribed proportion through mixing throat *5* into pipeline *6*. The mixed oil is heated at the delivery end in heat exchanger *7* and the solvent is evaporated in tower *8*. The evaporated fraction is cooled and liquefied in heat exchanger *9*, and returned by pump *10* through line *11* into tank *4*. — The mixing ratio of crude and solvent is to be matched to the changing ground temperature.

7.3.3. Chemical treatment

Flow parameters can be improved by adding certain chemicals to the crude. The chemicals in question belong to two distinct groups. The first includes compounds that enter and modify the paraffin lattice during the cooling of the crude. The compounds of the second group constitute solvates in the oil and, assuming a preferred orientation parallel to the main direction of flow, prevent the radial displacement of liquid particles in turbulent flow, thus reducing the energy dissipation due to impulse exchange. Lattice-modifying additives include the so-called P-inhibitors. Molecules of this inhibitor adhere to the edges or dislocations of the paraffin crystals formed during the cooling of the crude. This results in most of the paraffin forming minute crystal grains, owing to the reduction in the adhesion affinity between paraffin particles and in the tendency to form aggregates (Price 1971). — According to Price, the P-inhibitors found to be useful in practice belong to two groups of chemical compounds, notably ethylene copolymers and higher polymers. In this latter, the main chain of the polymer has side chains of n-paraffins. Both types of compounds are anchored by these side chains to the paraffin lattice.

Of the additives tried out in European practice, the best-known ones bear the code designations ECA 841 and ECA 5217 (Brod et al. 1971). ECA 841 is a paste at room temperature; it dissolves readily in aromatic solvents such as benzene and its homologues. ECA 5217 is jelly or paste-like at room temperature; it dissolves well in these solvents, too. Both compounds are presumably esters with methacrylate type polymer chains attached to them. The effects of the two compounds were investigated both in the laboratory and in the field. Table 7.3—2 (after Brod et al. 1971) shows how the flow behaviour of a 4 : 1 mixture of an African and a Middle-East crude was modified by the addition of 0.15 percent ECA 5217. The properties improved by the additive were retained even after storage in a tank for 25 days.

Table 7.3—2

Flow parameters of a crude, untreated and treated with a 0.15% solution of ECA-5217 (after Brod et al. 1971)

Flow parameters at 4 °C	Oil	
	untreated	treated
ASTM upper pour point, °C	21	6
Plastic viscosity, Ns/m²	0.112	0.058
Static shear stress, N/m², measured with		
rheoviscosimeter	32	0.43
TAL pipe model	35	1.0

The extent to which flow properties are improved by the additives depends on concentration and temperature. Figure 7.3—7 summarizes the results of experiments with ECA 841, performed at the Petroleum Engineering Department of the University of Heavy Industries, Miskolc (Hungary). The points plotted were established for Algyő crude at 4 °C temperature and 32 s^{-1} shear rate, using a Haake rotation viscosimeter. The parameter is the temperature at which the inhibitor was added. The shear stress at a given concentration of additive is seen to be the less (within the range of concentrations examined) the higher the temperature of addition. This is explained by the nature of the adsorption mechanism.

The additive is the more effective, the less the separated paraffin in the oil at the time when the additive is added and, hence, the smaller size the structures formed by the paraffin crystals. The difference in shear stress is considerable between 40 and 50 °C. Minimum shear stress was almost identical in the three cases. This value was ensured by adding just 0.02 weight percent ECA 841 at temperatures of 50 and 80 °C, whereas 0.05 percent was required to have the same effect at 40 °C. The flow curve

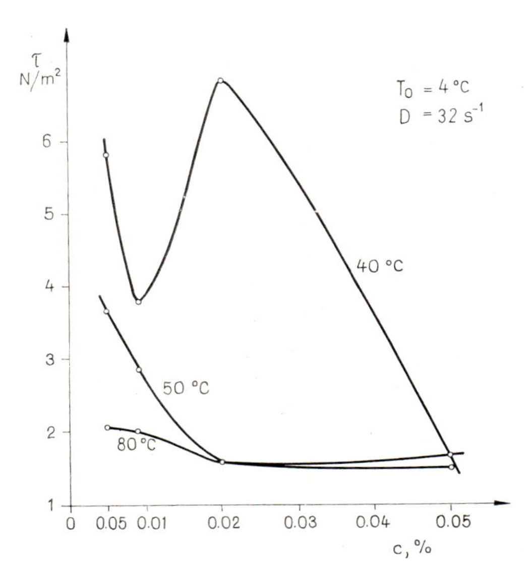

Fig. 7.3—7 Influence of P-inhibitor ECA 841 upon shear stress in Algyő crude

of the optimally treated oil can be determined by means of the rotation viscosimeter, in the knowledge of the optimum temperature of addition. Using these data the pressure gradients to be expected in the pipeline can be calculated, as explained in connection with Figs 7.3—4 and 7.3—5.

It was in 1954 that drag reduction in both Newtonian and non-Newtonian fluids under the influence of certain additives was first observed. Experience has shown the most effective drag reducers to be long straight-chained lyophilic polymers. In polar liquids, e.g. water, the most effective molecules are those which include polar groups such as the carboxy group. One such compound is CMC (carboxy-methyl-cellulose). These macromolecules form solvates in the fluid to which they are added (Elperin et al. 1966). In the oil industry, drag-reducing additives are added mainly to the water-based injection fluids of hydraulic formation fracturing. No such additives have so far been commercially employed, as far as I am aware, in the transmission of crudes and products. However, Lescarboura et al. (1970) state that the first laboratory and field tests into drag reduction in light crudes and diesel fuel have already been performed. The additives tested included the high-molecular-weight hydrocarbon polymer of code designation CDR and several variants of a polyisobutylene called Vistanex. At a given concentration, these additives were found to be the more effective the more readily they dissolved in the fluid to be treated. The presence of polymers may reduce the drag in turbulent flow in two ways. In the flowing liquid, the elastic macromolecules arranged parallel to the flow direction hinder the radial movement of the flowing liquid's particles, and adsorb and temporarily store the energy that would otherwise get dissipated by impulse exchange (Pruitt et al. 1965). The other effect involved is that some of the macromolecules will adhere to the pipe wall and reduce its relative roughness (White 1964). The presence of polymers augments the apparent viscosity of the liquid transported, and imparts non-Newtonian properties even to Newtonian fluids.

The pressure drop to be expected in a pipeline cannot be predicted on the basis of flow curves determined by means of the rotation viscosimeter. This is due to two factors. One, the rotation viscosimeter will not evaluate the influence of pipe roughness. Two, according to certain hypotheses (Elperin et al. 1966) flow in the pipe may arrange the polymer molecules so that their concentration will differ from the axis to the pipe wall. Lescarboura et al. (1970) give a procedure for predicting the pressure drop to be expected. Strictly, this procedure holds only for the fluids and additives tested, but it will give a rough idea of what to expect with other fluids and additives, and permits to estimate the extent of drag reduction. Laboratory experiments were performed in one-in. pipe of 3.7 m length, in a flow velocity range from 0.6 to 4.9 m. These were succeeded by field experiments in 45 km of 8-in. and 51.5 km of 12-in. line, respectively. It was found that the drag reduction $\Delta p_{dr}/\Delta p_o$ due to a given additive in a given fluid depends on additive concentration c_{dr}, flow velocity v and pipe diameter d_i. The drag reduction due to the additive termed CDR in the 8- and 12-in. pipes tested (and, in a fair approximation, all over the pipe size range from 4 to 16 in.) is

$$\frac{\Delta p_{dr}}{\Delta p_o} = \frac{2.19\, v - 1}{\dfrac{c_{dr}\, v}{0.32\, c_{dr} - 30.9} + \dfrac{1000}{76.6 + 0.098\, c_{dr}}} \qquad 7.3\text{—}1$$

where c_{dr} is the additive concentration in parts per million (10^{-6} m^3/m^3).

Example 7.3—1. Find the percent drag reduction in light oil flowing at 2.3 m/s velocity with CDR added in a proportion of 1000 ppm. — Equation 7.3—1 yields

$$\frac{\Delta p_{dr}}{\Delta p_o} = \frac{2.19 \times 2.3 - 1}{\dfrac{1000 \times 2.3}{0.32 \times 1000 - 30.9} + \dfrac{1000}{76.6 + 0.098 \times 1000}} = 0.30$$

that is, drag reduction due to the additive amounts to 30 percent.

By Eq. 7.3—1, drag reduction increases with increasing flow velocity. The maximum to be expected in practice is 50—60 percent. One thing to be given the proper attention in choosing an additive is that the 'strength' of the polymer molecule must be sufficient, that is, it must not be degraded by the buffeting it receives in the flowing liquid. Drag-reduction efficiency decreases as degradation proceeds.

7.3.4. Oil transport in a water bed

A fairly large number of publications since 1950 have contained theoretical considerations and experimental data concerning the common transport of oil and water in the same pipeline. These publications assert if during flow the phase in contact with the pipe is the water rather than the oil, and the oil remains enclosed in a sort of water sheath, then the drag of flow and the driving power required may be significantly less even though the throughput is increased by the addition of water. Figure 7.3—8, a plot of M. E. Charles' experimental data, shows the pressure needed to pump the oil alone, referred to the pressure demand of the oil–water mixture, v. the water content of the mixture (Szilas 1966). The pressure reduction factor is seen to attain a maximum of about 9 for a water cut of 0.3—0.4. The practical application of this idea was hampered by the fact that the two phases, flowing together, tend to form highly unstable flow patterns; it was especially in pipelines laid over hilly terrain that the two phases were least likely to flow the way they were expected to. — In the last few years, the common transport of water and oil was attempted in a different way, in the form of 'oil-in-water' emulsions prepared by adding surface-active substances to the water. The viscosity of this type of emulsion is independent of oil viscosity. Laboratory tests by Rose and Marsden (1970) showed that the viscosity of the emulsion is best described by the Richardson formula

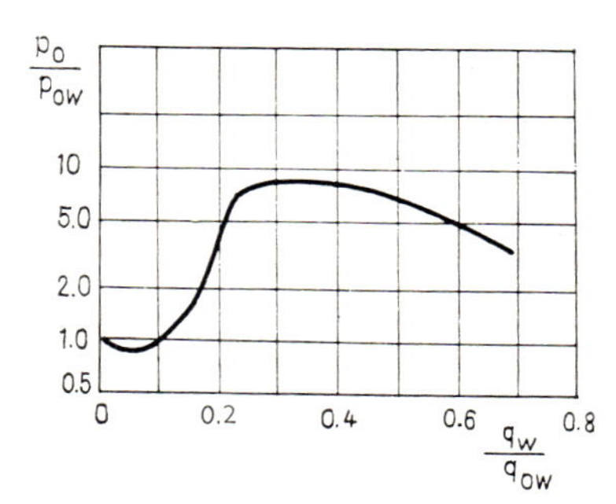

Fig. 7.3—8 Power to drive pure crude and crude in water bed v. water content, after M. E. Charles

$$\mu_{ow} = \mu_w \exp\left(a \frac{q_o}{q_{ow}}\right) \qquad 7.3\text{—}2$$

where a is a constant. — In the cases investigated, the viscosity of the emulsion was much lower than that of the oil. At $q_o/q_{ow} = 0.5$, drag was one-ninth of the drag of oil alone.

Important research in this field was carried out by Soviet workers, who experimented in long pipelines with Mangyshlak crude and heavy fuel oil, respectively, in emulsion with water treated with sulphonol. A relationship suitable for calculating a valid friction factor for the common flow of the two phases was given by Gubin and Stepanyugin (1970a) as

$$\lambda_{ow} = \frac{0.390 R_\sigma^{-1.13} \left(\frac{q_o}{100\, q_{ow}} \right)^{n_1}}{(245 A^{1.10} N_{\mathrm{Reow}})^{n_2}} \qquad 7.3-3$$

where

$$R_\sigma = \frac{\sigma_{ow}}{\sigma_o + \sigma_w}\,; \qquad A = \frac{\mu_w^2}{\mu_{ow}\,\sigma_{ow}\, d_i}\,;$$

$$N_{\mathrm{Reow}} = \frac{v_{ow}\, d_i}{\nu_w}\,;$$

$$n_1 = 3.50\,(1 - 3.82 R_\sigma)^2\,; \qquad n_2 = 1.75 \times (1 - 3.82 R_\sigma).$$

This relationship shows the friction factor to be independent indeed of the flow properties of the oil phase. The influence of temperature is very slight: it will merely change the viscosity of the additive-treated water. Investigations showed the active material to coat the water-oil interface and the pipe wall. Adsorption takes place in some 30 min. Using the flow equations thus derived, a calculation procedure was developed permitting the optimization of transmission cost (Gubin and Stepanyugin 1970b). Figure 7.3—9 shows optimal q_o/q_{ow} ratios for various length of pipeline v. annual throughput, calculated by this method. Clearly, the greater the throughput, the higher the optimal relative water content. For a given throughput of oil, $(q_o/q_{ow})_{\mathrm{opt}}$ is the less the longer the pipeline. Figure 7.3—10 shows optimal pipe size, likewise v. oil throughput. The optimal size for a given length is the larger the greater the throughput and, for a given throughput, it is the smaller the longer the pipeline. In both Figures, the numbers on the graphs indicate pipeline length in thousand-kilometre units. — At the delivery end, the emulsion must, of course, be rectified, and the water must be removed. In order to prevent pollution of the environment, it is necessary to add to the water a chemical that biologically neutralizes or decomposes the surface-active additive (Gubin et al. 1971).

Oil that is solid at the flow temperature in the pipeline *can be transported in suspension in water*. The oil grains have been shown to concentrate

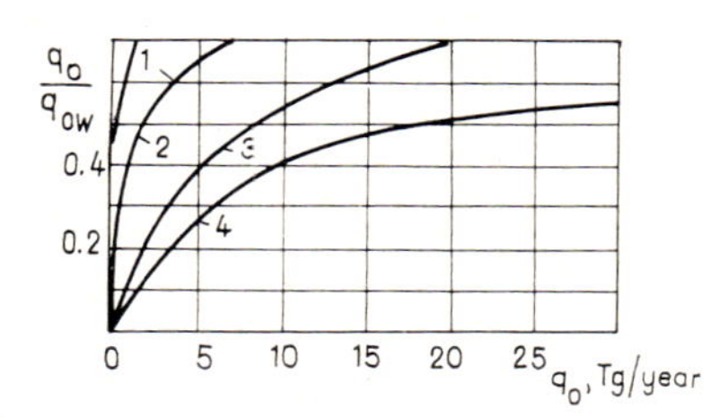

Fig. 7.3—9 Optimum O/W ratio of transmission in a water bed, after Gubin and Stepanyugin(1970b)

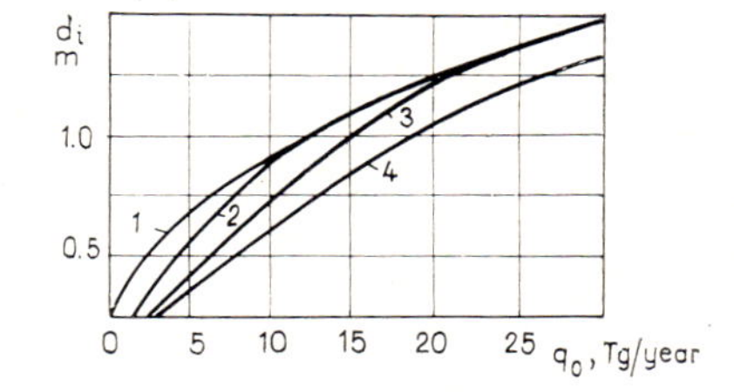

Fig. 7.3—10 Optimum pipe size of transmission in a water bed, after Gubin and Stepanyugin (1970b)

along the pipe axis, whereas the fluid next to the pipe wall is practically pure water. The drag is accordingly slight. The suspension is thixotropic-pseudoplastic. Flow in a pipe is determined by the slippage of the water 'sheath', shear rate, shear duration and flow temperature (Lamb and Simpson 1963). The transport of oil suspended in water was first employed in 1962 in Indonesia. The schematic diagram of the transmission system

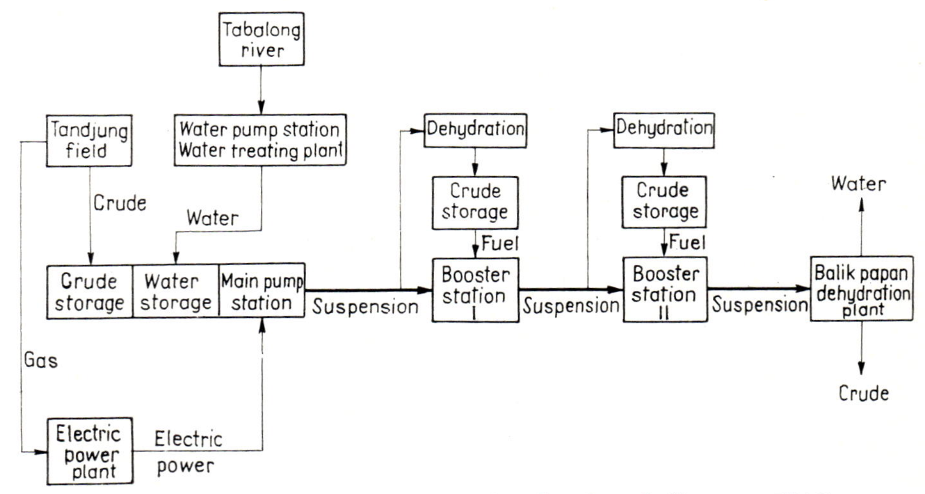

Fig. 7.3—11 Tandjung pipeline, after Lamb and Simpson (1963)

is shown as Fig. 7.3—11. The oil of about 41 °C pour point, containing 33 percent paraffin, is transported at a rate of 2 Tg per year in a 20-in. line of 238 km length from the Tandjung field to the Balik Papan refinery. This method permitted the decrease in apparent viscosity at a flow velocity of 0.7 m/s from 10^2 Ns/m^2 to 0.4 Ns/m^2; 70 percent of the fluid is gelled oil; the rest is water (Lamb and Simpson 1963).

Chapter 8

Pipeline transportation of natural gas*

8.1. Physical and physico-chemical properties of natural gas

In the following we shall be concerned only with those physical and physicochemical properties that affect the transmission of gas in pipelines; even those properties will be discussed only in so far as they enter into the relevant hydraulic theories.

8.1.1. Equation of state, compressibility, density, gravity

It is the gas laws or, in a broader scope, the equations of state concerning natural gas that describe the interrelationships of pressure, specific volume and temperature, all of which may be subsumed under the notion 'pVT behaviour' of the gas. Natural gas is not ideal, and the deviation of its behaviour from the ideal gas laws is seldom negligible and occasionally quite considerable. For the purposes of practical calculations, those equations describing the pVT behaviour of real gases are to be preferred that account for deviation from ideal-gas behaviour by a single correction factor, based on a consideration valid for any gas. Such a consideration is the *theorem of corresponding states*, expressed in terms of state parameters reduced to the critical state: this theorem states the existence of a function of the reduced state parameters, $f(p_r, V_r, T_r) = 0$, that is valid for any gas. The pVT behaviour of natural gas can be described in terms of an equation of state corrected by the compressibility factor z:

$$pV = z\frac{R}{M}T \qquad 8.1-1$$

where V is thespeci fic volume valid at pressure p and temperature T; R is the universal gas constant, whose rounded-off value is 8314 J/(kmole. K). Any given gas is characterized by its specific gas constant, or simply gas constant,

$$R' = \frac{R}{M}. \qquad 8.1-2$$

By the theorem of corresponding states, the compressibility factors z of two gases are equal if the reduced state parameters of the two gases are

* This Chapter has been written in co-operation with Ferenc Patch.

equal, that is, if the gases are in the same corresponding state. Reduced pressure is

$$p_r = p/p_c$$

and reduced temperature is

$$T_r = T/T_c.$$

In the case of gas mixtures, reduced parameters of state are to be replaced by the pseudo-reduced parameters p_{pr} and T_{pr} defined in terms of the pseudocritical pressure and temperature p_{pc} and T_{pc}, both depending on gas composition, as follows:

$$p_{pr} = p/p_{pc} \qquad 8.1-3$$

and

$$T_{pr} = T/T_{pc}. \qquad 8.1-4$$

If gas composition is known, p_{pc} and T_{pc} can be determined by applying the principle of additivity or—usually at a lower accuracy—using empirical diagrams. Additivity means that molar mass, pseudocritical pressure and pseudocritical temperature of a mixture can be added up from the respective parameters of the components, combined with their molar fractions or volume fractions:

$$M = \sum_{i=1}^{n} y_i M_i$$

$$p_{pc} = \sum_{i=1}^{n} y_i p_{ci}$$

$$T_{pc} = \sum_{i=1}^{n} y_i T_{ci}.$$

The accuracy of this calculation is usually impaired by the circumstance that, in the composition of the gas, the relative abundance of the heaviest components is given by a single figure denoted C_{n+}. The critical parameters are usually put equal to those of the component C_{n+1}; if e.g. the combined heavy fraction is C_{6+}, then it is taken into account with the critical parameters of heptane, C_7. This is, of course, an approximation whose accuracy depends on the actual composition of the combined heavy fraction.

Example 8.1—1. Find the molar mass and pseudocritical parameters of state at a pressure of 88 bars and 280 K temperature of the wet natural gas whose composition is given in Columns (1) and (2) of Table 8.1—1. The physical constants of the gas components are listed in Table 6.4—1. The calculation whose details are given in Columns (3), (4) and (5) of Table 8.1—1, furnishes M = 21.5 kg/kmole. By Eqs 8.1—3 and 8.1—4,

$$p_{pr} = \frac{8.8 \times 10^6}{4.59 \times 10^6} = 1.92$$

and

$$T_{pr} = \frac{280.0}{224.2} = 1.25.$$

Table 8.1—1

Components of natural gas	z_i	$M_i z_i$	$T_{ci} z_i$	$p_{ci} z_i$
		kg/kmol	K	10^5 N/m²
1	2	3	4	5
Methane	0.790	12.67	150.7	36.67
Ethane	0.100	3.01	30.5	4.89
Propane	0.055	2.43	20.3	2.34
i-Butane	0.010	0.58	4.1	0.36
n-Butane	0.015	0.87	6.4	0.57
n-Pentane	0.024	1.73	11.3	0.81
CO_2	0.001	0.04	0.3	0.08
N_2	0.005	0.14	0.6	0.17
Total	1.000	21.47	224.2	45.90

Figure 8.1—1 permits us to read off pseudocritical pressures v. pressure and pseudocritical temperatures v. temperature for hydrocarbon gases of various molar mass.

Example 8.1—2. Find using Fig. 8.1—1 pseudoreduced parameters of state for a gas of molar mass $M = 21.5$ at the pressure and temperature stated in the foregoing example. — Figure 8.1—1 furnishes $p_{pc} = 4.61$ MN/m² and $T_{pc} = 223$ K. — By Eqs 8.1—3 and 8.1—4,

$$p_{pr} = \frac{8.8 \times 10^6}{4.61 \times 10^6} = 1.91$$

and

$$T_{pr} = \frac{280}{223} = 1.26 .$$

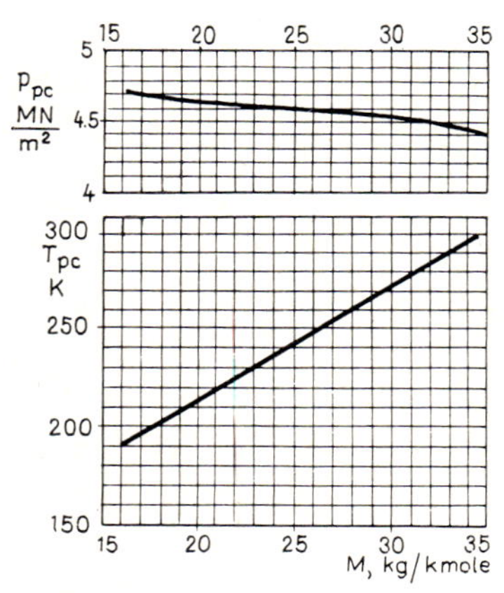

Fig. 8.1—1 Pseudocritical parameters of natural gas v. molar mass (from Katz 1959, p. 111; used with permission of McGraw-Hill Book Company)

The results furnished by the two procedures are in approximate agreement, but more substantial deviations may occur in other cases. Empirical diagrams other than Fig. 8.1—1 have been published (e.g. Stearns et al. 1951).

The compressibility factor z can be determined most accurately by laboratory experiments on pVT behaviour. If no $z = f(p, T)$ diagram based on experiment is available, then z is determined as a function of the reduced state parameters out of empirical diagrams or relationships. One of the best-known empirical diagrams is that of Standing and Katz (1942; Fig. 8.1—2). For the pseudoreduced parameters $p_{pr} = 1.92$ and $T_{pr} = 1.25$ of Example 8.1—1, it furnishes $z = 0.92$.

Natural gas often contains substantial amount of non-hydrocarbon gases, N_2 and CO_2 first of all. If the volume percentage of these is less than 8 percent for N_2 and 10 percent for CO_2, then the compressibility factor is

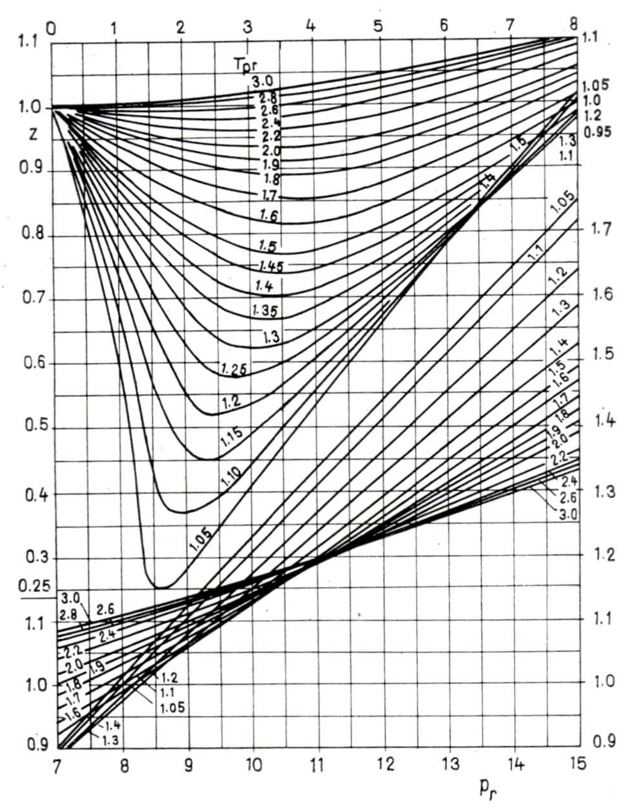

Fig. 8.1—2 Compressibility factor of natural gas v. pseudo-reduced parameters of state, after Standing (1952; reproduced by permission of the copyright owner—copyright © Chevron Research Company 1951; all rights reserved under the International Copyright Convention)

most readily determined by a procedure suggested by Gráf (Szilas 1967): the hydrocarbons are considered to be a single component whose pseudo-critical parameters and compressibility factors z are determined separately by some suitable method. The compressibility factors z_{CO2} and z_{N2} for CO_2 and N_2 are then read off Figures 8.1—3 and 8.1—4, and the value z' for the mixed gas as a whole is calculated using the mixing rule

$$z' = y_{N2}z_{N2} + y_{CO2}z_{CO2} + (1 - y_{N2} - y_{CO2})z.$$

In writing computer programs it is an advantage if the compressibility factor can be found by calculation, without having to resort to diagrams. Literature contains several calculation procedures. The relevant formulae are mathematical representations of more or less extensive domains of the families of curves constituting the empirical diagrams. The French gas

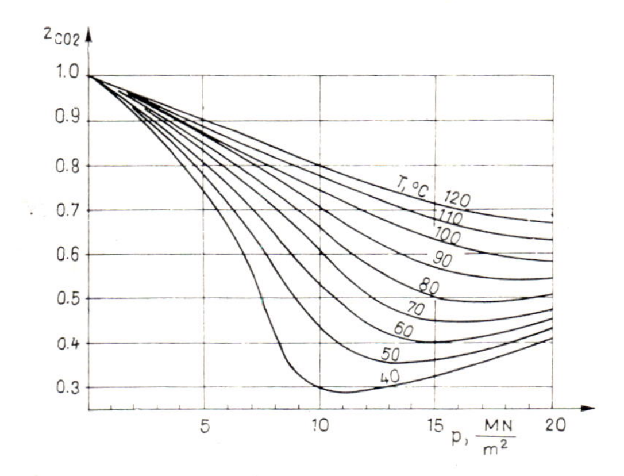

Fig. 8.1—3 Compressibility factor of CO_2 according to Reamer and Elters; after Török et al. (1966)

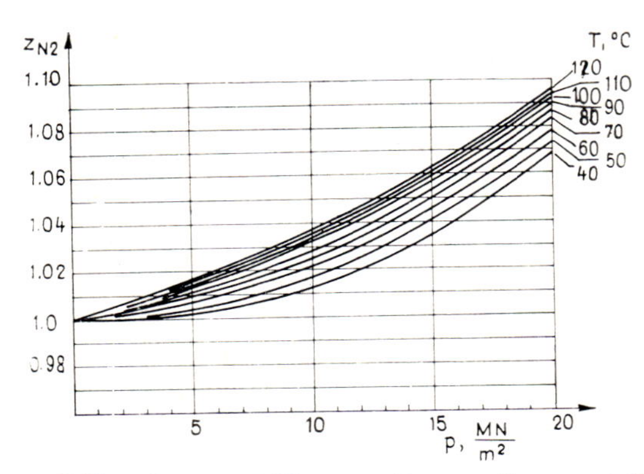

Fig. 8.1—4 Compressibility factor of N_2 according to Sage and Lacey; after Török et al. (1966)

industry uses for pressures of below 70 bars at soil-temperature flow the formula

$$z = 1 - 2\times 10^{-8}p. \qquad 8.1-5$$

(Société . . . Manuel 1968). A relationship accounting also for temperature, applicable below 60 bars, is

$$z = \frac{1}{1 + kp} \qquad 8.1-6$$

k is listed for certain temperatures in Table 8.1—2 (likewise from Société . . . Manuel 1968).

Table 8.1—2

T °C	k
0	2.65×10^{-8}
15	2.04×10^{-8}
30	1.65×10^{-8}

Relationships for calculating the pseudocritical parameters of state have been given by Thomas et al. (1970)

$$p_{\rm pc} = 4.894\times 10^6 - 4.050\times 10^5\, \varrho_r \qquad 8.1-7$$

$$T_{\rm pc} = 94.71 + 170.7\, \varrho_r. \qquad 8.1-8$$

Wilkinson (1964) gave for $p_r < 1.5$ the formula

$$z = 1 + 0.257 p_{\rm pr} - 0.533 \frac{p_{\rm pr}}{T_{\rm pr}} \qquad 8.1-9$$

where $p_{\rm pr}$ and $T_{\rm pr}$ are the values furnished by Eqs 8.1—3 and 8.1—4, respectively.

Gas density at pressure p and temperature T can be obtained putting $V = 1/\varrho$ in Eq. 8.1—1:

$$\varrho = \frac{pM}{zRT}\,. \qquad 8.1-10$$

If this equation is written up for the standard state, then $p = p_n$, $T = T_n$ and $z = z_n = 1$. Then

$$\varrho_n = \frac{p_n\, M}{R\, T_n}\,. \qquad 8.1-11$$

Let e.g. $p_n = 1.013$ bar and $T_n = 273.2$ K. Since furthermore, $R = 8314$ J/(kmole. K),

$$\varrho_n = \frac{M}{22.42}\,. \qquad 8.1-12$$

Rearranging we get, at the standard-state parameters stated above, the standard molar volume

$$V_{\text{mol}} = M \varrho_n = 22.42 \text{ m}^3/\text{kmole}. \qquad 8.1-13$$

Relative density is the standard-state density of the gas referred to the standard-state density of air:

$$\varrho_r = \frac{\varrho_n}{\varrho_{an}}. \qquad 8.1-14$$

Now by Eqs 8.1—12 and 8.1—14,

$$\varrho_r = \frac{M}{M_a} = \frac{M}{28.96}. \qquad 8.1-15$$

Gravity is

$$\gamma = \varrho g. \qquad 8.1-16$$

8.1.2. Viscosity

Gas viscosity, as distinct from liquid viscosity, increases as temperature increases, decreases as molecular weight increases, and is independent of pressure at medium pressures. At atmospheric pressure, viscosities of hydrocarbon gases vary linearly with temperature between 0 and 200 °C (Fig. 8.1—5). The viscosity of hydrocarbon mixtures at atmospheric pressures is readily calculated using a relationship published by Herning and Zipper:

$$\mu_a = \frac{\sum_{i=1}^{n} \mu_{ai} y_i \sqrt{M_i}}{\sum_{i=1}^{n} y_i \sqrt{M_i}}. \qquad 8.1-17$$

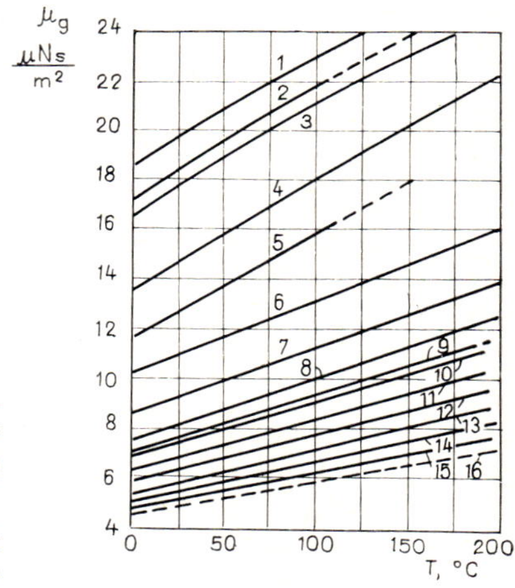

Fig. 8.1—5 Viscosities of natural gas components at atmospheric pressure; after Carr (1954). *1* helium, *2* air, *3* nitrogen, *4* carbon dioxide, *5* hydrogen sulphide, *6* methane, *7* ethane, *8* propane, *9* isobutane, *10* n-butane, *11* n-pentane, *12* n-hexane, *13* n-heptane, *14* n-octane, *15* n-nonane, *16* n-decane

Figure 8.1–6 is a plot of values furnished by this relationship, and, more generally, of measured viscosities of artificial hydrocarbon-gas mixtures. It permits us to find atmospheric pressure viscosity in terms of the molecular weight and relative gravity of the gas (Carr et al. 1954). The influence of non-hydrocarbon components may be taken into account by a viscosity-increasing correction depending on relative density, provided

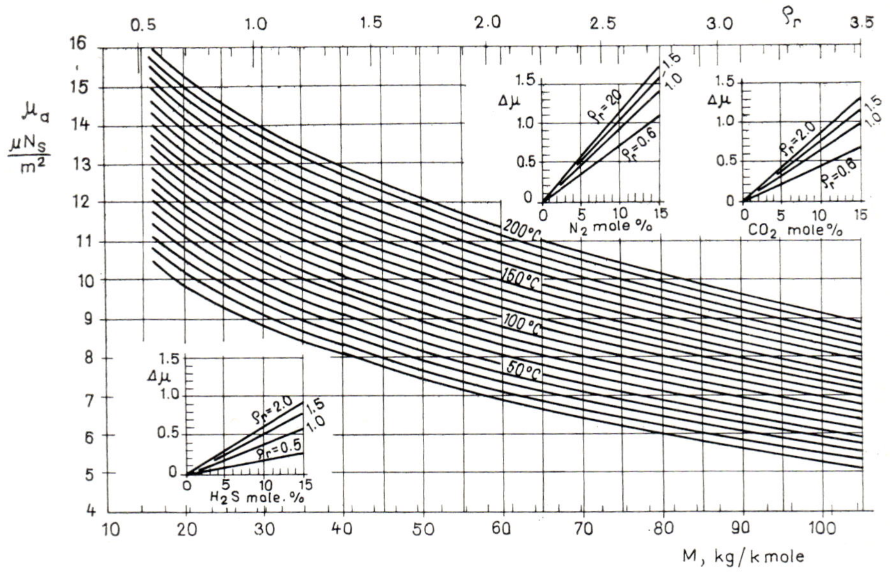

Fig. 8.1–6 Viscosity of natural gas at atmospheric pressure (Carr et al. 1954)

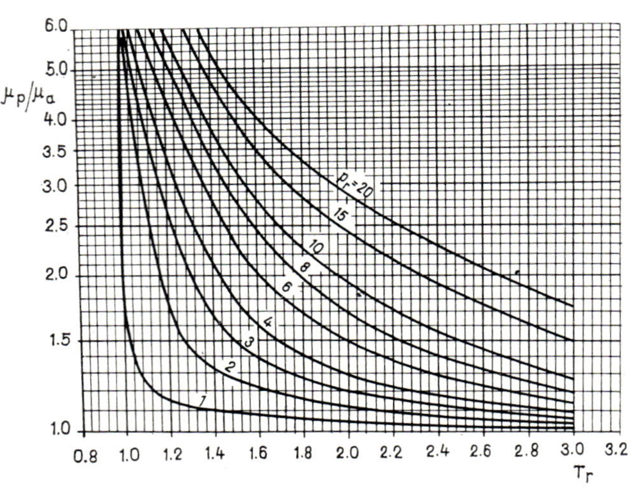

Fig. 8.1–7 Variation of $k = \mu_p/\mu_a$ v. the pseudo-reduced parameters of state (Carr et al. 1954)

the share of these components does not exceed 15 percent. — A relationship between viscosity and pressure may be set up making use of the theorem of corresponding states. Figure 8.1—7 allows us to find, in the knowledge of the pseudocritical pressure and temperature, the factor $k = \mu_p/\mu_a$ by which the viscosity at atmospheric pressure is to be multiplied in order to obtain the value that holds at pressure p (Carr et al. 1954). The relationship is valid in the gaseous state only, and it is therefore necessary in critical cases to check by phase examinations whether or not a liquid phase is present.

8.1.3. Specific heat, molar heat, adiabatic gas exponent, Joule—Thomson effect

Specific heat is the heat capacity of the unit mass or molar mass of a substance or, in the case we are discussing, the ratio of the heat dQ imparted to a unit mass of gas to the resulting temperature change dT, provided no phase change takes place during the temperature change. The usual cases investigated in a gas are temperature changes at constant pressure, on the one hand, and at constant volume, on the other, and accordingly, two distinct specific heats may be defined, the isobaric c_p

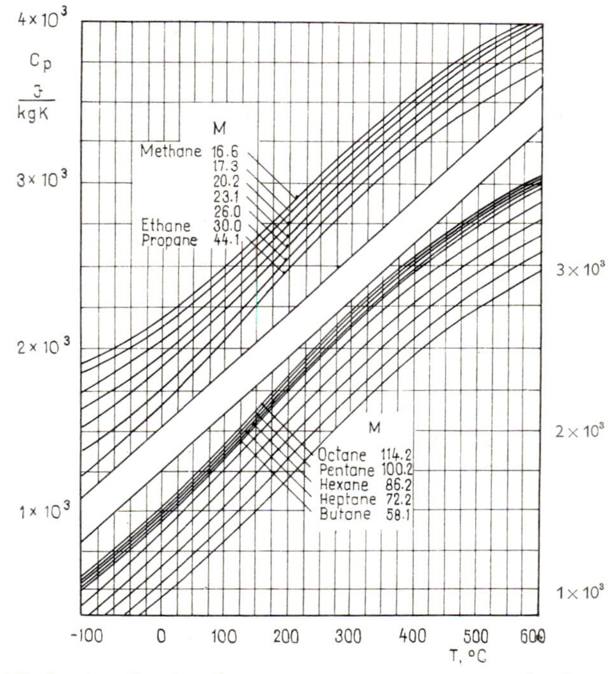

Fig. 8.1—8 Specific heats of natural-gas components at atmospheric pressure, after Brown (1945)

and the isochoric c_v. $c_p > c_v$, since part of the heat supplied to the system will expand the gas in the isobaric case. In ideal gases, the difference between the two constants equals the gas constant, that is,

$$c_p - c_v = R. \qquad 8.1-18$$

The difference between the characteristic specific heats in a real gas is not constant. Figure 8.1—8 after Brown (1945) shows isobaric molar heats v. temperature of hydrocarbon homologues at atmospheric pressure. The c_{pa} of gas mixtures can be determined on the basis of additivity, in terms of the components' specific heats and molar fractions, that is,

$$c_{pa} = \sum_{i=1}^{n} y_i \, c_{pai}.$$

The specific heat c_{pp} of a gas mixture at pressure p exceeds the value c_{pa} at atmospheric pressure by Δc_p. Figure 8.1—9 is a plot of Δc_p v. pseudo-

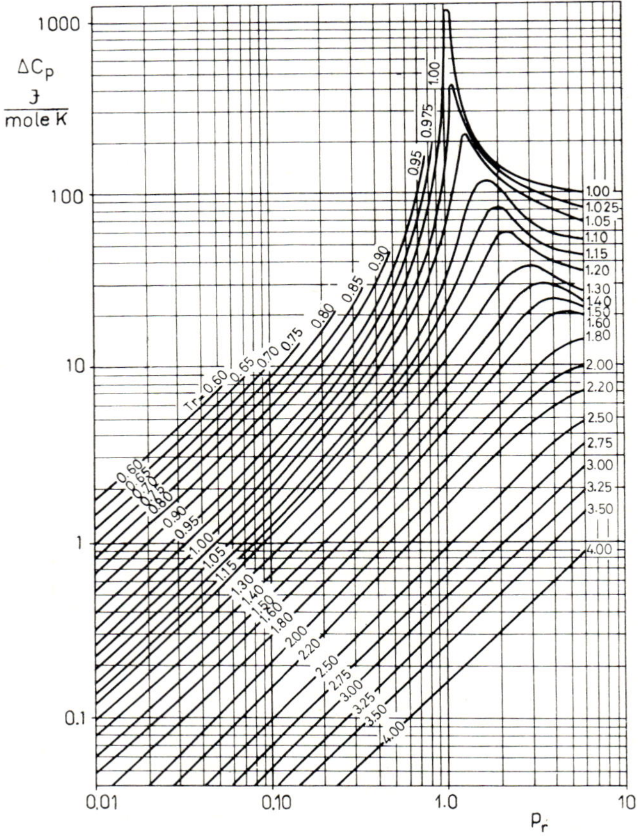

Fig. 8.1—9 Specific-heat correction, after Perry (1969)

reduced pressure p_{pr} for various pseudoreduced temperatures T_{pr} after Edmister (Perry 1969). Let us point out that Δc_p is stated in molar terms, and has to be divided by the molar M mass in order to transform it into a quantity having the nature of a specific heat.

The adiabatic gas exponent

$$\varkappa = \frac{c_p}{c_v} \qquad 8.1-19$$

is usually required in thermodynamic calculations. Its value can be determined, e.g., by reading the molar-heat difference $(c_p - c_v)$ off Fig.8.1—10

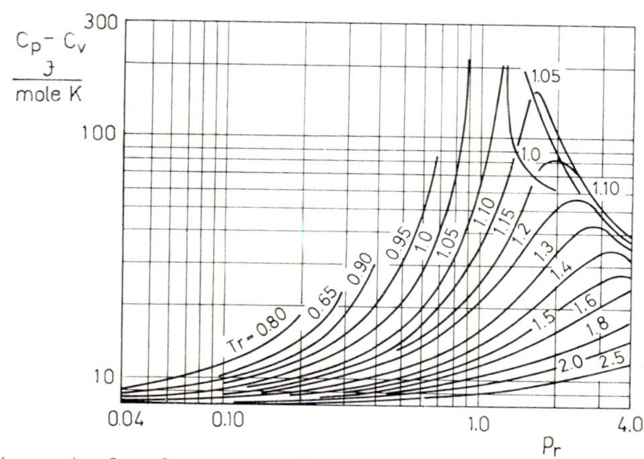

Fig. 8.1—10 $(c_p - c_v)$ of real gases v. the reduced parameters of state, after Perry (1969)

(after Perry 1969) and calculating c_v in the knowledge of c_p. The use of the figure presupposes knowledge of the pseudoreduced parameters of state. The molar heat or specific heat is calculated for a certain temperature range rather than for a single pair of pressure and temperature values. The mean molar heat c_p can be determined for instance by planimetering the specific heats calculated for various temperatures by the procedure outlined above. In a simpler procedure, one may read the enthalpy values h_1 and h_2 corresponding to the initial and terminal temperatures T_1 and T_2 off suitable diagrams or tables. Then

$$\bar{c}_p = \frac{h_2 - h_1}{T_2 - T_1}. \qquad 8.1-20$$

If the pressure of an ideal gas is lowered without the gas delivering energy, then, if the gas is ideal and the change of state is adiabatic, the total internal energy of the system remains unchanged, that is, the state change is isoenthalpic, and the temperature of the gas remains unchanged, too. If, however, the gas undergoing said change is real, then its volume change will differ from ideal gas behaviour. As a result, its internal energy and hence also its temperature will be affected (Joule—Thomson effect). Among the temperature changes taking place during gas flow, it is expedient to account for this effect by the Joule—Thomson coefficient μ_d, which is a measure of temperature change per unity pressure change. $\mu_d \gtreqless 0$, that is, expansion may increase, reduce or leave unchanged the

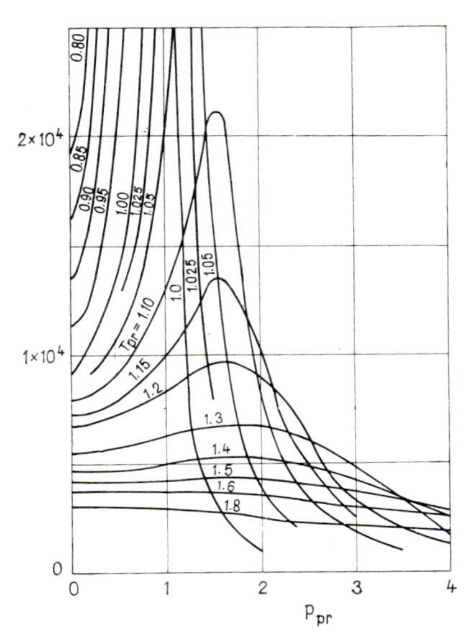

Fig. 8.1—11 Relationship for determining the choke effect, after Korchazhkin (1963)

temperature of the gas. Several relationships for determining μ_d have been derived. Figure 8.1—11 gives the values [in J/(K kmole)] of the expression

$$\frac{p_{pc}\, c_p\, \mu_d}{T_{pc}}$$

in terms of the pseudoreduced parameters of state, and this expression may be solved to yield μ_d (Korchazhkin 1963).

8.1.4. Hydrocarbon hydrates

Hydrocarbon gas hydrate is a solid granular substance resembling snow or ice. It is composed of water and the molecules of one or more hydrate-forming gases. The molecules of this gas enter cavities in the H_2O lattice, which is looser than the ice lattice, without entering into chemical bond with the water. The lattice thus forming may be one of two pentagonal dodecahedra. The conditions of hydrate formation and stability are: (i) sufficiently low temperature and high pressure; (ii) the hydrate-forming gas is held together by covalent bonds; its molecules are shorter than 8 Å;

and when liquid, it is immiscible with water; (iii) during hydrate formation, water is liquid; (iv) hydrate is resistant to water and no Van der Waals forces arise between its molecules.

Hydrates include besides water methane, ethane, propane or butane, alone or mixed together. In addition to the hydrocarbons, other, non-hydrocarbon gas components such as nitrogen, carbon dioxide or hydrogen sulphide may also be hydrate-forming. Hydrate composition depends on the nature of the hydrate-forming gas but is not governed by the rules of

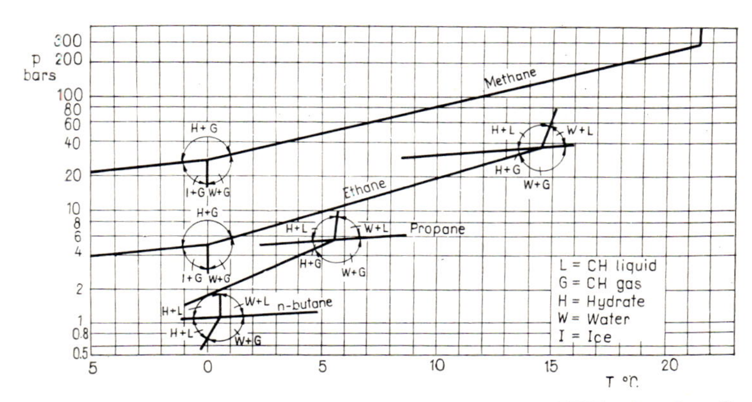

Fig. 8.1—12 State diagram of hydrocarbons according to Willard; after Orlicek and Pöll 1951, Table 118 (used with permission of Springer-Verlag, Wien/New York)

stoichiometry. The least water-to-methane ratio in methane hydrate would be 4.5, in view of the number of methane molecules that can be accommodated in the water lattice. However, methane-unsaturated hydrates with more than 4.5 moles of H_2O per mole of methane also occur. The least water content of ethane hydrate is about 7.7 moles H_2O per mole of ethane. The propane and butane molecules may enter but the largest cavities of the lattice, and hence, in propane hydrate, 17 moles at least of water are required per mole of propane.

Figure 8.1—12 shows state diagrams of various two-component hydrocarbon hydrates after Willard (Orlicek and Pöll 1951). The upper temperature limit e.g. of propane hydrate formation is seen to be 5.6 °C, with a corresponding pressure of 5.6 bars. The point defined by these parameters of state is an invariant of the propane-water system, a four-phase point with no degree of freedom, where propane hydrate as the solid phase is at equilibrium with gaseous propane saturated with water vapour, water saturated with liquid propane and propane saturated with water. The figure shows which phases may coexist in the individual regions. In reality, it is usual for hydrates to involve more than one hydrocarbon component. The critical pressure of hydrate formation is substantially reduced e.g. if methane is accompanied by some hydrocarbon of larger molar mass, propane or butane first of all. Even quite low concentrations of these may displace the phase diagram rather considerably. For approximate estimates one may use Fig. 8.1—13, which shows critical hydrate-formation pressures and temperatures for hydrocarbons of various relative gravities. The presence of CO_2 and H_2S at a given temperature may lower the critical pressure, whereas the presence of N_2 tends to raise it. The inset in Fig. 8.1—13 provides the correction factor C_{N2} which shows how many times the critical pressure of hydrate formation is higher in the presence than in the absence of a certain quantity of nitrogen.

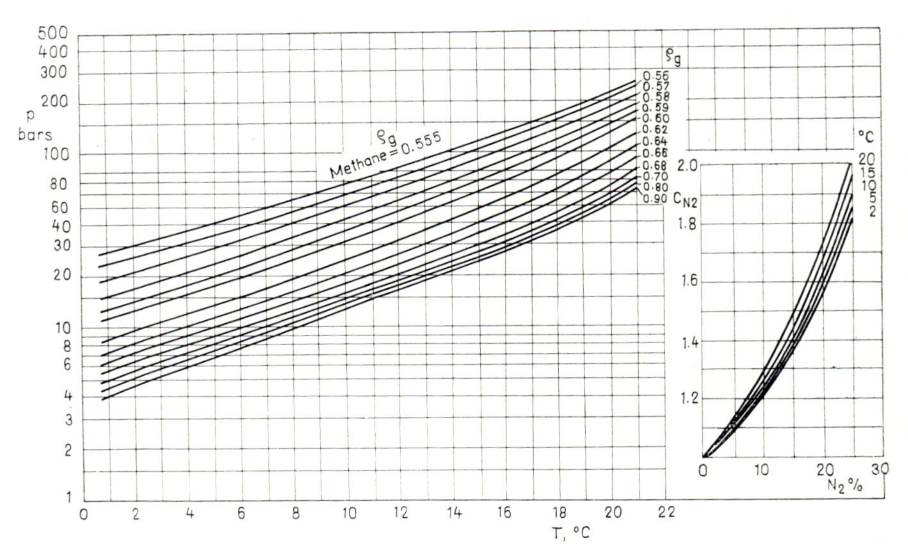

Fig. 8.1—13 Limits of gas hydrate formation, after Katz (1959, p. 213; used with permission of McGraw-Hill Book Company)

Several more accurate procedures have been devised. For nitrogenless natural gas, up to about 280 bars pressure, Katz' procedure involving equilibrium constants is best suited. The condition of hydrate formation is

$$\sum_{i=1}^{n} \frac{z_{hi}}{K_{hi}} = 1.$$

The K_{hi}s are to be read off the $K_h = f(T)_p$ diagrams of the hydrate-forming components (Katz 1959). — Heinze (1971) prefers the modified McLeod—Campbell procedure for determining hydrate formation temperatures (hydrate points) of natural gas containing nitrogen up to about 400 bars pressure. The hydrate point is calculated using the relationship

$$T = \sqrt{\frac{K_h}{0.445}}. \qquad 8.1\text{—}21$$

The values of K_h for various pressures are contained in Table 8.1—3. The K_h values falling between those given in the table may be found by

Table 8.1—3
Hydrate-equilibrium factors K_h for hydrate-forming natural-gas components (modified after Heinze 1971)

Components \ p, bars	50	100	150	200	250	300	350	390
CH_4	34,543	35,949	36,719	37,357	37,814	38,204	38,531	38,767
C_2H_6	45,535	47,101	48,078	48,704	49,316	49,772	50,140	50,435
C_3H_8	85,060	83,970	79,836	75,610	73,150	71,340	70,103	69,154
i-C_4H_{10}	102,096	94,310	89,319	82,481	78,791	75,569	74,533	73,304
n-C_4H_{10}	57,979	51,133	47,648	45,032	43,846	43,328	43,276	43,234
N_2	30,555	32,133	33,369	33,695	34,214	34,656	35,005	35,251
CO_2	38,788	43,504	44,812	46,773	50,371	51,660	52,269	54,018
H_2S	63,986	69,972	74,001	76,349	78,554	80,426	81,373	82,148

linear interpolation. The hydrate factor for multi-component natural gas of known molar ratios may be found by applying the principle of additivity.

Example 8.1—3 (Heinze 1971). Find the hydrate point at 147 bars pressure for the gas composition given in Column 2 of Table 8.1—4. Let us point out that, regardless of the quantity and nature of the longer-molecule non-hydrate-forming natural-gas components, it is assumed that

$$\sum_{i=1}^{n} z_{hi} = 1.$$

By the data in Column 4 of the Table, $K_h = 39{,}459$ and hence, hydrate point is at

$$T = \sqrt{\frac{39{,}459}{0.445}} = 297.8 \text{ K} = 24.6 \text{ °C}.$$

Hydrate point may be substantially reduced by adding to the natural gas a hydrate inhibitor such as calcium chloride, methanol, ethylene glycol, diethylene glycol.

Table 8.1—4
Finding the hydrate formation temperature of Thönse gas (after Heinze 1971)

Components	z_{hi}	K_{hi} at 147 bars	$z_{hi}K_{hi}$
CH_4	0.865	30,673	31,722
C_2H_6	0.073	48,020	3,505
C_3H_8	0.028	80,084	2,242
i-C_4H_{10}	0.013	89,618	1,165
N_2	0.010	33,295	333
CO_2	0.011	44,734	492

$\sum_{i=1}^{n} z_{hi}K_{hi} = 39{,}459$

8.2. Temperature of flowing gases

In most long uninsulated pipelines, the temperature of flowing gas approaches soil temperature after a travel sufficiently short for flow temperature to be identified for all practical purposes with soil temperature over the full length of the pipeline. In certain cases, however, the flow temperature of gas may significantly differ from the temperature of the surrounding soil, and it may then be important to determine temperature traverses for the pipeline. The cases in question include the following. (i) It is necessary to decide in designing where the flow temperature drops

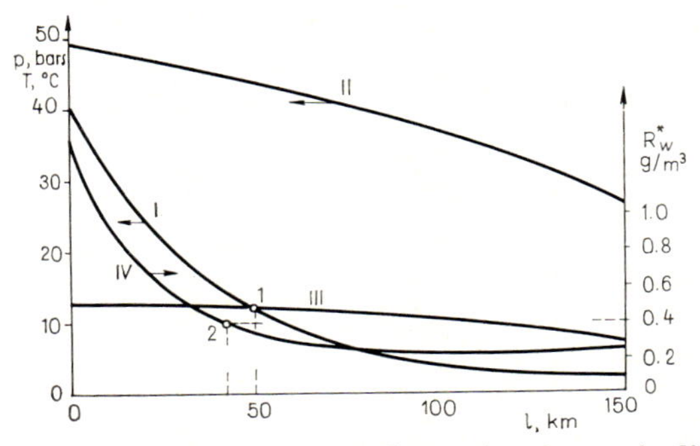

Fig. 8.2—1 Pinpointing hazard of hydrate formation in a pipeline, after Smirnov and Shirkovsky (1957) and Török et al. (1968)

below the hydrate point; (ii) it is desired to chill the gas by injecting liquefied gas, in order to increase the throughput capacity of the pipeline (Gudkov et al. 1970); (iii) in arctic regions, the gas may cause an undesirable warming up of the permafrost soil in which the pipeline is laid.

Figure 8.2—1 (after Smirnov and Shirkovsky 1957; and Török et al. 1968) is a temperature traverse of a given pipeline (Graph I). It permits us to delimit the line segment where there is a risk of hydrate formation. Graph II is the pressure traverse. The accurate calculation of the two traverses takes a successive-approximation procedure. In the knowledge of pressure, the hydrate-point traverse (Graph III) may be calculated in the manner explained in Section 8.1.4. At point *1*, where Graphs II and III meet ($l = 50$ km), the hydrate point T_h is just equal to the temperature T_g of gas flow. For hydrate to form, it is sufficient that there be some free water available at this line section. Graph IV is a water vapour saturation traverse along the pipeline. Points of this curve can be determined by means of auxiliary diagrams for corresponding pairs of p and T_g (e.g., Katz 1959). Assuming the water vapour content of the gas to be 0.4 g/m³, the dew point of the gas turns out to occur at point *2*; from there on, the pipeline does contain condensed water: that is, this condition of hydrate formation is also satisfied at and beyond $l = 50$ km.

The temperature of gas flowing in the pipeline depends, for a given inflow temperature T_1 and soil temperature T_s, on the following factors: (i) heat exchange with the environment, depending primarily on the heat transfer coefficient (cf. Section 7.2.3). The internal convection coefficient, α_1, is infinite in a fair approximation. (ii) The Joule—Thomson effect due to friction, velocity increase and altitude change. (iii) Phase changes (condensation, evaporation) due to pressure and temperature changes. (iv) The energy loss of flow, which end up as heat.

These effects are accounted for in steady-state flow by the following equation (Pápay 1970), stating flow temperature at a distance l_x from the head end of the line to be

$$T_{lx} = \frac{C_1^{\frac{C_2}{C_3}} \left[T_s + \frac{C_4}{C_2} - \frac{C_1 C_5}{C_2(C_2 + C_3)}\right]}{(C_1 + C_2 l_x)^{\frac{C_2}{C_3}}} - \frac{C_4 + C_5 l_x}{C_2} + \frac{C_5(C_1 + C_3 l_x)}{C_2(C_2 + C_3)} \qquad 8.2\text{—}1$$

where

$$C_1 = z_{V1} c_{pL} + (1 - z_{V1}) c_{pV}; \qquad C_2 = \frac{k}{q_m};$$

$$C_3 = \frac{z_{V2} - z_{V1}}{l} (c_{pL} - c_{pV})$$

$$C_4 = z_{V1} c_{pL} \mu_{dL} \frac{p_1 - p_2}{l} + (1 - z_{V1}) c_{pV} \mu_{dV} \frac{p_1 - p_2}{l} +$$

$$+ Q \frac{z_{V2} - z_{V1}}{l} + v_1 \frac{v_2 - v_1}{l} + g \frac{h}{l} - \frac{k\pi d_o}{q_m} T_1$$

$$C_5 = \frac{(z_{V2} - z_{V1})(p_1 - p_2)}{l^2} (c_{pL} \mu_{dL} - c_{pV} \mu_{dV}) + \left(\frac{v_2 - v_1}{l}\right).$$

In deriving this equation, Pápay assumed pressure, flow rate and phase transitions to be linear functions of distance from the head end. It is therefore recommended in problems where a high accuracy is required to calculate temperature changes for shorter line segments. Let us point out that suffix 1 invariably refers to the head end and 2 to the tail end of the line of length l, except, of course, in the numbering of the constants C.

In the case when the phase changes are left out of consideration—that is, in single-phase flow—Eq. 8.2—1 simplifies to

$$T_{lx} = T_s + (T_1 - T_s) e^{-al_x} - \frac{\mu_{dV}(p_1 - p_2)}{al} \times$$

$$\times (1 - e^{-al_x}) - \frac{gh}{al\, c_{pV}} (1 - e^{-al_x}) - \frac{v_2 - v_1}{al\, c_{pV}} \times$$

$$\times \left[\left(v_1 - \frac{v_2 - v_1}{al}\right)(1 - e^{-al_x}) + \frac{(v_2 - v_1) l_x}{l}\right] \qquad 8.2\text{—}2$$

where

$$a = \frac{k}{q_m c_{pV}}.$$

The first two terms of this equation describe heat exchange with the environment; the third one accounts for the Joule—Thomson effect, the fourth for the change in geodetic head and the fifth for the change in velocity head. In practical calculations, the last two terms may be neglected. The resulting error is usually less than the error due to uncertainties in the various parameters. If the pressure drop is small, then so is the temperature drop due to expansion, and the third term may also be neglected, in which case Eq. 8.2—2 simplifies to Eq. 7.2—15.

8.3. Steady-state flow in pipeline systems

The fundamental relationships of gas flow in a pipeline are stated in Section 1.2. That section refers to a single pipeline. Actual gas transmission systems, however, often form connected nets, in which flow is governed by relationships much more involved than those referring to a single line. Pipe nets may be high-pressure, with pressure changes entailing significant changes in specific volume, and low-pressure, with such changes negligible. The first type includes transmission systems of regional supply, the second includes gas nets supplying local consumer groups, most often community utilities and households. In the latter we shall concentrate on the first type.

Flow in transmission systems is almost invariably transient, but numerous design and operation control problems may be solved notwithstanding by assuming flow to be steady-state. Network models based on the assumption of steady flow permit the establishment of pressure contour maps of both radial and looped networks for periods of peak demand. This map permits us to pinpoint the critical segments of the net, where consumer demand cannot be fully met in peak demand periods. The pressure map is useful in designing new systems, and in expanding or checking the operation of existing ones.

8.3.1. Design fundamentals

The two basic elements of a pipeline network are the *nodes* and the *node connecting elements* (NCEs). Nodes include those points where a pipeleg ends, or where two or more NCEs join, or where there is injection or delivery of gas. The pressure map of the network is determined by node pressures. The most important NCEs are pipelegs, compressor stations, regulators, valves, and underground gas storages. Prior to constructing a model of a a complex system it is necessary to establish mathematical models for individual NCEs. These models are in effect pressure v. throughput relationships valid at given parameters.

The characteristic equation of a *high-pressure pipeleg* is, by Eq. 1.2—7,

$$p_1^2 - p_2^2 = k_1 q^2. \qquad 8.3-1$$

The gas flow expressed in standard volume units is

$$q = \left[\frac{p_1^2 - p_2^2}{k_1}\right]^{0.5} \tag{8.3—2}$$

where

$$k_1 = 1.95 \times 10^{-4} \left(\frac{p_n}{T_n}\right)^2 \frac{lM\,\bar{T}\bar{z}\,\bar{\lambda}}{d_i^5}. \tag{8.3—3}$$

In a *low-pressure pipeleg*, with pressure close to atmospheric, we have $\bar{z} = 1$ and

$$p_1^2 - p_2^2 = (p_1 + p_2)(p_1 - p_2) \approx 2p_n(p_1 - p_2)$$

and the above equations modify to

$$p_1 - p_2 = k_2 q^2 \tag{8.3—4}$$

and

$$q = \left[\frac{p_1 - p_2}{k_2}\right]^{0.5} \tag{8.3—5}$$

respectively, with

$$k_2 = 0.975 \times 10^{-4} \frac{p_n l M \bar{T} \bar{\lambda}}{T_n^2 d_i^5}. \tag{8.3—6}$$

Compressor characteristics are provided by the manufacturer. These may usually be approximated by a function of the type

$$q = \frac{P}{k_3 \left(\frac{p_2}{p_1}\right)^{k_4} + k_5} \tag{8.3—7}$$

where k_3, k_4 and k_5 are compressor constants.

Pressure regulators may be described by the flow equations of chokes (cf. Section 1.5—1). If the pressure drop is less than critical (flow is subsonic), then Eq. 1.5—2 will hold if the gas is liquidless, that is,

$$q = k_6 p_1 \sqrt{\left(\frac{p_2}{p_1}\right)^{\frac{2}{\varkappa}} - \left(\frac{p_2}{p_1}\right)^{\frac{\varkappa+1}{\varkappa}}} \tag{8.3—8}$$

where

$$k_6 = \sqrt{2R}\,\frac{\pi}{4}\,d_{ch}^2 \frac{T_n}{p_n}\,\alpha \sqrt{\frac{1}{MT_1}\,\frac{\varkappa}{\varkappa - 1}}. \tag{8.3—9}$$

If the pressure drop is above-critical (flow is sonic), then p_2/p_1 is to be replaced by the expression in Eq. 1.5—3, and the characteristic relationship is

$$q = k_7 p_1. \tag{8.3—10}$$

The production of wells tapping an *underground gas reservoir* can be described by the relationship

$$q = k_8(p_1^2 - p_2^2)^n \quad 8.3\text{—}11$$

where, as distinct from the usual productivity relationship (cf. Section 3.1) p_1 means formation pressure and p_2 means wellhead pressure; k_8 is a productivity index corresponding to this latter definition.

In the knowledge of the gas transmission system's elements, a mathematical-hydraulic model of the entire system may be constructed. In laying down the principles of modelling, the recognition of an analogy between gas flow in pipe networks and flow of electricity in electrical networks was extensively exploited. Kirchhoff's laws apply to gas flow, too. The first law applies to any node; the algebraic sum of gas flows entering and leaving the node is zero, that is

$$\sum_{i=1}^{m} q_i = 0 \quad 8.3\text{—}12$$

where m is the number of NCEs meeting at the node. Gas flowing into the node is given the positive sign. By Kirchhoff's second law, for any loop in the high-pressure system, the algebraic sum of pressure drops, taken with signs corresponding to a consistent sense of rotation around the loop, is zero, that is,

$$\sum_{i=1}^{n} (p_1^2 - p_2^2)_i = 0. \quad 8.3\text{—}13$$

where n is the number of NCEs in the loop, and p_1 and p_2 are, respectively, the head-end and tail-end pressures of said pipelegs, head and tail being taken with respect to the sense of rotation chosen. This relationship is also called the loop law. In low-pressure gas distribution networks, the compressibility of the gas is negligible and the loop law accordingly simplifies to

$$\sum_{i=1}^{n} (p_1 - p_2)_i = 0. \quad 8.3\text{—}14$$

There are two fundamental types of gas transmission systems, loopless and looped.

8.3.2. Loopless systems

(a) Numerical methods

In a loopless system, NCEs joined by nodes form no closed loops anywhere in the system.

Figure 8.3—1 shows a hypothetical loopless system. Gas enters through node I and leaves through Nodes II, III and IV. Pressures and throughputs in such a system, assuming all NCEs to be pipelegs, are calculated as follows. In the knowledge of the gas volumes respectively injected into and taken out of the nodes, Eq. 8.3—12 furnishes the gas flows q_i in the pipelegs. In possession of these latter, Eq. 8.3—1 yields pressure drops in the pipelegs and node pressures.

Example 8.3—1. Given the gas flows into and out of Nodes from I to IV of the pipeline shown in Fig. 8.3—1 and the parameters of pipelegs from *1* to *3* and the prescribed terminal pressure $p_{IV} = 18$ bars for Node IV; find the injection pressure p_I necessary to ensure the throughputs and the terminal pressure prescribed, and find the individual node pressures. The resistance factors calculated using Eq. 8.3—3 from the parameters of the pipelegs are listed in Column 3 of Table 8.3—1. The k_{1i}s have been replaced by k_is. Node throughputs are listed in Column 4, and the prescribed terminal pressure appears in the last row of column 9. In the possession of the node

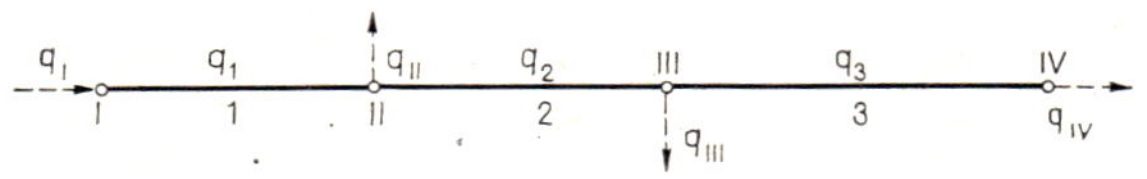

q_I q_1 q_{II} q_2 III q_3 IV
I 1 II 2 3 q_{IV}
q_{III}

Fig. 8.3—1

throughputs, the pipeleg throughputs listed in Column 5 were calculated using node law 8.3—12. Column 7 states the pressure drops in the pipelegs. Now with p_{IV}, and hence, p_{IV}^2, given, one may find the remaining node pressures using the relationship

$$p_j^2 = p_{IV}^2 + \sum_{i=j}^{3} k_i q_i^2 ; \qquad 8.3-15$$

$$j = \text{III; II; I.}$$

The calculation reveals that an injection pressure of $p_I = 54.9$ bars is required to ensure a terminal pressure of $p_{IV} = 18$ bars.

The situation is somewhat more complicated if the injection and terminal pressures are fixed, and so are the injection and delivery rates at the intermediate nodes, and the problem is to find the maximum gas output that

Table 8.3—1

Node j	Pipeleg i	k_i $10^{10}\frac{N^2s^2}{m^{10}}$	q_j $\frac{m^3}{s}$	q_i $\frac{m^3}{s}$	k_iq_i $10^{10}\frac{N^2s}{m^7}$	$k_iq_i^2$ $10^{10}\frac{N^2}{m^4}$	p_j^2 $10^{10}\frac{N^2}{m^4}$	p_j $10^5\frac{N}{m^2}$
1	2	3	4	5	6	7	8	9
I			+2.38				3008.5	54.9
	1	249.0		2.38	592.6	1410.4		
II			−0.38				1598.1	40.0
	2	145.0		2.00	290.0	580.0		
III			−0.80				1018.1	31.9
	3	482.0		1.20	578.4	694.1		
IV			−1.20				324.0	18.0

Table

Node	Pipeleg	k_i	q_j	$q_i^{(1)}$	$k_i q_i^{(1)}$	$k_i q_i^{(1)2}$
		$10^{10}\ \frac{N^2s^2}{m^{10}}$	$\frac{m^3}{s}$	$\frac{m^3}{s}$	$10^{10}\ \frac{N^2s}{m^7}$	$10^{10}\ \frac{N^2}{m^4}$
1	2	3	4	5	6	7
I						
	1	249.0		2.38	592.6	1410.4
II			−0.38			
	2	145.0		2.00	290.0	580.0
III			−0.8			
	3	482.0		1.20	578.4	694.1
IV						

estimation (1): $\Sigma k_i q_i^{(1)} = 1461.0 \times 10^{10}$

$q_{I} = 2.38\ m^3/s \rightarrow q_{IV} = 1.2\ m^3/s$

$$\Delta q = -\frac{(16\times10^5)^2 - (18.5\times10^5)^2}{2\times1461.0\times10^{10}} = 0.029 \approx 0.03$$

can be ensured at the delivery end of the line. The solution involves a successive approximation (Hain 1968) in the following steps: (i) Estimate the maximum throughput $q_I^{(1)}$ of the first pipeleg. (ii) Using Eq. 8.3—12, find the first-approximation throughputs $q_i^{(1)}$ of the individual pipelegs. (iii) In possession of these latter find the pressure drops in the pipelegs using Eq. 8.3—1. (iv) Using the relationship

$$p_j^2 = p_I^2 - \sum_{i=1}^{j-1} (p_1^2 - p_2^2)_i ; \qquad 8.3\text{—}16$$

$$j = \text{II};\ \text{III};\ \ldots,\ m$$

where m is the number of nodes, find the node pressures belonging to the $q_i^{(1)}$s calculated in the first approximation. (v) If the square of terminal pressure p_m deviates from the square of terminal pressure $p_m^{(1)}$ by more than the error permitted, then the throughputs determined in (ii) for the individual pipelegs are to be corrected using the expression $q_i^{(2)} = q_i^{(1)} + \Delta q$ where

$$\Delta q = -\frac{p_m^2 - p_m^{(1)2}}{2\sum_{i=1}^{n} k_i q_i^{(}}. \qquad 8.3\text{—}17$$

(vi) The procedure is repeated from Step (iii) on until the prescribed and calculated terminal pressures agree to within a prescribed tolerance.

Example 8.3—2. Find the maximum delivery rate at node IV in the pipeline characterized in the foregoing example, if $p_I = 55$ bars and $p_{IV} = 16$

8.3 – 2

p_i^2	p_j	$q_i^{(2)}$	$k_i q_i^{(2)}$	$k_i q_i^{(2)2}$	p_j^2	p_j
$10^{10}\ \frac{N^2}{m^2}$	$10^5\ \frac{N}{m^2}$	$\frac{m^3}{s}$	$10^{10}\ \frac{N^2 s}{m^7}$	$10^{10}\ \frac{N^2}{m^4}$	$10^{10}\ \frac{N^2}{m^4}$	$10^5\ \frac{N}{m^2}$
8	9	10	11	12	13	14
3025.0	55.0				3025.0	55.0
		2.41	600.1	1446.2		
1614.6					1578.8	39.7
		2.03	296.2	601.2		
1034.6					977.6	31.3
		1.23	592.9	729.2		
340.5	18.5				248.4	15.8

estimation (2): $\Sigma k_i q_i^{(2)} = 1489.2 \times 10^{10}$

$q_{\text{I}} = 2.41$ m/s $\rightarrow q_{\text{IV}} = 1.23$ m³/s

$$\Delta q = -\frac{(16 \times 10^5)^2 - (15.8 \times 10^5)^2}{2 \times 1489.2 \times 10^{10}} = -0.0025$$

bars. — The main data of the solution are listed in Table 8.3 – 2. It shows that, at the given offtakes at intermediate nodes, the maximum delivery rate attainable at the delivery end of the pipeline is 1.23 m³/s.

In the two approximations employed to solve the problem, the values of the k_is were unchanged although throughputs and tail-end pressures of the pipelegs were different. The reasons for this are, one, that flow is fully turbulent so that the friction factor is independent of the throughput-dependent Reynolds number and, two, the change in the mean pressures of the pipelegs is so slight that change in the compressibility factor z is negligible.

If there is a booster pump station installed somewhere along the pipeline, then the maximum throughput capacity of the pipeline can be calculated as follows (Hain 1968): Steps (i)—(iii) of the calculation are as above. (iv) In the knowledge of the gas throughput, intake pressure and installed compressor capacity, the output pressure of the pump can be determined for the node examined. (v) The tail-end pressure of the pipeline is calculated in the knowledge of the output pressure and of pressure drops in the individual pipelegs. (vi) If the calculated tail-end pressure differs from the prescribed one, pipeleg throughputs are once more corrected using the relationship $q_i^{(2)} = q_i^{(1)} + \Delta q$, but the correction itself is now calculated by means of the relationship

$$\Delta q = -\frac{p_m^2 - p_m^{(1)2}}{2\left[\frac{(p_2)_c^2 - (p_1)_c^2}{q_c} + \sum_{i=1}^{n} k_i q_i^{(1)}\right]} \qquad 8.3\text{—}18$$

where $(p_1)_c$ and $(p_2)_c$ are the intake and discharge pressures of the compressor, respectively, and q_c is compressor output.

(b) Graphical methods

A quick and simple graphical method for solving problems involving the throughput and pressure parameters of gas transmission lines, based on diagrams, has been proposed by van den Hende (1969).

The first step is to plot the family of curves $\Delta p = f(p_1)_q$, resembling Fig. 8.3 – 2 for the pipeline examined, with both axes of coordinates calibrated in the same units. The plots are calculated using the equations

$$\Delta p = p_1 - p_2 = p_1 - \sqrt{p_1^2 - kq^2} . \qquad 8.3-19$$

Here, p_2 is expressed using Eq. 8.3 – 1. The k factor in Eq. 8.3 – 19 is identical with the k_1 figuring in Eq. 8.3 – 3. The values of λ and z figuring there are functions of p and q. Plotting the family of curves may be simplified, however, by assuming for the purposes of approximate calculations that λ and z are constant for any value of p and q. Van den Hende has developed a procedure for the calculation of k out of a function $k = Cl$, where C can be read off a table as a function of pipe size d_i. Whichever way k is determined, the individual curves of the family may be constructed by the graphical procedure illustrated in Fig. 8.3 – 3. The value of p_o is furnished by Eq. 8.3 – 1 after the substitution $p_2 = 0$, that is,

$$\Delta p_o = q\sqrt{k} .$$

Plotting any point of the curve is performed in a manner similar to the construction starting from point A. Equation 8.3 – 1 holds for any point of this curve. For instance, it is clear on inspection that the hypotenuse of triangle OAB is precisely equal to the abscissa of point P, and that, in the triangle,

$$p_1^2 = p_2^2 + kq^2 .$$

In possession of a family of curves characterized by a given k, any one of the three parameters p_1, p_2 and q of the pipeline can be determined rapidly in possession of the other two.

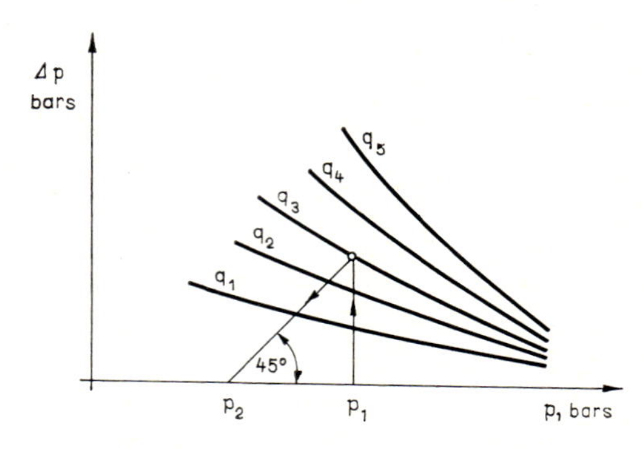

Fig. 8.3 – 2 Graphical procedure, after van den Hende (1969)

Example 8.3—3. Find the head-end pressure p_I required to ensure the tail-end pressure $p_{IV} = 18$ bars in the pipeline schematically shown as Fig. 8.3—1, under the conditions stated in Example 8.3—1. — Figure 8.3—4 illustrates the family of curves plotted using Eq. 8.3—19. The manner of constructing the head-end pressure is shown by the full line in the Figure. The resulting head-end pressure is $p_I = 55$ bars. This graphical method lends itself well to the solution of numerous other problems, too.

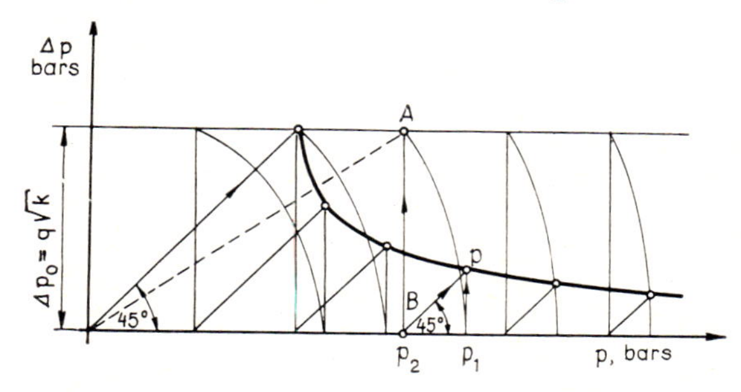

Fig. 8.3—3 Graph $\Delta p = f(p)$, after van den Hende (1969)

Example 8.3—4. In the pipeline characterized in the foregoing example, we want to raise the pressure at Node III from 31.9 to 35 bars. What is the pressure increment required at the head end, and what will the pressures at Nodes II and IV be? — The solution of the problem is shown in dashed line in Fig. 8.3—4. The result is $\Delta p_I = 1{,}5$ bar for the required pressure increment at the head end, and 42.7 and 24.3 bars, respectively for the resulting pressures at Nodes II and IV.

8.3.3. Looped systems

The first procedure for modelling a low-pressure looped network was developed by Cross (1936); it was adapted with some modifications also to high-pressure systems (Hain 1968). Let us illustrate the application of this method on the loop shown as Fig. 8.3—5. The gas flows into and out of the nodes are known, and so is pressure p_I at Node I. We are to find the gas throughputs of the indivual pipelegs, as well as the pressures in the remaining nodes. The solution is based on the following consideration. Taking clockwise flow as positive, let us assume a first-approximation value $q_1^{(1)}$ for the throughput of pipeleg *1*. Let us then use the node law to find the gas throughputs $q_i^{(1)}$, whose signs will be according as flow is clockwise or counterclockwise. In the steady state, the loop law (8.3—13 or 8.3—14) will have to apply. If it is assumed that the first-approximation values of throughputs in the individual pipelegs differ by Δq from the actual throughput, then

$$\sum_{i=1}^{n} k_i(q_i^{(1)} + \Delta q)\,|q_i^{(1)} + \Delta q| = 0\,,$$

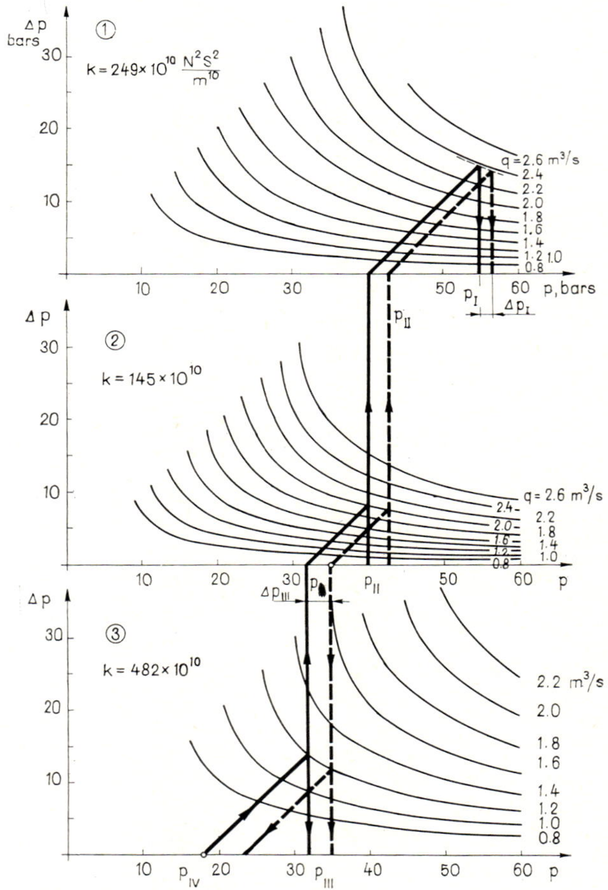

Fig. 8.3–4 Throughput and pressure distribution in the pipeline specified in Fig. 8.3–1

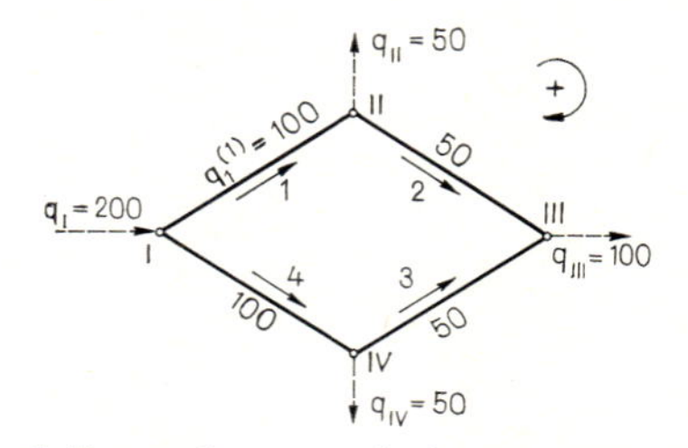

Fig. 8.3–5 Looped transmission system to Cross' method

where n is the number of node-connecting elements (pipelegs). By this relationship, the correction is

$$\Delta q = -\frac{\sum_{i=1}^{n} k_i \, |q_i^{(1)}| \, q_i^{(1)}}{2 \sum_{i=1}^{n} k_i \, |q_i^{(1)}|} \qquad 8.3\text{—}20$$

provided $|\Delta q| \ll q_i$; the second-approximation values of the gas throughputs in the individual pipelegs can now be calculated as

$$q_i^{(2)} = q_i^{(1)} + \Delta q. \qquad 8.3\text{—}21$$

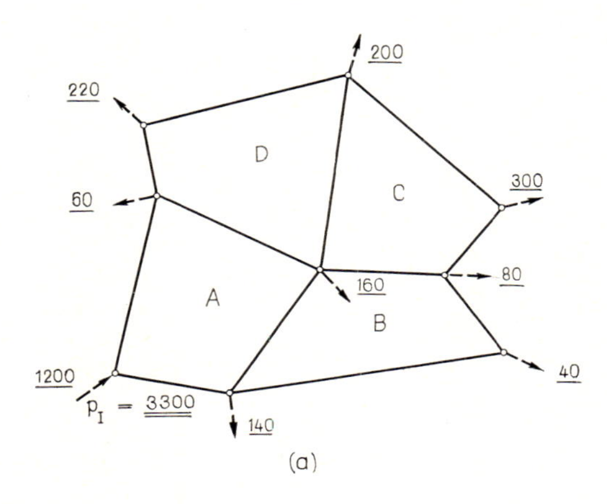

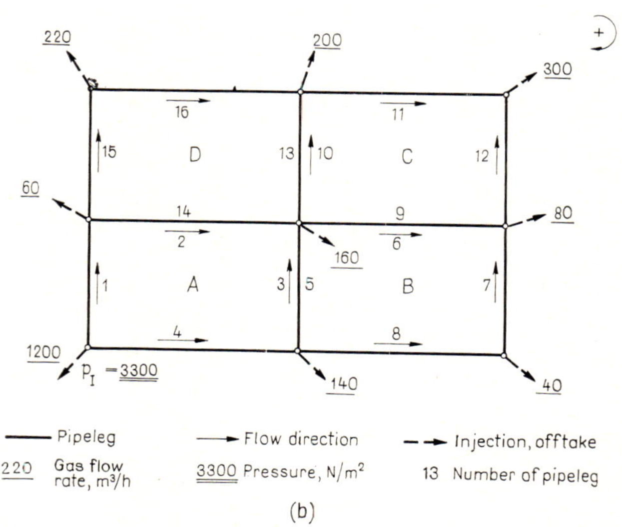

Fig. 8.3—6 Layout and hookup of low-pressure looped gas supply system

Table

Loop	Pipeleg	d_i	l_i	k_i	$q_i^{(1)}$	$q_i^{(1)}$	$k_i\|q_i\|^{(1)}$
		m	m	$\frac{10^4\ Ns^2}{m^8}$	m^3/h	$10^{-2}\ m^3/s$	$\frac{10^4\ Ns}{m^5}$
1	2	3	4	5	6a	6b	7
A	1	0.3071	450	0.1240	700	19.444	0.0241
	2(14)	0.1541	420	4.5759	200	5.556	0.2542
	3(5)	0.1541	370	4.0311	−200	− 5.556	0.2240
	4	0.2589	280	0.1917	−500	−13.889	0.0266
	$\sum_{i=1}^{4}$						0.5289
B	5(3)	0.1541	370	4.0311	200	5.556	0.2240
	6(9)	0.1023	290	28.087	200	5.556	1.5604
	7	0.1023	240	23.245	−120	−3.333	0.7748
	8	0.1023	660	63.923	−160	−4.444	2.8410
	$\sum_{i=5}^{8}$						5.4002
C	9(6)	0.1023	290	28.087	−200	−5.556	1.5604
	10(13)	0.1541	480	5.2296	40	1.111	0.0581
	11	0.1023	480	46.489	60	1.667	0.7748
	12	0.1023	220	21.308	−240	−6.667	1.4205
	$\sum_{i=9}^{12}$						3.8138
D	13(10)	0.1541	480	5.2295	− 40	−1.111	0.0581
	14(2)	0.1541	420	4.5759	−200	−5.556	0.2542
	15	0.3071	180	0.0496	440	12.22	0.0061
	16	0.2051	500	1.1859	220	6.111	0.0725
	$\sum_{i=13}^{16}$						0.3909

If after the kth successive approximation Δq is within the tolerance admitted, then the node pressures can be calculated using the relationship

$$p_j^2 = p_{\mathrm{I}}^2 - \sum_{i=1}^{j-1} k_i \, |q_i^{(k)}| \, q_i^{(k)} ; \qquad 8.3\text{—}22$$

$$j = \mathrm{I};\ \mathrm{II};\ \ldots$$

for a high-pressure network or

$$p_j = p_{\mathrm{I}} - \sum_{i=1}^{j-1} k_i \, |q_i^{(k)}| \, q_i^{(k)} \qquad 8.3\text{—}23$$

for a low-pressure network; k equals k_1 in Eq. 8.3—3 in the first case, and k_2 in Eq. 8.3—6 in the second.

If the system is composed of several loops, then, after a first-approximation estimation of the throughputs in the individual pipelegs, one calculates a Δq for each loop, and then performs the correction of the pipelegs'

8.3—3

$\Delta p_i = k_i q_i^{(1)} \lvert q_i^{(1)} \rvert$	Δq	$q_i^{(2)}$	. . .	$q_i^{(s)}$	$q_i^{(s)}$	$\Delta p = k_i q_i^{(s)} \lvert q_i^{(s)} \rvert$	p_{Ii}	p_{IIi}
$\frac{N}{m^2}$	$10^{-2}\ \frac{m^3}{s}$	$10^{-2}\ \frac{m^3}{s}$		$10^{-2}\ \frac{m^3}{s}$	m^3/h	$\frac{N}{m^2}$	$\frac{N}{m^2}$	$\frac{N}{m^2}$
8	9	10		11a	11b	12	13	14
46.87		19.192	. . .	20.960	754.6	54.47	3300	3246
141.23	−0.252	4.076	. . .	4.532	163.2	93.98	3246	3152
−124.42		−6.298	. . .	−5.486	−197.5	−121.30	3152	3273
−36.99		−14.141	. . .	−12.373	−445.4	−29.35	3273	3302
26.70						−2.20		
12.42		6.298	. . .	5.486	197.5	121.30	3273	3152
866.89	0.490	3.846	. . .	4.353	156.7	532.14	3152	2619
−258.27		−2.843	. . .	−1.888	−68.0	−82.81	2619	2702
1262.7		−3.954	. . .	−2.999	108.0	−574.75	2702	3277
−529.64						−4.12		
−866.89		−3.846	. . .	−4.353	−156.7	−532.14	2619	3152
6.46	2.200	2.083	. . .	1.220	43.9	7.79	3152	3144
129.14		3.867	. . .	4.315	155.3	865.76	3144	2278
−947.0		−4.466	. . .	−4.018	144.6	−343.98	2278	2622
−1678.3						−2.58		
−6.46		−2.083	. . .	−1.220	−43.9	−7.79	3144	3152
−142.23	1.228	−4.076	. . .	−4.532	−163.2	−93.98	3152	3246
7.41		13.450	. . .	14.762	531.4	10.81	3246	3235
44.29		7.339	. . .	8.651	311.4	88.74	3235	3146
95.99						−2.22		

throughputs loop after loop. The pipelegs common to two loops are corrected using the Δqs determined for both loops. Let us illustrate this procedure by an example referring to a low-pressure network.

Example 8.3—5. Given the gas flows into and out of the nodes of the network shown as Fig. 8.3—6a, and given the pressure $p_{I(1)} = 3300\ N/m^2$ of node $I_{(1)}$; find the gas throughputs of the individual pipelegs, and the individual node pressures. The loops are considered to be balanced if the condition $\left| \sum_{i=1}^{n} \Delta p_i \right| < 5\ N/m^2$ is satisfied. A working model of the network is shown as part (b) of Fig. 8.3—6. The numbering, sizes and length of the pipelegs composing the loops in the Figure are given in Columns 2—4 of Table 8.3—3. The pressure drops in the individual pipelegs are calculated using the relationship

$$\Delta p_i = k_i \lvert q_i \rvert q_i \qquad 8.3\text{—}24$$

derived from Eq. 8.3—4; k_i is furnished by Eq. 8.3—6, with $p_n = 1.014 \times 10^5$ N/m², $T_n = 288.2$ K, $M = 16.03$ kg/kmole, $\overline{T} = 283$ K, and λ is obtained using Eq. 1.2—5. After substitutions,

$$k_i = 5.079 \times 10^{-3} \frac{l_i}{d_i^{5.333}}.$$

The values calculated in this way are listed in Column 5. Column 6 of the Table lists the first-approximation throughputs of the individual pipelegs, with clockwise rotation regarded as positive. In estimating these throughputs, the circumstance that the condition implicit in Eq. 8.3—12 must hold for each node separately was taken into due account. The pressure drops in Column 8 were calculated using Eq. 8.3—24. The Δqs for the individual loops were determined from Eq. 8.3—20; for instance, in loop A,

$$\Delta q_A = -\frac{26.70}{2 \times 0.5289 \times 10^4} = -0.252 \times 10^{-2} \text{m}^3/\text{s}.$$

The summed data of Column 8 show that $\left| \sum_{i=1}^{4} \Delta p_i \right|$ exceeds in each loop the tolerance of 5 N/m², so that the values listed in Column 6b have to be corrected. In the pipelegs which belong to one loop only, the corrected throughputs are supplied by Eq. 8.3—21. In the pipelegs common to two loops, the throughputs must of course be the same (Column 10). The corrected throughputs are calculated as shown in the example below. By the values for pipelegs *2* and *14* in Column 6b, the first approximation throughput was 5.556×10^{-2} m³/s with signs according to the sense of rotation. The absolute value of the corrected throughput, calculated by means of Eq. 8.3—21 but not stated in the Table, is 5.304×10^{-2} m³/s in pipeleg *2*. The throughput of pipeleg *14* is equated with this value, and then corrected using the correction for loop D and Eq. 8.3—21. The value obtained is -4.076×10^{-2} m³/s, and accordingly the throughput in leg *2* is 4.076×10^{-2} m³/s. It is these values that are entered into the corresponding rows of Column 10. Iteration is pursued with the q_is of Column 10. After seven steps of iteration, not given in detail, one obtains the last-step data and the final results listed in Columns 11—14. Clearly, by the data in Column 12, $\left| \sum_{i=1}^{n} \Delta p_i \right|$ is in every case within the tolerance of 5 N/m². The data in Columns 13 and 14 are the p_{Ii}s and p_{IIi}s, the pressures at the head and tail ends, respectively, of the individual pipelegs, based on the pressure drop data in Column 12 and on the condition of $p_{I(1)} = 3300$ N/m² at the node common to pipelegs *1* and *4*.

The main advantage of the Cross method is its simplicity, whereas its main drawback is the slowness of the convergence, which renders this method uneconomical in many applications. In order to eliminate these drawbacks, Renouard developed a variant of the Cross method (Société . . . Manuel 1968). The Renouard method is suited for the modelling of steady-state operation in not-too-complicated looped networks. The method was generalized by Pernelle for networks of any size (Société . . . Manuel 1968)

The essence of this latter method is as follows. In the foregoing example we required a throughput correction Δp for each loop in each step of iteration. Let us denote the throughput correction to be calculated by Δq_A in the case of loop A, Δq_B in the case of loop B, etc. The throughput correction of the pipeleg(s) common to loops A and B is, then, $(\Delta q_A - \Delta q_B)$, that of the pipelegs common to loops B and C, $(\Delta q_B - \Delta q_C)$, etc. — Equation 8.3—20 permits us to write up for the n loops n linear equations in the n throughput corrections Δq_A, Δq_B, Δq_C, etc. The corrections are furnished by the solution of this system of equations. After applying the corrections to the throughputs, the values obtained are checked to see whether they satisfy loop law 8.3—13; if the aggregate pressure drops of the loops exceed the prescribed tolerance, then the procedure is repeated. This method furnishes, according to its authors, a result of sufficient accuracy in two or three steps even if the first estimates of the individual throughputs of the pipelegs are rather wide of the mark.

Stoner's method for solving looped networks is based on the node continuity equation (Stoner 1970). It has the advantage that, whereas the Cross method can be used to establish throughput and pressure maps of the network only, the Stoner method will furnish any parameter (pipe size in a leg, compressor horsepower required, number of storage wells, size of pressure-reducing choke, etc.) of the complex system. It is, however, significantly more complicated than the previously mentioned methods, and it requires much more computer time. — The way of constructing the model is illustrated in Fig. 8.3—7. Node *11*, selected as an example, receives gas

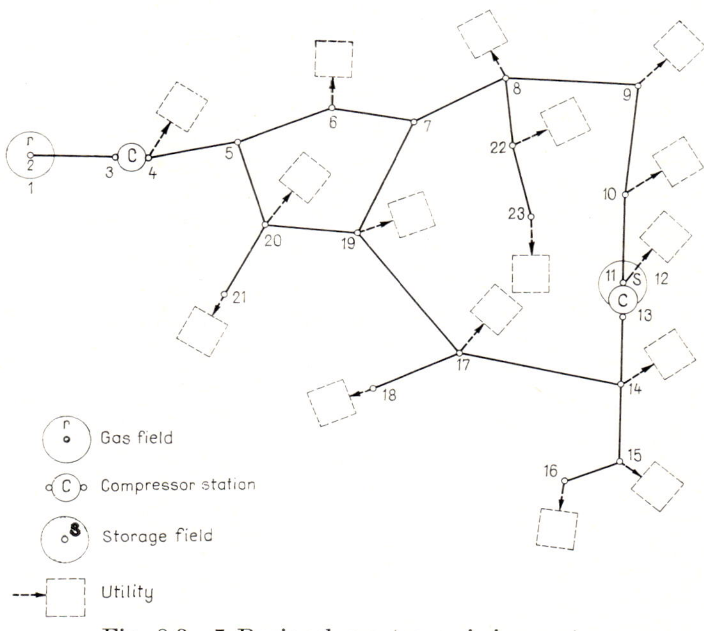

Fig. 8.3—7 Regional gas transmission system

from underground storage facility (*12—11*) and pipeleg (*10—11*), and delivers gas into the intake of pump (*13—11*), and into the consumer supply circuit directly attached to the node. By this model, the node equation 8.3—12 can be given the form

$$F_{11} = (q_{12-11})_s - (q_{13-11})_c + (q_{10-11})_p - q_{o11} = 0 \qquad 8.3-25$$

where suffixes s, p and c respectively refer to storage, pipeleg and compressor, and q_{o11} is the flow of gas out of node *11*. Flow into the node is positive. The measure of imbalance at the node is F_{11}; its value is zero if the node is balanced, that is, if the condition $|F_j| < \varepsilon$ is satisfied, where ε is the tolerance. Introducing into this Equation the Relationships 8.3—1, —7 and —11, we get

$$F_{11} = J_{12-11}(|\,p_{12}^2 - p_{11}^2\,|)_n S_{12-11} - \frac{p_{13-11}}{k_3\left(\dfrac{p_{13}}{p_{11}}\right)^{k_4} + k_5} + \qquad 8.3-26$$

$$+ \frac{(|\,p_{10}^2 - p_{11}^2\,|)^{0.5}}{(k_1)_{10-11}^{0.5}} S_{10-11} - q_{o11} = 0$$

where $S_{i,j}$ is a sign factor accounting for flow direction:

$$S_{i,j} = \text{sign}\,(p_i - p_j) = \begin{cases} +1 \text{ if } p_i \geq p_j \\ -1 \text{ if } p_i < p_j \end{cases}.$$

Writing up in a similar fashion n equations of continuity for the n nodes of the system, one obtains the non-linear system of Equations constituting the mathematical model of the system in the steady state.

The equations contain node pressures, inputs/outputs and the parameters of the NCEs (node connecting elements), altogether $(2n + m)$ parameters, where n is the number of nodes and m is the number of NCEs. The model of n equations will in principle yield any n unknowns of the $(2n + m)$ parameters, if the remaining $(n + m)$ parameters are given. These equations, similar to Eq. 8.3—26, can thus be written in the form

$$F_j(x_1, x_2, \ldots x_n) = 0 \qquad 8.3-27$$
$$j = 1, 2, \ldots, n.$$

The only criterion in choosing the n unknowns to be calculated is that the continuity equations of the type 8.3—26, written up for the nodes, must remain mutually independent. Since the value of the nodal gas throughput is independent in $(n - 1)$ equations only, at least one of the values q_{oj} must be known. It is likewise necessary to state at least one node pressure.

The solution of the non-linear system of Equations 8.3—27 constituting the mathematical model of the network may be achieved by the Newton—Raphson technique. The essence of this method is that it provides linear relationships for correcting the initial, estimated values of the unknowns, and said relationship of correction ensures that the successive steps of iteration make the system approach the solution. Let the value of the ith unknown, denoted x_i, be $x_i^{(k)}$ after the kth step of iteration; then

$$x_i^{(k+1)} = x_i^{(k)} + \Delta x_i^{(k+1)}; \; i = 1, 2, \ldots, n \qquad 8.3-28$$

where the Δx_is are furnished in each step of iteration by the solution of the linear system of equations

$$\sum_{i=1}^{n} \frac{\partial F_j}{\partial x_i} \Delta x_i = - F_j; \qquad j = 1, 2, \ldots, n. \qquad 8.3-29$$

The $\partial F_j/\partial x_i$s are the values of the derivatives of the node continuity equations taken at the x_is calculated in the foregoing step of iteration. The linear system of Equations 8.3—29 may be solved by direct elimination. The Newton—Raphson method requires that the initial values $x_i^{(1)}$ of the unknowns x_i be estimated. The convergence behaviour of the iteration will depend to a significant degree on the goodness of these estimates, even in a fairly simple system. A suitable rate of convergence may be ensured, according to Stoner's proposition, in the following way. Introducing the acceleration factor α_i, Eq. 8.3—28 can be written in the form

$$x_i^{(k+1)} = x_i^{(k)} + \Delta x_i^{(k+1)} \alpha_i . \qquad 8.3-30$$

The value of α_i can be expressed in terms of the actual and the foregoing corrections Δx_i as follows. Let

$$A_i = \frac{\Delta x_i^{(k+1)}}{\Delta x_i^{(k)}};$$

if $A_i \leq -1$,	then	$\alpha_i = 0.5 \mid A_i \mid$;
if $-1 < A_i < 0$,	then	$\alpha_i = 1.0 - 0.5 \mid A_i \mid$;
if $0 < A_i < 1$,	then	$\alpha_i = 1.0 + 2.0 \mid A_i \mid$;
if $A_i \geq 1$,	then	$\alpha_i = 3$.

In the first two steps of iteration, where divergence is most likely to occur, it is best to put $\alpha_i = 0.5$ to ensure convergence. In the subsequent steps, the values of α_i are determined as above in every other step; in the steps in between, α_i is put equal to 1.0. This method ensured convergence in every case and gave results of satisfactory accuracy after 6—10 steps.

Stoner (1971, 1972), in a development of the above method, gave a procedure for determining the 'sensitivity' of the system in steady-state operation. The purpose of the calculation is in this case to find out in what way some change(s) in some parameter(s) of the system affect the remaining parameters. For instance, what changes in input pressures and flow rates, or compressor horsepower, are to be effected in order to satisfy a changed consumer demand? Schematically the method can be outlined as follows. — Let y_i denote those n parameters whose changes we are interested in, after other m parameters of the system, denoted x_i in their turn, have been changed. In this notation, the non-linear system of equations resembling

Eq. 8.3—26, constituting the model of the system, can be written in the form

$$F_j(y_1, y_2, \ldots, y_n; x_1, x_2, \ldots, x_m) = 0 \qquad 8.3\text{—}31$$

$$j = 1, 2, \ldots; n.$$

The Taylor series expansion of the function F_j, with all but the linear terms dropped, is

$$\sum_{i=1}^{n} \frac{\partial F_j}{\partial y_i} \mathrm{d}y_i + \sum_{i=1}^{m} \frac{\partial F_j}{\partial x_i} \mathrm{d}x_i = 0 ; \quad j = 1, 2, \ldots, n. \qquad 8.3\text{—}32$$

Each one of the two sets of derivatives, $\partial F_j/\partial y_i$ and $\partial F_j/\partial x_i$, formally identical with the derivatives figuring in Eq. 8.3—29, can be regarded as forming a matrix:

$$\mathbf{J} = \begin{bmatrix} \frac{\partial F_1}{\partial y_1} & \cdots & \frac{\partial F_1}{\partial y_n} \\ \vdots & & \vdots \\ \frac{\partial F_n}{\partial y_1} & \cdots & \frac{\partial F_n}{\partial y_n} \end{bmatrix};$$

$$\mathbf{C} = \begin{bmatrix} \frac{\partial F_1}{\partial x_1} & \cdots & \frac{\partial F_1}{\partial x_m} \\ \vdots & & \vdots \\ \frac{\partial F_n}{\partial x_1} & \cdots & \frac{\partial F_n}{\partial x_m} \end{bmatrix}.$$

Using these identities, Eq. 8.3—32 may be rewritten in matrix notation as

$$\mathbf{J}\,\mathrm{d}y + \mathbf{C}\,\mathrm{d}x = 0 \qquad 8.3\text{—}33$$

which after rearranging becomes

$$\mathrm{d}y = -\mathbf{J}^{-1}\mathbf{C}\,\mathrm{d}x$$

where $\mathbf{J}^{-1}$ is the inverse of $\mathbf{J}$. The matric resulting from the multiplication $-\mathbf{J}^{-1}\mathbf{C}$ is the so-called sensitivity matrix of the system, to be denoted by the symbol $[\mathrm{d}_y/\mathrm{d}_x]$. It is a measure of change in the parameters y_i resulting from unity changes in the parameters x_i, provided that the node continuity equation 8.3—12 is satisfied for every node. It suffices to determine the sensitivity matrix just once to be able to determine by a simple matrix multiplication the change in the parameters y_i of the system, resulting from any change in the parameters x_i, represented by the vector Δx:

$$\Delta y = \left[\frac{\mathrm{d}_y}{\mathrm{d}_x}\right] \Delta x\,. \qquad 8.3\text{—}34$$

8.4. Transient flow in pipeline systems

Flow parameters in gas transmission pipelines are usually time-dependent, the main reason for which is the variation in demand, as a function of a variety of factors. If the fluid flowing in the pipe were incompressible, Newtonian, then the change in throughput would take place at the same instant and would be of the same magnitude at any pipeline section, including the head and the tail end. Such transient flow could, then, at any instant be described by the equations of steady-state flow over the whole length of the pipeline. If, however, demand at the delivery end of a pipeline conveying compressible gas changes then it takes a time Δt for the resulting pressure reduction to make itself felt at the head end of the pipeline. The equations of steady-state flow will, then, apply to infinitesimal lengths of pipeline only. Still, even the equations of flow in a complex system made up of pipelegs of considerable length may be derived from these fundamental relationships. It has been primarily thanks to the electronic computer that several increasingly more accurate procedures for modelling gas flow in complicated pipeline systems have been able to be developed in the last decade and a half.

8.4.1. Fundamental relationships

The relationships describing flow in pipelines of finite length may be derived from four fundamental relationships; any differences in these are merely matters of formulation. The equation of continuity is

$$\frac{\partial q_m}{\partial x} + \frac{\partial(\varrho A_i)}{\partial t} = 0. \qquad 8.4-1$$

The equation of energy or of motion is the transient form, accounting for the change of parameters in time, of Eq. 1.2—1:

$$\frac{\partial p}{\partial x} + \varrho g \sin \alpha + \frac{\lambda v^2 \varrho}{2d_i} + \varrho \frac{\partial v}{\partial t} = 0. \qquad 8.4-2$$

The equation of state for a gas flow regarded as isothermal is, by Eq. 8.1—1

$$\frac{p}{\varrho} = z \frac{R}{M} T .$$

The fourth fundamental relationship

$$z = \mathrm{f}(p)_T ,$$

has several solutions employed in practice, one of which is Eq. 8.1—9. If z is replaced by its average value and considered constant, then the number of fundamental equations reduces to three, and Eq. 8.1—1 may be written in the simpler form

$$\frac{p}{\varrho} = B^2 , \qquad 8.4-3$$

where B is the isothermic speed of sound. Eqs 8.4—1 and 8.4—3 imply

$$F_1 = \frac{B^2 \partial q_m}{A_i \partial x} + \frac{\partial p}{\partial t} = 0 \qquad 8.4\text{—}4$$

where mass flow is

$$q_m = \varrho A_i v = \frac{p}{B^2} A_i v \,.$$

By Eqs 8.4—2, 8.4—3 and the above definition of q_m,

$$F_2 = \frac{1 \partial p^2}{2 \partial x} + \frac{p \partial q_m}{A_i \partial t} + \frac{p^2 g}{B^2} \sin \alpha + \frac{\lambda\, B^2 q_m \,|q_m|}{2 d_i\, A_i^2} = 0. \qquad 8.4\text{—}5$$

Equations 8.4—4 and —5 constitute a system of non-linear partial differential equations; then, on the assumption that $\bar{z}$ = constant, describe transient flow in the pipeline system.

8.4.2. Flow in pipelines

(a) Matching the system to variable consumer demand

There are two frequently-employed ways of adapting flow in pipelines to the (usually daily) fluctuation of consumer demand. It holds for both cases that over the period of fluctuation (which we shall henceforth equate with one day) the gas quantity injected into the pipeline equals the gas offtake, that is, the consumption out of the line. In the first case, the hourly injection of gas into the pipeline is constant, say q_{in} in standard volume units. At the delivery end of the pipeline, pressure varies between the minimum required at the head end of the consumer supply network, p_{2min}, and the peak pressure p_{2max}, occurring when demand is at a low end. This setup has the advantage that the rate of injecting gas into the pipeline is constant, requiring no regulation within one day. The difference between the constant input and the fluctuating offtake is taken up by the pipeline acting as a buffer storage facility. The drawback of this setup is that a significant portion of the pressure energy in the gas is dissipated by a throttle at the delivery end, since over most of the day the tail-end pressure of the pipeline exceeds the p_{2min} required by the consumer supply network. Energy loss is minimized in certain instances by making the gas pass through a gas turbine rather than a pressure reducer, and letting the turbine drive an electric generator. The output of this latter is fed to the grid. An approximate designing of the pipeline may be performed by the following procedure of Smirnov and Shirkovsky, slightly modified (Szilas 1967).

Equation 1.2—7 may be restated in the following simple form:

$$q = k \sqrt{\frac{p_1^2 - p_2^2}{\bar{z}}} \,. \qquad 8.4\text{—}6$$

Let us point out that, as opposed to the factor k_1 in Eq. 8.3—2, the k occurring here does not include the compressibility factor $\bar{z}$. By Eq. 8.4—6,

$$p_1 = \sqrt{p_2^2 + \frac{q^2 \bar{z}}{k^2}}\,, \qquad 8.4\text{—}7$$

and, introducing the expression $p_1/p_2 = R_p$,

$$p_1 = \frac{q R_p \sqrt{\bar{z}}}{k\sqrt{R_p^2 - 1}}. \qquad 8.4\text{—}8$$

Likewise by Eq. 8.4—6,

$$p_2 = \sqrt{p_1^2 - \frac{q^2 \bar{z}}{k^2}} \qquad 8.4\text{—}9$$

and, putting $p_1/p_2 = R_p$,

$$p_2 = \frac{q\sqrt{\bar{z}}}{k\sqrt{R_p^2 - 1}}. \qquad 8.4\text{—}10$$

Introducing the expressions of p_1 and p_2 in Eqs 8.4—8–10 into the Eq. 1.2—26 for the mean pressure, and rearranging, we get

$$\frac{3\bar{p}k}{2q\sqrt{\bar{z}}} = \frac{R_p^2 + R_p + 1}{(R_p + 1)\sqrt{R_p^2 - 1}}. \qquad 8.4\text{—}11$$

Figure 8.4—1 is a plot of the expression $3\bar{p}k/2q\sqrt{\bar{z}}$ v. R_p as furnished by Eq. 8.4—11.

The above-given relationships permit us to find out whether the maximum pressure p_{1max} that can be ensured at the head end of the pipeline is sufficient to satisfy consumer demand, provided injection rate of gas into the pipe-

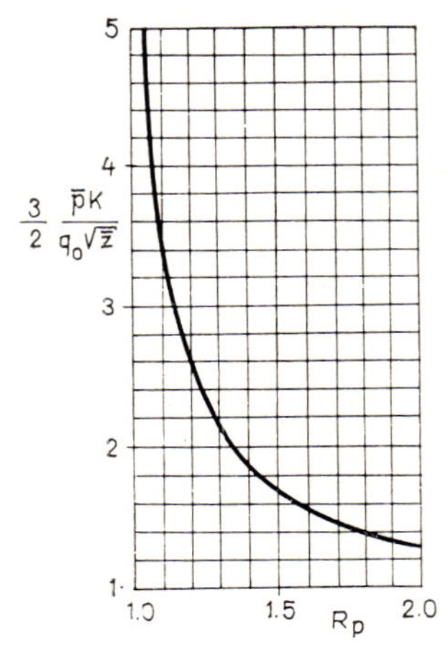

Fig. 8.4—1 For determining the buffer action of a pipeline (I), after Smirnov and Shirkovsky (1957) and Szilas (1967)

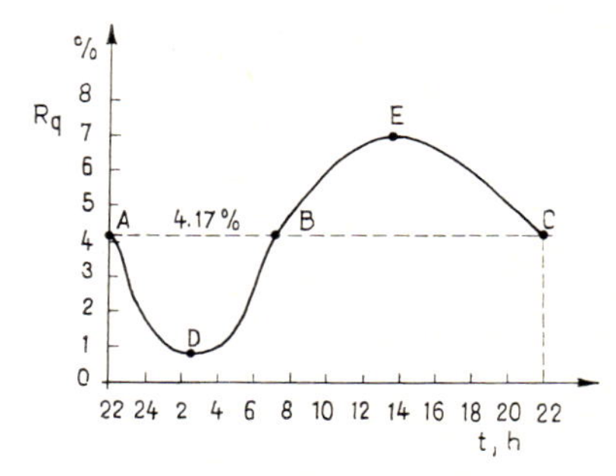

Fig. 8.4—2 For determining the buffer action of a pipeline (II), according to Smirnov and Shirkovsky (1957) and Szilas (1967)

line is uniform. Calculation proceeds as follows. (i) Gas consumption is plotted v. time t on the basis of daily consumption records. Figure 8.4—2 shows the daily fluctuation of R_q, the percentage hourly consumption referred to daily consumption. The input into the pipeline per hour is 1/24th of the daily consumption, that is,

$$\bar{q} = \frac{1}{24} \int_{t=0}^{24} q(t)\, dt\,.$$

The line parallel to the abscissa axis, having ordinate $\bar{q}$, intersects the curve at points A, B and C. In segment $A—B$, consumption is less than $\bar{q}$; that is, gas accumulates in the pipeline. In segment $B—C$, the gas thus accumulated is used to cover higher-than-average demand. (ii) Gas flow into the pipeline is precisely equal to demand at point A. The gas reserve in the pipeline has dropped to zero at that point; that is, pressure at the tail end of the pipeline must at that instant be $p_{2\min}$, which is the least pressure required at the head end of the consumer supply system. Applying in an approximation the relationship for steady-state flow, we get by Eq. 8.4—7 for pressure at the head end of the pipeline at this same instant

$$p_1 = \sqrt{p_{2\min}^2 + \frac{q^2 \bar{z}}{k^2}}$$

where $\bar{z}$ can be determined by iteration using the given $\bar{T}$ and the p calculated from p_1 and $p_{2\min}$, applying Eq. 1.2—26. The mean pressure in the pipeline is, then,

$$\bar{p} = \frac{2}{3}\left(p_1^2 + \frac{p_{2\min}^2}{p_1 + p_{2\min}}\right).$$

For the standard-state volume of gas in the pipeline, the general gas law furnishes

$$V_{OA} = \frac{T_n \bar{p} A_i l}{p_n \bar{z} T}. \qquad 8.4—12$$

(iii) The area embraced by curve segment $A—D—B$ and line segment $A—B$ is determined by planimetering. Multiplying this by the scale of the diagram, we get the volume V_{OAB} of gas stored up in the slack-demand period. (iv) Said stored-up volume will be a maximum at the instant corresponding to B, and hence, the head- and tail-end pressures p_1 and p_2, respectively, of the pipeline will also be maximal at that instant. The mean pressure in the pipeline, corrected by $\bar{z}$, is at that instant

$$\frac{\bar{p}}{\bar{z}} = \frac{p_n}{T_n} V_{Oal} \frac{T}{A_i l}, \qquad 8.4—13$$

where $V_{Oal} = V_{OA} + V_{OAB}$. — From $\bar{p}/\bar{z}$, $\bar{p}$ and $\bar{z}$ may be determined by successive approximation. (v) The next step is to calculate the value of the expression $3\bar{p}k/2q\sqrt{\bar{z}}$ and to read the corresponding R_p off Diagram

8.4—1. In the possession of this latter, Eq. 8.4—10 may be used to furnish p_2, whereupon $p_1 = R_p p_2$. If the technically feasible maximum pressure at the head end of the pipeline is p_{1max}, then the quantity of gas V_{OAB} stored up in the pipeline during the period $A-B$ will suffice to cover the excess demand in the period $B-C$, if $p_1 = p_{1max}$.

The second way to satisfy fluctuating consumer demand is to inject into the pipeline gas at varying pressures and rates of injection, so as to ensure an unvaried tail-end pressure p_2 in the pipeline, equal to what is required by the consumer supply network. This, of course, can be realized only if the variations in demand in the supply circuit can be predicted to a fair degree of accuracy. The problem can be solved making use of the relationships given in Section 8.1—4. In the following we shall describe the principle of the Batey—Courts—Hannah method of solution and discuss the conclusions that can be drawn from a numerical example (Batey et al. 1961).

Gas-consumption variations in time can be represented by a Fourier function. Figure 8.4—3 shows the graph of such a function, $q_2 = f(t)$. In possession of this graph, and of the constant pressure p_2 prescribed, we may, starting from fundamental relationships, calculate functions of gas flow rate and pressure v. time, step by step for various pipeline sections, proceeding backward along the line. Such functions referring to the head end of the pipeline are illustrated by graphs $q_1 = f'(t)$ and $p_1 = f''(t)$ shown in Fig. 8.4—3. The group of diagrams in Fig. 8.4—4 has been prepared on the basis of several similar diagrams. These permit us to conclude upon several characteristics of transient flow. The six top diagrams are plots of each variable v. the frequency of the demand wave. Each curve in parts (a), (c) and (e) of the Figure shows the output flow-rate wave to have a lower amplitude q_{ao} than the input flow-rate wave (whose ampli-

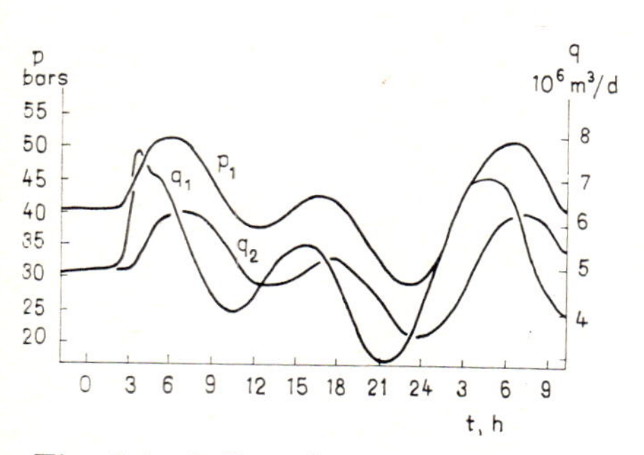

Fig. 8.4—3 Transient flow (I), after Batey et al. (1961)

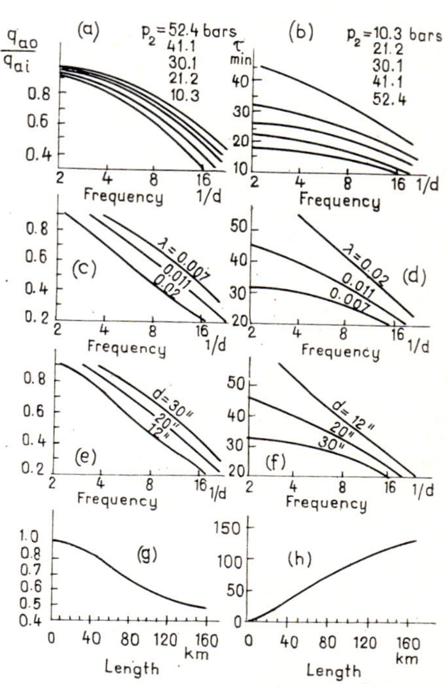

Fig. 8.4—4 Transient flow (II), after Batey et al. (1961)

tude is q_{ai}); damping increases as the frequency. For a given frequency, the amplitude ratio is the less, that is, damping is the stronger, the lower the tail-end pressure p_2 (part (a) of the Figure), the lower the friction factor λ, and the less the pipe size d_i. Each curve in parts (b), (d), and (f) of the Figure shows the phase shift of the demand wave to be the less, that is, phase velocity to be the higher, the higher the frequency. For a given frequency, phase shift is the less, the higher the tail-end pressure p_2, the less the friction factor λ, and the larger the pipe size. The diagram denoted (g) shows the amplitude ratio to decrease with pipeline length; the diagram denoted (h) shows the phase shift to increase with it. — This way of satisfying consumer demand has for its main advantage that no pressure energy need be dissipated by throttling at the tail end, that is the power consumption of compressor is reduced. A precondition of any application of this method is, of course, a sufficiently accurate foreknowledge of the demand wave, in the form of the relevant relationships $q_1 = f'(t)$ and $p_1 = f''(t)$; also, the compressor station should be operated in keeping with these relationships.

No analytic solution of general validity is known for the system of partial differential equations discussed as describing transient gas flow in Section 8.4.1. In special cases, however, connected with specific initial and boundary conditions, the system of equation can be solved. Literature contains descriptions of many such solutions (e.g. Komikova 1971, Wilkinson et al. 1964). One of the common traits of these solutions is that they permit the analysis of partial problems and that their computer time demand is not excessive. In practice, numerical solutions are often preferred in simulating transient flow conditions.

(b) Numerical solutions

In the approach to the numerical solution of the system of partial differential equations 8.4—4 and 8.4—5, the system is transformed into a system of algebraic equations using the method of finite differences. This algebraic system is capable of solution. For the transformation, the method of central finite differences can be used to advantage. It consists in essence of replacing the function, continuous in the interval under investigation, by a chord extending across a finite domain of the independent variable. The slope of said chord is approximately equal to the slope of the tangent to the curve at the middle of the domain. It is subsequently simple to calculate numerically the derivative of the curve.

For solving the system of differential equations, literature (e.g. Zielke 1971) usually cites three methods: *the implicit method, the explicit method and the method of characteristics*. A common trait to the three methods is that calculation proceeds step by step, deriving pressures and flow rates prevailing at various points of the pipeline at the instant $t + \Delta t$ on the basis of the known distribution of pressures and flow rates at the instant t. The differences are as follows. — In the explicit method, the partial differential equations are transformed into algebraic equations, so that the unknown pressures and flow rates at the instant $t + \Delta t$ depend only on the known pressures and flow rates of the preceding time step, which permits us to find their values one by one solving the individual equations for them. — In the implicit method, a system of algebraic equations results, which contains the unknown pressures and flow rates at the instant $t + \Delta t$ at the neighbouring points of the pipeline so as to be made available only by the solution of the entire simultaneous set of equations. The system of equations furnished by the transformation may, in both cases, be either linear or not. There is the fundamental difference that, whereas in the explicit system the time step is limited for reasons of stability, the only consideration that limits the time step in the implicit method is the accuracy required, but steps are usually significantly longer than what is admissible in the explicit

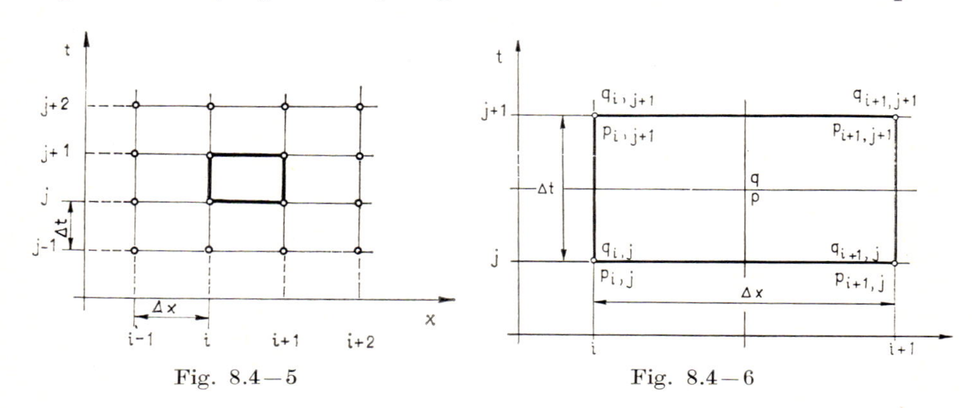

Fig. 8.4—5

Fig. 8.4—6

method. — The method of characteristics is essentially an explicit method whose essence is to seek in the $[x, t]$ plane such directions along which the partial differential equation can be reduced to a common differential equation. This latter can be solved numerically by the method of finite differences. The time step is rather restricted also in this method.

Let us now discuss the transforming of the system of partial differential equations into one of algebraic equations by the method of finite differences as performed in the implicit method (Streeter and Wylie 1970, Zielke 1971). The pipeleg under examination is divided up into segments of length Δx. The time-variable flow rates and pressures of the line sections thus obtained can be assigned to the nodes of the lattice in Fig. 8.4—5, with a distance step Δx and a time step Δt. Figure 8.4—6 is a blow-up of the cell bounded by the lattice points $(i;\ i + 1)$ in space and $(j;\ j + 1)$ in time. On the basis of this Figure, approximate values for the derivatives figuring in the systems of partial differential equations 8.4—4 and 8.4—5, relative to said cell, can be written up (with q_m replaced by q) as follows:

$$\frac{\partial p^2}{\partial x} = \frac{p^2_{i+1,j+1} + p^2_{i+1,j} - p^2_{i,j+1} - p^2_{i,j}}{2\Delta x} \qquad 8.4\text{—}14$$

$$\frac{\partial p}{\partial t} = \frac{p_{i,j+1} + p_{i+1,j+1} - p_{i,j} - p_{i+1,j}}{2\Delta t} \qquad 8.4\text{—}15$$

$$\frac{\partial q}{\partial x} = \frac{q_{i+1,j+1} + q_{i+1,j} - q_{i,j+1} - q_{i,j}}{2\Delta x} \qquad 8.4\text{—}16$$

$$\frac{\partial q}{\partial t} = \frac{q_{i,j+1} + q_{i+1,j+1} - q_{i,j} - q_{i+1,j}}{2\Delta t} . \qquad 8.4\text{—}17$$

Regarding pressure p and mass flow rate q figuring in Eqs 8.4—4 and 8.4—5 as time and space averages that are constant within the cell, we get

$$q = \frac{1}{4}(q_{i,j} + q_{i+1,j} + q_{i,j+1} + q_{i+1,j+1}) \qquad 8.4\text{—}18$$

$$p = \frac{1}{4}(p_{i,j} + p_{i+1,j} + p_{i,j+1} + p_{i+1,j+1}). \qquad 8.4\text{—}19$$

Resubstituting Eqs 8.4—14—8.4—19 into Eqs 8.4—4 and 8.4—5, and rearranging, we have a system of non-linear algebraic equations:

$$\left.\begin{aligned}
F_1 = {} & \frac{1}{\Delta t}(p_{i,j+1} + p_{i+1,j+1} - p_{i,j} - p_{i+1,j}) + \\
& + \frac{B^2}{A\Delta x}(q_{i+1,j+1} + q_{i+1,j} - q_{i,j+1} - q_{i,j}) = 0\,; \\
F_2 = {} & \frac{1}{\Delta x}(p^2_{i+1,j+1} + p^2_{i+1,j} - p^2_{i,j+1} - p^2_{i,j}) + \\
& + \frac{1}{2A\,\Delta t}(p_{i,j} + p_{i+1,j} + p_{i,j+1} + p_{i+1,j+1}) \times \\
& \times (q_{i,j+1} + q_{i+1,j+1} - q_{i,j} - q_{i+1,j}) + \\
& + \frac{g \sin \alpha}{4B^2}(p^2_{i,j} + p^2_{i+1,j} + p^2_{i,j+1} + p^2_{i+1,j+1}) + \\
& + \frac{\lambda B^2}{8d\,A^2}(q_{i,j} + q_{i+1,j} + q_{i,j+1} + q_{i+1,j+1}) \times \\
& \times |q_{i,j} + q_{i+1,j} + q_{i,j+1} + q_{i+1,j+1}| = 0.
\end{aligned}\right\} \qquad 8.4\text{—}20$$

If the values of the parameters $q_{i,j}$; $q_{i+1,j}$; $p_{i,j}$; $p_{i+1,j}$ at the instant j are known, either because they figure in the initial conditions or as a result of the calculation for a preceding time step, then the pair of equations contains four unknowns in all: the parameters $q_{i,j+1}$; $q_{i+1,j+1}$; $p_{i,j+1}$; $p_{i+1,j+1}$, belonging to the instant $t + \Delta t$. The pair of equations 8.4—20 may be written up in a similar manner for each one of the n cells. Thus, in any time step, we have to solve $2n + 2$ equations, including the two boundary conditions, for $2n + 2$ unknowns altogether. For solving the

$2n + 2$ non-linear algebraic equations, Streeter and Wylie (1970) have proposed the Newton—Raphson method of iteration. The procedure of solution is influenced by the way the two boundary conditions are stated; it will differ according to whether the two boundary conditions refer to the same or to opposite ends of the pipe segment examined. Said boundary conditions are most often time functions of node gas flow rate or pressure. — The number of steps of iteration required to solve the system of equations depends to a significant extent on the choice of initial values for the variables. In order to accelerate convergence it is to be recommended to estimate the initial values by extrapolating from the values found for the preceding time steps. A solution of sufficient accuracy of the system of equations may thus be achieved in just one or two steps. The implicit method has the advantage of being stable even if the time steps Δt exceed $\Delta x/B$, and that, consequently, time-step length is limited by accuracy considerations only. There is, however, the drawback that the values of the variables for the instant $t + \Delta t$ may occasionally be furnished by a non-linear system of equations of almost-unmanageable size.

The method of characteristics has also been employed (Streeter and Wylie 1970) for the solution of Eqs 8.4—4 and 8.4—5. The advantages and disadvantages of this method resemble those of the explicit method; all there is to do in order to find the pressure and mass flow rate in the next instant of time is the solution of a system of two quadratic equations in two unknowns, but the time step, for reasons of stability, must be quite small: $\Delta t < \Delta x/B$. There is an advantage in simultaneously using the characteristic and the implicit method. This will increase the largest admissible time step Δt rather significantly against what is admitted by the sole use of the method of characteristics. Furthermore, the method of characteristics permits the breakdown of complex gas transmission systems into simpler elements. The implicit method, applied to the individual elements, will yield a smaller number of non-linear equations per system, and accordingly, time needed to solve these equations will be reduced rather substantially.

8.4.3. Flow in pipeline systems

If there is injection or offtake of gas at certain intermediate points of the transmission line, then these points are to be regarded as nodes and the node law must apply to them. The system of non-linear algebraic equations, written up for the implicit cells of the individual pipeline segments and composed of pairs of equations resembling Eqs 8.4—20, is then complemented by node continuity equations of the form

$$q_n + \sum_{i=1}^{m} q_i = 0\,,$$

where q_n is the gas mass flow into or out of the node, and the q_is are the mass flow rates in the pipelegs meeting at the node. It is the solution of this extended non-linear system of algebraic equations that furnishes the time-dependence of pressures and flow rates in a transmission line with injections and offtakes at intermediate points. Describing in this way the transients taking place in the transmission line system is fairly complicated; this is why, despite its accuracy, it is used to solve simpler, radial systems only (Wylie et al. 1970). Modelling the transients of more complicated, looped nets is usually performed by some simpler method resulting from certain neglections. The most usual neglections are as follows (Guy 1967).

In a gas transmission line system, neglecting the altitude difference between the system's nodes does not usually introduce a significant error. The third-term on the right-hand side of Eq. 8.4—5 describing transient flow may therefore be dropped. It can further be shown that the term $(p/A_i)(\partial q/\partial t)$, describing the change per unit of time in the rate of mass flow on the right-hand side of Eq. 8.4—5, is in the majority of practical cases less by an order of magnitude than the friction term

$$\frac{\lambda B^2 q^2}{2d_i A_i^2},$$

and is therefore negligible, too. These simplifications reduce the system composed of Eqs 8.4—4 and 8.4—5 to the following, much simpler, form:

$$\frac{\partial p}{\partial t} = -\frac{B^2 \partial q}{A \partial x}, \qquad 8.4\text{—}21$$

$$\frac{\partial p^2}{\partial x} = \frac{\lambda B^2}{\mathrm{d} A^2} q^2, \qquad 8.4\text{—}22$$

where we have changed the notation concerning internal cross-sectional area and *ID* of the pipe ($A_i \to A$; $d_i \to d$). — Equation 8.4—21 states the pressure change per unit of time in an infinitesimal length of pipe dx, brought about by an infinitesimal change in the gas mass flow rate. The equation describes the capacitive property of the pipeline. — By Eq. 8.4—22, the flowing pressure drop in the infinitesimal length of pipe dx can be calculated using the relationship for steady-state flow. The equation expresses the resistance to flow of the pipeline. The physical content of these equations can be generalized to systems of pipelines as follows.

One assigns to any node half the length of each pipeleg tying in to that node, and the half-legs thus obtained are summed. Let the volume thus assigned to node j be V_j. The flows $q_{i,j}$ into and out of the node and the offtake $q_{o,j}$ at the node determine the change of the mass flow rate at the node. Equation 8.4—21 may, therefore, be rewritten for this node in the following form (Fincham 1971):

$$\frac{V_j \mathrm{d} p_j}{B^2 \mathrm{d} t} = \sum_{i=1}^{m} q_{i,j} - q_{o,j} \qquad 8.4\text{—}23$$

where m is the number of pipelegs tying in to the node. — By Eq. 8.4—22, mass flow rates in the pipelegs assigned to node j can be calculated using the relationship

$$q_{i,j} = \left[\frac{d_{i,j} A_{i,j}^2}{\lambda_{i,j} B^2} \frac{| p_i^2 - p_j^2 |}{l_{i,j}} \right]^{0.5} S_{i,j} \qquad 8.4-24$$

where the $l_{i,j}$s are the lengths of the individual pipelegs and

$$S_{i,j} = \text{sign}\,(p_i - p_j).$$

Introducing Eq. 8.4—24 into Eq. 8.4—23, and employing the notation

$$K_j = \frac{B}{V_j} \quad \text{and} \quad J_{i,j} = J_{j,i} = \left(\frac{\mathrm{d}A^2}{\lambda l} \right)_{i,j}^{0.5}$$

we get after rearranging the differential equation

$$\frac{\mathrm{d}p_j}{\mathrm{d}t} = K_j \sum_{i=1}^{m} [J_{i,j}(| p_i^2 - p_j^2 |)^{0.5} S_{i,j}] - q_{oj}. \qquad 8.4-25$$

Applying the method of finite differences,

$$\frac{\mathrm{d}p_j}{\mathrm{d}t} = \frac{p_j(t + \Delta t) + p_j(t)}{\Delta t}.$$

Using this, Eq. 8.4—25 assumes after rearranging the form

$$p_j(t + \Delta t) = \Delta t\, K_j \{ \sum_{i=1}^{m} [J_{i,j}(| p_i^2 - p_j^2 |)^{0.5} S_{i,j}] - q_{oj} \} + p_j t\,. \qquad 8.4-26$$

Writing up similar equations for the other nodes we obtain a system of non-linear algebraic equations concerning the transients in the complex system. The solution of this system of equations furnishes the pressures prevailing at the individual nodes at the instant $(t + \Delta t)$. Differential equation 8.4—25 can be solved using the implicit or explicit method, as follows.

Let us introduce the notation

$$C_j = \Delta t\, K_j \{ \sum_{i=1}^{m} J_{i,j}(| p_i^2 - p_j^2 |)^{0.5} S_{i,j} - q_{oj} \}\,.$$

Equation 8.4—26 may accordingly be written up in two ways:

$$p_j(t + \Delta t) = C_j(t) + p_j(t)$$

for the explicit method, and

$$p_j(t + \Delta t) = C_j(t + \Delta t) + p_j(t)$$

for the implicit method. If the node pressures at the instant t are fixed by some initial condition, then the explicit method will directly furnish the pressures prevailing at the instant $(t + \Delta t)$. If, on the other hand, the implicit method is adopted, then said pressures may be obtained only by simultaneously solving the system of non-linear algebraic equations, including an equation resembling 8.4–26 for each node. If the system incorporates other elements (compressor, choke, etc.) as well, then the models described so far are further complicated. These elements having no transient storage capacity, however, their transient behaviour will be characterized by the same mathematical models as are described in Section 8.3.1 for steady-state operation.

In connection with the application to concrete cases of the mathematical models outlined above, we have invariably pointed out the necessity of formulating suitable initial and boundary conditions. The alternatives arising in this respect were summarized by Batey et al. (1961).

In fixing initial conditions there are two options. (i) Flow rates and pressures are determined by simultaneous measurement at various points of the pipeleg examined, and the pressure and flow rate distribution functions thus obtained will fix the initial state of the system. This procedure is bound to run up against a number of difficulties, and it is therefore much more common that: (ii) flow is considered as steady-state at the instant $t = 0$ in the pipeleg examined. The initial pressure distribution required for the transient calculation can then be calculated using the steady-state model, and the mass flow rate is constant.

In defining boundary conditions, the following alternatives enter into consideration. One may fix the variation in time of the injection or delivery pressure, of the flow rate or of throughput. As these six parameters are not independent mathematically, it is necessary and sufficient to fix the time variations of two parameters. In a complex system it is usually necessary to start from time-variable consumer demand at the various nodes. The mathematical formulation of the relevant time functions may be based e.g. on the harmonic analysis of measurement results. Once the time functions have been established, one of the problems to be solved may be the adjustment, within the feasible limits, of the flow rates and pressures of the individual sources of gas, and possibly of compressor horsepower, so that the pressures at the consumer offtake points do not exceed the least supply pressure contracted for by more than a certain safety reserve. This is the way in which energy losses due to expansion can be minimized.

8.5. Computer modelling of gas transmission systems

8.5.1. Case of the digital computer

The practical mathematical modelling of gas transmission systems has been made possible by the advent of the high-speed electronic computer. This statement is amply illustrated by the computation time demand of the steady-state, let alone the transient, network models described above. Employing the digital computer for systems modelling requires the mathematical formulation, 'intelligible' to the computer, of the fundamental data (e.g. system configuration) and of the fundamental relationships describing the particular model envisaged. This formulation is something of a special problem. In solving it, the systems analyst is assisted by an independent branch of modern mathematics, graph theory (Haray 1969), which has been applied—e.g. by Szendy (1967)—to the topological characterization of electric networks, too.

(a) Application of the graph theory

The complex gas transmission system composed of nodes and NCEs may, with due attention to the known or assumed directions of flow, be regarded as a directed graph whose connection matrix **A** is rather simple to write up. Let the columns of **A** represent NCEs, that is, edges of graph, and let the rows represent nodes. Let element a_{ij} of the matrix be

$$a_{i,j} = \begin{cases} +1, & \text{if edge } j \text{ emerges from node } i, \\ -1, & \text{if edge } j \text{ ends in node } i, \\ 0, & \text{if edge } j \text{ and node } i \text{ are unconnected.} \end{cases}$$

The connection matrix of the graph in Fig. 8.5—1a, representing the network in Fig. 8.3—6, is accordingly

$$\mathbf{A} = \begin{array}{c} \\ 1 \\ 2 \\ 3 \\ 4 \\ 5 \\ 6 \\ 7 \\ 8 \\ 9 \end{array} \begin{array}{c} \begin{array}{cccccccccccc} 1 & 2 & 3 & 4 & 5 & 6 & 7 & 8 & 9 & 10 & 11 & 12 \end{array} \\ \left[\begin{array}{cccccccccccc} 1 & 0 & 0 & 1 & 0 & 0 & 0 & 0 & 0 & 0 & 0 & 0 \\ -1 & 1 & 0 & 0 & 0 & 0 & 0 & 0 & 0 & 0 & 0 & 1 \\ 0 & -1 & -1 & 0 & 0 & 0 & 0 & 0 & 0 & 1 & 1 & 0 \\ 0 & 0 & 1 & -1 & 1 & 0 & 0 & 0 & 0 & 0 & 0 & 0 \\ 0 & 0 & 0 & 0 & -1 & 1 & 0 & 0 & 0 & 0 & 0 & 0 \\ 0 & 0 & 0 & 0 & 0 & -1 & 1 & 0 & 0 & 0 & -1 & 0 \\ 0 & 0 & 0 & 0 & 0 & 0 & -1 & -1 & 0 & 0 & 0 & 0 \\ 0 & 0 & 0 & 0 & 0 & 0 & 0 & 1 & -1 & -1 & 0 & 0 \\ 0 & 0 & 0 & 0 & 0 & 0 & 0 & 0 & 1 & 0 & 0 & -1 \end{array}\right] \\ \text{Node-connecting elements} \end{array} \text{Nodes} \qquad (8.5\text{—}1)$$

This connection matrix uniquely defines system configuration. In network calculation, one requires in addition to the connection matrix also a definition of the loops—the senses of rotation—in the network, which

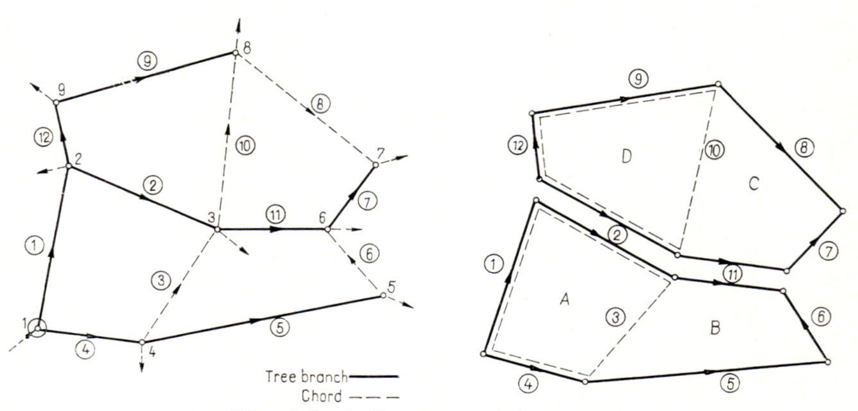

Fig. 8.5—1 Gas transmission system

may be performed with the aid of the so-called loop matrix. In order to derive the loop matrix from the connection matrix it is necessary to introduce the concept of a tree. This term denotes a connected graph in which there is one and only one trajectory between any two nodes. Thus, any loopless graph is a tree. If a graph is looped, it is possible to turn it into a tree by eliminating some of its edges. This may be performed automatically, by adding up the rows of the connection matrix. Let us designate on the tree chosen a so-called point of reference or base point, and let us drop the corresponding row from the connection matrix. Now rearranging the matrix so as to separate tree branches and chord branches (the latter are those which are to be eliminated to form the tree), we may write up the so-called matching matrix of the system. Let e.g. the reference point be node *1*, in the graph shown as Fig. 8.5—1a, and let us eliminate loops by dropping edges *3*, *6*, *8* and *10*. The matching matrix of the graph is, then, written in the form

$$\mathbf{B} = [\mathbf{B}_f \mathbf{B}_h] = \begin{array}{c} \\ 2 \\ 3 \\ 4 \\ 5 \\ 6 \\ 7 \\ 8 \\ 9 \end{array} \begin{array}{c} \begin{array}{cccccccccccc} 1 & 2 & 4 & 5 & 7 & 9 & 11 & 12 & 3 & 6 & 8 & 10 \end{array} \\ \left[\begin{array}{cccccccc|cccc} -1 & 1 & 0 & 0 & 0 & 0 & 0 & 1 & 0 & 0 & 0 & 0 \\ 0 & -1 & 0 & 0 & 0 & 0 & 1 & 0 & -1 & 0 & 0 & 1 \\ 0 & 0 & -1 & 1 & 0 & 0 & 0 & 0 & 1 & 0 & 0 & 0 \\ 0 & 0 & 0 & -1 & 0 & 0 & 0 & 0 & 0 & 1 & 0 & 0 \\ 0 & 0 & 0 & 0 & 1 & 0 & -1 & 0 & 0 & -1 & 0 & 0 \\ 0 & 0 & 0 & 0 & -1 & 0 & 0 & 0 & 0 & 0 & -1 & 0 \\ 0 & 0 & 0 & 0 & 0 & -1 & 0 & 0 & 0 & 0 & 1 & -1 \\ 0 & 0 & 0 & 0 & 0 & 1 & 0 & -1 & 0 & 0 & 0 & 0 \end{array} \right] \end{array} \qquad 8.5\text{—}2$$

Tree branches $\mathbf{B}_f$ (columns 1, 2, 4, 5, 7, 9, 11, 12); Chord branches $\mathbf{B}_h$ (columns 3, 6, 8, 10)

Clearly, in a graph of n nodes and m edges, the number of independent so-called basic loops is $k = m - n + 1$. It can be shown that the transpose $\mathbf{C}^T$ of the matrix $\mathbf{C}$ of these basic loops is defined by the relationship

$$\mathbf{C}^T = \begin{bmatrix} \mathbf{C}_f^T \\ \mathbf{C}_h^T \end{bmatrix} = \begin{bmatrix} -\mathbf{B}_f^{-1}\mathbf{B}_h \\ \mathbf{I} \end{bmatrix}, \qquad 8.5\text{—}3$$

where $\mathbf{I}$ is the unity matrix.

$\mathbf{B}_f^{-1}$ is the inverse of matrix $\mathbf{B}_f$. It can be produced either by inverting matrix $\mathbf{B}_f$, or by writing up directly as follows. The rows of $\mathbf{B}_f^{-1}$ are the tree branches; its columns are the nodes. Let element b_{fi}^{-1} of matrix $\mathbf{B}_j^{-1}$ be

$$b_{ji}^{-1} = \begin{cases} +1, & \text{if the trajectory from the base point to node } i \text{ includes branch } j \text{, with the branch directed towards the base point,} \\ -1, & \text{idem, with the branch directed towards the node,} \\ 0, & \text{if the trajectory from the base point to the node does not include branch } j. \end{cases} \qquad 8.5\text{—}4$$

For an example, the inverse of matrix $\mathbf{B}_f$ referring to the tree shown in continuous line in Fig. 8.5—1a, written up according to Eq. 8.5—4, is

$$
-\mathbf{B}_f^{-1} = \begin{array}{r} \\ 1 \\ 2 \\ 4 \\ 5 \\ 7 \\ 9 \\ 11 \\ 12 \end{array}
\begin{array}{c} \begin{array}{cccccccc} 2 & 3 & 4 & 5 & 6 & 7 & 8 & 9 \end{array} \\
\left[\begin{array}{cccccccc}
1 & 1 & 0 & 0 & 1 & 1 & 1 & 1 \\
0 & 1 & 0 & 0 & 1 & 1 & 0 & 0 \\
0 & 0 & 1 & 1 & 0 & 0 & 0 & 0 \\
0 & 0 & 0 & 1 & 0 & 0 & 0 & 0 \\
0 & 0 & 0 & 0 & 0 & 1 & 0 & 0 \\
0 & 0 & 0 & 0 & 0 & 0 & 1 & 0 \\
0 & 0 & 0 & 0 & 1 & 1 & 0 & 0 \\
0 & 0 & 0 & 0 & 0 & 0 & 1 & 1
\end{array}\right] \\ \text{Nodes} \end{array} \text{Tree branches}
$$

In possession of this inverse matrix, the transpose $\mathbf{C}^T$ of matrix $\mathbf{C}$ referring to the independent loops is obtained, after performing the matrix multiplication, written up in Eq. 8.5—3 in the form (cf. Fig. 8.5—1b):

$$
\mathbf{C}^T = \begin{bmatrix} -\mathbf{B}_f^{-1}\mathbf{B}_h \\ \mathbf{I} \end{bmatrix} = \begin{array}{r} \\ 1 \\ 2 \\ 4 \\ 5 \\ 7 \\ 9 \\ 11 \\ 12 \\ 3 \\ 6 \\ 8 \\ 10 \end{array}
\begin{array}{c} \text{Loops} \\ \begin{array}{cccc} A & B & C & D \end{array} \\
\left[\begin{array}{rrrr}
-1 & -1 & 0 & 0 \\
-1 & -1 & -1 & 1 \\
1 & 1 & 0 & 0 \\
0 & 1 & 0 & 0 \\
0 & 0 & -1 & 0 \\
0 & 0 & 1 & -1 \\
0 & -1 & -1 & 0 \\
0 & 0 & 1 & -1 \\
1 & 0 & 0 & 0 \\
0 & 1 & 0 & 0 \\
0 & 0 & 1 & 0 \\
0 & 0 & 0 & 1
\end{array}\right] \end{array} \text{NCEs}
$$

The system is uniquely defined by its connection matrix $\mathbf{A}$ and the loop matrix $\mathbf{C}$ derived from it. If in the following we agree to represent gas flow in the individual NCEs by m-dimensional column vector q, and the gas offtakes at the individual nodes by n-dimensional column vector q_0, then Kirchhoff's node law may be written in the form of a matrix equation

$$\mathbf{A}q = q_o,$$

or, in more detail, of the relationship

$$\sum_j a_{ij}q_j = q_{oi}\,.$$

Kirchhoff's second law may be written up in a similar fashion, by representing pressure changes $\Delta p^2 = p_1^2 - p_2^2$ across NCEs by column vectors ΔP. The loop law then assumes the form

$$\mathbf{C}\Delta P = 0,$$

or, in more detail,

$$\sum_j c_{kj}\Delta P_j = 0,$$

where k is the subscript of loops.

It emerges from the above considerations that modelling gas transmission systems by means of directed graphs is fairly simple; description using matrices of such systems affords a clear insight into the essence of the problem; and the calculation is readily performed by computer. It should be noted, however, that if the system is extensive, the procedure takes a considerable storage capacity. Another problem is that matrices **A**, **B** and **C** are usually highly sparse; that is, a high percentage (up to 90 or even 98 percent in some cases) of their elements may be zero. In order to reduce storage capacity demand and to simplify calculation, special sparse-matrix solution methods have been devised.

(b) Review of system-modelling programs

As a consequence of the fast-increasing popularity of the digital computer, numerous systems, simulation programs have been developed by the research teams involved with the problem. The most widely known programs were reviewed by Goacher (1969), who divided them in three main groups.

(i) General programs suitable also for the modelling of gas transmission systems. These are essentially programs suited for the solution of differential equations of various types. Several of these are included in the software of almost every medium and big general-purpose computer. The best-known such programs include CSMP (Continuous System Modelling Program (IBM 1130/360)); Digital Simulation Language (IBM 1130/7090/360); MIMIC; MIDAS (Modified Integration Digital Analog Simulation); KALDAS (Kidsrove Algol Digital Analog Simulation (ICL 1900 Series)); SLANG (Simulation Language (ICL 503/803/4120/4130/ATLAS)). These programs have the common drawback that the system of differential equations describing the process taking place in the system has to be formulated by the gas engineer, who must, in addition, bring the system to the most suitable form or, indeed, reduce it to the most fundamental operations (addition, subtraction), as the system of equations is fed to the computer as a basic data. Preparing the equations of the boundary conditions is not less cumbersome. Another disadvantage is that all programs named above employ the explicit method to solve the system of differential equations, and although the results for any time step are obtained rather fast, time steps must be quite short, which is a considerable disadvantage when handling transients of long duration.

(ii) Programs modelling steady states. These programs are used for two distinct purposes: first, independently, to investigate one of the fairly large class of steady-state or nearly-steady-state technical and engineering problems, and secondly, to furnish initial conditions for the dynamic models. The programs developed by workers of the Gas Council's London Research Station and their main characteristics are given in Table 8.5—1 (after Goacher 1969). These programs satisfy in their majority the requirement that the user should not have to know about the structure and operation of the program. The input data including the network configuration, the parameters of the pipelegs, the pressures and yields of the sources.

Table 8.5—1

Main features of steady-state network analysis programs (after Goacher 1969)

Program	Pipelegs	Nodes	Pressure-defined nodes	Loops	Compressors/Regulators
1	2	3	4	5	6
MANNA (8K)	150	150	12	15	None
MANNA (32K)	300	300	40	50	None
			50 altogether		
MANN 1 (32K)	600	500	100	200	None
			200 altogether		
SONIC (8K)	150	150	20	20	10
				20 altogether	
SONIC (32K)	300	300	40	50	20 (at least 1)
				50 altogether	
DEVIL (32K)	300	300	40	50	20 (at least 1)
				50 altogether	
SNAC (32K)	400	300	25	150	25

MANNA = Matrix Algebra for Non-linear Network Analysis (IBM 1130)
SONIC = Steady-state of Networks Including Compressors (IBM 1130)
SNAC = Steady-state Network Analysis with Compressors

and consumer demands is readily compiled with reference to a set of instructions. In order to solve a loop it is sufficient to estimate the throughput in one pipeleg included in the loop. From these data, the computer will calculate the steady-state conditions by an iteration procedure.

(iii) Programs modelling steady and transient states. In these programs, the part modelling the steady state serves to provide the initial conditions required for the transient calculations. The program is formulated also in this case so that the gas engineer in control of the system may use it as a 'black box' provided he observes certain instructions and rules of operation. Although the input data list differs from one program to the next, each program will require as a matter of course the data mentioned in connection with the programs simulating steady states. These must be complemented with the transient boundary conditions and with the parameters of the compressors, regulators, valves, etc. included in the system. The most widely known programs and their main characteristics are listed in Table 8.5—2. Not mentioned in the Table is General Electric's (USA) fairly successful GE Simulator. This latter, similarly to ENSMP and CAP, solves the differential equations by the implicit method, whereas PIPETRAN and SATAN use the explicit method. The time step is, therefore, much shorter in the latter. For an example as to the structure of a program simulating a complex gas transmission system, let us consider that of SATAN.

Table 8.5 – 2

A summary of the main features available in four dynamic analysis programs (after Goacher 1969)

Program	Pipelegs	Nodes	Flow-defined nodes	Pressure defined nodes	Compressors	Regulators	Loops	Valves	Flow pressure profiles	Lonble or parallel pipes
1	2	3	4	5	6	7	8	9	10	11
CAP	150	150 may have upper and lower pressure limits	No limit	5	Not available	Not available	10	10	No limit	150
ENSMP	300	300	No limit	30	20 PCO	Not available	10 Does not include compressor loops	Not available	5 standard type. No limit on specific type	300
				30 altogether						
PIPETRAN	105	106	30	5	10 PCO PCI FCO FCI HPC HPM	10 PCO PCI FCO FCI	No limit	Not available	30	105
SATAN	300	300 may have upper and lower pressure limits	No limit	25	PCO FCO HPM	50 altogether PCO FCO	150	No limit	50	Not available

PCO = Pressure-controlled outlet
PCI = Pressure-controlled inlet
FCO = Flow-controlled outlet
FCI = Flow-controlled inlet
HPC = Horsepower control
HPM = Horsepower maximum
ENSMP = Extended Network Systems Modelling Program; Engineering Research Station of Gas Council (England)
CAP = Control Advisory Program; Engineering Research Station of Gas Council (England)
PIPETRAN = Pipeline Transients, Electronic Associates Inc. (USA)
SATAN = Steady and Transient Analysis of Network Gas Council London Research Station

The program consists of three main units, which may be run separately if so desired. — *Phase 1.* Calculation of the steady state. Quite often, it is necessary to compare only the steady-state operation of various system configurations, or to furnish initial conditions for the transient calculation. In that case, running this program separately, one may examine up to 10 variants in succession. At the end of each run, in addition to the printout of the desired results, all data required for the transient calculation except the boundary conditions is stored in the background memory from where

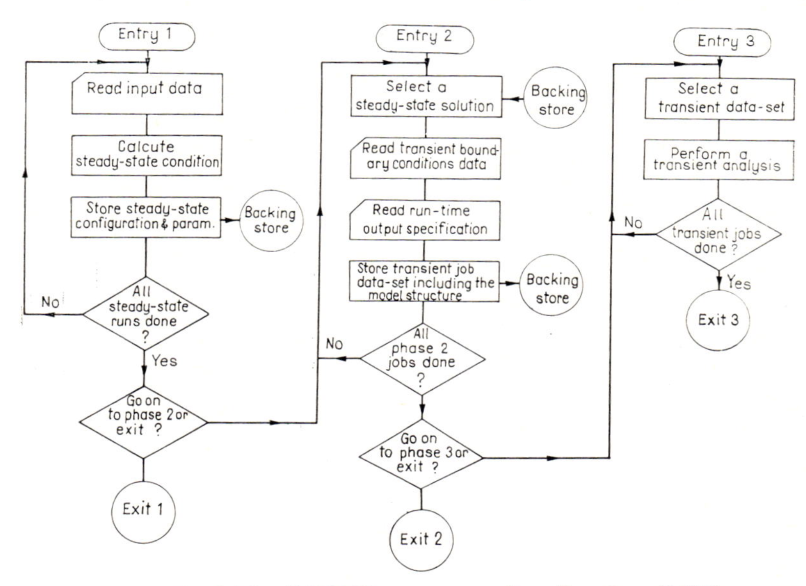

Fig. 8.5—2 The SATAN program, after Goacher (1969)

it may be called in as and when required. — *Phase 2.* This is the connecting phase in which the transient model is built up step by step out of the results of one or more variants of the previous phase, judged to be the most interesting, and out of the boundary conditions fed into the machine. The results are once more relegated to the background memory. Up to ten dynamic-model variants may be stored also in this case. — *Phase 3.* Transient analysis. The computer calls in the intermediate results, calculated in the foregoing phases, of one or several preferred variants, and calculates the transient flow rates and pressures.

The block diagram of the program is given after Goacher in Fig. 8.5—2. The program structure is such that routines providing higher accuracy, or a faster solution, or a more economical use of storage space can be introduced into the program without changing its structure, merely by exchanging certain segments.

8.5.2. Case of the analog computer

Analog computers incorporate cells simulating the four main types of elements of the gas transmission network: pipelegs, consumers, gas sources and compressor stations. The system's model can be composed of these cells by simple plug-in. In analog computers simulating steady-state flow, the so-called pipeleg cells simulating flow in a length of pipe represent the relationship

$$E_1 - E_2 = RI^n$$

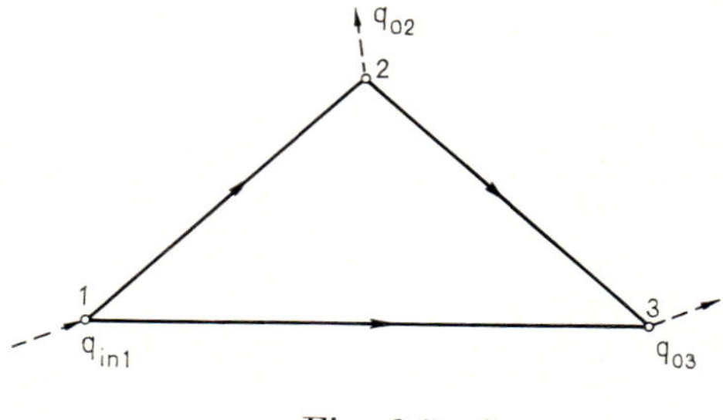

Fig. 8.5 – 3

analogous to the friction equation, where E is E.M.F., R is electric resistance and I is intensity of current. If the variation with pressure of the flowing medium's gravity can be neglected, then pressure p in the pipeline is proportional to E. This analogy applies to low-pressure community utilities. In high-pressure gas transmission systems, where gravity varies markedly with pressure, this latter is proportional to the square root of E. The extraction of the radical is performed by a function generator. In modern apparatuses, *pipe cells* are constructed so that, if the direction of flow in a given pipeleg is opposite to what has been assumed, an automatic reversal of flow takes place. The cell simulating the consumer is a current sink which, irrespective of any voltage fluctuation at the node, absorbs current at a constant rate out of the system. The so-called *source cell* modelling the injection of gas is a d.c. power supply unit of variable output voltage. The *compressor-station cell* is an electronic amplifier.

Determining the various constants is facilitated by auxiliary diagrams in which said constants are plotted v. certain characteristic parameters, for instance, resistance to flow of a pipeleg v. pipe size and length, etc. Setting these constants by means of calibrated potentiometers is fairly simple and quick.

The results of the modelling procedure appear at a central control console, where any node or cell of the system may be called in by pushing a single button. This makes node pressure and throughput, or, in the case of a cell, the difference between input and output pressure and the gas flow rate, appear on the readouts of digital voltmeters or as a printout of a printer hooked up to the system; each data is provided with a code permitting identification. The system can be instructed to scan all nodes in succession and to record results as the scan proceeds.

In an analog system, variants of the gas-transmission-system model take much less time to build than when using a digital computer. This is a substantial advantage both when designing new systems and when examining the operation of existing ones. The analog machine can be used to advantage also in the examination of transient behaviour. For an example let us consider one of the possible analog models of transient flow in the network section shown as Fig. 8.5—3. The principle of the procedure was outlined by Goacher (1969). The mathematical model of the network section, the system of first-degree differential equations 8.4—25, can be written in the form

$$\left.\begin{aligned}\frac{dp_1}{dt} &= K_1 [-J_{2,1}(p_1^2 - p_2^2)^{0.5} - J_{3,1}(p_1^2 - p_3^2)^{0.5} + q_{in1}] \\ \frac{dp_2}{dt} &= K_2 [J_{1,2}(p_1^2 - p_2^2)^{0.5} - J_{3,2}(p) - p_3^2)^{0.5} - q_{02}] \\ \frac{dp_3}{dt} &= K_3 [J_{1,3}(p_1 - p_3)^{0.5} + J_{2,3}(p_2 - p_3)^{0.5} - q_{03}]\end{aligned}\right\} \quad 8.5-5$$

The block diagram of the analog system solving this system of equations is shown as Fig. 8.5—4.

The analog computer has the considerable advantage that it will solve a variety of problems without need for mathematical abstraction, by simple manual intervention, and practically at once, with no time required for the calculation. This type of modelling is therefore of considerable interest when the problem to solve is the choice among a large number of variants in designing a new system or in optimizing an existing one. One of the applications it is best suited for is the training of transmission system operations engineers, as it lends itself very well to illustrating the way the gas supply system will function under various conditions (cf.

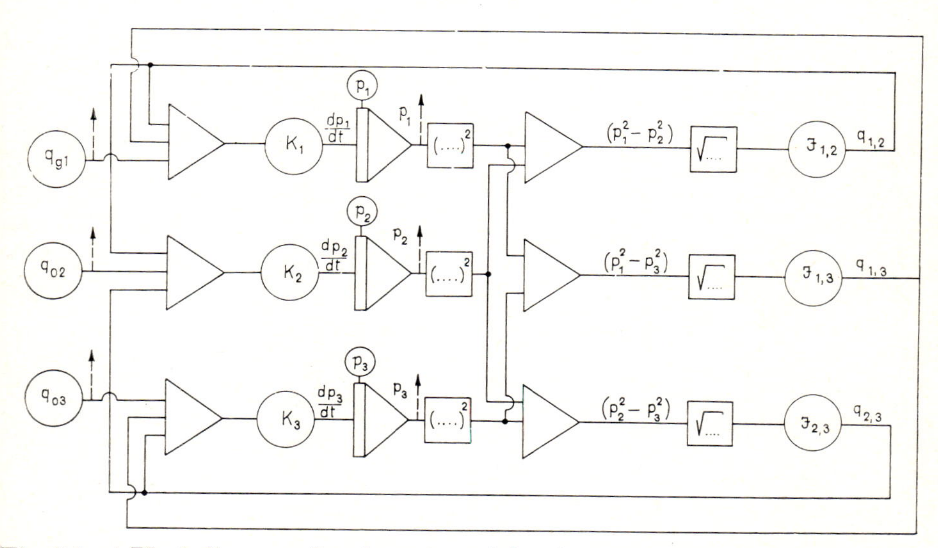

Fig. 8.5—4 Block diagram of analog setup solving the system of Equation 8.5—5, after Goacher (1969)

also Shephard and Williams 1965, de Brem and Tonnelier 1970). — The main drawback of the analog computer is that it is a special-purpose machine; a device designed for the simulation of gas supply systems, that can be used for hardly anything else. Machines suitable for representing complicated systems include a large number of cells and their price is accordingly high.

The largest-size analog computer suitable for modelling a regional gas supply system is operated by SNAM at Metanopoli, for simulating the Italian grid. After a publication in 1970, the total length of pipelines in the grid is 4000 km, and this is to be expanded by a further 5000 km. In 1968 it conveyed 10×10^9 m^3 of gas per year, and this value is expected to increase to 18 or 20×10^9 m^3 by 1980. The analog machine models 210 pipelegs, 22 injection and 70 offtake nodes, 4 compressor stations, 20 pressure regulators and 40 valves. The transient behaviour of the grid is checked every three hours. The purpose of the check is to find the most economical way of satisfying demand, in terms of minimum compressor power and optimum storage-capacity exploitation. In 1969, a digital machine was installed alongside the analog machine built in 1968. The digital machine serves to acquire measurement and situation data, to perform side calculations, to give alarm in the case of malfunctions, and to prepare reports (Bonfiglioli and Croce 1970).

8.6. Pipeline transportation of natural gas; economics

Finding the economically optimum dimensions and operating parameters of a gas transmission pipeline is a comparatively simple task if there is a single line to convey gas from the site of production to the site of consumption. If no intermediate boosting is required, then the optimum trace and pipe size of the line can be determined similarly to the method outlined for oil pipelines in Section 7.1.4. Moreover, Renauldon's procedure will furnish the most economical pipe size relatively simply even if booster stations are installed at one or more points along the pipeline (Société . . . Manuel 1968).

In practice, however, the typical problem arising is the economic optimization of transmission networks or, even more generally, of regional gas supply grids, and this is a very complicated problem indeed. As was shown in the previous section, the grid includes the gas field proper as one of its components. It was tacitly assumed that the gas flow rate out of the field is given or known. In reality, the gas reserve of the field, the envisaged rhythm of its exploitation, and its expected life will all affect economy. Measures taken in order to smooth daily and monthly production rates over the year also carry a considerable importance. We shall consider these viewpoints below.

The duration of developing a gas field (drilling the gas wells, choosing the number of producing wells, the preparation and transmission equipment to be matched to them, the capacities and the construction and installation periods of these sets of equipment) may vary rather widely.

Mayer-Gürr illustrates the economic importance of these factors on a simple example (Mayer-Gürr 1971). Figure 8.6—1a shows the first two sections of the typical production curve of a gas field, in three variants. The exploitable gas reserve of the field was assumed to equal 55 km³. Three rhythms of exploitation have been envisaged. The first production period takes five years in all three. It is during this period that production is run up to full rated capacity. This latter is 4.5 km³ in the first case, 3.0 in the second, and 2.25 in the third. This is the output that is desired to maintain over about 60 percent of the field's life. The period of level production is 10 years in the first case, 15 in the second and 20 in the third. Part (b) of the Figure shows the drilling rates and numbers of wells required by the individual variants. The reason for drilling more wells even after the run-up period of five years is that, by the assumptions made, reservoir pressure and hence the productivity of the wells will decline during production. Part (c) of the Figure shows the first cost of wells and of production equipment in the field. First cost is seen to be exactly twice as high in the first case as in the third. The rhythm of development and the prescribed sustained level of production of the gas field does, then, substantially affect the economics of the regional gas supply grid.

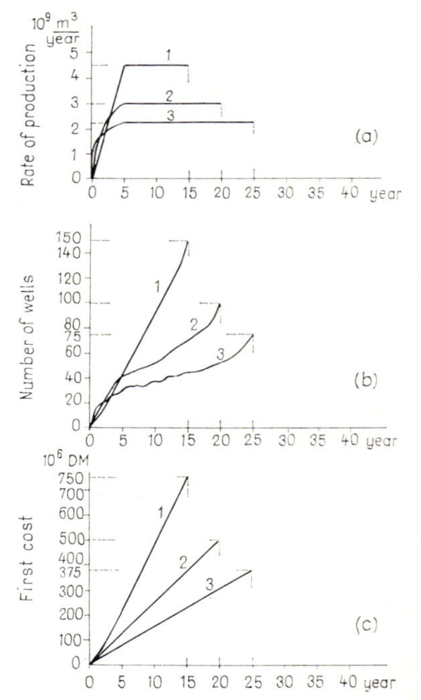

Fig. 8.6—1 Development costs of gas field, after Mayer-Gürr (1971)

The factors affecting the optimal operation of a gas supply system are, according to Graf (1971), as follows. (i) The load factor of the system should be as high as possible. (Load factor: ratio of mean to maximum hourly gas flow.) The load factor of the production system may differ from that of the supply system; the main reason for this is that one pipeline may convey gas coming from several gas fields. The gas supply company may, with a view to increasing the load factor, take the following measures: use the pipeline as a buffer storage facility (cf. Section 8.4—2a); establish an underground stratigraphic storage capacity (storage field); use a reserve of liquefied gas, or propane injection, or high-pressure gas storage to ensure an excess supply capacity for periods of peak demand. (ii) Of the above-enumerated measures, those are taken that ensure the most economical solution in any given case. (iii) Gas fields of various nature are to be produced in the most economical combination, possibly one after another. (iv) The system should ensure uninterrupted gas supply with a high degree of safety. The safety of supply can be measured by two factors.

One is availability, which is the ratio of the aggregate length of uninterrupted supply periods to total time. The other is the reserve factor. The higher the availability the less reserve is needed in the form of parallel lines, underground storage capacity or standby peak-demand supply systems.

Undergound storage in storage fields is a chapter of reservoir engineering. Here we shall just touch upon the essentials and the nature of the procedure. Natural gas may be stored in a gas reservoir, either exhausted or nearing exhaustion, in an exhausted crude oil reservoir, or in an aquifer. Requirements facing a reservoir are as follows. (i) It should have a cap-rock impervious to gas, (ii) it should have sufficiently high porosity and permeability, (iii) the storage wells should not establish communication betwen the formations traversed, that is, their casings should be cemented in so as to provide faultless packoff (isolation); (iv) it should be situated close to the area of consumption, (v) the reservoir rock should be chemically inert vis-à-vis the gas to be stored. — The gas reservoir may be closed, with no inflow of water from below or laterally. The storage space may in such cases be regarded in a fair approximation as a closed tank whose volume equals the pore volume. In open reservoirs, decrease of pressure entails the inflow of water from below or from the periphery towards the centre of the reservoir. On the injection of gas into the reservoir, the gas–water interface will sink. The gas-filled volume of this type of underground reservoir is, then, variable. In the USA, underground storage reservoirs for the storage of natural gas have been in use since 1915, and G. C. Grow (1965) predicted the volume of such reservoirs to attain 35 percent of country-wide annual consumption by 1980.

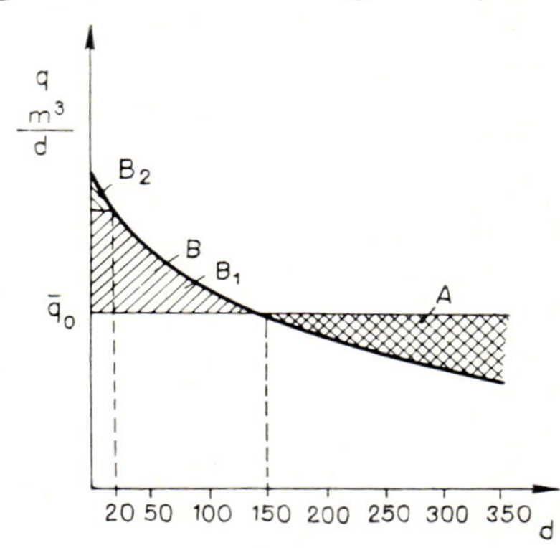

Fig. 8.6 – 2 Natural gas supply (after Kridner 1965)

The main purpose of an underground storage reservoir is to mitigate the economically harmful influence of seasonal fluctuations in demand. Such fluctuations are significant especially where one of the foremost uses of gas is in heating. In order to exploit the capacities of the production and transmission equipment more fully, a reserve is built up during the low-demand summer months in underground reservoirs (storage fields) close to consumption centres. This is where peak demand in the winter months is met from.

Figure 8.6 – 2 is a typical diagram of natural-gas supply. The bold line shows the number of days on which daily consumption exceeds some arbitrary daily consumption q_i. The area below the curve equals the annual gas consumption, provided the axes are suitably calibrated. The average height of the area below the curve equals the mean daily consumption, $\bar{q}$. Assuming that the production equipment produces, and the transmission system conveys, a gas flow q each day, the area A indicates the volume of gas that can be stored up in the low-demand period. Area B, equal to area A, is proportional to the volume of gas to be taken out of the reservoir on high-demand days. The diagram reveals that, in the case considered, gas will have to be taken out of the reservoir over 150 days. Area B is split in two parts. Gas volume B_1 is most expediently stored in a storage field, whereas the storage of volume B_2 by some other means may be more economical, because taking out B_2 of the storage field would need a fairly

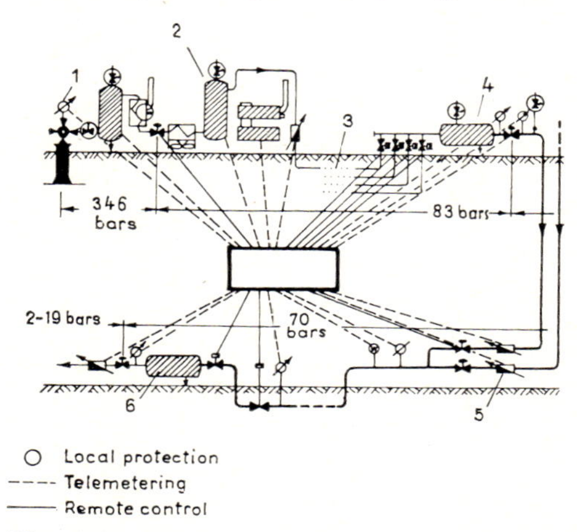

Fig. 8.6—3 The German Federal Republic's gas supply system, after Graf (1971)

considerable increase of the gas production capability of the producing field, which is rather a costly affair. It may therefore be expedient to supply volume B_2 from, for example, high-pressure undergound gas tanks, storage of liquefied gas, or of propane gas, etc. These methods have a copious literature.

Regional gas supply grids are usually too complicated to be operated optimally by unassisted men, in view of the continually changing demand and other conditions facing the system. This is why the process control of such systems is of so great an importance. Today's consensus is that joint off-line control by man and computer is to be preferred. One of the tasks of the computer is data acquisition and the presentation in due time of adequate information. In possession of this information, the dispatcher must be in a position to take optimal decisions and to implement them by remote control. The other main task of the computer is the on-line control of certain functions (Holland and Mix 1970). The first publication on off-line process control dates from 1966 (*Pipe Line News* 1966). Process control of gas supply systems has considerably spread since. Roberts (1970) describes the gas supply grid of Panhandle Eastern, controlled by a dispatcher assisted by an IBM-1800 computer. The grid incorporates pipelines of 2100 km aggregate length and supplies with natural gas 110 utilities catering to 22 million consumers. — Let us cast a brief glance on computer control, with the system realized at the Brigitta-Mobil Oil companies taken as an example. Figure 8.6—3 is a schematic sketch of the journey of natural gas from the gas wells to the consumers. Gas is produced from wells *1*, and led through the dehydrators *2* installed next to each well into flowline *3* and through it into well centres *4*. From there, gas flows through transmission line *5* and enters the low-pressure utility

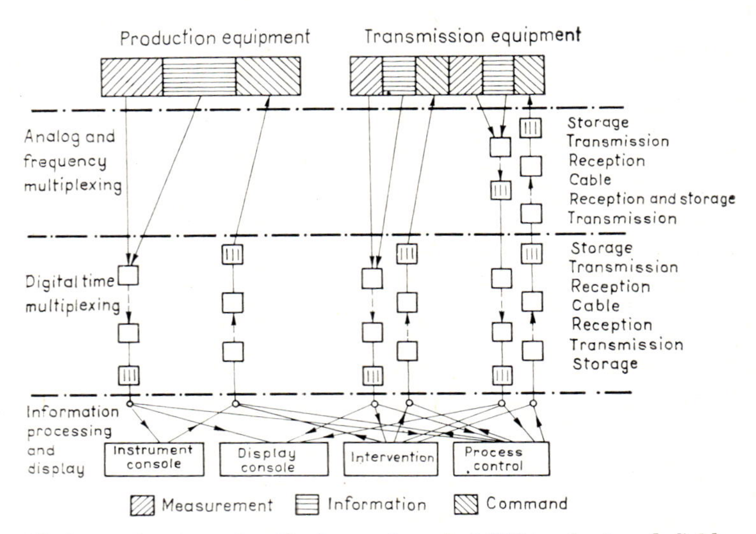

Fig. 8.6—4 Information transfer, display and control (Mittendorf and Schlemm 1971)

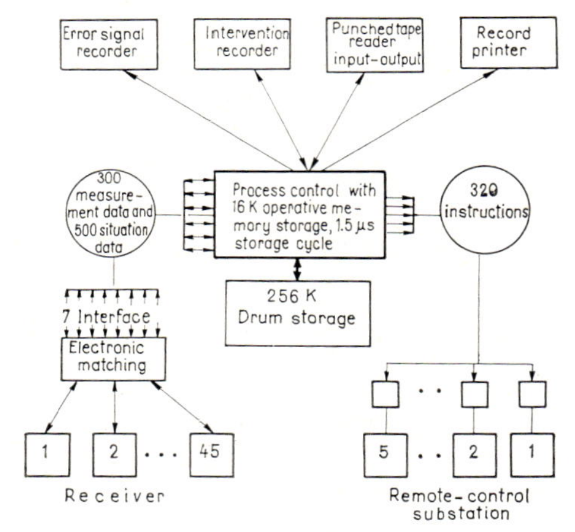

Fig. 8.6—5 Flow of information among telemetering, remote-control and program-control equipment (Mittendorf and Schlemm 1971)

at transfer station *6*. The data acquired all over the system is supplied to the central process control at Visbek. Critical points are provided with local automatic safety equipment (Graf 1971).

Figure 8.6—4 is a schematic diagram of information flow and control (Mittendorf and Schlemm 1971). It is seen that both the dispatcher and the computer may give commands controlling both the production and transmission equipment. — Figure 8.6—5 shows information flow between the computer, its peripheral units, and the remote-signalling and remote-control equipment.

Appendix

Gilbert's gradient curves for various d and q_0

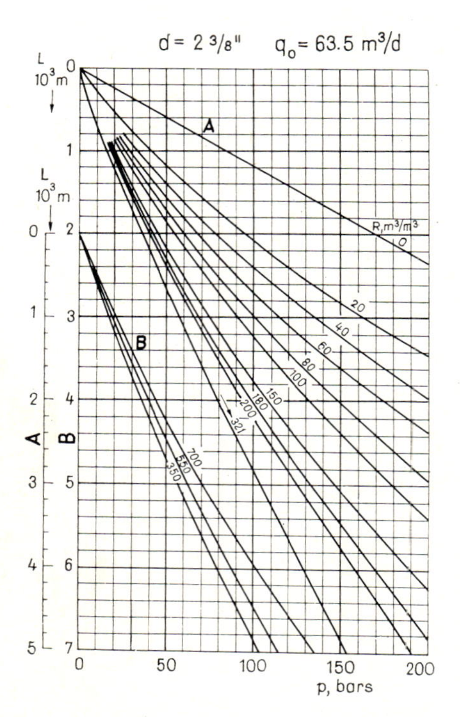

Fig. A—1

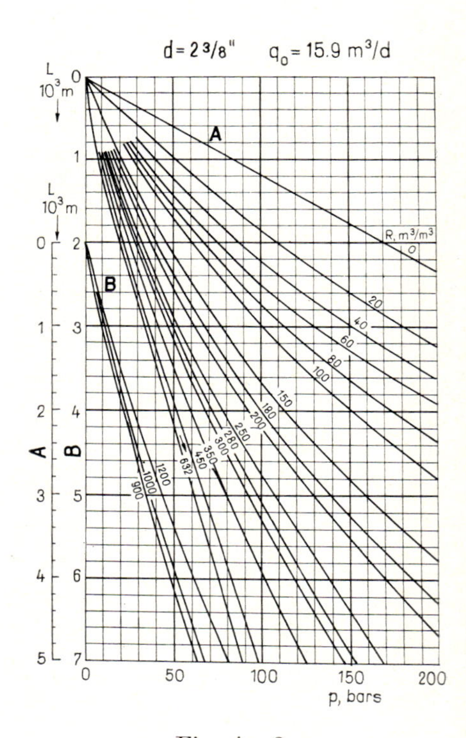

Fig. A—2

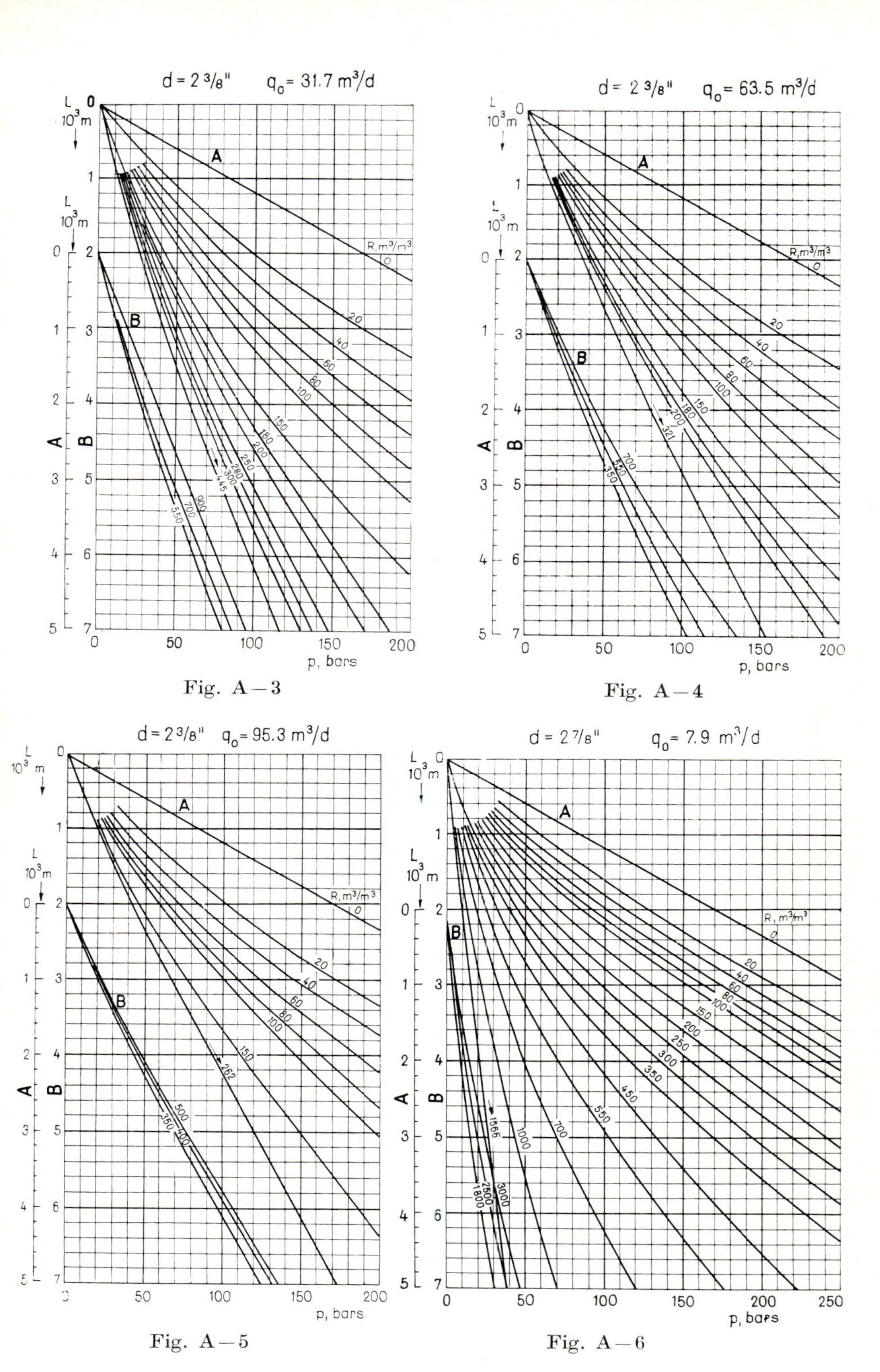

Fig. A—3

Fig. A—4

Fig. A—5

Fig. A—6

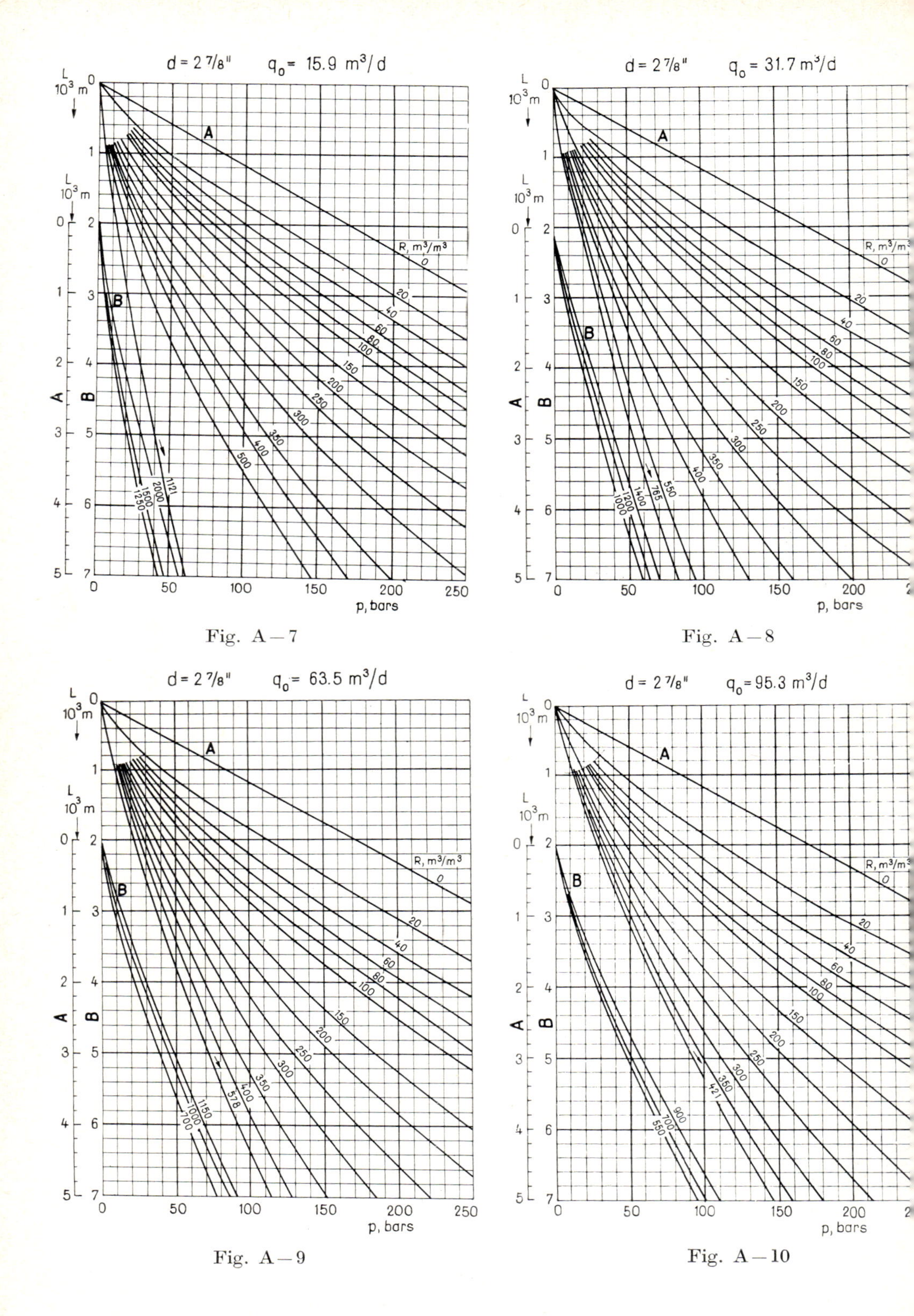

Fig. A—7

Fig. A—8

Fig. A—9

Fig. A—10

References

Abdurashitov, S. A., Avenesyan, V. G. (1964): Eksperimentalnoe issledovanie fizicheskikh svoistv emulsionnykh neftei. *Izvestia Neft i Gaz 1.*

Abramzon, L. S. (1968): Vybor temperatury podogreva nefti. *Transport i Khranenie Nefti i Nefteproduktov 4.*

Adonin, A. N. (1969): Vybor novogo oborudovaniya i rezhima otkachki dlya glubinnonasosnykh skazhin. *Neftyanoe Khozyaistvo 8.*

Agnew, B. G. (1956): Dynamometer stars in a new role. *Oil and Gas Journal. Oct. 22nd.*

Aliev, R. A., Bleikner, E. M., Kulkov, V. A., Loshchilin, A. E. (1969): Opredeleniye raschetnykh reologicheskikh parametrov smesei vysokozastivayushchei nefti s malovyazkimi razbavitelyami. *Transport i Khranenie Nefti i Nefteproduktov 8.*

American Gas Association (1955): Orifice metering of natural gas. *Gas measurement committee report 3.*

American Petroleum Institute (1961): *History of Petroleum Engineering.* Dallas and Houston.

Amyx, J. W. (1960): *Petroleum Reservoir Engineering.* McGraw-Hill Book Co. Inc., New York, Toronto, London.

Ando, M. (1970): Here's how the Japanese use electric pipeline heating. *Oil and Gas International 7.*

API RP 39 (1960): Standard procedure for the evaluation of hydraulic fracturing fluids. *7.*

Arutyunoff, S. (1965): Modern aspects of electrical submergible pumps. *ASME preprint 54-PET-28.*

Aziz, K. (1962—1963): Ways to calculate gas flow and static head. *Petroleum Management, Nov.—Aug.*

Aziz, K., Govier, G. W., Fogarasi, M. (1972): Pressure drop in wells producing oil and gas. *J. of Canadian Petroleum Technology. July—Sept.*

Baker, O. (1954): Simultaneous flow of oil and gas. *Oil and Gas Journal, July 26th.*

Balcke, H. (1949): *Die Wärmeschutztechnik.* Verlag von W. Knapp, Halle.

Bán, Á. (1962): *Vliyanie svoistv gornykh porod na dvizheniye v nikh zhidkosti.* Gostoptekhizdat, Moscow.

Barrett, R. (1970): Unique 'high efficiency' cyclone separator. *Europe and Oil 8.*

Barry, E. G. (1971): Pumping non-Newtonian waxy crude oils. *Journal of the Institute of Petroleum 3.*

Batey, E. H., Courts, H. R., Hannah, K. W. (1961): Dynamic approach to gas pipeline analysis. *Oil and Gas Journal, Dec. 18th.*

Baxendell, P. B. (1955): Pipeline flow of oil and gas mixtures. *Fourth WPC, Section II/E,* Paper 4.

Baxendell, P. B., Thomas, R. (1961): Calculation of pressure gradients in high-rate oil wells. *Journal of Petroleum Technology 10.*

Beadle, G., Harlan, J., Brown, K. E. (1963): Evaluation of surface back-pressure for continuous- and intermittent-flow gas lift. *Journal of Petroleum Technology 3.*

Bean, D. W. (1967): CIG tames two-phase flow with on-stream sphere system. *Pipe Line Industry 8, 9.*

Beeson, C. M., Knox, D. G., Stoddard, J. H. (1958): The plunger lift method of oil production. *Petroleum Engineer, June, July, August.*

Beggs, H. D., Brill, J. P. (1972): An experimental study of two-phase flow in inclined pipes. *SPE Preprint, No. 4007.*

Belházy, T. Z. (1970): Die Entwicklung der Absperrorgane. *Gas und Wasserfach 3.*
Belov, I. G. (1960): *Issledovanie raboty glubinnykh nasosov dinamografom.* Gostoptekhizdat, Moscow.
Bennett, E. N., Auvenshine, W. L. (1957): Dewatering of gas wells. *API Drilling and Production Practice, 1956.* Dallas.
Bertuzzi, A. F., Tek, M. R., Poettmann, F. H. (1956): Simultaneous flow of liquid and gas through horizontal pipe. *Trans. AIME. 207.*
Bilharz, H. L., Greenwell, H. E. (1953): Condensate well corrosion. *NGAA. Tulsa, Oklahoma.*
Binder, R. C. (1958): *Advanced Fluid Mechanics.* Vol. I. Prentice-Hall, Inc., New Jersey.
Bleakley, W. B. (1970a): Humble takes a look at man's role in automated production. *Oil and Gas Journal, Sept. 14th.*
Bleakley, W. B. (1970b): Pegasus automated to *n*th degree. *Oil and Gas Journal, Nov. 2nd.*
Bogdanov, A. A. (1968): *Pogruzheniye tsentrobezhniye elektronasosy dlya dobchi nefti.* NEDRA, Moscow.
Bogdanov, A. A., Rozantsev, V. P., Holodnyak, A. Yu. (1972): Napornaya kharakteristika pogruzhnogo tsentrobezhnogo nasosa pri rabote vyazkikh zhidkostyakh neftevodyanykh emul'siyakh. *Neftepromyslovoe Delo 12.*
Boldizsár, T. (1958): The distribution of temperature in flowing wells. *American Journal of Science 256,* 4.
Boley, W. (1967): Einsatzplanung und Betriebserfahrung mit elektrischen Tauchkreiselpumpen. *Erdoel-Erdgas-Zeitschrift 11.*
Bonfiglioli, G. P., Croce, L. (1970): Analog and numerical computers in the operation and design of gas networks. *International Gas Union Conf., Moscow, Preprint C9-70.*
Bonsall, J. G. (1960): Slim-hole completions reduce well costs 35%. *Oil and Gas Journal, July 18th.*
Bowers, J. H. (1970): Shell tries hydraulic lift for high-volume wells in West Texas. *Oil and Gas Journal, Jan. 19th.*
Boyd, M. C. (1960): Tandem rod pumps prove practical and economical in deep dual wells. *World Oil, June.*
Brameller, A., Chancellor, V. E., Hamam, Y. (1971): Recent advances in mathematical analysis for gas industry application. *IGE Journal 3.*
Brandt, H., Poynter, W. G., Hummel, J. D. (1965): Stimulating heavy oil reservoirs with downhole air-gas burners. *World Oil 9.*
de Brem, F. R., Tonnelier, J. (1970): Study of gas distribution nets with an analogic computer: application and performances of the new Gaz de France unit. *International Gas Union Conf., Moscow, Preprint D12—70.*
Brill, J. P., Doerr, T. C., Hagedorn, A. R. (1966): Practical use of recent research in multiphase vertical and horizontal flow. *Journal of Petroleum Technology 4.*
Brod, M., Deane, B. C., Rossi, F. (1971): Field experience with the use of additives in the pipeline transportation of waxy crudes. *Journal of the Institute of Petroleum 3.*
Broussard, W. F., Gravis, Ch. K. (1960): Three-phase separators. *World Oil 4.*
Brown, G. G. (1945): A series of enthalphy-entropy charts for natural gases. *Trans. AIME 160.*
Brown, K. E. (1967): *Gas Lift Theory and Practice, Including a Review of Petroleum Engineering Fundamentals.* Prentice Hall. Inc., Englewood Cliffs, New Jersey, USA.
Brüggemann, F., de Monyé, G. (1959): Neuerungen auf dem Gebiet der Erdölförderung mittels Tiefpumpen. *Erdoel Zeitschrift 5.*
Bucaram, S. M., Sullivan, J. H. (1971, 1972): A data gathering and processing system to optimize field operations. *SPE Preprint 3468; Journal of Petroleum Technology 2.*
Bummer, B. L. (1971): Improved paraffin prevention techniques reduce operating costs, Powder River Basin, Wyoming. *SPE Preprint 3352.*
Burrell, G. R., Cornett, D. E., Green, B. F. (1970): Computerized multified data acquisition and control system. *SPE Preprint 2933.*

Buthod, A. P., Castillo, G., Thompson, E. E. (1971): How to use computers to calculate heat, pressure in buried pipelines. *Oil and Gas Journal, March 18th.*
Byers, D. P. (1962–1963): Handbook for LACT operation. (Parts 1–4.) *Pipe Line Industry 11* 1962; *2* 1963; *4* 1963; *8* 1963.
Cabet, R. (1966): *L'économie du transport par conduite.* Société des Éditions Techniques, Paris.
Campbell, J. M. (1955): Elements of field processing. (Parts 2–5.) *Oil and Gas Journal, April 11th, May 30th, June 20th.*
Campbell, J. M. (1956): Separation of oil and water. *Oil and Gas Journal, Oct. 29th.*
Carr, N. L., Kobayashi, R., Burrows, D. B. (1954): Viscosity of hydrocarbon gases under pressure. *Trans. AIME 201.*
Carslaw, H. S., Jaeger, J. C. (1947): *Conduction of heat in solids.* Clarendon Press, Oxford.
Carter, R. D., Miller, S. C., Rilley, H. G. (1963): Determination of stabilized gas well performance from short flow tests. *Journal of Petroleum Technology 6.*
Chandrasekharan, K. P., Sikdar, P. K. (1970): Here's how waxy Indian crude is prepared for pipeline transit. *Oil and Gas International 10.*
Chapin, R. L., Woodhall, R. J. (1970): Analysis of an operating telemetering system at Elk Basin Field. *Journal of Petroleum Technology 4.*
Cheeseman, A. P. (1971): How to optimize gas pipeline design by computer. *Oil and Gas Journal, Dec. 20th.*
Cheeseman, A. P. (1972a): Here's how operations simulator works in gas-pipeline design. *Oil and Gas Journal, March 13th.*
Cheeseman, A. P. (1972b): Dynamic simulator an effective tool. *Oil and Gas Journal, April 24th.*
Chernikin, V. I. (1958): *Perekachka vyazkikh i zastyvayushchikh neftei.* Gostoptekhizdat, Moscow.
Chilingar, G. V., Beason, C. M. (1969): *Surface Operations in Petroleum Production.* American Elsevier P.C., New York.
Coberly, C. J. (1961): *Theory and Application of Hydraulic Oil Well Pumps.* KOBE Inc., Huntington Park, California.
Coltarp, E. D. (1971): Subsurface electrical pump well test analysis. *SPE Preprint 3548.*
Corley, C. J., Rike, J. L. (1959): How to save with tubingless completions. *Oil and Gas Journal, April 20th.*
Coulter, D. M. (1971): Cooled gas line from the Arctic? *Pipe Line Industry 10.*
Cox, J. B., Underrinner, C. F. (1970): Electronic-computer production management — a new era. *SPE Preprint 2934.*
Craft, B. C., Holden, W. R., Graves, E. D. (1962): *Well Design: Drilling and Production.* Prentice-Hall, Inc., New Jersey.
Crosby, G. E. (1969a): Miniaturized completions can be artificially lifted. *Petroleum Engineer, Febr.*
Crosby, G. E. (1969b): Fully rolled thread: breakthrough in rod couplings. *Petroleum Engineer, June.*
Cross, H. (1936): Analysis of flow in networks of conduits or conductors. *Bulletin of University of Illinois 286.*
Crumback, L. (1964): PVC not a substitute, but a first line material. *Canadian Petroleum Engineer 7.*
Cullender, M. H. (1955): The isochronal performance method of determining the flow characteristics of gas wells. *Trans. AIME 204.*
Davenport, T. C., Conti, V. J. (1971): Heat transfer problems encountered in the handling of waxy crude oils in large pipelines. *Journal of the Institute of Petroleum 5.*
Degtyarev, V. N., Didenko, V. S. (1972): Rezultaty polupromyshlennykh eksperimentov po termicheskoi obrabotke mangyshlakskikh neftei. *Neftyanoe Khozyaistvo 4.*
De Moss, E. E., Ellis, R. C., Kingsley, G. S. (1972): New gas lift concept — continuous flow production rates from deep low pressure wells. *SPE Preprint 4174.*
Dempsey, J. R., Patterson, J. K., Coats, K. H., Brill, J. P. (1971): An efficient model for evaluating gas field gathering system design. *Journal of Petroleum Technology, 9.*

DeVerteuil, L. R. (1953): The exercise of control of flowing wells and of input to gas lift installations by means of a differential flow control valve. *Journal of the Institute of Petroleum 39.*
Dobrinescu, D., Bulau, L. (1969): Contribuţii la calculul termic al conductelor îngropate. *Petrol şi Gaze 2.*
Dodge, D. W., Metzner, A. B. (1959): Turbulent flow of non-Newtonian systems. *A. I. Ch. E. Journal, June.*
Downer, L., Inkley, F. A. (1972): Thermal expansion coefficients need update to combat errors in crude oil measuring. *Petroleum and Petrochemical International 6.*
Drouin, A. H. (1970): Oil field valves for subzero temperatures. *Oil and Gas Journal, Febr. 16th.*
Dukler, A. E., Wicks, M., Cleveland, R. G. (1964): Frictional pressure drop in two-phase flow:
(A) A comparison of existing correlations for pressure loss and holdup.
(B) An approach through similarity analysis. *A. I. Ch. E. Journal, January.*
Duns, H. Jr., Ros, N. C. J. (1963): Vertical flow of gas and liquid mixtures in wells. *Sixth WPC, Section II/22.*
Ebert, M., Kalkhof, H. (1961): Flüssigkeitsmengenmessung wird von der Temperatur des Durchflussmediums unabhängig. *VDI-Nachrichten 16,* Nr. 41.
Eichberg, D. (1970): Zip-on jacketing insulates pipeline. *Oil and Gas Journal, July 27th.*
Ells, J. W., Brown, V. R. R. (1971): The design of pipelines to handle waxy crude oils. *Journal of the Institute of Petroleum, May.*
Elperin, I. T., Smolsky, B. M., Levental, L. I. (1966): K voprosu umensheniya gidrodinamicheskogo soprotivleniya truboprovodov. *Inzhinerno-Fizichesky Zhurnal 2.*
Espanol, J. H. (1968): Comparison of three methods for calculating a pressure traverse in vertical multiphase flow. *M. S. Thesis,* The University of Tulsa.
Espanol, J. H., Holmes, C. S., Brown, K. E. (1969): A comparison of existing multiphase flow methods for the calculation of pressure drop in vertical wells. *SPE Preprint 2553.*
Eubanks, J. M., Franks, B. L., Lawrence, D. K., Maxwell, T. E., Merryman, C. J. (1958): *Pumping Well Problem Analysis.* Joe Chastain, Midland, Texas.
Ewing, R. D. (1970): Long stroke pumping unit. *SPE Preprint 3186.*
Farrar, G. L. (1972): Computer control in the industry. *Oil and Gas Journal, Nov. 6th.*
Fincham, A. E. (1971): A review of computer programs for network analysis (developed at London Research Station). *GC (London) 189.*
Ford, P. E. (1955): Pipelines for viscous fuels. *Fourth WPC, Section VIII.*
Forsythe, V. L. (1964): Evaluation of a stabilizer charged gas lift valve for multiple-phase flow using graphical techniques. *Journal of Petroleum Technology 6.*
Franco, A. (1972): Polyurethane foam insulates 100-mile Brazil pipeline. *Oil and Gas Journal, July 10 th.*
Frick, Th. C. (1962): *Petroleum Production Handbook.* McGraw-Hill Book Co. Inc., New York, Toronto, London.
Garlyauskas, A. I. (1971): Diskretny printsip maksimuma v optimalnom upravlenii gazoprovodami. *Gazovaya Promyshlennost' 4.*
Gautier, M. (1970): L'expérience d'Elf Union sur le transport du fuel lourd par pipeline. *Revue de L'AFTP 11–12.*
Gazimov, M. G., Imamalikov, Yu. A., Makhyanova, K. I., Sattarov, U. G., Hisainov, B. H. (1971): Krivye i empiricheskie formuly dlya opredeleniya vyazkosti neftei i emulsii. *Neftepromyslovoe Delo 9.*
Geyman, M. A., Gadiev, S. M. (1960): Glubinnonasosniye privody dlya ekspluatatsii kustovykh dvukhstvolnikh skvazhin. *Azerbaidzhanskoe Neftyanoe Khozyaistvo 7.*
Gibbs, S. G., Nolen, K. B. (1972): Well site diagnosis of pumping problems using mini-computers. *SPE Preprint 4068.*
Gilbert, W. E. (1955): Flowing and gas-lift well performance. *API Drilling and Production Practice.* 1954, Dallas.
Goacher, P. S. (1969): Steady and transient analysis of gas flows in networks. *GC (London) 157.*

Govier, G. W., Aziz, K. (1972): *The flow of complex mixtures in pipes*. Van Nostrand Reinhold Company, New York.
Govier, G. W., Fogarasi, M. (1972): The interpretation of data on the rheological behaviour of thixotropic crude oils. *Journal of Canadian Petroleum Technology 4* (Oct.—Dec.)
Govier, G. W., Ritter, R. A. (1963): Pipeline flow characteristics of crude oils. *Sixth WPC, Section VII/1.*
Graf, H. G. (1957a): Vereinfachte Sammelstelle im Ölfeld. *Erdoel Zeitschrift 2.*
Graf, H. G. (1957b): Dual completion im Feld Eilte-West. *Erdoel Zeitschrift 7.*
Graf, H. G. (1970): Stand der Automatisierung in der Öl- und Gasproduktion. *Erdoel-Erdgas-Zeitschrift 1.*
Graf, H. G. (1971): Dispatching von Erdgas. *Erdoel-Erdgas-Zeitschrift 9.*
Graham, G. E. (1972): Optimizing gas-system design. *Oil and Gas Journal, Nov. 13.*
Gravis, Ch. K. (1960): The oil and gas separator. *World Oil, Jan.*
Greathouse, W. D., McGlasson, R. L. (1958): Progress in plastics for petroleum-production piping. *API Drilling and Production Practice*, 1957, Dallas.
Griffin, F. D. (1968): New API design calculations for sucker rod pumping systems. *API Preprint 875-22-F.*
Griffith, P. (1962): *Two-phase flow in pipes*. Special Summer Program, Massachusetts Institute of Technology, Cambridge, Mass.
Griffith, P., Wallis, G. B. (1961): Two-phase slug flow. *Journal of Heat Transfer, Trans. ASME 8.*
Grosse, L. (1951): *Arbeitsmappe für Mineralölingenieure*. VDI-Verlag GmbH, Düsseldorf.
Grow, G.C. Jr. (1965): Analysis of underground storage. *Pipe Line Industry 10.*
Gubin, V. E., Emkov, A. A., Protasova, L. A. (1971): Primenenie alkilsulfatov i alkilsulfatonov dlya gidrotransporta vysokovyazkikh i vysokozastyvayushchikh neftei. *Transport i Khranenie Nefti i Nefteproduktov 3.*
Gubin, V. E., Stepanyugin, V. N. (1970a): Gidravlichesky raschet truboprovodov dlya perekachki neftei v smesy s vodnym rastvorom PAV. *Neftyanoe Khozyaistvo 2.*
Gubin, V. E., Stepanyugin, V. N. (1970b): Optimalniye parametry dlya perekachki vysokovyazkikh i vysokozastyvayushckikh neftei v smesy s vodnym rastvorom PAV. *Neftyanoe Khozyaistvo 12.*
Gudkov, S. F., Benyaminovich, C. A., Odisahariya, G. E. (1970): Rational utilisation fields of technological schemes for the natural gas large volumes transport. *International Gas Union Conf. Moscow, Preprint C-16-70.*
Guy, J. J. (1967): Computation of unsteady gas flow in pipe networks. *Ind. Chem. Eng. Symposium. London, 23.*
Gyulay, Z. (1942): *Olajvezetékek tervezése és építése* (Designing and building oil pipelines). Mérnöki Továbbképző Intézet, Budapest.
Haarmann, K. (1970): Metallurgical factors in producing steels for large diameter pipe. *World Petroleum 10.*
Hacksma, J. D. (1972): How to predict plunger-lift performance. *Oil and Gas Journal, May 22nd.*
Haddenhorst, H. G. (1962): Die Verpumpung von schwereren Erdölen durch Rohrleitungen. *Erdoel Zeitschrift 5.*
Hagedorn, A. R., Brown, K. E. (1964): The effect of viscosity in two-phase vertical flow. *Journal of Petroleum Technology 2.*
Hagedorn, A. R., Brown, K. E. (1965): Experimental study of pressure gradients occurring during continuous two-phase flow in small-diameter vertical conduits. *Journal of Petroleum Technology 4.*
Hain, H. A. (1968): How to determine the maximum capability of a complex pipeline system. *Pipe Line News 9.*
Hangs, F. E. (1966): More insulated, heated-oil pipelines are in prospect. *Oil and Gas Journal, Oct. 3rd.*
Haray, F. (1969): *Graph Theory*. Addison Wesley, Massachusetts.
Harriger, J. W. (1971): Measurement trends in the 70s. *Pipe Line Industry 2.*
Harrison, O. R. (1970): Planning and implementing a computer production control system. *Journal of Petroleum Technology 2.*
Harvey, A. H., Briller, R., Arnold, M. H. (1971): Pipelining oils below their pour points. (I—II.) *Oil and Gas Journal, Aug. 23rd; Aug. 30th.*

Hedström, B. O. A. (1952): Flow of plastic materials in pipes. *Ind. and Eng. Chem. 3.*
Heideman, J. C. (1972): Optimal development of natural gas transmission system. *SPE Preprint 3980.*
Heinze, F. (1971): Hydratbildung. *Lehrbogen 3.3 von der Bergakademie Freiberg.*
Hende, van den, A. (1969): Résolution graphique des problèmes de perte de charge dans les canalisations. *L'Industrie Chimique 7.*
Hennig, J. (1970): Neue Messgeräte zur Erfassung von Erdgasmengen in Verbindung mit Messumformern für Dichte. *Erdoel-Erdgas-Zeitschrift 5.*
Herning, F. (1950): *Grundlagen und Praxis der Mengenstrommessung.* VDI-Verlag GmbH, Düsseldorf.
Holland, A. E., Mix, R. C. (1970): Computer aids gas dispatching. *Oil and Gas Journal, Oct. 5th.*
Hollis, R. G. (1968): Deep, high-volume hydraulic pumping Reno-field. *API Preprint 801-44A.*
Holzschuhmacher, W. (1972): Ausbauoptimierung von Gastransports-Systemen. *Erdöl-Erdgas-Zeitung 10.*
Hood, L. E. (1968): The flexible sucker rod—an innovation in pumping. *API Preprint 801-44F.*
Howell, J. K., Hogwood, E. E. (1962): *Electrified Oil Production.* The Petroleum Publishing Co., Box 1260, Tulsa, Oklahoma 74101.
Hudgins, Ch. M. (1970): Hydrogen sulfide corrosion can be controlled. *Petroleum Engineer, Dec.*
Hurst, W., Goodson, W. C., Leese, R. E. (1963): Aspects of gas deliverability. *Journal of Petroleum Technology 6.*
Hutlas, E. J., Granberry, W. R. (1972): A practical approach to removing gas well liquids. *Journal of Petroleum Technology 8.*
Ioachim, G. (1965): *Extracţia petrolului şi gazelor.* Editura Tehnica, Bucuresti.
Ives, O. G. (1971): Basic look at CPC end devices. *Petroleum Engineer, Aug.*
Jacobsen, R. S. (1970): Computerized calculation of orifice meter constants. *Pipe Line Industry 5.*
Jaeger, J. C., Clarke, M. (1942): *Proc. Roy. Soc. (Edinburgh) 61.*
Jernigan, W. F. (1971): New producers aid optimum design of high-volume pumping systems. *Oil and Gas Journal, June 28th.*
Jones, D. L., Brown, K. E. (1971): Application of system analysis can increase production by 200 percent in high volume gas lift wells. *SPE Preprint 3547.*
Joy, R. F., Coleman, G. H. (1968): Flexible sucker rod in the works. *Oil and Gas Journal, July 15th.*
Juch, A. H., Watson, R. J. (1969): New concepts in sucker-rod pump design. *Journal of Petroleum Technology 3.*
Karpov, S. V., Galiullin, Z. T., Khodanovich, I. E., Bikchentai, R. I. (1972): Transport bolshikh potokov gaza s uchetom teplovykh rezhimov gazoprovodov i okhlazhdeniya gaza na KS. *Gazovaya Promyshlennost' 5.*
Katz, D. L. (1959): *Handbook of Natural Gas Engineering.* McGraw-Hill Book Co. Inc., New York, Toronto, London.
Katz, D. L. (1972): Thermodynamic analysis of frictional heat effects in pipeline flow. *Oil and Gas Journal, March 6th.*
Klampfl, H. (1971): Neue Art des Schaltens von Tiefpumpenantrieben bei periodischen Betrieb. *Erdoel-Erdgas-Zeitschrift 7.*
Knudsen, J. G., Katz, D. L. (1958): *Fluid dynamics and heat transfer.* McGraw-Hill Book Co. Inc., New York, Toronto, London.
Kolodezny, P. A., Karachev, A. S. (1971): Issledovanie vikhrevoi kamery pri vysokikh davleniyakh i bolshikh raskhodakh. *Neft i Gaz 7.*
Komikova, V. G. (1971): Transport i khraneniye nefty i gaza. *Vypusk 97.* Nedra, Moscow.
Korchazhkin, M. T. (1963): Raschet protsessa drosselirovaniya prirodnogo gaza. *Gazovaya Promyshlennost' 7.*
Kornilov, G. G., Gallyamov, M. N. (1972): Dvizhenie po truboprovodam vysokovyazkikh plastichnyi zhidkostei sloem iz malovyazkoi zhidkosti. *Transport i Khranenie Nefti 6.*

Kridner, K. (1965): What are needs for gas storage? *Pipe Line Industry 10.*
Kriegel, E. (1967): Berechnung von Zweiphasenströmungen von Gas/Flüssigkeits-Systemen in Rohren. *Chemie-Ing. Techn. 22.*
Kruman, B. B., Geibovich, A. A. (1970): Otkachka vyazkikh zhidkostei s pomochyu vibratsionovo nasosa. *Neftyanoe Khozyaistvo 8.*
Kulizade, K. N. (1960): *Elektrooborudovaniye v neftedobyche.* Azneftneshr., Baku.
Kut, S. (1969): Internal epoxy coating natural gas lines. *Erdöl und Kohle, Erdgas, Petrochemie 3.*
Kut, S. (1972): Internal and external pipe coatings (carrying gas, liquid and solids). *Rohre-Rohrleitungsbau-Rohrleitungstransport 2.*
Laabs, H. (1969): Armaturen für den Transport von Erdöl, Erdgas und deren Produkte durch Pipelines. *Erdöl und Kohle, Erdgas, Petrochemie 12.*
Lamb, M. J., Simpson, W. C. (1963): Pipeline transportation of waxladen crude oil as water suspension. *Sixth WPC, Section VII,* Paper 13.
LeBaron Bowen, R. (1961): Scaleup for non-Newtonian fluid flow. (7 articles.) *Chem. Eng., Jun. 12th—Sept. 4th.*
Ledet, Cha. A., Leutwyler, K., Turner, R. G. (1968): Dewater gas wells automatically. *Petroleum Engineer, Dec.*
LeFur, E., Martin, M. (1967): Transport en conduite de liquides non newtoniens. *Revue de l'Institut Français du Pétrole 7., 8.*
Lescarboura, J. A., Culter, J. D., Wahl, H. A. (1970): Drag reduction with a polymeric additive in crude oil pipelines. *SPE Preprint 3087.*
Little, L. F. (1963): Stock tank vapor recovery pays off. *World Oil, June.*
Lobkov, A. M. (1955): *Sbor i transport nefty na promyslakh.* Gostoptekhizdat, Moscow.
Longwell, P. A. (1966): *Mechanics of fluid Flow.* McGraw-Hill, Book Co. Inc., New York, St. Louis, San Francisco, Toronto, London, Sidney.
Lubinski, A., Blenkarn, K. A. (1957): Buckling of tubing in pumping wells, its effects and means for controlling it. *Petroleum Transaction AIME 210.*
MacMillan, J. I. (1971): Economics of remote data processing for oil and gas production. *ASME Preprint Pet. -39.*
Maddox, R. N. (1963): Some properties of mixtures. *Oil and Gas Journal, March 4th.*
Maher, J. L., Coggins, R. W. (1969): How to estimate size, cost of producing equipment. (Part 1.) Separators and treaters. *World Oil, Aug.*
Makowski, M. M., Mochlinski, K. (1956): An evaluation of two rapid methods of assessing the thermal resistivity of soil. *Proceedings of the Institution of Electrical Engineers, Oct.*
Maksutov, R. A., Gazimov, M. G. (1971): Vliyanie elektricheskikh svoistv materialov na otlozhenie parafina. *Neftyanoe Khozyaistvo 7.*
Mamedov, Z. I., Mustafaev, S. D., Bairamov, A. A. (1971): Prinuditelnoe izmenenie vremeni opredeleniya plastovogo davleniya glubinonasosnykh skvazhin. *Neft i Gaz 6.*
Mapes, G. J. (1960): The low temperature separation unit. *World Oil, Jan.*
Martch, H. B., McCall, N. J. (1972): Optimization of the design and operation of natural gas pipeline systems. *SPE Preprint 4006.*
Martin, J. S. (1961): Stop pounding that pumping well. *Petroleum Engineer, May.*
Martin, J. T. (1970): Low level automation for marginal leases. *Petroleum Engineer, May.*
Mayer-Gürr, A. (1971): Erdgas-Produzenten und Erdgas-Abnehmer. *Gasverwendung 10.*
McAfee, R. V. (1961): The evaluation of vertical-lift performance in producing wells. *Journal of Petroleum Technology 4.*
McCurdy, R. C., Elkins, L. E. (1967): *Impact of New Technology on the US Petroleum Industry 1946—1965.* National Petroleum Council, Washington.
McDannold, G. R. (1960): How to determine pumping loads with hollow sucker rods. *Oil and Gas Journal, Aug. 1st.*
McDonald, E., Baker, O. (1964): Spheres up capacity 30—70%. *Oil and Gas Journal, June 15th, 22nd, 29th.*
McGhee, E. (1957): How to get cheap daily tests on every well. *Oil and Gas Journal, April 8th.*
McLeod, W. R., Anderson, D. L., Day, J. J. (1972): A comparison of vertical two-phase computation techniques. *ASME Preprint 72-Pet-38.*

Metcalf, J. P., Landry, R. W. (1971): Turbine meters offer benefits for the big oil producers. *Oil and Gas International 4.*

Metters, E. W. (1970): A new concept in pumping unit technology. *SPE Preprint 3193.*

Metzner, A. B., Reed, J. C. (1955): Flow of non-Newtonian fluids — correlation of the laminar, transition, and turbulent flow regions. *A. I. Ch. E. Journal 12.*

Michie, T. W., Page, W. A., Kidd, A. N. (1970): Program for maintaining equipment reliability and data integrity in a large-scale computer production control project. *SPE Preprint 2935.*

Mikhalkov, P. V., Vosrikova, V. P., et al. (1967): Termoobrabotka nefty kak odin iz faktorov uvelicheniya proizvoditel'nosty nefteprovodov. *Neftyanoe Khozyaistvo 1.*

Milinsky, V. M., Kharlamenko, V. I., Lutfullin, A. Kh., Buskunov, A. A. (1970): Puti snizheniya energeticheskikh zatrat glubinnonasosnykh ustanovok. *Neftyanoe Khozyaistvo 8.*

Milley, Gy. (1970): Nevezéktani problémák a kőolajok magas hőmérsékleten dermedő természetes komponenseivel kapcsolatban, különös tekintettel a reológiai tulajdonságokat befolyásoló komponensekre (Problems of terminology concerning the high-pour-point natural components of crudes, with special regard to those components affecting rheological behaviour). *Research Report,* Petroleum Engineering Department, University of Heavy Industries, Miskolc, Hungary.

Milton, L. H. (1972): PCP. Humble computerized data system for manually operated oil field. *Oil and Gas Journal, March 13th.*

Mirzadzhanzade, A. H., Mingaraev, R. Sh., Zaitsev, Yu. V., Gurbanov, R. S., Rasizade Ya. M. (1969): A nem newtoni folyadékok áramlása csövekben. Néhány hidrodinamikai hőcserefeladat (Flow of non-Newtonian fluids in pipes. Some problems of heat exchange in hydrodynamics). *Kőolaj és Földgáz 3.*

Mittendorf, H., Schlemm, F. (1971): Prozessrechnereinsatz zur Leitungsüberwachung in Ferngasnetzen. *Erdoel-Erdgas-Zeitschrift 11.*

Mokhanov, V. I. (1962): *Izmerenie razkhoda i kolichestva zhidkosty, gaza i para.* Gosenergoizdat, Moscow.

Monyé, G. (1959): Regelung des Tiefpumpenantriebes durch den Ölzufluss. *Erdöl und Kohle 1.*

Muradov, A., Mametklychev, K. H. (1970): Opredelenie optimalnoi temperatury podogreva parafinovykh neftei pered perekachkoi. *Transport i Khraneniye Nefty i Nefteproduktov 6.*

Muraviev, I. M. (1959): *Spravochnik po dobyche nefty.* (Vol. II.) Gostoptekhizdat, Moscow.

Muraviev, I. M., Knisenko, G. N., Mishchenko, I. T., Knyazev, N. S., Olifer, S. L. (1971): Vliyanie reologicheskikh svoistv nefti na rabotu pogruzhnogo tsentrobezhnogo nasosa. *Neftepromyslovoe Delo 2.*

Muraviev, I. M., Krylov, A. P. (1949): *Ekspluatatsiya neftyanykh mestorozhdeniy.* Gostoptekhizdat, Moscow.

Natural Gasoline Association of America (1957): *Equilibrium Ratio Data Book.* Tulsa.

Nematizadeh, F. (1969): Silicone injection boosts separator capacity 40%. *World Oil, May.*

Nemudrov, A. G., Kulikov, V. A. (1971): Optimalnoe rezhimy rabotu nasosnykh stantsii "goryachego" nefteprovoda. *Neftyanoe Khozyaistvo 6.*

Nichols, R. (1968): Dewatering gas wells. *API Preprint 851-42-A.*

Nind, T. E. W. (1964): *Principles of Oil Well Production.* McGraw-Hill Book Co., New York, Toronto, London.

Nomisikov, Yu. P., Khalichin, A. G., Samilev, M. A. (1970): Oborudovanie skvazhiny rekomenduemoe dlya mestorozhdenii soderzhashchikh serovodorodov. *Gazovaya Promyshlennost' 10.*

November, M. H. (1972): Electronic density-measuring system is accurate, versatile. *Oil and Gas Journal, Febr. 21st.*

Novoselov, V. F., Kandurov, A. A. (1971): Uchet inertsii pri rashchete neustanovivshegosya dvizheniya zhidkosti v truboprovode. *Neftyanoe Khozyaistvo 9.*

Oates, J. A. (1962): *Pipes and Pipelines Manual and Directory.* (1st ed.) Scientific Surveys Ltd., London.

O'Brien, H. L. (1951): *Petroleum Tankage and Transmission.* Graver Tank and Mfg. Co. Inc., Houston, Texas.
O'Donnell, U. P. (1968): Getty's new heated crude line is probably world's largest. *Oil and Gas Journal, April 8th.*
O'Donnell, J. P. (1969): TGT makes valuable use of densitometers. *Oil and Gas Journal, Nov. 10th.*
Oil and Gas Journal (1956): Automatic custody transfer. — *June 11th.*
Oil and Gas Journal (1963): Treating plant will make high-wax crude transportable by pipeline. — *Febr. 25th.*
Oil and Gas Journal (1964): $ 500.000 automation investment. — *July 27th.*
Oil and Gas Journal (1969): Chemicals solve tank shortage. — *Jan. 27th.*
Oil Well Supply (1957): Selection and application of subsurface pumps. *Bulletin 6—956.*
Omana, R., Houssiere, Jr. C., Brown, K. E., Brill, J. P., Thompson, R. E. (1969): Multiphase flow through chokes. *SPE Preprint 2682.*
O'Neil, R. K. (1972): Operating costs and performance of large-volume submergible pumps. *Journal of Petroleum Technology 12.*
Orkiszewski, J. (1967): Predicting two-phase pressure drops in vertical pipe. *Journal of Petroleum Technology 6.*
Orlicek, A. F., Pöll, H. (1951, 1955): *Hilfsbuch für Mineralöltechniker* (I—II). Springer Verlag, Wien/New York.
Palm, C. D. (1962): NGPL 'insured' with automatic valve. *Pipeline Engineer 6.*
Panteleev, G. V. (1970): Turbulizator potoka primenyaemy dlya uvelicheniya davleniya na uste skvazhin. *Neftepromyshlovoe Delo 9.*
Pápay, J. (1970): Termelőkutak és vezetékek hőmérséklet viszonyai stacionér állapotban (Steady temperature distributions in producing wells and pipelines). *Kőolaj és Földgáz 11.*
Pápay, J. (1971): Gáztelep, -kút és -vezeték teljesítményegyenlete (Performance equations of oil reservoir, well, and flowline). *Kőolaj és Földgáz 8.*
Pasqualini, M. (1950): Le rechauffage des produits petroliers lors de leur transport dans les pipe-lines. *Bulletin de l'Association Française des Techniciens du Pétrole 8.*
Patsch, F. (1971): Függőleges kétfázisú áramlás nyomásviszonyainak számítása (Calculation of pressure traverses in vertical two-phase flow). *Bányászati és Kohászati Lapok, Kőolaj és Földgáz 12.*
Patton, L. D. (1970): Continuous rod design. *Petroleum Engineer, Aug.*
Pearson, W. G. (1969): Is there a computer in your oil fields future? *Oil and Gas Journal, May 26th.*
Pektyemirov, G. A. (1951): *Ásványolajtelepi kézikönyv* (Petroleum tankage handbook). NIK, Budapest. (Translation from Russian.)
Perkins, T. K., Turner, J. B. (1971): Starting behaviour of the gathering lines and pipelines when filled with gelled Prudhoe Bay oil. *Journal of Petroleum Technology 3.*
Perry, C. W. (1964): The modern ball valve. *Petroleum Times, May 15th.*
Perry, J. H. (1969): *Vegyészmérnökök kézikönyve* (Chemical engineer's handbook). Műszaki Könyvkiadó, Budapest. (Translation from English.)
Petroleum Engineer (1958): Sonic pump bows in. *Sept.*
Petroleum Extension Service (1955): *Oil Pipe Line Measurement and Storage Practices.* The University of Texas, Texas.
Petroleum Extension Service (1956): *Field Handling of Natural Gas.* — The University of Texas, Dallas.
Phoenix-Rheinrohr AG (1960): *Tabellen für Tiefbohrtechnik.* Düsseldorf.
Pipeline News (1966): Annual automation symposium. — *10.* Bayonne, N. J.
Pipeline News (1972): Shell moves viscous oil through unheated lines. — *5.* Bayonne, N. J.
Pisarik, M. N. (1972): Raspredelenie temperatur po stvolu skvazhiny pri rabote nagrevatelei. *Neftepromyslovoe Delo 9.*
PLI Staff (1970): New triple action syphon pig leaves pipe lines clean and dry. *Pipe Line Industry 1.*
Poettmann, F. H., Beck, R. L. (1963): New charts developed to predict gas-liquid flow through chokes. *World Oil, March.*

Poettmann, F. H., Carpenter, P. G. (1953): The multiphase flow of gas, oil and water through vertical flow strings. *API Drilling and Production Practice, 1952*. Dallas.
Powers, M. L. (1972): Optimizing sucker rod replacement. *Oil and Gas Journal, Febr. 7th.*
Price, R. C. (1971): Flow impovers for waxy crudes. *Journal of the Institute of Petroleum 3.*
Pruitt, G. T., Simmons, C. M., Neil, G. H., Grawford, H. R. (1965): A method to minimize the cost of pumping fluids containing friction-reducing additives. *Journal of Petroleum Technology 5.*
Prutzman, F. G. (1955): Problems and economics of dual completions. *Petroleum Engineer, July.*
Puscoiu, N. (1970): Debitmetri de masa folosite la măsurarea producției de țiței și gaze. *Petrol și Gaze 4.*
Pylant, H. S. (1961): Plastic pipe in oil and gas fields. *Oil and Gas Journal, Febr. 27th.*
Raabe, J. (1970): Pumpen. *Hydraulische Maschinen und Anlagen* (Teil 3). VDI-Verlag GmbH, Düsseldorf.
Rabinovitch, E. Z., Kuznetsov, P. B. (1970): Gidravlicheskie soprotivleniya magistralnykh nefteprovodov. *Transport i Khranenie Nefti i Nefteproduktov 7.*
Rachford, H. H. (1972): Transient-pipeline calculations improve design. *Oil and Gas Journal, Oct. 30th.*
Raulins, G. M. (1972): Safety by down-hole well control. *Journal of Petroleum Technology 3.*
Reher, E. O., Mylius, E. (1967): Eine Übersicht über einige Methoden zur Berechnung des Druckverlustes beim Transport zeitunabhängiger nicht-Newton'scher Flüssigkeiten in Rohrleitungen. *Chem. Techn. 3.*
Reid, A. M. (1969): Valve design and selection. *Pipes and Pipelines International 4.*
Reneau, G. (1953): Design and development of improved annular survey equipment and techniques. *API Drilling and Production Practice, 1952*. Dallas.
Reppisch, I. (1958): Mengenmessung mit Ovalrad-Flüssigkeitzählern in der Erdölindustrie. *Erdoel Zeitschrift 8.*
Resen, L. (1957): Humble tries L.A.C.T., gives it stamp of approval. *Oil and Gas Journal, March 4th.*
Ritter, R. A., Baticky, J. P. (1967): Numerical prediction of the flow characteristics of thixotropic liquids. *Society of Petroleum Engineers Journal, Dec.*
Ritter, R. A., Govier, G. W. (1970): The development and evaluation of theory of thixotropic behaviour. *Canadian Journal of Chemical Engineering 48.*
Roberts, B. J. (1970): Panhandle extends its automation. *Oil and Gas Journal, July 23rd.*
Ros, N. C. J. (1961): Simultaneous flow of gas and liquid as encountered in well tubing. *Journal of Petroleum Technology 10.*
Rose, S. C., Marsden, S. S. (1970): The flow of North Slope crude oil and its emulsions at low temperatures. *SPE Preprint 2996.*
Sanders, J. M. (1969): Basic criteria for the sizing and selection of control valves. *Pipes and Pipelines International 10, 12.*
Saye, H. A. (1958): Automatic well testing. *Oil and Gas Journal, Jan. 6th.*
Schlichting, P. (1970): Der Transport von Öl-Wasser-Gas Gemischen in Erdölfernleitungen. *Erdoel-Erdgas-Zeitschrift 6.*
Schmoe, W. P. (1959): Bottom hole separators increase production. *World Oil, April.*
Scott, R. W. (1965): Vacuum stabilizers boosts income minimize production problems. *World Oil, Jan.*
Scott, J. (1967): Production automation breaks barriers. *Petroleum Engineer, June.*
Scott, J. (1969): The revolution in artificial lift. *Petroleum Engineer, Febr.*
Scott, J. (1971): CPC new way of oil country life. *Petroleum Engineer, Aug.*
Scott, J., Crosby, G. (1970): What does industry expect of computer production control. *Petroleum Engineer, May.*
Sens, M. (1969): Evolution of gas measurement in France. *Pipe Line Industry 2.*
Shaver, R. G., Merrill, E. W. (1959): Turbulent flow of pseudoplastic polymer solutions in straight cylindrical tubes. *A. I. Ch. E. Journal 6.*
Shamir, U. (1971): Optimal route for pipelines in two-phase flow. *Society of Petroleum Engineers Journal, March.*

Shaw, S. F. (1939): *Gas-lift Principles and Practices.* Gulf P. C., Houston, Texas.
Shephard, P. A., Williams, R. W. (1965): The application of analogue computer techniques to the distribution system of the North Thames Gas Board. *The Institution of Gas Engineers, Communication 700 (London), 11.*
Sherman, P. (1963): Changes in the rheological properties of emulsions on aging, and their dependence on the kinetics of globule coagulations. *I. Phys. Chem. 12.*
Shirkovsky, A. I. (1970): Estimating minimum flow rate required for continuous removal of condensate from gas wells. *SPE Preprint 2937.*
Shishchenko, R. Y., Apriesov, K. A. (1952): *A kőolaj szállítása és tárolása* (Transport and storage of petroleum). NIK, Budapest. (Translation from Russian.)
Simmons, W. E. (1972): Optimizing continuous flow gas lift wells. (Parts I–II.) *Petroleum Engineer, Aug., Sept.*
Slonegger, J. C. (1961): *Dynagraph Analysis of Sucker Rod Pumping.* Gulf P. C., Houston, Texas.
Smirnov, A. S., Shirkovsky, A. I. (1957): *Dobycha i transport gaza.* Gostoptekhizdat, Moscow.
Smith, R. S. (1968): Submergible pump completion in two-phase flowing wells. *Petroleum Engineer, Dec.*
Snyder, R. E. (1970): New pumping methods boost oil production. *World Oil, May.*
Société du Journal des Usines à Gaz (1968): *Manuel pour le transport et la distribution du gaz.* Paris.
Sommer, H. A. (1972): An approach to oil field automation. *SPE Preprint 3971.*
Speel, L. (1967): High pressure gas well completions. *Seventh WPC Proceedings 3.*
Spink, L. K. (1958): *Principles and Practice of Flow Meter Engineering.* The Foxboro Co., Massachusetts.
Standing, M. B. (1952): *Volumetric and Phase Behaviour of Oil Field Hydrocarbon Systems.* Reinhold P. C., New York.
Standing, M. B., Katz, D. L. (1942): Density of crude oil saturated with natural gases. *Trans. AIME 146.*
Stearns, R. F., Johnson, R. R., Jackson, R. M., Larson, C. A. (1951): *Flow Measurement With Orifice Meters.* D. Van Nostrand Co. Inc., Toronto, New York, London.
Stephens, M. M., Spencer, O. F. (1950): *Natural Gas Engineering.* The Pennsylvania State College. Pennsylvania.
Stewart, T. L. (1971a): Computer speeds surge calculations. *Oil and Gas Journal, Nov. 22nd.*
Stewart, T. L. (1971b): Modelling liquid pipelines in transient and steady state. *Pipeline News 12.*
Stoner, M. A. (1970): A new way to design natural gas systems. *Pipe Line Industry 2.*
Stoner, M. A. (1971, 1972): Sensitivity analysis applied to a steady-state model of natural gas transportation systems. *SPE Preprint 3056;* and *Society of Petroleum Engineers Journal, April.*
Stradtmann, F. H. (1961): *Stahlrohr Handbuch.* (6. Aufl.) Vulkan-Verlag, Essen.
Streeter, V. L., Wylie, E. E. (1970): Natural gas pipeline transients. *Society of Petroleum Engineers Journal, Dec.*
Swinney, G. W. (1972): Gas measurement with on-line computers. *Pipe Line Industry 7.*
Szendy, K. (1967): *Korszerű hálózatszámítási módszerek* (Modern network analysis methods). Akadémiai Kiadó, Budapest.
Szilas, A. P. (1955): Betriebsverhältnisse gasfreies Öl fördernder Sonden. *Mitteilungen der Fakultäten für Bergingenieure und Geo-Ingenieure (XVIII).* Techn. Universitätsfakultäten, Sopron.
Szilas, A. P. (1957): A gazdaságos termelőberendezés kiválasztása (Choosing the most economical production equipment). *Bányászati Lapok 6.*
Szilas, A. P. (1959): Bestimmung des Druckkoeffizienten aus den Betriebsdaten der Ölsonde. *Mitteilungen der Fakultäten für Bergingenieure und Geo-Ingenieure (XX).* Techn. Universitätsfakutäten, Sopron.
Szilas, A. P. (1962): Intermittende Gaslift- und Plungerliftförderung – zwei rationelle Förderverfahren für geringproduktive Sonden. *Freiberger Forschungshefte A233.*

Szilas, A. P. (1964): Kontinuierlicher oder intermittierender Betrieb von Pumpsonden? *Erdoel-Zeitschrift 3.*

Szilas, A. P. (1966): Erdölgewinnung. *Fernstudium.* Bergakademie, Freiberg.

Szilas, A. P. (1967): *A földgáz termelése, előkészítése és szállítása* (Production, handling and transportation of natural gas). Műszaki Könyvkiadó, Budapest.

Szilas, A. P. (1968): Startdruckänderung eines in einer Erdölfernleitung abkühlenden strukturviskosen Öles. *Erdoel-Erdgas-Zeitschrift 9.*

Szilas, A. P. (1971): Tixotrop pszeudoplasztikus kőolaj nyomásveszteségszámításra alkalmas folyási görbéinek meghatározása (Determining the flow curves permitting head loss calculation for thixotropic-pseudoplastic crudes). *Kőolaj és Földgáz 4.*

Szilas, A. P., Falk, R. (1959): Kőolaj- és földgáztermelés (Oil and gas production). *Mining Handbook.* II. Műszaki Könyvkiadó, Budapest.

Taylor, C. J. (1960): How and why tubing anchors reduce operating costs of rods. *Canadian Oil and Gas Journal 6.*

Terris, J. (1965): How and when to automate oil producing facilities. *World Oil, June.*

Thielen, H. (1971): Die Druckwellengeschwindigkeit in Pipelines unter dem Einfluss von Temperatur und Druck. *Rohre-Rohrleitungsbau-Rohrleitungstransport, Oct.*

Thomas, L. K., Hankinson, R. W., Phillips, K. A. (1970): Determination of acoustic velocities for natural gas. *Journal of Petroleum Technology 7.*

Thomas, N. L. (1965): Polyurethane foam insulates Australian heavy oil line. *Pipe Line Industry 5.*

Thompson, D. D., Nicksic, S. W. (1970): Frequency can help improve *BS&W* monitor's accuracy. *Oil and Gas Journal, Nov. 23rd.*

Török, J., Haraszti, E., Kassai, L. (1966): *Földgázok PVT tulajdonságai* (PVT relationships of natural gases). NIMDOK, Budapest.

Török, J., Pápay, J., Haraszti, E., Berecz, E., Kassai, L. (1968): *Fázisegyensúlyok és olajmérnöki alkalmazásuk* (Phase equilibria and their application in petroleum engineering). II. NIMDOK, Budapest.

Tube Turns Division of Chemotron Corp. (1969): *Piping Engineering.* Louisville, Kentucky.

Tugunov, P. Y. (1968): Opredelenie optimalnoy tolshchiny teplovoi izolatsii dlya magistralnykh truboprovodov. *Neftyanoe Khozyaistvo 5.*

Turner, M. C. (1954): When do dual completions pay? *Oil and Gas Journal, May 17th.*

Turner, R. G., Hubbard, M. G., Dukler, A. E. (1969): Analysis and prediction of minimum flow rate for the continuous removal of liquids from gas wells. *Journal of Petroleum Technology 11.*

Uhde, A., Kopp, G. (1971): Pipelining problems resulting from the handling of waxy crudes. *Journal of the Institute of Petroleum, March.*

Uhl, A. E. (1967): Steady flow in gas pipe lines. (Parts 3—4.) *Pipe Line Industry 1, 3.*

Upp, E. L. (1969): New developments in gas flow measurement. *Pipe Line Industry 10.*

USI (1959): *Handbook of Gas Lift.* Garrett Oil Tools. US Industries Inc., New York.

Vance, R. L. (1971): Dens + computer = faster volume finding. *Oil and Gas Journal, June 14th.*

Vermeersch, F. (1968): Pr ms and techniques in producing gas wells in South-West France. *Journal of Institute of Petroleum, Sept.*

Verschuur, E., Den Hartog, A. P., Verheul, C. M. (1971): The effect of thermal shrinkage and compressibility on the yielding of gelled waxy crude oils in pipelines. *Journal of the Institute of Petroleum, May.*

Versluys, J. (1930): Mathematical development of the theory of flowing wells. *Trans. AIME (Petroleum Development and Technology) 86.*

VNIIOEHG (1969): *Sostoyaniye i puty sovershenstvovaniya sbora i transporta nefti gaza na promyslakh vostochnikh rayonov SSSR.* Moscow.

Vodolaga, V. S. (1971): Issledovaniya dinamiki temperaturnykh potei vokrug podzemnogo gazoprovoda. *Gazovaya Promyshlennost' 11.*

Wafelman, H. R. (1969): Displacer operated level gauges most specified throughout Europe. *Europe and Oil 3.*

Walker, E. W. (1959): Heavy crude is more attractive now. *Oil and Gas Journal, Sept. 28th.*

Walton, T. A. (1972): Correlations for design of gas transmission lines. *SPE Preprint 4010.*
Werner, K., Becker, J. (1968): Zweizonenförderung in Hochdruck-Gasbohrungen. *Erdöl und Kohle, Erdgas, Petrochemie 12.*
Westphal, K. (1952): Rohrleitungstransport von zähflüssigen Rohölen mit hohem Stockpunkt. *Erdoel und Kohle 12.*
Whinery, K. F., Campbell, J. M. (1958): A method for determining optimum second stage pressure in three stage separation. *Journal of Petroleum Technology 4.*
White, G. L. (1964): Friction pressure reducers in well stimulation. *Journal of Petroleum Technology 8.*
White, G. W., O'Connell, B. T., Davis, R. C., Stacha, L. A. (1963): An analytical concept of the static and dynamic parameters of intermittent gas lift. *Journal of Petroleum Technology 3.*
White, J. E. (1969): Pipeline cost and investment analyses reveal big-inch economies. *Oil and Gas International 3.*
Wieland, D. R. (1961a): Wire-line installations are a boost to gas lift. *Oil and Gas Equipment 8.*
Wieland, D. R. (1961b): How gas-lift valves aid in dual multiple completions. *Oil and Gas Equipment 9.*
Wijdiecks, J. (1971): Water hammer in large oil transmission lines. *Rohre, Rohrleitungsbau, Rohrleitungstransport 4.*
Wilkinson, J. V., Holliday, D. V., Batey, E. H. (1964): Analytic solution for gas flow. *Pipe Line Industry 11.*
Wilson, P. M. (1973): Deep well tests show jet pump advantage. *World Oil, May.*
Winkler, H. W., Camp, G. F. (1956): Down-hole chambers increase gas-lift efficiency. *Petroleum Engineer, May, June.*
Winkler, H. W., Smith, S. (1962): *CAMCO Gas Lift Manual.* CAMCO Inc., Houston, Texas.
Withers, V. R. (1970): The expansion of an oil storage tank under the liquid head of its contents. *Journal of the Institute of Petroleum, Jan.*
Woods, J. B. (1961): How to improve your hydraulic pumping operations. *World Oil, Jan.*
Wood, R. S. (1958): Capacitance-type *BS&W* recorder. *Oil and Gas Journal, Dec. 8th.*
World Oil (1970): Automation ideas improve marginal lease operation. — *June.*
Wylie, E. B., Stoner, M. A., Streeter, V. L. (1970): Network system transient calculations by implicit method. *SPE Preprint 2963.*
Wylie, E. B., Streeter, V. L., Stoner, M. A. (1972): Unsteady natural gas calculations in complex piping systems. *SPE Preprint 4004.*
Zaba, J., Doherty, W. T. (1956): *Practical Petroleum Engineers' Handbook.* (4th ed.) Gulf Publishing Co., Houston, Texas.
Zaitsev, Yu. B. (1968): Tekhnika i tekhnologiya podema nefty iz skvazhin i puty ikh dalneyshego sovershenstvovaniya. *Neftepromyshlovoe Delo 7.*
Zeinalov, T. K., Sarkisov, E. J., Bagiryan, G. G. (1968): Teploizolatsiya neftegazoprovodov penopoliuretanom. *Gazovoe Delo 6.*
Zielke, W. (1971): Einige Betrachtungen über die Berechnung instationärer Strömungsvorgänge in Gasfernleitungen. *Das Gas- und Wasserfach 11.*
Zongker, F. (1969): Testing program indicates advantages of polyurethane pipeline cleaning tools. *Europe and Oil 3.*
Zorkóczy, B. (1968): *Metallográfia és anyagvizsgálat* (Metallography and material testing). Tankönyvkiadó, Budapest.
Zotov, A. G., Kand, A. G. (1967): Nagruzki i napryazheniya v polykh nasosnykh stangakh iz osteklovannykh trub. *Izv. V.U.Z. Nefty i Gaz 8.*

Subject index